Short-Circuit Currents in Three-Phase Systems

Gerd Balzer

Short-Circuit Currents in Three-Phase Systems

Springer

Gerd Balzer
Department of Electrical Power System
TU Darmstadt
Darmstadt, Germany

ISBN 978-3-658-50709-1 ISBN 978-3-658-50710-7 (eBook)
https://doi.org/10.1007/978-3-658-50710-7

This Springer imprint is published by the registered company Springer Fachmedien Wiesbaden GmbH, part of Springer Nature.
The registered company address is: Abraham-Lincoln-Str. 46, 65189 Wiesbaden, Germany

Preface

The main purpose of this book is to explain the background and prerequisites that led to the specifications in the international standard (IEC 60909-0), "Short-Circuit Current Calculation in Three-Phase Systems." The first edition of the IEC standard was published in 1988; therefore, the technical content of this IEC standard is very strongly influenced by the earlier German VDE 0102 standards, so these older German standards are referred to in places in the explanations, derivations, and references.

The first independent VDE regulation on the subject of short-circuit current calculation was published in 1962 (high voltage) and 1964 (low voltage), respectively, and it has developed considerably in the following years, as new considerations had to be added. This book describes some of the processes in which the author has been involved as a member of the German VDE-0102 commission (later UK 121.1) from 1972 to 2024 and as Convenor of IEC 73.1, "Short Circuit Current Calculation," since 1998.

The content is therefore not a complete description of the procedure for calculating short-circuit currents in electrical networks, for which the relevant IEC standard is of course authoritative, but is intended to provide explanations of individual specifications, e.g., peak short-circuit current, breaking current, and steady-state short-circuit current. In addition, many examples are given to help the reader understand the short-circuit current calculation. Furthermore, the comments refer to the status of the second edition of IEC Standard 60909-0 (2016).

An IEC standard should primarily describe the result of how a value is to be determined or how a process is to be represented; in many cases, the physical background cannot be explained in detail. For this reason, the book attempts to close this gap by presenting some physical relationships. This also means that, for example, in the context of long-term network planning, the components must be dimensioned over the entire service life in such a way that they can withstand the operational load conditions with sufficient safety. In contrast, the implementation of standard specifications regarding representation in computer programs remains the task of appropriate reference books.

Some of the contents of this book have already been published by the author in books, journals, and conference papers over the past decades, so this is a summarized presentation on the subject of short-circuit current calculation.

The author has been a member of the corresponding VDE commission since 1972 and additionally the Convenor of the relevant IEC working group since 1998; he had to leave these organizations at the request of the "German Electrotechnical Commission" (DKE) due to his age in 2024. The content of the book therefore summarizes the activities in the field of short-circuit current calculation over this period.

The author would like to thank his doctoral supervisor, Prof. Dr.-Ing. Gerhard Hosemann, for entrusting him with the topic of "Short-Circuit Current Calculation" as a member of staff at the Institute for Electrical Power Supply at the former TH Darmstadt (now TU Darmstadt) in 1972, thus giving him the opportunity to work on his Ph.D. in the field of short-circuit current calculation. Special thanks also go to the author's son, Dr. C. Balzer, for reviewing some of the chapters.

Darmstadt, Germany
April 2026

Gerd Balzer

Formula Symbol

Symbol

I_k''	Initial symmetrical short-circuit current
Λ	Magnetic conductance
a	Complex operator
A	Cross section
b	Internal inductance, hollow conductor
B	Magnetic flux density (inductance)
$\mathbf{C}$	Transformation matrix
c	Voltage factor
C, C'	Capacity, -per unit
c_E	Earth fault factor
$\cos \varphi$	Power factor
$\mathbf{D}$	Determinant
d	Distance, diameter
D	Electric flux density
d_A	Radius of a hollow conductor
d_E	Penetration depth of the return current in the earth
E	Active voltage
E	Earth
E	Electrical field
F	Failure, deviation
f	Frequency
f_0	Natural frequency
f_c	Equivalent frequency
f_n	Nominal frequency
G	General value (matrix)
G, G'	Admittance, per unit
h	Height

H	Magnetic field strength
I, $\mathbf{I}$	Current, matrix
I_1	Stator current
I_2	Rotor current
I_{AC}	AC component
I_b	Symmetrical short-circuit breaking current
I_d	Steady-state current
I_k	Steady-state short-circuit current
I_{LR}	Symmetrical locked-rotor current
i_p	Peak short-circuit current
I_{th}	Thermal equivalent short-circuit current
j	Imaginary quantity
k	Amplification factor
$\mathbf{K}$	Chain matrix
K	Impedance correction factor
l	Length
L, L'	Inductivity, per unit
L_b	Operating inductance
L_H	Magnetizing inductance
m	Factor for the heat effect of the DC component
m	Load factor
M	Mutual inductance
n	Factor for the heat effect of the AC component
n	Number
n_1	Synchronous rotation speed
n_2	Rotating field of the rotor
P	Active power
p	Laplace operator
p	Ratio of the maximum excitation voltage to the rated excitation voltage of the generator
p	Voltage range (transformer, generator)
P_{krT}	Short-circuit losses (rated value)
P_{nat}	Natural power
Q	Charge
q	Factor calculation of breaking current of asynchronous motors
Q	Reactive power
r	Radius
r	Specific electrical resistance
R, R'	Resistance, per unit
r_B	Equivalent radius bundled conductor
R_E	Resistance of field winding
R_{Fe}	Iron loss
R_{GS}	Resistance of the armature winding

S	Apparent power
s	Distance
s	Slip
t	Time
T	Time constant
$\mathbf{T}$	Transformation matrix
t	Transformation ratio
T_{d}	Time constant (d axis)
T_{d0}	Open-circuit time constant
T_{DC}	DC time constant
T_{k}	Short-circuit duration
t_{min}	Minimum time delay
U	Voltage
U_1	Stator voltage
U_2	Rotor voltage
U_{c}	Continuous operating voltage (surge arrester)
U_{f}	Excitation voltage of synchronous machine
u_{k}	Rel. short-circuit voltage
U_{m}	Highest voltage of equipment
U_{n}	Nominal system voltage
U_{r}	Rated voltage
U_{ref}	Reference voltage
U_{rest}	Residual voltage (motors)
u_{Rr}	Resistive component of the short-circuit voltage (rated value)
w	Winding
X, x	Reactance, related
X_2	Leakage reactance (motor)
X_{d}, x_{d}	Synchronous reactance (d axis), related
X_{H}	Magnetizing reactance (transformer, motor)
X_{Hd}	Magnetizing reactance (synchronous machine, d axis)
X_{Hq}	Magnetizing reactance (synchronous machine, q axis)
X_{σ}	Leakage reactance
$X_{\sigma\mathrm{A}}$	Armature leakage reactance (synchronous machine)
$X_{\sigma\mathrm{Dd}}$	Armature leakage reactance (damper winding, d axis)
$X_{\sigma\mathrm{Dq}}$	Armature leakage reactance (damper winding, q axis)
$X_{\sigma\mathrm{E}}$	Leakage reactance (field winding)
$\underline{Y}$, $\underline{\mathbf{Y}}$	Admittance, -matrix
Y_{q}	Shunt admittance
Z	Impedance
$\underline{Z}$, $\underline{\mathbf{Z}}$	Impedance, -matrix
Z_{l}	Longitudinal impedance
Z_{W}	Surge impedance

ΔU	Voltage drop
Φ	Flux (magnetic)
Φ_H	Magnetizing flux
Φ_σ	Leakage flux
ϑ_E	Conductor temperature at the end of the short circuit
ϑ, φ	Phase angle
Λ	Magnetic conductance
Θ	Ampere turns (magnetic)
α	Damping constant
α	Potential coefficient
α	Temperature coefficient
α, β	Correction factor
β	Phase constant
ε_0	Dielectric constant
ε_r	Related dielectric constant
γ	Propagation constant
η	Efficiency factor
φ	Electric potential
κ	Factor calculation of the peak short-circuit current
λ	Eigen value
λ	Factor calculation of the steady-state short-circuit current
μ	Factor calculation of the symmetrical short-circuit breaking current
μ	Permeability
μ_0	Permeability constant
ρ_E	Specific soil resistivity
ω	Angular frequency
ω_0	Natural frequency

Subscripts

(0)	Zero sequence system
(1)	Positive sequence system
(2)	Negative sequence system
+	With
0	Open circuit, without
0, α, β	Components
A	Armature (synchronous machine)
A	Line conductor/fictitious outer sheath
AC	Alternating current (AC)
asym	Asymmetrical
B	Bundled conductor
B	Load
b	Operation

C	Capacitance
C	Converter
CS	Current source
d	d axis
D	Damper winding
D	Earth wire
D	Short-circuit limiting reactor
D	Throughput power
DC	Direct current (DC)
E	Earth
E	End
E	Excitation
E	Total winding power
EE	Double line-to-earth short-circuit
F	Fault location
f	Fictitious
F	Overhead line
G	Generator
GA	Asynchronous generator
GD	Doubly fed induction generator
GS	Synchronous machine
HV	High voltage
i	Inside
Im	Imaginary
K	Cable
K	Compensation
K	Correction
k	Short circuit
k1	Line-to-earth
k2	Line-to-line
k3	Three phase
kE2E	Line-to-line with earth contact
kEE	Double line-to-earth short-circuit
L	Conductor, line
LV	Low voltage
M	Motor
M	Sheath
max	Maximum
min	Minimum
N	Network
n	Nominal value
n	Reference number
PA	Power station (asynchronous generator + transformer)

PD	Power station (doubly fed induction generator + transformer)
PS	Power station (synchronous generator + transformer)
q	q axis
r	Rated value
R, S, T	Conductor identification
real	Real
res	Resultant
S	Switch-gear
S	System
T	Subconductor
t	Transferred value
T	Transformer
th	Thermal
U	Converter
V	Consumer
VS	Voltage source
α	Index
σ	Leakage

Superscripts

$'$	Transient
$''$	Subtransient
*	Changed node voltage

Competing Interests The author has no competing interests to declare that are relevant to the content of this manuscript.

Contents

Chapter 1
Introduction

The short-circuit current calculation is a "safety calculation" that applies to the design of the systems, devices, and the network and personal protection. For this reason, it makes sense that suitable methods have been derived in the past for calculating the currents, which are described in more detail in the following chapters.

1.1 Basics

The short-circuit current calculation according to the standard [1] deals with a calculation method to ensure the security resp. safety of electrical networks and thus also of persons, and it has therefore always been understood as a "safety standard" for the design of equipment, which applies in particular to the maximum short-circuit current. For this reason, the aim has not primarily been the accuracy of the results, but the results should be on the "safer" side. This basic approach can be seen in particular in the determination of the short-circuit breaking current: The application of the factors μ and q (Sect. 10.2) represents the envelope of the possible breaking currents of three-phase machines, so that the exact simulation of a specific machine always leads to a possibly significantly smaller current during the short-circuit current waveform. However, this should not be critical, as the breaking current is used to design the circuit breakers. As a result, the equations for calculating the short-circuit currents in three-phase systems always describe the conditions under which the short-circuit current must be calculated. In contrast to this, the calculation methods are not described in detail, e.g. setting up or inverting an impedance matrix, as these currents can be determined in extended networks with several voltage levels. In contrast, this description of the mathematical implementation can be found in the appropriate specialist books [2].

It must be pointed out that the existing standard [1] can only be used to determine the total short-circuit currents at the fault locations and not the partial short-circuit

G. Balzer, *Short-Circuit Currents in Three-Phase Systems*,
https://doi.org/10.1007/978-3-658-50710-7_1

currents that flow to a fault location. The fact that these partial short-circuit currents cannot be easily determined becomes apparent when calculating the peak short-circuit currents in the individual network branches. In contrast to this, however, the partial short-circuit currents of generator and motor equipment must be calculated in order to determine the total short-circuit breaking current at the fault location.

For the reasons mentioned above, two different questions are thus covered, namely:

- Design of the installations and components to ensure proper operation of a network. In this context, it must always be ensured that the mechanical forces occurring in the event of a fault are controlled and that equipment can fulfill its function, e.g. the interruption of the short-circuit current by switching devices. To solve this problem, it is necessary to determine the maximum short-circuit current at the relevant network nodes.
- Design of the system protection in order to disconnect a network part or a faulty piece of equipment from the network in the event of a fault selectively. To solve this problem, the minimum short-circuit current is required so that the protection can be set accordingly. This means that the starting current should be lower than the actual current in order to ensure reliable tripping of the protection.

When using the equations given in this book, it is expressly pointed out that only the formulas of the applicable standard are to be used for the calculations of short-circuit currents. In some places, the equations and boundary conditions given here go beyond the representation in the standard [1].

1.2 Definition of Currents

For this reason, different short-circuit currents are determined for the technical problems described above, which are listed below. The terms required to calculate the short-circuit currents are also listed completely in Sect. 5.1.

- Initial symmetrical short-circuit current I''_k (rms value): AC component of the short-circuit current at the beginning of the short-circuit occurrence.
- Peak short-circuit current i_p (peak value): Maximum value of a short-circuit current that occurs after a maximum of $t = 10$ ms, depending on the R/X ratio of the short-circuit loop, at an operating frequency of $f = 50$ Hz. The peak short-circuit current is thus determined from the AC and the DC component of the initial symmetrical short-circuit current.
- Short-circuit breaking current I_b (rms value): AC component of a short-circuit current during the mechanical contact separation of a switching device. The IEC regulation specifies values for the following minimum time delays: $t_{min} = 0.02$; 0.05, 0.10, 0.25 s.
- Steady state short-circuit current I_k (rms value): Short-circuit current after all transient phenomena, e.g. from generators, have finished.

- Thermal equivalent short-circuit current I_{th} (rms value): Short-circuit current over a certain time range, e.g. $t = 1$ s, whose thermal effect is the same as the actual short-circuit current, including the existing DC component and a decaying AC component of the current.

In the past, the short-circuit power was often used as the input value instead of the initial short-circuit current. The disadvantage, however, is that when specifying the short-circuit power, the nominal system voltage U_n must be known which was used to determine the power, as the short-circuit current calculation uses the nominal system voltage as the basis. While the highest voltage U_m for equipment according to [3] is always clearly defined for a voltage range, in contrast to this two different nominal system voltages are sometimes defined (e.g. $U_m = 245$ kV: $U_n = 220$ kV and $U_n = 230$ kV) or for voltages $U_m > 300$ kV no nominal system voltages U_n are defined.

This is also the reason why the term short-circuit power is no longer used in the calculation equations of the IEC regulation [1], so that the conversion into a short-circuit current is no longer necessary. In addition, the design variable of equipment and the setting value of the network protection is always the calculated short-circuit current.

1.3 Procedure

The application of the standard relates to the planning of electrical systems that are operated over a long period, e.g. approx. >40 years. This means that only data should be used that should not lose its validity over this period. For this reason, operational values (current, tap position of the transformers) should not be used that result from the load flow, for example. However, the internal voltage of the feeders is essential for the level of the short-circuit current, so that it is essential to consider the voltage in the calculation. For this reason, the maximum permissible or operational voltage in the system is used as the basis for the short-circuit current calculation for simplification purposes. For the maximum short-circuit current, the related voltage factor c_{max} is used or taken into account by an additional impedance correction factor K, if necessary. The voltage factor c thus indirectly describes the operational condition of the network or equipment.

In the currently valid IEC regulation, the method of the equivalent voltage source at the short-circuit location is used. In this method, a load flow is not considered, but is taken into account by a voltage factor c. It is assumed that there is only a maximum deviation of 5% to the unsafe side between this calculation result and the current according to more precise methods (e.g. the superposition method taking into account the load flow that leads to the maximum short-circuit current).

Due to the increased use of full-size converters that are operated in a current-carrying mode, it is necessary to take their contribution to the short-circuit current into account using the simulation as a current source, if they contribute to the short-circuit current based on the control strategy. In this case, these currents must be superimposed on the currents calculated from the equivalent voltage source at the short-circuit location to determine the resulting short-circuit current.

In principle, arcs or arc resistances at the short-circuit location are not taken into account when calculating the short-circuit current, so that the metallic short-circuit is always assumed. In practice, the arc generally leads to a lower current than the calculated current, but this should only have a particular effect in low-voltage networks directly at the infeed.

In general, it should be possible to calculate the short-circuit currents using a computer or by hand and achieve the same result. However, this was not always the case in the past, as estimates or topology simplifications were sometimes permitted with regard to manual calculations.

The existing standards [1] for calculating short-circuit currents only deal with short-circuits as shown in Figs. 3.4 and 3.5, which means that earth faults in isolated or compensated networks are not covered.

1.4 Fault Statistics

Various boundary conditions are responsible for the occurrence of short-circuits and their fault type (single or multi-phase with and without earth contact), e.g.:

- Operating voltage,
- equipment (overhead line, cable, transformer, etc.),
- neutral point treatment,
- ambient conditions.

In Germany, different fault types are published annually in the failure and availability statistics of the FNN (VDE) [4]. Table 1.1 shows the percentage distribution of fault types for the period 2013–2016 depending on the voltage ranges. The line-to-earth faults include both earth faults and short-circuits to earth.

While line-to-earth short-circuits or earth faults are clearly predominant in the high-voltage and extra-high-voltage grids (HV, EHV), multi-phase faults also occur more frequently in medium-voltage grids (MV), which is a consequence of the small distances between the conductors compared to the high-voltage grids, for example, in the event of a lightning strike to an overhead line. Compensated neutral point treatment predominates in medium and high-voltage grids in Germany (MV approx.

Table 1.1 Percentage distribution of short-circuit faults for the years 2013 through 2016

Type of fault	MV	HV	EHV
Line-to-earth	35.4	91.3	89.7
Line-to-line	16.5	1.8	8.4
Three-phase	9.6	0.7	1.3
Unknown	38.5	6.2	0.6

MV medium voltage (>1 kV—72.5 kV), *HV* high voltage (>72.5 kV—125 kV), *EHV* extra high voltage (>125 kV)

72%; HV approx. 81%), which is also a consequence of the proportion of overhead lines in the grids. In contrast, only low-impedance (solidly) earthing is common in the extra-high-voltage grids, which primarily leads to high current, line-to-earth short circuits.

1.5 Short-Circuit Times

During a short-circuit process until the short-circuit current is switched off by the selective protection, various time steps are run through, which are shown in Fig. 1.1.

The following steps are defined in detail:

Fault initiation	Initiation of the short circuit
Energize	The set response value of the system protection is exceeded
Tripping	Trip command to the circuit-breaker
Opening of contacts (mechanic)	Opening of the contacts of a circuit breaker
Opening of contacts (electric)	Interruption of the short-circuit current, due to the different current zero crossings of the individual phases, the electrical interruption by the breaker poles cannot occur simultaneously, which depends on the type of fault
Actual value lower than operating value	The set response value of the system protection is undercut
Release	The system protection returns to its output mode

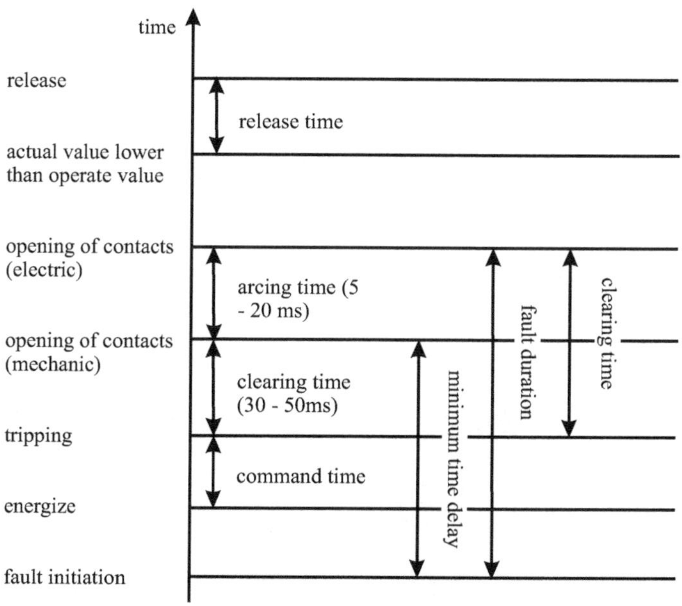

Fig. 1.1 Terms and sequences of the system protection (the time axe is not full-scaled)

In the short-circuit current calculation, the minimum time delay t_{min} is used to determine short-circuit breaking current, which is determined according to Fig. 1.1 from the time between the initiation of the fault and the mechanical contact opening of the circuit-breaker.

References

1. IEC 60909-0:2016 Edition 2.0 (2016) Short-circuit current calculations in three-phase a.c. systems—Part 0: calculation of currents. IEC, Geneva
2. Das J C (2018) Power systems handbook—Volume 1: Short-circuits in AC and DC systems. CRC Press Taylor & Francis Group, New York
3. IEC 60038:2009+AMD1:2021 CSV (2021) IEC Standard Voltages. IEC, Geneva
4. FNN Forum Netztechnik/Netzbetrieb im VDE (2014, 2015, 2016, 2017) Störungs-und Verfügbarkeitsstatistik, Berichtsjahre 2013/2014/2015/2016, Berlin

Chapter 2
Standards for Short-Circuit Current Calculation

This section presents the national (DKE/VDE) and international (IEC) standards in the field of short-circuit current calculation. In addition, the development of the standard is presented, the origin of which lies in the preparation of the VDE regulation (Germany) at the beginning of the 1960s and was developed by the UK 121.1 committee and the VDE 0102 commission.

This is the reason why many publications or references dealing with the subject of short-circuit current calculation are written in German. This applies in particular to the derivation of the method "equivalent voltage source at the short-circuit location".

2.1 Development of Standardization

After discussions within the "International Electrotechnical Commission" (IEC), the first meeting of Committee 73 "Short-circuit currents" took place in Solna/Stockholm in 1972. It was decided to set up a series of two working groups (WG), or today Maintenance Teams (MT), to which the following activities were assigned:

- MT 1: Calculation of short-circuit currents,
- MT 2: Heating and mechanical effects.

The contents of these working groups coincide with the VDE commissions for the preparation of the standards of VDE 0102 and VDE 0103 and the DKE commissions UK 121.1 and UK 121.2. The valid standards prepared by the IEC working group MT 1 are listed in Sect. 2.2.

The first VDE regulations on short-circuit current calculation were published between 1962 and 1964, whereby the considerations for these documents had already begun in 1954. Due to the different areas of application and groups of people, two separate standards were drawn up at that time:

G. Balzer, *Short-Circuit Currents in Three-Phase Systems*,
https://doi.org/10.1007/978-3-658-50710-7_2

- Guidelines for the calculation of short-circuit currents, VDE 0102 Part 1/9.62 Three-phase systems with rated voltages above 1 kV (consisting of 29 pages)
- Guidelines for the calculation of short-circuit currents, VDE 0102 Part 2/4.64 Three-phase systems with rated voltages up to 1000 V (consisting of 11 pages)

The main difference between the two guidelines, apart from the boundary conditions tailored to the different requirements, was that in Part 1 (high voltage) the short-circuit current was calculated using the complex calculation, whereas this was not the case in Part 2 (low voltage). In addition, the active voltage in the method using the equivalent voltage source at the short-circuit location (Sect. 3.6.6) was $1.1 \cdot U_n/\sqrt{3}$ (U_n nominal system voltage), while in the low-voltage section the value was set at $1.0 \cdot U_{rT}/\sqrt{3}$ (U_{rT} rated voltage of the transformer on the low-voltage side).

Since the short-circuit current calculation in high-voltage and low-voltage networks cannot differ for physical reasons, it made sense to abandon this subdivision later on, also from the point of view of the shift from manual to computer calculation. This was also because the short-circuit current calculation is carried out across several voltage limits in one calculation process.

Until the first VDE standard, the short-circuit current calculation was a part of VDE standard 0670/1.54 (AC switchgear). In the first standard (1962), the following types of faults were dealt with in Part 1:

- Three-phase (symmetrical) short circuit, § 5,
- unbalanced faults, § 6:
  - Line-to-line short circuit with and without earth contact,
  - double line-to-earth short-circuit,
  - line-to-earth short-circuit,
- partial short-circuit currents through earth, § 7:
  - partial short-circuit currents in installations,
  - partial short-circuit currents on overhead line towers,
- influence of loads, § 8.

In the case of multiple-fed short-circuits, the method of the equivalent voltage source at the short-circuit location with $1{,}1 \cdot U_n$ was introduced.

Part 2 of the VDE standard referred exclusively to the three-phase and line-to-earth short-circuit current for far-from-generator. For near-to-generator faults, reference was made to Part 1.

The revision of the existing guidelines was published in 1971, with the following parts:

- Guidelines for the calculation of short-circuit currents, VDE 0102 Part 1/11.71 Three-phase systems with rated voltages above 1 kV (consisting of 32 pages)
- Guidelines for the calculation of short-circuit currents, VDE 0102 Part 2/11.75 Three-phase systems with rated voltages up to 1000 V (consisting of 36 pages)

The main changes in the first part were that the equivalent voltage source method can be used at the short-circuit location up to a nominal system voltage of $U_n = 380$ kV. As load currents are not taken into account in this case, it is also possible to use the superposition method for a more precise calculation [1]. It is pointed out that "the user is free to apply calculation methods that are more accurate or better adapted to the existing operating conditions". In addition, the equivalent frequency method was introduced for meshed networks, when calculating the peak short-circuit current [2]. Furthermore, the chapter on the contribution of motors has been significantly expanded.

The changes in the second part relate, for example, to the fact that the equivalent voltage source method is now also used in the low-voltage range, with $1.0 \cdot U_{rT}$ (secondary side of the low-voltage transformer). Since a large number of asynchronous motors are used in industrial low-voltage networks in particular, the conditions to be taken into account for motors are described in detail. This is based on extensive measurements, which are described in [3]. For the calculation of line-to-earth short-circuit currents in particular, extensive tables are provided for the use of $R_{(0)}/R_{(1)}$ and $X_{(0)}/X_{(1)}$-ratios for different low-voltage cable types, which were determined according to [4]. In addition, the curves used in the standard are approximated by equations to allow their use in calculation programs [5].

Based on these basic principles, IEC Standard 909:1988 (Short-circuit current calculation in three-phase a.c. systems) was published, which is characterized by the following significant changes compared to the earlier VDE version:

- Summary of part 1 (high voltage) and part 2 (low voltage) and application of the equivalent voltage source method at the fault location for all voltage levels.
- Introduction of impedance correction factors for transformers.
- Two methods are available for the consideration of power stations:
  - Individual correction method: Here, the impedances of equipment (generator and transformer) are corrected differently.
  - Total correction method: In this case, the impedance of the power station is given a single factor.
- The breaking current in meshed grids must be calculated using a new method [6].
- The application of the λ-factor when determining the steady state short-circuit current relates exclusively to single-fed and multiple single-fed short circuits.

In 2001, a revision of the main standard was published (IEC 60909-0:2001–07) which, in addition to a new structure, resulted in the following significant changes

- Revision of the correction factors for power stations.
- Introduction of correction factors for network transformers.
- New definition of the voltage factors for the low-voltage range due to the new voltage values of low-voltage transformers.
- The short-circuit power is no longer used as a calculation variable.

The new main standard, which was published in 2016 (IEC 60909-0, Ed. 2.0, 2026-01), also deals with the short-circuit contribution

- of wind power station units, and the
- contribution of full-size converters

as a result of technical developments and the increased use of this installations in the generation of electrical energy from renewable energy sources.

Future revisions will deal with the improved consideration of wind and PV installations and the application of HVDC (high voltage direct current) systems.

2.2 Existing Standards

The calculation of short-circuit currents is defined internationally by the working group of IEC TC 73 [Short-circuit current-calculation in three-phase a.c. systems]. In addition to the calculation standards [7, 8], there are also supplementary reports which describe the calculation process, examples and also data from equipment [9–11] and are therefore considered to be application guidelines. The following standards and supplements are currently available:

- IEC 60909-0:2016: Short-circuit current calculations in three-phase a.c. systems—Part 0: Calculation of currents [7]
- IEC TR 60909-1:2002: Short-circuit currents in three-phase a.c. systems—Part 1: Factors for the calculation of short-circuit currents according to IEC 60909-0 [9]
- IEC TR 60909-2:2008: Short-circuit current calculations in three-phase a.c. systems—Part 2: Data of electrical equipment for short-circuit current calculations [10]
- IEC 60909-3:2009: Short-circuit currents in three-phase a.c. systems—Part 3: Currents during two separate simultaneous line-to-earth short circuits and partial short-circuit currents flowing through earth [8]
- IEC TR 60909-4:2021: Short-circuit currents in three-phase a.c. systems—Part 4: Examples for the calculation of short-circuit currents [11].

In the following sections, some information on the application of the currently valid IEC standard [7] is taken as a basis, but the content of this standard is not reproduced in full. In any case, the wording of the valid IEC standard is always decisive for the short-circuit current calculation and not the equations listed here.

The regulations described above deal with the calculation of short-circuit currents in three-phase systems; in addition, the IEC Commission also generally covers the calculation in DC auxiliary power systems, which includes the following publications:

- IEC 61660-1:1997: Short-circuit currents—Short-circuit currents in d.c. auxiliary installations in power plants and substations—Part 1: Calculation of short-circuit currents [12]
- IEC TR 61660-3:2000: Short-circuit currents in d.c. auxiliary installations in power plants and substations—Examples of calculations [13]

In the following, however, reference is made exclusively to the calculation of short-circuit currents in three-phase systems.

2.3 Conclusions

VDE standards for the calculation of short-circuit currents have existed since 1962, from which the IEC standards emerged in later years. In the meantime, different editions have been developed, in which the calculation principles and general equipment data are also presented in more detail.

References

1. Funk G (1962) Der Kurzschluß im Drehstromnetz. R. Oldenbourg Verlag, München
2. Koglin H-J (1971) Der abklingende Gleichstrom beim Kurzschluß in Energieversorgungsnetzen. Dissertation TH Darmstadt, p D17
3. Webs A (1972) Einfluß von Asynchronmotoren auf die Kurzschlußstromstärken in Drehstromanlagen. VDE-Fachberichte 27:86–92
4. Balzer G (1977) Impedanzmessung in Niederspannungsnetzen zur Bestimmung der Kurzschlussströme. Dissertation TH Darmstadt, p D17
5. Hosemann G, Balzer G (1975) Impedanzwerte zur Berechnung des Kurzschlußstroms nach DIN 57 102 Teil 2/VDE 0102 Teil 2. Ein Beispiel für die Anpassung einer Norm an die Rechenmaschinentechnik. Elektronorm Jg 29(3):118–121
6. Hosemann G, Balzer G (1984) Der Ausschaltwechselstrom beim dreipoligen Kurzschluss im vermaschten Netz. ETZ-Archiv 6(2):51–56
7. IEC 60909– 0:2016 Edition 2.0 (2016) Short-circuit current calculations in three-phase a.c. systems—Part 0: calculation of currents. IEC, Geneva
8. IEC 60909–3:2009 Edition 3.0 (2009) Short-circuit currents in three-phase a.c. systems—Part 3: currents during two separate simultaneous line-to-earth short circuits and partial short-circuit currents flowing through earth. IEC, Geneva
9. IEC/TR 60909–1:2002 Edition 2.0 (2002) Short-circuit currents in three-phase a.c. systems—Part 1: Factors for the calculation of short-circuit currents according to IEC 60909–0
10. IEC/TR 60909–2:2008 Edition 3.0 (2008) Short-circuit current calculations in three-phase a.c. systems—Part 2: data of electrical equipment for short-circuit current calculations. IEC, Geneva
11. IEC/TR 60909–4:2021 Edition 2.0 (2021) Short-circuit currents in three-phase a.c. systems—Part 4: examples for the calculation of short-circuit currents. IEC, Geneva
12. IEC 60660–1:1997 Edition 1.0 (1997) Short-circuit currents—short-circuit currents in d.c. auxiliary installations in power plants and substations—Part 1: calculation of short-circuit currents. IEC, Geneva
13. IEC/TR 60660-3: 2000 Edition 1.0 (2000) Short-circuit currents in d.c. auxiliary installations in power plants and substations—examples of calculations. IEC, Geneva

Chapter 3
Requirements and Calculation Methods

The specified standards for short-circuit current calculation according to Chap. 2 require various boundary conditions, which are explained in more detail in this chapter. These conditions are partly necessary in order to use the method of the equivalent voltage source at the short-circuit location, which simplifies the calculation considerably [1].

Due to the use of converters with VSC technology, which are intended to supply a defined inductive current during the short-circuit period, it is also necessary to consider the influence of these current sources separately, as shown in Sect. 3.6.7.

3.1 General Information on Short-Circuit Current Calculation

The scope of this standard IEC 60909-0 [2] refers to three-phase systems with operating frequencies of 50 Hz and 60 Hz. Networks with voltages $U_m > 420$ kV are not covered by the described method of the equivalent voltage source at the short-circuit location, but must always be calculated using the superposition method. The reason for this is that the calculation using the equivalent voltage source at the short-circuit location requires a voltage factor c, which depends on the usual topology, the mode of operation and equipment at this voltage level. As calculations of the voltage factor were not carried out comprehensively for networks with voltages $U_m > 420$ kV, it is not possible to make a statement about this. However, as shown later, the fundamental problem with more precise calculation using the superposition method is that the load flow that leads to the largest short-circuit current is not easy to determine.

In general, it is possible to determine the calculation of short-circuit currents in three-phase systems using two methods if current sources are not taken into account, e.g.:

G. Balzer, *Short-Circuit Currents in Three-Phase Systems*,
https://doi.org/10.1007/978-3-658-50710-7_3

- Superposition method,
- equivalent voltage source at the short-circuit location.

Both options, which are also presented in Sect. 3.6, are permissible for calculating the short-circuit currents. In addition, it is permitted to use other methods or simplifications if they provide results that are demonstrably on the safe side. This means that the calculation of the maximum short-circuit current should result in values that are too high and the calculation of the minimum short-circuit current should result in values that are too low.

While the two calculation methods listed above are used to determine steady-state values, e.g. the initial symmetrical short-circuit current, it is of course also possible to determine the short-circuit current curve using programs that show the entire transient current curve as a result. This includes, for example, the EMTP/ATP calculation program. In these cases, however, it is necessary to consider the complete simulation, including all resistances and time constants, in the case of generators and motors. In addition, it is of course also possible to determine the short-circuit current from network tests or with the aid of network models.

The superposition method can always be used to determine the exact short-circuit current for an operating case. However, the requirement for this calculation is that a load flow is known which takes into account all voltages at the feeders, the transformer tap positions, the consumer loads, etc.

In contrast, the calculation using the equivalent voltage source at the short-circuit location provides a result that is independent of the existing load flow in the nedwork in individual cases, as these parameters are generally taken into account by the voltage factor c or by impedance correction factors. The purpose of this method is to determine the design parameters of equipment to be installed and operated in the coming decades for the planning calculation of a nedwork. However, the planning engineer is at liberty to use the superposition method for nedwork expansion planning as well. However, it must be ensured that a load flow is used for the end of the expansion planning that leads to the maximum or minimum short-circuit currents. However, this is likely to be quite difficult in most cases, especially as the planning engineer would have to take responsibility for an inappropriate design of equipment if an "unsuitable" load flow is assumed as a prerequisite. The main advantages of the simplified procedure can be summarized as follows:

- The input variables should result from the standard values of equipment (rated values, length, etc.).
- Simple simulation of equipment, e.g. transformer tap changers in main-position.
- Easy to apply to meshed networks.
- It should be possible to simulate unbalanced faults.
- On average, the calculations should provide results that are on the safe side.

The application of the equivalent voltage source method at the short-circuit location is a method that is adapted to the usual technical operating modes of networks and is therefore particularly suitable for planning calculations. In addition, the exact data of the components (e.g. the subtransient reactance of a generator) are not yet precisely defined for many nedwork expansion plans. For this reason, it goes without saying that a precise calculation based on exact data using the superposition method provides different results than the simplified calculation using the equivalent voltage source method at the short-circuit location.

The different results are particularly evident when the contributions of motors or generators to the short-circuit current are to be determined. While decay factors μ and q (Sect. 5.6) derived from an envelope curve of different measurements are used to calculate the short-circuit breaking current according to the IEC standard, it would be a random coincidence, if the same breaking alternating current were calculated for the exact simulation of a generator (with time constants and reactances).

When using the equivalent voltage source method at the short-circuit location, the actual short-circuit current should be determined with a maximum deviation of 5% (in relation to the unsafe side). As this tolerance limit is already inherent in the application of the method, it is therefore not permissible to determine a short-circuit current and then deduct a safety margin of 5%, e.g. when determining the maximum short-circuit current.

When the first VDE standard for short-circuit current calculation was drawn up (1962), the method of the equivalent voltage source at the short-circuit location was developed taking into account the networks and their operating modes at the time. In the meantime, however, it has become apparent that this is no longer the case in many cases today, so that a change in the calculation was unavoidable. This led to the introduction of impedance correction factors, which are discussed in more detail in Chap. 8.

3.2 System Topology

When calculating the short-circuit current, different network topologies are taken into account, which also have an influence on the calculation method and therefore on the result. The network structures relevant for the short-circuit current calculation are:

- Single fed (radial system), one branch, Fig. 3.1a,
- Single fed (radial system), parallel branches, Fig. 3.1b,
- Multiple single-fed, Fig. 3.1c,
- Multiple fed via a common impedance, Fig. 3.1d,
- Multiple/single fed (meshed system), Fig. 3.1e.

Figure 3.1 shows the various system topologies and short-circuit locations.

For systems according to Fig. 3.1a, the calculation always provides a clear result for all different types of short-circuit current, whereas in Fig. 3.1b–e, the 20 Hz method must be used to calculate the peak short-circuit current, for example, if there are different R/X ratios of the two networks. In the case of multiple single-fed, the calculation according to [2] allows the individual amounts to be added together, so

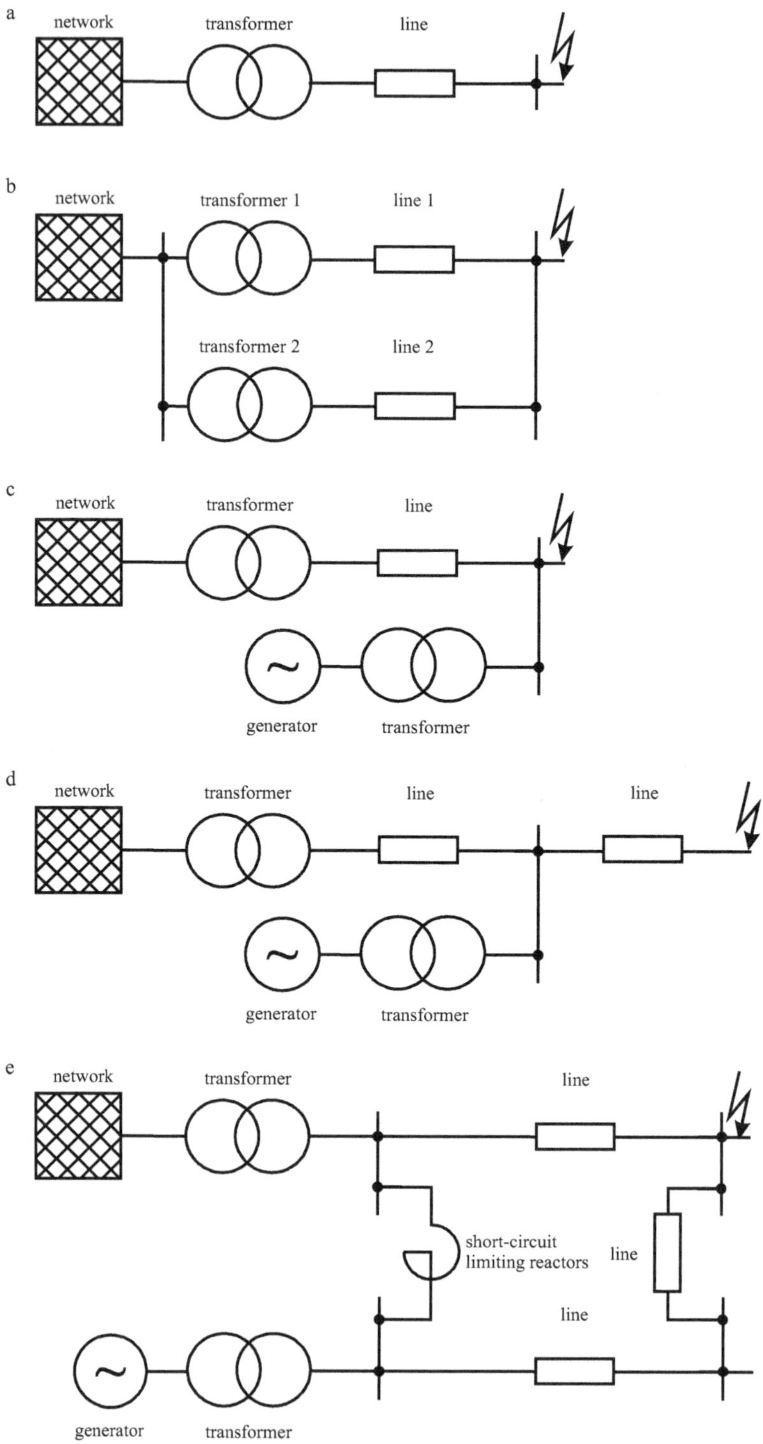

Fig. 3.1 Different network topologies and short-circuit locations, (**a**) single fed (radial system), one branch, (**b**) single fed (radial system), parallel branches, (**c**) multiple single-fed, (**d**) multiple fed via a common impedance, (**e**) multiple/single fed (meshed system)

that this result should be on the safe side compared to the complex calculation. The reason for this is that the time of the maximum peak value is not identical depending on the *R*/*X* ratio of the individual branches. For networks according to Fig. 3.1d, e, the corresponding procedures (Eq. (5.59)) must be considered, when calculating the short-circuit breaking current in order to avoid currents that are on the unsafe side.

3.3 Selection of Short-Circuit Currents and Short-Circuit Cases

The maximum and minimum short-circuit currents must be determined to dimension equipment and design the protection, so that specific boundary conditions are decisive for this and must be considered when determining the maximum and minimum short-circuit currents:

- Maximum short-circuit current:
  - voltage factor c_{max},
  - system configuration and power generation units, which lead to the maximum short-circuit current,
  - system topology, which lead to the maximum short-circuit current,
  - the impedance correction factors have to be used,
  - maximum motor use,
  - the temperture of cables and overhead lines is 20°C.
- Minimum short-circuit current
  - voltage factor c_{min},
  - system configuration and power generation units, which lead to the minimum short-circuit current (e.g. wind power station units and photovoltaic station units should be neglected),
  - system topology, which lead to the minimum short-circuit current,
  - the impedance correction factors have to be used with the value $K = 1$,
  - contributions of motors shall be neglected,
  - the resistances of lines are to be used according to the conductor temperature at the end of the short circuit, Eq. (3.1).

The following equation applies to the calculation of the effective resistance at the cable end temperature ϑ_E:

$$R_{L\vartheta} = [1 + \alpha \cdot (\vartheta_E - 20\ °C)] \tag{3.1}$$

Equation (3.1) can be used to determine the corresponding resistance at different temperatures based on the resistance at a temperature of 20 °C. For the temperature coefficient factor, the value $\alpha = 0.004/K$ can be set with sufficient accuracy for copper and aluminum. For a cable end temperature of e.g. $\vartheta_E = 80\ °C$ follows from this:

$$R_{L80} = [1 + 0.004 \cdot (80\ °C - 20\ °C)] \cdot R_{L20} = 1.24 \cdot R_{L20} \tag{3.2}$$

This means that at the final temperature after a short circuit, the resistance is 24% higher than at 20 °C. Due to the *R*/*X* ratio, the increase in resistance will have a particular effect in low-voltage networks. In this context, it must be pointed out that a maximum limitation of the cable end temperature to e.g. $\vartheta_E = 80\ °C$ in low-voltage networks, is no longer included in the regulations, as was previously the case. This will have a particular impact when using polymer cables, as these cable types can be expected to have a permissible end temperature of e.g. $\vartheta_E = 250\ °C$ in the event of a short circuit.

The conditions listed above also mean that in the case of a network feeder consisting of two parallel transformers; only one transformer should take into account when calculating the minimum short-circuit current, if operation with one transformer is possible [$(n-1)$ security]. In contrast, when determining the maximum short-circuit current, the system condition that is technically possible and leads to the maximum current values must be taken into account. Figure 3.2 illustrates the use of the network topology to determine the minimum and maximum short-circuit current.

- *Maximum short-circuit current:* If it is operationally possible for both transformers T1 and T2 as well as generator G to feed into the 10 kV busbar to supply the motor group (design of the 10 kV system corresponds to the maximum short-circuit current), all equipment (T1, T2, M and G) must be taken into account in the short-circuit current calculation.

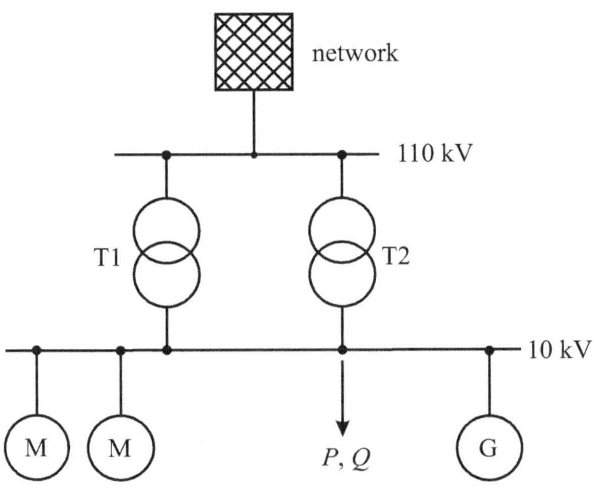

Fig. 3.2 Example for selecting the maximum and minimum short-circuit current

- *Minimum short-circuit current:* As the motor group can be out of operation, it does not contribute to the short-circuit current and therefore does not need to be taken into account, as the system protection must also be functional for this operating condition. In addition, it must be considered whether the network feeder (consisting only of a transformer for low-load operation) or the generator feeder (if island operation is possible) provides the minimum short-circuit current. This contribution must then be taken into account when dimensioning the system protection.

The short-circuit current calculation always assumes a metallic short-circuit; this means that an arc at the fault location is not taken into account. However, this type of fault will rarely occur in practice if the closing of a circuit-breaker to a closed earthing switch is not taken into account. The arc voltage that arises at the fault location reduces the fault current, as the arc voltage counteracts the active voltage of the network. In extreme cases, this current will go to zero if this arc voltage corresponds to the active system voltage. Due to the magnitude of this arc voltage, which generally has no phase shift to the short-circuit current, it depends in a first approximation on the length of the arc and therefore this process has less effect on arc faults in high-voltage and extra-high-voltage networks than in low-voltage networks. As it increases less than the active voltage, as is possible, for example, immediately after the feed-in from the superimposed network into the low-voltage network. According to [3], the arc voltage related to the length is almost independent of the current and therefore practically constant. It is approx. 1000 to 1600 V/m for an arc burning freely in air.

When determining the minimum short-circuit current, it must therefore be taken into account that only currents that occur due to the topology and operating mode of the network are determined, so that further reductions, e.g. arcing faults, are not taken into account and must be considered accordingly depending on the application. In addition, it will not be possible to specify generally valid voltages at the fault location, as they are very dependent on the local conditions or constructions. For example, the corresponding IEC standard (IEC 61439:2020: Low-voltage switchgear and controlgear assemblies) and, where applicable, special measures for personal protection must be observed when manufacturing low-voltage switchgear and controlgear assemblies with regard to protection from arcing faults. If necessary, the document IEEE 1584–2018 (Guide for Performing Arc-Flash Hazard Calculations) can be referred to if the short-circuit current and the energy conversion in a switch panel are to be determined taking an arc into account.

When using selective protection in low-voltage public supply networks, fuses are used whose tripping characteristic is dependent on the rated current, so that these two characteristic values (short-circuit current, rated current) must be matched in each individual case. For this reason, the minimum short-circuit current according to IEC standards provides an upper limit value for the design of the selective protection.

When determining the maximum short-circuit current, an arcing fault therefore leads to a reduction in the actual short-circuit current, so that the result of the calculation is on the safe side. However, this is no longer the case when determining the minimum short-circuit current. The reduction of the minimum short-circuit current is not taken into account, particularly in the case of faults in low-voltage systems due to the reduced voltage factor c_{min}, as the reduced voltage factor c_{min} corresponds, for example, to a changed load flow including a lower voltage level in the system. The calculation of the minimum short-circuit current using the boundary conditions specified above therefore always leads to the upper limit values for the "minimum short-circuit current" in networks. Particularly in the case of short circuits in low-voltage systems at the terminals of the infeed transformer, a reduction in the calculated value of the minimum short-circuit current will occur, which applies especially to the line-to-line fault, as it should be the minimum short-circuit current in these cases. To take the arc into account, extensive short-circuit tests are generally necessary in test fields that lie outside a short-circuit current calculation standard. For this reason, reference is made in this context to the specifications of the IEEE standard [4].

In the calculation, it is generally assumed that the short circuit occurs when there is a voltage zero crossing between the two base points of the short circuit, which has a particular effect on the peak short-circuit current and the direct current component, Sects. 5.4 and 5.5. In general, however, an insulation flashover should not occur at this time, but at a voltage value different from zero, so that this assumption leads to a result that is on the safe side.

In principle, various currents are used for the dimensioning of devices and installations, which must of course also be provided by a standard; these are the following currents in detail (Sect. 1.2):

- Initial symmetrical short-circuit current I''_{k}
- Peak short-circuit current i_{p},
- Short-circuit breaking current I_{b},
- Steady state short-circuit current I_{k},
- Thermal equivalent short-circuit current I_{th}.

The main principle of the standard for calculating the short-circuit currents is that the initial symmetrical short-circuit current is used as the calculation variable (input variable) for determining the other currents, using different factors shown in Fig. 3.3. The calculation steps for deriving the different short-circuit currents are summarized in Chap. 5.

Various faults can occur in a three-phase system, which are classified depending on the conductor affected by the short circuit. Figure 3.4 shows the various possible short circuits. The currents marked with a closed arrow are determined in the IEC 60909-0 [2]. Standard for short-circuit current calculation. Index 3 is not used in the illustration of the three-phase short-circuit if no confusion is possible. In Fig. 3.4, the short circuits are each entered "symmetrically" to phase R so that real transformation matrices of the components are created when using the symmetrical components to calculate the short-circuit currents (Sect. 9.2.2).

The standard [2] also covers the line-to-earth short-circuit current (Fig. 3.4d), but not the capacitive current to earth in an isolated or compensated network, as these

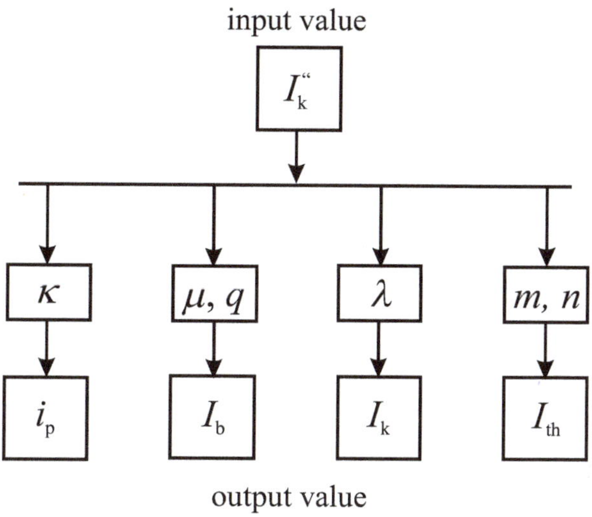

Fig. 3.3 Relationship between the various short-circuit currents

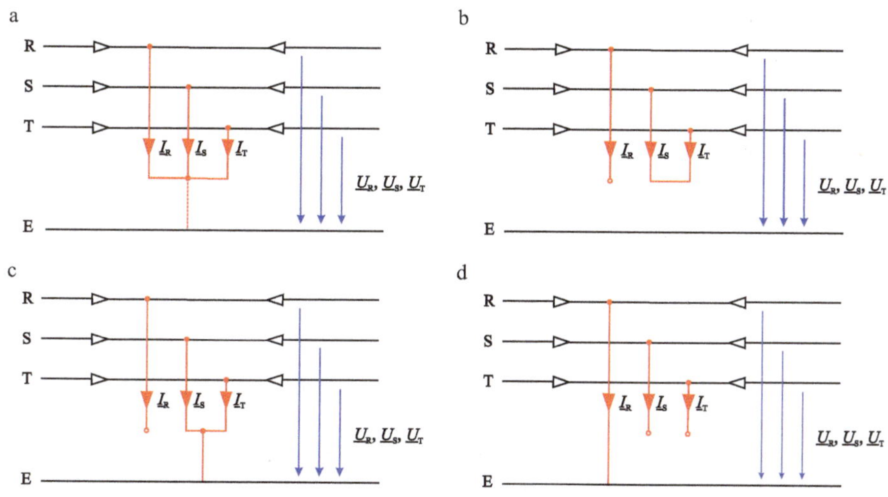

Fig. 3.4 Types of short circuit in the three-phase system. (**a**) three phase short circuit (with and without earth contact), (**b**) line-to-line short circuit, (**c**) line-to-line short circuit with earth contact, (**d**) line-to-earth short circuit, ➔ total current, ▷ partial current

currents are not relevant to the design. In this case, suitable models of the neutral system must be used.

Not included in the current main document of IEC 60909-0 [2] is the calculation of the double line-to-earth short-circuit current according to Fig. 3.5, to which a separate standard [5] is dedicated. The calculation of double line-to-earth short-circuit current is described in detail in Chap. 6.

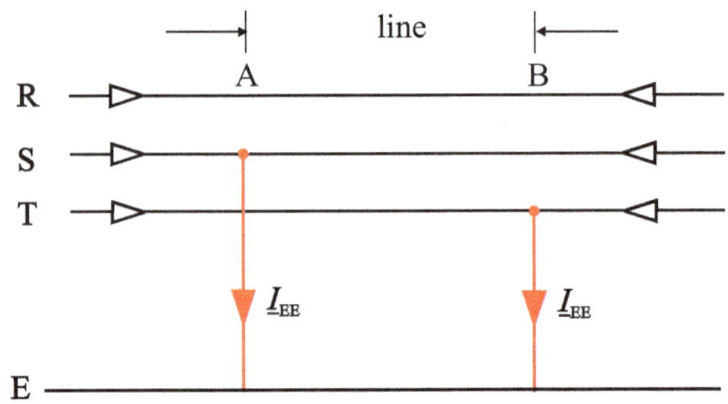

Fig. 3.5 Simultaneous double earth short circuit in two different conductors

Table 3.1 Selection of short-circuit currents during planning

Short-circuit currents	Equipment	Relevant currents k3	k2	k1
Maximum currents Stress/influence:				
– Dynamic	Components of installations	i_p	i_p	–
– Switching on	Switching devices	i_p	–	i_p
– Switching off	Switching devices	I_b	–	I_b
– Thermal	Components of installations, lines	I_{th}	–	I_{th}
– Inductive influence	Overhead lines			I_k''
– Earth voltage, touch voltage	Installation			I_k''
Minimum currents				
– Excitation, tripping	System protection	–	I_k'',I_k	I_k'',I_k

i_p peak short-circuit current, I_b short-circuit breaking current, I_{th} thermal equivalent short-circut current, I_k'' initial symmetrical short-circuit current, I_k steady state short-circuit, k3, k2, k1 short-circuit (three phase, line-to-line, line-to-earth)

When dimensioning and designing electrical installations, both the maximum and minimum short-circuit currents must be calculated. The maximum possible short-circuit current is important for the dimensioning of the components with regard to mechanical and thermal strength, while the minimum short-circuit current is important for the setting of the system protection. The planner of electrical installations and networks must therefore have suitable calculation methods that calculate the short-circuit currents with sufficient certainty.

Table 3.1 provides an overview of the currents and short-circuit types to be considered in detail. When determining the maximum short-circuit current, the peak short-circuit current i_p (Sect. 5.4), the short-circuit breaking current I_b (Sect. 5.6 and Chap. 10) and the thermal equivalent short-circuit current I_{th} (Sect. 5.8 and Chap. 11) are important. The current I_{th} is calculated from the initial symmetrical short-circuit current I_k'', the steady state short-circuit current I_k and the peak factor κ using the factors m (thermal effect of the direct current component) and n (thermal effect of the alternating current component). In addition, for unbalanced short circuits with earth contact (line-to-earth, double earth short circuit), the earth and touch voltages and the inductive influence of parallel lines must be taken into account.

In contrast, when setting the protective devices, the minimum initial symmetrical short-circuit current I''_{k} is important for excitation and the steady state short-circuit current I_{k} for tripping.

The short circuits shown in Fig. 3.4 can be calculated using the symmetrical components, see Chap. 9. The following section provides an overview of the calculation of the short-circuit current, taking into account the positive and zero sequence systems. It is assumed that the negative sequence impedance $\underline{Z}_{(2)}$ is equal to the positive sequence impedance $\underline{Z}_{(1)}$, which applies for the calculation of the initial symmetrical short-circuit current I''_{k}. The influence of full-size converters is not taken into account in this case.

- Line-to-earth short circuit, Fig. 3.4d:

$$\underline{I}''_{\mathrm{k1}} = \frac{\sqrt{3} \cdot cU_{\mathrm{n}}}{2 \cdot \underline{Z}_{(1)} + \underline{Z}_{(0)}} \tag{3.3}$$

- Line-to-line short circuit, without earth contact, Fig. 3.4b:

$$\underline{I}''_{\mathrm{k2}} = \frac{cU_{\mathrm{n}}}{2 \cdot \underline{Z}_{(1)}} \tag{3.4}$$

- Line-to-line short circuit, with earthh contact, part via earth, Fig. 3.4c:

$$\underline{I}''_{\mathrm{kE2E}} = \frac{\sqrt{3} \cdot c \cdot U_{\mathrm{n}}}{\underline{Z}_{(1)} + 2 \cdot \underline{Z}_{(0)}} \tag{3.5}$$

- Three phase short circuit, Fig. 3.4a:

$$\underline{I}''_{\mathrm{k3}} = \frac{cU_{\mathrm{n}}}{\sqrt{3} \cdot \underline{Z}_{(1)}} \tag{3.6}$$

When determining the maximum or minimum short-circuit current, it is not necessary to determine all currents first, but the following simplifications can be made using Eqs. (3.3)–(3.6), if different impedances $\underline{Z}_{(1)}$ and $\underline{Z}_{(2)}$ are not taken into account due to the control strategies of full-size converters.

- Maximum short-circuit current
  - $\underline{I}''_{\mathrm{k3}} > \underline{I}''_{\mathrm{k2}}$ because of $\underline{I}''_{\mathrm{k2}} = \frac{\sqrt{3}}{2} \cdot \underline{I}''_{\mathrm{k3}}$ (1)
  - $\underline{I}''_{\mathrm{k3}} > \underline{I}''_{\mathrm{k1}}$ if $\underline{Z}_{(0)} > \underline{Z}_{(1)}$ resp. (2)
  - $\underline{I}''_{\mathrm{k1}} > \underline{I}''_{\mathrm{k3}}$ if $\underline{Z}_{(0)} < \underline{Z}_{(1)}$ (3)

- Minimum short-circuit current
  - $\underline{I}''_{k2} < \underline{I}''_{k3}$ because of $\underline{I}''_{k2} = \frac{\sqrt{3}}{2} \cdot \underline{I}''_{k3}$ (1)
  - $\underline{I}''_{k3} < \underline{I}''_{k1}$ if $\underline{Z}_{(0)} < \underline{Z}_{(1)}$ resp. (3)
  - $\underline{I}''_{k1} < \underline{I}''_{k3}$ if $\underline{Z}_{(0)} > \underline{Z}_{(1)}$ (2)

1. The relationship results from Eqs. (3.4) and (3.6)
2. Applies to ratios $\underline{Z}_{(0)}/\underline{Z}_{(1)}$ of cables and overhead lines
3. In the vicinity of transformers Yd, Dy due to decoupling of the zero systems

The relationships illustrate that the line-to-line short-circuit current is always lower than the three-phase one. The distinction between three-phase and line-to-earth short-circuit current depends on the ratio $\underline{Z}_{(0)}/\underline{Z}_{(1)}$. In general, the zero sequence impedance of overhead lines and cables is greater than the positive sequence impedance, so that the following applies to extended networks

$$\underline{I}''_{k3} > \underline{I}''_{k1} \tag{3.7}$$

In contrast to this, however, the line-to-earth short-circuit current at the terminals of an infeed transformer with vector group Dy or Yd is generally greater. In this case, due to the neutral point earthing of the transformer, the flux caused by the zero sequence currents is compensated by the leakage flux caused by the windings in the delta winding of the transformer, which is designed for full transmission power. As a result, the resulting zero sequence impedance may be lower than the positive sequence impedance due to the relationship $X_{(0)T} \leq X_{(1)T}$ of the transformer. In addition, the zero sequence of the upstream network is not transferred. This should generally be the case for power stations due to vector group Yd of the step-up transformer.

3.4 Transient Short-Circuit Current Curve

A symmetrical or asymmetrical short-circuit current curve, e.g. for a three-phase short circuit, depends on the time at which the short circuit is initiated. In an exclusively inductive circuit, the maximum direct current component and therefore the greatest shift in the short-circuit current occurs, when the short-circuit takes place at zero voltage, as shown below.

If a short-circuit is calculated based on Fig. 3.6, the short-circuit current curve is generally determined according to Eq. (3.8). Here, the short-circuit impedance will be mainly inductive, since both transformers and lines etc. have an inductive behavior in the event of a short-circuit.

In the Laplace range, the impedance $Z(p) \approx p \cdot L$:

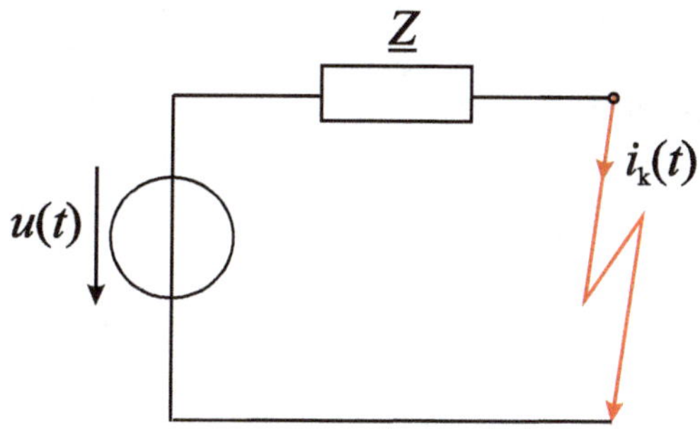

Fig. 3.6 Single line diagram for determining the short-circuit current using the short-circuit impedance $\underline{Z}$

$$I(p) = \frac{U(p)}{Z(p)} \tag{3.8}$$

The two extreme values apply to the voltage $U(p)$ at the time the short circuit occurs:

$$U(p) = \widehat{U} \cdot \frac{\omega}{p^2 + \omega^2} \quad \text{with} \quad u(t) = \widehat{U} \cdot \sin(\omega \cdot t) \tag{3.9}$$

$$U(p) = \widehat{U} \cdot \frac{p}{p^2 + \omega^2} \quad \text{with} \quad u(t) = \widehat{U} \cdot \cos(\omega \cdot t) \tag{3.10}$$

If it is assumed in a first step that the short-circuit occurs at voltage zero crossing $[t = 0 \rightarrow \sin(\omega \cdot t) = 0]$, then follows:

$$I(p) = \frac{U(p)}{Z(p)} = \frac{\omega \cdot \widehat{U}}{p^2 + \omega^2} \cdot \frac{1}{p \cdot L} \tag{3.11}$$

After the transformation, the following results for the short-circuit current curve or for the maximum value:

$$i_k(t) = \frac{\widehat{U}}{\omega L} \cdot [1 - \cos(\omega t)] \quad i_{kmax} = 2 \cdot \frac{\widehat{U}}{\omega L} \tag{3.12}$$

If, on the other hand, the short circuit occurs at the voltage maximum $[t = 0 \rightarrow \cos(\omega \cdot t) = 1]$, then follows:

$$I(p) = \frac{U(p)}{Z(p)} = \frac{\widehat{U}}{L} \cdot \frac{p}{p^2 + \omega^2} \tag{3.13}$$

After transformation, the following applies:

$$i_k(t) = \frac{\widehat{U}}{\omega L} \cdot \sin(\omega t) \qquad i_{kmax} = \frac{\widehat{U}}{\omega L} \tag{3.14}$$

As Eq. (3.12) shows, in the event of a short circuit at voltage zero crossing, a shift in the power frequency short-circuit current occurs, as both voltage and current are in phase at time $t = 0$, although a phase shift of 90° should occur as a result of the inductive circuit. This is compensated for by the direct current component that occurs. In contrast, the phase shift is always present when a short circuit occurs at the voltage maximum.

Due to the current displacement according to Eq. (3.12), a short-circuit that is initiated at voltage zero crossing leads to the maximum peak value, so that this boundary condition is assumed when determining the peak short-circuit current and is independent of the *R*/*X* ratio of the short-circuit impedance.

3.5 Near-to-Generator and Far-From-Generator Short Circuit

The changes in current over time during the short circuit depend on the electrical data of the synchronous machine (Sect. 4.2.1.2), the behavior of the voltage regulator, the type of fault, and possible grid changes (protective switch-offs).

In addition to the type of fault, a distinction is generally made between the short-circuit locations, so that short-circuits near-to-generator and far-from-the generator are defined. The following distinction applies here:

- Far-from-generator short circuit: In this case, the AC component remains constant during the short-circuit time, which means that the different, time-dependent short-circuit currents are identical, Eq. (3.15).

$$\underline{I}''_{\mathrm{k}} = \underline{I}_{\mathrm{b}} = \underline{I}_{\mathrm{k}} \tag{3.15}$$

Fig. 3.7 shows the typical current curve $i_{\mathrm{k}}(t)$ of a short-circuit far from the generator with a constant AC component over the entire time curve, when a short circuit occurs at voltage zero crossing.

- Near-to-generator short-circuit: In this case, the AC component decays during the short-circuit duration. This means that the RMS values of the short-circuit current become smaller, Eq. (3.16).

$$\underline{I}''_{\mathrm{k}} > \underline{I}_{\mathrm{b}} > \underline{I}_{\mathrm{k}} \tag{3.16}$$

Figure 3.8 shows the typical current curve $i_{\mathrm{k}}(\mathrm{t})$ of a near-to-generator short circuit with a decreasing AC component, when a short-circuit occurs at voltage zero crossing.

According to the IEC 60909-0 Standard [2], a short-circuit is defined as being near to the generator, if the short-circuit current contribution of a generator exceeds

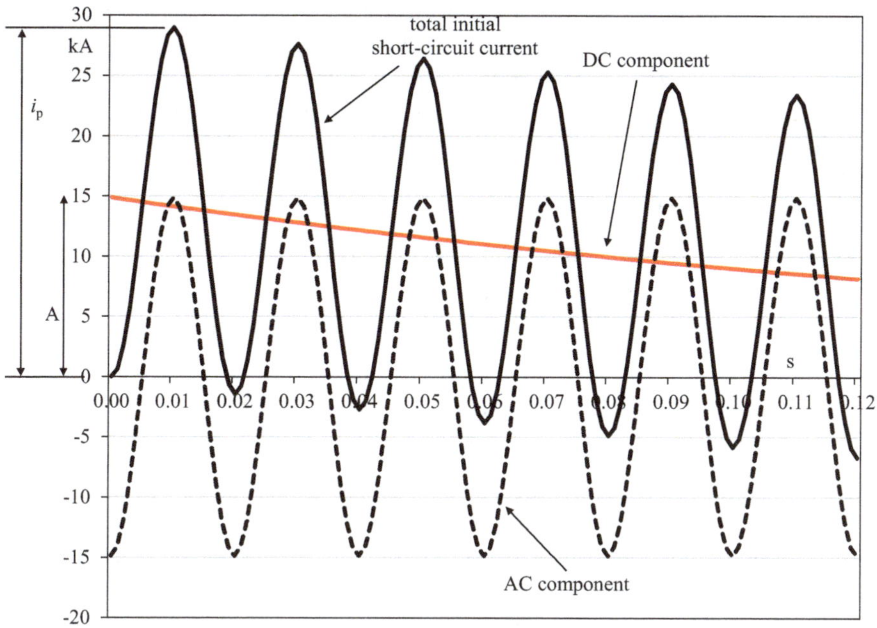

Fig. 3.7 Current curve for a short circuit far from the generator. *A* maximum DC component i_p peak short-circuit current

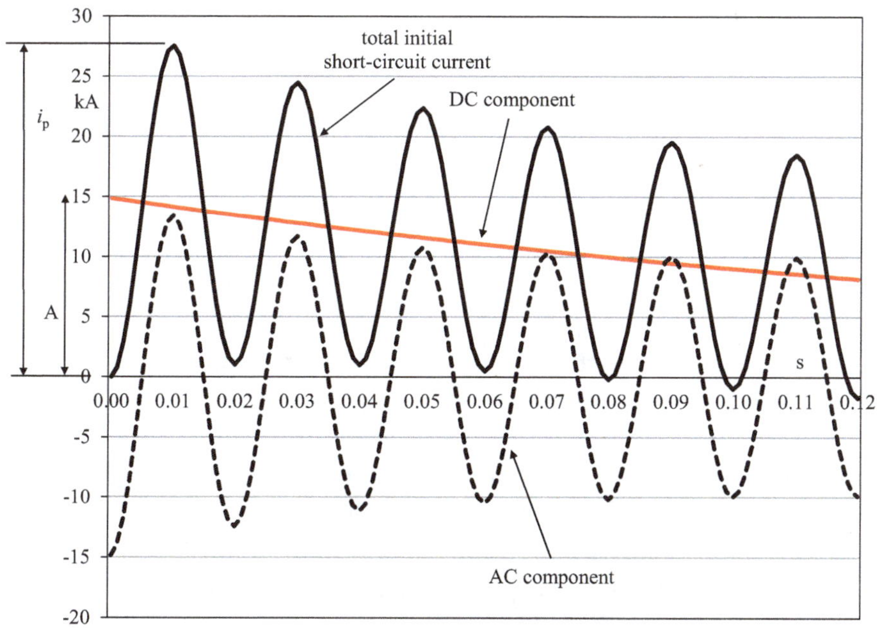

Fig. 3.8 Current curve for a short circuit near the generator with missing current zero crossings. *A* maximum DC component, i_p peak short-circuit current

twice its rated current. This can be taken from the μ-curves, when calculating the breaking short-circuit current, Sect. 5.6.1.

In the case of short circuits near to the generator, it is possible that the AC component decays faster than the associated DC component, which can occur particularly with large generators. In these cases, it is possible that the short-circuit current has missing zero crossings for some periods. The following boundary conditions influence the occurrence of missing zero crossings (Fig. 3.8):

- Load conditions before the fault occurs (inductive, capacitive, no-loaded),
- Short-circuit occurrence: Fault sequence,
- Fault type: Three-phase,
- Opening sequence of the breaker poles,
- Interruption time of the circuit-breaker,
- Neutral point treatment of the network,
- Arc resistance (fault location, switching chamber).

The calculation of missing current zero crossings must always be carried out using an appropriate three-phase calculation program, as missing current zero crossings only occur in one conductor in most cases. Further explanations can be found at ([5], Sect. 5.3).

3.6 Basic Methods for Determining Steady-State Short-Circuit Currents

Methods and boundary conditions are required to determine the short-circuit currents, which are explained in more detail in the following section, whereby only the options for calculating the steady-state short-circuit current are considered, in contrast to options that determine the flow of current over time.

This section also shows the influence of the load on the result, depending on the different calculation methods. In contrast to this, Sect. 3.7 also shows this influence on the calculation of the three-phase and line-to-earth short-circuit current.

3.6.1 *Calculation with the Real Distribution of Source Voltages*

In electrical networks, currents and voltages can always be determined using Kirchhoff's laws, which are characterized by the following relationships:

- At each node, the sum of all currents is zero,
- the sum of all voltages is zero for each mesh in a closed circuit.

The calculation with the real voltage distribution can be carried out both with and without a load impedance, which has a different influence on the short-circuit current and is considered in the following sections. In any case, the actual short-circuit currents or current distributions in the network are determined using this method.

3.6.1.1 With Load

According to Fig. 3.9, a short-circuit current $\underline{I}_k$ is fed in by two voltages $\underline{U}_1$ and $\underline{U}_2$. In addition, a load $\underline{Z}_B$ is taken into account in the network.

Taking into account the known voltages $\underline{U}_1$ and $\underline{U}_2$, the currents can be determined using the mesh and node equations so that Eq. (3.17) can be set up. In total, two meshes and three node equations are required, so that five unknown currents can be determined. As the example, Fig. 3.9 is easy to understand and shows a multiple, simple short-circuit location, the short-circuit current can of course also be derived directly from the diagram.

$$\begin{pmatrix} \underline{U}_1 \\ \underline{U}_2 \\ 0 \\ 0 \\ 0 \end{pmatrix} = \begin{pmatrix} \underline{Z}_1 & 0 & 0 & \underline{Z}_B & 0 \\ 0 & \underline{Z}_2 & 0 & 0 & 0 \\ 0 & 0 & \underline{Z}_3 & -\underline{Z}_B & 0 \\ -1 & 0 & 1 & 1 & 0 \\ 0 & 1 & 1 & 0 & -1 \end{pmatrix} \cdot \begin{pmatrix} \underline{I}_1 \\ \underline{I}_2 \\ \underline{I}_3 \\ \underline{I}_B \\ \underline{I}_k \end{pmatrix} \tag{3.17}$$

The various currents can be determined using Cramer's rule according to Eq. (3.18), so that the following applies:

$$\underline{I}_i = \frac{\mathbf{D}_i}{\mathbf{D}} \tag{3.18}$$

With

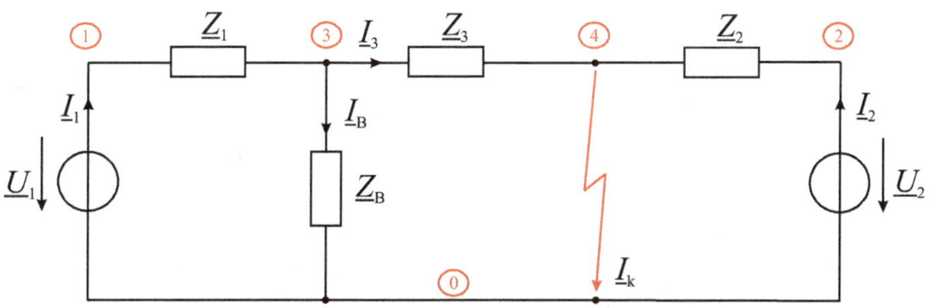

Fig. 3.9 Calculation of the short-circuit current $\underline{I}_k$ with the real source voltages, taking into account a load $\underline{Z}_B$

D result of the determinant according to Eq. (3.17),
$\mathbf{D}_i$ result of the partial determinant.

For the determinant **D** according to Eq. (3.17) applies by solving the determinant according to any rows or columns.

$$\mathbf{D} = \begin{vmatrix} \underline{Z}_1 & 0 & 0 & \underline{Z}_B & 0 \\ 0 & \underline{Z}_2 & 0 & 0 & 0 \\ 0 & 0 & \underline{Z}_3 & -\underline{Z}_B & 0 \\ -1 & 0 & 1 & 1 & 0 \\ 0 & 1 & 1 & 0 & -1 \end{vmatrix} = (-1) \cdot \begin{vmatrix} \underline{Z}_1 & 0 & 0 & \underline{Z}_B \\ 0 & \underline{Z}_2 & 0 & 0 \\ 0 & 0 & \underline{Z}_3 & -\underline{Z}_B \\ -1 & 0 & 1 & 1 \end{vmatrix} =$$
$$(-\underline{Z}_2) \cdot \begin{vmatrix} \underline{Z}_1 & 0 & \underline{Z}_B \\ 0 & \underline{Z}_3 & -\underline{Z}_B \\ -1 & 1 & 1 \end{vmatrix} = (-\underline{Z}_2 \cdot \underline{Z}_1) \cdot \begin{vmatrix} \underline{Z}_3 & -\underline{Z}_B \\ 1 & 1 \end{vmatrix} + \underline{Z}_2 \cdot \begin{vmatrix} 0 & \underline{Z}_B \\ \underline{Z}_3 & -\underline{Z}_B \end{vmatrix} =$$
$$-\underline{Z}_2 \cdot [\underline{Z}_1 \cdot (\underline{Z}_3 + \underline{Z}_B) + \underline{Z}_3 \cdot \underline{Z}_B] \tag{3.19}$$

To calculate the current $\underline{I}_1$, apply the partial determinant $\mathbf{D}_1$ by replacing the corresponding column (1) with the left vector Eq. (3.17):

$$\mathbf{D}_1 = \begin{vmatrix} \underline{U}_1 & 0 & 0 & \underline{Z}_B & 0 \\ \underline{U}_2 & \underline{Z}_2 & 0 & 0 & 0 \\ 0 & 0 & \underline{Z}_3 & -\underline{Z}_B & 0 \\ 0 & 0 & 1 & 1 & 0 \\ 0 & 1 & 1 & 0 & -1 \end{vmatrix} = (-1) \cdot \begin{vmatrix} \underline{U}_1 & 0 & 0 & \underline{Z}_B \\ \underline{U}_2 & \underline{Z}_2 & 0 & 0 \\ 0 & 0 & \underline{Z}_3 & -\underline{Z}_B \\ 0 & 0 & 1 & 1 \end{vmatrix} =$$
$$(-\underline{Z}_2) \cdot \begin{vmatrix} \underline{U}_1 & 0 & \underline{Z}_B \\ 0 & \underline{Z}_3 & -\underline{Z}_B \\ 0 & 1 & 1 \end{vmatrix} = (-\underline{Z}_2 \cdot \underline{U}_1) \cdot \begin{vmatrix} \underline{Z}_3 & -\underline{Z}_B \\ 1 & 1 \end{vmatrix} = -\underline{Z}_2 \cdot \underline{U}_1 \cdot (\underline{Z}_3 + \underline{Z}_B) \tag{3.20}$$

Taking into account the results from Eq. (3.19) and (3.20), the current $\underline{I}_1$ according to Eq. (3.18) is as follows:

$$\underline{I}_1 = \frac{-\underline{Z}_2 \bullet \underline{U}_1 \bullet (\underline{Z}_3 + \underline{Z}_B)}{-\underline{Z}_2 \bullet [\underline{Z}_1 \bullet (\underline{Z}_3 + \underline{Z}_B) + \underline{Z}_3 \bullet \underline{Z}_B]} = \frac{1}{\underline{Z}_1 + \frac{\underline{Z}_3 \bullet \underline{Z}_B}{\underline{Z}_3 + \underline{Z}_B}} \bullet \underline{U}_1 \tag{3.21}$$

The current $\underline{I}_2$ is determined directly from the second line of the system of Eq. (3.17) and is independent of the load impedance $\underline{Z}_B$.

$$\underline{I}_2 = \frac{1}{\underline{Z}_2} \bullet \underline{U}_2 \tag{3.22}$$

The current $\underline{I}_3$ is determined in the same way according to Eq. (3.18), resulting in:

$$\underline{I}_3 = \frac{-\underline{Z}_2 \bullet \underline{Z}_B \bullet \underline{U}_1}{-\underline{Z}_2 \bullet [\underline{Z}_1 \bullet (\underline{Z}_3 + \underline{Z}_B) + \underline{Z}_3 \bullet \underline{Z}_B]} = \frac{1}{\underline{Z}_1 + \underline{Z}_3 + \frac{\underline{Z}_1 \bullet \underline{Z}_3}{\underline{Z}_B}} \bullet \underline{U}_1 \tag{3.23}$$

According to Eq. (3.23), the short-circuit current contribution $\underline{I}_3$, which feeds directly into the fault location, is also dependent on the load impedance $\underline{Z}_B$, so that the following applies:

$$\underline{Z}_B \Rightarrow \infty \qquad \underline{I}_{3(\infty)} = \frac{1}{\underline{Z}_1 + \underline{Z}_3} \bullet \underline{U}_1 \tag{3.24}$$

$$\underline{Z}_B = 0 \qquad \underline{I}_{3(0)} = 0 \tag{3.25}$$

This means that the short-circuit impedance changes when a load is taken into account and therefore has an influence on the short-circuit current contribution. The short-circuit current $\underline{I}_k$ is calculated from the sum of the two currents $\underline{I}_2$ and $\underline{I}_3$, Eq. (3.17).

$$\underline{I}_k = \frac{1}{\underline{Z}_2} \bullet \underline{U}_2 + \frac{1}{\underline{Z}_1 + \underline{Z}_3 + \frac{\underline{Z}_1 \bullet \underline{Z}_3}{\underline{Z}_B}} \bullet \underline{U}_1 \tag{3.26}$$

With equal voltages $\underline{U}_1 = \underline{U}_2 = \underline{U}_0$ the result is:

$$\underline{I}_k = \left[\frac{1}{\underline{Z}_2} + \frac{1}{\underline{Z}_1 + \underline{Z}_3 + \frac{\underline{Z}_1 \bullet \underline{Z}_3}{\underline{Z}_B}}\right] \bullet \underline{U}_0 \tag{3.27}$$

In this case, the short-circuit current $\underline{I}_k$ is reduced by the load impedance $\underline{Z}_B$. This applies under the condition that the voltage at the connection node of the load is not kept constant, but decreases with the load.

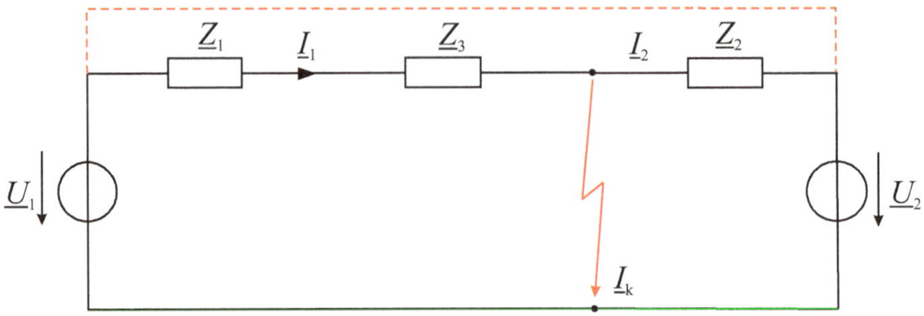

Fig. 3.10 Equivalent circuit diagram without load, dashed line: Equipotential connection

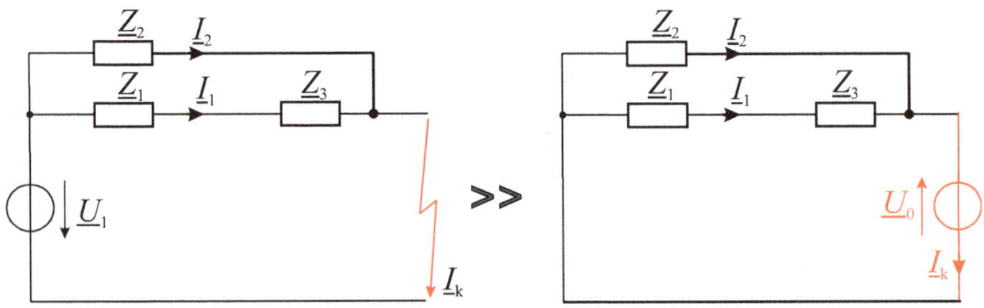

Fig. 3.11 Equivalent circuit diagram for the short-circuit current calculation with an active voltage. Left: summarized representation of the resulting voltages at the feed locations, right: arrangement of the resulting voltages at the short-circuit location

3.6.1.2 Without Load

If, in contrast to Sect. 3.6.1.1, the load is not taken into account in the short-circuit current calculation, the short-circuit current is calculated according to Eq. (3.28) by using $\underline{Z}_B \rightarrow \infty$. A 3×3 system of equations for the unknown currents $\underline{I}_2$, $\underline{I}_k$ and $\underline{I}_1 = \underline{I}_3$ can also be set up according to Eq. (3.17).

$$\underline{I}_k = \frac{1}{\underline{Z}_2} \cdot \underline{U}_2 + \frac{1}{\underline{Z}_1 + \underline{Z}_3} \cdot \underline{U}_1 \tag{3.28}$$

The resulting short-circuit current is made up of the two current components and depends essentially on the two voltages $\underline{U}_1$ and $\underline{U}_2$. If these voltages are identical, they can be combined as shown in Fig. 3.10, as the two network nodes are at the same potential, as indicated by the dashed connection.

The advantage for the short-circuit current calculation with the same voltage sources in this case is that only one system voltage source is used for the calculation, which can be used at the short-circuit location according to Fig. 3.11 with the opposite voltage direction in order to obtain the same current flow.

The voltage $\underline{U}_0$ depends on the voltage at the fault location before the short circuit occurs. It is therefore dependent on the voltages of the feeders, the network topology, the components and the load flow. This voltage is determined by introducing the voltage factor c in accordance with Chap. 7. The method of the equivalent voltage source at the short-circuit location can therefore be derived from the simplifications shown, Section 3.6.6.

3.6.2 Node Admittance Matrix

In principle, the admittance matrix $\underline{\mathbf{Y}}_{\mathrm{N}}$ of the network can be set up to calculate the short-circuit currents in order to determine the currents $\underline{I}_i$ using the impedance matrix $\underline{\mathbf{Z}}_{\mathrm{N}}$ and voltage distribution $\underline{U}_i$. A node with zero potential (earth) is defined as the reference node. The following applies in general:

$$\underline{\mathbf{I}}_i = \underline{\mathbf{Y}}_{\mathrm{N}} \cdot \underline{\mathbf{U}}_i \qquad \underline{\mathbf{U}}_i = \underline{\mathbf{Z}}_{\mathrm{N}} \cdot \underline{\mathbf{I}}_i \tag{3.29}$$

The following generally applies to the admittance matrix of a network:

$$\underline{\mathbf{Y}}_{\mathrm{N}} = \begin{pmatrix} \underline{Y}_{11} & \cdots & \underline{Y}_{1i} & \cdots & \underline{Y}_{1\mathrm{m}} \\ \vdots & \ddots & \vdots & \ddots & \vdots \\ \underline{Y}_{i1} & \cdots & \underline{Y}_{ii} & \cdots & \underline{Y}_{i\mathrm{m}} \\ \vdots & \ddots & \vdots & \ddots & \vdots \\ \underline{Y}_{\mathrm{m}1} & \cdots & \underline{Y}_{\mathrm{m}i} & \ddots & \underline{Y}_{\mathrm{mm}} \end{pmatrix} \tag{3.30}$$

The symmetrical admittance matrix according to Eq. (3.30) is set up according to the following rules:

- The elements of the main diagonal ii consist of the sum of the admittances $\underline{Y}_{ij}$ of all nodes j that are connected to node i.
- The elements of the secondary diagonals ij consist of the negative sum of the admittances between the two nodes i and j.

For the example shown in Fig. 3.9, the following complete admittance matrix can be derived, Eq. (3.31), corresponding to the node numbering from 0 to 4, where node 0 is the reference node:

$$\mathbf{Y}_{\mathrm{N}} = \begin{pmatrix} \underline{Y}_1 & 0 & -\underline{Y}_1 & 0 \\ 0 & \underline{Y}_2 & 0 & -\underline{Y}_2 \\ -\underline{Y}_1 & 0 & \underline{Y}_1 + \underline{Y}_{\mathrm{B}} + \underline{Y}_3 & -\underline{Y}_3 \\ 0 & -\underline{Y}_2 & -\underline{Y}_3 & \underline{Y}_2 + \underline{Y}_3 \end{pmatrix} \tag{3.31}$$

All voltages and currents, including the short-circuit current $\underline{I}_{\mathrm{k}}$, can be determined, as the voltages $\underline{U}_1$ and $\underline{U}_2$ are known according to Eq. (3.32) and the other variables ($\underline{U}_3$, $\underline{I}_1$, $\underline{I}_2$ and $\underline{I}_{\mathrm{k}}$) can be determined using the four equations. The currents of the feeders are counted as positive, if they feed into the node, otherwise they are counted as negative. The result for the short-circuit current corresponds to the result of Eq. (3.26). All branch currents between neighboring nodes can be determined from the node voltages to earth.

$$\begin{pmatrix} \underline{I}_1 \\ \underline{I}_2 \\ 0 \\ -\underline{I}_\mathrm{k} \end{pmatrix} = \begin{pmatrix} \underline{Y}_1 & 0 & -\underline{Y}_1 & 0 \\ 0 & \underline{Y}_2 & 0 & -\underline{Y}_2 \\ -\underline{Y}_1 & 0 & \underline{Y}_1 + \underline{Y}_\mathrm{B} + \underline{Y}_3 & -\underline{Y}_3 \\ 0 & -\underline{Y}_2 & -\underline{Y}_3 & \underline{Y}_2 + \underline{Y}_3 \end{pmatrix} \cdot \begin{pmatrix} \underline{U}_1 \\ \underline{U}_2 \\ \underline{U}_3 \\ 0 \end{pmatrix} \tag{3.32}$$

According to Fig. 3.9, voltage sources are provided as feed-in sources, resulting in a number of four nodes (without reference nodes). If these voltage sources are converted into equivalent current sources, the number of nodes is reduced to two.

3.6.3 Meshed Impedanz Matrix

In addition to the node admittance method, the mesh impedance method can also be used, so that the following applies accordingly:

$$\underline{\mathbf{U}}_i = \underline{\mathbf{Z}}_\mathrm{N} \cdot \underline{\mathbf{I}}_i \tag{3.33}$$

Eq (3.34) generally applies to the impedance matrix of a network.

$$\underline{\mathbf{Z}}_\mathrm{N} = \begin{pmatrix} \underline{Z}_{11} & \cdots & \underline{Z}_{1i} & \cdots & \underline{Z}_{1\mathrm{m}} \\ \vdots & \ddots & \vdots & \ddots & \vdots \\ \underline{Z}_{i1} & \cdots & \underline{Z}_{ii} & \cdots & \underline{Z}_{i\mathrm{m}} \\ \vdots & \ddots & \vdots & \ddots & \vdots \\ \underline{Z}_{\mathrm{m}1} & \cdots & \underline{Z}_{\mathrm{m}i} & \ddots & \underline{Z}_{\mathrm{mm}} \end{pmatrix} \tag{3.34}$$

The impedance matrix according to Eq. (3.34) can be derived, for example, by inversion of the admittance matrix, using the mesh method, graph theory or open circuit measurements [6].

The following shows how to set up the $\underline{\mathbf{Z}}$ -matrix using mesh analysis. According to Fig. 3.12, the example shown in Fig. 3.9 is examined. Three loops I, II and III with independent mesh currents $\underline{I}_1$, $\underline{I}_2$ and $\underline{I}_3$ can be defined. If a branch has two mesh currents, the algebraic sum must be used in each case. To set up the equations, a direction of rotation must be defined for each mesh, e.g. clockwise.

The following equations result for the three mesh circuits with currents $\underline{I}_1$, $\underline{I}_2$ and $\underline{I}_3$.

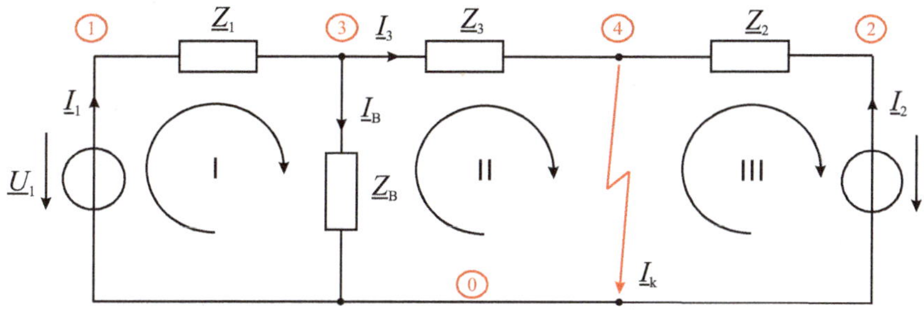

Fig. 3.12 Equivalent circuit for determining the impedance matrix using mesh analysis

$$\begin{aligned}\underline{U}_1 &= \underline{Z}_1 \cdot \underline{I}_1 + \underline{Z}_B \cdot (\underline{I}_1 - \underline{I}_3)\\ 0 &= \underline{Z}_3 \cdot \underline{I}_3 - \underline{Z}_B \cdot (\underline{I}_1 - \underline{I}_3)\\ \underline{U}_2 &= \underline{Z}_2 \cdot \underline{I}_2\end{aligned} \tag{3.35}$$

The system of Eq. (3.36) can also be set up directly, if it is constructed using the following rules:

- Definition of the meshes and determination of the independent currents.
- The elements of the main diagonals *ii* consist of the sum of the impedances of a mesh.
- The elements of the secondary diagonals *ij* form the coupling impedances between two meshes. The sign is positive if the two mesh currents in the coupling impedance have the same sign.
- The voltage sources of the meshes appear as a voltage vector, whereby the sign is negative, if it is aligned with the sign of the mesh current.

$$\begin{pmatrix}\underline{U}_1\\ \underline{U}_2\\ 0\end{pmatrix} = \begin{pmatrix}\underline{Z}_1 + \underline{Z}_B & 0 & -\underline{Z}_B\\ 0 & \underline{Z}_2 & 0\\ -\underline{Z}_B & 0 & \underline{Z}_3 + \underline{Z}_B\end{pmatrix} \cdot \begin{pmatrix}\underline{I}_1\\ \underline{I}_2\\ \underline{I}_3\end{pmatrix} \tag{3.36}$$

With the current constraint $\underline{I}_k = \underline{I}_3 + \underline{I}_2$ at node 4, the following results from Eq. (3.35):

$$\begin{pmatrix}\underline{U}_1\\ \underline{U}_2\\ 0\end{pmatrix} = \begin{pmatrix}\underline{Z}_1 + \underline{Z}_B & \underline{Z}_B & -\underline{Z}_B\\ 0 & \underline{Z}_2 & 0\\ -\underline{Z}_B & -(\underline{Z}_3 + \underline{Z}_B) & \underline{Z}_3 + \underline{Z}_B\end{pmatrix} \cdot \begin{pmatrix}\underline{I}_1\\ \underline{I}_2\\ \underline{I}_k\end{pmatrix} \tag{3.37}$$

The short-circuit current $\underline{I}_k$ can be determined according to Fig. 3.12 based on this system of equations, e.g. using Cramer's rule (Sect. 3.6.1.1), resulting in:

$$\underline{I}_k = \frac{\underline{U}_2}{\underline{Z}_2} + \frac{\underline{Z}_B}{\underline{Z}_1 \cdot (\underline{Z}_3 + \underline{Z}_B) + \underline{Z}_B \cdot \underline{Z}_3} \cdot \underline{U}_1 \tag{3.38}$$

The result of Eq. (3.38) corresponds to the derivation according to Eq. (3.26).

3.6.4 Superposition Method

If components in the network has a linear behavior in the event of a short circuit, the current in any branch can be determined in such a way that *n* partial currents are determined, which are caused by *n* feeders, whereby (*n*-1) voltage sources are short-circuited or, in the case of current sources, interrupted. The total current results by superimposing the individual partial currents.

3.6.4.1 General Application of the Superposition Method

As an example, the superposition method for the network circuit shown in Fig. 3.9 is applied in two steps by assuming one voltage source as the active voltage source, while the other is short-circuited. This results in the two equivalent circuits shown in Fig. 3.13.

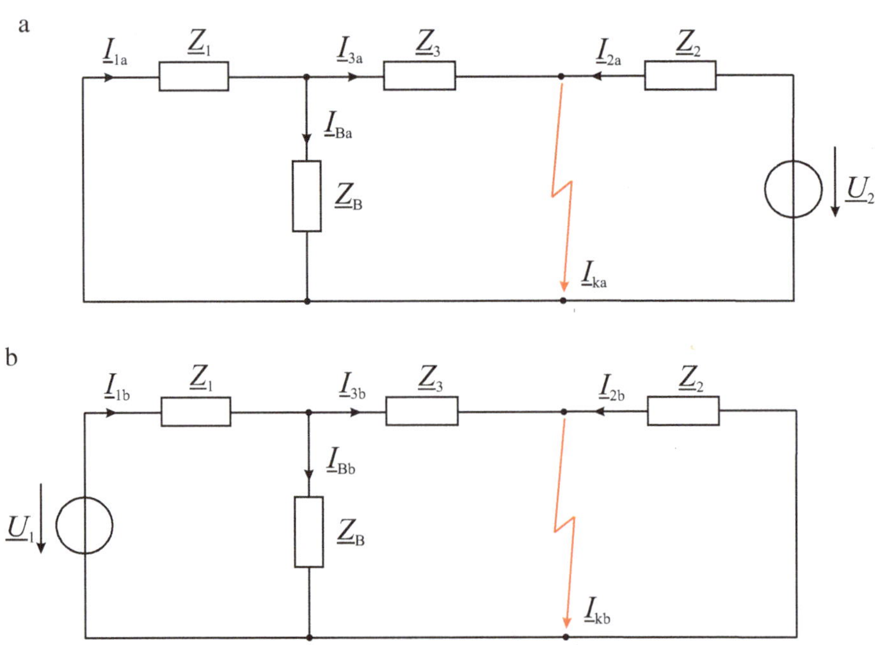

Fig. 3.13 Equivalent circuit diagrams for determining the short-circuit current using the superposition method, (**a**) voltage $\underline{U}_1$ short-circuited, (**b**) voltage $\underline{U}_2$ short-circuited

The currents in Fig. 3.13 can be derived directly, resulting in the following values:

$$\underline{I}_{ka} = \underline{I}_{2a} = \frac{\underline{U}_2}{\underline{Z}_2} \qquad \underline{I}_{1a} = \underline{I}_{3a} = \underline{I}_{Ba} = 0 \tag{3.39}$$

$$\underline{I}_{kb} = \underline{I}_{3b} = \frac{\underline{Z}_B}{\underline{Z}_1 \cdot (\underline{Z}_3 + \underline{Z}_B) + \underline{Z}_B \cdot \underline{Z}_3} \cdot \underline{U}_1 \tag{3.40}$$

$$\underline{I}_{1b} = \frac{\underline{U}_1}{\underline{Z}_1 + \dfrac{\underline{Z}_B \cdot \underline{Z}_3}{\underline{Z}_3 + \underline{Z}_B}} \qquad \underline{I}_{2b} = 0 \tag{3.41}$$

$$\underline{I}_{Bb} = \frac{\underline{Z}_3}{\underline{Z}_1 \cdot (\underline{Z}_3 + \underline{Z}_B) + \underline{Z}_B \cdot \underline{Z}_3} \cdot \underline{U}_1 \tag{3.42}$$

During the short circuit, the resulting currents are obtained by superimposing the various partial currents $\underline{I}_{ka}$ and $\underline{I}_{kb}$, resulting in the short-circuit current:

$$\underline{I}_k = \underline{I}_{ka} + \underline{I}_{kb} = \frac{\underline{Z}_B}{\underline{Z}_1 \cdot (\underline{Z}_3 + \underline{Z}_B) + \underline{Z}_B \cdot \underline{Z}_3} \cdot \underline{U}_1 + \frac{\underline{U}_2}{\underline{Z}_2} \tag{3.43}$$

3.6.4.2 Using the Superposition Method to Derive the Equivalent Voltage Source at the Short-Circuit Location

In the following, the superposition method is applied by first deriving a resulting voltage source at the final short-circuit location, which is the result of the various source voltages and load conditions in a network [7, 8]. In a second step, this voltage is used to determine a current distribution that is to be superimposed on the original load flow in order to obtain the final short-circuit current.

If it is assumed that the load and voltage conditions are known before the short circuit occurs, the exact values of the short-circuit current at the short-circuit location and in the branches can be determined using the superposition method. This is performed in three steps according to Fig. 3.14, as already shown above, which is exemplified by the network circuit according to Fig. 3.14a.

- 1. Step
  In the first step, the voltage U_{bk} at the short-circuit location, the current distribution I_b and the internal voltages of the active elements in the network before the short-circuit occurs are determined (load flow calculation, Fig. 3.14b). For the calculation, the load condition before the short-circuit occurs, i.e. the node loads and feeders, the impedances in the network and the subtransient reactances of the

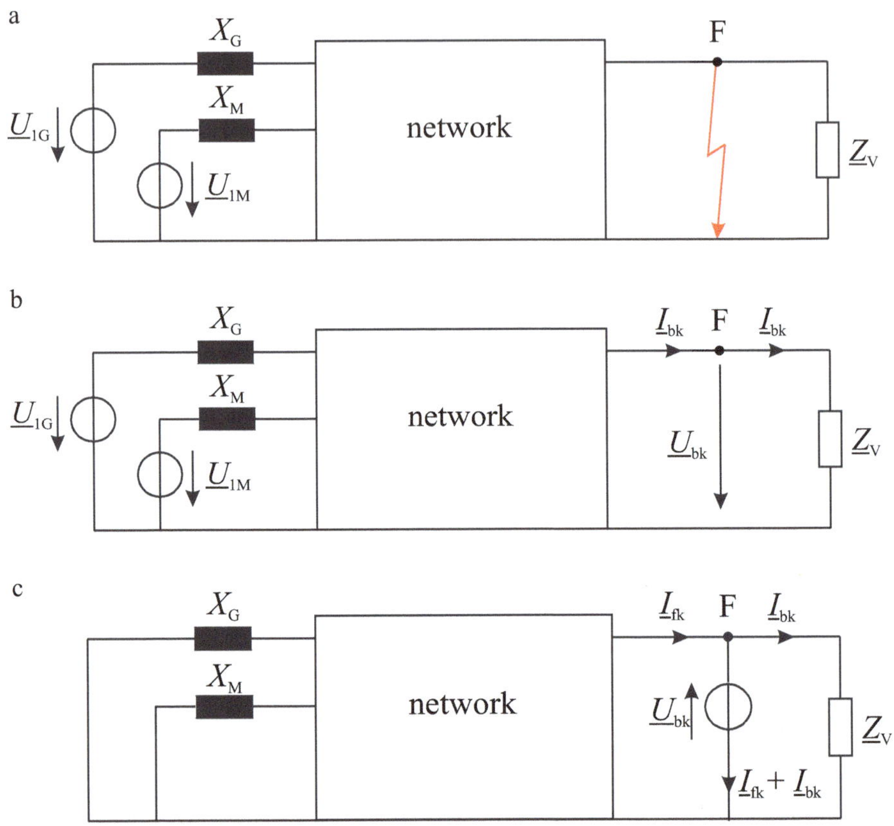

Fig. 3.14 Equivalent circuit for deriving the superposition method, (**a**) network circuit diagram in the positive sequence system, (**b**) determination of the operating voltage $\underline{U}_{bk}$ at the short-circuit location F, (**c**) determination of the current distribution using the equivalent voltage source $\underline{U}_{bk}$; X_G, X_M reactances of the active elements (generator, motor), $\underline{Z}_V$ load impedance

active elements are taken into account, as well as the actual tap position of the transformers.

- 2. Step
  According to Fig. 3.14c, the currents $\underline{I}_f$ and voltages $\underline{U}_f$ in the network are calculated using an equivalent voltage source $\underline{U}_{fk} = -\underline{U}_{bk}$ at the short-circuit location, whereby this voltage source is the only voltage source in the network. The equivalent voltage U_{fk} has the opposite polarity so that the voltage at the fault location is zero after superposition. The internal voltages of the active elements are short-circuited.
- 3. Step
  A superposition of the conditions Fig. 3.14b, c corresponds to the current distribution in the network after the short circuit has occurred. The voltage at the short-circuit location is zero according to Eq. (3.44).

$$\underline{U}_{\mathrm{fk}} = \underline{U}_{\mathrm{bk}} + (-\underline{U}_{\mathrm{bk}}) = 0 \tag{3.44}$$

For the load branch at the fault location, the superposition of the currents also results in the value zero (the load impedance is short-circuited). The resulting current at the short-circuit location is calculated as follows

$$\underline{I}''_{\mathrm{k}} = 0 + (\underline{I}_{\mathrm{fk}} + \underline{I}_{\mathrm{bk}}) \tag{3.45}$$

The disadvantage of this method is that the voltage U_{bk} at the short-circuit location, before the short-circuit occurs, and thus also the calculated short-circuit current, is strongly dependent on the relevant load condition and the "most unfavorable" load flow must first be found, which, for example, leads to the maximum short-circuit current in the calculation. Sect. 3.6.6 therefore explains a method with which the maximum and minimum short-circuit current can be calculated independently of the relevant load condition, if certain boundary conditions are met. The results should be on the safe side with this method, which is called the equivalent voltage source at the short-circuit location.

While in this section the superposition is carried out using voltage sources, this method can of course also be used with current sources, as shown, for example, in the application of converters that feed-in a constant current during the short-circuit phase, Chap. 13. An example of the combination of voltage and current sources is shown in Sect. 3.6.7. The procedure described above is illustrated using the example in Fig. 3.9.

- 1. Step: Feed-in of $\underline{U}_2$ and short-circuit of voltage $\underline{U}_1$.
  The current $\underline{I}_{\mathrm{k(I)}}$ is calculated as follows:

$$\underline{I}_{\mathrm{k(I)}} = \frac{\underline{U}_2}{\underline{Z}_2} \tag{3.46}$$

- 2. Step: Feed-in of $\underline{U}_1$ and short-circuit of voltage $\underline{U}_2$.
  The current $\underline{I}_{\mathrm{k(II)}}$ is calculated as follows:

$$\underline{I}_{\mathrm{k(II)}} = \frac{1}{\underline{Z}_1 + \underline{Z}_3 + \dfrac{\underline{Z}_1 \cdot \underline{Z}_3}{\underline{Z}_{\mathrm{B}}}} \cdot \underline{U}_1 \tag{3.47}$$

- 3. Step: Superposition of the two partial currents $\underline{I}_{\mathrm{k(I)}}$ and $\underline{I}_{\mathrm{k(II)}}$ to get the resulting short-circuit current.

$$\underline{I}_{\mathrm{k}} = \underline{I}_{\mathrm{k(I)}} + \underline{I}_{\mathrm{k(II)}} = \frac{\underline{U}_2}{\underline{Z}_2} + \frac{\underline{U}_1}{\underline{Z}_1 + \underline{Z}_3 + \dfrac{\underline{Z}_1 \cdot \underline{Z}_3}{\underline{Z}_{\mathrm{B}}}} \tag{3.48}$$

The result according to Eq. (3.48) corresponds to the result determined using Kirchhoff's laws according to Eq. (3.26), so that these methods are equivalent.

The superposition method is therefore an exact method for determining the short-circuit currents. In principle, the calculation according to the superposition method will lead to results that are better adapted to the actual conditions, so that the application of this method is covered by the IEC Standard [2]. However, it has the disadvantage that the switching conditions and the voltage conditions must be known before the short circuit occurs. This also applies in particular to the tap-position of the transformers, which are taken into account in the calculation according to IEC in the middle position. However, the most unfavorable load condition of the network and thus the voltage conditions are often not known before the short circuit occurs, especially for the system design in the planning phase.

3.6.5 Equivalent Voltage Source, Equivalent Current Source

In principle, it is possible to separate any network and carry out a network reduction. In this case, the relationship $\underline{U}(\underline{I})$ at the terminals must be linear, if a network consists of linear elements. This linear relationship can be determined by carrying out two tests at the terminals:

- Open circuit test and
- short-circuit test.

From these tests, the necessary quantities of an equivalent circuit can be determined, namely:

- Equivalent voltage source: Open-circuit voltage $\underline{U}_0$ and internal impedance $\underline{Z}_i$.
- Equivalent current source: Source current $\underline{I}_0$ and internal impedance $\underline{Z}_i$

Figure 3.15 shows an example of the conversion of a simple network circuit consisting of a voltage source $\underline{U}_1$ and two impedances $\underline{Z}_1$ and $\underline{Z}_2$ into an equivalent voltage or current source.

In Fig. 3.15 (left), a single voltage $\underline{U}_1$ is assumed for simplification. In the case of extended networks with different feeders and thus voltages, there is a resulting open-circuit voltage at the output terminals, which results from the different voltages $\underline{U}_1$ … $\underline{U}_n$ of the feeders and the impedances of the network.

The two tests described above provide the following results for the network circuit shown in Fig. 3.15 (left):

$$\text{Open-circuit } (I_2 = 0), \text{index } (0): \quad \underline{U}_{2(0)} = \frac{\underline{Z}_2}{\underline{Z}_1 + \underline{Z}_2} \cdot \underline{U}_1 \tag{3.49}$$

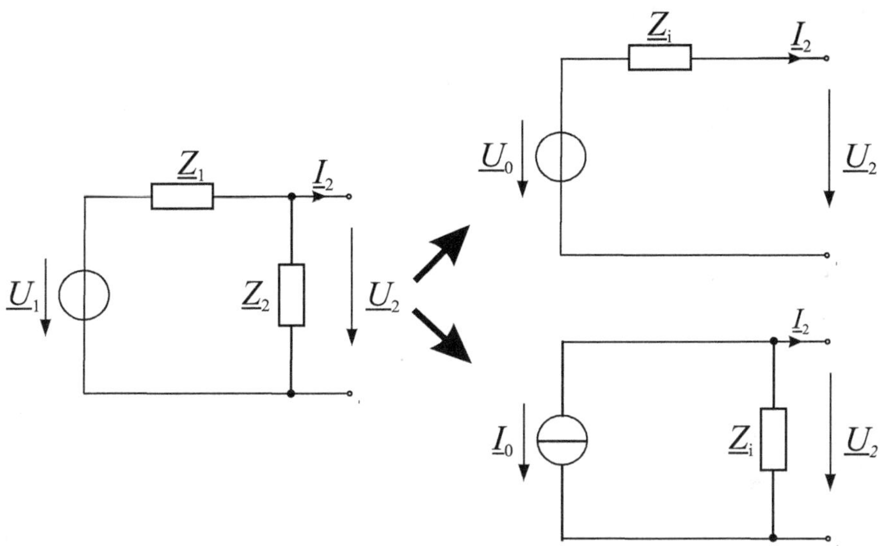

Fig. 3.15 Example of a conversion to a voltage or current source

$$\text{Short-circuit } (U_2 = 0), \text{index } (k) : \underline{I}_{2(k)} = \frac{\underline{U}_1}{\underline{Z}_1} \tag{3.50}$$

From this, the following characteristic values of the equivalent circuits according to Fig. 3.15 (right) can be derived so that the behavior at the terminals is identical:

- Equivalent voltage source:
  Open-circuit voltage:

$$\underline{U}_0 = \underline{U}_{2(0)} = \frac{\underline{Z}_2}{\underline{Z}_1 + \underline{Z}_2} \cdot \underline{U}_1 \tag{3.51}$$

Internal impedance:

$$\underline{Z}_\text{i} = \frac{\underline{U}_0}{\underline{I}_{2(\text{k})}} = \frac{\underline{Z}_1 \cdot \underline{Z}_2}{\underline{Z}_1 + \underline{Z}_2} \tag{3.52}$$

- Equivalent current sourve:
  Source current:

$$\underline{I}_0 = \underline{I}_{2(\text{k})} = \frac{\underline{U}_1}{\underline{Z}_1} \tag{3.53}$$

Internal impedance:

$$\underline{Z}_\text{i} = \frac{\underline{U}_{2(0)}}{\underline{I}_0} = \frac{\underline{Z}_1 \cdot \underline{Z}_2}{\underline{Z}_1 + \underline{Z}_2} \tag{3.54}$$

The internal impedance $\underline{Z}_i$ can also be determined directly by the value of the impedance that results at the terminals of the original network, when the voltage $\underline{U}_1$ is short-circuited. This impedance is identical for both equivalent circuits. Using the voltage equation $\underline{U}_0 = \underline{Z}_i \cdot \underline{I}_0$, the voltage source can be converted into a current source.

The principle of the equivalent voltage source is used, when calculating short-circuit currents in accordance with IEC [2].

3.6.6 Procedure Using the Equivalent Voltage Source at the Short-Circuit Location

The calculation of the short-circuit currents in accordance with IEC [2] is carried out using the "equivalent voltage source at the short-circuit location" method. This method is derived from the superposition method (Sect. 3.6.4), whereby the load condition of the network is taken into account approximately via the voltage factor c (equivalent EMF method, [3]).

According to Fig. 3.16, the internal voltages in the network are short-circuited (from generators and motors) and all shunt admittances are disregarded. The single voltage source $c \cdot U_n/\sqrt{3}$ is used at the short-circuit location; this corresponds to step 3 according to Sect. 3.6.4 or Fig. 3.14c).

From the linear system of equations (admittance matrix), Eq. (3.30), the corresponding impedance matrix can be determined by inversion, Eq. (3.55).

$$\underline{\mathbf{Z}}_N = \begin{pmatrix} \underline{Z}_{11} & \cdots & \underline{Z}_{1i} & \cdots & \underline{Z}_{1m} \\ \vdots & \ddots & \vdots & \ddots & \vdots \\ \underline{Z}_{i1} & \cdots & \underline{Z}_{ii} & \cdots & \underline{Z}_{im} \\ \vdots & \ddots & \vdots & \ddots & \vdots \\ \underline{Z}_{m1} & \cdots & \underline{Z}_{mi} & \ddots & \underline{Z}_{mm} \end{pmatrix} \tag{3.55}$$

If the method of the equivalent voltage source is used at the short-circuit location F, the following applies to the voltage at node i (at the short-circuit location):

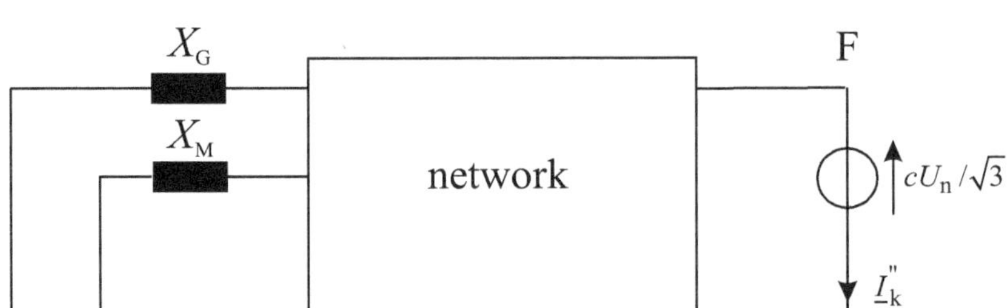

Fig. 3.16 Equivalent circuit diagram in the positive sequence system for calculating of the short-circuit current $\underline{I}_k$ using the equivalent voltage source method at the short-circuit location F

$$\underline{U}_1^* = -c \cdot U_\mathrm{n}/\sqrt{3} \tag{3.56}$$

The node voltages to earth are marked with an asterisk (*) to distinguish these voltages from the actual node voltages during the short circuit. The voltages are counted as positive if they point from a node to a reference node. This also applies to currents that feed into a node (Sect. 3.6.2). As this does not apply to the equivalent voltage and the short-circuit current as shown in Fig. 3.16, these are given a negative sign.

Using the impedance matrix, it is possible to calculate the short-circuit current at the short-circuit location directly. Here, the equivalent voltage at node i is used according to Eq. (3.57), while all node currents have the value $\underline{I}_i = 0$, except for the fault current $\underline{I}''_{\mathrm{k}i}$. The following applies in general:

$$\begin{pmatrix} \underline{U}_1^* \\ \vdots \\ -c \cdot U_\mathrm{n}/\sqrt{3} \\ \vdots \\ \underline{U}_m^* \end{pmatrix} = \begin{pmatrix} \underline{Z}_{11} & \cdots & \underline{Z}_{1i} & \cdots & \underline{Z}_{1\mathrm{m}} \\ \vdots & \ddots & \vdots & \ddots & \vdots \\ \underline{Z}_{i1} & \cdots & \underline{Z}_{ii} & \cdots & \underline{Z}_{i\mathrm{m}} \\ \vdots & \ddots & \vdots & \ddots & \vdots \\ \underline{Z}_{\mathrm{m}1} & \cdots & \underline{Z}_{\mathrm{m}i} & \ddots & \underline{Z}_{\mathrm{mm}} \end{pmatrix} \cdot \begin{pmatrix} 0 \\ \vdots \\ -\underline{I}''_{\mathrm{k}i} \\ \vdots \\ 0 \end{pmatrix} \tag{3.57}$$

From Eq. (3.57), the short-circuit current $\underline{I}''_{\mathrm{k}i}$ results:

$$c \cdot U_\mathrm{n}/\sqrt{3} = \underline{Z}_{ii} \cdot \underline{I}''_{\mathrm{k}i} \qquad \underline{I}''_{\mathrm{k}i} = \frac{c \cdot U_\mathrm{n}/\sqrt{3}}{\underline{Z}_{ii}} \tag{3.58}$$

All node voltages against reference earth can be derived from the system of Eq. (3.57). When determining the node voltages, it is assumed that there is no-load flow before the short circuit occurs and that the voltages of all feeders correspond to the voltage value of the equivalent voltage source at the fault location. The following applies to the voltage $\underline{U}_1$ at grid node 1 in the event of a short circuit at node i:

$$\underline{U}_1^* = -\underline{Z}_{1i} \cdot \underline{I}''_{\mathrm{k}i} = -\frac{\underline{Z}_{1i}}{\underline{Z}_{ii}} \cdot c \cdot U_\mathrm{n}/\sqrt{3} \tag{3.59}$$

It should be noted that the voltage according to Eq. (3.59) results from the changed voltage profile due to the equivalent voltage source at the fault location. The actual voltages $\underline{U}_1$ during the short circuit result in:

$$\underline{U}_1 = c \cdot U_\mathrm{n}/\sqrt{3} - \frac{\underline{Z}_{1i}}{\underline{Z}_{ii}} \cdot c \cdot U_\mathrm{n}/\sqrt{3} = \left(1 - \frac{\underline{Z}_{1i}}{\underline{Z}_{ii}}\right) \cdot c \cdot U_\mathrm{n}/\sqrt{3} \tag{3.60}$$

If the method of the equivalent voltage source at the short-circuit location is used for the example in Fig. 3.17, the admittance $\underline{Y}_B$ of the load is taken into account in a first step due to comparability, although this is not permitted for the method with the equivalent voltage source at the short-circuit location. This results in two nodes 1, 2 and the reference node 0, to which the node voltages are related.

For Fig. 3.17, the following admittance matrix results according to Eq. (3.61), which can also be derived from Eq. (3.31) by deleting nodes 1 and 2.

$$\begin{pmatrix} 0 \\ -\underline{I}_k \end{pmatrix} = \begin{pmatrix} \underline{Y}_1 + \underline{Y}_B + \underline{Y}_3 & -\underline{Y}_3 \\ -\underline{Y}_3 & \underline{Y}_2 + \underline{Y}_3 \end{pmatrix} \cdot \begin{pmatrix} \underline{U}_1^* \\ -\underline{U}_0 \end{pmatrix} \tag{3.61}$$

This results in the short-circuit current:

$$\underline{I}_k = \underline{Y}_3 \cdot \underline{U}_1^* + (\underline{Y}_2 + \underline{Y}_3) \cdot \underline{U}_0 \tag{3.62}$$

With:

$$\underline{U}_1^* = -\frac{\underline{Y}_3}{\underline{Y}_1 + \underline{Y}_B + \underline{Y}_3} \cdot \underline{U}_0 \tag{3.63}$$

$$\underline{I}_k = \left\{ (\underline{Y}_2 + \underline{Y}_3) - \frac{\underline{Y}_3^2}{\underline{Y}_1 + \underline{Y}_B + \underline{Y}_3} \right\} \cdot \underline{U}_0 = \left\{ \underline{Y}_2 + \frac{(\underline{Y}_1 + \underline{Y}_B) \cdot \underline{Y}_3}{\underline{Y}_1 + \underline{Y}_B + \underline{Y}_3} \right\} \cdot \underline{U}_0 \tag{3.64}$$

$$\underline{I}_k = \left\{ \frac{1}{\underline{Z}_2} + \frac{1 + \frac{\underline{Z}_1}{\underline{Z}_B}}{\underline{Z}_1 + \underline{Z}_3 + \frac{\underline{Z}_1 \cdot \underline{Z}_3}{\underline{Z}_B}} \right\} \cdot \underline{U}_0 \tag{3.65}$$

A comparison of this result with Eq. (3.27) shows that the results are not identical, which is a consequence of the shunt admittance $\underline{Y}_B$. This means that using the

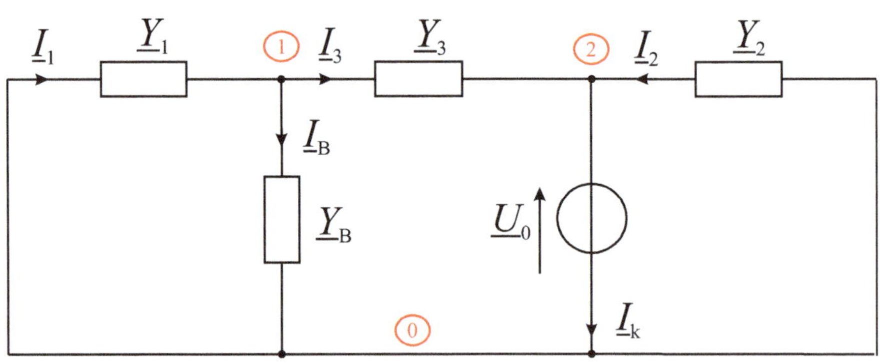

Fig. 3.17 Equivalent circuit diagram for the application of the admittance matrix

method of the equivalent voltage source at the short-circuit location, shunt admittances distort the result and should therefore not be taken into account. Whereas with $\underline{Y}_B = 0$, the same values result.

The reason for the incorrect calculation is that the voltage profile is reversed due to the use of the equivalent voltage source at the short-circuit location. This means that $\underline{U}_1$ is the voltage difference to the active voltage and results in a voltage that is the reverse of the actual distribution in the event of a grid fault. This means that the voltage at the short-circuit location is maximum, while in reality it has a value of zero.

This can be easily interpreted using Fig. 3.17. If, for example, a load impedance were taken into account in parallel with the fault location F, the calculation would result in a load current due to the equivalent voltage source, but this would not actually be present because the load is short-circuited due to its connection point and is therefore currentless. In this case, the lack of superposition has an effect, as it is carried out according to the exact procedure in Fig. 3.14c.

By applying the equivalent voltage source method at the short-circuit location, the short-circuit currents can be calculated in a simple way using the inverse of the nodal admittance matrix. The main diagonal elements correspond to the short-circuit impedances of the individual nodes.

3.6.7 Short-Circuit Current Calculation Using the Equivalent Voltage Source and Current Sources

The calculation process resulting from the superposition of voltage and current sources in a network is shown below. According to Fig. 3.18, a network (N) feeds in via a transformer and line to a short circuit at the busbar F. In addition, a converter (U) feeds in the busbar A. The behavior of the converter can be regarded as a constant current source during the short-circuit according to Chap. 13. The positive sequence impedance of the current source is assumed to be infinite in accordance with Sect. 13.2.1. The consideration of current sources in the short-circuit current calculation is described in detail repeatedly in Chap. 13.

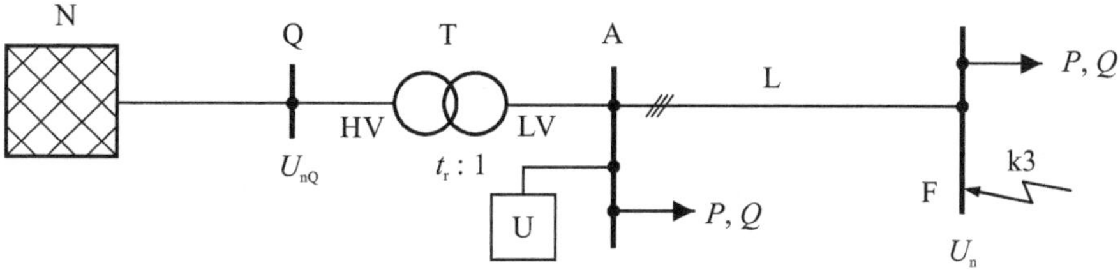

Fig. 3.18 Network circuit diagram for determining the three-phase short-circuit current

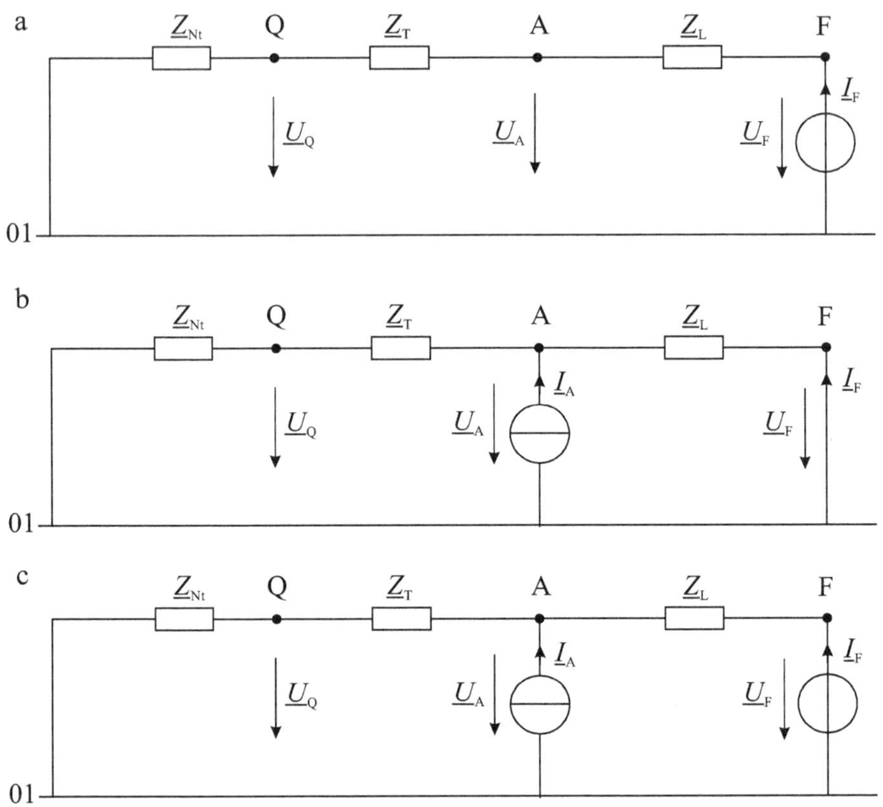

Fig. 3.19 Equivalent circuit for the network circuit according to Fig. 3.18. (**a**) equivalent voltage source at the short-circuit location, (**b**) current sources, (**c**) resulting equivalent circuit

While the network voltage is simulated by the equivalent voltage source at the short-circuit location, each individual converter must be simulated by a separate current source. This results in Fig. 3.19. The network impedance is converted to the underlaid voltage level.

When calculating the short-circuit current, it is assumed that the resulting current according to Fig. 3.19 can be fed in from various sources:

- Equivalent voltage source at the fault location using the equivalent voltage source method at the short-circuit location, Fig. 3.19a.
- Generation units that can be modeled as a current source, Fig. 3.19b.

The two processes shown in Fig. 3.19a, b can generally be combined into a resulting equivalent circuit as shown in Fig. 3.19c under the condition that the impedance of the current source is assumed to be infinite.

To determine the system behavior at the network nodes Q, A and F, the $\underline{\mathbf{Y}}$-matrix can be set up according to Eq. 3.66, whereby in a first step the system of equations is only considered for the equivalent voltage source at the short-circuit location. According to Eq. (3.56), the node voltages caused by the equivalent voltage source are marked with an asterisk (*) to distinguish them from the actual voltages during the short circuit.

$$\begin{pmatrix} \underline{I}_{\mathrm{Q}} \\ \underline{I}_{\mathrm{A}} \\ \underline{I}_{\mathrm{F}} \end{pmatrix} = \begin{pmatrix} \underline{Y}_{\mathrm{Nt}} + \underline{Y}_{\mathrm{T}} & -\underline{Y}_{\mathrm{T}} & 0 \\ -\underline{Y}_{\mathrm{T}} & \underline{Y}_{\mathrm{T}} + \underline{Y}_{\mathrm{L}} & -\underline{Y}_{\mathrm{L}} \\ 0 & -\underline{Y}_{\mathrm{L}} & \underline{Y}_{\mathrm{L}} \end{pmatrix} \cdot \begin{pmatrix} \underline{U}^*_{\mathrm{Q}} \\ \underline{U}^*_{\mathrm{A}} \\ \underline{U}^*_{\mathrm{F}} \end{pmatrix} \tag{3.66}$$

By converting the $\underline{\mathbf{Y}}$-matrix, the resulting $\underline{\mathbf{Z}}$-matrix can be determined:

$$\begin{pmatrix} \underline{U}^*_{\mathrm{Q}} \\ \underline{U}^*_{\mathrm{A}} \\ \underline{U}^*_{\mathrm{F}} \end{pmatrix} = \begin{pmatrix} (\underline{Z}_{\mathrm{Nt}}) & (\underline{Z}_{\mathrm{Nt}}) & (\underline{Z}_{\mathrm{Nt}}) \\ (\underline{Z}_{\mathrm{Nt}}) & (\underline{Z}_{\mathrm{Nt}} + \underline{Z}_{\mathrm{T}}) & (\underline{Z}_{\mathrm{Nt}} + \underline{Z}_{\mathrm{T}}) \\ (\underline{Z}_{\mathrm{Nt}}) & (\underline{Z}_{\mathrm{Nt}} + \underline{Z}_{\mathrm{T}}) & (\underline{Z}_{\mathrm{Nt}} + \underline{Z}_{\mathrm{T}} + \underline{Z}_{\mathrm{L}}) \end{pmatrix} \cdot \begin{pmatrix} \underline{I}_{\mathrm{Q}} \\ \underline{I}_{\mathrm{A}} \\ \underline{I}_{\mathrm{F}} \end{pmatrix} \tag{3.67}$$

Since it is first assumed that the equivalent voltage source at the short-circuit location F is the only source in the network, the system of equations for calculating the short-circuit current and the node voltages is given by Eq. (3.68). The direction of the voltage and current at the short-circuit location must be taken into account as shown in Fig. 3.16.

$$\begin{pmatrix} \underline{U}^*_{\mathrm{Q}} \\ \underline{U}^*_{\mathrm{A}} \\ -cU_{\mathrm{n}}/\sqrt{3} \end{pmatrix} = \begin{pmatrix} (\underline{Z}_{\mathrm{Nt}}) & (\underline{Z}_{\mathrm{Nt}}) & (\underline{Z}_{\mathrm{Nt}}) \\ (\underline{Z}_{\mathrm{Nt}}) & (\underline{Z}_{\mathrm{Nt}} + \underline{Z}_{\mathrm{T}}) & (\underline{Z}_{\mathrm{Nt}} + \underline{Z}_{\mathrm{T}}) \\ (\underline{Z}_{\mathrm{Nt}}) & (\underline{Z}_{\mathrm{Nt}} + \underline{Z}_{\mathrm{T}}) & (\underline{Z}_{\mathrm{Nt}} + \underline{Z}_{\mathrm{T}} + \underline{Z}_{\mathrm{L}}) \end{pmatrix} \cdot \begin{pmatrix} 0 \\ 0 \\ -\underline{I}''_{\mathrm{kVS}} \end{pmatrix} \tag{3.68}$$

The short-circuit current at the fault location F is determined directly from this:

$$\underline{I}''_{\mathrm{kVS}} = \frac{cU_{\mathrm{n}}/\sqrt{3}}{\underline{Z}_{\mathrm{Nt}} + \underline{Z}_{\mathrm{T}} + \underline{Z}_{\mathrm{L}}} \tag{3.69}$$

The additional index "VS" (Voltage Source) is introduced to indicate that the portion of the short-circuit current according to Eq. (3.69) is caused by a voltage source.

If, contrary to the representation according to Eq. (3.68), the feeder of a current source according to Fig. 3.19 is also taken into account, the system of equations expands according to Eq. (3.70). The direction of the currents must be taken into account, since the current fed in by the current source flows in the opposite direction to the short-circuit current.

$$\begin{pmatrix} \underline{U}_{\mathrm{Q}}^{*} \\ \underline{U}_{\mathrm{A}}^{*} \\ -cU_{\mathrm{n}}/\sqrt{3} \end{pmatrix} = \begin{pmatrix} (\underline{Z}_{\mathrm{Nt}}) & (\underline{Z}_{\mathrm{Nt}}) & (\underline{Z}_{\mathrm{Nt}}) \\ (\underline{Z}_{\mathrm{Nt}}) & (\underline{Z}_{\mathrm{Nt}}+\underline{Z}_{\mathrm{T}}) & (\underline{Z}_{\mathrm{Nt}}+\underline{Z}_{\mathrm{T}}) \\ (\underline{Z}_{\mathrm{Nt}}) & (\underline{Z}_{\mathrm{Nt}}+\underline{Z}_{\mathrm{T}}) & (\underline{Z}_{\mathrm{Nt}}+\underline{Z}_{\mathrm{T}}+\underline{Z}_{\mathrm{L}}) \end{pmatrix} \cdot \left[\begin{pmatrix} 0 \\ 0 \\ -\underline{I}_{\mathrm{k}}'' \end{pmatrix} + \begin{pmatrix} 0 \\ \underline{I}_{\mathrm{CS}} \\ 0 \end{pmatrix} \right] \tag{3.70}$$

This leads to the last line for the resulting short-circuit current:

$$\underline{I}_{\mathrm{k}}'' = \underline{I}_{\mathrm{kVS}}'' + \underline{I}_{\mathrm{kCS}}'' = \frac{cU_{\mathrm{n}}/\sqrt{3}}{\underline{Z}_{\mathrm{Nt}} + \underline{Z}_{\mathrm{T}} + \underline{Z}_{\mathrm{L}}} + \frac{\underline{Z}_{\mathrm{Nt}} + \underline{Z}_{\mathrm{T}}}{\underline{Z}_{\mathrm{Nt}} + \underline{Z}_{\mathrm{T}} + \underline{Z}_{\mathrm{L}}} \cdot \underline{I}_{\mathrm{CS}} \tag{3.71}$$

The second part of Eq. (3.71) means that the current of the current source $\underline{I}_{\mathrm{A}} = \underline{I}_{\mathrm{CS}}$ (CS: Current Source) is divided based on the impedance ratio. The system of Eq. (3.70) can be used to determine the voltages at the nodes that are caused by the partial short-circuit currents according to Eq. (3.71). This results in:

$$\begin{aligned} \underline{U}_{\mathrm{Q}}^{*} &= \underline{Z}_{\mathrm{Nt}} \cdot \underline{I}_{\mathrm{CS}} - \underline{Z}_{\mathrm{Nt}} \cdot \underline{I}_{\mathrm{k}}'' \\ \underline{U}_{\mathrm{A}}^{*} &= (\underline{Z}_{\mathrm{Nt}} + \underline{Z}_{\mathrm{T}}) \cdot \underline{I}_{\mathrm{CS}} - (\underline{Z}_{\mathrm{Nt}} + \underline{Z}_{\mathrm{T}}) \cdot \underline{I}_{\mathrm{k}}'' \end{aligned} \tag{3.72}$$

After inserting the resulting short-circuit current according to Eq. (3.71), the voltages are obtained under these conditions:

$$\begin{aligned} \underline{U}_{\mathrm{Q}}^{*} &= \frac{\underline{Z}_{\mathrm{Nt}}}{\underline{Z}_{\mathrm{Nt}} + \underline{Z}_{\mathrm{T}} + \underline{Z}_{\mathrm{L}}} \cdot \left(-\frac{cU_{\mathrm{n}}}{\sqrt{3}} + \underline{Z}_{\mathrm{L}} \cdot \underline{I}_{\mathrm{CS}} \right) \\ \underline{U}_{\mathrm{A}}^{*} &= \frac{\underline{Z}_{\mathrm{Nt}} + \underline{Z}_{\mathrm{T}}}{\underline{Z}_{\mathrm{Nt}} + \underline{Z}_{\mathrm{T}} + \underline{Z}_{\mathrm{L}}} \cdot \left(-\frac{cU_{\mathrm{n}}}{\sqrt{3}} + \underline{Z}_{\mathrm{L}} \cdot \underline{I}_{\mathrm{CS}} \right) \\ \underline{U}_{\mathrm{F}}^{*} &= -\frac{cU_{\mathrm{n}}}{\sqrt{3}} \end{aligned} \tag{3.73}$$

With the calculated voltages, it should be noted that the voltage profile changes because of the equivalent voltage source. This means that the actual voltage values during the short circuit result from the difference between the active voltage $cU_{\mathrm{n}}/\sqrt{3}$ and those calculated according to Eq. (3.73).

3.7 Consideration of Loads or Shunt-Impedances for Line-to-Earth and Three Phase Short-Circuit Currents

When using the method of the equivalent voltage source at the short-circuit location, shunt impedance must generally be neglected in order to avoid incorrect results. In this section, the influence of the ignoring on the result of the calculation is derived, depending on the type of short-circuit. The neglect of transverse impedances does not apply to motor branches, as they themselves provide a power frequency contribution to the short-circuit current similar to a synchronous machine.

3.7.1 Three Phase Short-Circuit Current

According to the IEC regulation [2], shunt impedances in the positive sequence system do not have to be taken into account. In the following, the influence of capacitances X_C and inductances X_D (conductor to earth in each case) on the magnitude of the three phase short-circuit current is determined. Here, the method of the equivalent voltage source at the short-circuit location is compared with the calculation using the real feed-in of the source voltages, as the latter is the exact method.

3.7.1.1 Calculation with the Real Feeder

The calculation with the real feeder is based on the equivalent circuit according to Fig. 3.20.

The capacitance C in Fig. 3.20 is assumed to be at different locations for the further calculation, with the limit values:

- $\alpha = 0$: Capacitance at the voltage source;
- $\alpha = 1$: Capacitance at the short-circuit location.

For the three-phase short-circuit current $\underline{I}_{k0}$ at the fault location F without taking into account the capacitance C or the reactance X_C, the following results:

$$\underline{I}_{k0} = \frac{\underline{E}}{jX} \tag{3.74}$$

The short-circuit reactance X_{kC} is calculated using the capacitance C and the factor α for the location A of the capacitance:

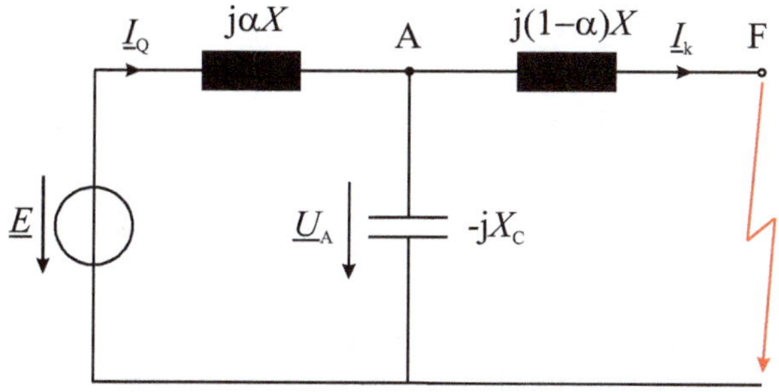

Fig. 3.20 Equivalent circuit for determining the influence of a capacitance on the three phase short-circuit current $\underline{I}_k$

$$X_{kC} = jX \cdot \left[\alpha + \frac{(1-\alpha) \cdot X_C}{X_C - (1-\alpha) \cdot X}\right] \quad (3.75)$$

The total current $\underline{I}_Q$ can be derived from this in accordance with Fig. 3.20. For the three phase short-circuit current $\underline{I}_{kCa}$ at the fault location F, taking into account the capacitance and constant source voltage $\underline{E}$ (additional index 1):

$$\underline{I}_{kC1} = \frac{X_C}{X_C - (1-\alpha) \cdot X} \cdot \underline{I}_Q = \frac{X_C}{jX} \cdot \frac{\underline{E}}{X_C - \alpha \cdot (1-\alpha) \cdot X} \quad (3.76)$$

If the two currents with and without capacitance are set in relation to each other, the following results:

$$\frac{\underline{I}_{k0} - \underline{I}_{kC1}}{\underline{I}_{kC1}} = -\alpha \cdot (1-\alpha) \cdot \frac{X}{X_C} \quad (3.77)$$

With

$\underline{I}_{k0}$ short-circuit current without capacitance,
$\underline{I}_{kC1}$ short-circuit current with capacitance and constant voltage $\underline{E}$.

If a maximum reactance of $|X| = 0.1\ |X_C|$ is assumed - this corresponds to a maximum voltage increase of approx. 10% above the reactance X in the no-loaded case—this results in the following simplification

$$F1 = \frac{\underline{I}_{k0} - \underline{I}_{kC1}}{\underline{I}_{kC1}} = -0.1 \cdot \alpha \cdot (1-\alpha) \quad (3.78)$$

Figure 3.21 shows the deviation of the short-circuit current calculation depending on the connection point of capacitance C.

It can be seen that, taking into account a constant voltage $\underline{E}$ (factor $F1$), the actual short-circuit current $\underline{I}_{kC1}$ at the fault location is always greater than the short-circuit current $\underline{I}_{k0}$ without capacitance. In this case, the deviation is a maximum of 2.5%.

If, on the other hand, the voltage before the short circuit occurs at the connection point of capacitance C is kept constant, the voltages $\underline{U}_A$ and $\underline{E}_1$ (changed source voltage as a function of the capacitance) result as a function of the source voltage $\underline{E}$.

$$\underline{U}_A = \frac{X_C}{X_C - \alpha \cdot X} \cdot \underline{E} \quad \text{resp.} \quad (3.79)$$

$$\underline{E}_A = \left(1 - \alpha \cdot \frac{X}{X_C}\right) \cdot \underline{U}_A = \left(1 - \alpha \cdot \frac{X}{X_C}\right) \cdot \underline{E} \quad (3.80)$$

To derive Eq. (3.79), it is assumed that the voltage $\underline{U}_A$ (with capacitor) is equal to the voltage $\underline{E}$. The new source voltage $\underline{E}_1$ can be determined from this. According to Eq. (3.80), this voltage is always less than $\underline{E}$ if no capacitive impedance to earth is

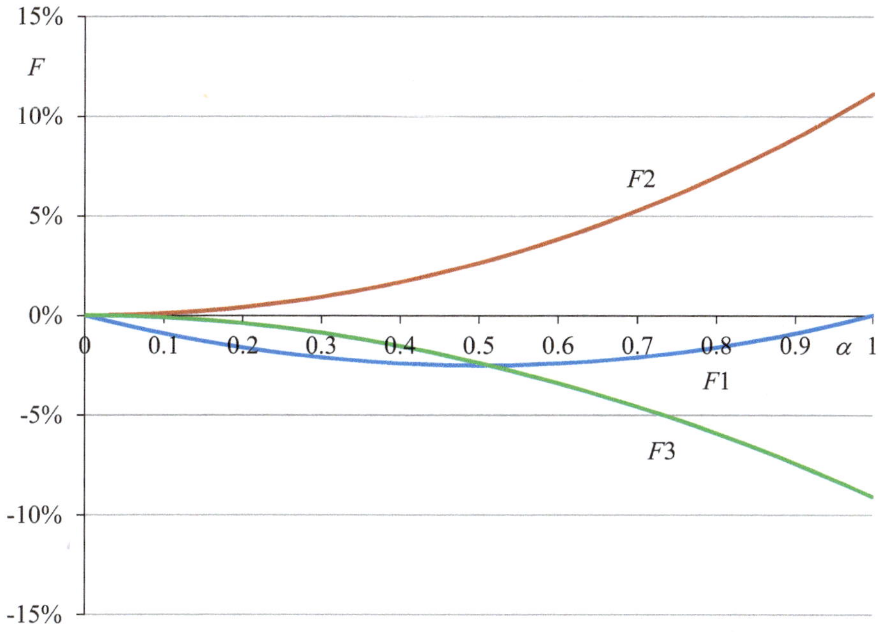

Fig. 3.21 Deviations when neglecting shunt impedances in %. *F*1 circuit without/with capacitance *C* and constant source voltage *E* (Eq. 3.78), *F*2 circuit without/with capacitance *C* and constant voltage $\underline{U}_A$, at the capacitance (Eq. 3.81), *F*3 circuit without/with inductance *L* and constant voltage $\underline{U}_A$, at the inductance (Eq. 3.83)

assumed. This means that the original source voltage $\underline{E}$ must be reduced as a result of the voltage increase at the capacitance connection point in order to achieve a constant network voltage $\underline{U}_A$ at the capacitance. Taking this voltage change into account (Eq. 3.80), the current ratio is determined by the changed source voltage $\underline{E}_A$ (index 2):

$$F2 = \frac{\underline{I}_{k0} - \underline{I}_{kC2}}{\underline{I}_{kC2}} = \frac{\frac{\underline{E}}{jX} - \frac{X_C}{jX} \cdot \frac{\underline{E}_A}{X_C - \alpha \cdot (1-\alpha) \cdot X}}{\frac{X_C}{jX} \cdot \frac{\underline{E}_A}{X_C - \alpha \cdot (1-\alpha) \cdot X}} = \frac{\alpha^2 \cdot X}{X_C - \alpha \cdot X} \tag{3.81}$$

With

$\underline{I}_{k0}$ short-circuit current without capacitance,
$\underline{I}_{kC2}$ short-circuit current with capacitance and constant voltage $\underline{U}_A$,
$\underline{E}_A$ source voltage at constant voltage $\underline{U}_A$ at the connection point of the capacitance.

With $|X| = 0.1\ |X_C|$, the curve *F*2 is as shown in Fig. 3.21. The calculations show, when taking capacitance and constant voltage $\underline{U}_A$ into account, the calculation of the short-circuit currents $\underline{I}_{k0}$ without taking capacitance into consideration always leads to greater values compared to the determination of $\underline{I}_{kC2}$. The reason for this is that the capacitive effect causes an increase in voltage at connection point A, which leads to a reduction in the internal voltage $\underline{E}_A$ compared to the voltage $\underline{E}$.

If, in contrast to the illustration in Fig. 3.20, the capacitance is replaced by an inductance ($|X|_C = |X|_L$), the current ratio is determined as follows

- Constant source voltage E

$$\frac{\underline{I}_{k0} - \underline{I}_{kL1}}{\underline{I}_{kL1}} = \alpha \cdot (1 - \alpha) \cdot \frac{X}{X_L} \tag{3.82}$$

- Constant reactor voltage $U_A = E$

$$F3 = \frac{\underline{I}_{k0} - \underline{I}_{kL2}}{\underline{I}_{kL2}} = \frac{-\alpha^2 \cdot X}{X_L + \alpha \cdot X} \tag{3.83}$$

With

$\underline{I}_{k0}$ short-circuit current without inductance,
$\underline{I}_{kL1}$ short-circuit current with inductance and constant voltage $\underline{E}$,
$\underline{I}_{kL2}$ short-circuit current with inductance and constant voltage $\underline{U}_A$,
$\underline{E}_A$ source voltage at constant $\underline{U}_A$ at the connection point of the inductance.

The deviation curve for the case of a constant voltage $\underline{U}_A$ under the condition $X_L = 10\ X$ is also plotted in Fig. 3.21 (factor $F3$). It can be seen, if the inductance is neglected, currents are calculated that are lower than those actually present. The maximum deviation depends on the size of the shunt impedance. In this example, there is a maximum deviation of 9.09% to the unsafe side, if inductive shunt impedances (loads) are not taken into account. This is the reason that a short-circuit current calculation without loads is only possible, if a voltage factor c according to Table 7.1, Sect. 7.1, is also taken into account, as an inductive load flow is generally assumed.

3.7.1.2 Calculation with the Equivalent Voltage Source at the Short-Circuit Location

The calculations specified in Sect. 3.6.1.1 take into account the real distributions of the source voltages in the network so that the current distribution (short-circuit current at the fault location and load current) is calculated correctly. With the method of the equivalent voltage source at the short-circuit location, the voltage distribution in the network deviates from this because the maximum voltage is present at the short-circuit location. If the equivalent voltage source is used at the short-circuit location, this results in the equivalent circuit shown in Fig. 3.22. The following applies to the short-circuit reactance X_{kC}:

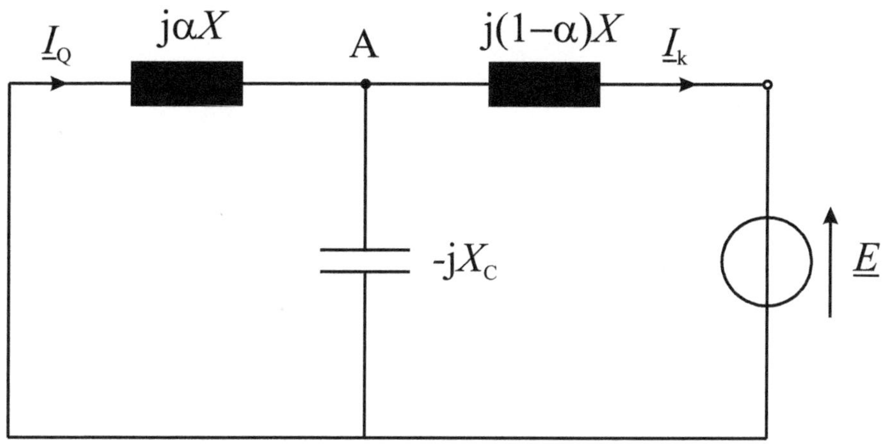

Fig. 3.22 Equivalent voltage source at the short-circuit location according to the example in Fig. 3.20

$$X_{kC} = j(1-\alpha)\cdot X - \frac{j\alpha\cdot X\cdot X_C}{\alpha\cdot X - X_C} = jX\cdot\left(1 - \frac{\alpha^2\cdot X}{\alpha\cdot X - X_C}\right) \tag{3.84}$$

The short-circuit current can be determined from this. In this case, the values are given the index "3", as a different voltage injection is used compared to Sect. 3.7.1.1.

$$\underline{I}_{kC3} = \frac{\underline{E}}{jX\cdot\left(1 - \frac{\alpha^2\cdot X}{\alpha\cdot X - X_C}\right)} \tag{3.85}$$

In order to specify the deviation between the two methods, when calculating the short-circuit currents at the fault location, the current according to Eq. (3.76) is used, which is indicated by $\underline{I}_{kC1}$, as this current is determined using an exact calculation method. The following applies to the deviation:

$$\frac{\underline{I}_{kC3} - \underline{I}_{kC1}}{\underline{I}_{kC1}} = \frac{X_C - \alpha\cdot(1-\alpha)\cdot X}{\left(1 - \frac{\alpha^2\cdot X}{\alpha\cdot X - X_C}\right)\cdot X_C} - 1 \tag{3.86}$$

With

$\underline{I}_{kC1}$ short-circuit current with capacitance and constant voltage $\underline{E}$,
$\underline{I}_{kC3}$ short-circuit current with capacitance and equivalent voltage $\underline{E}$ at the short-circuit location.

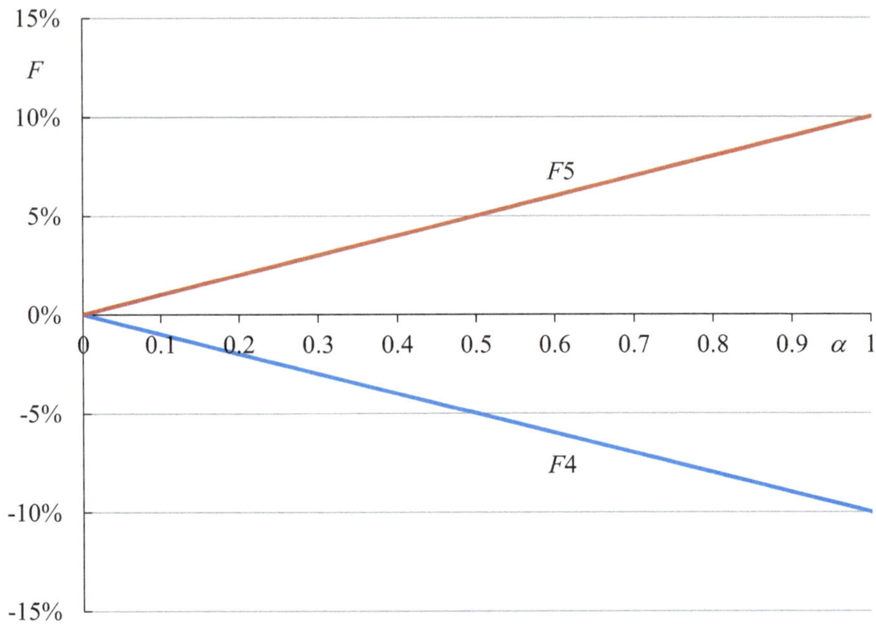

Fig. 3.23 Deviations between the calculation methods. $F4$ circuit with capacitance C and the equivalent voltage source/source voltage $\underline{E}$ (Eq. 3.87), $F5$ circuit with inductance L and the equivalent voltage source/source voltage $\underline{E}$ (Eq. 3.87)

If $|X| = 0.1\ |X_C|$ is set again, the following applies:

$$F4 = \frac{\underline{I}_{kC3} - \underline{I}_{kC1}}{\underline{I}_{kC1}} = \frac{1 - \alpha \cdot (1 - \alpha) \cdot 0.1}{\left(1 - \frac{\alpha^2 \cdot 0,1}{\alpha \cdot 0.1 - 1}\right)} - 1 \tag{3.87}$$

Figure 3.23 shows the deviation $F4$ as a function of the value α. An inductive shunt impedance for the equivalent circuit according to Fig. 3.22 can be taken into account by inserting $X_L = -X_C$ into Eq. (3.84)–(3.87). The factor $F5$ determined in this way is also shown in Fig. 3.23.

According to Fig. 3.23, the error in the short-circuit current calculation increases with increasing value α, whereby the absolute value depends on the power factor of the impedance. If the capacitance or other shunt impedances are taken into account, the result is that for a load directly at the fault location (Fig. 3.22, $\alpha = 1$), the calculation produces a capacitive current even though the capacitance is short-circuited. The error results from the fact that in the method with the equivalent voltage source at the short-circuit location, the superposition of the currents before the short-circuit is missing, which is particularly noticeable with this arrangement of the load or the shunt impedance.

3.7.2 Line-to-Earth Short-Circuit Current

In contrast to the calculation according to Sect. 3.7.1, the shunt impedance is only taken into account in the zero sequence system according to IEC 60909-0 [2], when determining the line-to-earth short-circuit current. The influence of a shunt impedance on the calculation of the line-to-earth short-circuit current is determined below.

According to Sect. 9.2.2.3, the line-to-earth short-circuit current is calculated from the sum of the three component currents, which are equal to each other at the short-circuit location, so only the behavior of the current $\underline{I}_{(1)}$ of the positive sequence system is considered in the following sections.

3.7.2.1 Calculation with the Real Feeder

For the calculation, the component equivalent circuit according to Fig. 3.24 applies, with a capacitance C or a reactance X_C as shunt impedance. It is assumed that the capacitance is the same in all component systems ($X_C = X_{(1)C} = X_{(2)C} = X_{(0)C}$), as is the case with a single-core cable, for example.

The following relationships apply to the impedances of the negative and zero sequence system (index C: with capacitance):

$$\underline{Z}_{(0)C} = \mathrm{j}(1-\alpha) \cdot X_{(0)} + \mathrm{j}\frac{\alpha \cdot X_{(0)} \cdot X_C}{X_C - \alpha \cdot X_{(0)}} = \mathrm{j}X_{(0)} \cdot \left(1 + \frac{\alpha^2 \cdot X_{(0)}}{X_C - \alpha \cdot X_{(0)}}\right) \quad (3.88)$$

$$\underline{Z}_{(2)C} = \mathrm{j}(1-\alpha) \cdot X_{(1)} + \mathrm{j}\frac{\alpha \cdot X_{(1)} \cdot X_C}{X_C - \alpha \cdot X_{(1)}} = \mathrm{j}X_{(1)} \cdot \left(1 + \frac{\alpha^2 \cdot X_{(1)}}{X_C - \alpha \cdot X_{(1)}}\right) \quad (3.89)$$

By inserting the impedances $\underline{Z}_{(2)C}$ and $\underline{Z}_{(0)C}$, the equivalent circuit is simplified to Fig. 3.25. If the impedance $\underline{Z}_A$ is introduced for simplification, the resulting short-circuit impedance $\underline{Z}_k$ follows:

$$\underline{Z}_k = \mathrm{j}\alpha \cdot X_{(1)} + \mathrm{j}\frac{X_C \cdot \underline{Z}_A}{\mathrm{j}X_C - \underline{Z}_A} \quad (3.90)$$

With

$$\underline{Z}_A = \mathrm{j}(1-\alpha) \cdot X_{(1)} + \underline{Z}_{(2)C} + \underline{Z}_{(0)C} \quad (3.91)$$

The following applies to the total current $\underline{I}_{(1)Q}$ in the positive sequence system according to Fig. 3.25, taking the impedances into account:

$$\underline{I}_{(1)Q} = \frac{\underline{E}}{\underline{Z}_k} = \frac{-\mathrm{j}\underline{E}}{\alpha \cdot X_{(1)} + \left(\frac{X_C \cdot \underline{Z}_A}{\mathrm{j}X_C - \underline{Z}_A}\right)} \quad (3.92)$$

or for the component current $\underline{I}_{(1)}$ at the short-circuit location.

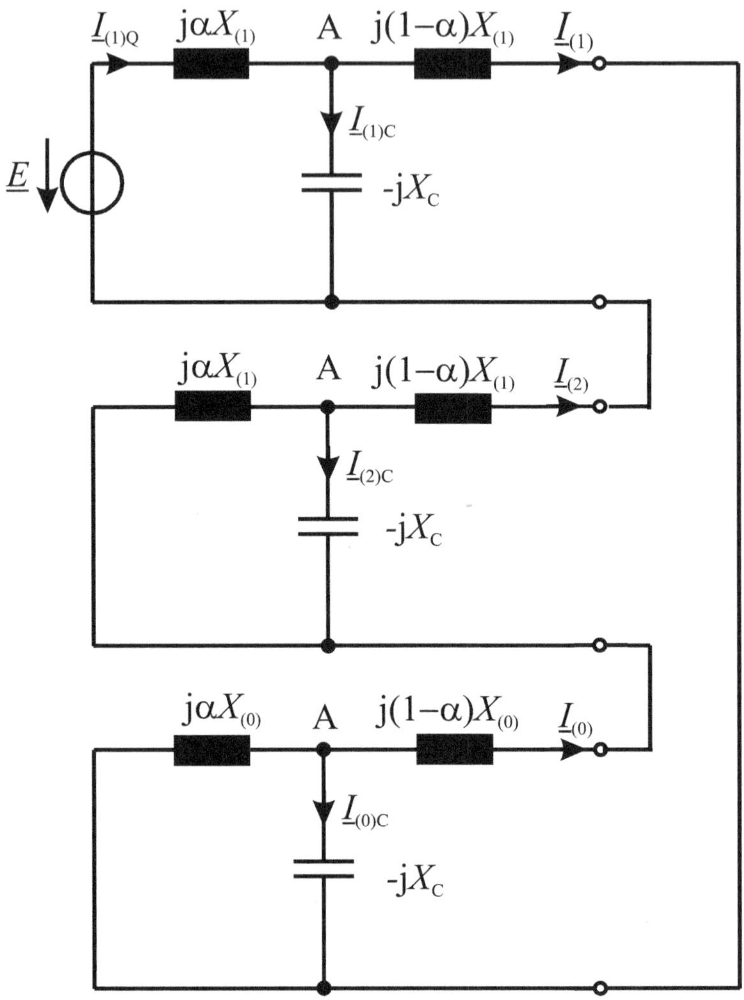

Fig. 3.24 Component equivalent circuit for determining the line-to-earth short-circuit current. $\underline{I}_{(1,2,0)\mathrm{k}}$ current in positive, negative and zero sequence system

$$\underline{I}_{(1)} = \frac{-\mathrm{j}X_\mathrm{C}}{\underline{Z}_\mathrm{A} - \mathrm{j}X_\mathrm{C}} \cdot \underline{I}_{(1)\mathrm{Q}} = \frac{\mathrm{j}X_\mathrm{C}}{\mathrm{j}X_\mathrm{C} - \underline{Z}_\mathrm{A}} \cdot \frac{-\mathrm{j}\underline{E}}{\alpha \cdot X_{(1)} + \left(\dfrac{X_\mathrm{C} \cdot \underline{Z}_\mathrm{A}}{\mathrm{j}X_\mathrm{C} - \underline{Z}_\mathrm{A}}\right)}$$

$$= \frac{X_\mathrm{C}}{X_\mathrm{C} \cdot \underline{Z}_\mathrm{A} + \alpha \cdot X_{(1)} \cdot (\mathrm{j}X_\mathrm{C} - \underline{Z}_\mathrm{A})} \cdot \underline{E} = \frac{X_\mathrm{C}}{\underline{Z}_\mathrm{A} \cdot \left(X_\mathrm{C} - \alpha \cdot X_{(1)}\right) + \mathrm{j}\alpha \cdot X_{(1)} \cdot X_\mathrm{C}} \cdot \underline{E} \tag{3.93}$$

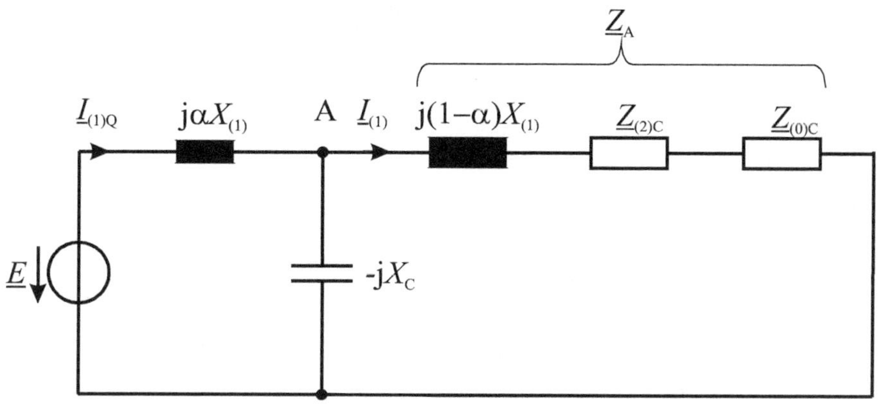

Fig. 3.25 Reduced equivalent circuit according to Fig. 3.24

To simplify matters, the same assumption is made for the maximum value of X_{C} in accordance with Sect. 3.7.1.1, Eq. (3.78).

$$|X_{\mathrm{C}}| = 10 \cdot |X_{(1)}| \tag{3.94}$$

In addition, the following should apply to the zero reactance in this example, which corresponds approximately to the zero reactance ratio of an overhead line with an earth wire (Sect. 4.7.2.4):

$$X_{(0)} = 3 \cdot X_{(1)} \tag{3.95}$$

After inserting these relationships, the following applies to the single impedances of the component systems:

$$\underline{Z}_{(0)\mathrm{C}} = \mathrm{j}\ 3 \cdot X_{(1)} \cdot \left[1 + \frac{3 \cdot \alpha^2}{10 - 3 \cdot \alpha}\right] \tag{3.96}$$

$$\underline{Z}_{(2)\mathrm{C}} = \mathrm{j}X_{(1)} \cdot \left[1 + \frac{\alpha^2}{10 - \alpha}\right] \tag{3.97}$$

$$\begin{aligned}\underline{Z}_{\mathrm{A}} &= \mathrm{j}X_{(1)} \cdot \left[(1 - \alpha) + 1 + \frac{\alpha^2}{10 - \alpha} + 3 + \frac{9 \cdot \alpha^2}{10 - 3 \cdot \alpha}\right] \\ &= \mathrm{j}X_{(1)} \cdot \left[5 - \alpha + \alpha^2 \cdot \left(\frac{1}{10 - \alpha} + \frac{9}{10 - 3 \cdot \alpha}\right)\right]\end{aligned} \tag{3.98}$$

For the component current $\underline{I}_{(1)}$, Eq. (3.93), the following results under these conditions

$$\underline{I}_{(1)} = \frac{\underline{E}}{jX_{(1)}} \bullet \frac{10}{\left[5 - \alpha + \alpha^2 \bullet \left(\frac{1}{10-\alpha} + \frac{9}{10-3\bullet\alpha}\right)\right] \bullet (10-\alpha) + 10\bullet\alpha} \tag{3.99}$$

If, on the other hand, the component current $\underline{I}_{(1)0}$ is calculated without taking into account the capacitances in the positive sequence, negative sequence and zero sequence systems, the result is as follows:

$$\underline{I}_{(1)0} = \frac{\underline{E}}{\underline{Z}_{(1)0} + \underline{Z}_{(2)0} + \underline{Z}_{(0)0}} = \frac{\underline{E}}{j5 \bullet X_{(1)}} \tag{3.100}$$

The line-to-earth short-circuit current at the fault location is generally determined from the sum of the component currents $\underline{I}_{(1)}$, $\underline{I}_{(2)}$ and $\underline{I}_{(0)}$, which are equal to each other. These currents $\underline{I}_{(1)}$ and $\underline{I}_{(1)0}$ can be compared directly in order to determine the deviations or the influence of the loads on the line-to-line short-circuit current at the fault location in the original system. The following applies to the ratio $F1$:

$$\begin{aligned} F1 &= \frac{\underline{I}_{(1)0} - \underline{I}_{(1)}}{\underline{I}_{(1)}} \\ &= \frac{1}{50} \bullet \left[10\bullet\alpha + (10-\alpha)\bullet\left\{5 - \alpha + \alpha^2\left(\frac{1}{10-\alpha} + \frac{9}{10-3\bullet\alpha}\right)\right\}\right] \\ &- 1 \end{aligned} \tag{3.101}$$

With

$\underline{I}_{(1)}$ component current with capacitances, constant source voltage $\underline{E}$,
$\underline{I}_{(0)0}$ component current without capacitances, constant source voltage $\underline{E}$.

Depending on the value α (connection point of the capacitance), the curve $F1$ is shown in Fig. 3.26.

With a constant source voltage $\underline{E}$ (curve $F1$), the maximum deviation is 17.14% (the calculation without capacitance results in too high a current in the positive sequence system) in the event that the capacitance is directly at the fault location ($\alpha = 1$). The reason for this is that the deviation in the calculation of the zero sequence impedance is greatest due to the parallel circuit. With a larger ratio of zero sequence reactance/positice sequence reactance $X_{(0)}/X_{(1)}$, this error will be even greater.

If, in contrast to the calculation for the deviation $F1$, the voltage $\underline{U}_A$ is kept constant before the fault occurs, depending on the connection point A, the active voltage $\underline{E}$ must be multiplied by the factor ($1 - \alpha{\cdot}X/X_C$), according to Eq. (3.80) if the capacitance is taken into account. This means that as the distance α increases, the voltage is reduced due to the capacitive effect. Of course, this only has an effect when calculating the short-circuit current, considering a load. After insertion, the result for the deviation is $F2$.

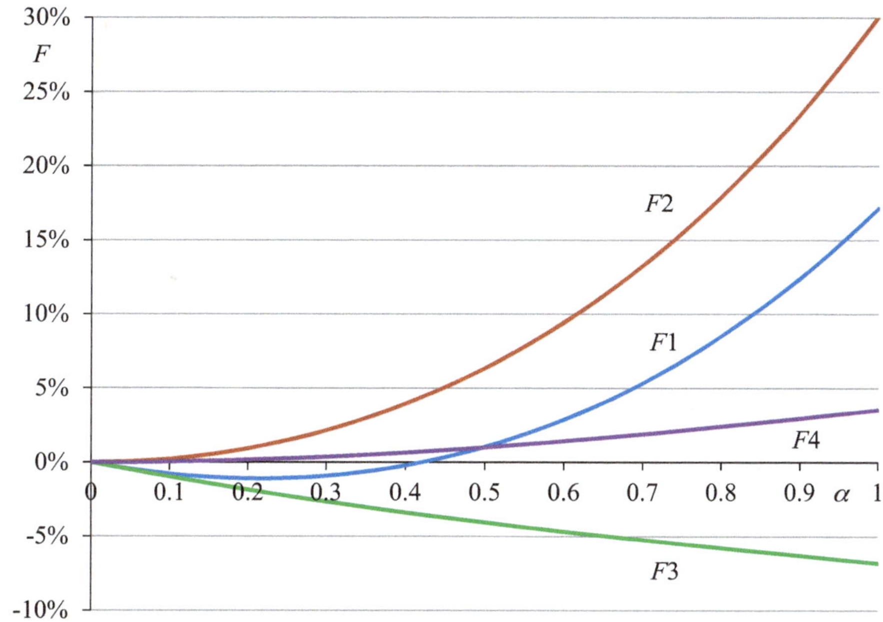

Fig. 3.26 Deviations when neglecting capacitive shunt impedances in %. *F*1 circuit without/with capacitances in the positive sequence and zero sequence system and constant source voltage $\underline{E}$, *F*2 circuit with capacitances in the positive sequence and zero sequence system and constant voltage $\underline{U}_A$, *F*3 circuit with zero sequence capacitance and constant source voltage $\underline{E}$, *F*4 circuit with zero sequence capacitance and constant voltage $\underline{U}_A$

$$F2 = \frac{\underline{I}_{(1)} - \underline{I}_{(1)1}}{\underline{I}_{(1)1}} = \frac{1}{50 \cdot (1 - 0.1 \cdot \alpha)} \left[10 \cdot \alpha + (10 - \alpha) \cdot \left\{ 5 - \alpha + \alpha^2 \cdot \left(\frac{1}{10 - \alpha} + \frac{9}{10 - 3 \cdot \alpha} \right) \right\} \right] - 1 \tag{3.102}$$

With

$\underline{I}_{(1)}$ component current with capacitances, constant source voltage $\underline{E}$,
$\underline{I}_{(1)1}$ component current with capacitances, constant voltage $\underline{U}_A$.

The evaluation shows that at a constant voltage $\underline{U}_A$, the deviations according to Fig. 3.26 can be even greater at a maximum of 30.16%.

According to IEC Standard 60909–0, shunt impedances are only to be taken into account in the zero sequence system, so that this calculation is compared below with the determination of the current according to Eq. (3.99). In this case, where the impedance is only used in the zero sequence system, the component current $\underline{I}_{(1)kIEC}$ is calculated as follows:

$$\underline{I}_{(1)\mathrm{IEC}} = \frac{\underline{E}}{\underline{Z}_{(1)0} + \underline{Z}_{(2)0} + \underline{Z}_{(0)\mathrm{C}}} \tag{3.103}$$

with

$$\underline{Z}_{(1)0} = \underline{Z}_{(2)0} = \mathrm{j}X_{(1)} \tag{3.104}$$

and

$$\underline{Z}_{(0)\mathrm{C}} = \mathrm{j}3 \cdot X_{(1)} \cdot \left(1 + \frac{3 \cdot \alpha^2}{10 - 3 \cdot \alpha}\right) \tag{3.105}$$

according to Eq. (3.96), it follows for $\underline{I}_{(1)\mathrm{IEC}}$

$$\underline{I}_{(1)\mathrm{IEC}} = \frac{\underline{E}}{\mathrm{j}X_{(1)} \cdot \left(5 + \frac{9 \cdot \alpha^2}{10-3 \cdot \alpha}\right)} \tag{3.106}$$

Assuming a constant voltage $\underline{E}$, this results in the factor $F3$:

$$F3 = \frac{\underline{I}_{(1)\mathrm{IEC}} - \underline{I}_{(1)}}{\underline{I}_{(1)}} = \frac{10 \cdot \alpha + (10 - \alpha) \cdot \left\{5 - \alpha + \alpha^2 \cdot \left(\frac{1}{10 - \alpha} + \frac{9}{10 - 3 \cdot \alpha}\right)\right\}}{5 + \frac{9 \cdot \alpha^2}{10 - 3 \cdot \alpha}} - 1 \tag{3.107}$$

With

$\underline{I}_{(1)}$ component current with capacitances, constant source voltage $\underline{E}$.
$\underline{I}_{(1)\mathrm{IEC}}$ capacitances only in zero sequence system, constant source voltage $\underline{E}$.

The deviation $F3$ in relation to the actual current can be determined at a constant voltage $\underline{E}$, which is also shown in Fig. 3.26. If the voltage at the connection point A of the capacitance is also kept constant in this case, the deviation $F4$ is obtained (Fig. 3.26). The calculation is similar to the determination of $F2$, Eq. (3.108).

$$F4 = \frac{10 \cdot \alpha + (10 - \alpha) \cdot \left\{5 - \alpha + \alpha^2 \cdot \left(\frac{1}{10-\alpha} + \frac{9}{10-3 \cdot \alpha}\right)\right\}}{\left(5 + \frac{9 \cdot \alpha^2}{10-3 \cdot \alpha}\right) \cdot (1 - 0.1 \cdot \alpha)} - 1 \tag{3.108}$$

It can be seen that under the conditions, the line-to-earth short-circuit current is determined to be a maximum of 3.5% too high. In this case, the capacitance in the positive and negative sequence system is not taken into account, but is taken into

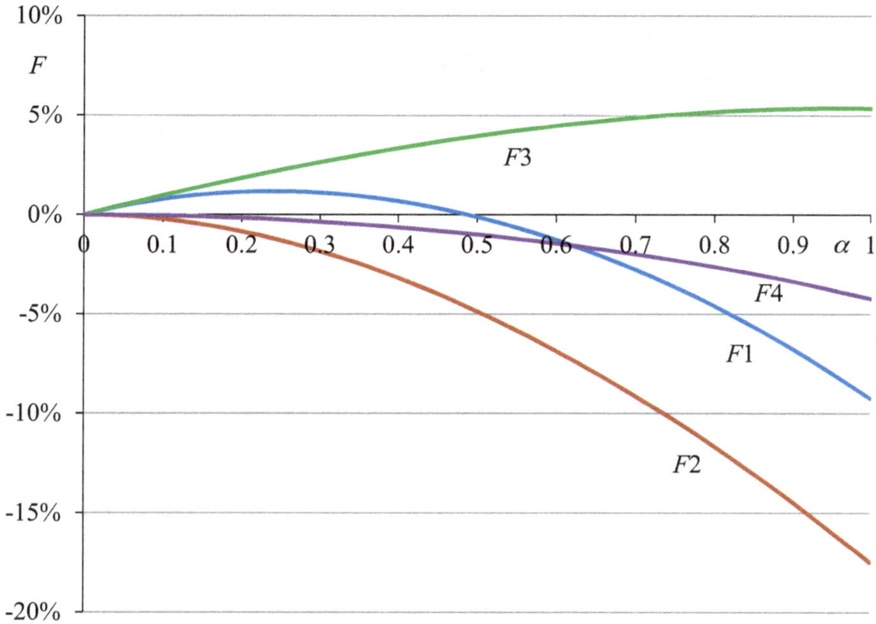

Fig. 3.27 Deviations when neglecting inductive shunt impedances in %. *F*1 circuit without/with inductances and constant source voltage $\underline{E}$, *F*2 circuit without inductances and constant voltage $\underline{U}_A$, *F*3 circuit with zero sequence inductance and constant source voltage $\underline{E}$, *F*4 circuit with zero sequence inductance and constant voltage at the connection point of the capacitance $\underline{U}_A$

consideration in the zero sequence system and the voltage at the connection point of the capacitance is kept constant, factor *F*4.

If the component current $\underline{I}_{(1)}$ is determined using the same procedure, assuming that the capacitance is replaced by an inductance, the deviations according to Fig. 3.27 result. These cases also show that the calculation of the line-to-earth short-circuit current leads to errors that are considerably smaller, if the shunt impedances are taken into account in the zero sequence system.

3.7.2.2 Calculation with the Equivalent Voltage Source at the Short-Circuit Location

If, in contrast to the illustration in Sect. 3.7.2.1, the calculation is carried out with the equivalent voltage source at the short-circuit location, Fig. 3.28 must be used as a basis, provided that the shunt impedances in the positive and negative sequence system are not taken into account.

Since, according to Fig. 3.28, the positive and negative sequence systems have no shunt impedances, in this case the currents $\underline{I}_{(1)\mathrm{IEC}}$, $\underline{I}_{(2)\mathrm{IEC}}$ and $\underline{I}_{(0)\mathrm{IEC}}$ correspond to the

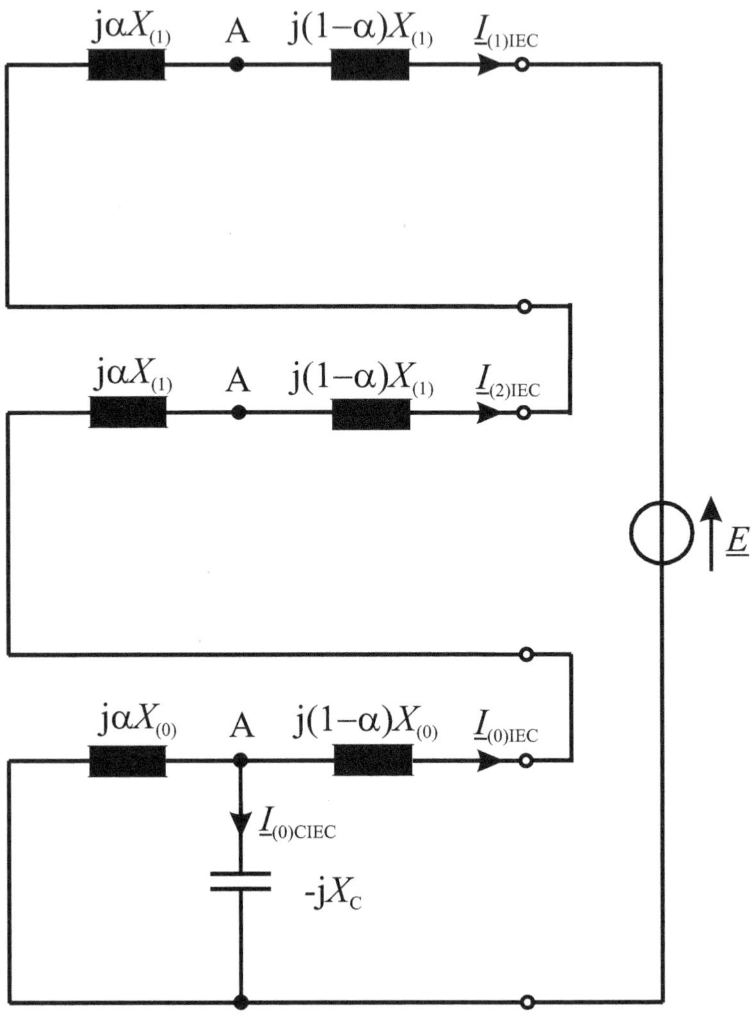

Fig. 3.28 Component equivalent circuit for calculating the line-to-earth short-circuit current using the equivalent voltage source at the short-circuit location

current distribution according to Eq. (3.103) to (3.106). This means that the deviations in the calculation of the currents correspond to the representation of faults *F*3 and *F*4, according to Figs. 3.26 and 3.27.

The deviations for calculating the line-to-earth short-circuit current with the specification that the shunt impedance is not taken into account in the positive and negative sequence systems, but in the zero sequence system, can therefore be tolerated.

3.7.2.3 Determining the Capacitor Current

Due to the consideration of the zero sequence capacitance, a current through the capacitance is determined as a consequence of the changed voltage profile when applying the equivalent voltage source method at the short-circuit location, which is not physically present in this case either. The magnitude of this current is determined below and compared with the exact current.

Eqs. (3.92) and (3.93) with the currents $\underline{I}_{(1)\mathrm{Q}}$ and $\underline{I}_{(1)}$ according to Fig. 3.24, taking into account the actual source voltage $\underline{E}$, apply to the calculation of the exact current of the shunt capacitance according to Sect. 3.7.2.1. From this, the capacitance currents for the component systems can be determined, with the impedance $\underline{Z}_{\mathrm{A}}$ according to Eq. (3.91).

$$\underline{I}_{(1)\mathrm{C}} = \frac{\underline{Z}_{\mathrm{A}}}{\underline{Z}_{\mathrm{A}} - \mathrm{j}X_{\mathrm{C}}} \bullet \underline{I}_{(1)\mathrm{Q}} \tag{3.109}$$

$$\underline{I}_{(2)\mathrm{C}} = -\frac{\alpha \bullet X_{(1)}}{\alpha \bullet X_{(1)} - X_{\mathrm{C}}} \bullet \underline{I}_{(1)} = \frac{\alpha \bullet X_{(1)}}{\alpha \bullet X_{(1)} - X_{\mathrm{C}}} \bullet \frac{\mathrm{j}X_{\mathrm{C}}}{\underline{Z}_{\mathrm{A}} - \mathrm{j}X_{\mathrm{C}}} \bullet \underline{I}_{(1)\mathrm{Q}} \tag{3.110}$$

$$\underline{I}_{(0)\mathrm{C}} = -\frac{\alpha \bullet X_{(0)}}{\alpha \bullet X_{(0)} - X_{\mathrm{C}}} \bullet \underline{I}_{(1)} = \frac{\alpha \bullet X_{(0)}}{\alpha \bullet X_{(0)} - X_{\mathrm{C}}} \bullet \frac{\mathrm{j}X_{\mathrm{C}}}{\underline{Z}_{\mathrm{A}} - \mathrm{j}X_{\mathrm{C}}} \bullet \underline{I}_{(1)\mathrm{Q}} \tag{3.111}$$

According to Sect. 9.2.2.3, the currents in the original system can be determined from the component currents using Eq. (3.112).

$$\begin{pmatrix} \underline{I}_{\mathrm{RC}} \\ \underline{I}_{\mathrm{SC}} \\ \underline{I}_{\mathrm{TC}} \end{pmatrix} = \begin{pmatrix} 1 & 1 & 1 \\ 1 & \underline{a}^2 & \underline{a} \\ 1 & \underline{a} & \underline{a}^2 \end{pmatrix} \bullet \begin{pmatrix} \underline{I}_{(0)\mathrm{C}} \\ \underline{I}_{(1)\mathrm{C}} \\ \underline{I}_{(2)\mathrm{C}} \end{pmatrix} \tag{3.112}$$

For the capacitor current $\underline{I}_{\mathrm{RC}}$ of the faulty conductor R, the equations given above are taken into account:

$$\underline{I}_{\mathrm{RC}} = \frac{\underline{I}_{(1)\mathrm{Q}}}{\underline{Z}_{\mathrm{A}} - \mathrm{j}X_{\mathrm{C}}} \bullet \left[\underline{Z}_{\mathrm{A}} + \frac{\mathrm{j}\alpha \bullet X_{\mathrm{C}} \bullet X_{(1)}}{\alpha \bullet X_{(1)} - X_{\mathrm{C}}} + \frac{\mathrm{j}\alpha \bullet X_{\mathrm{C}} \bullet X_{(0)}}{\alpha \bullet X_{(0)} - X_{\mathrm{C}}}\right] \tag{3.113}$$

With

$$\underline{I}_{(1)\mathrm{Q}} = \frac{-\mathrm{j}\underline{E}}{\alpha \bullet X_{(1)} + \left(\frac{X_{\mathrm{C}} \bullet \underline{Z}_{\mathrm{A}}}{\mathrm{j}X_{\mathrm{C}} - \underline{Z}_{\mathrm{A}}}\right)} \tag{3.114}$$

Taking into account the specifications $|X_{\mathrm{C}}| = 10 \cdot |X_{(1)}|$ and $X_{(0)} = 3 \cdot X_{(1)}$, the following values result.

$$\underline{I}_{\mathrm{RC}} = \frac{\underline{I}_{(1)\mathrm{Q}}}{\frac{\underline{Z}_{\mathrm{A}}}{\mathrm{j}X_{(1)}} - 10} \bullet \left[\frac{\underline{Z}_{\mathrm{A}}}{\mathrm{j}X_{(1)}} + \frac{10 \bullet \alpha}{\alpha - 10} + \frac{30 \bullet \alpha}{3 \bullet \alpha - 10}\right] \tag{3.115}$$

With

$$\frac{\underline{Z}_{\mathrm{A}}}{\mathrm{j}X_{(1)}} = 5 - \alpha + \alpha^2 \bullet \left(\frac{1}{10 - \alpha} + \frac{9}{10 - 3 \bullet \alpha}\right) \tag{3.116}$$

$$\underline{I}_{(1)\mathrm{Q}} = \frac{-\mathrm{j}\underline{E}}{X_{(1)}} \bullet \frac{1}{\alpha + \left(\frac{10 \bullet \underline{Z}_{\mathrm{A}}/\mathrm{j}X_{(1)}}{10 - \underline{Z}_{\mathrm{A}}/\mathrm{j}X_{(1)}}\right)} \tag{3.117}$$

Figure 3.28 applies to the capacitor current $\underline{I}_{\mathrm{RCIEC}}$, if the equivalent voltage source method is used at the short-circuit location:

$$\underline{I}_{(1)\mathrm{CIEC}} = \underline{I}_{(2)\mathrm{CIEC}} = 0 \tag{3.118}$$

$$\underline{I}_{(0)\mathrm{CIEC}} = -\frac{\alpha \bullet X_{(0)}}{\alpha \bullet X_{(0)} - X_{\mathrm{C}}} \bullet \underline{I}_{(1)\mathrm{IEC}} \tag{3.119}$$

With

$$\underline{I}_{(1)\mathrm{IEC}} = \frac{-\mathrm{j}\underline{E}}{2 \bullet X_{(1)} + \frac{\alpha \bullet X_{\mathrm{C}} \bullet X_{(0)}}{X_{\mathrm{C}} - \alpha \bullet X_{(0)}} + (1 - \alpha) \bullet X_{(0)}} \tag{3.120}$$

Under the same boundary conditions, the capacitor current is given by Eq. (3.121):

$$\underline{I}_{(0)\mathrm{CIEC}} = \frac{-\mathrm{j}\alpha \bullet X_{(0)} \bullet \underline{E}}{\left[2 \bullet X_{(1)} + (1 - \alpha) \bullet X_{(0)}\right] \bullet \left(X_{\mathrm{C}} - \alpha \bullet X_{(0)}\right) + \alpha \bullet X_{\mathrm{C}} \bullet X_{(0)}} \tag{3.121}$$

resp.

$$\underline{I}_{\mathrm{RCIEC}} = \underline{I}_{(0)\mathrm{CIEC}} = \frac{-\mathrm{j}3 \bullet \alpha \bullet \underline{E}}{X_{(1)}} \bullet \frac{1}{9 \bullet \alpha^2 - 15 \bullet \alpha + 50} \tag{3.122}$$

Table 3.2 shows the capacitor currents as a function of the two calculation methods and the installation location of the capacitance (α). In each case, the current of the faulty conductor is shown and the short circuit is at the end of the line.

The results in Table 3.2 show that currents are determined for both fault cases line-to-earth, three-phase) if the equivalent voltage source is used at the short-circuit location that are not physically present ($\alpha = 1.0$). For this reason, it makes sense not to take the shunt impedances into account, when calculating the three-phase short-circuit current. In contrast, the deviation when determining the line-to-earth short-circuit current is too large, if shunt impedances in the zero sequence system are not taken into account, especially as the deviations or errors are smaller compared to using the calculation with the actual voltage distribution.

Table 3.2 Capacitor current of the faulty conductor R as a function of the installation location of the capacitance α, related to 1 p.u. $= E/X_{(1)}$

Procedure/error	Amplitude / p.u.		
	$\alpha = 0$	$\alpha = 0.5$	$\alpha = 1$
Line-to-earth			
– Real voltage source	0.1000	0.0495	0.0000
– Equivalent voltage source	0.0000	0.0335	0.0682
Three-phase			
– Real voltage source	0.1000	0.0513	0.0000
– Equivalent voltage source	0.0000	0.0513	0.1000

3.8 Conclusions

In principle, there are several methods for calculating a system short-circuit. To simplify matters, the standard [2] uses the method of the equivalent voltage source at the short-circuit location, which has the great advantage that the load flow before the short circuit occurs, and therefore the tap position of the transformers, does not have to be taken into account. The consequence is that the voltage at the short-circuit location is generally simulated with a uniform voltage factor. In addition, this method does not take into account the shunt impedances, especially when calculating the three-phase short-circuit current, as otherwise currents are determined that do not exist in reality.

References

1. Balzer G, Nelles D, Tuttas C (2009) Kurzschlussstromberechnung nach IEC und DIN EN 60909-0 (VDE 0102):2002–07. VDE-Schriftreihe 77. Aktualisierte Auflage, VDE Verlag, Berlin, p 2
2. IEC 60909–0:2016 Edition 2.0 (2016) Short-circuit current calculations in three-phase a.c. systems—Part 0: calculation of currents. IEC, Geneva
3. Balzer G, Neumann C, Balzer C (2025) Schalt-und Ausgleichsvorgänge in elektrischen Netzen, 2nd edn. Springer, Heidelberg
4. IEEE Std 1584™-2018 (2018) IEEE guide for performing Arc-Flash hazard calculations. IEEE Industry Applications Society, New York
5. IEC 60909–3:2009 Edition 3.0 (2009) Short-circuit currents in three-phase AC systems—Part 3: Currents during separate simultaneous line-to-earth short circuit and partial short-circuit currents flowing through earth. IEC, Geneva
6. Das JC (2018) Power Systems Handbook: Short-circuits an AC and DC systems, vol 1. CRC Press, Taylor & Francis Group, New York
7. Funk G (1962) Der Kurzschluß im Drehstromnetz. R Oldenbourg, München
8. Glover JD, Sarma MS (2002) Power system analysis and design, 3rd edn. Brooks/Cole, Pacific Grove

Chapter 4
Calculation Variables for Equipment

In the following sections, the impedances and other variables of different equipment that are relevant for a short-circuit current calculation are determined. In this context, the positive and zero sequence impedances of overhead lines and cables are derived in detail. In addition, active elements (voltage sources, current sources) are also considered. In general, the quantities in the positive sequence, negative sequence and zero sequence systems are designated with the subscripts (1), (2) and (0). If no confusion is possible, the index is omitted for the representation in the positive sequence system.

4.1 Network Feeders

A network feeder is the combination of various active elements that contribute to the short-circuit current (e.g. generators, motors) and passive elements (e.g. transformers, lines). To avoid having to perform a detailed simulation of various equipment for each calculation, a substitute representation can be selected that consists of a voltage source and an internal impedance.

To calculate the short-circuit current in the network according to Fig. 4.1, the impedance of the network infeed must be determined using the following Eq. (4.1), with reference to the busbar Q.

$$Z_{\mathrm{N}} = \frac{c \cdot U_{\mathrm{nQ}}}{\sqrt{3} \cdot I''_{\mathrm{kN}}} \quad X_{\mathrm{N}} = \frac{Z_{\mathrm{N}}}{\sqrt{1 - (R_{\mathrm{N}}/X_{\mathrm{N}})^2}} \tag{4.1}$$

G. Balzer, *Short-Circuit Currents in Three-Phase Systems*,
https://doi.org/10.1007/978-3-658-50710-7_4

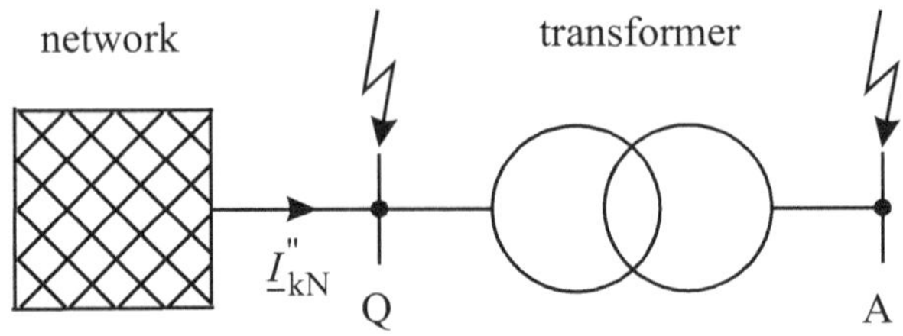

Fig. 4.1 System diagram for determining the three-phase short-circuit current at the busbars Q and A with a network feeder N

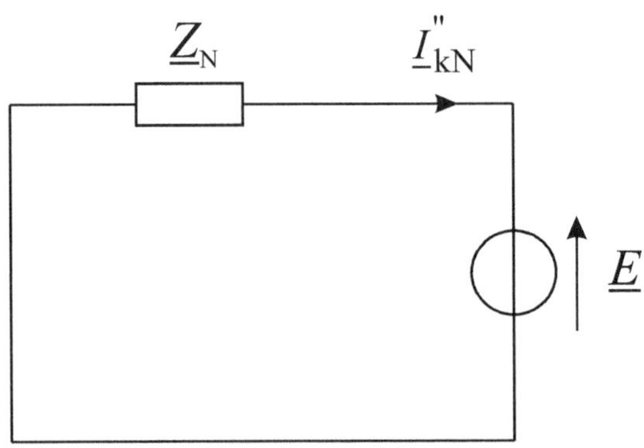

Fig. 4.2 Equivalent circuit for determining the three-phase short-circuit current for a network feeder on busbar Q using the equivalent voltage source at the short-circuit location

With

I''_{kN} initial symmetrical short-circuit current of the network N,
U_{nQ} nominal system voltage (busbar Q),
c voltage factor, Table 7.1.

In many cases, the ratio R_N/X_N is not known, so that simplifications can be made, e.g. for high-voltage networks:

$$R_N/X_N = 0,1 \quad \Rightarrow \quad X_N = 0,995 \cdot Z_N$$

In low-voltage networks, in contrast, smaller cable cross-sections are used so that the resistance will have a greater influence on the result. Information on the impedance ratios of low-voltage lines is given in [1], so that a general impedance ratio for low-voltage networks can be derived from these values.

The equivalent circuit for calculating the short-circuit current using the equivalent voltage source at the short-circuit location as shown in Fig. 4.2 can be specified from the information in Eq. (4.1), $\underline{E} = c \cdot U_{nQ}/\sqrt{3}$.

If a short circuit is assumed on busbar A, Fig. 4.1, the impedance of the network $\underline{Z}_N$ must be converted to the new voltage level using the square of the transformation ratio:

$$\underline{Z}_{\mathrm{A}} = \underline{Z}_{\mathrm{Nt}} = \frac{\underline{Z}_{\mathrm{N}}}{t_{\mathrm{r}}^2} \tag{4.2}$$

With

t_r rated transformation ration of the transformer (main position of the tap changer $t_r = U_{rTOS}/U_{TUS}$).

Depending on the power station feeder and the network topology, a distinction must be made between the minimum and maximum short circuit of a network, Sect. 3.3. The short-circuit power for calculating the impedance is often specified for the initial symmetrical short-circuit current I''_k, Sect. 1.2. In practice, however, it has proved useful to avoid this designation, as only the short-circuit currents are decisive for the selection of equipment (e.g. circuit-breakers). In addition, it is not always known which voltage is used to determine the short-circuit power, since, for example, a system voltage is no longer standardized for the highest voltages for equipment $U_m > 300$ kV and, for example, the voltage designations $U_n = 380$ kV and $U_n = 400$ kV are commonly used in Europe. For this reason, the term short-circuit power should not be used.

The short-circuit current contribution of a network feeder is constant over the entire time range, so that no decay of the AC component takes place. This means that the initial symmetrical short-circuit current $\underline{I}''_{kN}$ is equal to the breaking current $\underline{I}_b$ and the steady state short-circuit current $\underline{I}_k$, Eq. (4.3). A network feeder is therefore a far-from-generator short circuit, Sect. 3.5.

$$\underline{I}''_{\mathrm{kN}} = \underline{I}_{\mathrm{bN}} = \underline{I}_{\mathrm{kN}} \tag{4.3}$$

The zero sequence reactance of the network can be determined, for example, from the line-to-earth short-circuit current and the $R_{(0)}/X_{(0)}$ ratio; the three-phase short-circuit current must be specified in this case. According to Eqs. (4.4) and (4.5), the following applies to the absolute values:

$$\sqrt{\left(2 \cdot R_{(1)} + R_{(0)}\right)^2 + \left(2 \cdot X_{(1)} + X_{(0)}\right)^2} = \frac{\sqrt{3} \cdot c \cdot U_{\mathrm{n}}}{I''_{\mathrm{k1}}} \tag{4.4}$$

$$\sqrt{R_{(1)}{}^2 + X_{(1)}{}^2} = \frac{\sqrt{3} \cdot c \cdot U_{\mathrm{n}}}{I''_{\mathrm{k3}}} \tag{4.5}$$

Summarizing the equations follows:

$$\frac{\left(2 \cdot \frac{R_{(1)}}{X_{(1)}} + \frac{R_{(0)}}{X_{(1)}}\right)^2 + \left(2 + \frac{X_{(0)}}{X_{(1)}}\right)^2}{\left(R_{(1)}/X_{(1)}\right)^2 + 1} = 9 \cdot \left(\frac{I''_{\mathrm{k3}}}{I''_{\mathrm{k1}}}\right)^2 \tag{4.6}$$

$X_{(0)}/X_{(1)}$ the following results after conversion:

$$X_{(0)}/X_{(1)} = -A + \sqrt{A^2 - B} \tag{4.7}$$

With

$$A = 2 \cdot \frac{\frac{R_{(0)}}{X_{(0)}} \cdot \frac{R_{(1)}}{X_{(1)}} + 1}{\left(R_{(0)}/X_{(0)}\right)^2 + 1} \tag{4.8}$$

$$B = \frac{4 \cdot \left(\frac{R_{(1)}}{X_{(1)}}\right)^2 + 4 - 9 \cdot \left(\frac{I''_{k3}}{I''_{k1}}\right)^2 \cdot \left(\frac{R_{(1)}}{X_{(1)}} + 1\right)}{\left(R_{(0)}/X_{(0)}\right)^2 + 1} \tag{4.9}$$

The ratio $R_{(0)}/X_{(0)}$ depends on the structure of a network and can be, for example, a value of $R_{(0)}/X_{(0)} = 0.2$ for an overhead line network.

4.2 Generators

In principle, synchronous and asynchronous generators can be used to generate electrical energy. While synchronous generators are used in thermal and hydraulic power stations, asynchronous generators are used, for example, to convert wind into electrical energy and doubly-fed asynchronous generators are used in pumped storage power stations. In contrast, static converters are used to convert solar energy into electrical energy (photovoltaic systems).

To differentiate between the various generator types, the electrical variables are given the following indices:

- Synchronous generator: GS
- Asynchronous generator: GA
- Doubly-fed asynchronous generator: GD

4.2.1 Synchronous Generators

In accordance with the IEC Standard [2], synchronous generators are represented in the short-circuit current calculation exclusively with the aid of the subtransient reactance and the effective resistance of the generator, so that the initial symmetrical short-circuit current can be determined from this. To simplify matters, the decay behavior during the short-circuit process is simulated using factors μ, λ, Sects. 5.6.1 and 5.7.1. If, on the other hand, the temporal behavior during a short circuit is to be simulated using appropriate calculation programs (transient curves), the corresponding reactances and time constants must be used (Sect. 4.2.1.2).

4.2.1.1 Representation According to the IEC

For a generator that feeds directly into a short circuit, the electrical conditions described in [3, 4] result, so that the subtransient generator impedance is used to calculate the initial symmetrical short-circuit current, Eq. (4.10). The following equations are only to be used if a generator feeds directly into a voltage level, e.g. industrial network. If, on the other hand, a generator with a step-up transformer is considered, the "power station unit" equipment must be used in this case (Sect. 4.3).

$$\underline{Z}_{\mathrm{GS}} = R_{\mathrm{GS}} + \mathrm{j}X_{\mathrm{GS}} = R_{\mathrm{GS}} + \mathrm{j}X''_{\mathrm{d}} \tag{4.10}$$

In order to take into account the increased internal voltage due to a preloading before the short circuit occurs, the impedance $\underline{Z}_{\mathrm{GS}}$ is corrected with a factor K_{GS}, Sect. 8.2. This results in the impedance $\underline{Z}_{\mathrm{GSK}}$, which is used in the short-circuit current calculation.

$$\underline{Z}_{\mathrm{GSK}} = K_{\mathrm{GS}} \cdot \underline{Z}_{\mathrm{GS}} = K_{\mathrm{GS}} \cdot \left(R_{\mathrm{GS}} + \mathrm{j}X''_{\mathrm{d}}\right) \tag{4.11}$$

$$X''_{\mathrm{d}} = \frac{x''_{\mathrm{d}}}{100\%} \cdot \frac{U^2_{\mathrm{rGS}}}{S_{\mathrm{rGS}}} \tag{4.12}$$

With

- R_{GS} resistance of the armature winding,
- X''_{d}, x''_{d} subtransient reactance resp. relative value in %,
- S_{rGS} rated power of the generator,
- U_{rGS} rated voltage of the generator,
- K_{GS} correction factor of the generator impedance, Sect. 8.2.

Since the short-circuit usually takes place at the rated voltage of the generator, the saturated values (X''_{d}) must be used as reactances, which lead to higher short-circuit currents. In contrast, the unloaded synchronous machine may be energized to a reduced voltage in case of peak short-circuit tests in order to keep the mechanical stress low, so that the unsaturated reactances apply in these cases.

The correction factor K_{GS}, the derivation of which is shown in Sect. 8.2, results in:

$$K_{\mathrm{GS}} = \frac{U_{\mathrm{n}}}{U_{\mathrm{rGS}}} \cdot \frac{c_{\max}}{1 + x''_{\mathrm{d}} \cdot \sqrt{1 - \cos^2\varphi_{\mathrm{rGS}}}} \tag{4.13}$$

With

- U_{n} nominal system voltage,
- φ_{rGS} power factor under rated conditions ($I_{\mathrm{rGS}}/U_{\mathrm{rGS}}$),
- $c_{\max}$ voltage factor Table 7.1.

For the maximum, three-phase short-circuit current, this results in:

$$\underline{I}''_{\mathrm{k}} = \frac{cU_{\mathrm{n}}/\sqrt{3}}{\underline{Z}_{\mathrm{GS}} \cdot K_{\mathrm{GS}}} = \frac{U_{\mathrm{rGS}}/\sqrt{3}}{R_{\mathrm{GS}} + \mathrm{j}X''_{\mathrm{d}}} \left(1 + x''_{\mathrm{d}} \cdot \sqrt{1 - \cos^2 \varphi_{\mathrm{rGS}}}\right) \tag{4.14}$$

When calculating the minimum short-circuit current, the value of c_{min} is used in Eq. (4.14), while the correction factor is always formed with c_{max}. The effective resistance R_{GS} used in Eq. (4.11) is the resistance of the armature winding, so that the armature resistance R_{GS} must always be used when calculating the initial symmetrical short-circuit current.

In contrast, when calculating the K factor to determine the peak short-circuit current, a fictitious resistance R_{GSf} is used, which also takes into account the decay process of the subtransient component of the short-circuit current within the first half period. This is due to the reduction of the generator voltage and is particularly dependent on compensating currents in the damper winding of the synchronous machine, Eq. (4.18). For this reason, the values for R_{GSf} are higher than the corresponding armature resistances R_{GS} and, based on experience, should be used as follows:

- $R_{\mathrm{GSf}} = 0.05 \cdot X''_{\mathrm{d}}$ with $U_{\mathrm{rGS}} > 1\,\mathrm{kV}$, $S_{\mathrm{rGS}} \geq 100\,\mathrm{MVA}$
- $R_{\mathrm{GSf}} = 0.07 \cdot X''_{\mathrm{d}}$ with $U_{\mathrm{rGS}} > 1\,\mathrm{kV}$, $S_{\mathrm{rGS}} < 100\,\mathrm{MVA}$
- $R_{\mathrm{GSf}} = 0.15 \cdot X''_{\mathrm{d}}$ with $U_{\mathrm{rGS}} < 1\,\mathrm{kV}$

For the reason given above, the value R_{GSf} is used exclusively for calculating the peak short-circuit current i_{p}. In contrast, the actual armature resistance R_{GS} must be used when determining the DC time constant T_{a} or the DC element I_{DC} for the design of the current transformer.

For the generator voltage U_{rGS} used in Eq. (4.13), it is assumed that the actual terminal voltage U_{GS} does not deviate from the rated voltage U_{rGS}. If, on the other hand, the terminal voltage is changed so that U_{rGS} is exceeded, then $U_{\mathrm{GSmax}} = U_{\mathrm{rGS}}$ $(1 + p_{\mathrm{GS}})$ must be used for U_{rGS} in Eq. (4.13). The factor p_{G} represents the control range of the voltage regulator.

The correction factor specified in Eq. (4.13) also applies to the calculation of unbalanced short-circuit currents, i.e. the generator impedances in the positive sequence, negative sequence and zero sequence systems must be multiplied by the same factor K_{GS}. In the case of an impedance between the generator neutral point and earth, as is possible with generators in low-voltage systems, for example, the neutral point impedance is not multiplied by the correction factor. The low-impedance treatment of the generator neutral point is a necessity here if the associated network has the "TN network" (terra-neutral) protective measure, so that any line-to-earth short circuit immediately leads to a tripping so that no unacceptable touch voltages can arise.

Table 4.1 provides an overview of typical data for generators with different rated power.

Table 4.1 Data of synchronous generators, exemplary

Variables	Turbo-generator	Salient-pole-generator
$x''_d/\%$	10–30	12–30
$x'_d/\%$	14–40	20–45
$x_d/\%$	140–300	80–140
$x_{(2)}/\%$	10–30	12–30
$x_{(0)}/\%$	3–10	5–20
T''_d/s	0.02–0.04	0.02–0.1
T'_{d0}/s	5–15	4–10
T_a/s	0.05–0.4	0.1–0.4

x''_d subtransient reactance (saturated, *d* axis)
x'_d transient reactance (saturated, *d* axis)
x_d synchronous reactance (unsaturated, *d* axis)
$x_{(2)}$ negative sequence reactance
$x_{(0)}$ zero sequence reactance
T''_d subtransient time constant
T'_{d0} transient open-circuit time constant
T_a armature (DC) time constant

4.2.1.2 Representation of the Reactances and Time Constants (Operating Frequency)

In general, the modeling of a synchronous machine is based on Park's two-axis theory, whereby a three-phase system is broken down into the diagonal components of the rotor [4]. Due to the structure, the stator has a three-phase winding, whereas the field winding and possibly the damper winding are located on the rotor. In principle, the reactances are different in the direct axis and quadrature axis, which is particularly pronounced in a salient pole rotor; in addition, the field winding is located exclusively in the direct axis. Figure 4.3 shows the schematic arrangement of the windings in the direct (*d*) and quadrature (*q*) axis of the synchronous machine. While the damper windings are short-circuited in both axis in order to short-circuit transient phenomenon, excitation takes place exclusively in the *d*-axis.

In principle, different status areas can be defined after the occurrence of a short circuit, which result from the linking of the various windings:

- Subtransient area,
- transient area,
- steady state area.

The causes of these areas lie in the magnetic interactions of the different windings (armature, field and damper). The compensating currents in the damper winding decay quickly (subtransient area), while the decay behavior in the field winding is slower (transient area). This results in the following definitions:

- Subtransient reactance X''_d: This reactance consists of the leakage reactances of the armature, rotor and damper windings and the magnetizing reactance of the direct axis, Figs. 4.4 and 4.5.

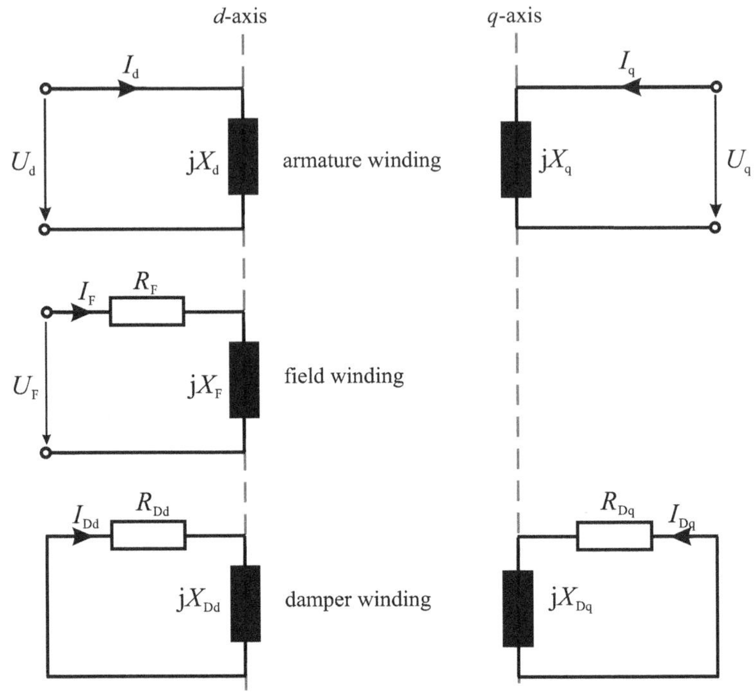

Fig. 4.3 Representation of the synchronous machine in *d*, *q*-axis

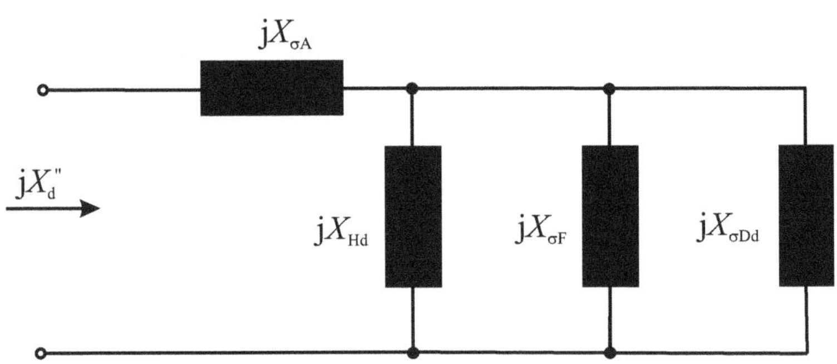

Fig. 4.4 Equivalent circuit diagram of the synchronous machine (subtransient, direct axis), $X_{\sigma A}$ armature leakage reactance, X_{Hd} magnetizing reactance, $X_{\sigma E}$ leakage reactance of the field winding, $X_{\sigma Dd}$ leakage reactance of the damper winding

- Transient reactance X'_d: Corresponds to the subtransient reactance after the decay process in the damper winding.
- Synchronous reactance X_d: Steady-state reactance after all oscillations have decayed.

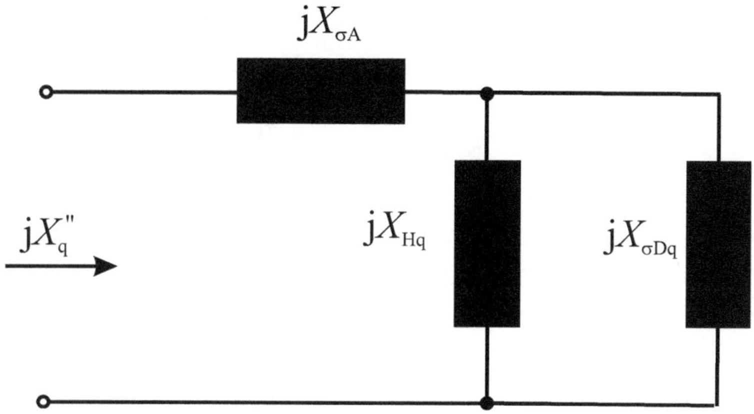

Fig. 4.5 Equivalent circuit diagram of the synchronous machine (subtransient, quadrature axis). $X_{\sigma A}$ armature leakage reactance, X_{Hq} magnetizing reactance, $X_{\sigma Dq}$ leakage reactance of the damper winding

In machines, the 0, *d*, *q* components, which rotate with the rotor, are often used instead of the 0, 1, 2 components. Here, the direct axis (*d*-axis) represents the magnetic axis of the magnet wheel, while the quadrature axis (*q*-axis) is offset by 90°. In turbo-generators, the differences in the reactances are small due to the symmetrical design of the rotor, whereas they are significant in salient-pole generators.

In the subtransient case, the input reactance in the direct direction consists of the leakage reactance of the armature $X_{\sigma A}$, the magnetizing reactance X_{Hd} and the leakage reactances of the field winding ($X_{\sigma F}$) and damper winding ($X_{\sigma Dd}$), so that according to Eq. (4.15) or Fig. 4.4:

$$X''_{d} = X_{\sigma A} + \frac{1}{\frac{1}{X_{Hd}} + \frac{1}{X_{\sigma F}} + \frac{1}{X_{\sigma Dd}}} \tag{4.15}$$

To simplify matters, the resistances of the various windings are neglected in Fig. 4.4.

The equivalent circuit diagram in the quadrature axis is shown in Fig. 4.5 without a field winding; the subtransient quadrature impedance is calculated according to Eq. (4.16).

$$X''_{q} = X_{\sigma A} + \frac{1}{\frac{1}{X_{Hq}} + \frac{1}{X_{\sigma Dq}}} \tag{4.16}$$

In the event of a transient phenomenon, e.g. a short circuit at the generator terminals, the resistances of the field and damper windings, which are in series with the corresponding leakage reactances, have an effect. Since the resistance R_F of the field winding is smaller than that of the damper winding R_D, it can be neglected and

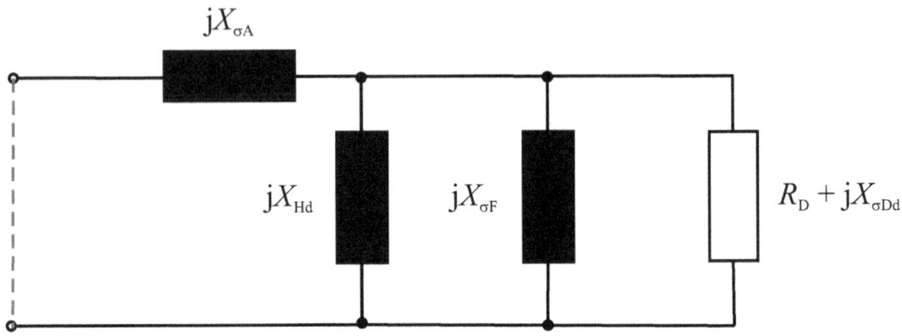

Fig. 4.6 Equivalent circuit diagram for a terminal short circuit or open circuit, decay process in the damper winding. R_D resistance of the damper winding

the current in the damper winding decays quickly, so that, for simplification, Fig. 4.6 applies by short-circuited input terminals as shown in Fig. 4.4.

Taking Fig. 4.6 into account, the following values result if a terminal short-circuit is assumed and all reactances are short-circuited, as seen from the damper winding:

$$X_D = X_{\sigma Dd} + \frac{1}{\frac{1}{X_{\sigma A}} + \frac{1}{X_{Hd}} + \frac{1}{X_{\sigma F}}} \tag{4.17}$$

The subtransient time constant T''_d is used to describe the decay process in the damper winding, taking into account the resistance R_D of the damper winding.

$$T''_d = \frac{X_D}{\omega \cdot R_D} \tag{4.18}$$

A time constant T''_{d0} can also be determined under open-circuit conditions by leaving the connection open as shown in Fig. 4.6, so that the following applies:

$$T''_{d0} = \frac{X_{D0}}{\omega \cdot R_D} \quad X_{D0} = X_{\sigma Dd} + \frac{1}{\frac{1}{X_{Hd}} + \frac{1}{X_{\sigma F}}} \tag{4.19}$$

The time constants for the q-axis are determined accordingly, assuming that the q-axis does not have an excitation winding.

$$T''_q = \frac{X_Q}{\omega \cdot R_Q} \quad X_Q = X_{\sigma Dq} + \frac{1}{\frac{1}{X_{\sigma A}} + \frac{1}{X_{Hq}}} \tag{4.20}$$

and

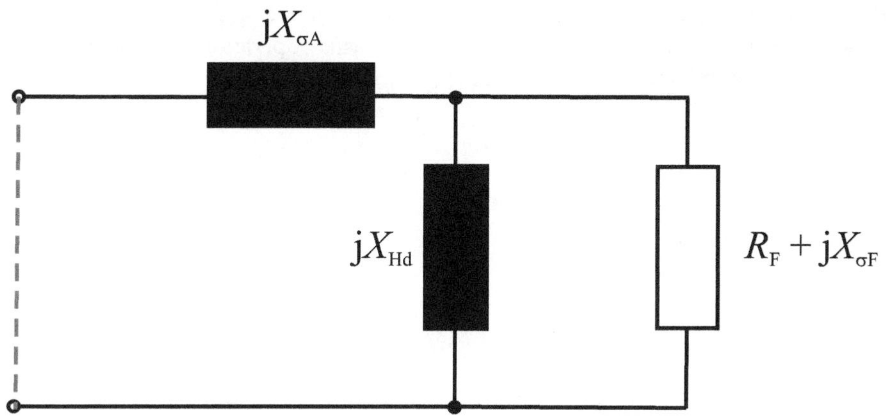

Fig. 4.7 Equivalent circuit diagram for a terminal short circuit or open-circuit, decay process in the field winding. R_F Resistance of the field winding

$$T''_{q0} = \frac{X_{Q0}}{\omega \cdot R_Q} \quad X_{Q0} = X_{\sigma Dq} + X_{Hq} \tag{4.21}$$

The transient reactances X'_d and X'_q are determined on the basis of Figs. 4.4 and 4.5 and Eqs. (4.15) and (4.16) under the condition that the damper winding is assumed to be interrupted:

$$X'_d = X_{\sigma A} + \frac{1}{\frac{1}{X_{Hd}} + \frac{1}{X_{\sigma F}}} \tag{4.22}$$

$$X'_q = X_{\sigma A} + X_{Hq} \tag{4.23}$$

After the decay process, the damper winding can be regarded as interrupted, so that the transient time constant T'_d can be determined according to Fig. 4.7.

$$T'_d = \frac{X_F}{\omega \cdot R_F} \quad X_F = X_{\sigma F} + \frac{1}{\frac{1}{X_{\sigma A}} + \frac{1}{X_{Hd}}} \tag{4.24}$$

As in the case of the subtransient time constant, open-circuit time constant can also be determined in the transient range.

$$T'_{d0} = \frac{X_{F0}}{\omega \cdot R_F} \quad X_{F0} = X_{Hd} + X_{\sigma F} \tag{4.25}$$

Depending on the time constants, the transient processes described above transfer to steady-state operation, which corresponds to normal operation, so that the input impedance is equal to the synchronous reactance (Figs. 4.4 and 4.5).

$$X_{\text{d}} = X_{\sigma\text{A}} + X_{\text{Hd}} \quad X_{\text{q}} = X_{\sigma\text{A}} + X_{\text{Hq}} \tag{4.26}$$

Taking into account Eqs. (4.22) and (4.23), the relationship between the transient time constants can be summarized, resulting in:

$$T'_{\text{d}} = T'_{\text{d0}} \cdot \frac{X'_{\text{d}}}{X_{\text{d}}} \tag{4.27}$$

In the event of a short-circuit at the terminals of the synchronous machine at voltage zero crossing, a DC component occurs as a result of the inductive short-circuit current, the armature time constant T_{a} (DC time constant) of which is determined from the subtransient reactances and the armature resistance $R_{\text{A}} = R_{\text{GS}}$ according to Eq. (4.28).

$$T_{\text{a}} = \frac{X''_{\text{d}} + X''_{\text{q}}}{2 \cdot \omega \cdot R_{\text{GS}}} \approx \frac{X''_{\text{d}}}{\omega \cdot R_{\text{GS}}} \tag{4.28}$$

System Fault

If, in contrast to the above presentation, the contribution of a generator to a system short-circuit is considered, the effective short-circuit impedance increases by the impedance $\underline{Z}_{\text{N}}$ between the generator terminal and the fault location, resulting in the following values:

$$\text{subtransient}: \underline{Z}''_{\text{k}} = R_{\text{GS}} + R_{\text{N}} + \text{j}\left(X''_{\text{d}} + X_{\text{N}}\right) \tag{4.29}$$

$$\text{transient}: \underline{Z}'_{\text{k}} = R_{\text{GS}} + R_{\text{N}} + \text{j}\left(X'_{\text{d}} + X_{\text{N}}\right) \tag{4.30}$$

$$\text{synchronous}: \underline{Z}_{\text{k}} = R_{\text{GS}} + R_{\text{N}} + \text{j}(X_{\text{d}} + X_{\text{N}}) \tag{4.31}$$

The time constants (additional index SF, system fault) can be determined approximately by adding the network reactance X_{N} to the armature leakage reactance $X_{\sigma\text{A}}$, neglecting the resistance R_{N} [5].

$$T''_{\text{dSF}} \approx T''_{\text{d}} \cdot \frac{1 + \dfrac{X_{\text{N}}}{X''_{\text{d}}}}{1 + \dfrac{X_{\text{N}}}{X'_{\text{d}}}} \quad T'_{\text{dSF}} = T'_{\text{d}} \cdot \frac{1 + \dfrac{X_{\text{N}}}{X'_{\text{d}}}}{1 + \dfrac{X_{\text{N}}}{X_{\text{d}}}} \tag{4.32}$$

$$T_{\text{aF}} = \frac{X''_{\text{d}} + X_{\text{N}}}{\omega \cdot (R_{\text{GS}} + R_{\text{N}})} \quad \text{with } X''_{\text{d}} = X''_{\text{q}} \tag{4.33}$$

Negative Sequence Reactance

Due to the rotating field, which rotates at twice the synchronous speed in the opposite direction to the rotor, the negative sequence reactance $X_{(2)GS}$ is equal to the mean value of the subtransient reactances of the direct and quadrature axis [5].

$$X_{(2)GS} = \frac{X''_d + X''_q}{2} \tag{4.34}$$

Zero Sequence Reactance

In principle, the magnetic fields should cancel each other out due to the physical arrangement of the winding, so that the zero sequence reactance is generally low. According to [5], the following guide values result:

$$X_{(0)GS} = \left(\frac{1}{3} \cdots \frac{1}{6}\right) \cdot X'_d \tag{4.35}$$

The neutral points of high-voltage generators are usually insulated from earth, so that zero sequence reactance is not required for the short-circuit current calculation. This is in contrast to low-voltage generators, which are earthed directly or via a reactance as a result of the system protection used (TN network). In such cases, earthing is often carried out via a reactance in order to suppress the harmonics that occur.

Short-Circuit Current Profile

The subtransient and transient behavior of synchronous machines is clearly evident in the event of a terminal or system short circuit and load shedding. In the event of a terminal short circuit, the subtransient reactances act in the first moment and are then replaced by the transient reactances. After all transient phenomena have subsided, the steady-state short-circuit current follows. The transient phenomena can be described for a three-phase terminal short-circuit $i_k(t)$ according to Eq. (4.36), with the terminal (rated) voltage of the generator U_{rGS}:

$$i_k(t) \approx \sqrt{\frac{2}{3}} \cdot U_{rGS} \cdot \left\{ \left[\frac{1}{X''_d} - \frac{1}{X'_d}\right] \cdot e^{-t/T''_d} + \left[\frac{1}{X'_d} - \frac{1}{X_d}\right] \cdot e^{-t/T'_d} + \frac{1}{X_d} \right\} \cdot \cos(\omega \cdot t) + \sqrt{\frac{2}{3}} \cdot U_{rGS} \cdot \left\{ -\frac{1}{2} \cdot \left[\frac{1}{X''_d} + \frac{1}{X''_q}\right] \cdot e^{-t/T_a} - \frac{1}{2} \cdot \left[\frac{1}{X''_d} - \frac{1}{X''_q}\right] \cdot e^{-t/T_a} \cdot \cos(2 \cdot \omega \cdot t) \right\} \tag{4.36}$$

Since the following applies approximately for turbo-generators:

$$X''_d \approx X''_q \tag{4.37}$$

The simplified result for the short-circuit current is:

$$i_\mathrm{k}(t) \approx \sqrt{\frac{2}{3}} \cdot U_\mathrm{rGS} \cdot \left(\begin{array}{l} \left\{ \left[\frac{1}{X''_\mathrm{d}} - \frac{1}{X'_\mathrm{d}} \right] \cdot \mathrm{e}^{-t/T''_\mathrm{d}} + \left[\frac{1}{X'_\mathrm{d}} - \frac{1}{X_\mathrm{d}} \right] \cdot \mathrm{e}^{-t/T'_\mathrm{d}} + \frac{1}{X_\mathrm{d}} \right\} \cdot \cos(\omega \cdot t) \\ - \frac{1}{X''_\mathrm{d}} \cdot \mathrm{e}^{-t/T_\mathrm{a}} \end{array} \right) \tag{4.38}$$

Equations (4.36) and (4.38) apply to the case where the synchronous machine is at no-load before the short-circuit occurs. The bracketed expression associated with the cosine term represents the decay of the AC component, while the remaining part represents the DC component, Eqs. (4.39) and (4.40).

- AC component:

$$i_\mathrm{kAC}(t) \approx \sqrt{\frac{2}{3}} \cdot U_\mathrm{rGS} \cdot \left\{ \left[\frac{1}{X''_\mathrm{d}} - \frac{1}{X'_\mathrm{d}} \right] \cdot \mathrm{e}^{-t/T''_\mathrm{d}} + \left[\frac{1}{X'_\mathrm{d}} - \frac{1}{X_\mathrm{d}} \right] \cdot \mathrm{e}^{-t/T'_\mathrm{d}} + \frac{1}{X_\mathrm{d}} \right\} \cdot \cos(\omega \cdot t) \tag{4.39}$$

- DC component:

$$i_\mathrm{kDC}(t) \approx \sqrt{\frac{2}{3}} \cdot U_\mathrm{rGS} \cdot \left(- \frac{1}{X''_\mathrm{d}} \cdot \mathrm{e}^{-t/T_\mathrm{a}} \right) \tag{4.40}$$

Depending on the time constants, it is possible that the AC component decays faster than the DC component. This is particularly the case for large generators with large time constants T_a according to Eq. (4.28). The consequence is that in a time range, e.g. $t \sim 100$ ms, the current curve over time has no zero crossings. Since three-phase circuit-breakers are dependent on current zero crossings during interruption, missing zero crossings should be avoided in the case of short circuits, Sect. 3.5.

In the event that the synchronous machine feeds into a short circuit under load conditions, the following generally applies to the AC component of the short-circuit current:

$$I_\mathrm{kAC}(t) = \left(I''_\mathrm{k} - I'_\mathrm{k} \right) \cdot \mathrm{e}^{-t/T''_\mathrm{d}} + \left(I'_\mathrm{k} - I_\mathrm{k} \right) \cdot \mathrm{e}^{-t/T'_\mathrm{d}} + I_\mathrm{k} \tag{4.41}$$

With the currents and voltages (resistances are neglected):

$$I''_\mathrm{k} = \frac{E''}{X''_\mathrm{d}} \qquad E'' \approx U_\mathrm{rGS}/\sqrt{3} + I_\mathrm{GS} \cdot X''_\mathrm{d} \cdot \sin \varphi_\mathrm{GS} \tag{4.42}$$

$$I'_\mathrm{k} = \frac{E'}{X'_\mathrm{d}} \qquad E' \approx U_\mathrm{rGS}/\sqrt{3} + I_\mathrm{GS} \cdot X'_\mathrm{d} \cdot \sin \varphi_\mathrm{GS} \tag{4.43}$$

$$I_k = \frac{E}{X_d} \quad E \approx U_{rGS}/\sqrt{3} + I_{GS} \cdot X_d \cdot \sin \varphi_{GS} \tag{4.44}$$

With

U_{rGS} rated voltage (generator),
I_{GS} load current of the generator before the short circuit occurs,
sin φ_{GS} power factor between voltage U_{rGS} and current I_{GS}, φ is positive for overexcitation, negative for underexcitation of the generator,
E internal generator voltage (subtransient, transient, synchronous).

In the event of a system fault, the reactance between the generator terminals and the fault location must be added to the relevant generator reactance in Eqs. (4.42) to (4.44), Eqs. (4.29) to (4.31), as shown in Sect. 11.3 using Fig. 11.1.

4.2.2 *Asynchronous Generator*

According to IEC [2], the calculation of the contribution to the short-circuit current is considered both for asynchronous generators and for doubly-fed asynchronous generators, which are used, for example, in the conversion of wind energy or in pumped storage power stations. In the analysis, the generator and the step-up transformer are considered as a single unit, so that each is referred to as a power station unit.

4.2.2.1 Asynchronous Generators (e.g. Squirrel Cage Motor, Slip Ring Motor)

In the following, the asynchronous generator and the step-up transformer are referred to as a power station unit, which is used, for example, to convert wind energy into electrical energy and feed it into the higher-level high-voltage network. Fig. 4.8 shows the corresponding equivalent circuit. When calculating the short-circuit currents, the electrical data is referred to the high-voltage level.

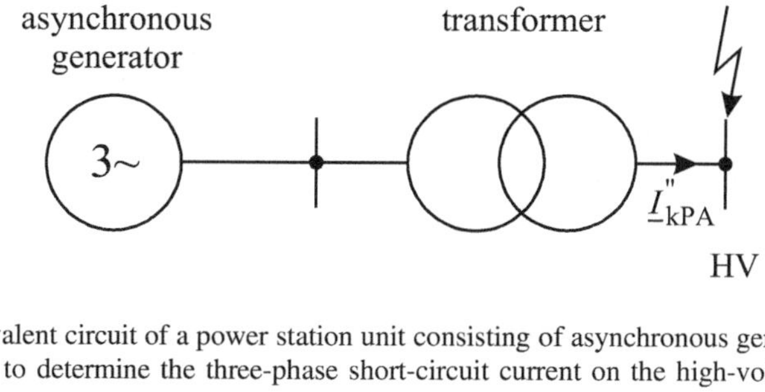

Fig. 4.8 Equivalent circuit of a power station unit consisting of asynchronous generator and step-up transformer to determine the three-phase short-circuit current on the high-voltage side of the transformer

The impedance Z_{GA} of asynchronous generators is determined in a similar way to the consideration of motors, Sect. 4.6.1.

$$Z_{GA} = \frac{1}{I_{LR}/I_{rGA}} \cdot \frac{U_{rGA}}{\sqrt{3} \cdot I_{rGA}} = \frac{1}{I_{LR}/I_{rGA}} \cdot \frac{U_{rGA}^2}{S_{rGA}} \tag{4.45}$$

With

$$\underline{Z}_{GA} = R_{GA} + jX_{GA} \tag{4.46}$$

U_{rGA} rated voltage of the asynchronous generator,
I_{rGA} rated current of the asynchronous generator,
S_{rGA} rated power of the asynchronous generator,
L_{LR}/I_{rGA} ratio of the symmetrical locked-rotor current to the rated current,
R_{GA} resistance of the asynchronous generator,
X_{GA} saturated reactance of the asynchronous generator.

If no other values are known, the complex generator impedance can be determined for simplification purposes using a ratio of $R_{GA}/X_{GA} = 0.1$. The total impedance of the power station unit (asynchronous generator + step-up transformer) $\underline{Z}_{PA}$ in the positive sequence system is determined according to Eq. (4.47), with reference to the high-voltage side of the transformer.

$$\underline{Z}_{PA} = t_r^2 \cdot \underline{Z}_{GA} + \underline{Z}_{THV} \tag{4.47}$$

With

$\underline{Z}_{THV}$ impedance of the transformer, related to the high-voltage side,
t_r rated transformation ratio of the transformer.

As the asynchronous generator takes an inductive load current from the network to cover its excitation, the correction of the impedance of the step-up transformer according to Sect. 4.8.1 is not necessary, as the correction factor K is only necessary for transformers if the inductive load current before the short-circuit occurs and the subsequent short-circuit current have the same phase angle. This is therefore not the case with an asynchronous generator (squirrel-cage motor).

To calculate the breaking current of asynchronous machines, the fictitious R_{GAf}/X_{GAf} ratios must generally be used, which also take into account the decay of the AC component. These values are given in Sect. 4.6.1.

When calculating unbalanced short circuits, the negative sequence system is equal to the positive sequence system, while the zero sequence system depends on the type of earthing of the transformer on the high-voltage side.

4.2.2.2 Doubly Fed Induction Generator

In a doubly fed induction generator, the stator and rotor windings are connected to network voltage so that only slip ring rotors can be used [6]. While the stator is

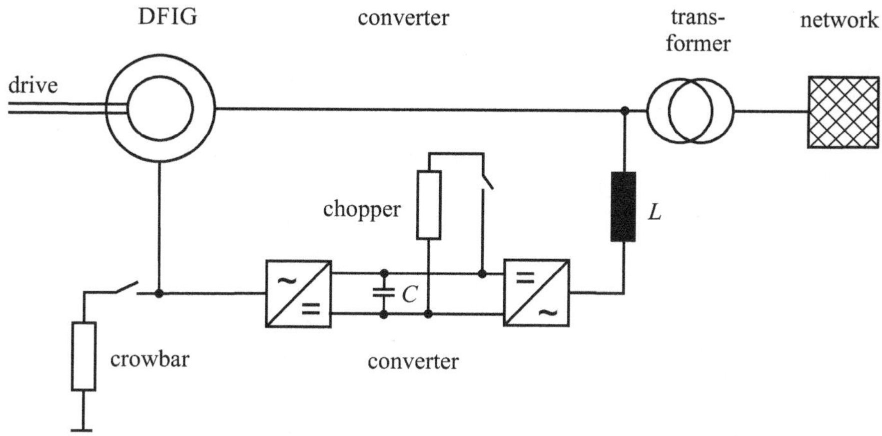

Fig. 4.9 Power station unit consisting of a doubly fed induction generator (DFIG) including step-up transformer

connected directly to the network, the rotor circuit is connected to the network via a converter (IGBT technology) so that reactive power can be both supplied and absorbed. The basic structure is shown in Fig. 4.9. In this case, too, all values refer to the high-voltage side of the transformer.

The "crowbar" and "chopper" elements shown in Fig. 4.9 are used to reduce the current contribution during a short-circuit so that the maximum peak short-circuit current is not reached [7, 8]. For example, by using the crowbar, the rotor circuit is short-circuited so that the high rotor current does not flow via the converter in the event of a short circuit. In these cases, the doubly-fed asynchronous generator acts as a squirrel cage motor. In order to continue to calculate with the equivalent voltage source at the short-circuit location, a fictitious positive sequence impedance $\underline{Z}_{PD}$ of the power station unit is determined according to [2] from the peak short-circuit current measured in practice, taking a factor κ_{PD} into account. This factor describes the reduction of the peak short-circuit current based on a fictitious initial symmetrical short-circuit current through the protection system of the installation (crowbar or chopper). The following therefore applies to the impedance:

$$Z_{PD} = \frac{\sqrt{2} \cdot \kappa_{PDf} \cdot U_{rTOS}}{\sqrt{3} \cdot i_{PDmax}} \tag{4.48}$$

$$\underline{Z}_{PD} = R_{PD} + jX_{PD} \tag{4.49}$$

With

U_{rTOS} rated voltage of the transformer on the high-voltage side,
κ_{PDf} factor (fictitious) for calculating the peak short-circuit current specified by the manufacturer,
i_{PDmax} peak short-circuit current for a three-pole short-circuit,

R_{PD} resistance of the asynchronous generator and transformer,
X_{GA} reactance of the asynchronous generator and transformer.

If the factor κ_{PD} is not specified by the manufacturer, a value of $\kappa_{PDf} = 1.7$ can be used. To calculate the initial symmetrical short-circuit current, a ratio of $R_{PD}/X_{PD} = 0.1$ can be taken into account to determine the complex impedance $\underline{Z}_{PD}$; this value κ is then also used to determine the DC component in the system. The two variables $\kappa_{PD}\{f(R_{PD}/X_{PD})\}$ and κ_{PDf} $\{f(R_{PDf}/X_{PD})\}$, Eq. (4.48), are therefore not correlated, but are comparable with the actual resistance R_{GS} of a synchronous generator and the fictitious resistance R_{GSf}.

The negative sequence and zero sequence impedances of the power station unit depend on the control strategy and the earthing and type of transformer and must therefore be obtained from the manufacturer.

The determination of the impedance from a short-circuit test, Eq. (4.48), corresponds to the IEC Standard from 2016 [2], which is not always clear in practice. Since the behavior of a doubly fed induction motor basically corresponds to that of a generator, it is intended to carry out the short-circuit impedance according to Eqs. (4.10) to (4.12), including a transformer correction, so that the manufacturer has to provide the necessary information. A proposal is shown in [9, 10], using the parameters of a doubly fed induction motor. This means that the impedance of the asynchronous generator is determined from the resistance and the saturated reactance according to Sect. 4.6.2, Eq. (4.57), and the transformer impedance is determined from the manufacturer's specifications. Measurements of the short-circuit current curve are listed in [11]. In these cases, the generator and transformer equipment are simulated individually.

4.3 Power Station Unit with Synchronous Generator

The synchronous generator and the step-up transformer of a power station are combined into one generation unit for the calculation of short-circuits in the network, so that a short circuit on the generator busbar is not covered by this simulation. The consequence is that, according to the valid IEC Standard [2], it is not permissible to split a power station into its two components for the short-circuit current calculation: synchronous generator—step-up transformer (or network transformer) and to apply the individual impedance correction factors of equipment (Sects. 4.2.1.1 and 4.7.1) if short-circuit currents are calculated on the high-voltage side of the transformer.

In principle, a short-circuit current calculation with a maximum voltage factor c_{max} leads to values that can be on the unsafe side if the load current before the short-circuit occurs is large and inductive. For this reason, it is necessary to apply an impedance correction for the entire power station unit. A distinction is made between two possibilities:

- Power station unit with transformer tap changer,
- Power station unit with transformer tap changer.

In this classification, it is assumed that in the case of a tap changer, voltage regulation on the HV side of the power station unit is performed by the transformer and the generator voltage is kept at its rated value. In the other case, the voltage is controlled by the voltage regulator of the generator and the step-up transformer does not have automatic voltage regulation, but taps that can be changed manually. The derivation of the listed correction factor K_{PS} is shown in [12, 13].

The equations given in the following sections are always used to determine the contribution of the power station to the network short circuit. In the event of a fault on the generator busbar, special considerations must be made, which are listed in the specification [2]. However, as these calculations are seldom (only when designing a new power station), they are not shown in this context.

The following applies to the calculation of the impedance of the $\underline{Z}_{PSK}$ power station (in relation to the HV side) according to [13]:

$$\underline{Z}_{PSK} = K_{PS} \cdot \left(t_r^2 \cdot \underline{Z}_{GS} + \underline{Z}_{TOS}\right) \tag{4.50}$$

With K_{PS} for the transformer with tap changer.

$$K_{PS} = \frac{U_{nQ}^2}{U_{rGS}^2} \cdot \frac{1}{t_r^2} \cdot \frac{c_{max}}{1 + \left|x_d'' - x_T\right| \cdot \sqrt{1 - \cos^2\varphi_{rGS}}} \tag{4.51}$$

With K_{PS} for the transformer without tap-changer.

$$K_{PS} = \frac{U_{nQ}}{U_{rGS} \cdot (1 + p_{GS})} \cdot \frac{1}{t_r} \cdot \frac{(1 \pm p_T) \cdot c_{max}}{1 + x_d'' \cdot \sqrt{1 - \cos^2\varphi_{rGS}}} \tag{4.52}$$

This means:

$\underline{Z}_{GS}$ impedance of the generator ($\underline{Z}_{GS} = R_{GS} + jX_d''$), Sect. 4.2.1.1,
$\underline{Z}_{TOS}$ impedance of the transformer on the HV side,
U_{nQ} nominal system voltage at connection point Q,
U_{rGS} rated voltage of the synchronous generator,
t_r rated transformation ratio of the transformer (main position, tap-changer),
x_d'' relative subtransient reactance of the synchronous generator,
x_T relative short-circuit reactance of the transformer in main position,
p_{GS} range of the synchronous generator voltage regulator,
p_T tap-changer range of the transformer.

The derivation of Eq. (4.51) assumes that the voltage at the generator busbar is kept constant by the voltage regulator. If, in contrast to this, the terminal voltage of the generator can also be changed, the maximum generator voltage must be determined in accordance with Sect. 4.2.1 (generator).

In the case of an unbalanced fault, the impedance correction factor is also used in the negative sequence and zero sequence systems. If a generator is operated in the underexcited range, higher short-circuit currents are generally only possible for unbalanced faults with earth contact, so that the superposition method should be used as an example.

4.4 Power Station Unit with Self-Commutated Converter

In connection with the generation of renewable energies (wind/sun or photovoltaics), self-commutated converters have been increasingly used in recent years [8, 14]. Figure 4.10 shows an example of the basic arrangement of a wind feed with an asynchronous generator and a PV feeder. The simulation is generally carried out in a power station unit, in which the electrical specifications refer to the high voltage side of the transformer as shown in Fig. 4.10. This arrangement is defined as a generation unit or power station unit (PF).

The units shown in Fig. 4.10 are modeled as a current source according to [2], as these generation units have the ability to feed in a constant current with any phase angle during the short-circuit duration depending on the regulation. As the grid operators may require the feed-in of an inductive current due to the voltage level in the event of a short circuit [15], this can be implemented accordingly in the short-circuit current calculation, as shown in Chap. 13. The simulation of the converter including the transformer is carried out as follows, whereby the index "CS" (current source) is used for the elements of the current source:

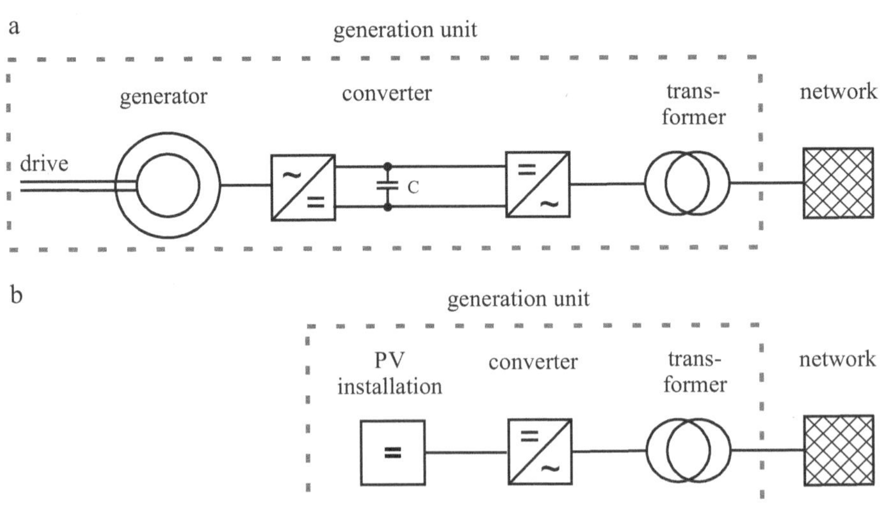

Fig. 4.10 Application of self-commutated converters in a generation unit (power station unit) (**a**) Wind feeder, (**b**) PV feeder

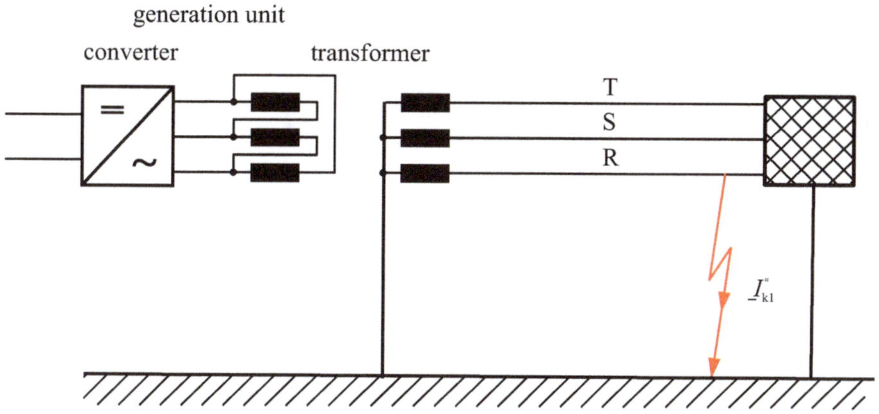

Fig. 4.11 Consideration of the zero sequence system or a line-to-earth fault

- Positive sequence system: The current of the current source is specified by the manufacturer (on the HV side). The shunt impedance $\underline{Z}_{(1)CS}$ (Fig. 13.6) is assumed to be infinite.
- Negative sequence system: The current of the current source generally depends on the control strategy and the value of the current in the positive sequence system. The size of the shunt impedance $\underline{Z}_{(2)CS}$ in the negative sequence system depends on the type of converter and the control strategy and must be specified by the manufacturer or can be assumed to be infinite.
- Zero sequence system: To determine the zero sequence impedance, the vector group, the design and the neutral point treatment of the transformer must be taken into account so that the zero sequence impedance $\underline{Z}_{(0)CS}$ of the generating unit is equal to the zero sequence impedance of the transformer, Fig. 4.11. In the case of an earthed neutral point of the transformer with vector group Yd5, only the zero sequence impedance of the transformer has an effect on the high-voltage side, as the zero sequence impedance on the secondary side is short-circuited by the delta winding. If, in contrast, a converter is used for a PV feeder without a transformer, then the zero sequence impedance $\underline{Z}_{(0)CS} \rightarrow \infty$, as is the case when used in low-voltage grids.

4.5 Generation Units (Wind and Photovoltaic Installations)

Different generation units (wind, PV) in accordance with Sects. 4.2.2.1, 4.2.2.2 and 4.4 can be combined into one generation unit. In accordance with VDE application rule 4120 [15], two different basic types of generating units can be defined, depending on the network connection points NP1 and NP2:

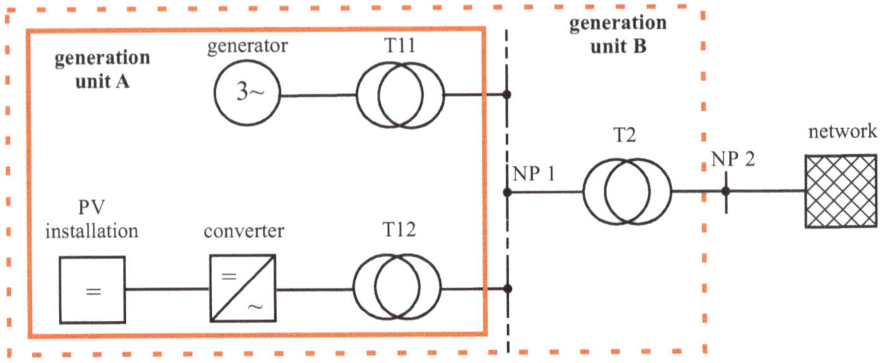

Fig. 4.12 Network example for generation units. (**a**) Without step-up transformer T2, (**b**) with step-up transformer T2

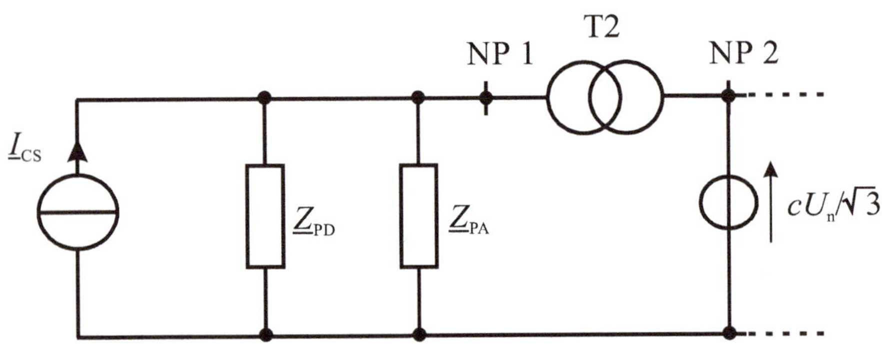

Fig. 4.13 Equivalent circuit diagram of a generation unit (wind, photovoltaic) incl. step-up transformer T2, simulation using a current source and the equivalent voltage source at fault location NP 2

- Generation unit without step-up transformer T2, type A, Fig. 4.12.
- Generation unit with step-up transformer T2, type B, Fig. 4.12.

For simplification, the internal cable connections can be neglected in the short-circuit current calculation due to the cross-sections, so that a reduced system model can be used. This model consists exclusively of current sources for feeders with full-size converters (Sect. 4.4) and impedances of the asynchronous generators (Sect. 4.2.2.1) or doubly fed asynchronous generators (Sect. 4.2.2.2). The influence of the asynchronous generators is determined with the aid of the equivalent voltage source at the short-circuit location. Figure 4.13 shows the reduced equivalent circuit diagram, depending on the network connection points NP 1 and NP 2, according to Fig. 4.12.

According to Fig. 4.13, the following applies to the individual variables, in each case in relation to the nominal system voltage U_{nNP1} of the generating system A:

$\underline{I}_{CS}$ sum of the feeder currents of the existing generating units in the event of a short circuit with a full-size converter in accordance with Sect. 4.4.

$\underline{Z}_{PA}$ parallel circuit of the existing generating units with asynchronous generators in accordance with Sect. 4.2.2.1.

$\underline{Z}_{PD}$ parallel circuit of the existing generating units with doubly fed asynchronous generators in accordance with Sect. 4.2.2.2.

The step-up transformer T2 (Fig. 4.13) is not provided with an impedance correction factor (Sect. 4.8.1), as it can be assumed that the power factor of the load flow before the short circuit occurs should be between cos $\varphi = 0.9$ and cos $\varphi = 1.0$.

The resulting positive sequence impedance and zero sequence impedance of a generating system A depend on the impedances of the individual generating units.

4.6 Motors

Both asynchronous motors (slip ring motors, squirrel cage motors) and converter-fed motors are used in industrial networks. The converter-fed motors can also be DC motors that are supplied from the three-phase network via a converter and can feed a short-circuit current into the three-phase network during the short-circuit period.

4.6.1 *Impedances of Asynchronous Motors According to IEC*

Asynchronous motors contribute to the short-circuit current depending on the short-circuit location. The following applies to the calculation of the impedance $\underline{Z}_M$:

$$Z_M = \frac{1}{I_{LR}/I_{rM}} \cdot \frac{U_{rM}}{\sqrt{3} \cdot I_{rM}} = \frac{1}{I_{LR}/I_{rM}} \cdot \frac{U_{rM}^2}{S_{rM}} \tag{4.53}$$

and

$$\underline{Z}_M = R_M + jX_M \tag{4.54}$$

$$S_{rM} = \frac{P_{rM}}{\eta \cos \varphi} \tag{4.55}$$

With

I_{LR}/I_{rM} symmetrical locked-rotor current of the motor in relation to the rated current,

U_{rM} rated voltage,

I_{rM} rated current,

S_{rM} rated apparent power,

P_{rM} rated active power,
R_M resistance of the asynchronous motor,
X_M reactance of the asynchronous motor.

In principle, the saturated motor reactance that leads to the largest short-circuit current should be used to calculate the short-circuit current. However, in the event of a short circuit, the active motor voltage is lower than the assumed nominal system voltage, e.g. $0.9 \cdot U_n/\sqrt{3}$, so that the saturated reactance not taken into account is compensated for by the increased voltage.

The ratio *R*/*X* can be derived from the electrical data of the asynchronous motor. According to [1], the *R*/*X* ratio can be determined from the DC time constant T_{DC}, so that, for example, a value of $R/X \sim 0.06$ results for a time constant of $T_{DC} = 50$ ms. This means that the resistance R_M can be neglected when calculating the initial symmetrical short-circuit current, especially as the result is on the safe side.

If, in contrast to this, the peak short-circuit current is determined, fictitious R_{Mf}/X_M ratios are to be used, comparable with the synchronous machine, Sect. 4.2.1.1, which take into account the decay of the alternating current element during the first half-period of the short-circuit. The following values can be used for the R_{Mf}/X_M ratio:

$R_{Mf} = 0.10\ X_M$ with $U_{rM} > 1$ kV and active power P_{rM} per pair of poles ≥ 1 MW,
$R_{Mf} = 0.15\ X_M$ with $U_{rM} > 1$ kV and active power P_{rM} per pair of poles < 1 MW,
$R_{Mf} = 0.42\ X_M$ with $U_{rM} < 1000$ V with connection cables.

While each motor must be considered individually when calculating the contribution to the short-circuit current of high-voltage motors due to the generally different characteristics, low-voltage motors can be combined into a motor group for simplification. The following then applies to the motor group:

- Locked-rotor current: $I_{LR}/I_{rM} = 5$
- Rated apparent power: $S_{rM} = S_{rT}$

In this case, S_{rT} is the rated power of the low-voltage transformer that feeds this motor group. In practice, the locked-rotor current of low-voltage motors will be $I_{LR}/I_{rM} > 5$, but this does not take into account connections from the terminals to the busbar, which reduce the short-circuit contribution of the motors. This means that no additional cable connections between the low-voltage busbar and the motor terminals need to be taken into account. The entire motor group is then connected directly to the busbar.

4.6.2 Equivalent Circuit Diagram of an Asynchronous Motor

For an asynchronous motor, the equivalent circuit diagram according to Fig. 4.14 can be specified for the positive sequence system, which essentially corresponds to a transformer [16]. The following variables are shown in Fig. 4.14:

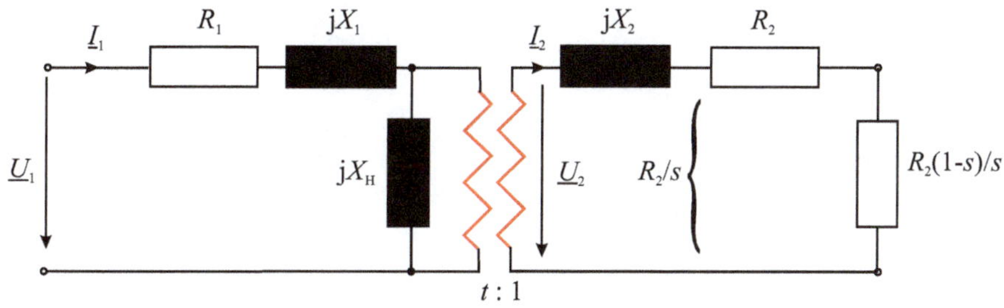

Fig. 4.14 Equivalent circuit of an asynchronous motor in the positive sequence system

$\underline{U}_1$ stator voltage (voltage to earth),
$\underline{I}_1$ stator current,
R_1 resistance of the stator,
X_1 leakage reactance,
X_H magnetizing reactance,
$\underline{U}_2$ rotor voltage,
$\underline{I}_2$ rotor current,
R_2 resistance of the rotor,
X_2 leakage reactance of the rotor,
s slip.

The following applies to the slip s in the positive sequence system according to Eq. (4.56):

$$s = \frac{n_1 - n_2}{n_1} \tag{4.56}$$

With

n_1 rotating field of the stator (synchronous rotation speed),
n_2 rotating field of the rotor.

In the circuit diagram of Fig. 4.14, the resulting resistance of the rotor R_2/s can be interpreted as the final resistance of a transformer. Here, R_2 represents the power loss converted in the rotor, while the value $R_2(1 - s)/s$ represents the value of the mechanical power delivered to the motor shaft. Two characteristic states can be defined depending on the slip:

- $s = 1$, stationary of the motor,
- $s = 0$, no-load.

During starting (from stationary of the motor), where the slip has the value $s = 1$, the series connection of the leakage reactances X_1 and X_2 acts as shown in Fig. 4.14. This reactance X_M also acts during a terminal short-circuit and is used as a short-circuit reactance according to Eq. (4.55) [17].

$$X_M = X_1 + \frac{X_H \cdot X_2}{X_H + X_2} \approx X_1 + X_2 \tag{4.57}$$

The subtransient time constant T'' with which the rotor current decays in the event of a short circuit is determined according to Eq. (4.58) and the angular frequency ω, since the stator resistance can be neglected compared to the resistance of the rotor.

$$T'' \approx \frac{X_1 + X_2}{\omega \cdot R_2} \tag{4.58}$$

The DC time constant T_{DC} results from the decay process in the stator according to Eq. (4.59):

$$T_{DC} = \frac{X_1 + X_2}{\omega \cdot R_1} \tag{4.59}$$

These variables can be used to determine the AC and DC components in the event of a short circuit, taking into account the *R*/*X* ratio of the motor, whereby the fictitious ratio R_{Mf}/X_{Mf} according to Sect. 4.6.1 represents the combination of the two time constants specified above.

- AC component:

$$i_{AC}(t) = \frac{\sqrt{2} \cdot U_1}{Z_M} \cdot e^{-t/T''} \tag{4.60}$$

- DC component:

$$i_{DC}(t) = \frac{\sqrt{2} \cdot U_1}{Z_M} \cdot e^{-t/T_{DC}} \tag{4.61}$$

For the derivation of the equivalent circuit diagram in the negative sequence system, the slip $s_{(2)}$, is determined by replacing the speed n_1 with $-n_1$ in Eq. (4.56), as the rotating field in the negative sequence system behaves in the opposite direction to the field in the positive sequence system. This results in $s_{(2)}$:

$$s_{(2)} = \frac{n_1 + n_2}{n_1} \tag{4.62}$$

This results in the following characteristic states for the behavior in the negative sequence system:

- $s = 0$: stationary of the motor,
- $s = 1$: no-load.

The value for $s_{(2)}$ is generally determined by adding Eqs. (4.56) and (4.62):

$$s_{(1)} + s_{(2)} = \frac{n_1 - n_2}{n_1} + \frac{n_1 - n_2}{n_1} = 2 \rightarrow s_{(2)} = 2 - s_{(1)} \tag{4.63}$$

This results in the equivalent circuit diagram of the asynchronous motor by replacing the value s in Fig. 4.14 with (2—s). The result is that the negative sequence system is equal to the positive sequence system and the line-to-line short-circuit current can be derived from the three-phase one, Sect. 12.4.1.

Since the neutral point of the asynchronous motors is always isolated, i.e. the neutral point is not connected to earth with a low impedance, the zero sequence system is infinite.

4.6.3 Static Converter-Fed Drives

Depending on the technology, different static converter-fed motors are used, which make a different contribution to the short-circuit current.

4.6.3.1 Reversible Static Converter-Fed Drives

The first VDE Standard on short-circuit current calculation in three-phase networks from 1962 [18] already specified the influence of converter-fed DC drives on the three-phase short circuit in the three-phase network when the converters are in inverter mode. The basis for the evaluation is an investigation presented in [19]. In this standard, the contribution to the peak short-circuit current i_{pDC} could be estimated as follows, if the short-circuit current contribution is limited by a high-speed overcurrent switch after 10 ms and switched off after 15 ms (with delta connection of the primary winding of the converter transformer):

$$i_{\mathrm{pDC}} = 2.9 \cdot \frac{P_{\mathrm{DC}}}{U_{\mathrm{n}}} \tag{4.64}$$

The following generally applies to the peak short-circuit current i_{pN} of the network at the connection point, without the contribution of the DC drive:

$$i_{\mathrm{pN}} = \sqrt{2} \cdot \kappa_{\mathrm{N}} \cdot I''_{\mathrm{kN}} = \sqrt{2} \cdot \kappa_{\mathrm{N}} \cdot \frac{S''_{\mathrm{kN}}}{\sqrt{3} \cdot U_{\mathrm{n}}} \tag{4.65}$$

By combining Eqs. (4.64) and (4.65), the maximum contribution of the peak short-circuit current of the drive can be specified as a function of the network parameters:

$$i_{\mathrm{pDC}} = \frac{2.9 \cdot \sqrt{3}}{\sqrt{2} \cdot \kappa_{\mathrm{N}}} \cdot \frac{P_{\mathrm{DC}}}{S''_{\mathrm{kN}}} \cdot i_{\mathrm{pN}} \tag{4.66}$$

With a maximum peak factor in an industrial network of $\kappa_{\mathrm{N}} = 1.8$, the following results after insertion into Eq. (4.66):

$$i_{\mathrm{pDC}} = 1.973 \cdot \frac{P_{\mathrm{DC}}}{S''_{\mathrm{kN}}} \cdot i_{\mathrm{pN}} \approx 2 \cdot \frac{P_{\mathrm{DC}}}{S''_{\mathrm{kN}}} \cdot i_{\mathrm{pN}} \tag{4.67}$$

With

P_{DC} rated power of the DC drive in MW,
S''_{kN} initial symmetrical short-circuit power at the connection point of the DC drive,
i_{pN} peak short-circuit current corresponding to the three-phase network without drive.

Since the stated Eq. (4.64) referred to the application of a DC drive in a balance choke circuit, it was necessary to make a change for DC drives as a result of the changed converter technology, which was implemented in 1990 [20].

These DC motors are connected to the network via converters using thyristor technology in a three-phase bridge circuit. In principle, these drives are no longer installed today, but they are still in operation. These motors can contribute to the three-phase short-circuit current if inverter operation is possible. Figure 4.15 shows an example of the circuit with a three-phase short circuit and with two conductive thyristors, which is assumed in [21] for the following derivation.

Under the same conditions as shown in [19], this results in a contribution to the peak short-circuit current of:

$$i_{\mathrm{pDC}} = 17.5 \cdot I_{\mathrm{rM}} = \sqrt{2} \cdot \kappa \cdot \frac{I_{\mathrm{LR}}}{I_{\mathrm{rM}}} \cdot I_{\mathrm{rM}} \tag{4.68}$$

This results in a maximum locked-rotor current ratio of:

$$\frac{I_{\mathrm{LR}}}{I_{\mathrm{rM}}} = \frac{17.5}{\sqrt{2} \cdot \kappa} \tag{4.69}$$

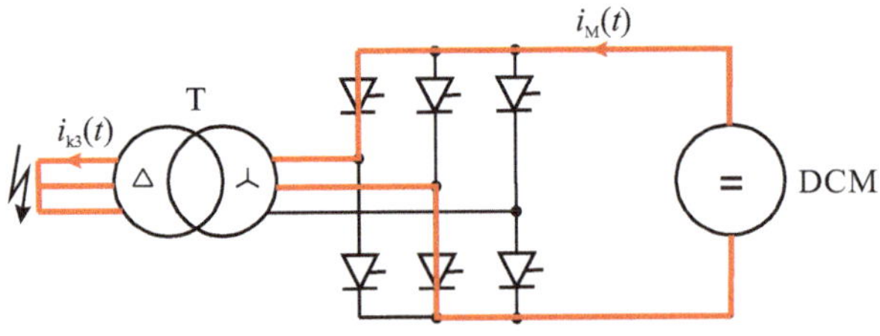

Fig. 4.15 DC drive DCM in the event of a three-phase network short circuit

Taking into account the skin effect in the resistors and additional inductances, the current contribution according to Eq. (4.69) is reduced at a value of $\kappa = 1.8$ to:

$$\frac{I_{LR}}{I_{rM}} = \frac{9}{\sqrt{2} \cdot \kappa} \approx 3.5 \tag{4.70}$$

From this information, a DC drive can be simulated by a motor impedance according to Eq. (4.53). The following values or relationships then apply:

U_{rM} rated voltage of the converter transformer,
I_{rM} rated current of the converter transformer,
I_{LR}/I_{rM} symmetrical locked-rotor current ratio (= 3),
R_M/X_M resistance/reactance ratio (= 0.1).

The values of the symmetrical locked-rotor current ratio according to Eq. (4.70) are shown in detail at [21], whereby the following general conditions apply:

- At the time of the short circuit on the network side, the refeeding motor has the maximum speed so that the induced motor voltage is equal to the network voltage.
- The motor-side inductance is represented exclusively by the motor inductance.
- The interruption of the motor contribution begins approx. 10 ms after the fast switch is activated, which becomes active when the motor current exceeds two and a half times the rated current.
- Saturation of the transformer is not reached.

4.6.3.2 Motors with Current Source Inverter (CSI)

This technology is used, for example, as converter-fed drives in power station auxiliary power networks or as marine diesel engines and the basic behavior corresponds to that of the HVDC LCC, so that only a contribution to the peak short-circuit current is made [22], Sect. 13.3. Fig. 4.16 shows an example of a system with a current source inverter using thyristor technology.

According to [22, 23], current source inverters contribute to the peak short-circuit current, and the inductance L in the DC link has a significant influence on the short-

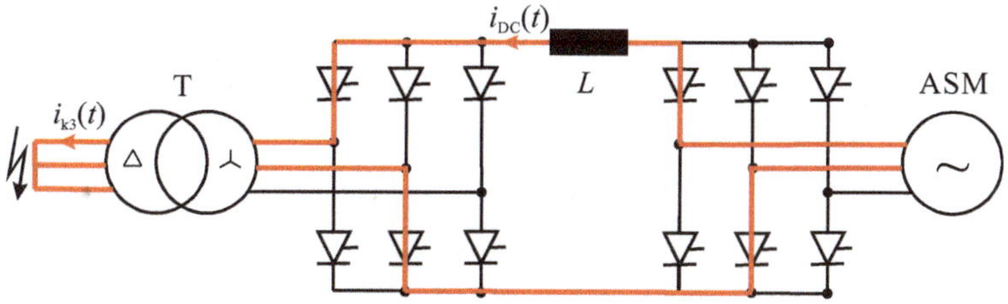

Fig. 4.16 Equivalent circuit of a current source inverter with an asynchronous machine (ASM)

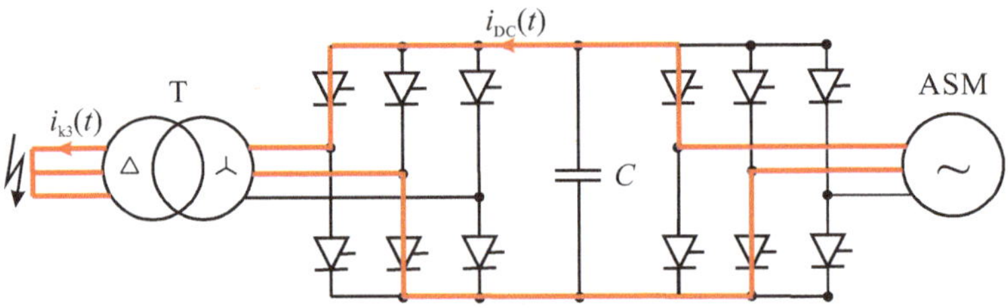

Fig. 4.17 Equivalent circuit of a voltage source inverter with an asynchronous machine (ASM)

circuit current contribution in the event of a three-phase short circuit. This drive is not currently covered in the IEC Standard [2].

4.6.3.3 Motors with Voltage Source Inverter (VSI)

Motors connected via voltage source inverter in IGBT technology basically correspond to the behavior of full-size converters for use in the generation of electrical energy by wind/PV systems (Sect. 4.4), but without the requirement of an inductive current, as is required, for example, in HVDC systems with connection to the public power supply, Sect. 13.2. Fig. 4.17 shows an example of a system with a voltage source inverter.

If the inverter system is not to contribute to the short-circuit current on the three-phase side, the short-circuit current can be interrupted directly by controlling the IGBTs so that there is no contribution to the peak short-circuit and breaking current. In this case, the drive is not included in the short-circuit current calculation.

If the installation is to provide an inductive contribution throughout the short-circuit duration, the simulation is carried out as a current source and the maximum reactive current must be specified by the manufacturer. In general, consideration is only necessary in the event of a short circuit at the voltage level of the converter transformer.

4.7 Overhead Lines and Cables

The positive and zero sequence impedances of overhead lines and cables are essential for calculating the short-circuit current, which is particularly true for the zero sequence system. The determination of the zero sequence impedance of the cable also depends on the type of earthing of the shield or sheath.

4.7.1 *Simplified Equivalent Circuit Diagram in the Positive and Zero Sequence System of Overhead Lines*

In general, a line is described by the line equations in which the initial values (U_S, I_S) are represented by the final values (U_R, I_R) [3]. These equations therefore apply to overhead lines, cables and gas-insulated lines (GIL). Here, the equation is expressed in the positive sequence system, index 1.

$$\begin{pmatrix} \underline{U}_{(1)S} \\ \underline{I}_{(1)S} \end{pmatrix} = \begin{pmatrix} \cosh\left(\underline{\gamma}_{(1)}\cdot\ell\right) & \underline{Z}_{(1)W}\cdot\sinh\left(\underline{\gamma}_{(1)}\ell\right) \\ \underline{Z}_{(1)W}^{-1}\cdot\sinh\left(\underline{\gamma}_{(1)}\cdot\ell\right) & \cosh\left(\underline{\gamma}_{(1)}\cdot\ell\right) \end{pmatrix}\cdot\begin{pmatrix} \underline{U}_{(1)R} \\ \underline{I}_{(1)R} \end{pmatrix} \tag{4.71}$$

The following applies to the surge impedances $\underline{Z}_{(1)W}$ and propagation constants $\gamma_{(1)}$ in the positive sequence system:

$$\underline{Z}_{(1)W} = \sqrt{\frac{R'_{(1)} + j\omega\cdot L'_{(1)}}{G'_{(1)} + j\omega\cdot C'_{(1)}}} \tag{4.72}$$

$$\underline{\gamma}_{(1)} = \alpha_{(1)} + j\beta_{(1)} = \sqrt{\left(R'_{(1)} + j\omega\cdot L'_{(1)}\right)\cdot\left(G'_{(1)} + j\omega\cdot C'_{(1)}\right)} \tag{4.73}$$

With

- ℓ length of the line,
- α damping constant,
- β phase constant (angular phase-shift).

Based on the quadripole Eq. (4.71), the elements of the π equivalent circuit (Fig. 4.18), which are described by Eq. (4.74), can be determined by comparing the coefficients.

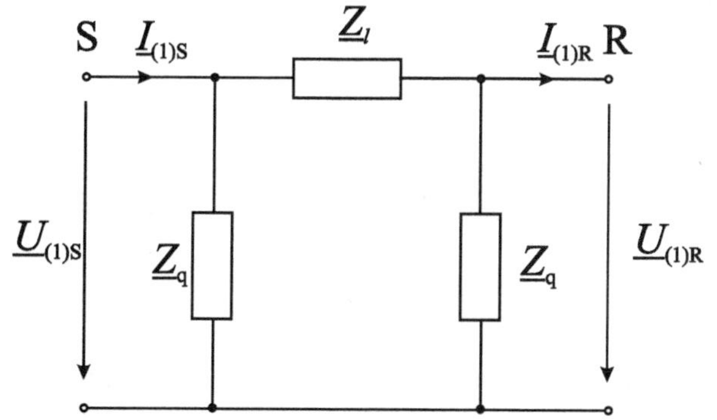

Fig. 4.18 π-equivalent circuit of a line

$$\begin{pmatrix} \underline{U}_{(1)\mathrm{S}} \\ \underline{I}_{(1)\mathrm{S}} \end{pmatrix} = \begin{pmatrix} 1+\underline{Z}_\ell \cdot \underline{Y}_\mathrm{q} & \underline{Z}_\ell \\ \underline{Y}_\mathrm{q}(2+\underline{Z}_\ell \cdot \underline{Y}_\mathrm{q}) & 1+\underline{Z}_\ell \cdot \underline{Y}_\mathrm{q} \end{pmatrix} \cdot \begin{pmatrix} \underline{U}_{(1)\mathrm{R}} \\ \underline{I}_{(1)\mathrm{R}} \end{pmatrix} \tag{4.74}$$

A comparison of the matrices (4.71) and (4.74) provides the values of the π equivalent circuit diagram for homogeneous lines.

$$\underline{Z}_\ell = \underline{Z}_{(1)\mathrm{W}} \cdot \sinh\left(\underline{\gamma}_{(1)} \cdot \ell\right) \tag{4.75}$$

$$\underline{Y}_\mathrm{q} = \frac{\cosh\left(\underline{\gamma}_{(1)} \cdot \ell\right) - 1}{\underline{Z}_{(1)\mathrm{W}} \cdot \sinh\left(\underline{\gamma}_{(1)} \cdot \ell\right)} = \frac{\tanh\left(\underline{\gamma}_{(1)} \cdot \ell/2\right)}{\underline{Z}_{(1)\mathrm{W}}} \tag{4.76}$$

In electrical power supply, it is generally permissible to calculate with "short" lines, whereby lines are described as electrically short if the condition $(\gamma\ell) < 1$ is fulfilled, i.e. the hyperbolic functions can be expressed by the first members of the Taylor series expansion, so that the following applies:

$$\sinh\left(\underline{\gamma} \cdot \ell\right) = \left(\underline{\gamma} \cdot \ell\right) + \frac{\left(\underline{\gamma} \cdot \ell\right)^3}{3!} + \ldots + \frac{\left(\underline{\gamma} \cdot \ell\right)^{2 \cdot n+1}}{(2 \cdot n+1)!} + \ldots \tag{4.77}$$

$$\cosh\left(\underline{\gamma} \cdot \ell\right) = l + \frac{\left(\underline{\gamma} \cdot \ell\right)^2}{2!} + \ldots + \frac{\left(\underline{\gamma} \cdot \ell\right)^{2 \cdot n}}{(2 \cdot n)!} + \ldots \tag{4.78}$$

If the series expansion is interrupted after the first or second member, the result is as follows:

$$\sinh\left(\underline{\gamma} \cdot \ell\right) \approx \left(\underline{\gamma} \cdot \ell\right) \quad \cosh\left(\underline{\gamma} \cdot \ell\right) \approx 1 + \frac{\left(\underline{\gamma} \cdot \ell\right)^2}{2} \tag{4.79}$$

For the relative error F_rel or the associated length of the overhead line, the estimate applies if the second element is neglected in Eqs. (4.77) and (4.78):

$$\left| \frac{\left(\underline{\gamma} \cdot \ell\right)^3 / 3!}{\underline{\gamma} \cdot \ell} \right| \le F_\mathrm{rel} \tag{4.80}$$

$$\ell \le \sqrt{6 \cdot F_\mathrm{rel}} \cdot \frac{1}{\gamma} = \sqrt{6 \cdot F_\mathrm{rel}} \cdot \frac{\lambda}{2 \cdot \pi} \tag{4.81}$$

With a wavelength of $\lambda = 6000$ km, this results in a maximum line length of $\ell = 405$ km if an error of 3% is permitted. This length increases to $\ell = 523$ km with an error accuracy of 5%. This means that lines up to the length specified above can be modeled as one π element, assuming a frequency of $f = 50$ Hz, taking into account an accepted error.

The elements of the π equivalent circuit (Fig. 4.17) can be derived for the short line from Eqs. (4.75) and (4.76), resulting in the following:

$$\underline{Z}_{\ell} = \underline{Z}_{(1)\mathrm{W}} \cdot \left(\underline{\gamma} \cdot \ell\right) = \left(R'_{(1)} + \mathrm{j}\omega L'_{(1)}\right) \cdot \ell \tag{4.82}$$

$$\underline{Y}_{\mathrm{q}} = \frac{1 + \left(\underline{\gamma} \cdot \ell\right)^2 / 2 - 1}{\underline{Z}_{(1)\mathrm{W}} \cdot \left(\underline{\gamma} \cdot \ell\right)} = \left(G'_{(1)} + \mathrm{j}\omega \cdot C'_{(1)}\right) \cdot \frac{\ell}{2} \tag{4.83}$$

In the event of a short circuit at the receiving end of the line ($\underline{U}_{(1)\mathrm{R}} = 0$), the short-circuit current $\underline{I}_{(1)\mathrm{R}}$ at the fault location is given by Eqs. (4.74) and (4.82):

$$\underline{I}_{(1)\mathrm{R}} = \frac{\underline{U}_{(1)\mathrm{S}}}{\underline{Z}_{\ell}} = \frac{\underline{U}_{(1)\mathrm{S}}}{\left(R'_{(1)} + \mathrm{j}\omega L'_{(1)}\right) \cdot \ell} \tag{4.84}$$

This means that when calculating the short-circuit current in the event of a fault at the end of a line, it is sufficient to consider only the longitudinal impedance according to the equivalent circuit, Fig. 4.18. The short-circuit impedances of the overhead lines and cables are generally calculated for the positive sequence and zero sequence systems as follows.

$$\underline{Z}_{(1)} = R_{(1)} + \mathrm{j}X_{(1)} \qquad \underline{Z}_{(0)} = R_{(0)} + \mathrm{j}X_{(0)} \tag{4.85}$$

with

$R_{(1)}$, $R_{(0)}$ resistance of the positive and zero sequence system,
$X_{(1)}$, $X_{(0)}$ reactance of the positive and zero sequence system.

The impedances of overhead lines can be calculated from the geometric data of the line configuration and, as an example, Eqs. (4.86) and (4.87) give the positive and zero sequence impedance of a symmetrical overhead line (the distances between the conductors are identical) without an earth wire.

$$\underline{Z}_{(1)} = R_{(1)} + \mathrm{j}\omega \cdot \frac{\mu_0}{2 \cdot \pi} \cdot \ell \cdot \left(0,25 + \ln \frac{d}{r}\right) \tag{4.86}$$

$$\underline{Z}_{(0)} = R_{(1)} + 3 \cdot \omega \cdot \frac{\mu_0}{8} \cdot \ell + \mathrm{j}\omega \cdot \frac{\mu_0}{2 \cdot \pi} \cdot \ell \cdot \left(0,25 + 3 \cdot \ln \frac{d_{\mathrm{E}}}{\sqrt[3]{r \cdot d^2}}\right) \tag{4.87}$$

With

μ_0 permeability ($= 4\pi \cdot 10^{-7}$ Vs/Am),
ω angular frequency,
d distance between conductors,

Table 4.2 Characteristic values of selected overhead lines with two circuits per tower (double-circuit line), with an earth wire, $\rho_E = 100\ \Omega m$, data per circuit

Term	Dimension	123 kV	420 kV	765 kV
Number of subconductors	–	1	4	4
Cross section of subconductors	mm^2	240/40	240/40	560/50
$R'_{(1)}$	Ω/km	0.12	0.03	0.013
$X'_{(1)}$	Ω/km	0.393	0.251	0.276
$R'_{(0)}$	Ω/km	0.34	0.24	0.20
$X'_{(0)}$	Ω/km	1.62	1.40	1.31

$R'_{(1)}; R'_{(0)}$ resistance per unit (positive and zero sequence system)
$X'_{(1)}; X'_{(0)}$ reactance per unit (positive and zero sequence system)

Table 4.3 Characteristic values of cables of different voltage levels in the positive sequence system, prerequisite: installation in earth, load factor $m = 0.7$; delta installation

U_0/U	kV	0.6/1		12/20		64/110	230/400
Q	mm^2	120		150		400	1200RMS
Type		NAYY	NA2XY	N2XS2Y	NA2XS2Y	NA2XS(FL)2Y	2XS(FL)2Y
$R'_{(1)}$	Ω/km	0.253	0.253	0.124	0.206	0.078	0.015
$X'_{(1)}$	Ω/km	0.080	0.077	0.123	0.123	0.129	0.173

$R'_{(1)}$ resistance per unit (positive sequence system), DC resistance 20 °C
$X'_{(1)}$ reactance per unit (positive sequence system)

r radius of conductor,
ℓ line length,
d_E depth of penetration of the return current through earth.

Examples of data for high-voltage overhead lines are listed in Table 4.2. The corresponding positive sequence impedances of standard cable types can be taken from the manufacturer's specifications, while the zero sequence impedances can be determined from measurements or calculations. Example values are listed in Table 4.3. The calculations of the positive sequence and zero sequence impedance of cables and overhead lines are described in detail in Sects. 4.7.2 and 4.7.3.

Further exemplary data for overhead lines and cables can be found at [1].

4.7.2 Detailed Derivation of the Positive and Zero Sequence Impedances of Overhead Lines

In this subsection, the equations for the positive sequence and zero sequence impedances (longitudinal and shunt impedances) of lines in the event of a short circuit are derived based on the geometric arrangement. While the shunt impedances in the positive sequence system must not be taken into account when using the

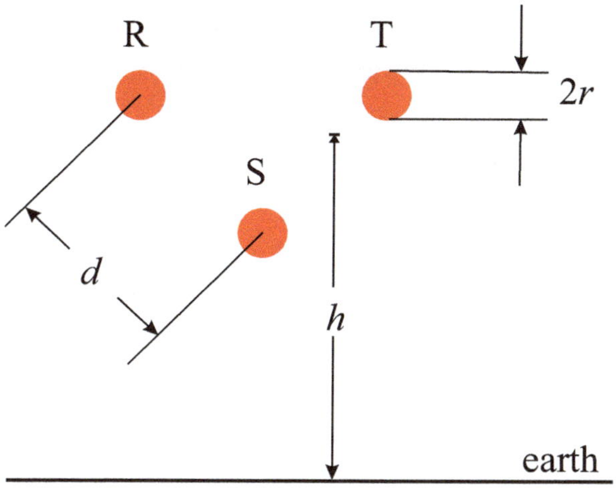

Fig. 4.19 Conductor arrangement for determining the inductance and capacitance of a three-phase line

calculation method of the equivalent voltage source at the fault location, this is appropriate when using the superposition method, for example.

4.7.2.1 Positive Sequence Impedance

In general, the positive sequence impedance of a three-phase overhead line is considered taking into account the geometric arrangement shown in Fig. 4.19, where d is the distance between the conductors and h is the height of a conductor above earth.

To calculate the positive sequence impedance, the capacitances of a line are not taken into account in the first step, so that only the resistance and inductance are determined. The capacitance of a line is determined according to Sect. 4.7.2.5. For the line-to-earth voltages $\underline{U}_{\mathrm{R}}$, $\underline{U}_{\mathrm{S}}$ and $\underline{U}_{\mathrm{T}}$ in matrix notation, the following applies for an overhead line with a short-circuited line end according to Fig. 4.20:

$$\begin{pmatrix} \underline{U}_{\mathrm{R}} \\ \underline{U}_{\mathrm{S}} \\ \underline{U}_{\mathrm{T}} \end{pmatrix} = \begin{pmatrix} \underline{Z}_{11} & \underline{Z}_{12} & \underline{Z}_{13} \\ \underline{Z}_{21} & \underline{Z}_{22} & \underline{Z}_{23} \\ \underline{Z}_{31} & \underline{Z}_{32} & \underline{Z}_{33} \end{pmatrix} \cdot \begin{pmatrix} \underline{I}_{\mathrm{R}} \\ \underline{I}_{\mathrm{S}} \\ \underline{I}_{\mathrm{T}} \end{pmatrix} \tag{4.88}$$

With

$$\underline{Z}_{\nu\nu} = R_{\nu} + \mathrm{j}\omega L_{\nu\nu} \quad \underline{Z}_{\nu\mu} = \mathrm{j}\omega M_{\nu\mu} \tag{4.89}$$

Since conductor loops must always be considered in the inductance calculation, it is assumed for the arrangement in Fig. 4.21 that the return conductor for each

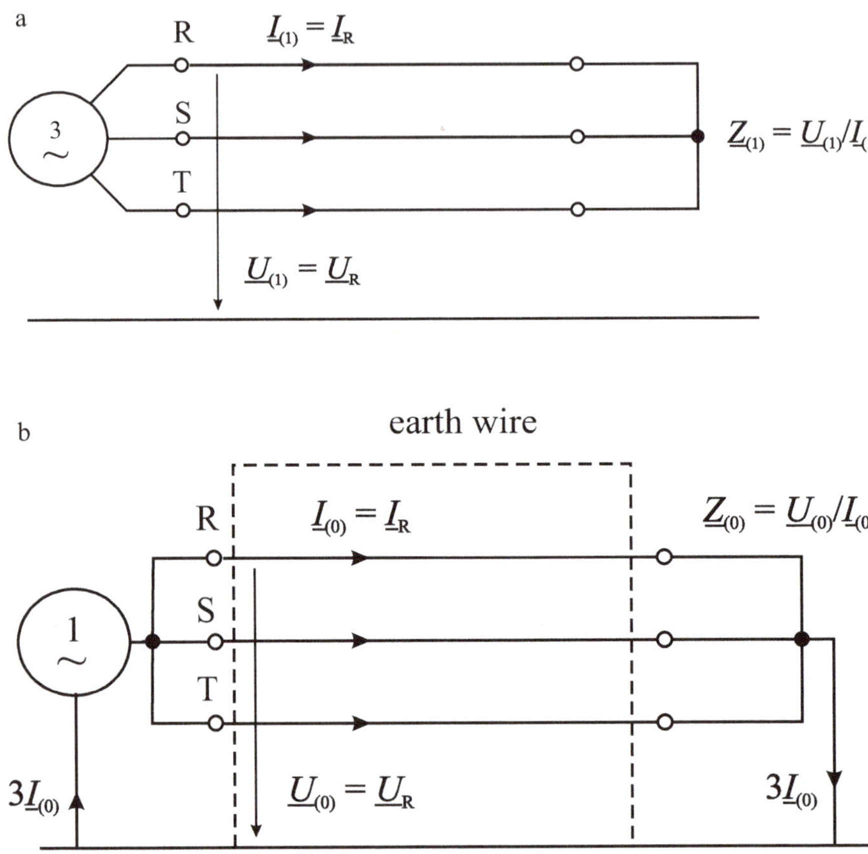

Fig. 4.20 Positive sequence and zero sequence impedance of a three-phase overhead line (without capacitance). (**a**) Positive sequence system, (**b**) zero sequence system

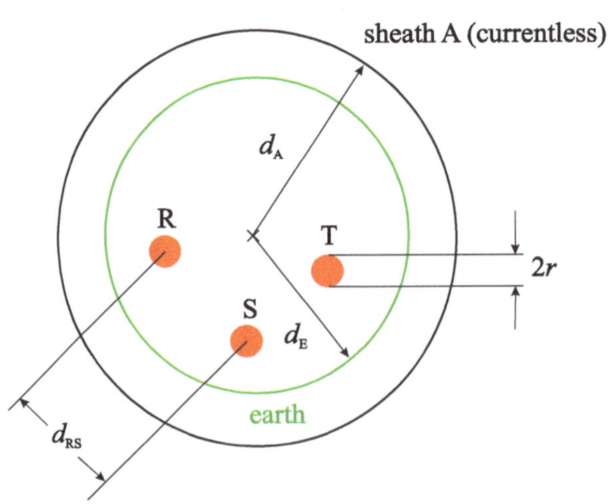

Fig. 4.21 Conductors R, S and T of a three-phase line within the (imaginary) common sheath A

conductor R, S and T is the common sheath with the radius d_A. The advantage of this arrangement is that the sum of all currents flowing in the common sheath conductor is zero, so that the radius of the imaginary sheath no longer appears in the result of the impedance calculation.

In this case, the self-inductance of the conductor ν in the sheath A according to Sect. 2.1.3 of [3] applies:

$$L_{\nu\nu} = \frac{\mu_0}{2\pi} \cdot \ell \cdot \left(0.25 + \ln\frac{d_A}{r_\nu}\right) \tag{4.90}$$

with

d_A radius of the sheath,
r_ν radius of the conductor ν.

According to Chap. 2 of [3], the coupling $M_{\nu\mu}$ of two conductors R and S within the sheath with the radius d_A results if the return conductor is the common sheath in each case:

$$M_{\nu\mu} = \frac{\mu_o}{2\pi} \cdot \ell \cdot \ln\frac{d_A}{d_{\nu\mu}} \tag{4.91}$$

The first line of the system of Eq. (4.88) can be transformed using the following current conditions:

$$\underline{I}_R + \underline{I}_S + \underline{I}_T = 0 \tag{4.92}$$

$$\underline{I}_T = \underline{a} \cdot \underline{I}_R \quad \underline{I}_S = \underline{a}^2 \cdot \underline{I}_R \tag{4.93}$$

$$\underline{a} = -\frac{1}{2} + j\frac{\sqrt{3}}{2} \quad \underline{a}^2 = -\frac{1}{2} - j\frac{\sqrt{3}}{2} \tag{4.94}$$

Due to the symmetrical operation of the line, the impedance of the conductor R is generally as follows:

$$\begin{aligned} \frac{\underline{U}_R}{\underline{I}_R} = \underline{Z}_R &= \left\{\underline{Z}_{11} + \underline{Z}_{12} \cdot \left[-\frac{1}{2} - j\frac{\sqrt{3}}{2}\right] + \underline{Z}_{13} \cdot \left[-\frac{1}{2} + j\frac{\sqrt{3}}{2}\right]\right\} \cdot \\ &= \left\{\underline{Z}_{11} - \frac{1}{2} \cdot [\underline{Z}_{12} + \underline{Z}_{13}] + j\frac{\sqrt{3}}{2} \cdot [\underline{Z}_{13} - \underline{Z}_{12}]\right\} \end{aligned} \tag{4.95}$$

After inserting Eqs. (4.89) to (4.95) into the system of Eq. (4.88), the impedances of the three conductors R, S and T are determined:

$$\underline{Z}_R = R_R + \omega \cdot \frac{\mu_0}{4\pi} \cdot \sqrt{3} \cdot \ell \cdot \ln \frac{d_{RT}}{d_{RS}} + j\omega \cdot \frac{\mu_0}{2\pi} \cdot \ell \cdot \left[0.25 + \ln \frac{\sqrt{d_{RS} \cdot d_{RT}}}{r_R}\right] \quad (4.96)$$

$$\underline{Z}_S = R_S + \omega \cdot \frac{\mu_0}{4\pi} \cdot \sqrt{3} \cdot \ell \cdot \ln \frac{d_{RS}}{d_{ST}} + j\omega \cdot \frac{\mu_0}{2 \cdot \pi} \cdot \ell \cdot \left[0.25 + \ln \frac{\sqrt{d_{RS} \cdot d_{ST}}}{r_S}\right] \quad (4.97)$$

$$\underline{Z}_T = R_T + \omega \cdot \frac{\mu_0}{4\pi} \cdot \sqrt{3} \cdot \ell \cdot \ln \frac{d_{ST}}{d_{RT}} + j\omega \cdot \frac{\mu_0}{2\pi} \cdot \ell \cdot \left[0.25 + \ln \frac{\sqrt{d_{RT} \cdot d_{ST}}}{r_T}\right] \quad (4.98)$$

with

d_{RS}, d_{RT}, d_{ST}: distance between two conductors,
r_R, r_S, r_T: radii of the conductors.

If the following characteristic values are assumed for a horizontal configuration, i.e. all conductors are in one plane (horizontal configuration of a 110 kV overhead line):

$d_{12} = d_{23} = 4$ m; $d_{13} = 8$ m; $r_\nu = 10.95$ mm (240/40 mm^2)
$R'_R = R'_S = R'_T = 0.119\ \Omega/\text{km}$; $\mu_0 = 4\pi \cdot 10^{-7} \cdot \frac{\text{Vs}}{\text{Am}}$

this results in the following impedance per unit:

$$\underline{Z}'_R = [0.119 + 0.038 + j0.130] \cdot \Omega/\text{km}$$

$$\underline{Z}'_S = [0.119 + j0.123] \cdot \Omega/\text{km}$$

$$\underline{Z}'_T = [0.119 - 0.038 + j0.130] \cdot \Omega/\text{km}$$

The results of the impedance calculation make it clear that the resistances and inductances of the individual conductors are different in an unsymmetrical arrangement of the conductors and therefore lead to a voltage unbalance at the end of the line if symmetrical currents are assumed. If, in contrast to the arrangement shown in Fig. 4.19, the distances between the conductors are the same, $\underline{Z}_{12} = \underline{Z}_{13}$, for the first line of the system of Eq. (4.88) or for Eq. (4.95) follows:

$$\underline{U}_R = (\underline{Z}_{11} - \underline{Z}_{12}) \cdot \underline{I}_R = \left[R_R + j\omega \cdot \frac{\mu_0}{2\pi} \cdot \ell \cdot \left(0.25 + \ln \frac{d}{r}\right)\right] \cdot \underline{I}_R = [R_R + j\omega \cdot L_b] \cdot \underline{I}_R \quad (4.99)$$

The term L_b is referred to as the operating inductance of a line and represents the mean value of the impedances according to Eqs. (4.96) to (4.98), which is also referred to as the positive sequence impedance of the overhead line.

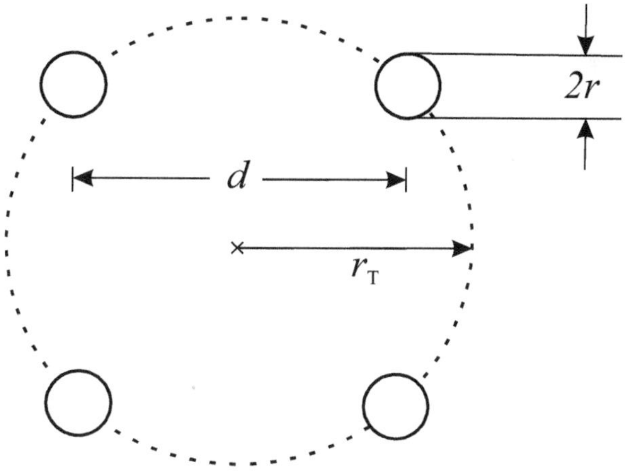

Fig. 4.22 Equivalent circuit diagram for determining the equivalent radius of a bundled conductor

4.7.2.2 Inductances of Bundled Conductors

To reduce the border field strength on the surface of conductor wires, bundled conductors (Table 4.2) are used at the extra-high voltage level, Fig. 4.22. Each of the n subconductors of the bundle carries the current I/n, thus determining the operating inductance L_b:

$$L_b = \frac{\mu_0}{2\pi} \cdot \ell \cdot \left(\frac{0.25}{n} + \ln \frac{d}{r_B} \right) \tag{4.100}$$

The equivalent radius of the bundled conductor r_B is calculated according to Eq. (4.101):

$$r_B = \sqrt[n]{n \cdot r \cdot r_T^{n-1}} \tag{4.101}$$

with

r radius of a subconductor,
r_T radius of a reference circle (Fig. 4.22),
n number of bundled conductors.

Equation (4.100) shows that a larger equivalent radius r_B results in a smaller operating inductance. If an approximately equal ampacity is to be assumed for two systems, the following possibilities arise (I_d continuous ampacity):

- System I: 2 × 120/20 mm^2 $I_d = 2 \times 410$ A
- System II: 1 × 380/50 mm^2 $I_d = 840$ A

For the equivalent radius r_B ($n = 2$) of system I, the Eq. (4.101) gives $r = 7.75$ mm for $r_T = 200$ mm:

$$r_B = \sqrt{2 \cdot 7.75 \cdot 200}\,\text{mm} = 55.68\,\text{mm}$$

A value of $r = 13.5$ mm results for the radius r of system II. With a conductor spacing of $d = 5$ m, the ratio of the impedances is as follows:

$$\frac{L_{BI}}{L_{BII}} = \frac{0.25/2 + \ln(5000/55.68)}{0.25 + \ln(5000/13.50)} = 0.75$$

The use of the bundled conductor system reduces the operating inductance to approx. 75%. This means that the short-circuit current increases accordingly.

4.7.2.3 Zero Sequence Impedance

While the calculation of the operational inductance is based on the field between the conductors, the field between the conductors and earth is relevant for the calculation of the zero sequence inductance. This is based on the loop impedance of an "infinitely" long line with return via earth as shown in Fig. 4.23. The earth resistances R_A, R_B can be neglected in this calculation.

It should be noted, as shown in Fig. 4.23, that the depth of penetration of the return current in earth d_E is considerably greater than the overhead line height h, which applies to the range of the usual operating frequencies of overhead lines.

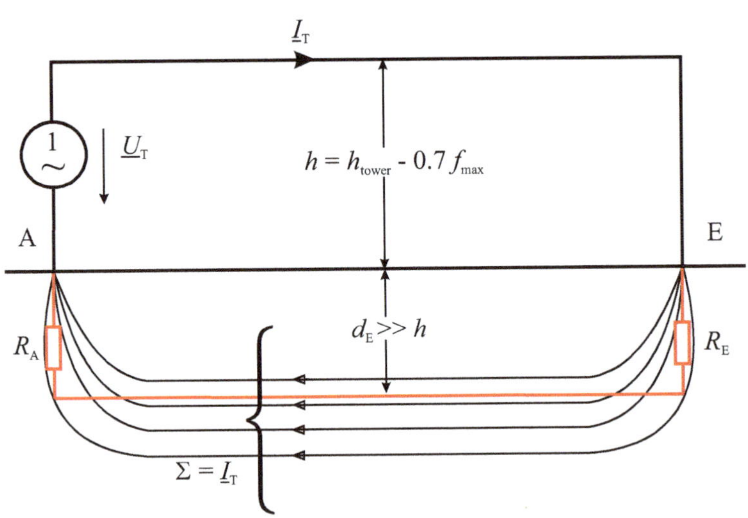

Fig. 4.23 Equivalent circuit diagram for calculating the loop impedance conductor– earth. d_E earth current depth of the return current, R_A, R_E resistances of the earthing systems, h average conductor height, R' resistance per unit of the conductor, ρ_E specific earth resistance, h_{tower} conductor wire height at the tower, f_{max} maximum sag

For the calculation of the zero sequence impedance, the earth return conductor is assumed to be an additional cylinder with the radius d_E within the sheath A (Fig. 4.21). The system of Eq. (4.102) results for the induced longitudinal voltages 1–3 (conductor) and 4 (earth):

$$\begin{pmatrix} \underline{U}_1 \\ \underline{U}_2 \\ \underline{U}_3 \\ \underline{U}_4 \end{pmatrix} = \begin{pmatrix} \underline{Z}_{11} & \underline{Z}_{12} & \underline{Z}_{13} & \underline{Z}_{14} \\ \underline{Z}_{21} & \underline{Z}_{22} & \underline{Z}_{23} & \underline{Z}_{24} \\ \underline{Z}_{31} & \underline{Z}_{32} & \underline{Z}_{33} & \underline{Z}_{34} \\ \underline{Z}_{41} & \underline{Z}_{42} & \underline{Z}_{43} & \underline{Z}_{44} \end{pmatrix} \cdot \begin{pmatrix} \underline{I}_1 \\ \underline{I}_2 \\ \underline{I}_3 \\ \underline{I}_4 \end{pmatrix} \tag{4.102}$$

With (according to Fig. 4.20b)

$$\underline{I}_1 = \underline{I}_2 = \underline{I}_3 = \underline{I}_{(0)} \quad \text{and} \quad \underline{I}_4 = -3 \cdot \underline{I}_{(0)}$$

With

$\underline{I}_1, \underline{I}_2, \underline{I}_3$ conductor currents of the three-phase system acc. to Fig. 4.20,
$\underline{U}_1, \underline{U}_2, \underline{U}_3$ longitudinal voltages of the three-phase system acc. to Fig. 4.20.

If the longitudinal voltages U_1, U_2 and U_3 are related to the voltage U_4, the first line of the system of Eq. (4.102) is obtained:

$$\begin{aligned} &\underline{U}_1 - \underline{U}_4 = \underline{U}_{(0)} = \\ &\underline{I}_{(0)} \cdot (\underline{Z}_{11} + \underline{Z}_{12} + \underline{Z}_{13} - \underline{Z}_{41} - \underline{Z}_{42} - \underline{Z}_{43}) + \underline{I}_4 \cdot (\underline{Z}_{14} - \underline{Z}_{44}) \end{aligned} \tag{4.103}$$

Equation (4.104) only applies to symmetrical overhead lines, as in principle three different conductor equations can be derived for an asymmetrical equipment. In this case, all currents $\underline{I}_1$, $\underline{I}_2$ and $\underline{I}_3$ are identical using the system of Eq. (4.102). The following applies to the individual impedances ν, $\mu = 1, 2, 3$, according to Eq. (4.90), self-inductance, and (4.91), mutual inductance:

$$\underline{Z}_{\nu\nu} = R_\nu + \mathrm{j}\omega \cdot \frac{\mu_0}{2\pi} \cdot \ell \cdot \left(0.25 + \ln \frac{d_A}{r_\nu}\right) \quad \text{(self – inductance : Conductor)} \tag{4.104}$$

$$\underline{Z}_{\nu\mu} = \mathrm{j}\omega \cdot \frac{\mu_0}{2\pi} \cdot \ell \cdot \ln \frac{d_A}{d_{\nu\mu}} \quad \text{(mutual inductance : Conductor-Conductor)} \tag{4.105}$$

$$\underline{Z}_{4\mu} = \underline{Z}_{\nu 4} = \mathrm{j}\omega \cdot \frac{\mu_0}{2\pi} \cdot \ell \cdot \ln \frac{d_A}{d_E} \quad \text{(mutual inductance : Conductor-Earth)} \tag{4.106}$$

$$\underline{Z}_{44} = \frac{\omega \cdot \mu_0 \cdot \ell}{8} + \mathrm{j}\omega \cdot \frac{\mu_0}{2\pi} \cdot \ell \cdot \ln \frac{d_A}{d_E} \quad \text{(self-inductance : Earth)} \tag{4.107}$$

With the earth current depth

Table 4.4 Reference values for the earth current depth d_E as a function of the conductivity κ_E of the soil are as follows at $f = 50$ Hz [7, 18, 24]

Type of soil according to	Alluvial soil, marl, moorland	Clay	Prous sand-lime brick, clay slate, loam, clay, farmland soil	Quartz, solid lime		Granite, gneiss, clayey slate	
VDE and CCITT				Wet sand	Wet gravel	Dry sand and gravel	Stony soil
$\rho_E/\Omega{\cdot}m$	30	50	100	200	500	1000	3000
$\kappa_E/\mu S/cm$	333	200	100	50	20	10	3,3
d_E/m	510	660	930	1320	2080	2940	5100

$$d_E = \frac{1.85}{\sqrt{\mu_0 {\cdot} \omega / \rho_E}} \tag{4.108}$$

and the radius d_A of the currentless sheath and the distances $d_{\nu\mu}$ between the conductors.

The resistance of the return conductor ($\omega{\cdot}\mu_0/8 \approx 50\ \Omega$ m/km at $f = 50$ Hz) is independent of the conductivity of the soil. The earth current depth d_E can be interpreted as the distance of a fictitious earth return conductor from the earth's surface, which causes the same inductance of the conductor loop as the actual current distribution. For Al/St ropes, the relative permeability μ_r must be used to determine the internal inductance $\mu_r \approx 5$–10 (single layer). For multi-layer Al/St ropes and for Al and Cu conductors, $\mu_r \approx 1$ applies. The specific conductivity κ_E of the soil can be found in Table 4.4.

Is set for simplification for the distances $d_{\nu\mu}$.

$$d_{\nu\mu} = \sqrt[3]{d_{RS} \cdot d_{RT} \cdot d_{ST}} = d \tag{4.109}$$

it follows from Eq. (4.103)

$$\underline{U}_{(0)} = (\underline{Z}_{11} + 2{\cdot}\underline{Z}_{12} - 3{\cdot}\underline{Z}_{14}){\cdot}\underline{I}_{(0)} + (\underline{Z}_{14} - \underline{Z}_{44}){\cdot}\underline{I}_4 \tag{4.110}$$

or

$$\frac{\underline{U}_{(0)}}{\underline{I}_{(0)}} = \underline{Z}_{(0)} = (\underline{Z}_{11} + 2{\cdot}\underline{Z}_{12} - 6{\cdot}\underline{Z}_{14} + 3{\cdot}\underline{Z}_{44}) \tag{4.111}$$

After insertion follows for the zero sequence impedance $\underline{Z}_{(0)}$:

$$\underline{Z}_{(0)} = R_{(1)} + 3\cdot\omega\cdot\frac{\mu_0}{8}\cdot\ell + \mathrm{j}\omega\cdot\frac{\mu_0}{2\pi}\cdot\ \ell\ \cdot\left(0.25 + 3\cdot\ln\frac{d_\mathrm{E}}{\sqrt[3]{r\cdot d^2}}\right) \tag{4.112}$$

When determining the zero sequence impedance of an overhead line with an additional earth wire, the system of Eq. (4.103) is extended by a fifth column or row. The result is:

$$\begin{pmatrix} \underline{U}_1 \\ \underline{U}_2 \\ \underline{U}_3 \\ \underline{U}_4 \\ \underline{U}_5 \end{pmatrix} = \begin{pmatrix} \underline{Z}_{11} & \underline{Z}_{12} & \underline{Z}_{13} & \underline{Z}_{14} & \underline{Z}_{15} \\ \underline{Z}_{21} & \underline{Z}_{22} & \underline{Z}_{23} & \underline{Z}_{24} & \underline{Z}_{25} \\ \underline{Z}_{31} & \underline{Z}_{32} & \underline{Z}_{33} & \underline{Z}_{34} & \underline{Z}_{35} \\ \underline{Z}_{41} & \underline{Z}_{42} & \underline{Z}_{43} & \underline{Z}_{44} & \underline{Z}_{45} \\ \underline{Z}_{51} & \underline{Z}_{52} & \underline{Z}_{53} & \underline{Z}_{54} & \underline{Z}_{55} \end{pmatrix} \cdot \begin{pmatrix} \underline{I}_1 \\ \underline{I}_2 \\ \underline{I}_3 \\ \underline{I}_4 \\ \underline{I}_5 \end{pmatrix} \tag{4.113}$$

In this system of equations, lines 1–3 represent the longitudinal voltages of the conductors R, S and T, while $\underline{U}_4$ is the voltage of the earth wire D and $\underline{U}_5$ is the longitudinal voltage above the earth. If the values are related to the conductor-to-earth voltage in the same way as in Eq. (4.103), the result for the first line is as follows, taking the relationships into account:

$$3\cdot\underline{I}_{(0)} + \underline{I}_4 + \underline{I}_5 = 0 \tag{4.114}$$

$$\begin{gathered}\underline{U}_1 - \underline{U}_5 = \underline{U}_{(0)} = (\underline{Z}_{11} + 2\cdot\underline{Z}_{12} - 6\cdot\underline{Z}_{15} + 3\cdot\underline{Z}_{55})\cdot\underline{I}_{(0)} + \\ (\underline{Z}_{14} - 2\cdot\underline{Z}_{15} + \underline{Z}_{55})\cdot\underline{I}_4\end{gathered} \tag{4.115}$$

The same applies to the voltage along the earth wire:

$$\begin{gathered}\underline{U}_4 - \underline{U}_5 = \\ (3\cdot\underline{Z}_{14} - 6\cdot\underline{Z}_{15} + 3\cdot\underline{Z}_{55})\cdot\underline{I}_{(0)} + (\underline{Z}_{44} + \underline{Z}_{55} - 2\cdot\underline{Z}_{15})\cdot\underline{I}_4\end{gathered} \tag{4.116}$$

Since the earth wire is regularly earthed (at each tower), the voltage $\underline{U}_4$—$\underline{U}_5 = 0$, so that the current flowing in the earth wire can be calculated from this.

$$\underline{I}_4 = -\ \frac{3\cdot(\underline{Z}_{55} + \underline{Z}_{14} - 2\cdot\underline{Z}_{15})}{\underline{Z}_{44} + \underline{Z}_{55} - 2\cdot\underline{Z}_{15}}\cdot\underline{I}_{(0)} \tag{4.117}$$

After insertion into Eq. (4.116), the following results:

$$\underline{Z}_{(0)\mathrm{E}} = (\underline{Z}_{11} + 2\cdot\underline{Z}_{12} - 6\cdot\underline{Z}_{15} + 3\cdot\underline{Z}_{55}) - \frac{3\cdot(\underline{Z}_{55} + \underline{Z}_{14} - 2\cdot\underline{Z}_{15})^2}{\underline{Z}_{44} + \underline{Z}_{55} - 2\cdot\underline{Z}_{15}} \tag{4.118}$$

The relationships apply to the individual expressions, taking into account conditions (4.104) to (4.107) (plus a fifth conductor).

$$\underline{Z}_{11}+2\cdot\underline{Z}_{12}+3\cdot\underline{Z}_{55}-6\cdot\underline{Z}_{15}=R_{(1)}+3\cdot\omega\cdot\frac{\mu_0}{8}\cdot\ell + \mathrm{j}\omega\cdot\frac{\mu_0}{2\pi}\cdot\ell\cdot\left[0.25+3\cdot\ln\frac{d_{\mathrm{E}}}{\sqrt[3]{r\cdot d^2}}\right] \tag{4.119}$$

$$\underline{Z}_{55}+\underline{Z}_{14}-2\cdot\underline{Z}_{15}=\omega\cdot\frac{\mu_0}{8}\cdot\ell+\mathrm{j}\omega\cdot\frac{\mu_0}{2\pi}\cdot\ell\cdot\ln\frac{d_{\mathrm{E}}}{d_{\mathrm{D}}} \tag{4.120}$$

$$\underline{Z}_{44}+\underline{Z}_{55}-2\cdot\underline{Z}_{15}=R_{\mathrm{D}}+\omega\cdot\frac{\mu_0}{8}\cdot\ell+\mathrm{j}\omega\cdot\frac{\mu_0}{2\pi}\cdot\ell\cdot\left[0.25+\ln\frac{d_{\mathrm{E}}}{r_{\mathrm{D}}}\right] \tag{4.121}$$

with

d_{D} mean geometric distance between phase and earth wire.

The above considerations can also be extended to the zero sequence impedances of double circuit lines with one or more earth wires.

4.7.2.4 Positive and Zero Sequence Impedance Ratios

To calculate short-circuit currents with earth contact, the zero sequence impedance of overhead lines is required, so that the ratios $R_{(0)}/R_{(1)}$ and $X_{(0)}/X_{(1)}$ must be known. Using Eqs. (4.99) and (4.112) or (4.118), these ratios can be determined depending on the topology of the overhead line (with and without an earth wire). Two examples are determined below.

Without Earth Wire

As an example, the following geometric values are used for a medium-voltage overhead line in a horizontal configuration ($U_{\mathrm{n}} = 20$ kV):

Conductor radius: $r = 6.25$ mm (cross section 95 mm^2)
Average conductor spacing: $d = 1.26$ m (spacing $d_{1,2} = 1$ m or 2 m)
Earth current depth: $d_{\mathrm{E}} = 930$ m (spec. soil resistivity $\rho_{\mathrm{E}} = 100\ \Omega$ m)
Resistance of the conductor: $R'_{(1)} = 0.308\,\Omega/\mathrm{km}$ aluminium
Permeability: $\mu_0 = 4\pi \cdot 10^{-7}$Vs/Am.

The impedance ratios can be determined from this

$$R_{(0)}/R_{(1)}=\frac{R'_{(1)}+3\cdot\omega\cdot\frac{\mu_0}{8}}{R'_{(1)}}=1.48$$

$$X_{(0)}/X_{(1)}=\frac{\left(0.25+3\cdot\ln\frac{d_{\mathrm{E}}}{\sqrt[3]{r\cdot d^2}}\right)}{\left[0.25+\ln\frac{d}{r}\right]}=4.57$$

With Earth Wire

The following geometric values are used as an example for a high-voltage overhead line in a horizontal configuration ($U_n = 110$ kV):

Conductor radius: $r = r_D = 10.95$ mm (cross section 240/40 mm^2)
Average conductor spacing: $d = d_D = 5.04$ m (spacing $d_{1,2} = 4$ m or 8 m)
Earth current depth: $d_E = 930$ m (spec. soil resistivity $\rho_E = 100\ \Omega$ m)
Resistance of the conductor: $R'_{(1)} = 0.119\,\Omega/\text{km}$ (aluminum, conductor-, earth wire)

The positive sequence impedance is determined according to Eq. (4.99), while the zero sequence impedance is calculated according to Eq. (4.118).

$$\underline{Z}'_{(1)} = (0.119 + \text{j}0.401)\,\Omega/\text{km} \quad \underline{Z}'_{(0)} = (0.225 + \text{j}1.116)\,\Omega/\text{km}$$

This results in the following zero sequence impedance ratios:

$$R_{(0)}/R_{(1)} = 1.89 \quad X_{(0)}/X_{(1)} = 2.79$$

The zero sequence impedance ratios given in Table 4.2 for the 110 kV overhead line differ from the values given above, as the example there considers a double-circuit line with an earth wire and therefore deviates from the example calculation.

4.7.2.5 Capacities (Positive and Negative Sequence System)

Three-phase overhead lines have capacitances between the conductors and to earth. To calculate the capacitances, it is assumed that the extended line conductor with the length ℓ according to Fig. 4.24 carries the charge $+Q$ [25]. The counter charge $-Q$ is at infinity.

For the electric flux density D and the electric field strength E, the following applies at a distance s:

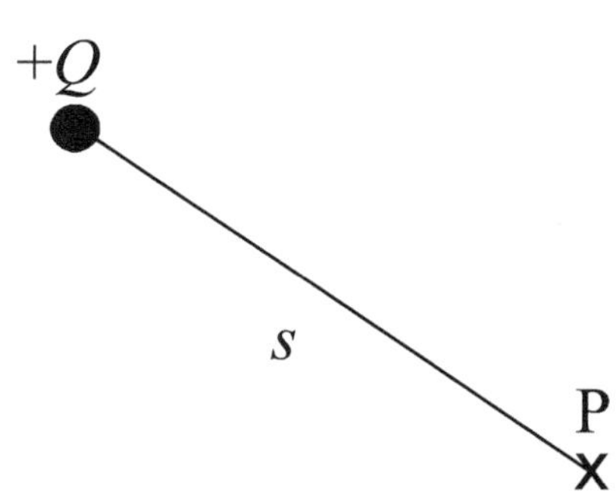

Fig. 4.24 Calculation of the electric field for a line conductor with the charge $+Q$

$$D = \frac{Q}{2\pi \cdot \ell} \cdot \frac{1}{s} \quad E = \frac{D}{\varepsilon_r \cdot \varepsilon_0} = \frac{Q}{2\pi \cdot \varepsilon_r \cdot \varepsilon_0 \cdot \ell} \cdot \frac{1}{s} \tag{4.122}$$

With the dielectric constant $\varepsilon_0 = 8.854$ pF/m or As/Vm (vacuum) and the relative constant ε_r. Based on the electric field strength, the potential φ can be determined according to Eq. (4.123), taking into account the integration constant K.

$$\varphi = -\int E \cdot ds = \frac{Q}{2\pi \cdot \varepsilon_r \cdot \varepsilon_0 \cdot \ell} \cdot \ln(s) + K \tag{4.123}$$

For different conductors with different charges (Q_1, Q_2, …) acting on a point P, the resulting potential results from the superposition.

$$\varphi = -\frac{1}{2\pi \cdot \varepsilon_r \cdot \varepsilon_0 \cdot \ell} \cdot \{Q_1 \cdot \ln(s_1) + Q_2 \cdot \ln(s_2) + \ldots\} + K \tag{4.124}$$

If, in contrast to the above illustration (Fig. 4.24), the earth is assumed to be the conducting surface (Fig. 4.25), the principle of mirroring is applied, so that the following applies to the potential under these conditions:

$$\varphi = -\frac{1}{2\pi \cdot \varepsilon_r \cdot \varepsilon_0 \cdot \ell} \cdot Q_1 \cdot \ln\left(\frac{s_1'}{s_1}\right) + K \tag{4.125}$$

Or if several line charges act on one point P, Eq. (4.126).

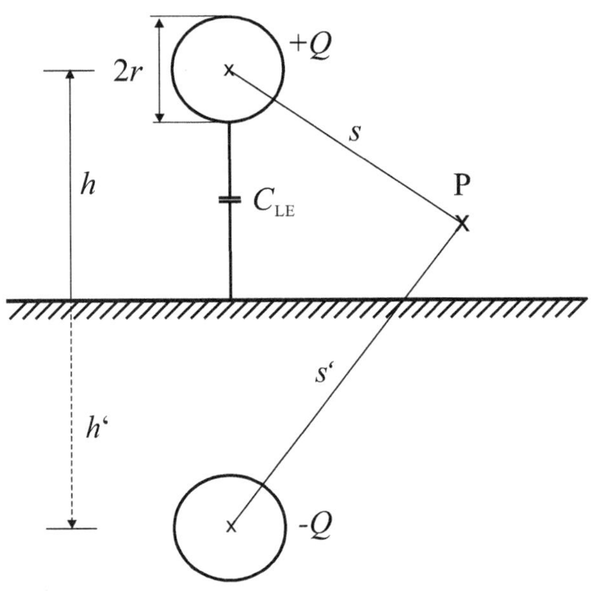

Fig. 4.25 Equivalent circuit diagram for capacitance calculation, conductor above the earth's surface (mirroring method)

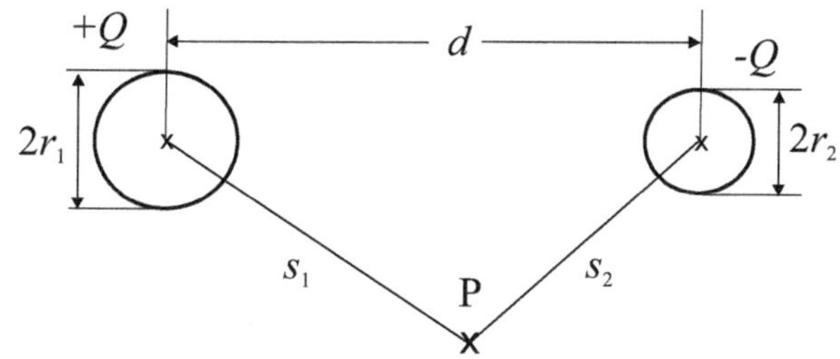

Fig. 4.26 Capacitance of a conductor consisting of two line charges (without earth)

$$\varphi = -\frac{1}{2\pi \cdot \varepsilon_r \cdot \varepsilon_0 \cdot \ell} \cdot \left\{ Q_1 \cdot \ln\left(\frac{s_1'}{s_1}\right) + Q_2 \cdot \ln\left(\frac{s_2'}{s_2}\right) + \ldots \right\} + K \tag{4.126}$$

For the calculation of a loop (consisting of a forward and return conductor, without earth) according to Fig. 4.26, the following generally applies to the potential φ of a point P with $Q_1 = -Q_2$, derived from G (4.125):

$$\varphi = -\frac{Q}{2\pi \cdot \varepsilon_r \cdot \varepsilon_0 \cdot \ell} \cdot \ln\left(\frac{s_2}{s_1}\right) + K \tag{4.127}$$

The following boundary conditions result for the potentials φ_1 and φ_2 on the conductors:

- Surface of conductor 1: $s_1 = r_1$ and $s_2 = d$
- Surface of conductor 2: $s_1 = d$ and $s_2 = r_2$

After insertion, the voltage U between the two conductors is obtained:

$$U = \varphi_1 - \varphi_2 = \frac{Q}{2\pi \cdot \varepsilon_r \cdot \varepsilon_0 \cdot \ell} \cdot \left\{ \ln\left(\frac{d}{r_1}\right) + \ln\left(\frac{r_2}{d}\right) \right\} = \frac{Q}{\pi \cdot \varepsilon_r \cdot \varepsilon_0 \cdot \ell} \cdot \ln\left(\frac{d}{\sqrt{r_1 \cdot r_2}}\right) \tag{4.128}$$

This results in the capacitance of the conductor loop, or between two conductors, according to Fig. 4.26 with $r = r_1 = r_2$:

$$C = \frac{Q}{U} = \frac{\pi \cdot \varepsilon_r \cdot \varepsilon_0 \cdot \ell}{\ln \frac{d}{r}} \tag{4.129}$$

In contrast, the capacitance of a conductor loop with earth is determined on the basis of Eq. (4.125), so that the potential of point P can be derived from the superposition according to Fig. 4.27. In this case, too, the surfaces of the conductors are used to determine their potential, Eq. (4.131):

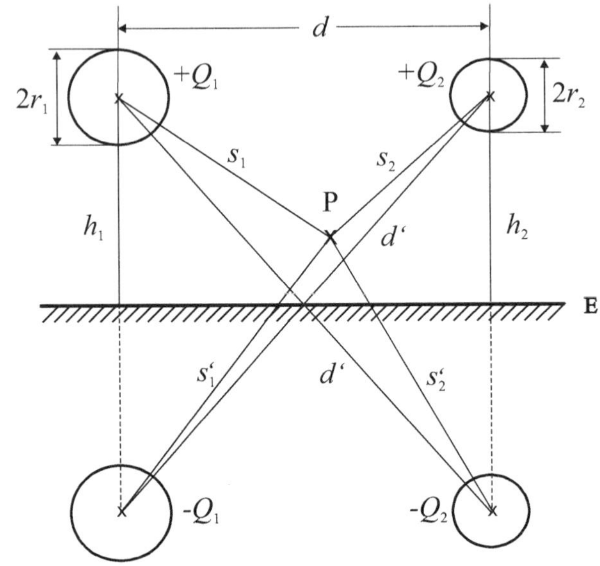

Fig. 4.27 Capacitance of a conductor consisting of two line charges (with earth)

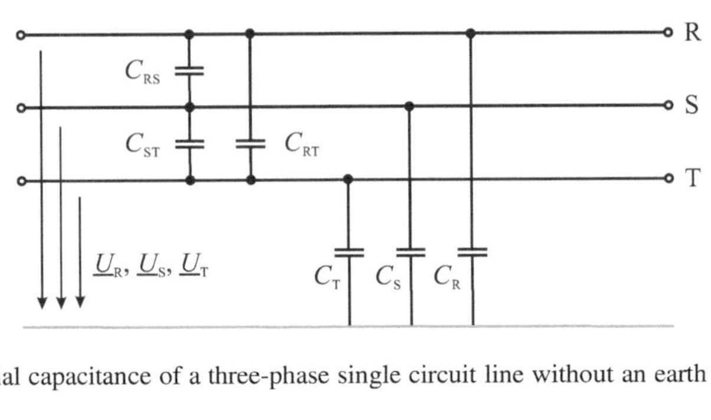

Fig. 4.28 Partial capacitance of a three-phase single circuit line without an earth wire

$$\varphi_1 = -\frac{1}{2\pi \cdot \varepsilon_r \cdot \varepsilon_0 \cdot \ell} \cdot \left\{ Q_1 \cdot \ln\left(\frac{2h_1}{r_1}\right) + Q_2 \cdot \ln\left(\frac{d'}{d}\right) \right\} \tag{4.130}$$

A comparable equation can be derived for the potential φ_2. The capacitance of a single conductor can be determined from Eq. (4.127) if the point P lies on the surface of the conductor, so that the relationships $s_1' = 2h - r$ and $s_1 = r$ apply.

$$C_{LE} = \frac{Q}{\phi_{L0}} = \frac{2\pi \cdot \varepsilon_r \cdot \varepsilon_0 \cdot \ell}{\ln\left(\frac{2h-r}{r}\right)} \approx \frac{2\pi \cdot \varepsilon_r \cdot \varepsilon_0 \cdot \ell}{\ln\left(\frac{2h}{r}\right)} \tag{4.131}$$

Positive Sequence Capacitance of a Three-Phase Overhead Line

If a three-phase overhead line is fed by a symmetrical voltage system as shown in Fig. 4.28, the following applies:

$$\underline{U}_R = \underline{U}_R \quad \underline{U}_S = \underline{a}^2 \cdot \underline{U}_R \quad \underline{U}_T = \underline{a} \cdot \underline{U}_R \tag{4.132}$$

$$\underline{Q}_R = \underline{Q}_R \quad \underline{Q}_S = \underline{a}^2 \cdot \underline{Q}_R \quad \underline{Q}_T = \underline{a} \cdot \underline{Q}_R \tag{4.133}$$

Based on Eq. (4.129), the system of equations can be developed according to (4.134) if no ground wire is assumed.

$$\begin{pmatrix} \underline{U}_R \\ \underline{U}_S \\ \underline{U}_T \end{pmatrix} = \begin{pmatrix} \alpha_{11} & \alpha_{12} & \alpha_{13} \\ \alpha_{21} & \alpha_{22} & \alpha_{23} \\ \alpha_{31} & \alpha_{32} & \alpha_{33} \end{pmatrix} \cdot \begin{pmatrix} \underline{Q}_R \\ \underline{Q}_S \\ \underline{Q}_T \end{pmatrix} \tag{4.134}$$

The variables α represent the potential coefficients of the system of equations and can be derived from Eq. (4.128).

$$\alpha_{\nu\nu} = \frac{1}{2\pi \cdot \varepsilon_0 \cdot \ell} \cdot \ln\left(\frac{2h_\nu}{r_\nu}\right) \quad \text{for} \quad \nu = \text{R, S, T} \tag{4.135}$$

$$\alpha_{\nu\mu} = \frac{1}{2\pi \cdot \varepsilon_0 \cdot \ell} \cdot \ln\left(\frac{d'}{d}\right) \quad \text{for} \quad \nu, \mu = \text{R, S, T}; \quad \mu \neq \nu \tag{4.136}$$

If the voltage and charge conditions according to Eq. (4.128) are taken into account, the first line of the system of Eq. (4.132) follows if the distances of a symmetrical overhead line according to Fig. 4.27 are used:

$$\underline{U}_R = (\alpha_{11} - \alpha_{12}) \cdot \underline{Q}_R \tag{4.137}$$

The following results for the positive sequence capacitance:

$$C_{(1)} = \frac{\underline{Q}_R}{\underline{U}_R} = \frac{1}{\alpha_{11} - \alpha_{12}} = \frac{2\pi \cdot \varepsilon_r \cdot \varepsilon_0 \cdot \ell}{\ln\left(\frac{2h}{r}\right) - \ln\left(\frac{d'}{d}\right)} \tag{4.138}$$

With

$$d' \approx \sqrt{(2h)^2 + d^2} \quad h = \sqrt[3]{h_R \cdot h_S \cdot h_T} \tag{4.139}$$

follows

$$C_{(1)} = \frac{2\pi \cdot \varepsilon_r \cdot \varepsilon_0 \cdot \ell}{\ln \frac{d}{r \cdot \sqrt{1 + \left(\frac{d}{2h}\right)^2}}} \tag{4.140}$$

The positive sequence capacitance is independent of the conductor height h if twice the conductor height 2 h is significantly greater than the distance d between the conductors, which is the case for high-voltage overhead lines. The following then applies for simplification:

$$C_{(1)} \approx \frac{2\pi \cdot \varepsilon_0 \cdot \ell}{\ln(d/r)} \tag{4.141}$$

Zero Sequence Capacitance of a Three-Phase Overhead Line

To calculate the zero sequence or earth capacitance, the three conductors must be connected in parallel as shown in Fig. 4.27 and supplied with a voltage $\underline{U}_{\mathrm{R}} = \underline{U}_{\mathrm{S}} = \underline{U}_{\mathrm{T}}$. Under this assumption follows:

$$\underline{Q}_{\mathrm{R}} = \underline{Q}_{\mathrm{S}} = \underline{Q}_{\mathrm{T}} \tag{4.142}$$

For a symmetrical line, the following results according to Eq. (4.134)

$$\underline{U}_{\mathrm{R}} = (\alpha_{11} + 2\alpha_{12}) \cdot \underline{Q}_{\mathrm{R}} \tag{4.143}$$

After inserting the potential coefficients, the results are as follows:

$$C_{(0)} = \frac{\underline{Q}_{\mathrm{R}}}{\underline{U}_{\mathrm{R}}} = \frac{1}{\alpha_{11} + 2\alpha_{12}} = \frac{2\pi \cdot \varepsilon_{\mathrm{r}} \cdot \varepsilon_0 \cdot \ell}{\ln\left(\frac{2h}{r}\right) + \ln\left(\frac{\sqrt{(2h)^2 + d^2}}{d}\right)} = \frac{2\pi \cdot \varepsilon_{\mathrm{r}} \cdot \varepsilon_0 \cdot \ell}{3 \cdot \ln\left(\frac{2h}{\sqrt[3]{r \cdot d^2}}\right) \cdot \sqrt[3]{1 + \left(\frac{d}{2h}\right)^2}} \tag{4.144}$$

If the condition $d << 2\ h$ is fulfilled, the following applies by simplification:

$$C_{(0)} \approx \frac{2\pi \cdot \varepsilon_{\mathrm{r}} \cdot \varepsilon_0 \cdot \ell}{3 \cdot \ln\left(\frac{2h}{\sqrt[3]{r \cdot d^2}}\right)} \tag{4.145}$$

For three-phase overhead lines, the earth capacitance is in the range of $C_{(0)} \approx 0.45$–$0.6\ C_{(1)}$.

Capacitance Ratio

Taking into account the following geometric data of a 110 kV overhead line (horizontal configuration), the capacitances in the positive und zero sequence system result.

Conductor radius: $r = 10.95$ mm (cross section 240/40 mm^2)
Average conductor spacing: $d = 5.04$ m (spacing $d_{1,2} = 4$ m or 8 m)
Average conductor height: $h = 21$ m
Positive sequence capacitance:

$$C_{(1)} = \frac{2\pi \cdot 1 \cdot 8.854}{\ln \frac{5.04}{10.95 \cdot 10^{-3} \cdot \sqrt{1 + \left(\frac{5.04}{2 \cdot 24}\right)^2}}} \text{nF/km} = 9.094 \text{ nF/km} \qquad \text{acc. to Eq.(4.140)}$$

$$C_{(1)} = \frac{2\pi \cdot 1 \cdot 8.854}{\ln \frac{5.04}{10.95 \cdot 10^{-3} \cdot}} \text{nF/km} = 9.073 \text{ nF/km} \qquad \text{acc. to Eq.(4.141)}$$

- Zero sequence capacitance:

$$C_{(0)} = \frac{2\pi \cdot 1 \cdot 8.854 \cdot \text{nF/km}}{3 \cdot \ln\left(\frac{2 \cdot 24}{\sqrt[3]{10.95 \cdot 10^{-3} \cdot 5.04^2}}\right) \cdot \sqrt[3]{1 + \left(\frac{5.04}{2 \cdot 24}\right)^2}}$$

$$= 4.311 \text{nF/km} \quad \text{acc. to Eq.} \qquad (4.144)$$

$$C_{(0)} = \frac{2\pi \cdot 1 \cdot 8.854}{3 \cdot \ln\left(\frac{2 \cdot 24}{\sqrt[3]{10.95 \cdot 10^{-3} \cdot 5.04^2}}\right)} \text{nF/km} = 4.315 \text{ nF/km} \qquad \text{acc. to Eq.(4.145)}$$

The capacitance values result in a capacitance ratio $C_{(0)}/C_{(1)} = 4.315$ nF/km/ 9.073 nF/km = 0.476.

4.7.3 *Impedances of Cables*

The structure of a cable (three-core, single-core cable, shield, and medium-voltage and low-voltage cable) differs significantly from that of an overhead line when determining the impedances, so that the cable impedances are determined separately in this section.

4.7.3.1 Positive Sequence Impedance of Medium-Voltage Cables

The resistance of a cable is determined according to Eq. (4.146).

$$R = R_{20} \cdot [1 + \alpha_{20} \cdot (\theta_L - 20\,°C)] \quad (4.146)$$

with

R_{20} DC resistance at 20 °C,
α_{20} temperature coefficient ($\alpha_{20} \approx 0.004$ 1/K),
θ_L cable temperature.

The resistance according to Eq. (4.146) is generally increased by skin and proximity effects. At the usual technical frequencies, however, the increase can be neglected for small and medium cross-sections (e.g. <240 mm^2).

In principle, however, the operation of the cable with alternating current results in frequency-dependent additional losses, which are expressed by an increase in effective resistance. The value R given in Eq. (4.146) is influenced by additional resistances, which increase the losses in the metal sheath and armoring, resulting in the total effective resistance R_K of the cable:

$$R_K = R + \Delta R_M + \Delta R_B \quad (4.147)$$

with

ΔR_M additional resistance for induction losses in the sheath,
ΔR_B additional resistance for the hysteresis losses in the steel band.

To determine the additional resistance ΔR_M of a conductor through a metal sheath, it is assumed according to Fig. 4.29 that three single-core cables are arranged in a triangle.

To calculate the positive sequence impedance, the system of Eq. (4.88) is extended by three rows and columns to represent the sheath voltages and currents, so that the following applies:

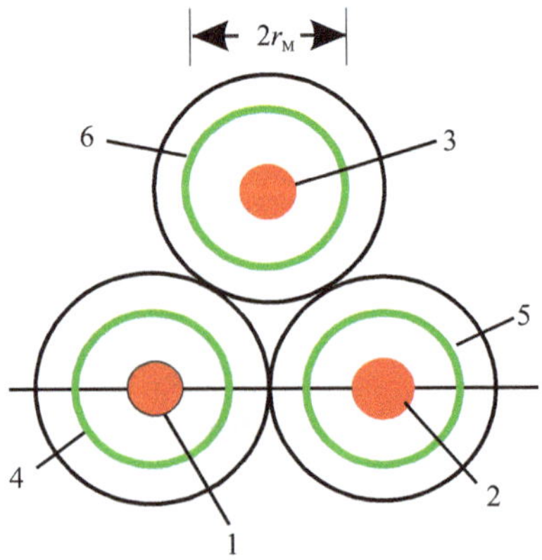

Fig. 4.29 Delta configuration of single-core cables. 1, 2, 3 conductors, 4, 5, 6 sheaths, r_M radius of the sheath

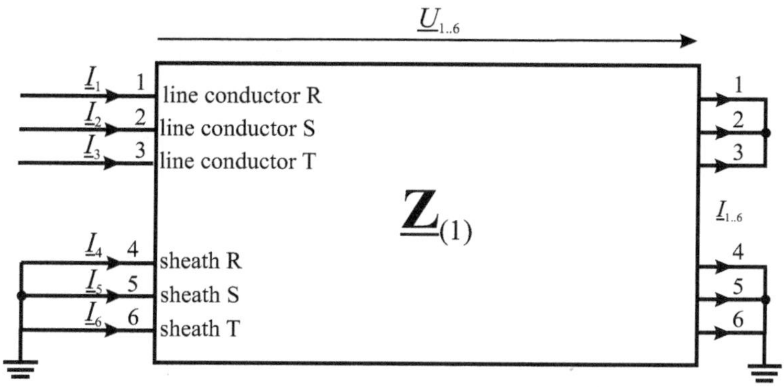

Fig. 4.30 Equivalent circuit for determining the cable impedance in the positive sequence system

$$\begin{pmatrix} \underline{U}_1 \\ \underline{U}_2 \\ \underline{U}_3 \\ \underline{U}_4 \\ \underline{U}_5 \\ \underline{U}_6 \end{pmatrix} = \begin{pmatrix} \underline{Z}_{11} & \underline{Z}_{12} & \underline{Z}_{13} & \underline{Z}_{14} & \underline{Z}_{15} & \underline{Z}_{16} \\ \underline{Z}_{21} & \underline{Z}_{22} & \underline{Z}_{23} & \underline{Z}_{24} & \underline{Z}_{25} & \underline{Z}_{26} \\ \underline{Z}_{31} & \underline{Z}_{32} & \underline{Z}_{33} & \underline{Z}_{34} & \underline{Z}_{35} & \underline{Z}_{36} \\ \underline{Z}_{41} & \underline{Z}_{42} & \underline{Z}_{43} & \underline{Z}_{44} & \underline{Z}_{45} & \underline{Z}_{46} \\ \underline{Z}_{51} & \underline{Z}_{52} & \underline{Z}_{53} & \underline{Z}_{54} & \underline{Z}_{55} & \underline{Z}_{56} \\ \underline{Z}_{61} & \underline{Z}_{62} & \underline{Z}_{63} & \underline{Z}_{64} & \underline{Z}_{65} & \underline{Z}_{66} \end{pmatrix} \cdot \begin{pmatrix} \underline{I}_1 \\ \underline{I}_2 \\ \underline{I}_3 \\ \underline{I}_4 \\ \underline{I}_5 \\ \underline{I}_6 \end{pmatrix} \tag{4.148}$$

The positive sequence impedance is determined according to Fig. 4.30, comparable to the representation of overhead lines according to Fig. 4.20a, in that the cable sheaths are connected to earth at both ends and the sheath and conductor currents add up to zero. From this, the system of Eq. (4.148) can be simplified.

Taking into account the symmetries and current and voltage conditions, the system of Eq. (4.149) can be specified. The conductor currents $\underline{I}_1$, $\underline{I}_2$ and $\underline{I}_3$ are applied as injected currents of a symmetrical three-phase current system.

$$\begin{pmatrix} \underline{U}_1 \\ 0 \\ 0 \\ 0 \end{pmatrix} = \begin{pmatrix} (\underline{Z}_{11} - \underline{Z}_{12}) & \underline{Z}_{14} & \underline{Z}_{12} & \underline{Z}_{12} \\ (\underline{Z}_{14} - \underline{Z}_{12}) & \underline{Z}_{44} & \underline{Z}_{12} & \underline{Z}_{12} \\ (\underline{Z}_{14} - \underline{Z}_{12}) \cdot \underline{a}^2 & \underline{Z}_{12} & \underline{Z}_{44} & \underline{Z}_{12} \\ (\underline{Z}_{14} - \underline{Z}_{12}) \cdot \underline{a} & \underline{Z}_{12} & \underline{Z}_{12} & \underline{Z}_{44} \end{pmatrix} \cdot \begin{pmatrix} \underline{I}_1 \\ \underline{I}_4 \\ \underline{I}_5 \\ \underline{I}_6 \end{pmatrix} \tag{4.149}$$

In Eq. (4.149) means:

$$\underline{I}_4 + \underline{I}_5 + \underline{I}_6 = 0 \qquad \underline{U}_4 = \underline{U}_5 = \underline{U}_6 = 0 \tag{4.150}$$

$$\underline{Z}_{11} = R_1 + \mathrm{j}\omega \cdot \frac{\mu_0}{2\pi} \cdot \ell \cdot \left[0.25 + \ln \frac{d_\mathrm{A}}{r} \right] \tag{4.151}$$

$$\underline{Z}_{44}=\underline{Z}_{55}=\underline{Z}_{66}=R_{M}+j\omega\cdot\frac{\mu_0}{2\pi}\cdot\ell\cdot\ln\frac{d_A}{r_M} \tag{4.152}$$

$$\underline{Z}_{14}=\underline{Z}_{41}=\underline{Z}_{52}=\underline{Z}_{25}=\underline{Z}_{63}=\underline{Z}_{36}=j\omega\cdot\frac{\mu_0}{2\pi}\cdot\ell\cdot\ln\frac{d_A}{r_M} \tag{4.153}$$

$$\begin{aligned}&\underline{Z}_{12}=\underline{Z}_{15}=\underline{Z}_{13}=\underline{Z}_{16}=\underline{Z}_{42}=\underline{Z}_{45}=\underline{Z}_{43}=\underline{Z}_{46}=\underline{Z}_{53}=\underline{Z}_{54}=\underline{Z}_{56}\\&=\underline{Z}_{61}=\underline{Z}_{62}=\underline{Z}_{64}=\underline{Z}_{65}=j\omega\frac{\mu_0}{2\pi}\cdot\ell\cdot\ln\frac{d_A}{d}\end{aligned} \tag{4.154}$$

With

R_1 resistance of the conductor,
R_M resistance of the sheath,
d_A radius of the fictitious outer sheath,
r radius of the conductor,
r_M inner radius of the sheath,
d distance between conductors.

The sheath currents are determined from this, taking into account Eq. (4.150), which in this case (delta configuration) form a symmetrical three-phase current system.

$$\begin{aligned}\underline{I}_4&=-\frac{\underline{Z}_{14}-\underline{Z}_{12}}{\underline{Z}_{44}-\underline{Z}_{12}}\cdot\underline{I}_1\\ \underline{I}_5&=-\frac{\underline{Z}_{14}-\underline{Z}_{12}}{\underline{Z}_{44}-\underline{Z}_{12}}\cdot\underline{a}^2\cdot\underline{I}_1\\ \underline{I}_6&=-\frac{\underline{Z}_{14}-\underline{Z}_{12}}{\underline{Z}_{44}-\underline{Z}_{12}}\cdot\underline{a}\cdot\underline{I}_1\end{aligned} \tag{4.155}$$

Substituting the impedance of the cable into Eq. (4.149) gives the following result:

$$\frac{\underline{U}_1}{\underline{I}_1}=\underline{Z}_{(1)}=(\underline{Z}_{11}-\underline{Z}_{12})-\frac{(\underline{Z}_{14}-\underline{Z}_{12})^2}{\underline{Z}_{44}-\underline{Z}_{12}} \tag{4.156}$$

$$\underline{Z}_{(1)}=R_{(1)}+j\omega\cdot\frac{\mu_0}{2\pi}\cdot\ell\cdot\left[0.25+\ln\frac{d}{r}\right]-\frac{\left[j\omega\cdot\frac{\mu_0}{2\pi}\cdot\ell\cdot\ln\frac{d}{r_M}\right]^2}{R_M+j\omega\cdot\frac{\mu_0}{2\pi}\cdot\ell\cdot\ln\frac{d}{r_M}} \tag{4.157}$$

$$\mathrm{Re}\left\{\underline{Z}_{(1)}\right\}=R_{(1)}+R_M\cdot\frac{\left[\omega\cdot\frac{\mu_0}{2\pi}\cdot\ell\cdot\ln\frac{d}{r_M}\right]^2}{R_M^2+\left[\omega\cdot\frac{\mu_0}{2\pi}\cdot\ell\cdot\ln\frac{d}{r_M}\right]^2} \tag{4.158}$$

$$\mathrm{Im}\left\{\underline{Z}_{(1)}\right\}=\omega\cdot\frac{\mu_0}{2\pi}\cdot\ell\cdot\left[0.25+\ln\frac{d}{r}\right]-\frac{\left[\omega\cdot\frac{\mu_0}{2\pi}\cdot\ell\cdot\ln\frac{d}{r_M}\right]^3}{R_M^2+\left[\omega\cdot\frac{\mu_0}{2\pi}\cdot\ell\cdot\ln\frac{d}{r_M}\right]^2} \tag{4.159}$$

While the resistance of the cable R_1 is increased by an earthed sheath, the inductance is reduced. If the sheath resistance is not taken into account in Eq. (4.157) ($R_M \to \infty$), the result is Eq. (4.160), which corresponds to the positive sequence impedance of a symmetrical overhead line, Eq. (4.99).

$$\underline{Z}_{(1)} = R_{(1)} + j\omega \cdot \frac{\mu_0}{2\pi} \cdot \ell \cdot \left[0.25 + \ln\frac{d}{r}\right] = R_{(1)} + jX_{(1)} \tag{4.160}$$

The following values result for a cable 3 × 1 × 150 mm²/25 mm² NA2XS2Y (20 kV) (resistance per unit):

$$R'_{(1)} = 0.206\,\Omega/\text{km} \quad R'_M = 1.236\,\Omega/\text{km}$$

$$r = 6.91\,\text{mm} \quad r_M = 12.25\,\text{mm}$$

$$d = 36\,\text{mm} \quad X'_{(1)} = 0.104\,\Omega/\text{km}$$

With

d cable outer diameter.

After insertion, additional impedances per unit are determined in each case:

$$\Delta R'_M = +0.0024\ \Omega/\text{km} \quad \Delta X'_M = -0.0001\ \Omega/\text{km}$$

If the cables are not laid directly next to each other, but the center-to-center spacing of the conductors (Fig. 4.29) corresponds to twice the outer diameter of the cable, the additional values increase under these conditions to:

$$\Delta R'_M = +0.0072\ \Omega/\text{km} \quad \Delta X'_M = -0.0006\ \Omega/\text{km}$$

The additional resistance or the additional inductance caused by a cable armoring can be calculated in the same way, whereby a value of $\mu_r \approx 300$ is generally used for a steel band armoring.

Based on Eq. (4.155), the sheath current in symmetrical operation is calculated in relation to the conductor currents by inserting the impedances, resulting in:

$$\underline{I}_4 = -\frac{j\omega \cdot \frac{\mu_0}{2\pi} \cdot \ell \cdot \ln\frac{d}{r_M}}{R_M + j\omega \cdot \frac{\mu_0}{2\pi} \cdot \ell \cdot \ln\frac{d}{r_M}} \cdot \underline{I}_1 \tag{4.161}$$

Taking into account the geometric and electrical parameters of a 20 kV NA2XS2Y cable, the following results:

$$\underline{I}_4 = -(2.994 + j54.635) \cdot 10^{-3} \cdot \underline{I}_1 \quad \text{and} \quad I_4 = 0.0547 \cdot I_1$$

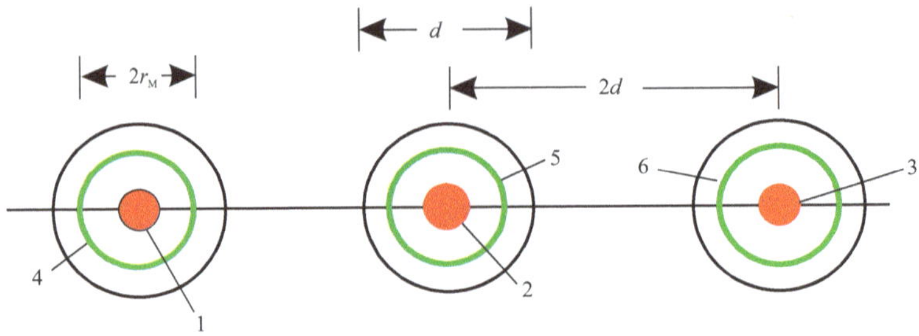

Fig. 4.31 Single-core cable in horizontal configuration, legend Fig. 4.29. *d* outer cable diameter

This means that in this case the sheath current is 5.47% of the associated conductor current. If the cable center-to-center distance is increased to twice this value, the sheath current increases to approx. 8.37%. If, based on Fig. 4.29, the distance between the conductors is increased, the sheath current $\underline{I}_4$ approaches the negative value of the conductor current $\underline{I}_1$ with $d \rightarrow \infty$, Eq. (4.160), which means that the entire conductor current flows in the sheath, but in the opposite direction.

If, in contrast to this example, the sheath is only earthed on one side, e.g. to avoid additional losses in the sheath, the current $\underline{I}_4 = 0$ according to Eq. (4.155), resulting in the induced voltage $\underline{U}_4$ of the sheath to ground plane at the end of the cable.

$$\underline{U}_4 = (\underline{Z}_{14} - \underline{Z}_{12}) \cdot \underline{I}_1 = \left(\mathrm{j}\omega \cdot \frac{\mu_0}{2\pi} \cdot \ell \cdot \ln \frac{d}{r_\mathrm{M}} \right) \cdot \underline{I}_1 \tag{4.162}$$

Taking the geometric data into account, the induced voltage $\underline{U}_4$ is obtained:

$$\underline{U}_4 = 0.0677 \cdot \frac{\underline{I}_1}{\mathrm{kA}} \cdot \left(\frac{\mathrm{kV}}{\mathrm{km}} \right)$$

This means that with a three-phase short-circuit current of $I_1 = 25$ kA, an induced voltage of $U_4 = 1.693$ kV/km is calculated at the unearthed sheath (cable end). As a result, surge arresters may have to be used on the cable sheath in order not to damage the cable insulation. This has a particular effect in the event of a line-to-earth short circuit, as in this case the conductor currents do not partially cancel each other out.

In contrast to the illustration in Fig. 4.29, single-core cables are laid in the horizontal configuration shown in Fig. 4.31, so that the distances between the individual conductors are different.

To determine the positive sequence impedance, the same current, voltage and impedance conditions are used as listed in Eqs. (4.147) to (4.152). Changes refer to some mutual impedances due to the changed distances.

$$\underline{Z}_{12} = \underline{Z}_{21} = \underline{Z}_{15} = \underline{Z}_{51} = \underline{Z}_{23} = \underline{Z}_{32} = \underline{Z}_{24} = \underline{Z}_{42} = \underline{Z}_{45} = \\ \underline{Z}_{54} = \underline{Z}_{53} = \underline{Z}_{35} = \underline{Z}_{56} = \underline{Z}_{65} = \underline{Z}_{26} = \underline{Z}_{62} = \mathrm{j}\omega \frac{\mu_0}{2\pi} \cdot \ell \cdot \ln \frac{d_\mathrm{A}}{2d} \tag{4.163}$$

$$\underline{Z}_{13} = \underline{Z}_{31} = \underline{Z}_{16} = \underline{Z}_{61} = \underline{Z}_{34} = \underline{Z}_{43} = \underline{Z}_{46} = \underline{Z}_{64} = \mathrm{j}\omega \frac{\mu_0}{2\pi} \cdot \ell \cdot \ln \frac{d_\mathrm{A}}{4d} \tag{4.164}$$

After insertion, the system of equations is simplified according to Eq. (4.165), assuming that the conductor currents are injected and represent a symmetrical three-phase system.

$$\begin{pmatrix} \underline{U}_1 \\ \underline{U}_2 \\ \underline{U}_3 \\ 0 \\ 0 \\ 0 \end{pmatrix} = \begin{pmatrix} \underline{Z}_{11} & \underline{Z}_{12} & \underline{Z}_{13} & \underline{Z}_{14} & \underline{Z}_{12} & \underline{Z}_{13} \\ \underline{Z}_{12} & \underline{Z}_{11} & \underline{Z}_{12} & \underline{Z}_{12} & \underline{Z}_{14} & \underline{Z}_{12} \\ \underline{Z}_{13} & \underline{Z}_{12} & \underline{Z}_{11} & \underline{Z}_{13} & \underline{Z}_{12} & \underline{Z}_{14} \\ \underline{Z}_{14} & \underline{Z}_{12} & \underline{Z}_{13} & \underline{Z}_{44} & \underline{Z}_{12} & \underline{Z}_{13} \\ \underline{Z}_{12} & \underline{Z}_{14} & \underline{Z}_{12} & \underline{Z}_{12} & \underline{Z}_{44} & \underline{Z}_{12} \\ \underline{Z}_{13} & \underline{Z}_{12} & \underline{Z}_{14} & \underline{Z}_{13} & \underline{Z}_{12} & \underline{Z}_{44} \end{pmatrix} \cdot \begin{pmatrix} \underline{I}_1 \\ \underline{a}^2 \cdot \underline{I}_1 \\ \underline{a} \cdot \underline{I}_1 \\ \underline{I}_4 \\ \underline{I}_5 \\ \underline{I}_6 \end{pmatrix} \tag{4.165}$$

After solving the system of equations, the following conductor impedances, Eqs. (4.166) to (4.168), and sheath currents, Eqs. (4.169) to (4.171), result:

$$\frac{\underline{U}_1}{\underline{I}_1} = \underline{Z}_{(1)} = \underline{Z}_{11} + \underline{a}^2 \cdot \underline{Z}_{12} + \underline{a} \cdot \underline{Z}_{13} + (\underline{Z}_{14} - \underline{Z}_{13}) \cdot \frac{\underline{I}_4}{\underline{I}_1} + (\underline{Z}_{12} - \underline{Z}_{13}) \cdot \frac{\underline{I}_5}{\underline{I}_1} \tag{4.166}$$

$$\frac{\underline{U}_2}{\underline{a}^2 \cdot \underline{I}_1} = \underline{Z}_2 = (\underline{Z}_{11} - \underline{Z}_{12}) - \frac{(\underline{Z}_{14} - \underline{Z}_{12})^2}{\underline{Z}_{44} - \underline{Z}_{12}} \tag{4.167}$$

$$\frac{\underline{U}_3}{\underline{a} \cdot \underline{I}_1} = \underline{Z}_3 = \underline{Z}_{11} + \underline{a} \cdot \underline{Z}_{12} + \underline{a}^2 \cdot \underline{Z}_{13} + (\underline{Z}_{13} - \underline{Z}_{14}) \cdot \frac{\underline{I}_4}{\underline{I}_1} + (\underline{Z}_{12} - \underline{Z}_{14}) \cdot \frac{\underline{I}_5}{\underline{I}_1} \tag{4.168}$$

$$\frac{\underline{I}_4}{\underline{I}_1} = -\frac{\underline{Z}_{14} + \underline{a}^2 \cdot \underline{Z}_{12} + \underline{a} \cdot \underline{Z}_{13}}{\underline{Z}_{44} - \underline{Z}_{13}} - \frac{(\underline{Z}_{12} - \underline{Z}_{13})}{(\underline{Z}_{44} - \underline{Z}_{13})} \cdot \frac{\underline{I}_5}{\underline{I}_1} \tag{4.169}$$

$$\frac{\underline{I}_5}{\underline{I}_1} = -\frac{\underline{Z}_{14} - \underline{Z}_{12}}{\underline{Z}_{44} - \underline{Z}_{12}} \cdot \underline{a}^2 \tag{4.170}$$

$$\frac{\underline{I}_6}{\underline{I}_1} = -\left(\frac{\underline{I}_4}{\underline{I}_1} + \frac{\underline{I}_5}{\underline{I}_1} \right) \tag{4.171}$$

Taking into account the electrical and geometric specifications of the single-core cable, the following values can be calculated for the individual conductor impedances and sheath currents, based on the conductor current $\underline{I}_1$:

$$\underline{Z}_1 = (0.2198 + j0.0497)\cdot\Omega/\text{km} \quad Z_1 = 0.2253\cdot\Omega/\text{km}$$
$$\underline{Z}_2 = (0.2070 - j0.0518)\cdot\Omega/\text{km} \quad Z_2 = 0.2134\cdot\Omega/\text{km}$$
$$\underline{Z}_3 = (0.1925 + j0.0684)\cdot\Omega/\text{km} \quad Z_3 = 0.2043\cdot\Omega/\text{km}$$
$$\underline{I}_4 = -0.1778 - j0.0402\cdot\underline{I}_1 \quad I_4 = 0.1823\cdot I_1$$
$$\underline{I}_5 = -0.0244 + j0.0150\cdot\underline{I}_1 \quad I_5 = 0.0286\cdot I_1$$
$$\underline{I}_6 = 0.2022 + j0.0252\cdot\underline{I}_1 \quad I_6 = 0.2038\cdot I_1$$

In contrast to delta installation with a sheath current of approx. 5.5% in relation to the conductor current, a horizontal configuration with approx. 20.4% results in significantly higher currents in the sheath and therefore higher losses.

4.7.3.2 Zero Sequence Impedance of Low-Voltage Cables

While the zero sequence impedance of overhead lines is determined with an earth wire in Sect. 4.7.2.3, the zero sequence impedance of a low-voltage cable can be determined using the measurement circuit shown in Fig. 4.32, so that the neutral conductor, sheath/shield and earth act as a return conductor [26]. The zero sequence impedance of a three-core medium-voltage cable can also be derived from this by omitting the neutral conductor.

The following cable types or their cable structures are considered in detail, according to Table 4.5.

Due to the cable geometry, the impedances of all conductors to the sheath and all conductors to earth are identical. The difference between the mutual inductances of the outer conductors and the mutual inductance of the neutral conductor to the outer conductors is only slight. In addition, the self-inductances of the outer conductors and usually also the neutral conductors are the same. The following simplifications for the length-related impedances therefore apply to the self- and mutual impedances:

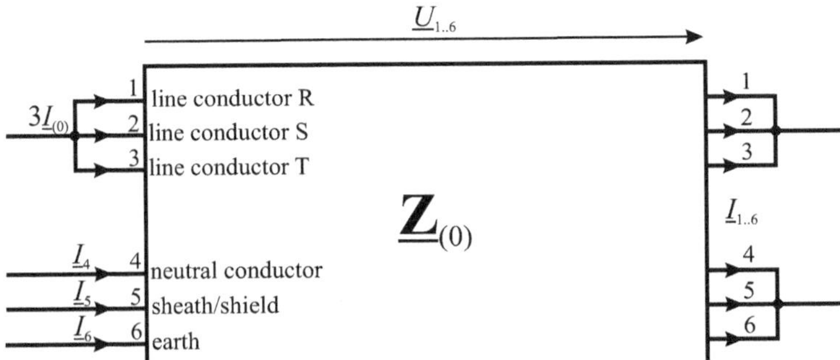

Fig. 4.32 Equivalent circuit for the zero sequence system of a four-core cable (low-voltage cable) with sheath and earth return in the event of a short circuit

Table 4.5 Structure of typical low-voltage cables (conductor identification according to Fig. 4.32)

Identification	Cable structure/cross-section
Four-core cable with neutral conductor/sheath as return conductor; N(A)YY	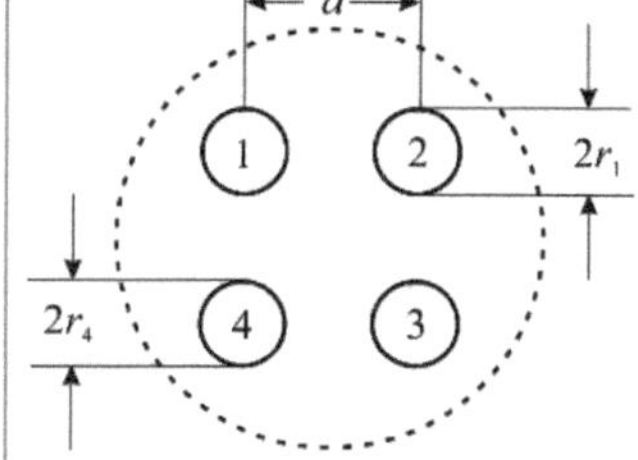
Three-core cable with sheath/earth as return conductor NAKLEY resp. N(A)YCWY	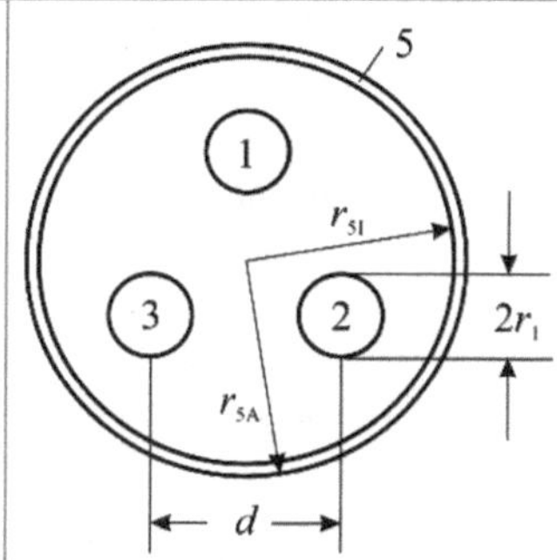
Four-core cable with neutral conductor/sheath/earth as return conductor; N(A)KBA resp. N(A)YCWY	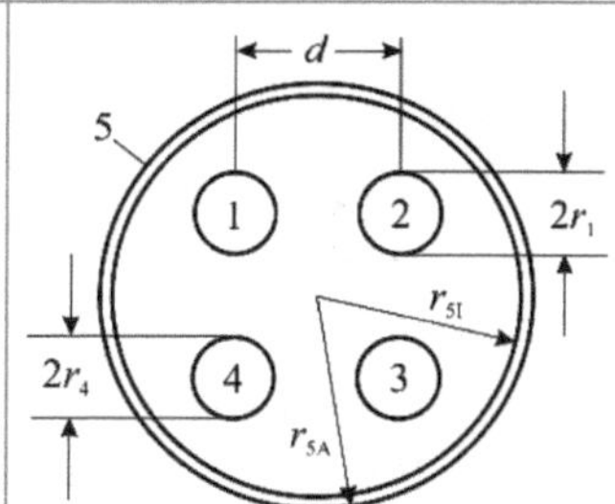

$$\underline{Z}_{11} = \underline{Z}_{22} = \underline{Z}_{33} = R'_1 + j\omega \cdot \frac{\mu_0}{2\pi} \cdot \left[0.25 + \ln\left(\frac{d_A}{r_1}\right)\right] \tag{4.172}$$

$$\underline{Z}_{44} = R'_4 + j\omega \cdot \frac{\mu_0}{2\pi} \cdot \left[0.25 + \ln\left(\frac{d_A}{r_4}\right)\right] \tag{4.173}$$

$$\underline{Z}_{55} = R'_5 + j\omega \cdot \frac{\mu_0}{2\pi} \cdot \left[b + \ln\left(\frac{d_A}{r_5}\right)\right] \tag{4.174}$$

$$\underline{Z}_{66} = R'_6 + j\omega \cdot \frac{\mu_0}{2\pi} \cdot \ln\left(\frac{d_A}{d_E}\right) \tag{4.175}$$

$$b = \frac{1}{\left(r_{5A}^2 - r_{5I}^2\right)^2} \cdot \left[\ln\left(\frac{r_{5A}}{r_{5I}}\right) - \frac{3 \cdot r_{5I}^2 - r_{5A}^2}{4} \right] \tag{4.176}$$

$$\underline{Z}_{12} = \underline{Z}_{23} = \underline{Z}_{13} = \underline{Z}_{14} = \underline{Z}_{24} = \underline{Z}_{34} = \mathrm{j}\omega \cdot \frac{\mu_0}{2\pi} \cdot \ln\left(\frac{d_A}{d}\right) \tag{4.177}$$

$$\underline{Z}_{16} = \underline{Z}_{26} = \underline{Z}_{36} = \underline{Z}_{46} = \underline{Z}_{56} = \mathrm{j}\omega \cdot \frac{\mu_0}{2\pi} \cdot \ln\left(\frac{d_A}{d_E}\right) \tag{4.178}$$

With

R'_1, R'_4	resistance phase, neutral conductor,
R'_5	resistance sheath,
R'_6	resistance earth ($= \omega\, \mu_0/8$),
r_1, r_4	radius phase, neutral conductor,
b	inner inductance of a hollow conductor [27],
d	mean distance phase, neutral conductor,
r_{5I}, r_{5A}	radius hollow conductor (inner, outer),
d_E	penetration depth of the return current in the earth,
d_A	radius of the fictitious sheath conductor.

To derive the zero sequence impedance, the system of Eq. (4.113) is extended by one row and column (center conductor), so that Eq. (4.179) generally results, taking into account the simplifications given above:

$$\begin{pmatrix} \underline{U}_1 \\ \underline{U}_2 \\ \underline{U}_3 \\ \underline{U}_4 \\ \underline{U}_5 \\ \underline{U}_6 \end{pmatrix} = \begin{pmatrix} \underline{Z}_{11} & \underline{Z}_{12} & \underline{Z}_{12} & \underline{Z}_{12} & \underline{Z}_{15} & \underline{Z}_{16} \\ \underline{Z}_{12} & \underline{Z}_{11} & \underline{Z}_{12} & \underline{Z}_{12} & \underline{Z}_{15} & \underline{Z}_{16} \\ \underline{Z}_{12} & \underline{Z}_{12} & \underline{Z}_{11} & \underline{Z}_{12} & \underline{Z}_{15} & \underline{Z}_{16} \\ \underline{Z}_{12} & \underline{Z}_{12} & \underline{Z}_{12} & \underline{Z}_{44} & \underline{Z}_{15} & \underline{Z}_{16} \\ \underline{Z}_{15} & \underline{Z}_{15} & \underline{Z}_{15} & \underline{Z}_{15} & \underline{Z}_{55} & \underline{Z}_{16} \\ \underline{Z}_{16} & \underline{Z}_{16} & \underline{Z}_{16} & \underline{Z}_{16} & \underline{Z}_{16} & \underline{Z}_{66} \end{pmatrix} \cdot \begin{pmatrix} \underline{I}_1 \\ \underline{I}_2 \\ \underline{I}_3 \\ \underline{I}_4 \\ \underline{I}_5 \\ \underline{I}_6 \end{pmatrix} \tag{4.179}$$

Based on the system of Eq. (4.179), the zero sequence impedances of some variants of the different cable types are determined according to Table 4.5.

Four-Core Cable only with Neutral Conductor as Return Conductor (Cable Type NAYY)

Deleting rows and columns 5 and 6 results in the system of Eq. (4.179):

$$\begin{pmatrix} \underline{U}_1 \\ \underline{U}_2 \\ \underline{U}_3 \\ \underline{U}_4 \end{pmatrix} = \begin{pmatrix} \underline{Z}_{11} & \underline{Z}_{12} & \underline{Z}_{12} & \underline{Z}_{12} \\ \underline{Z}_{12} & \underline{Z}_{11} & \underline{Z}_{12} & \underline{Z}_{12} \\ \underline{Z}_{12} & \underline{Z}_{12} & \underline{Z}_{11} & \underline{Z}_{12} \\ \underline{Z}_{12} & \underline{Z}_{12} & \underline{Z}_{12} & \underline{Z}_{44} \end{pmatrix} \cdot \begin{pmatrix} \underline{I}_1 \\ \underline{I}_2 \\ \underline{I}_3 \\ \underline{I}_4 \end{pmatrix} \tag{4.180}$$

Taking into account the measuring circuit for determining the zero sequence voltage, the following results:

$$\underline{U}_1 - \underline{U}_4 = \underline{U}_{(0)} = (\underline{Z}_{11} - \underline{Z}_{12}) \cdot \underline{I}_1 + (\underline{Z}_{12} - \underline{Z}_{44}) \cdot \underline{I}_4 \tag{4.181}$$

According to Fig. 4.32, the current condition applies:

$$\underline{I}_1 + \underline{I}_2 + \underline{I}_3 + \underline{I}_4 = 3 \cdot \underline{I}_{(0)} + \underline{I}_4 = 0 \tag{4.182}$$

The zero sequence impedance then results in:

$$\begin{gathered} \frac{\underline{U}_{(0)}}{\underline{I}_{(0)}} = \underline{Z}_{(0)} = (\underline{Z}_{11} - \underline{Z}_{12}) + 3 \cdot (\underline{Z}_{44} - \underline{Z}_{12}) = \\ R'_1 + \mathrm{j}\frac{\mu_0}{2\pi}\left[0.25 + \ln\left(\frac{d}{r_1}\right)\right] + 3 \cdot \left\{R'_4 + \mathrm{j}\frac{\mu_0}{2\pi}\left[0.25 + \ln\left(\frac{d}{r_4}\right)\right]\right\} \end{gathered} \tag{4.183}$$

If the phase and neutral conductors of the cable are identical, Eq. (4.183) is simplified accordingly, resulting in the zero sequence impedance according to Eq. (4.184).

$$\underline{Z}_{(0)} = 4 \cdot R'_1 + \mathrm{j}4 \cdot \frac{\mu_0}{2\pi}\left[0.25 + \ln\left(\frac{d}{r_1}\right)\right] \tag{4.184}$$

Four-Core Cable with Neutral Conductor and Earth as Return Conductor (Cable Type NAYY)

In this case, an additional column and row are introduced for the earth (indicator 6) based on the system of Eq. (4.180), so that the following applies:

$$\begin{pmatrix} \underline{U}_1 \\ \underline{U}_2 \\ \underline{U}_3 \\ \underline{U}_4 \\ \underline{U}_6 \end{pmatrix} = \begin{pmatrix} \underline{Z}_{11} & \underline{Z}_{12} & \underline{Z}_{12} & \underline{Z}_{12} & \underline{Z}_{16} \\ \underline{Z}_{12} & \underline{Z}_{11} & \underline{Z}_{12} & \underline{Z}_{12} & \underline{Z}_{16} \\ \underline{Z}_{12} & \underline{Z}_{12} & \underline{Z}_{11} & \underline{Z}_{12} & \underline{Z}_{16} \\ \underline{Z}_{12} & \underline{Z}_{12} & \underline{Z}_{12} & \underline{Z}_{44} & \underline{Z}_{16} \\ \underline{Z}_{16} & \underline{Z}_{16} & \underline{Z}_{16} & \underline{Z}_{16} & \underline{Z}_{66} \end{pmatrix} \cdot \begin{pmatrix} \underline{I}_1 \\ \underline{I}_2 \\ \underline{I}_3 \\ \underline{I}_4 \\ \underline{I}_6 \end{pmatrix} \tag{4.185}$$

The following current and voltage conditions apply for determining the zero sequence impedance:

$$\begin{aligned} &\underline{I}_1 + \underline{I}_2 + \underline{I}_3 + \underline{I}_4 + \underline{I}_6 = 3 \cdot \underline{I}_{(0)} + \underline{I}_4 + \underline{I}_6 = 0 \\ &\underline{U}_4 = \underline{U}_6 \\ &\underline{U}_1 - \underline{U}_4 = \underline{U}_{(0)} \end{aligned} \tag{4.186}$$

After inserting the conditions according to Eq. (4.186), the zero sequence impedance results from the system of Eq. (4.185).

$$\frac{\underline{U}_{(0)}}{\underline{I}_{(0)}} = \underline{Z}_{(0)} = (\underline{Z}_{11} - \underline{Z}_{12}) + 3 \cdot (\underline{Z}_{44} - \underline{Z}_{12}) \cdot \frac{\underline{Z}_{66} + \underline{Z}_{12} - 2 \cdot \underline{Z}_{16}}{\underline{Z}_{66} + \underline{Z}_{44} - 2 \cdot \underline{Z}_{16}} \tag{4.187}$$

If the phase and neutral conductors have the same cross-section, Eq. (4.187) gives the zero sequence impedance:

$$\begin{aligned} \underline{Z}_{(0)} = {} & R_1' + \mathrm{j}\frac{\mu_0}{2\pi}\left[0.25 + \ln\left(\frac{d}{r_1}\right)\right] + \\ & 3 \cdot \frac{\left\{R_1' + \mathrm{j}\frac{\mu_0}{2\pi} \cdot \left[0.25 + \ln\left(\frac{d}{r_1}\right)\right]\right\} \cdot \left\{R_6' + \mathrm{j}\frac{\mu_0}{2\pi} \cdot \ln\left(\frac{d}{r_1}\right)\right\}}{R_6' + R_1' + \mathrm{j}\frac{\mu_0}{2\pi}\left[0.25 + \ln\left(\frac{d_\mathrm{E}}{r_1}\right)\right]} \end{aligned} \tag{4.188}$$

In [1, 26], data on cable impedances are listed, which were determined according to the equations given above for different cross-sections and cable types. Eq. (4.184) is again obtained for $R_6' \rightarrow \infty$ that means that the return conductor via earth is not present.

4.7.3.3 Positive Sequence and Zero Sequence Capacitance

Due to the small distances and the larger dielectric constants, the positive sequence capacitance $C_{(1)}$ is significantly greater in a cable than in an overhead line; as a result of the structure, a distinction is made between a single-core and three-core cable as shown in Fig. 4.33. Whereas in the case of three single-core cables, the mutual capacitances C_g between the conductors are short-circuited, these capacitances are present in a three-core cable. The corresponding equations for the calculation of the capacitance per unit can be found in Table 4.6.

The earth or zero sequence capacitance differs from the operating capacitance only in the case of the belted cable due to the capacitance between the conductors.

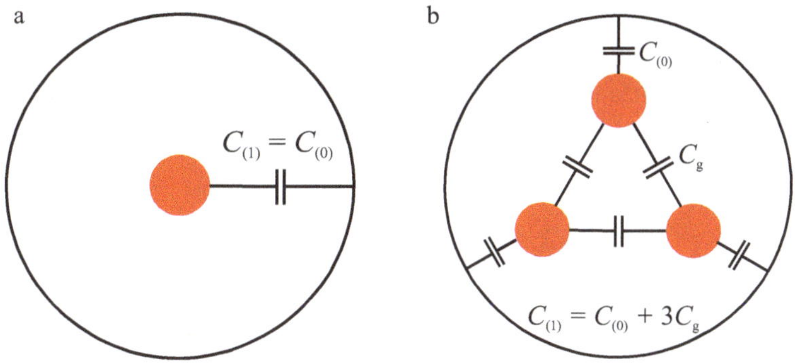

Fig. 4.33 Cable capacitances depending on the cable type. (**a**) Single-core cable (radial field cable), (**b**) Three-core cable (belted cable)

Table 4.6 Equations for determining the cable capacitances

Quantity	Cable type		
	Single-core, radial field cable	Three-core, belted cable	
Positive sequence capacitance $C_{(1)}$	$\frac{2\pi \cdot \varepsilon_0 \cdot \varepsilon_r}{\ln \frac{r_I}{r}} \cdot \ell$	$\frac{2\pi \cdot \varepsilon_0 \cdot \varepsilon_r}{\ln \sqrt{\frac{3 \cdot c^2 \cdot \left(r_I^2 - c^2\right)^3}{r^2 \cdot \left(r_I^6 - c^6\right)}}} \cdot \ell$	Eq. (4.190)
Zero sequence capacitance $C_{(0)}$	$C_{(0)} = C_{(1)}$	$\frac{2\pi \cdot \varepsilon_0 \cdot \varepsilon_r}{\ln \left[\frac{r_I^6 - c^6}{3 \cdot c^2 \cdot r \cdot r_I^3}\right]} \cdot \ell$	Eq. (4.190)

r_I inner radius of the conductive layer (sheath/shield)
r outer radius of the conductor
c distance of the conductor from the cable center point

4.8 Power Transformers

In the following, two different equivalent circuit diagrams for the short-circuit current calculation are derived in addition to the transformer impedance values:

- Simplified equivalent circuit diagram: In this case, the simulation is carried out using a longitudinal impedance; the calculation to a different voltage level is carried out by multiplying by the transformation ratio t.
- Integrated equivalent circuit diagram: The transformation ratio is integrated into a quadripole representation, resulting in a π equivalent circuit diagram.

The short-circuit current calculation can generally be carried out with both equivalent circuit diagrams.

This section deals exclusively with separate winding transformers, whereas autotransformers are dealt with in Chap. 16 and their application in the short-circuit current calculation. These are transformers whose windings are also electrically connected to each other.

4.8.1 Transformer Impedance according to IEC

For the short-circuit current calculation, the value according to Eq. (4.191) is used for the impedance Z_T of transformers.

$$Z_T = \frac{u_{kr}}{100\%} \cdot \frac{U_{rT}^2}{S_{rT}} \quad (4.191)$$

With

U_{rT} rated voltage of the transformer,
S_{rT} rated apparent power of the transformer,
u_{kr} rated short-circuit voltage (%).

The impedance $\underline{Z}_T$ of the transformer is determined using Eq. (4.211), taking into account the ohmic short-circuit losses. In addition to the transformer impedance, an impedance correction K_T must be inserted for the calculation according to [3], in each case in the positive sequence, negative sequence and zero sequence system, so that the expression according to Eq. (4.192) results for transformers.

$$\underline{Z}_{TK} = K_T \cdot \underline{Z}_{TK} \quad K_T = 0.95 \cdot \frac{c_{max}}{1 + 0.6 \cdot x_T} \quad (4.192)$$

With

c_{max} voltage factor,
x_T relative reactance of the transformer (X_T).

When applying the impedance correction factor K_T, an inductive load flow is assumed before the short-circuit occurs, which has the same direction as the short-circuit current. The derivation of the correction factor is discussed in more detail in Sect. 8.3. When applying Eq. (4.190), the following conditions are assumed:

- Only the middle position of the tap changer (principal tapping) is assumed for the impedance, as an impedance correction is provided which takes into account an inductive preload.
- If transformers with different transformation ratios are connected in parallel, the arithmetic mean of the individual values is to be taken as the resulting transformation ratio, Sect. 15.1.

The $Z_{(0)}/Z_{(1)}$ ratios of the transformers for various designs can be found in Sect. 4.8.2 and 4.8.4. Transformers with a vector group (phase shifting) have no effect when determining the transformer impedance, so that the short-circuit current at the fault location or in the same voltage level is not taken into account, even in the case of unbalanced short-circuits. However, if short-circuit currents, e.g. on the high-voltage side of a transformer, are converted, the vector group must be taken into account. The conversion of component currents and voltages for these cases is shown in Sect. 9.2.2.2, Table 9.1.

4.8.2 Simplified Equivalent Circuit Diagram in the Positive Sequence System

Based on the induction and Ampere's law, a T equivalent circuit diagram can be derived for the transformer, which is based on the representation shown in Fig. 4.34, which does not take into account the ohmic losses that occur [3]. The following relationships result for the terminal voltages:

$$\underline{u}_1(t) = w_1 \cdot \frac{d\Phi_1}{dt} = w_1 \cdot \frac{d}{dt}(\Phi_H + \Phi_{\sigma 1}) \tag{4.193}$$

$$\underline{u}_2(t) = w_2 \cdot \frac{d\Phi_2}{dt} = w_2 \cdot \frac{d}{dt}(\Phi_H + \Phi_{\sigma 2}) \tag{4.194}$$

With

w_1, w_2 turns of the windings 1, 2.

The magnetic flux Φ is calculated according to Eq. (4.195):

$$\Phi = A \cdot B = A \cdot \mu_0 \cdot \mu_r \cdot H = A \cdot \mu_0 \cdot \mu_r \cdot \frac{\Theta}{\ell} = \mu_0 \cdot \mu_r \cdot \frac{w \cdot i}{\ell} \cdot A \tag{4.195}$$

With

A cross-sectional area,
B magnetic induction,
H magnetic field strength,
μ_0, μ_r permeability,
Θ ampere turns,
ℓ length of magnetic flux.

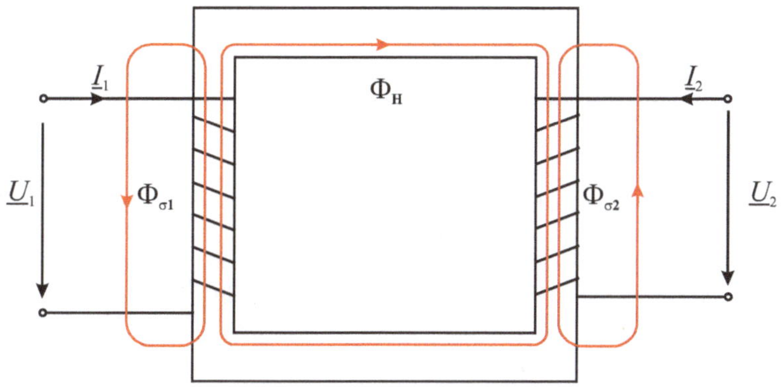

Fig. 4.34 Transformer diagram for determining the equivalent circuit. Φ_H main flux, $\Phi_{\sigma 1,2}$ leakage flux winding 1, 2

Using the abbreviation Λ for the magnetic conductance results in the following simplification:

$$\Phi = \Lambda \cdot w \cdot i \quad \Lambda = \mu_0 \cdot \mu_r \cdot \frac{A}{\ell} \tag{4.196}$$

This results in the following relationships for the fluxes according to Fig. 4.34:

$$\Phi_H = \Lambda_H \cdot (w_1 \cdot i_1 + w_2 \cdot i_2) \tag{4.197}$$

$$\Phi_{\sigma 1} = \Lambda_{\sigma 1} \cdot w_1 \cdot i_1 \quad \Phi_{\sigma 2} = \Lambda_{\sigma 2} \cdot w_2 \cdot i_2 \tag{4.198}$$

Substituting into Eqs. (4.193) and (4.194) results in.

$$\begin{aligned} u_1(t) &= w_1 \cdot \frac{\mathrm{d}}{\mathrm{dt}} [\Lambda_H \cdot (w_1 \cdot i_1 + w_2 \cdot i_2) + \Lambda_{\sigma 1} \cdot w_1 \cdot i_1] \\ u_2(t) &= w_2 \cdot \frac{\mathrm{d}}{\mathrm{dt}} [\Lambda_H \cdot (w_1 \cdot i_1 + w_2 \cdot i_2) + \Lambda_{\sigma 2} \cdot w_2 \cdot i_2] \end{aligned} \tag{4.199}$$

or

$$\begin{aligned} u_1(t) &= w_1^2 \cdot (\Lambda_H + \Lambda_{\sigma 1}) \cdot \frac{\mathrm{d}i_1}{\mathrm{dt}} + w_1 \cdot w_2 \cdot \Lambda_H \cdot \frac{\mathrm{d}i_2}{\mathrm{dt}} \\ u_2(t) &= w_1 \cdot w_2 \cdot \Lambda_H \cdot \frac{\mathrm{d}i_1}{\mathrm{dt}} + w_2^2 \cdot (\Lambda_H + \Lambda_{\sigma 1}) \cdot \frac{\mathrm{d}i_2}{\mathrm{dt}} \end{aligned} \tag{4.200}$$

If sinusoidal currents are assumed, the following results

$$\frac{\mathrm{d}i(t)}{\mathrm{dt}} = \frac{\mathrm{d}}{\mathrm{dt}}\left(\hat{i} \cdot e^{\mathrm{j}\omega t}\right) = \mathrm{j}\omega \cdot \hat{i} \cdot e^{\mathrm{j}\omega t} = \mathrm{j}\omega \cdot \sqrt{2} \cdot I \cdot e^{\mathrm{j}\omega t} \tag{4.201}$$

$$\begin{pmatrix} \underline{U}_1 \\ \underline{U}_2 \end{pmatrix} = \mathrm{j}\omega \cdot \begin{pmatrix} w_1^2 \cdot (\Lambda_H + \Lambda_{\sigma 1}) & w_1 \cdot w_2 \cdot \Lambda_H \\ w_1 \cdot w_2 \cdot \Lambda_H & w_2^2 \cdot (\Lambda_H + \Lambda_{\sigma 2}) \end{pmatrix} \cdot \begin{pmatrix} \underline{I}_1 \\ \underline{I}_2 \end{pmatrix} \tag{4.202}$$

If, in contrast to Eq. (4.202), the impedances are related to a voltage level (e.g. side 1), the current $\underline{I}_2$ or the voltage $\underline{U}_2$ according to Eq. (4.203) or Fig. 4.35 with the transformation ratio t applies.

$$\underline{I}_2 = \frac{w_1}{w_2} \cdot \underline{I}_{2t} = t \cdot \underline{I}_{2t} \text{ and } \underline{U}_2 = \frac{w_2}{w_1} \cdot \underline{U}_{2t} = \underline{U}_{2t}/t \tag{4.203}$$

After insertion into the system of Eq. (4.202), the following results after transformation.

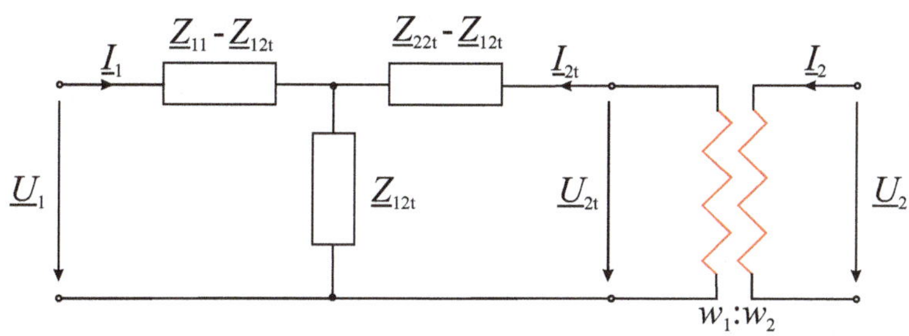

Fig. 4.35 T equivalent circuit diagram of the transformer, impedances according to Eq. (4.205)

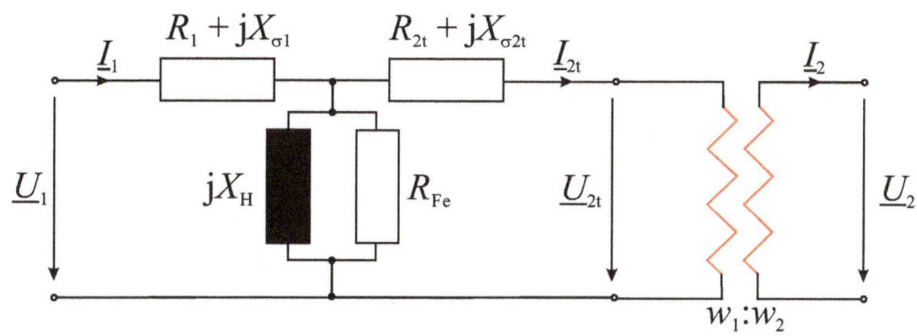

Fig. 4.36 Single-phase equivalent circuit diagram of a transformer (T equivalent circuit diagram) in the positive sequence system. R_1 winding resistance of the primary winding (1), R_{2t} winding resistance of the secondary winding (2) in relation to the winding (1), $X_{\sigma 1}$ leakage reactance of the primary winding (1), $X_{\sigma 2t}$ leakage reactance of the secondary winding (2) in relation to the winding (1), X_H magnetizing reactance, R_{Fe} iron losses (hysteresis, eddy current), t transformation ratio

$$\begin{pmatrix} \underline{U}_1 \\ \underline{U}_{2t} \end{pmatrix} = j\omega \cdot \begin{pmatrix} w_1^2 \cdot (\Lambda_H + \Lambda_{\sigma 1}) & w_1^2 \cdot \Lambda_H \\ w_1^2 \cdot \Lambda_H & w_1^2 \cdot (\Lambda_H + \Lambda_{\sigma 2}) \end{pmatrix} \cdot \begin{pmatrix} \underline{I}_1 \\ \underline{I}_{2t} \end{pmatrix} \tag{4.204}$$

The equivalent circuit diagram derived from this is shown in Fig. 4.35 with an ideal transformer, the impedances result from Eq. (4.205):

$$\begin{aligned} \underline{Z}_{11} - \underline{Z}_{12t} &= j\omega \cdot w_1^2 \cdot \Lambda_{\sigma 1} = j\omega \cdot L_{\sigma 1} \\ \underline{Z}_{22t} - \underline{Z}_{12t} &= j\omega \cdot w_1^2 \cdot \Lambda_{\sigma 2} = j\omega \cdot L_{\sigma 2} \\ \underline{Z}_{12t} &= j\omega \cdot w_1^2 \cdot \Lambda_H = j\omega \cdot L_H \end{aligned} \tag{4.205}$$

Figure 4.36 shows the equivalent circuit diagram derived from Eq. (4.203), including the ohmic losses. If, in contrast to this, a simplified equivalent circuit diagram with an ideal transformer is assumed, which only uses the short-circuit impedance according to Fig. 4.36 (this means that the leakage reactance and the ohmic losses are zero and the magnetizing reactance is almost infinite), this results from Eq. (4.202) under the condition $\underline{U}_1 = \underline{U}_2$, assuming that the transformer does not have a phase shifting vector group:

$$\underline{I}_2 = \frac{w_1 \cdot (w_1 - w_2)}{w_2 \cdot (w_2 - w_1)} \cdot \underline{I}_1 = -\frac{w_1}{w_2} \cdot \underline{I}_1 \tag{4.206}$$

From Eq. (4.206) it makes more sense to change the sign of the current indicator arrows, which is taken into account in the following equivalent circuits.

If, in contrast to Eq. (4.203), the transformation ratio is assumed to be complex, e.g. for transformers with a vector group Yd5 (phase shifting), the following applies for the transformation ratio with $t = \underline{t}$. The currents and voltages are determined from this:

$$\underline{I}_{2t} = \left(\frac{w_2}{w_1}\right) \cdot e^{j\delta} \cdot \underline{I}_2 = \frac{\underline{I}_2}{\underline{t}^*} \quad \text{und} \quad \underline{U}_{2t} = \left(\frac{w_1}{w_2}\right) \cdot e^{j\delta} \cdot \underline{U}_2 = \underline{t} \cdot \underline{U}_2 \tag{4.207}$$

The following applies to the impedance in relation to the transformed side:

$$\underline{U}_{2t} = \underline{Z}_{2t} \cdot \underline{I}_{2t} \quad \rightarrow \quad \underline{t} \cdot \underline{U}_2 = \underline{Z}_{2t} \cdot \frac{\underline{I}_2}{\underline{t}^*} \tag{4.208}$$

This results in the impedance, which must be referred to side 2 according to Fig. 4.35:

$$\underline{U}_2 = \frac{\underline{Z}_{2t}}{\underline{t} \cdot \underline{t}^*} \cdot \underline{I}_2 = \frac{\underline{Z}_{2t}}{t^2} \cdot \underline{I}_2 = \underline{Z}_2 \cdot \underline{I}_2 \tag{4.209}$$

According to Eq. (4.209), the advantage is that the vector group of a transformer does not have to be taken into account in the short-circuit current calculation if the impedance of the high-voltage side is to be converted to the low-voltage side (Sect. 4.1).

The impedance $\underline{Z}_T = R_T + jX_T = R_1 + R_{2t} + j(X_{\sigma 1} + X_{\sigma 2t})$ is also referred to as the short-circuit impedance of the transformer. In principle, it is possible to convert the T equivalent circuit diagram into an equivalent π equivalent circuit, but in this case there is no physical assignment of the elements, as is possible with the T equivalent circuit. For short-circuit calculations, the equivalent π equivalent circuit is generally used, as the number of nodes is not increased compared to the T equivalent circuit, see Sect. 4.8.3. For the value of the impedance in the positive sequence system, the following applies according to Eq. (4.210), corresponding to Sect. 4.8.1:

$$Z_T = \frac{u_{kr}}{100\%} \cdot \frac{U_{rT}^2}{S_{rT}} \tag{4.210}$$

Table 4.7 Zero sequence impedances of transformers and their neutral point load capacity in relation to the rated current I_r (reference values)

Vector group	Type	Earthed side	$X_{(0)}/X_{(1)}$	Neutral point load capacity %
Yy	3 single-phase, five-limb transformer	Y or y	100	0
	Three-limb transformer	Y or y	≈ 7	10[a]
	3 single-phase, five-limb transformer	Y and y	1	100
	Three-limb transformer	Y and y[b]	≈ 0.8[c]	100
Yd	3 single-phase transformer	Y	0,8–1	100
	Three-limb transformer	Y	≈ 0.8[c]	
Yz, Dz		z	0.15[c]	100
Yy(d)[e]		Y or y	1.5–2.4—4.5[f]	≤ 100
		Y and y	<1.5	100

[a]30% during 2 hours (tank heating)
[b]Also autotransformer
[c]Similar to the Yy vector group, a tank flow is formed, but to a much lesser extent. The induced tank currents act like a parallel short circuit in the zero sequence system, which reduces the expected value from $X_{(0)}/X_{(1)} = 1$ to $X_{(0)}/X_{(1)} \approx 0.8$
[d]The leakage channel between the zig and zag windings can be very narrow at low voltage, so the leakage flux is low
[e]With stabilizing winding
[f]Depending on the rated power and position of the stabilizing winding

Since the short-circuit voltage u_{kr} takes into account both the inductive and the ohmic voltage drop, the short-circuit reactance is as follows.

$$X_T = \sqrt{Z_T^2 - R_T^2} \qquad R_T = \frac{P_{krT}}{S_{rT}} \cdot \frac{U_{rT}^2}{S_{rT}} = \frac{u_{Rr}}{100\%} \cdot \frac{U_{rT}^2}{S_{rT}} \tag{4.211}$$

With

P_{krT} short-circuit losses (rated value),
u_{Rr} rated resistive component of the short-circuit voltage (%).

The following therefore applies to the transformer impedance.

$$\underline{Z}_T = R_T + jX_T \tag{4.212}$$

The values for the determination of Eqs. (4.210) and (4.211) can be taken from the manufacturer's specifications. For the zero sequence reactance ratios, the following data can generally be used as a simplification depending on the vector group, Table 4.7.

The ratios $R_{(0)}/R_{(1)}$ can be assumed to be approx. 1, with the exception of Yz, Dz transformers (≈0.4). The neutral point load capacity is required for low-voltage transformers to connect single-phase loads and for high-voltage transformers to connect an arc suppression coil (Petersen coil) and relates to the rated current.

4.8.3 Integrated Equivalent Circuit Diagram in the Positive Sequence System

Due to the size relationships of the individual elements (X_H, R_{Fe} >> X_T, R_T, Fig. 4.36), only the longitudinal impedances are taken into account in the short-circuit current calculation, so that Fig. 4.37 is used for simplification (an impedance correction is not taken into account in this case). In contrast, the shunt admittance must also be taken into account in load flow calculations, especially for low-voltage transformers.

The individual chain matrices (**A**) can be used to determine the overall matrix for the transformer as shown in Fig. 4.37. The relationships apply:

$$\text{Longitudinal impedance}: \begin{pmatrix} \underline{U}_1 \\ \underline{I}_1 \end{pmatrix} = \begin{pmatrix} 1 & \underline{Z}_T \\ 0 & 1 \end{pmatrix} \cdot \begin{pmatrix} \underline{U}_{2t} \\ \underline{I}_{2t} \end{pmatrix} \tag{4.213}$$

$$\text{Ideal transformer}: \begin{pmatrix} \underline{U}_{2t} \\ \underline{I}_{2t} \end{pmatrix} = \begin{pmatrix} t & 0 \\ 0 & 1/t \end{pmatrix} \cdot \begin{pmatrix} \underline{U}_2 \\ \underline{I}_2 \end{pmatrix} \tag{4.214}$$

With the transformation ratio $t = w_1/w_2$ of the transformer, this results for the series circuit:

$$\begin{pmatrix} \underline{U}_1 \\ \underline{I}_1 \end{pmatrix} = \begin{pmatrix} 1 & \underline{Z}_T \\ 0 & 1 \end{pmatrix} \cdot \begin{pmatrix} t & 0 \\ 0 & 1/t \end{pmatrix} \cdot \begin{pmatrix} \underline{U}_2 \\ \underline{I}_2 \end{pmatrix} = \begin{pmatrix} t & \underline{Z}_T/t \\ 0 & 1/t \end{pmatrix} \cdot \begin{pmatrix} \underline{U}_2 \\ \underline{I}_2 \end{pmatrix} \tag{4.215}$$

An analogous π equivalent circuit can be derived from the chain matrix according to Eqs. (4.213) and (4.214), Fig. 4.38.

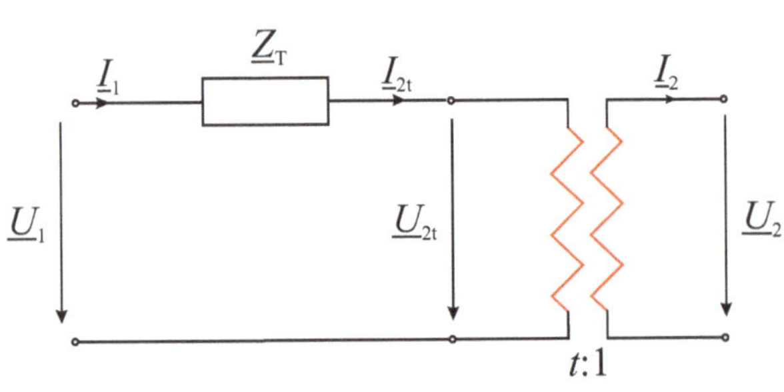

Fig. 4.37 Simplified transformer equivalent circuit diagram

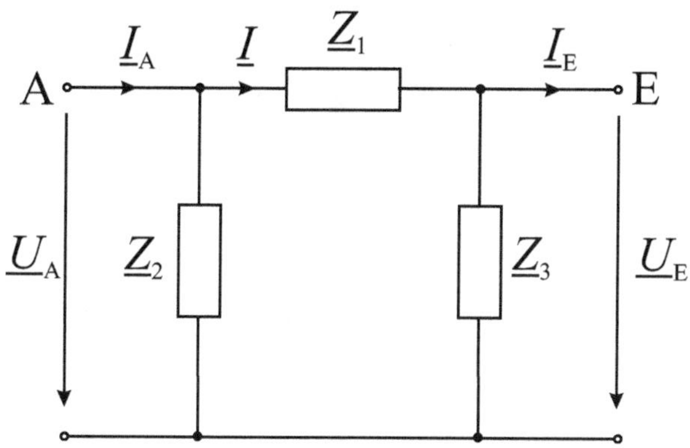

Fig. 4.38 π Equivalent circuit diagram of a transformer with ideal transformer

The following applies to the elements of the π equivalent circuit:

$$\underline{Z}_1 = \frac{\underline{Z}_T}{t} \quad \underline{Z}_2 = \frac{\underline{Z}_T}{1-t} \quad \underline{Z}_3 = \frac{\underline{Z}_T}{t^2 - t} \tag{4.216}$$

If the two individual elements (longitudinal impedance $\underline{Z}_{T2}$ and ideal transformer) are changed instead, the result is the same for the entire chain matrix:

$$\begin{pmatrix} \underline{U}_1 \\ \underline{I}_1 \end{pmatrix} = \begin{pmatrix} t & \underline{Z}_{T2} \cdot t \\ 0 & 1/t \end{pmatrix} \cdot \begin{pmatrix} \underline{U}_2 \\ \underline{I}_2 \end{pmatrix} \tag{4.217}$$

A comparison of the coefficients of the two Eqs. (4.214) and (4.217) leads to the result that the two equivalent circuits are only identical if $\underline{Z}_{T1} = t^2 \cdot \underline{Z}_{T2}$. This means that the impedance $\underline{Z}_T$ must be transformed with t^2 to the other voltage side.

The short-circuit current calculation using the equivalent voltage source at the short-circuit location requires that shunt impedances in the positive sequence system are not taken into account. In this case, the π equivalent circuit requires the use of shunt impedances, so that this is contradictory. For this reason, Sect. 15.1 shows the short-circuit current calculation with the different representations of the transformers and their results.

4.8.4 Equivalent Circuit Diagram in the Zero Sequence System

The value of the zero sequence impedances of transformers depends on the connection of the windings (e.g. delta, star), on the neutral point treatment, and on the design. In principle, a simplified T equivalent circuit diagram can be given according to Fig. 4.39, which contains the essential elements of this representation.

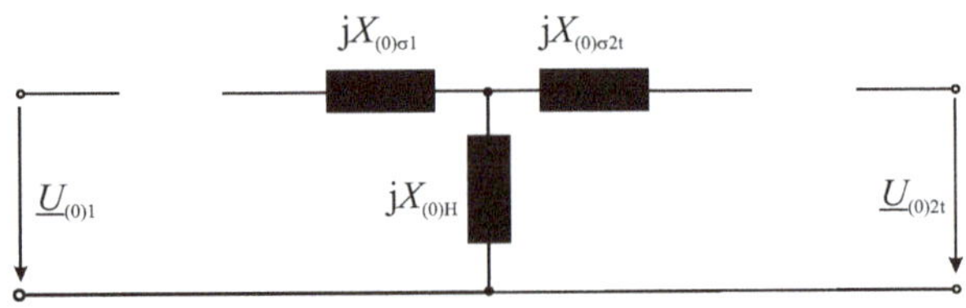

Fig. 4.39 General T equivalent circuit diagram of a transformer in the zero sequence system

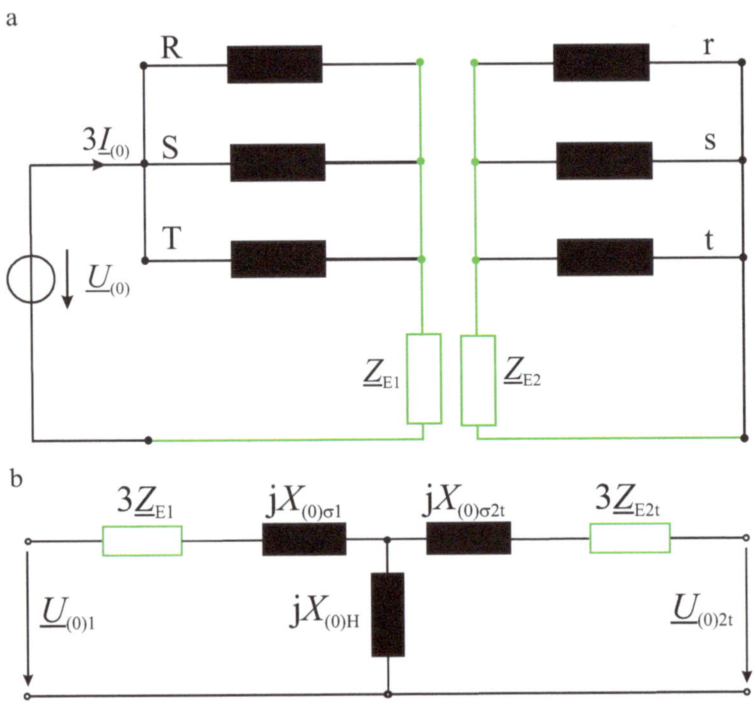

Fig. 4.40 Equivalent circuit diagram of a Y(N)y(n) transformer in the zero sequence system. (**a**) network circuit diagram, (**b**) equivalent circuit diagram

The open connections in Fig. 4.39 are used according to the treatment of the transformer neutral point. Fig. 4.40 applies to a Y(N)y(n) transformer in the zero sequence system with earthing on both sides via an impedance.

Since three times the zero sequence current flows through the earthing impedances, $3\underline{Z}_E$ must be used in the equivalent circuit of the zero sequence system (Fig. 4.40b). The following limit values result for $\underline{Z}_E$:

- $\underline{Z}_E = 0$: solidly earthed,
- $\underline{Z}_E \to \infty$: isolated.

With a Yy transformer, an ampere balance can only be formed for the windings if a current can also flow in the corresponding winding on the secondary side. Using the equivalent circuit diagram, the following statements can be made for the limit cases specified above [28] or Sect. 15.2:

- $\underline{Z}_{E1} = \underline{Z}_{E2} = 0$
  In the event of an unbalanced fault with earth contact, the zero sequence voltages and currents are transferred to the parallel system. For this reason, such operation is not permitted or does not make sense.
- $\underline{Z}_{E1} = 0$; $\underline{Z}_{E2} \to \infty$ (or vice versa)
  Earthing the high-voltage side has no effect, as theoretically no line-to-earth short-circuit current flows due to the high-reactance X_H, so this circuit is not used to reduce the earth-fault factor in networks. In practice, a line-to-earth short-circuit current will nevertheless flow due to the leakage reactances present, so that the reactance ratio can assume a value of $X_{(0)T}/X_{(1)T}$ up to 100 depending on the transformer design.

The equivalent circuit of a Y(N)d transformer can be seen in Fig. 4.41. Due to the delta arrangement of the low-voltage winding, a zero sequence current can flow on the primary winding, which does not affect the secondary winding. For this reason, the delta winding for the zero sequence system represents a decoupling of the two networks. As the magnetizing reactance in the zero sequence system is considerably greater than the corresponding small leakage reactances, the zero sequence current in the primary winding is mainly limited by the earthing impedance $\underline{Z}_{E1}$.

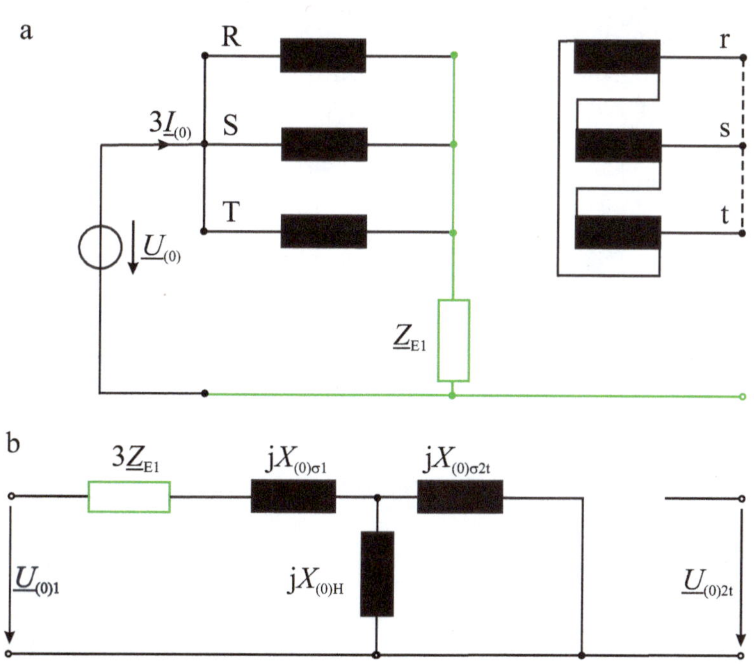

Fig. 4.41 Equivalent circuit diagram of a Y(N)d transformer in the zero sequence. (**a**) network circuit diagram, (**b**) equivalent circuit diagram

4.8.5 *Equivalent Circuit Diagram of a Three-Winding Transformer*

Transformers with a stabilizing winding (three-winding transformers) are frequently used in high-voltage networks when the HV and LV windings are connected in a star to save insulation. This is particularly necessary if a transformer neutral point is to be earthed, as the use of the delta winding as a stabilizing winding decouples the neutral systems. In the case of a line-to-earth short circuit, fault voltages are limited to one side of the system.

As a rule, the power of the stabilizing winding is only rated at 1/3 of the other windings. If, on the other hand, the power of the stabilizing winding is approximately equal to the secondary winding and the compensation winding is used to supply a medium-voltage network, it is referred to as a tertiary winding. These three-winding transformers (with tertiary winding) are used, for example, when consumers with strongly fluctuating loads are to be separated from the others. In this case, for example, the high-voltage winding is placed in the middle, unless there are technical insulation reasons for not doing so.

4.8.5.1 Positive Sequence System

In a three-winding transformer, a total of three different short-circuit tests must be carried out in order to clearly determine the different reactances. In each case, one winding is short-circuited, one remains unloaded, while voltage is applied to the last winding. The elements of the equivalent circuit shown in Fig. 4.42 can be determined using Table 4.8.

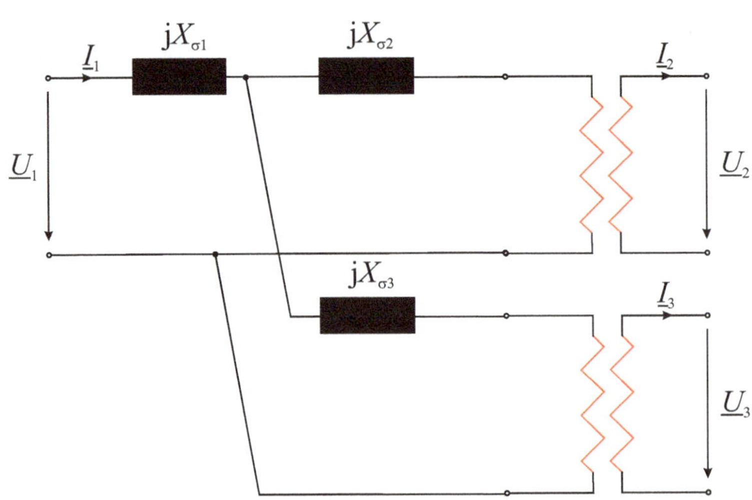

Fig. 4.42 Equivalent circuit of a three-winding transformer in the positive sequence system (neglecting the magnetizing reactance X_H)

Table 4.8 Short-circuit tests to determine the reactances

Winding			Reactances	
Test	Short circuit	No-load	Measurement	Calculation
1	2	3	$X_{12} = X_1 + X_2$	$2X_1 = X_{31} + X_{12} - X_{23}$
2	3	1	$X_{23} = X_2 + X_3$	$2X_2 = X_{12} + X_{23} - X_{31}$
3	1	2	$X_{31} = X_3 + X_1$	$2X_3 = X_{23} + X_{31} - X_{12}$

As a rule, one of the reactances X_1, X_2, X_3 is negative or significantly smaller in amount than the others. This is the reactance of the winding that lies between the other two. It should be noted that the negative reactance is not a capacitance, as the value increases proportionally with the frequency. In contrast, the reactances of the inner and outer windings are always positive, as seen from the core.

If only the positive sequence impedance from windings 1 and 2 is required, the three-winding transformer can be modeled as a two-winding transformer (winding 3 is the stabilizing winding without network connection). In this case, the first two lines (measurement 1 and 2) are added together, resulting in

$$X_1 + X_2 = \frac{(X_{31} + X_{12} - X_{23} + X_{12} + X_{23} - X_{31})}{2} = X_{12} \tag{4.218}$$

4.8.5.2 Zero Sequence System

The zero sequence system for a three-winding transformer can be derived from the considerations in Sects. 4.8.4 and 4.8.5.1. While the use of a Yy transformer is only suitable for coupling networks to a limited extent, the use of a transformer with an stabilizing winding eliminates these disadvantages. Figure 4.43 shows the zero sequence of a Ydy transformer. It can be seen that the zero sequence systems of both networks 1 and 2 are decoupled if, for example, $\underline{Z}_{E1} = 0$ and $\underline{Z}_{E2} \rightarrow \infty$. In this operating case, the transformer is a two-winding transformer (Yd), although the secondary side can still be realized in a star configuration. In principle, it is also possible to earth both transformer neutral points, as the neutral systems are always decoupled. However, this should be determined on a case-by-case basis for low-power transformers.

Examples of earthing on both sides of transformers and their limits are listed in Sect. 15.2.

4.8.6 Characteristic Values of Transformers

The characteristic values of transformers commonly used in Germany are listed in Table 4.9. Further data can be found at [1].

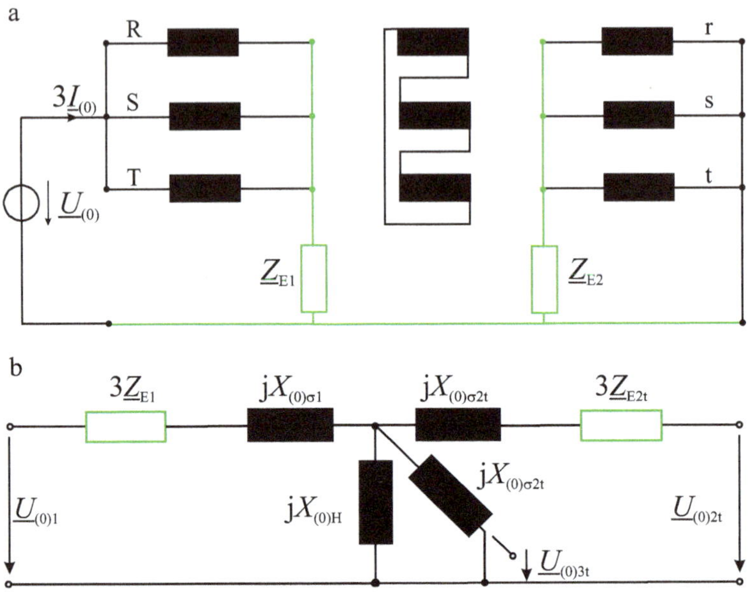

Fig. 4.43 Equivalent circuit diagram in the zero sequence for a Ydy transformer. (**a**) Network circuit diagram, (**b**) equivalent circuit diagram

Table 4.9 Characteristic values of transformers used in Germany

U_r/kV	u_{kr}/%	p_{kr}/%	$p_{\ell r}$/%	i_{0r}/%
≤30	6–10	0.4–0.9	0.06–0.17	0.5–1.3
$30 < U_{rTOS} < 110$	10–14	0.3–0.9	0.05–0.18	0.5 (0.05[a])—0.9
$110 < U_{rTOS} < 380$	12–20	0.2–0.4	0.03–0.07	0.45 (0.05[a])

U_r rated voltage of the HV side
S_r rated power
u_{kr} short-circuit voltage (rated value)
p_{kr} short-circuit losses (rated value)
$p_{\ell r}$ open-circuit losses (rated value)
I_{0r} open-circuit current (rated value)
[a]The lower values represent low-noise transformers with a rated flux density < 1.4 T

The zero sequence impedance values depend on the vector group, the neutral point earthing, and the design; typical values are listed in [1] or in Sect. 4.8.2, Table 4.7.

4.9 HVDC Installations

At present, the contribution of HVDC systems to a short circuit on the three-phase side is not yet taken into account in the IEC Standard. Basically, two different technologies must be assumed here, which also make a different contribution to a fault. These technologies are:

- Thyristor technology (LCC: Line Commutated Converter),
- IGBT technology, full-size converter (VSC: Voltage Source Converter).

For systems with thyristors, a contribution is only possible if there is inverter operation, i.e. the DC side feeds into the three-phase grid. The contribution relates exclusively to the initial symmetrical short-circuit current and the peak short-circuit current and consists of the discharge of the DC system. In contrast to this, the HVDC system in IGBT design can be regarded as a current source in the event of a short-circuit, which supplies a short-circuit current contribution depending on the voltage drop at the inverter terminals, which is modeled as an inverter as shown in [2].

A detailed description of the short-circuit current contribution of HVDC systems is given in Chap. 13.

4.10 Reactor

Reactors for short-circuit current limitation are designed as air-core reactors so that there is no reduction in reactance as a result of saturation by the short-circuit current. This means that the installation of a reactor in the short-circuit loop increases the positive, negative and zero sequence impedance of a network and thus causes a reduction in the short-circuit current.

The impedance of the reactor $\underline{Z}_R$ is determined according to Eq. (4.219), where the resistance is considerably smaller than the reactance ($R_R << X_R$; e.g. $R_R = 0.03\, X_R$).

$$Z_R = \frac{u_{kR}}{100\%} \cdot \frac{U_n/\sqrt{3}}{I_{rR}} \tag{4.219}$$

With

u_{kR} relative short-circuit voltage,
I_{rD} rated current of the reactor,
U_n nominal system voltage.

The positive sequence impedance of a short-circuit current limiting reactor calculated according to Eq. (4.219) is also valid for the negative sequence and zero sequence systems.

The use of a short-circuit current limiting reactor results in a voltage drop during normal operation depending on the operating current, so that the voltage downstream of the reactor is reduced. This occurs particularly when the operating current is inductive, e.g. inrush current of an asynchronous motor. In these cases, it makes sense to use an I_s limiter for normal operation in parallel with the reactor, which interrupts before the peak short-circuit current is reached in the event of a short-circuit, so that the short-circuit current can be limited by the reactor.

4.11 Capacitors

In principle, capacitors (line-to-earth) will participate in a short circuit due to their discharge. In general, this contribution to the peak short-circuit current should be smaller, as the amplitude will be lower than the operating frequency component and the maximum value of the discharge current will have decayed quickly as a result of the discharge time constant. In addition, the maximum value of the capacitor current occurs when the short circuit happens at the voltage maximum. In contrast, when calculating the operating-frequency peak short-circuit current, a short-circuit initiation at voltage zero crossing is assumed so that the two current components are decoupled in terms of time and cannot be superimposed.

However, these considerations no longer necessarily apply to capacitor banks and filter systems of HVDC installations as a result of the design, so that in these cases the discharge current can have an influence on the strength of the installations. The results of a calculation are shown in Sect. 14.1.

The use of series capacitors does not have to be taken into account when calculating the short-circuit current if they are fitted with an overvoltage protection device, Sect. 14.2.

4.12 Conclusions

In this section, the essential data of equipment is given in order to calculate the short-circuit currents. It describes in detail how the positive and zero sequence impedances of overhead lines and cables can be determined from the geometric values.

References

1. IEC/TR 60909–2:2008 Edition 3.0 (2008) Short-circuit current calculations in three-phase a.c. systems—Part 2: Data of electrical equipment for short-circuit current calculations. IEC, Geneva
2. IEC 60909–0:2016 Edition 2.0 (2016) Short-circuit current calculations in three-phase a.c. systems—Part 0: Calculation of currents. IEC, Geneva
3. Balzer G, Neumann C, Balzer C, (2025) Schalt- und Ausgleichsvorgänge in elektrischen Netzen. 2. AuflageSpringer, Heidelberg
4. Nelles D (2009) Netzdynamik. VDE, Berlin
5. Funk G (1962) Der Kurzschluß im Drehstromnetz. R. Oldenbourg, München
6. Binder A (2012) Elektrische Maschinen und Antriebe – Grundlagen, Betriebsverhalten. Springer, Heidelberg
7. IEC 60909–0:2001 Edition 1.0 (2001) Short-circuit current calculations in three-phase a.c. systems—Part 0: Calculation of currents. IEC, Geneva
8. Gursoy E, Walling RA (2011) Representation of variable speed wind turbine generators for short circuit analysis. In: IEEE Electrical Power and Energy Conference (EPEC), pp 444–449

9. Lager T, Hofmann L (2019) Improvement of the consideration of short-circuit current contributions from doubly-fed induction generator based wind turbines for short-circuit current calculations according to IEC 60909. In: 2019 IEEE Milan PowerTech, Milano
10. Schürhuber R, Oswald BR, Fickert L, Fortmann J (2020) Verhalten von Windkraftanlagen mit doppelt speisenden Asynchrongeneratoren (DFIG) bei Kurzschlüssen und anderen Netzfehlern. In: e&i elektrotechnik und informationstechnik, pp 415–424
11. Ruben M, Laubrock M (2017) Short-circuit currents of wind turbines with doubly-fed induction generators. In: Conference for Wind Power Drives CWD2017, March, Aachen
12. IEC/TR 60909–1:2002 Edition 2.0 (2002) Short-circuit currents in three-phase a.c. systems—Part 1: Factors for the calculation of short-circuit currents according to IEC 60909–0. IEC, Geneva
13. Oeding D, Waider G (1988) Maximale Teilkurzschlußströme von Kraftwerksblöcken ohne Stufenschalter. ETZArchiv 10:173–180
14. Walling RA, Gursoy E, English B (2011) Current contributions from Type 3 and Type 4 wind turbine generators during faults. In: IEEE Power and Energy Society General Meeting, San Diego
15. VDE-AR-N 4120 Anwendungsregel:2018–11 (2018) Technische Regeln für den Anschluss von Kundenanlagen an das Hochspannungsnetz und deren Betrieb (TAR Hochspannung). VDE, Berlin
16. Hosemann G, Boeck W (1991) Grundlagen der elektrischen Energietechnik, 4th edn. Springer, Berlin
17. Das JC (2018) Power systems handbook—Volume 1: short-circuits an AC and DC systems. CRC Press, Taylor & Francis Group, New York
18. VDE 0102, Teil 1/9.62 (1962) Leitsätze für die Berechnung der Kurzschlussströme, Teil 1 Drehstromanlagen mit Nennspannungen von 1 kV und darüber. VDE-Verlag GmbH, Berlin
19. Meyer M (1959) Einspeisung eines stromrichtergespeisten Gleichstromantriebes auf einen drehstromseitigen Sammelschienenkurzschluss. ETZ-A 80:784–787
20. DIN VDE 0102:1990–01 (1990) Berechnung von Kurzschlussströmen in Drehstromnetzen. VDE-Verlag GmbH, Berlin
21. Grotstollen H (1979) Der Beitrag stromrichtergespeister Gleichstromantriebe zum Stoßkurzschlussstrom im Drehstromnetz. ETZ Archiv 11:321–326
22. IEC 61363–1, 1998–02 (1998) Electrical installations of ships and mobile and fixed offshore units—Part 1: procedures for calculating short-circuit currents in three-phase a.c. IEC, Geneva
23. Haake D, Malsch M, Pfeiffer K (2009) Kurzschlussstromanteile Umrichter- bzw. Stromrichtergespeister Antriebe in Netzen des Kraftwerkseigenbedarfs. ETG-Kongress 2009 FT 1 + 2, 27.-28. Okt. 2009, Düsseldorf
24. Oeding D, Oswald RB (2011) Elektrische Kraftwerke und Netze, 7th edn. Springer, Berlin
25. Brüderlink R (1954) Induktivität und Kapazität der Starkstrom-Freileitungen. Verlag G. Braun, Karlsruhe
26. Balzer G (1977) Impedanzmessung in Niederspannungsnetzen zur Bestimmung der Kurzschlussströme. Dissertation, TH Darmstadt, p D17
27. Küpfmüller K, Mathis W, Reibiger A (2006) Theoretische Elektrotechnik, 17th edn. Springer-Verlag, Berlin, Heidelberg, p 357
28. Balzer G, Remde H (1985) Probleme bei der beidseitigen Erdung von Transformatoren. Brown Boveri Technik 7:349–354

Chapter 5
Calculation of the Short-Circuit Currents

In this section, the equations for calculating the various currents in accordance with IEC 60909-0 [1] are given, based on the data in Chap. 4, taking into account the basic procedure and the boundary conditions in Chap. 3. The following equations refer primarily to the calculation of the three-phase short-circuit current. It is expressly pointed out that the valid standards must always be observed when applying the equations.

Furthermore, the following explanations refer exclusively to the representation of the short-circuit current calculation based on the equivalent voltage source at the short-circuit location. Due to the use of full size converters with a specified current contribution during the short circuit, an additional simulation with current sources is necessary. This current contribution must always be superimposed on the result from the calculation using the voltage source. The procedure in these cases is described in Sect. 3.6.7.

5.1 Definitions

The terms used in the calculation of short-circuit currents are defined below.

- Initial symmetrical short-circuit current I_k''
  This is the rms value of the short-circuit current at time $t = 0$, Sect. 5.3.
- Short-circuit breaking current (symmetrical) I_b
  RMS value of the AC component of a short-circuit current at the time of contact separation by switching devices. Factors (μ, q) are taken into account in the calculation, Sect. 5.6.
- Rated voltage U_r
  Voltage value to which the characteristic values of a piece of equipment are related.

G. Balzer, *Short-Circuit Currents in Three-Phase Systems*,
https://doi.org/10.1007/978-3-658-50710-7_5

- Rated current I_r
  Design current of equipment to which all current-relevant variables are related.
- Steady state short-circuit current $\underline{I}_k$
  RMS value of the AC component after all transient phenomena of synchronous and asynchronous machines have decayed, Sect. 5.7.
- Equivalent frequency f_c
  Frequency for calculating the peak factor κ; for a system frequency of $f = 50$ Hz, the equivalent frequency is $f_c = 20$ Hz, Sect. 5.4.4.
- Equivalent voltage source
  Ideal voltage source of the equivalent voltage source at the short-circuit location as the only voltage in the positive sequence system, together with the voltage factor c.
- Far-from-generator short circuit
  The AC component of the short-circuit current remains constant over almost the entire time course, Sect. 5.2.
- Near-to-generator short circuit
  The short-circuit current curve includes a decaying AC component caused by synchronous and asynchronous machines, Sect. 5.2.
- Reactances of generators X''_d, X'_d, X_d
  Effective reactances (subtransient, transient and synchronous) of a generator as a function of time during a short circuit, Sect. 5.6. The subtransient reactance of a synchronous machine is required to calculate the initial symmetrical short-circuit current.
- DC component I_{DC}
  The DC component is responsible for the displacement of the short-circuit current. It occurs if the expected short-circuit current does not start at the zero crossing when the short-circuit occurs, e.g. when it occurs at the voltage zero crossing in an inductive circuit, Sect. 5.6.2.
- DC time constant T_{DC}
  The DC time constant describes the decay behavior of the DC component and depends on the *R*/*X* ratio of the short-circuit path, Sect. 5.4.1.
- Short-circuit duration
  Time range between the occurrence of the fault and the electrical interruption of the short-circuit current.
- Short-circuit impedance Z_k
  The short-circuit impedance is the resulting impedance for calculating the short-circuit currents. It is specified as a function of the short-circuit type for the positive sequence, negative sequence and zero sequence system, Sect. 5.3.
- Minimum time delay t_{min}
  The minimum time delay is the time between the start of the short circuit and the mechanical contact opening of a switching device. The value is used to determine the short-circuit breaking current, Sect. 5.6.
- Nominal system voltage U_n

RMS value of the voltage between the conductors of a network. In general, the nominal system voltages are defined by IEC 60038:2009 [2]. For the highest voltages for equipment $U_{\mathrm{m}} > 300$ kV, there are no normative specifications for U_{n}.

- Voltage factor c
  The voltage factor c describes the ratio of the equivalent voltage source at the short-circuit location in relation to the nominal voltage U_{n} of the network divided by $\sqrt{3}$, Table 7.1.
- Peak factor κ
  The peak factor κ is used to determine the maximum peak value of the short-circuit current, depending on the maximum value of the initial short-circuit current. The peak factor is influenced by the R/X ratio of the short-circuit loop, Sect. 5.4.1.
- Peak short-circuit current i_{p}
  Peak value of the maximum short-circuit current after the short-circuit occurs. At an operating frequency of $f = 50$ Hz, the value occurs at $t = 10$ ms ($R/X = 0$), Sect. 5.4.
- Thermal equivalent short-circuit current I_{th}
  RMS value of a symmetrical short-circuit current whose thermal effect is identical to the actual short-circuit current, taking into account an existing DC component and a possibly decaying AC component, Sect. 5.8.
- AC component I_{AC}
  The AC component is the short-circuit current that oscillates symmetrically around the zero axis with the operating frequency of the network.

5.2 Short-Circuit Current Propagation

Basically, two different short circuits are distinguished according to Sect. 3.5 on the basis of their profile:

- Far-from-generator, and
- near-to-generator.

According to the definition, a short-circuit is described as a near-to-generator short-circuit if, in the event of a short circuit, at least one synchronous machine supplies an expected initial short-circuit current that is greater than twice the rated current of the machine. In the case of asynchronous motors, it is a short circuit near-to-generator if the asynchronous motors contribute more than 5% to the initial short-circuit current without motors. This assumption that a decay of the short-circuit contribution of generators and motors is only considered from a value of $I''_{\mathrm{kG,M}} = 2 \cdot I_{\mathrm{rG,M}}$ is also reflected in the fact that the factors μ up to twice the rated current have the value $\mu = 1$.

When calculating the short-circuit current, it would be desirable to determine the time characteristic of the total short-circuit current (see Sect. 3.4 and 3.5). However, since only certain short-circuit current values are required for the design of equipment or the protection (Table 3.1), it is perfectly sufficient to calculate the values that occur at fixed times after the short-circuit occurs. For this reason, the following currents are determined in accordance with the existing IEC Standard, based on the initial short-circuit current, Sect. 5.3, as shown in Fig. 3.3:

- Peak short-circuit current, Sect. 5.4,
- Short-circuit breaking current, Sect. 5.6,
  - Generators,
  - Motors,
  - Full-size converter,
- Steady state short-circuit current, Sect. 5.7,
- Thermal equivalent short-circuit current, Sect. 5.8.

5.3 Initial Symmetrical Short-Circuit Current

The initial short-circuit current is used as an initial value to calculate the short-circuit currents required to dimension electrical equipment or to set the network protection. The following generally applies to the calculation of the three-phase initial short-circuit current I_k'':

$$\underline{I}_k'' = \frac{c \cdot U_n/\sqrt{3}}{\underline{Z}_k} \tag{5.1}$$

With

c voltage factor according to Table 7.1,
U_n nominal system voltage,
$\underline{Z}_k$ short-circuit impedance.

When determining the short-circuit impedance, only equipment that the short-circuit current flows through must be taken into account. The comment in Sect. 5.4.1 must be observed when considering the *R*/*X* ratio to determine the short-circuit impedance.

5.4 Peak Short-Circuit Current

The level of the peak short-circuit current influences both the thermal heating of equipment and the magnetic forces between the conductors [3], so that this characteristic value is important for the design of the installations.

5.4.1 General Calculation

Due to electrical equipment (cables, transformers, machines, etc.), which mainly have an inductive effect during a short circuit, there is a phase shift of almost $\varphi = 90°$ between the active voltage and the short-circuit current during a short circuit. Since the current in an inductive network branch generally starts at the zero crossing, a short circuit must result in a transient phenomenon at the zero crossing of the voltage, so that a DC component occurs. With an exclusively inductive networkbranch ($R/X = 0$), the maximum peak value of the current occurs after $t = 10$ ms (at 50 Hz), in relation to the occurrence of the short circuit. With larger R/X ratios, the maximum value shifts to shorter times.

The highest current occurring during a short circuit is called the peak short-circuit current and is used for the mechanical design of equipment. There are various methods for calculating this, depending on the network topology, which result in different calculation efforts. The various methods for calculating the peak short-circuit current are listed at [4].

To simplify matters, the calculation of the factor κ, Eq. (5.2), the ratio R/X of the positive sequence system, can be used for all fault types. Determining the peak short-circuit current is only of interest for calculating the largest short-circuit current, which is usually the three-phase short-circuit current. In special cases (in the vicinity of Yd transformers), the line-to-line short-circuit current may be greater than the three-phase short-circuit current, but it can be assumed that $R_{(1)}/X_{(1)} < R_{(0)}/X_{(0)}$ applies to the ratios, so that the use of $R_{(1)}/X_{(1)}$ to calculate the κ factor is on the safe side.

In the short-circuit current calculation according to [1], two different R/X ratios are used, namely

- The real ratio R/X of equipment for calculating the initial short-circuit current. The resistance R is determined from the ohmic resistances of equipment.
- The fictitious ratio R_f/X for determining the peak short-circuit current. In this case, not only the reduction or damping of the DC component because of the resistance is taken into account, but also the reduction of the AC component in the case of short-circuits near to the generator. In this case, the real ratio R/X must be used to determine the initial symmetrical short-circuit current, Eq. (5.2), while the fictitious ratio R_f/X must be taken into account to calculate the value κ, if required by the network topology.

Further explanations regarding the application of the fictitious ratio R_f/X are provided in Sect. 5.10.

The peak short-circuit current i_p is generally calculated from the initial short-circuit current I''_k as follows

$$i_p = \sqrt{2} \cdot \kappa \cdot I''_k \tag{5.2}$$

The peak short-circuit current i_p represents the maximum value of the current waveform $i_k(t)$, which can be expressed by Eq. (5.2). The short-circuit current is determined from an *RL* series resonant circuit using the active voltage *U* by initiating the short-circuit at the zero crossing of the voltage. It is assumed that the current was zero before the short circuit occurred. According to Eq. (5.3), this results in the Laplace domain:

$$I(p) = \frac{U(p)}{p \cdot L + R} = \frac{\omega \cdot \sqrt{2} \cdot U}{p^2 + \omega^2} \cdot \frac{1}{p \cdot L + R} = \frac{\omega \cdot \sqrt{2} \cdot U}{L} \cdot \frac{1}{p + R/L} \cdot \frac{1}{p^2 + \omega^2} \tag{5.3}$$

The reverse transformation results:

$$i_k(t) = \sqrt{2} \cdot U \cdot \frac{\omega}{L} \cdot \frac{1}{\omega^2 + (R/L)^2} \left[e^{-t \cdot R/L} - \cos(\omega t) + \frac{R/L}{\omega} \cdot \sin(\omega t) \right] \tag{5.4}$$

Using the trigonometric function according to Eq. (5.5), Eq. (5.3) can be transformed so that the relationship according to Eq. (5.6) results:

$$\cos(\omega t + \alpha) = \cos(\omega t) \cdot \cos(\alpha) - \sin(\omega t) \cdot \sin(\alpha) \tag{5.5}$$

$$i_k(t) = \frac{\sqrt{2} \cdot U}{X} \cdot \frac{1}{1 + (R/X)^2} \cdot \left[e^{-t/T_{DC}} - \sqrt{1 + (R/X)^2} \cdot \cos(\omega t + \alpha) \right] \tag{5.6}$$

resp.

$$i_k(t) = \frac{\sqrt{2} \cdot U}{X \cdot \sqrt{1 + (R/X)^2}} \cdot \left[\frac{e^{-t/T_{DC}}}{\sqrt{1 + (R/X)^2}} - \cos(\omega t + \alpha) \right] \tag{5.7}$$

With the DC time constant

$$T_{DC} = \frac{X}{\omega \cdot R} = \frac{X}{2\pi \cdot f \cdot R} \qquad \alpha = \arctan(R/X) \tag{5.8}$$

The maximum value of the short-circuit current $i_k(t)$ at time t_m can always be determined using the first derivative by setting it to zero.

$$\frac{di_k(t)}{dt} \stackrel{!}{=} 0 = \frac{-1}{T_{DC}} e^{-t_m/T_{DC}} + \sqrt{1 + (R/X)^2} \cdot \omega \cdot \sin(\omega t_m + \alpha) \tag{5.9}$$

resp.

$$e^{-t_m/T_{DC}} = T_{DC}\sqrt{1 + (R/X)^2} \cdot \omega \cdot \sin(\omega t_m + \alpha) \tag{5.10}$$

The time t_m of the maximum current cannot be specified explicitly, but can only be determined by approximation. Figure 5.1 shows the short-circuit current curve over time within the first 12 ms after the short-circuit occurs, assuming a nominal system voltage of $U_n = 110$ kV and an initial symmetrical short-circuit current of $I_k'' = 20\,\text{kA}$. It can be seen that the peak value of the short-circuit current occurs at shorter times as the *R*/*X* ratio increases. In addition, the amplitude decreases so that smaller DC components and thus factors κ are created. A value of $R/X \rightarrow \infty$ results in a minimum value of $t = 5$ ms.

According to [4], the maximum value of the short-circuit current always occurs at the zero crossing of the voltage before the fault occurs, regardless of the phase angle. This also results from the calculation of the derivative according to Eqs. (5.9) and (5.10). The peak factor κ_3 is determined for a three-phase short circuit that is initiated simultaneously at the zero crossing of a phase as follows

$$\kappa_3 = 1 + e^{-(\pi/2+\beta)\cdot R/X} \cdot \sin\beta \tag{5.11}$$

with

β arctan(*X*/*R*).

Depending on the *R*/*X* ratio, the following κ values result for the simultaneous three-phase short circuit according to Table 5.1.

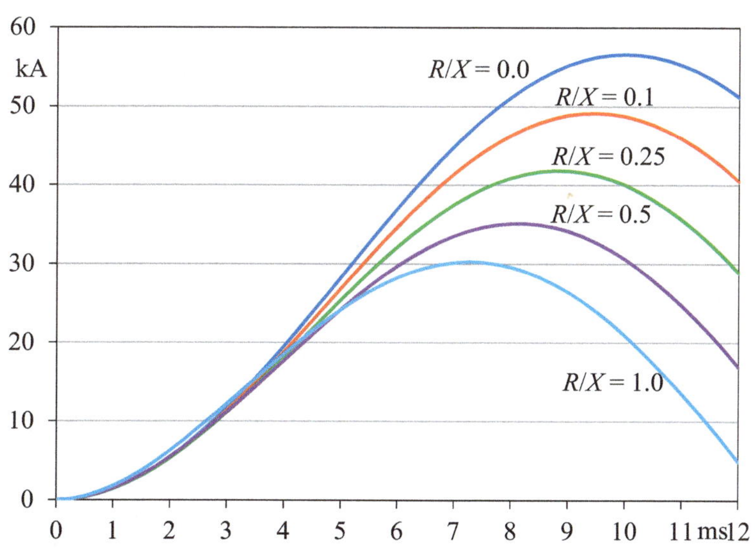

Fig. 5.1 Behavior of the DC component as a function of the *R*/*X* ratio

Table 5.1 Values for the peak factor κ for a three-phase short circuit (κ_3, simultaneous), for a line-to-line short circuit with transition to a three-phase short circuit (κ_{2-3}) and the approximation formula according to IEC

R/X	0.0	0.1	0.2	0.3	0.4	0.6	0.8	1.0	1.2
κ_3	2.000	1.734	1.544	1.407	1.308	1.180	1.109	1.067	1.042
κ_{3IEC}	2.000	1.746	1.558	1.418	1.315	1.182	1.109	1.069	1.047
κ_{2-3}	2.366	2.011	1.760	1.580	1.450	1.283	1.187	1.130	1.094

κ_3 according to Eq. (5.11)
κ_{3IEC} according to Eq. (5.12)
κ_{2-3} non-simultaneous short circuit after Eq. (5.23)

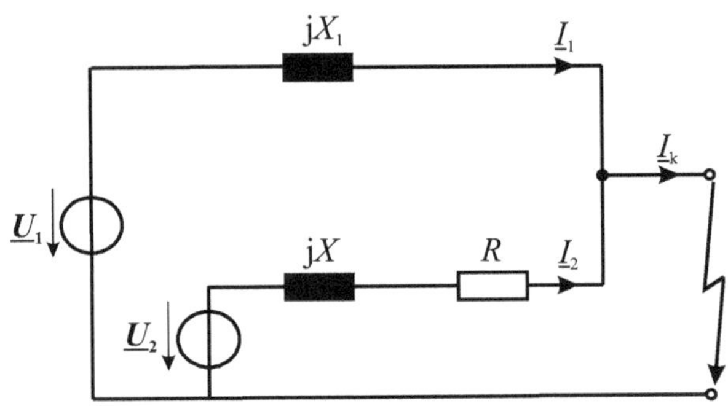

Fig. 5.2 Determination of the peak short-circuit current i_p for parallel branches

The factor κ_3 according to Eq. (5.11) has been approximated by a simplified equation in [1, 5]:

$$\kappa_{3IEC} \approx 1.02 + 0.98 \cdot e^{-3 \cdot R/X} \tag{5.12}$$

The author of this book determined the original numerical approximation of Eq. (5.12), based on the $\kappa = f(R/X)$ representation according to an edition of the old VDE standard for the calculation of short-circuit currents in low-voltage systems from 1975. To simplify matters, this equation was changed to the version valid today, Eq. (5.12). This procedure also applies to the μ and q curves according to Eqs. (5.47)–(5.50) and (5.52)–(5.55).

When determining the peak short-circuit current in a single short-circuit branch, this calculation of the factor κ always leads to a correct result if the R/X ratio of the branch is used. However, this does not apply if the short circuit is fed via parallel branches or the short circuit occurs in a meshed network. This is shown using the following example.

If the peak short-circuit current i_p is multiple single-fed at the short-circuit location according to Fig. 5.2, there are several methods for determining the factor κ. This figure also applies to the case where a short circuit is connected to a voltage source via parallel branches, as the voltages $\underline{U}_1 = \underline{U}_2 = \underline{U}$ are identical.

The exact value of the peak short-circuit current results from the superposition of the time curves of the partial currents $i_1(t)$ and $i_2(t)$. It should be assumed that the amounts of the individual currents are equal, so that the following applies:

$$X_1 = \sqrt{R^2 + X^2} = X\sqrt{1 + (R/X)^2} \tag{5.13}$$

The following applies to the superposition of the two partial short-circuit currents at the short-circuit location:

$$i_k(t) = i_1(t) + i_2(t) \tag{5.14}$$

Substituting Eq. (5.7) into (5.14) gives the total current for the example in Fig. 5.2:

$$i_k(t) = \frac{\sqrt{2} \cdot U}{X \cdot \sqrt{1 + (R/X)^2}} \cdot \left[1 - \cos(\omega t) \text{-} \cos(\omega t + \alpha) + \frac{e^{-t/T_{DC}}}{\sqrt{1 + (R/X)^2}} \right] \tag{5.15}$$

The short-circuit current $i_k(t)$ Eq. (5.15) has a DC component as well as an AC component, so that the total current results from the addition of the time curves. The actual peak short-circuit current i_p can be determined using the time curve of the short-circuit current $i_k(t)$. If the peak short-circuit current i_p is determined using the ratio R/X at the short-circuit location rather than by superimposing the time curves, the result is the total short-circuit impedance $\underline{Z}_k$.

$$\underline{Z}_k = \frac{jX_1 \cdot (R + jX)}{R + j(X_1 + X)} = X_1 \cdot \frac{R \cdot X_1 + j(X^2 + X \cdot X_1 + R^2)}{R^2 + (X_1 + X)^2} \tag{5.16}$$

$$R_k = X_1 \cdot \frac{R \cdot X_1}{R^2 + (X_1 + X)^2} \qquad X_k = X_1 \cdot \frac{(X^2 + X \cdot X_1 + R^2)}{R^2 + (X_1 + X)^2} \tag{5.17}$$

This results in the R/X ratio:

$$R_k/X_k = \frac{X_1 \cdot R}{R^2 + X \cdot (X_1 + X)} \tag{5.18}$$

After inserting the condition from Eq. (5.12), the result is as follows:

$$R_k/X_k = \frac{R/X}{1 + \sqrt{1 + (R/X)^2}} \tag{5.19}$$

Using the variables X_k and R_k/X_k, the peak short-circuit current i_{p1} can be calculated from the total short-circuit impedance at the fault location according to Eq. (5.7).

$$i_{p1} = \kappa_1 \frac{\sqrt{2} \cdot U}{Z_k} \tag{5.20}$$

The ratio f_1, i.e. the error according to Eq. (5.21), can be determined from the variables i_p according to Eq. (5.15) and i_{p1} according to Eq. (5.20). Here, the peak short-circuit current i_p is considered to be the actual value.

$$f_1 = \frac{i_{p1}}{i_p} \tag{5.21}$$

Figure 5.3 shows the peak short-circuit current ratio f_1 according to Eq. (5.21), determined from the value of the total short-circuit impedance.

Calculating the peak short-circuit current i_{p1} using the *R*/*X* ratio at the short-circuit location always results in smaller values than the currents that actually occur ($f_1 < 1$). This is particularly the case when the *R*/*X* ratios of parallel branches differ considerably. For this reason, IEC Standard 60,909-0 [6] suggests a total of three different ways of calculating the peak short-circuit current, Sects. 5.4.2–5.4.4.

According to Eq. (5.12), the factor κ can assume a maximum factor of 2, which means that a maximum of twice the peak value of the short-circuit current can occur. This value applies in the event that the three-phase short circuit occurs simultaneously between the conductors. For the theoretical case that a fault first occurs between two phases in the voltage zero crossing of the phase-to-phase voltage, which expands to a three-phase short circuit exactly after 5 ms, this results in a peak factor $\kappa > 2$, corresponding to the approximation Eq. (5.22) according to [7].

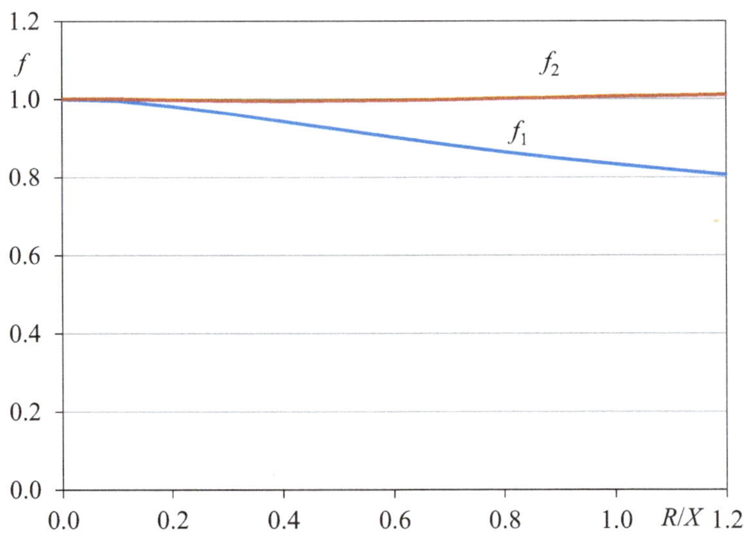

Fig. 5.3 Ratio of the peak short-circuit currents $f_1 = i_{p1}/i_p$ resp. $f_2 = i_{p2}/i_p$. i_{p1} calculation using the ratio *R*/*X* at the fault location, Eq. (5.20), i_p actual short-circuit current, Eq. (5.15), i_{p2} calculation using the equivalent frequency $f_c = 20$ Hz, Sect. 5.4.4

$$\kappa_{2-3} \approx 1 + e^{-(\pi/6+\beta)\cdot R/X} \cdot \left[\sin(\beta - \pi/3) + \frac{\sqrt{3}}{2} \cdot \cos(\beta)\right] + e^{-(2\pi/3+\beta)\cdot R/X} \cdot \frac{\sqrt{3}}{2} \cdot \sin(\beta) \tag{5.22}$$

The factors κ_{2-3} are also listed in Table 5.1 for different *R*/*X* values, whereby the maximum value is $\kappa_{2-3} = 2.366$. As the failure sequence described above is very unlikely in practice, the standard does not include a corresponding κ curve in the specifications.

For short circuits far from the generator, the κ value takes into account the decay of the DC component within the first 10 ms or half the period. In the case of short circuits near the generator, the reduction of the AC component in generators and motors is also taken into account. For this purpose, fictitious R_f/X values are specified by the IEC regulations for various equipment (see Sects. 4.2.1.1 and 4. 6.1). The specified fictitious resistances R_{Gf} of the generators differ significantly from the actual armature resistances, so that the fictitious values R_{Gf} must not be used for the decay of the DC component in the event of a short circuit. In these cases, the actual values, i.e. the armature resistances, must be used. The fictitious effective resistances have been derived from measurements and calculations. To simplify the calculation for motors, the actual resistances can be set equal to the fictitious resistances or neglected, as the result is on the safe side.

Calculations according to [4] have shown that the deviation $\Delta R/X''_d$, which reflects the influence between the decay time constant based on the armature resistance R_G and the actual decay process of the generator, is greater for small generators than for large generators. In addition, this influence is only relevant for short circuits near to the generator, so that the fictitious resistance R_{Gf} does not have to be taken into account for processes far from the generator. However, the results are on the safe side and the influence is small because of the *R*/*X* ratio of the network.

In the following Sects. 5.4.2 to 5.4.4, various methods for calculating the peak short-circuit current according to [1] are described in more detail. As a rule, the method using the equivalent frequency (Sect. 5.4.4) will correspond to the actual value, especially if the *R*/*X* ratios in a network are very different. The two other methods (Sects. 5.4.2 and 5.4.3) are much better suited for a manual calculation of the short-circuit currents. If the ratios $R/X \leq 0.3$, all methods provide almost the same results.

The initial short-circuit alternating current is always calculated using the actual *R*/*X* ratio of the total short-circuit impedance $\underline{Z}_k$.

5.4.2 *Uniform Ratio* R/X *or* X/R

The value κ is calculated according to the smallest ratio *R*/*X* (or *X*/*R*) of a branch in the network, whereby the impedances of a series connection can be combined. Only the network branches must be taken into account, which carry the partial short-circuit current.

Assessment This method leads to κ values that are always on the safe side, as small *R*/*X* ratios according to Eq. (5.2) lead to larger κ values and thus to larger peak short-circuit currents. The κ value is characterized by the smallest value of the *R*/*X* ratios, even if only a small short-circuit current will pass via this network branch. The calculation method is particularly suitable for manual calculation.

5.4.3 Ratio R/X or X/R at the Short-Circuit Location

The peak short-circuit current is determined according to Eq. (5.23) by adding a safety factor to the calculated value according to Eq. (5.2) if the peak factor κ is determined from the *R*/*X* (or *X*/*R*) ratio at the short-circuit location:

$$i_{\mathrm{pb}} = 1.15 \cdot \kappa \cdot \sqrt{2} \cdot I_k'' \tag{5.23}$$

In this case, the factor 1.15 takes into account the deviation according to Fig. 5.3, as it can be shown that the exact solution for parallel branches with different *R*/*X* ratios always leads to a result that is on the uncertain side [4]. Taking into account a tolerance in the calculation of $\pm 5\%$, the factor of 1.15 therefore covers a difference up to the *R*/*X* ratio of approx. 1.2, which should be sufficient in practice. In this case, this results in a ratio of $f \sim 0.8$ according to Fig. 5.3.

If the *R*/*X* ratio of all network branches is $R/X \leq 0.3$, the factor 1.15 does not need to be taken into account. If the value of the network branches is $R/X \leq 0.3$, it can be assumed from Fig. 5.3 that the calculation remains within the fault tolerance of $\pm 5\%$ specified above. As the calculation of the peak short-circuit current using the factor 1.15 will generally be on the safe side, the resulting factor $1.15{\cdot}\kappa$ is limited to 1.8 in low-voltage networks and to 2.0 in medium and high-voltage networks, as otherwise values higher than 2.0 could result.

Assessment In principle, this is a simple method, as the *R*/*X* ratio of the total short-circuit impedance is available, which is therefore also suitable for manual calculation. The disadvantage is that in many cases the safety factor of 1.15 is too high, as the relevant *R*/*X* range only occurs in exceptional cases in practice. The restriction that this factor does not have to be taken into account for values of *R*/*X* ratios ≤ 0.3 requires a topology query of the network, which should be difficult with computer programs, especially for branches with different equipment connected in series.

5.4.4 Procedure for the Equivalent Frequency f_c

In [4], the method is derived using the equivalent frequency. Extensive investigations have shown that two different methods for determining the peak short-circuit current produce different results, namely:

- Calculation with operating frequency: With this method, larger errors can occur if the *R*/*X* ratios of parallel network branches are very different, as shown as an example in Fig. 5.3. In addition, these deviations are on the uncertain side, which means that the actual peak short-circuit currents are greater than the calculated ones.
- Procedure with very low frequencies: In this method, the resulting impedance of a network is determined for the case that all inductances are almost short-circuited ($f = 0.1$ Hz). Under these conditions, calculations show that the results are primarily on the safe side.

These two fundamental considerations led to the conclusion that there must obviously be a frequency at which the calculated results correspond on average to the real currents, taking into account a statistical deviation. Investigations with sample networks have shown that this frequency, which is referred to as the equivalent frequency, can be found as a function of the operating frequency:

- Operating frequency $f = 50$ Hz: equivalent frequency $f_c = 20$ Hz,
- Operating frequency $f = 60$ Hz: equivalent frequency $f_c = 24$ Hz.

According to [4], at an equivalent frequency of $f_c = 20$ Hz, for an operating frequency of $f = 50$ Hz, the deviations in 3000 calculated cases are between -3.5% (uncertain) and + 3.1% (certain).

The calculation of the initial symmetrical short-circuit current must be determined in a first step using the actual *R*/*X* ratio at the operating frequency. The R_c/X_c ratio at the fault location is then determined using the equivalent frequency $f_c = 20$ Hz (at 50 Hz) or $f_c = 24$ Hz (at 60 Hz). The current *R*/*X* ratio for determining the peak factor κ is then determined using Eq. (5.24):

$$\frac{R}{X} = \frac{R_c}{X_c} \cdot \frac{f_c}{f} \tag{5.24}$$

The following applies to the equivalent impedance:

$$\underline{Z}_c = R_c + jX_c \tag{5.25}$$

Figure 5.3 shows the ratio $f_2 = i_{p2}/i_p$. In this case, the current i_{p2} or the factor κ has been determined using the equivalent frequency. A comparison with the actual peak short-circuit current shows that the deviations remain $\leq 1\%$ with this method.

Assessment This method leads to κ values that correspond on average to the actual value. The disadvantage is that the impedance matrix must also be set up again for the equivalent frequency. For this reason, this method is definitely suitable for calculations with a computer program.

5.4.5 Multiple Single-Fed

For simplification (manual calculation), the resulting peak short-circuit current at the busbar can be determined from the sum of the partial peak short-circuit currents for a multiple single-fed according to Fig. 3.1b, Eq. (5.26).

$$i_{\mathrm{p}} = i_{\mathrm{p}1} + i_{\mathrm{p}2} + \ldots i_{\mathrm{p}i} \ldots + i_{\mathrm{p}n} \tag{5.26}$$

The result of Eq. (5.26) is always on the safe side, as the maximum values in the partial branches occur at different times depending on the *R*/*X* ratio. Thus, the resulting peak short-circuit current is always smaller if the *R*/*X* ratios are different. This is illustrated by the example in Sect. 5.4.6.1, method 1 compared to method 3, as in this case the addition of the individual values results in a higher amount than the addition of the time-based values.

The consequence is that a manual calculation of the short-circuit current may provide a different result compared to a calculation using a computer program, since in the latter case the method in Sect. 5.4.4 (equivalent frequency) is generally used in many cases.

5.4.6 Example

Since according to Sects. 5.4.2 to 5.4.4 there are several methods for calculating the peak short-circuit current, these calculation options are explained using two examples.

5.4.6.1 Multiple Single-Fed Short Circuit

In the following, the result of Eq. (5.15) is examined with the aid of Fig. 5.2 using different calculation methods. The time curves for the individual partial currents $i_{\mathrm{k}1}(t)$ and $i_{\mathrm{k}2}(t)$ are shown in Fig. 5.2, taking into account Eq. (5.6) and the corresponding *R*/*X* ratios:

$$i_{\mathrm{k}1}(t) = \frac{\sqrt{2} \cdot U}{X_1} \cdot [1 - \cos(\omega t)] = \frac{\sqrt{2} \cdot U}{X \cdot \sqrt{1 + (R/X)^2}} \cdot [1 - \cos(\omega t)] \tag{5.27}$$

$$i_{\mathrm{k}2}(t) = \frac{\sqrt{2} \cdot U}{X} \cdot \frac{1}{\sqrt{1 + (R/X)^2}} \cdot \left[\frac{\mathrm{e}^{-t/T_{\mathrm{DC}}}}{\sqrt{1 + (R/X)^2}} - \cos(\omega t + \alpha)\right] \tag{5.28}$$

With the values according to Eq. (5.8) and X_1 according to Eq. (5.13). If a value of *R*/*X* = 1.0 (branch 2) is assumed, the following currents result at a frequency of $f = 50$ Hz:

$$i_{k1}(t) = \frac{\sqrt{2}\cdot U}{X\cdot\sqrt{2}}\cdot[1 - \cos(\omega t)] \tag{5.29}$$

$$i_{k2}(t) = \frac{\sqrt{2}\cdot U}{X\cdot\sqrt{2}}\cdot\left[\frac{e^{-100\cdot\pi\cdot t}}{\sqrt{2}} - \cos(\omega t + 45^{\circ})\right] \tag{5.30}$$

Equations (5.2) and (5.12) are used to determine the resulting peak short-circuit current i_p, so that the following values are obtained, with

$$\sqrt{2}\cdot I''_{k1} = \sqrt{2}\cdot I''_{k2} = \sqrt{2}\cdot I''_{k} = \frac{U}{X} \tag{5.31}$$

This means that the amount of the partial currents is identical in each case.

- Method 1: Sum of the peak short-circuit currents, as there is a multiple single-fed short circuit, Sect. 5.4.5.

  The following peak short-circuit currents result for the partial currents

$$i_{p1} = \sqrt{2}\cdot 2\cdot I''_k \qquad i_{p2} = \sqrt{2}\cdot 1.0688\cdot I''_k$$

The addition of the currents in relation to the total current results in:

$$i_p = i_{p1} + i_{p2} = \sqrt{2}\cdot 1.5344\cdot\left(2\cdot I''_k\right)$$

- Method 2: *R*/*X* ratio at the short-circuit location (Sect. 5.4.3)

  The ratio R_k/X_k at the short-circuit location can be determined from the short-circuit impedance according to Eq. (5.19).

$$R_k/X_k = \frac{1}{1+\sqrt{1+1^2}} = 0.41421 \quad \kappa = 1.3028$$

$$i_p = 1.15\cdot 1.3028\cdot\sqrt{2}\cdot\left(2\cdot I''_k\right) = \sqrt{2}\cdot 1.4983\cdot\left(2\cdot I''_k\right)$$

If the factor 1.15 according to Eq. (5.23) is not taken into account, the following applies to the resulting peak short-circuit current:

$$i_p = 1.3028\cdot\sqrt{2}\cdot\left(2\cdot I''_k\right)$$

- Method 3: Simulation of the equivalent frequency (Sect. 5.4.4)

With an equivalent frequency of $f_c = 20$ Hz, the resulting ratio R/X is determined using the short-circuit impedance according to Eq. (5.19):

$$R/X = \frac{R_c}{X_c} \cdot \frac{f_c}{f} = \frac{\frac{R}{0.4 \cdot X}}{1 + \sqrt{1 + \left(\frac{R}{0.4 \cdot X}\right)^2}} \cdot 0.4 = 0.2708 \qquad \kappa = 1.4549$$

$$i_p = 1.4549 \cdot \sqrt{2} \cdot \left(2 \cdot I_k''\right)$$

- Method 4: Simulation of the current curve over time
  The addition of the currents based on the time curves represents the correct result and is shown in Fig. 5.4; for the currents according to Eqs. (5.29) and (5.30), only the bracketed term is taken into account in each case. The total current is also shown. Figure 5.4 shows the maximum peak values resp. the κ values of the partial currents and the total current (in relation to the sum of the partial currents) with the corresponding times:

Partial current 1	$\kappa_1 = 2.0000$	$t_{1max} = 10.000$ ms
Partial current 2	$\kappa_2 = 1.0694$	$t_{2max} = 7.250$ ms
Sum	$\kappa_{1+2} = 2.8934$	$t_{max} = 8.625$ ms
Total current	$\kappa = 1.4467.$	

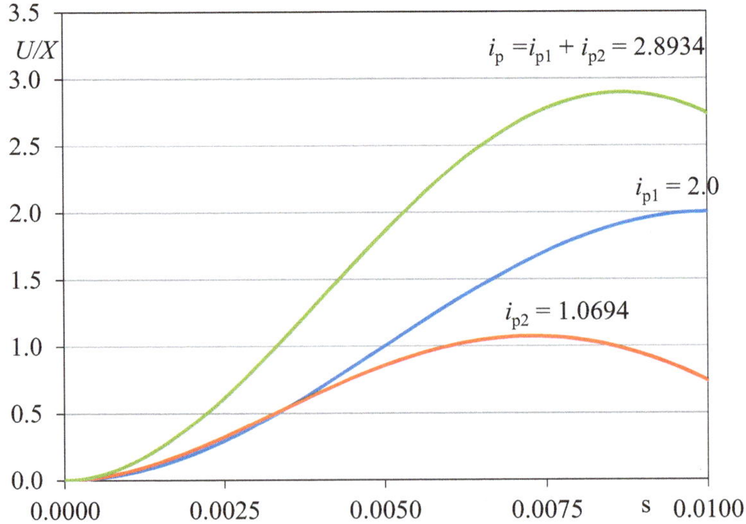

Fig. 5.4 Time curve of the short-circuit currents according to Fig. 5.2 (example). i_{p1} partial short-circuit current in branch 1, i_{p2} partial short-circuit current in branch 2, i_p total current

Assessment Adding the partial peak short-circuit currents for multiple single-fed short circuit according to Eq. (5.26) leads to a result that is on the safe side. In contrast, the calculation using the *R*/*X* ratio of the short-circuit impedance at the fault location produces a current value that is on the unsafe side (without a factor of 1.15). The calculation of the peak short-circuit current taking into account the equivalent frequency only results in a deviation of 0.57% and is on the safe side.

5.4.6.2 Network Example

As a second example, Fig. 5.5 shows a parallel overhead line/cable connection that is connected to network N. A three-phase short circuit occurs at busbar BB2, the peak short-circuit current of which must be calculated.

The following data is used for the individual components:

- Network: $U_{\mathrm{nN}} = 110\ \mathrm{kV}\ I''_{\mathrm{kN}} = 63\ \ \mathrm{kA}$ $R_{\mathrm{N}}/X_{\mathrm{N}} = 0.1$
- Overhead line: $\underline{Z}'_{\mathrm{OHL}} = (0.119 + \mathrm{j}0.401)\,\Omega/\mathrm{km}$ $\ell = 50\ \mathrm{km}$
- Cable: $\underline{Z}'_{\mathrm{C}} = (0.047 + \mathrm{j}0.123)\,\Omega/\mathrm{km}$ $\ell = 50\ \mathrm{km}$

The following applies to the calculation of the three-phase initial short-circuit current:

$$\underline{I}''_{\mathrm{k}} = \frac{cU_{\mathrm{n}}/\sqrt{3}}{\underline{Z}_{\mathrm{k}}} \tag{5.32}$$

The short-circuit impedance $\underline{Z}_{\mathrm{k}}$ is calculated as follows

$$\underline{Z}_{\mathrm{k}} = \underline{Z}_{\mathrm{N}} + \frac{\underline{Z}_{\mathrm{OHL}} \cdot \underline{Z}_{\mathrm{C}}}{\underline{Z}_{\mathrm{OHL}} + \underline{Z}_{\mathrm{C}}} \tag{5.33}$$

With the values based on the component:

$$Z_{\mathrm{N}} = \frac{cU_{\mathrm{nN}}}{\sqrt{3} \cdot I''_{\mathrm{kN}}} = \frac{1.1 \cdot 110\ \ \mathrm{kV}}{\sqrt{3} \cdot 63\ \ \mathrm{kA}} = 1.1089\ \ \Omega$$

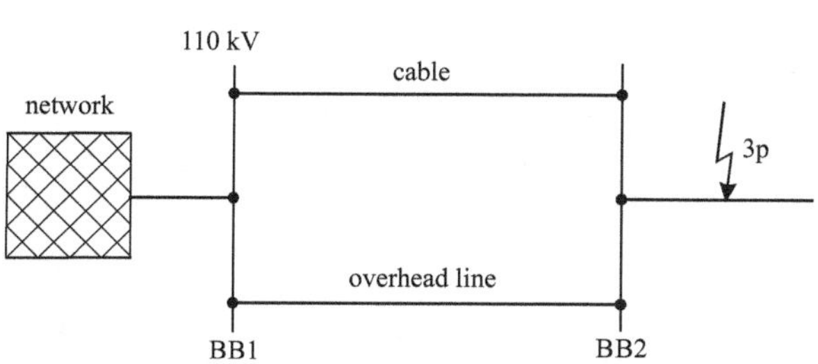

Fig. 5.5 Equivalent circuit for determining the peak short-circuit current at the fault location busbar SS2 in the event of a three-phase short circuit

$$X_N = 1.1034\ \Omega \qquad R_N = 0.1103\ \Omega$$

$$\underline{Z}_{OHL} = (5.95 + j20.05)\ \Omega \qquad R_{OHL}/X_{OHL} = 0.2968$$

$$\underline{Z}_C = (2.35 + j6.15)\ \Omega \qquad R_C/X_C = 0.3821$$

The short-circuit impedance is calculated according to Eq. (5.33):

$$\underline{Z}_k = 0.1103\ \Omega + j1.1034\ \Omega + \frac{(2.35 + 6.15) \cdot (5.95 + j20.05)}{8.30 + j26.20}\ \Omega$$

$$= (1.8126 + j5.8154)\Omega$$

$$Z_k = 6.0913\,\Omega \qquad R_k/X_k = 0.3117$$

Equation (5.32) applies accordingly for the three-phase short-circuit current:

$$I_k'' = \frac{1.1 \cdot 110\ \text{kV}/\sqrt{3}}{6.0913\ \Omega} = 11.4687\ \text{kA}$$

Different methods are used to calculate the peak short-circuit current:

(a) Uniform Ratio R/X

The smallest ratio *R*/*X* is calculated from the network feeder to $R_N/X_N = 0.1$. This results in the peak factor $\kappa_a = 1.746$, Eq. (5.12).

$$i_p = \kappa_a\sqrt{2} \cdot I_k'' = 1.7460 \cdot \sqrt{2} \cdot 11.4687\ \text{kA} = 28.3187\ \text{kA}$$

(b) Ratio *R/X* or *X/R* at the Short-Circuit Location

As the three network branches have an *R*/*X* ratio greater than 0.3, the value 1.15 must be taken into account when calculating the peak factor. The κ factor is determined from Eq. (5.12) from the *R*/*X* ratio at the short-circuit location. The following applies:

$$i_p = 1.15 \cdot \kappa_b\sqrt{2} \cdot I_k'' = 1.15 \cdot 1.4047 \cdot \sqrt{2} \cdot 11.4687\ \text{kA} = 26.2004\ \text{kA}$$

If, in contrast, the factor 1.15 is not taken into account, the resulting peak short-circuit current $i_p = 22.7830$ kA.

(c) Equivalent Frequency

The various network branches must be converted to an equivalent frequency ($f_c = 20$ Hz) at an operating frequency of $f = 50$ Hz. This results in the following values:

$$\underline{Z}_k = 0.1103\ \Omega + j0.4414\ \Omega + \frac{(2.35 + 2.46) \cdot (5.95 + j8.02)}{8.30 + j10.48}\ \Omega$$
$$= (1.8069 + j2.3334)\Omega$$

The calculation of the impedance $\underline{Z}_c$ is only used to determine the peak factor κ. The initial symmetrical short-circuit current is determined according to methods (a) or (b) using the short-circuit impedance $\underline{Z}_k$. The following applies to the ratio *R*/*X*

$$\frac{R}{X} = \frac{R_c}{X_c} \cdot \frac{f_c}{f} = \frac{1.8069}{2.3334} \cdot \frac{20}{50} = 0.3098$$

This results in the following for the peak short-circuit current:

$$i_p = \kappa_c \cdot \sqrt{2} \cdot I_k'' = 1.4069 \cdot \sqrt{2} \cdot 11.4687\ \text{kA} = 22.8187\ \text{kA}$$

Assessment A comparison of the different methods shows that the calculation using the equivalent frequency provides the smallest value compared to the other two. However, this requires more computational effort, as the short-circuit impedance must be determined twice. In this case, determining the peak short-circuit current using the impedance at the short-circuit location (method b) requires multiplication by a factor of 1.15, as the condition that all network branches have an *R*/*X* ratio of less than 0.3 is not met. If this factor is not taken into account, the result deviates from the actual value (method c) by only approx. 0.16%. It can be seen that the factor 1.15 is only justified if the *R*/*X* ratios of the network are between 0.0 and 1.2.

5.5 DC Component

In principle, a short-circuit current, especially in high-voltage systems, has an inductive behavior due to the *R*/*X* ratio ($R/X \approx 0.1$). As a result, a DC component $i_{DC}(t)$ occurs in the short-circuit current curve as shown in Fig. 3.7 due to the short-circuit occurrence at voltage zero crossing. The absolute value of the DC component can be determined according to Eq. (5.7), so that the following applies:

$$i_{\mathrm{DC}}(t) = \frac{\sqrt{2}\cdot U}{X\cdot\sqrt{1+(R/X)^2}}\cdot\frac{\mathrm{e}^{-t/T_{\mathrm{DC}}}}{\sqrt{1+(R/X)^2}} = \sqrt{2}\cdot I''_{\mathrm{k}}\cdot\frac{\mathrm{e}^{-t/T_{\mathrm{DC}}}}{\sqrt{1+(R/X)^2}}\ \text{resp.} \quad (5.34)$$

$$i_{\mathrm{DC}}(t) \approx \sqrt{2}\cdot I''_{\mathrm{k}}\cdot \mathrm{e}^{-t/T_{\mathrm{DC}}} = I_{\mathrm{DC}}\cdot \mathrm{e}^{-t/T_{\mathrm{DC}}} \quad (5.35)$$

With

U Voltage (line—earth),
T_{DC} DC time constant, Eq. (5.8),
I''_{k} initial symmetrical short-circuit current.

To determine the DC time constant, the *R*/*X* ratio at the short-circuit location must be determined. In the case of a single-fed network (without branches) or meshed networks, this ratio can be determined directly from the short-circuit impedance, as compared to calculating the maximum amplitude of the peak short-circuit current.

The safe side can be estimated by adding the partial DC components, e.g. for two parallel branches with different *R*/*X* ratios. The following relationships result for the individual DC components i_{DC1}, i_{DC2} and the addition i_{DC} if the notations according to Fig. 5.2 are used:

$$\begin{aligned} i_{\mathrm{DC1}}(t) &= \frac{\sqrt{2}\cdot U}{X_1\cdot\sqrt{1+(R_1/X_1)^2}}\cdot\frac{\mathrm{e}^{-t/T_{\mathrm{DC1}}}}{\sqrt{1+(R_1/X_1)^2}} \\ i_{\mathrm{DC2}}(t) &= \frac{\sqrt{2}\cdot U}{X\cdot\sqrt{1+(R/X)^2}}\cdot\frac{\mathrm{e}^{-t/T_{\mathrm{DC2}}}}{\sqrt{1+(R/X)^2}} \end{aligned} \quad (5.36)$$

On the right-hand side of Eq. (5.35) is the absolute value of the initial symmetrical short-circuit current and an expression that describes the curve of the DC component. If the *R*/*X* ratio is small (e.g. ≤ 0.3), the absolute value of the DC component is almost determined by the second part of Eq. (5.35). The deviation is shown below as a function of the *R*/*X* ratio.

The maximum direct current component results from the amounts of the individual components at the time of the short-circuit occurrence $t = 0$, so that the following generally applies to the maximum value:

$$i_{\mathrm{DC}}(0) = \sqrt{2}\cdot U\cdot\left[\frac{1}{X_1\cdot\left[1+(R_1/X_1)^2\right]} + \frac{1}{X\cdot\left[1+(R/X)^2\right]}\right] \quad (5.37)$$

According to Fig. 5.2, $R_1/X_1 = 0$, so the DC constant is simplified using Eq. (5.13).

$$i_{\mathrm{DC}}(0) = \sqrt{2} \cdot U \cdot \left[\frac{1}{X_1} + \frac{1}{X \cdot \left[1 + (R/X)^2 \right]} \right] \tag{5.38}$$

The following applies to the absolute value of the initial symmetrical short-circuit current, taking into account Eqs. (5.17) and (5.18):

$$I_{\mathrm{k}}'' = \frac{\sqrt{2} \cdot U}{\sqrt{R_{\mathrm{k}}^2 + X_{\mathrm{k}}^2}} = \frac{\sqrt{2} \cdot U}{X_1} \cdot \frac{R^2 + (X_1 + X)^2}{\sqrt{(R \cdot X_1)^2 + \left(X^2 + X \cdot X_1 + R^2\right)^2}} \tag{5.39}$$

The absolute value of the reactance X_1 is determined by the reactance X and the *R*/*X* ratio according to Eq. (5.40) and is also changed as a function of the factor $n = 2$, 1, 0.5.

$$X_1 = n \cdot X \cdot \sqrt{1 + (R/X)^2} \tag{5.40}$$

In Fig. 5.6, the ratio of the calculations according to Eq. ((5.34), $k = \sqrt{2} \cdot I_{\mathrm{k}}''/I_{\mathrm{DC}}$) is plotted as a function of the ratio n. It can be seen that the deviations remain below 2% for an *R*/*X* ratio ≤ 0.3, so that the simplification according to Eq. (5.35) can be used to calculate the DC component.

For larger *R*/*X* ratios, the calculation of the DC component, taking into account the simplification (initial symmetrical short-circuit current), produces a value that is greater than the DC component that actually occurs, so that the result is on the safe side.

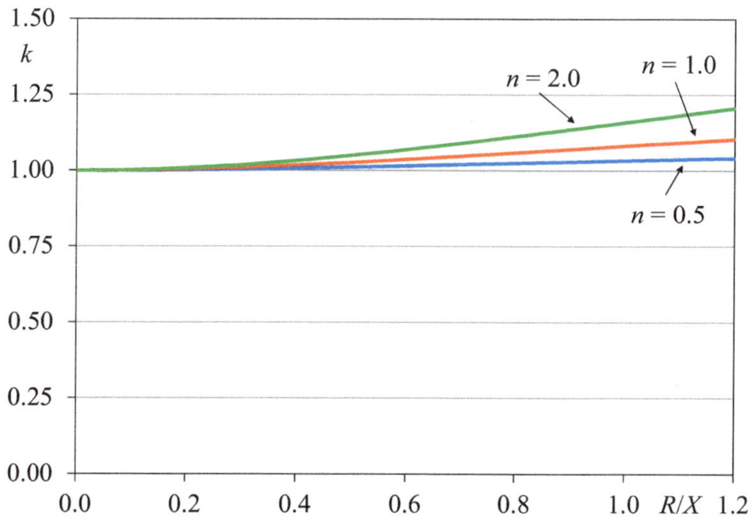

Fig. 5.6 Deviations in the calculation of the DC constant according to Eq. (5.34)

5.6 Short-Circuit Breaking Current I_b

The breaking current takes into account the decay processes of the short-circuit current in the event of short-circuits near to synchronous generators and motors. This current is necessary for the dimensioning of circuit-breakers, as the breaking process is only activated after a few milliseconds (Fig. 1.1). The short-circuit current has already decayed during this time, so that the switching capacity of a switching device can be lower than specified by the initial symmetrical short-circuit current. In general, a distinction is made between the symmetrical (AC component only) and the asymmetrical (AC and DC components).

In principle, the calculation of the breaking current is complex due to the electromagnetic transient phenomenon in the machines and in the network, so that according to [1], μ and q factors are specified for simplification, which should be on the safe side for determining the dimensioning currents for switching devices.

5.6.1 *Method for Calculating the Symmetrical Breaking Current*

In the case of near-to-generator short-circuits, the short-circuit current decays due to the time constants of the generators or motors [8, 9], which is described by a factor μ. This decay process of the short-circuit current propagation can be expressed, for example, for a synchronous generator in no-load operation before the short-circuit occurs according to Sect. 4.2.1, Eq. (4.39) by the AC element $i_{\mathrm{AC}}(t)$ at the time $t = t_{\min}$.

$$i_{\mathrm{kAC}}(t) \approx \sqrt{\frac{2}{3}} \cdot U_{\mathrm{rGS}} \cdot \left\{ \left[\frac{1}{X''_{\mathrm{d}}} - \frac{1}{X'_{\mathrm{d}}} \right] \cdot \mathrm{e}^{-t_{\min}/T''_{\mathrm{d}}} + \left[\frac{1}{X'_{\mathrm{d}}} - \frac{1}{X_{\mathrm{d}}} \right] \cdot \mathrm{e}^{-t_{\min}/T'_{\mathrm{d}}} + \frac{1}{X_{\mathrm{d}}} \right\} \cdot \cos(\omega \cdot t_{\min}) \tag{5.41}$$

with

U_{rGS} rated voltage of the generator (line-to-line),
$t_{\min}$ minimum time delay,
X''_{d} subtransient reactance,
X'_{d} transient reactance,
X_{d} synchronous reactance.

The effective value of the AC component according to Eq. (5.41) can be related to the initial symmetrical short-circuit current for the time $t = t_{\min}$, so that the following applies if a network reactance X_N is assumed in series with the synchronous machine (the time constants are extended by the index N for this case):

$$\mu = \frac{I_b}{I''_k} = \frac{I''_k(t = t_{min})}{I''_k(t = 0)} \approx \frac{\sqrt{\frac{2}{3}} \cdot U_{rGS}}{\sqrt{\frac{2}{3}} \cdot U_{rGS} \cdot \left[\frac{1}{X''_d + X_N}\right]} \cdot$$

$$\left\{ \left[\frac{1}{X''_d + X_N} - \frac{1}{X'_d + X_N}\right] \cdot e^{-t_{min}/T''_{dF}} + \left[\frac{1}{X'_d + X_N} - \frac{1}{X_d + X_N}\right] \cdot e^{-t_{min}/T'_{dF}} + \frac{1}{X_d + X_N} \right\} \tag{5.42}$$

or

$$\mu \approx \left[1 - \frac{X''_d + X_N}{X'_d + X_N}\right] \cdot e^{-t_{min}/T''_{dN}} + \left[\frac{X''_d + X_N}{X'_d + X_N} - \frac{X''_d + X_N}{X_d + X_N}\right] \cdot e^{-t_{min}/T'_{dN}} + \frac{X''_d + X_N}{X_d + X_N} \tag{5.43}$$

According to Table 4.1, the following mean values are assumed for the reactances and time constants:

Subtransient reactance: $x''_d = 20\%$.
Transient reactance: $x'_d = 30\% = 1.5 \cdot x''_d$.
Synchronous reactance: $x_d = 220\% = 11 \cdot x'_d$.
Subtransient time constant: $T''_d = 30\,\text{ms}$.
Open-circuit transient time constant: $T'_{d0} = 10\,\text{s}$ resp. $T'_d = 1.364\,\text{s}$.

Since a system short circuit is considered, a system reactance X_N or x_N between the generator terminals and the fault location is assumed to be a multiple of the subtransient reactance x''_d, in each case in relation to the generator impedance Z_{rG}.

$$x_N = 10 \cdot x''_d = 200\%$$

From these values, the relevant time constants for the evaluation of Eq. (5.43) can be determined according to Sect. 4.2.1.2.

$$T''_{dN} \approx T''_d \cdot \frac{1 + \frac{x_N}{x''_d}}{1 + \frac{x_N}{x'_d}} = 43.04\,\text{ms} \qquad T'_{dN} \approx T'_d \cdot \frac{1 + \frac{x_N}{x'_d}}{1 + \frac{x_N}{x_d}} = 1.0363\,\text{s}$$

Taking into account a time period for the minimum time delay of $t_{min} = 0.02$ to 0.25 s, the following approximations can be used, Sect. 10.2:

$$e^{-t_{min}/T'_{dN}} \sim 1 \qquad e^{-t/T''} \sim (1 - t/T'') \tag{5.44}$$

$$\mu \approx \left[\frac{X'_d - X''_d}{X'_d + X_N}\right] \cdot e^{-t_{min}/T''_{dN}} + \frac{X''_d + X_N}{X'_d + X_N} \approx 1 - \frac{X'_d - X''_d}{X'_d - X_N} \cdot \frac{t}{T''} \tag{5.45}$$

According to Eq. (10.10), the short-circuit reactance X_N can be represented by generator variables for simplification. For subtransient reactance x''_d, it can also be written $I''_{kGS}/I_{rGS} = 1/x''_d$ so that the factor μ according to Eq. (5.45) can be expressed by the current ratio. In addition, the subtransient time constant T'' also influences the factor μ. According to [1], this decay process is therefore simplified by the factor μ. This μ curve represents the envelope of the actual, possible symmetrical breaking currents of electrical machines. The result is therefore on the safe side, as the maximum currents are necessary for dimensioning the switchgear. For this reason, it is always likely that, for example, the actual breaking current of a specific generator should be lower than that calculated using the μ factor. This results in the following for the symmetrical breaking current of the synchronous generator I_{bGS}:

$$I_{bGS} = \mu \cdot I''_{kGS} \tag{5.46}$$

The factor μ is determined according to Eqs. (5.47) to (5.50) by specifying the dependence of the factor μ as a function of the current ratio I''_{kGS}/I_{rGS} [1, 5].

$$\mu = 0.84 + 0.26 \cdot e^{-0.26 I''_{kGS}/I_{rGS}} \qquad t_{min} = 0.02\,\mathrm{s} \tag{5.47}$$

$$\mu = 0.71 + 0.51 \cdot e^{-0.30 I''_{kGS}/I_{rGS}} \qquad t_{min} = 0.05\,\mathrm{s} \tag{5.48}$$

$$\mu = 0.62 + 0.72 \cdot e^{-0.32 I''_{kGS}/I_{rGS}} \qquad t_{min} = 0.10\,\mathrm{s} \tag{5.49}$$

$$\mu = 0.56 + 0.94 \cdot e^{-0.38 I''_{kGS}/I_{rGS}} \qquad t_{min} = 0.25\,\mathrm{s} \tag{5.50}$$

The μ curves are specified as a function of the minimum time delay t_{min}. This is the time between the occurrence of the short circuit and the opening of the circuit-breaker contacts. When applying the equations for calculating the μ curves, Eqs. (5.47) to (5.50), it must be ensured that the value $\mu = 1$ is always used for values $I''_{kGS}/I_{rGS} \leq 2$. The definition for a short-circuit far-from-generator according to Sect. 3.5 can also be derived from this limit, i.e. in this case no synchronous machine delivers a short-circuit current that exceeds twice the value of the rated current.

Low-voltage synchronous generators usually have voltage-regulated compound excitation (AVR Automatic Voltage Regulation). This means that the excitation voltage has a current-dependent voltage component both as a function of the terminal voltage and via a current transformer, so that the terminal voltage can be kept constant in the event of load changes. This therefore also applies to the short circuit, so that the factor μ should only be used for these generators up to a minimum time delay of $t_{min} = 0.1$ s. For a longer minimum time delay, the manufacturer's specifications must be requested, which also applies to the determination of the steady state short-circuit current, Sect. 5.7.1.

As an example, Fig. 5.7 shows the current propagation of a low-voltage generator with compound excitation (AVR) with a three-phase terminal short circuit from no-load operation and at rated speed. Here, a distinction is made between a symmetrical and asymmetrical breaking current in accordance with the definition, Sect. 5.6.2. It can be clearly seen that in this case, the short-circuit current increases again from a short-circuit duration of $t_k > 0.3$ s.

Figure 5.7 shows a low-voltage generator with a rated power of $S_{rGS} = 1000$ kVA and a voltage of $U_{rGS} = 400$ V, so that a rated current of $I_{rGS} = 1443$ A can be determined from this. According to Fig. 5.7, a value of $I''_{kGS} = 7790$ A can be read as the initial magnitude of the short-circuit current, resulting in a ratio of the locked-rotor current of $I''_{kGS}/I_{rGS} \approx 5,4$. Up to a short-circuit time of $t_k \sim 0.3$ s, the terminal short-circuit decreases to a value of approx. 3810 A and then increases again, in this case up to a current of $I_k \sim 6000$ A ($t_k = 1$ s).

The factor μ is also used to calculate the breaking current of asynchronous machines. In addition, however, the factor q is introduced according to Eqs. (5.52) to (5.55), as the short-circuit current of the asynchronous machine decays faster than the short-circuit current of a synchronous machine. The reason for this is that the excitation is taken from the motor terminal, whereas in the case of a synchronous machine it is an external excitation. For this reason, the following applies to the breaking current of motors:

$$I_{bM} = q \cdot \mu \cdot I''_{kM} \quad (5.51)$$

The factor q can be taken from Eqs. (5.52) to (5.55). The factors are specified as a function of the rated active power per number of pairs of poles $P_{rM}/p = m$ [1, 5].

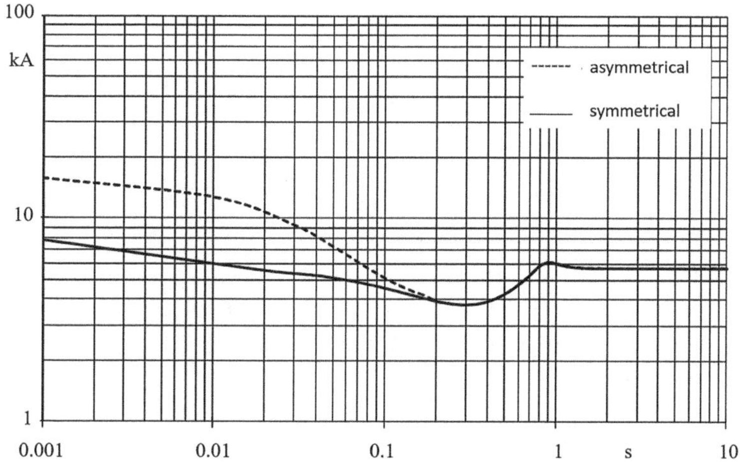

Fig. 5.7 Short-circuit current curve for a low-voltage generator with compound excitation. (Source: ABB)

$$q = 1.03 + 0.12 \cdot \ln(m) \qquad \text{bei}\, t_{\min} = 0.02\,\text{s} \tag{5.52}$$

$$q = 0.79 + 0.12 \cdot \ln(m) \qquad \text{bei}\, t_{\min} = 0.05\,\text{s} \tag{5.53}$$

$$q = 0.57 + 0.12 \cdot \ln(m) \qquad \text{bei}\, t_{\min} = 0.10\,\text{s} \tag{5.54}$$

$$q = 0.26 + 0.10 \cdot \ln(m) \qquad \text{bei}\, t_{\min} \geq 0.25\,\text{s} \tag{5.55}$$

The factor q does not take into account any short-circuit location, but is to be used independently of this for asynchronous motors when calculating the breaking current. In general, the current contribution depends on the residual voltage at the terminals in the event of a short-circuit, so that the relative contribution of the asynchronous motors to the breaking current is greater in the event of a network short-circuit than in the event of a terminal short-circuit. In the old VDE Standard [10], this behavior was taken into account by two q curves (a, b) for each minimum time delay. From 1971 [11], the distinction between terminal and network short-circuit is omitted by selecting an average value between the two q curves. Table 5.2 shows the values for a minimum time delay of $t_{\min} = 0.1$ s.

Although the values according to [1] are smaller for the network short-circuit compared to [10], so that the amount is on the uncertain side, in these cases the absolute current value of the asynchronous motors in relation to the total short-circuit current in the network is low, so that the deviation should be less than 5%. In addition, the distinction between terminal short circuit—and network short circuit is difficult to implement in a calculation.

The calculation of breaking current according to Eqs. (5.56) and (5.57) assumes that the contributions of the motors and generators to the short-circuit current for short-circuits near to the generator decrease during the short-circuit duration and do not influence each other. In contrast to this, a parallel network component $I_{\text{bN}} = I''_{\text{kN}}$ is assumed to be constant in the time range under consideration, so that $\underline{I}_{\text{b}}$ applies to the total breaking current:

Table 5.2 q factor for $t_{\min} = 0{,}1$ s for network and terminal short-circuits

		q_{62}	
P_{rM}/p MW	q_{16}	a	b
0.1	0.29	0.26	0.43
0.4	0.46	0.42	0.57
1.0	0.57	0.53	0.67
4.0	0.74	0.69	0.80

q_{16} q factor according IEC 60909-00102, 2016 [1]
q_{62} q factor according VDE 0102, 1962 [10]
a terminal short circuit
b network short circuit

$$I_b = I_{bM} + I_{bGS} + I''_{kN} \quad \text{resp.} \tag{5.56}$$

$$I_b = \mu_M \cdot q_M \cdot I''_{kM} + \mu_{GS} \cdot I''_{kGS} + I''_{kN} \tag{5.57}$$

With

I''_{kM} initial symmetrical short-circuit current of motors,
I''_{kGS} initial symmetrical short-circuit current of generators,
I''_{kN} initial symmetrical short-circuit current of the network.

However, the calculation of the breaking current according to Eq. (5.57) only applies if the individual partial short-circuit currents are decoupled during the entire short-circuit duration. In contrast to the example in Fig. 5.8, this means that in this case, there is mutual interference between the generator component and the network component, as the fault F is not located directly at the 6 kV busbar and therefore there is no multi single-fed infeed.

For the meshed network according to Fig. 5.8, it is necessary to determine the breaking current according to Eq. (5.58), since in these cases a calculation according to Eqs. (5.56) and (5.57) is not possible in these applications. The derivation of Eq. (5.58) according to [12] is described in detail in Chap. 10 and is based on the consideration of the initial voltage differences $\Delta\underline{U}''$ of the connected generators and motors, which are generally characterized by the indices G and M.

$$\underline{I}_b = \underline{I}''_k - \sum_i \frac{\Delta\underline{U}''_{Gi}}{cU_n/\sqrt{3}} \cdot (1-\mu_i) \cdot \underline{I}''_{kGi} - \sum_j \frac{\Delta\underline{U}''_{Mi}}{cU_n/\sqrt{3}} \cdot (1-\mu_j q_j) \cdot \underline{I}''_{kMj} \tag{5.58}$$

With

$$\Delta\underline{U}''_{Gi} = jX''_{di} \cdot \underline{I}''_{kGi} \tag{5.59}$$

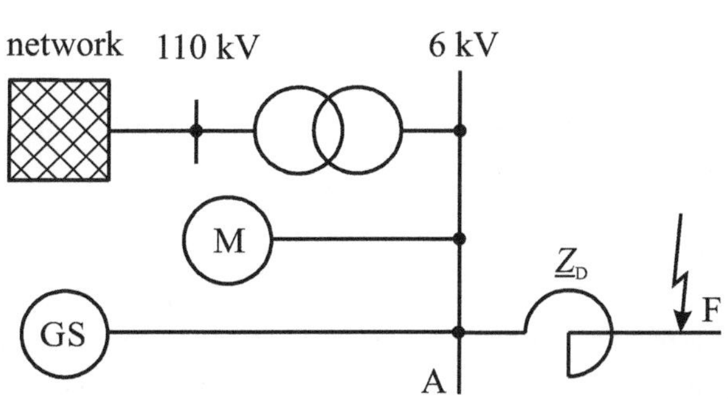

Fig. 5.8 Calculation of the breaking current I_b for a three-phase short circuit behind a common impedance $\underline{Z}_D$

$$\Delta \underline{U}''_{\mathrm{M}i} = \mathrm{j}X''_{\mathrm{M}j} \cdot I''_{\mathrm{kM}j} \tag{5.60}$$

This includes

$cU_{\mathrm{n}}/\sqrt{3}$	equivalent voltage source at the short-circuit location,
$\underline{I}''_{\mathrm{k}}, \underline{I}_{\mathrm{b}}$	initial symmetrical short-circuit current and breaking current, taking into account all network feeders, generators, and motors,
$\Delta \underline{U}''_{\mathrm{G}i}, \Delta \underline{U}''_{\mathrm{M}j}$	initial voltage differences at the connection point of the generators i and the motors j,
$\underline{I}''_{\mathrm{kG}i}, \underline{I}''_{\mathrm{kM}j}$	partial short-circuit currents of the generators i and the motors j.

For simplification, only the decay behavior of synchronous machines and asynchronous motors is taken into account in Eq. (5.58). [1] also includes power station feed-ins and wind power station units (asynchronous generator, doubly fed induction generator), so that the decay behavior according to Eq. (5.58) must also be taken into account in these cases.

If a calculation according to Eq. (5.58) is not possible or appears too complex, the following applies to the breaking current in meshed networks:

$$I_{\mathrm{b}} = I''_{\mathrm{k}} \tag{5.61}$$

This means that the result is always on the safe side, as the decay of the AC component of the short-circuit current is not taken into account.

According to [1], only the total value of the breaking current is determined in meshed networks and not the partial currents. In principle, only the current that is interrupted by the switchgear is decisive for the design of the switchgear. This generally means that the maximum alternating breaking current of the switchgear is less than or equal to the total current, e.g. at the busbar. As equipment is only available in defined rated values (31.5 kA, 40 kA), it is perfectly appropriate to use the same switchgear for all outgoing circuits of a busbar. In addition, replacement between different installation locations is possible without restriction.

5.6.2 *Method for Calculating the Asymmetrical Breaking Current*

While in Eqs. (5.46) and (5.51), only the AC component of the breaking current is determined, it is also necessary to take the DC component into account when designing circuit-breakers, as this causes a shift in the short-circuit current and results in a higher arc energy when switching off. The starting point is the calculation of the DC component in the short-circuit current curve, which is determined according to Eq. (5.62).

$$i_{DC} = \sqrt{2} \cdot I''_k \cdot e^{-2\pi \cdot f \cdot t \cdot R/X} \tag{5.62}$$

With

I''_k initial symmetrical short-circuit current,
f operating frequency,
R/X ratio of the short-circuit branch at the short-circuit location,
t time of calculation of the DC component.

The actual effective resistance of the short-circuit current branch must always be used for the resistance and not the fictitious resistance, as is the case with synchronous generators, for example (difference between R_G and R_{Gf}). The R/X ratio is determined from the impedance at the short-circuit location. The asymmetrical breaking current is calculated according to Eq. (5.63).

$$I_{b(asym)} = \sqrt{{I_b}^2 + i_{DC}^2} = I_b \cdot \sqrt{1 + \left(\frac{i_{DC}}{I_b}\right)^2} \tag{5.63}$$

With

$I_{b(asym)}$ asymmetrical breaking current (rms value),
I_b symmetrical breaking current (rms value) according Eq. (5.46) resp. (5.51),
i_{DC} DC component of the breaking current according Eq. (5.62).

5.7 Steady State Short-Circuit Current

The steady state short-circuit current I_k occurs after the subtransient and transient processes in the electrical machines have decayed and, as a result of the control behavior of the synchronous machines, it is less accurate to calculate than the initial symmetrical short-circuit current I''_k and the breaking current I_b. For this reason, the calculation of the steady state short-circuit current should be regarded as an estimate. In addition to the influences already described that affect the calculation of the breaking current, the following points become significant for the calculation of the steady state short-circuit current:

- saturated synchronous reactance in the direct axis,
- saturation of the iron, especially in the rotor,
- effectiveness of the voltage regulation of the generators,
- highest possible excitation voltage U_{fmax},
- clearing processes of the network protection.

The steady state short-circuit current is required, for example, to determine the thermally equivalent short-circuit current, which is necessary for the design of the cable cross-sections.

When calculating the steady state short-circuit current, a distinction must be made between two different network topologies, which also have an influence on the mathematical determination of the steady state short-circuit current. These are

- Isolated network with only one synchronous machine or multiple feeders in a non-meshed network,
- meshed network with several synchronous machines or network feeders.

A distinction is made between these options in Sects. 5.7.1 and 5.7.3 below.

5.7.1 *Isolated Network with a Synchronous Machine or Multiple Feeders in a Non-Meshed Network*

In this case, the steady state short-circuit current is determined in a network that is only fed by a synchronous machine. When calculating the continuous short-circuit current, a distinction is made between the minimum and maximum current:

$$I_{\mathrm{kmax}} = \lambda_{\mathrm{max}} \cdot I_{\mathrm{rGS}} \tag{5.64}$$

$$I_{\mathrm{kmin}} = \lambda_{\mathrm{min}} \cdot I_{\mathrm{rGS}} \tag{5.65}$$

With

I_{rGS} rated current of the generator,
λ factors according to the IEC Standard [1].

As turbogenerators and salient-pole generators differ considerably in the size of the synchronous reactances X_{d} and are also equipped with different excitation devices ($u_{\mathrm{fmax}} = U_{\mathrm{fmax}}/U_{\mathrm{fr}}$ is different), various values result for determining the factors λ. The factors λ are listed in [1] for different generator types and excitations.

For low-voltage generators with compound excitation (AVR Automatic Voltage Regulation), the contribution to the steady state short-circuit current can be equal to the initial symmetrical short-circuit current, so the manufacturer's specifications must be requested in order to determine the steady state short-circuit current for a three-phase short circuit. Figure 5.7 shows the current curve for a terminal short-circuit of a low-voltage generator. After the short-circuit current has decayed in the first time range up to approx. 0.3 s, the short-circuit current then increases due to voltage regulation, Sect. 5.6.1.

5.7.2 *Example: Single Fed resp. Multiple Single-Fed*

For the calculation of the steady state short-circuit current of a synchronous machine, it is assumed that in the event of a short circuit in F1 according to Fig. 5.9, only the

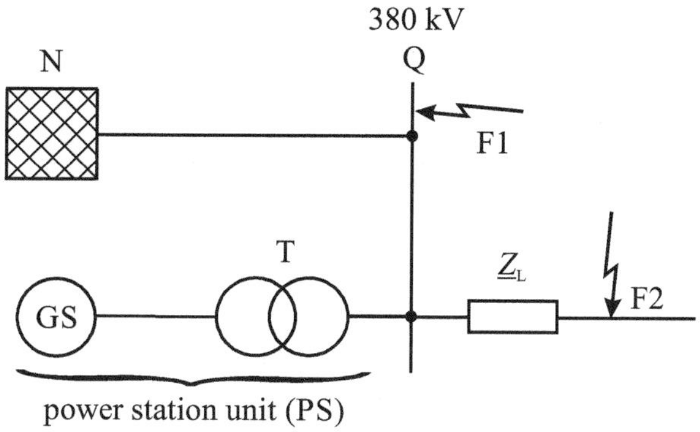

Fig. 5.9 Equivalent circuit diagram for determining the current profile in the event of a three-phase short circuit with parallel operation of generator and network

power station feeds into the 380 kV busbar. The network feeder is not taken into account for the time being.

The following data are assumed for the power station unit:

Generator GS	$S_{rGS} = 940$ MVA	$U_{rGS} = 27$ kV
	$x''_d = 23\%$	$R_{Gf} = 0.05 \cdot X''_d$
	$u_{fmax} = 1.3$	$x_d = 2.0$ p.u.
	$\sin\varphi_{rG} = 0.8$	
Transformer T	$S_{rT} = 850$ MVA	$t_r = 420$ kV/27 kV
	$u_{kr} = 17\%$	$u_{Rr} = 0.4\%$

The power station unit has a step-up transformer with a tap changer, so that the procedure in Sect. 4.3 must be applied in detail. The individual impedances are calculated from the specified data as follows:
Generator GS:

$$X''_d = \frac{x''_d}{100\%} \cdot \frac{U^2_{rGS}}{S_{rGS}} = 0.23 \cdot \frac{(27\ \text{kV})^2}{940\ \text{MVA}} = 178.4\ \text{m}\Omega$$

$$R_{GSf} = 0.05 \cdot X''_d = 8.9\ \text{m}\Omega$$

Transformer T (high voltage side):

$$Z_{TOS} = \frac{u_{kr}}{100\%} \cdot \frac{U^2_{rT}}{S_{rT}} = 0.17 \cdot \frac{(420\ \text{kV})^2}{850\ \text{MVA}} = 35.28\ \Omega$$

$$R_{TOS} = \frac{u_{Rr}}{100\%} \cdot \frac{U^2_{rT}}{S_{rT}} = 0.004 \cdot \frac{(420\ \text{kV})^2}{850\ \text{MVA}} = 830.1\ \text{m}\Omega$$

$$X_{TOS} = \sqrt{Z^2_{TOS} - R^2_{TOS}} = 35.27\ \Omega$$

The following applies to the correction factor of the power station unit according to Eq. (4.50):

$$K_{PS} = \frac{U_{nQ}^2}{U_{rGS}^2} \cdot \frac{U_{rTUS}^2}{U_{rTOS}^2} \cdot \frac{c_{max}}{1 + \left|x_d'' - x_T\right| \cdot \sqrt{1 - \cos^2\varphi_{rGS}}} =$$

$$\left(\frac{380\ \text{kV}}{27\ \text{kV}}\right)^2 \left(\frac{27\ \text{kV}}{420\ \text{kV}}\right)^2 \frac{1,1}{1 + |0.23 - 0.17| \cdot 0.8} = 0.8592$$

The impedance of the power station unit in relation to the HV side is determined as follows:

$$\begin{aligned} \underline{Z}_{PSK} &= K_{PS} \cdot \left(t_r^2 \cdot \underline{Z}_{GS} + \underline{Z}_{TOS}\right) \\ &= 0.8592 \cdot \left[\left(\frac{420}{27}\right)^2 \cdot (8.9 + j178.4)\text{m}\Omega + (0.83 + j35.27)\Omega\right] \\ &= (2.56 + j67.40)\Omega \end{aligned}$$

The three-phase short-circuit current on the HV side of the step-up transformer is formed taking into account Eq. (5.1), busbar Q:

$$\underline{I}_{kPS}'' = \frac{cU_{nQ}/\sqrt{3}}{\underline{Z}_{PSK}} = \frac{1.1 \cdot 380\ \text{kV}/\sqrt{3}}{(2.56 + j67.40)\ \Omega} = (0.14 - j3.58)\ \text{kA} \quad \Rightarrow \quad I_{kPS}'' = 3.58\ \text{kA}$$

To determine the steady state short-circuit current of the synchronous machine using the λ factors, the ratio I_{kGS}''/I_{rGS} must be determined. The following applies to the rated current of the generator in relation to the HV side of the transformer:

$$I_{rGS} = \frac{S_{rGS}}{\sqrt{3} \cdot U_{rGS}} \cdot \frac{1}{t_r} = \frac{940\ \text{MVA}}{\sqrt{3} \cdot 27\ \text{kV}} \cdot \frac{1}{420/27} = 1.29\ \text{kA}$$

With an initial symmetrical short-circuit current of $I_{kGS}'' = 3.58\ \text{kA}$, a current ratio is determined:

$$\frac{I_{kGS}''}{I_{rGS}} = \frac{3.58\ \text{kA}}{1.29\ \text{kA}} = 2.78$$

From this follows according [1]:

$$\lambda_{max} = 1.55 \quad \text{and} \quad \lambda_{min} = 0.45$$

The following results for the minimum and maximum three-phase steady state short-circuit current

$$I_{\mathrm{kPSmax}} = 1.55 \cdot 1.29\,\mathrm{kA} = 2.0\,\mathrm{kA} \qquad I_{\mathrm{kPSmin}} = 0.45 \cdot 1.29\,\mathrm{kA} = 0.58\,\mathrm{kA}$$

If the steady state short-circuit current is calculated according to Fig. 5.9 for the short-circuit location F1 (busbar Q), taking into account the network feeder, the following relationship applies for the steady state short-circuit current for the multi-sided single-feed in a non-meshed network:

$$I_{\mathrm{kF1}} = \lambda \cdot I_{\mathrm{rGS}} + I''_{\mathrm{kN}} = I_{\mathrm{kPS}} + I''_{\mathrm{kN}} \tag{5.66}$$

As the individual components do not influence each other in this case, the partial short-circuit currents can be added. If the following data is assumed for the network supply:

$$I''_{\mathrm{kN}} = 15\ \mathrm{kA} \quad U_{\mathrm{nQ}} = 380\,\mathrm{kV} \quad R/X = 0.1$$

results for the total maximum steady state short-circuit current i_{kF1max}:

$$I_{\mathrm{kF1\,max}} = 2..0\,\mathrm{kA} + 15.0\,\mathrm{kA} = 17.0\,\mathrm{kA}.$$

5.7.3 *Mesh Network with Synchronous Machines*

In meshed networks or in networks in which synchronous machines feed into a short-circuit via a common impedance, a simple calculation of the steady state short-circuit current in accordance with Sect. 5.7.1 is not possible, as the following example shows.

According to Fig. 5.9, a 940 MVA power station unit feeds parallel to a 380 kV network to the short-circuit location F2. In contrast to short-circuit location F1, short-circuit location F2 has a common impedance $\underline{Z}_{\mathrm{L}}$ between the short-circuit location and busbar A.

In the case of a short-circuit at the short-circuit location F1, the total current (Fig. 5.9c) contains beats frequencies after a time range $t \geq 0.5$ s, while the individual partial short-circuit currents (Fig. 5.10a, b) show the expected, unaffected course. In contrast, in the case of a short circuit in F2, all currents contain superimposed beat frequencies, but these only occur after a time of $t > 1$ s. Figures 5.10 and 5.11 show the transient current characteristics (qualitative).

The generator is accelerated by the short-circuit occurrence due to its lack of active power output, which results in an increase in speed and frequency. The generator therefore runs asynchronously with the network. In contrast, the network component retains its constant frequency, so that the total current is made up of two partial currents with two different frequencies. For the partial short-circuit currents, Eqs. (5.67) and (5.68) apply to the partial short-circuit currents for a three-phase short circuit F1 according to Fig. 5.9, if only the reactances are considered. Here, the

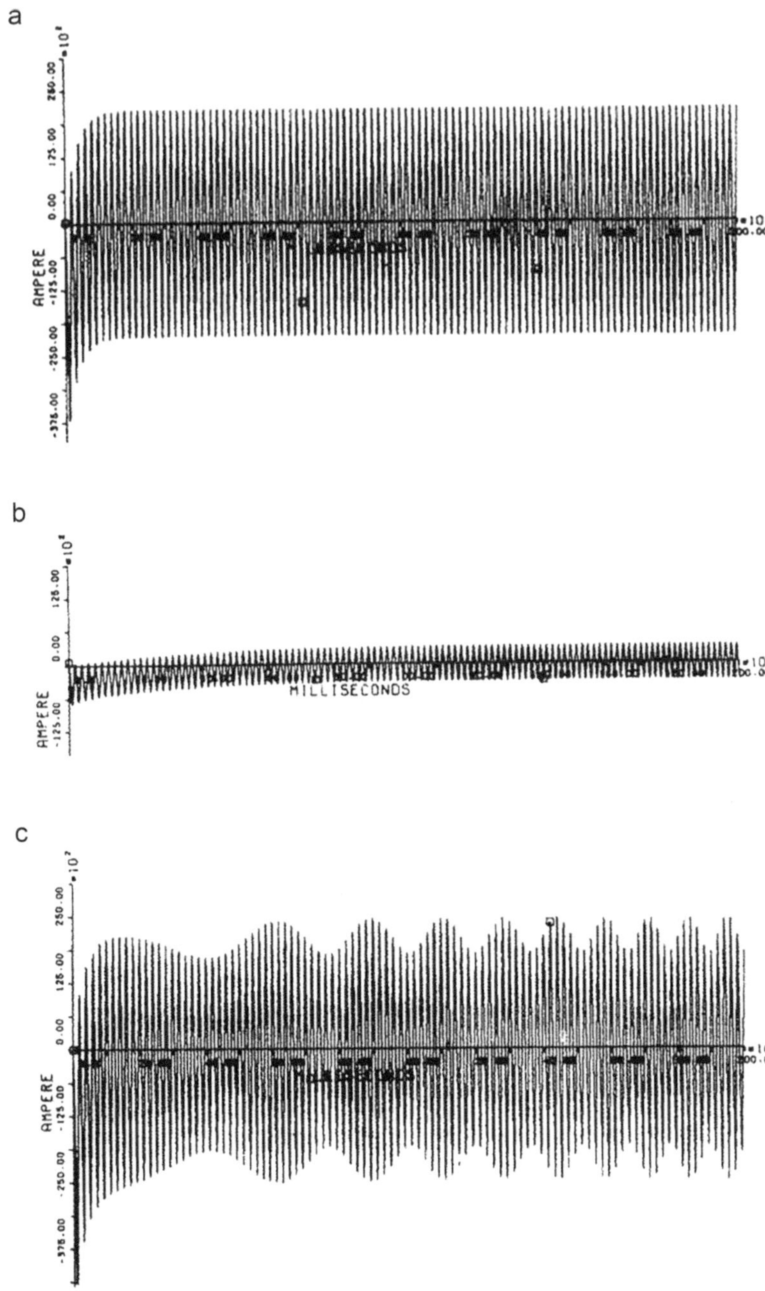

Fig. 5.10 Three-phase short-circuit in the 380 kV network according to Fig. 5.9, short-circuit location F1 [13]. (**a**) $i_{kN}(t)$ network current; (**b**) $i_{kGS}(t)$ generator current of the power station unit, (**c**) $i_k(t)$ total current

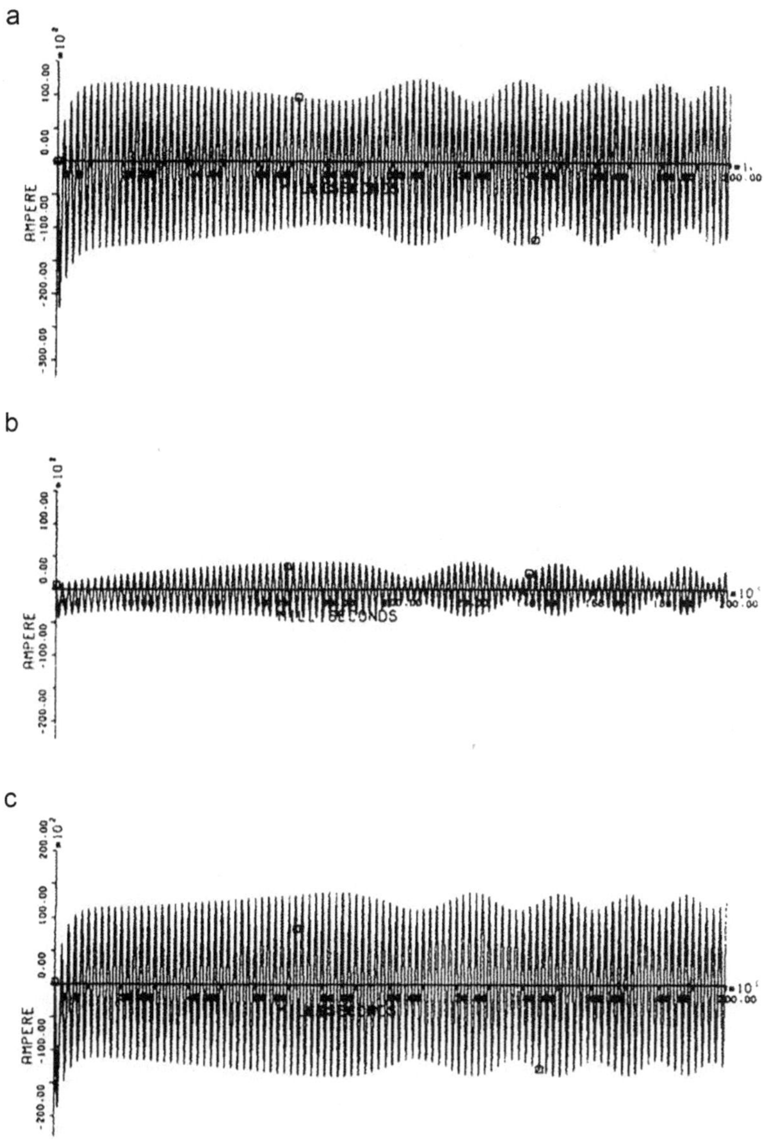

Fig. 5.11 Three-phase short-circuit in the 380 kV network according to Fig. 5.9, short-circuit location F2 [13]. (**a**) $i_{kN}(t)$ network current, (**b**) $i_{kGS}(t)$ generator current of the power stationunit, (**c**) $i_k(t)$ total current

reactance of the power station unit X_{PSK} is determined taking into account the synchronous reactance of the synchronous generator.

$$i_{kN}(\mathrm{t}) = \frac{\sqrt{2} \cdot cU_n/\sqrt{3}}{X_N} \cdot \sin(\omega t) \tag{5.67}$$

$$i_{kPS}(t) = \frac{\sqrt{2} \cdot cU_n/\sqrt{3}}{X_{PSK}} \cdot \sin\{\omega t + \Delta\omega(t) \cdot t\} \tag{5.68}$$

The following applies to the total current:

$$i_k(t) = \frac{\sqrt{2} \cdot cU_n/\sqrt{3}}{X_{PSK} \cdot X_N} \cdot [X_{PSK} \cdot \sin \omega t + X_N \cdot \sin\{\omega t + \Delta\omega(t) \cdot t\}] \tag{5.69}$$

The minima and maxima of the beat frequencies are determined by

$$i_{kmax} = \frac{X_N + X_{PSK}}{X_N \cdot X_{PSK}} \cdot \sqrt{2} \cdot cU_n/\sqrt{3} \tag{5.70}$$

$$i_{kmin} = \frac{X_N - X_{PSK}}{X_N \cdot X_{PSK}} \cdot \sqrt{2} \cdot cU_n/\sqrt{3} \tag{5.71}$$

It can be seen that the total current according to Eq. (5.70) is determined from the superposition of two components with different frequencies. The addition can be expressed by a sinusoidal function to which a beat frequency f_0 is superimposed.

$$|f_0| = \frac{|\omega\text{-}\omega\text{-}\Delta\omega|}{2\pi} = \frac{\Delta\omega}{2\pi} \tag{5.72}$$

The determination of the maximum according to Eq. (5.70) corresponds to the result according to Eq. (5.66) if the steady state short-circuit current is determined taking into account a multiple single-fed according to Sect. 5.7.2.

Based on the current curves shown, it can be seen that determining the steady state short-circuit current for near-to-generator short circuits in a meshed network using the factors λ does not lead to a sufficiently accurate result if generators or power station units are generally operated in parallel with a network. In addition to the described effect of the transient process, an additional error is to be expected in the calculation of the steady state short-circuit current, which occurs due to the neglect of the coupling by a common impedance, similar to the processes according to Sect. 5.4, when determining the short-circuit breaking current. This can be seen from the current curve in Fig. 5.10 if the short-circuit location is after a common impedance. For this reason, the steady state short-circuit current I_k in meshed networks must be calculated using the initial symmetrical short-circuit current, but without motors. A second short-circuit current calculation must therefore be carried out in networks with motor feeders.

A calculation of the steady state short-circuit current in meshed networks is determined according to Eqs. (5.73) and (5.74):

$$\text{maximum} : I_{\text{kmax}} = I''_{\text{kmax}} \tag{5.73}$$

$$\text{minimum} : I_{\text{kmin}} = I''_{\text{kmin}} \tag{5.74}$$

In this case, the maximum initial symmetrical short-circuit current I''_{kmax} is determined without taking motors into account, as their contribution to the steady state short-circuit current decays very quickly. This means that the initial symmetrical short circuit current must be calculated twice. In contrast, the connected motors are not taken into account when determining the minimum symmetrical short-circuit current for all short-circuit types.

5.7.4 Contribution of Asynchronous Machines

According to [1, 14], asynchronous machines do not contribute to the steady state short-circuit current in the event of a three-phase short-circuit at the motor terminals, as the excitation is lost as a result of the voltage drop at the motor terminals, in contrast to unbalanced short-circuits, Sect. 12.4.

Similar to the consideration in Sect. 5.6.2, the contribution of asynchronous machines is also a consequence of the voltage drop at the motor terminals in the event of a three-phase steady state short circuit in the network. According to [15], the steady state short-circuit current in the event of a network fault therefore depends on the residual voltage U_{Rest}, so that the following applies:

$$I_{\text{kM}} \approx \frac{U_{\text{Rest}}}{U_{\text{n}}} \cdot I''_{\text{kM}} \tag{5.75}$$

With

U_{n} nominal system voltage,
I''_{kM} initial symmetrical short-circuit current of the asynchronous motor in the event of a network short circuit.

Due to the low absolute contribution to the total short-circuit current, the motor contribution can be neglected in the case of a three-phase short circuit according to [1].

In the case of a line-to-line short circuit, residual voltages occur at the terminals of the motor so that, in contrast to a three-phase short circuit, there is a contribution to the short-circuit current, Sect. 5.9.

5.8 Thermal Equivalent Short-Circuit Current

The calculation of the thermally equivalent short-circuit current is necessary for the thermal design of equipment, as is the case for the dimensioning of cables, for example. The derivation of the factors m and n according to [6] is described in detail in Chap. 11. The basic idea for determining the thermally equivalent short-circuit current is to convert each short-circuit current transient curve, e.g. near-to-generator with DC displacement or decaying short-circuit current contribution of a synchronous machine, into a current curve that consists exclusively of a constant AC component. The thermal effect of this current curve then corresponds to the thermal effect of the actual short-circuit current during the specified short-circuit duration.

5.8.1 *Calculation Method*

Equation (5.76) applies to the calculation of the current I_{th}. This means that the initial symmetrical short-circuit current is taken as the initial value using the factors m and n, which correspond to the heat effect of the DC and AC components.

$$I_{\mathrm{th}} = I_{\mathrm{k}}'' \cdot \sqrt{m+n} \tag{5.76}$$

or

$$I_{\mathrm{th}}^2 = \frac{1}{T_{\mathrm{k}}} \int_0^{T_{\mathrm{K}}} i^2(t)\mathrm{d}t = \left(I_{\mathrm{k}}''\right)^2 \cdot (m+n) \tag{5.77}$$

With

I_{k}'' initial symmetrical short-circuit current,
m heat effect of the DC component,
n heat effect of the AC component.

The factors m and n can be taken from Eqs. (5.78) to (5.80).

$$m = \frac{e^{4 \cdot f \cdot T_{\mathrm{k}} \cdot \ln(\kappa - 1)} - 1}{2 \cdot f \cdot T_{\mathrm{k}} \cdot \ln(\kappa - 1)} \tag{5.78}$$

When determining the n-curves, a distinction must be made between two different current ratios $I_{\mathrm{k}}''/I_{\mathrm{k}}$, as the detailed Eq. (5.82) does not provide the physically reasonable curve according to Eq. (5.81).

$$\text{For } \frac{I''_k}{I_k} = 1 \quad \Rightarrow \quad n = 1 \tag{5.79}$$

$$\text{For } \frac{I''_k}{I_k} \geq 1,25$$

$$n = \frac{1}{\left(I''_k/I_k\right)^2} \cdot \left\{ \begin{aligned} &1 + \left(\frac{I''_k}{I_k} - \frac{I'_k}{I_k}\right)^2 \cdot \frac{T'_d}{20 \cdot T_k} \cdot \left(1 - e^{-20 \cdot T_k/T'_d}\right) + \left(\frac{I'_k}{I_k} - 1\right)^2 \cdot \frac{T'_d}{2 \cdot T_k} \cdot \left(1 - e^{-2 \cdot T_k/T'_d}\right) \\ &+ \left(\frac{I''_k}{I_k} - \frac{I'_k}{I_k}\right) \cdot \frac{T'_d}{5 \cdot T_k} \cdot \left(1 - e^{-10 \cdot T_k/T'_d}\right) + \left(\frac{I'_k}{I_k} - 1\right) \cdot 2 \cdot \frac{T'_d}{T_k} \cdot \left(1 - e^{-T_k/T'_d}\right) \\ &+ \left(\frac{I''_k}{I_k} - \frac{I'_k}{I_k}\right) \cdot \left(\frac{I'_k}{I_k} - 1\right) \cdot \frac{T'_d}{5.5 \cdot T_k} \cdot \left(1 - e^{-11 \cdot T_k/T'_d}\right) \end{aligned} \right\} \tag{5.80}$$

With

$$\frac{I'_k}{I_k} = \frac{I''_k/I_k}{0.88 + 0.17 \cdot I''_k/I_k} \qquad T'_d \approx \frac{3.1\,\text{s}}{I'_k/I_k} \tag{5.81}$$

Equation (5.82) applies to the thermal short-circuit strength of current conductors in the event that the short-circuit duration deviates from the rated value $T_{kr} = 1$ s. A value of $\eta = 1$ is generally used for this calculation, which means that no thermal energy is radiated during the short-circuit duration T_k, so that the results are on the safe side.

$$S_{th} = S_{thr} \cdot \sqrt{\frac{T_{kr}}{T_k} \cdot \frac{1}{\eta}} \tag{5.82}$$

With

- S_{thr} rated short-time current density,
- S_{th} short-time current density in case of a short-circuit time T_k,
- T_{kr} rated short-circuit duration (= 1 s),
- η factor to take into account the heat dissipation during the short-circuit duration, adiabatic: $\eta = 1$.

The use of the thermally equivalent short-circuit current is important when the required cable cross-section or the protection time must be determined depending on the short-circuit current curve over time.

5.8.2 Example

In the following, the required cross-section of a cable is determined under the following boundary conditions:

$$I_k'' = 40\ \text{kA} \quad \kappa = 1.8 \quad I_k''/I_k = 2 \quad T_k = 0.5\ \text{s}$$

The permissible short-term current density S_{thr} of a XLPE cable at an operating temperature of $\vartheta_B = 90\ °C$ and a short-circuit end temperature of $\theta_B = 250\ °C$ is for a short-circuit duration $T_{kr} = 1$ s:

$$S_{thr} = 143 \frac{\text{A}}{\text{mm}^2} \quad \text{Cu conductor}$$

$$S_{thr} = 94 \frac{\text{A}}{\text{mm}^2} \quad \text{Al conductor}$$

From the short-circuit duration $T_k = 0.5$ s and the factor $\kappa = 1.8$, the factor m is determined to be $m = 0.09$ according to Eq. (5.78). A value of $n = 0.712$ results for the factor n, Eq. (5.80), if a decay of the short-circuit AC component or a short-circuit current ratio of $I_k''/I_k = 2$ at a short-circuit duration of $T_k = 0.5$ s is taken into account. The current I_{th} is then calculated under these conditions as follows:

$$I_{th} = 40\,\text{kA} \cdot \sqrt{0.09 + 0.712} = 35.82\,\text{kA}$$

The short-circuit duration T_k and the rated values are used to determine the short-time current densities S_{th} at $\eta = 1$ for the various materials:

$$S_{th} = 143\ \frac{\text{A}}{\text{mm}^2} \cdot \sqrt{\frac{1}{0.5}} = 202.2\ \frac{\text{A}}{\text{mm}^2} \quad \text{(Cu)}$$

$$S_{th} = 94\ \frac{\text{A}}{\text{mm}^2} \cdot \sqrt{\frac{1}{0.5}} = 132.9\ \frac{\text{A}}{\text{mm}^2} \quad \text{(Al)}$$

This results in the required cable cross-section A:

$$A = \frac{I_{th}}{S_{th}} = \frac{35.82\ \text{kA}}{202.2\ \text{A}}\ \text{mm}^2 = 177.2\ \text{mm}^2 \quad \text{(Cu)}$$

$$A = \frac{I_{th}}{S_{th}} = \frac{35.82\ \text{kA}}{132.9\ \text{A}}\ \text{mm}^2 = 269.5\ \text{mm}^2 \quad \text{(Al)}$$

A reduction in the cable cross-sections can be achieved by shortening the tripping time of the selective protection. The cross-sections determined must always be compared with the cross-section dimensioning based on the load caused by the operating current. The larger of the two cross-sections is then decisive when selecting the cable cross-sections.

5.9 Unbalanced Short Circuits

While only the symmetrical, three-phase short-circuit current is considered in Sect. 5.2 to 5.8, Table 5.3 shows the calculation rules for the currents for line-to-earth and line-to-line short circuits in addition to Sect. 3.3.

When determining the peak factor κ, the value for the three phase short-circuit current can always be used, which applies in any case for the short circuit far from the generator. While in the case of a short-circuit near to the generator, the AC component in the negative sequence system always decays differently from the positive sequence system. This does not have a significant effect in the time range $t < 10$ ms. This assumes that the *R*/*X* ratio of the zero sequence system is greater than that of the positive sequence system, which is likely to be the case in practice, so that in principle $\kappa_{1\mathrm{p}} < \kappa_{3\mathrm{p}}$ applies. In these cases, the sum of the series connection of positive, negative and zero sequence impedance always leads to an *R*/*X* ratio that is greater than the ratio of the positive sequence system and therefore the result is on the safe side.

When calculating the breaking current for unbalanced short-circuits, e.g. for a line-to-line short circuit, the change in the generator reactance and thus the time constant during a short circuit only has an effect in the positive sequence system. This means that the reactance of a synchronous generator in the negative sequence system can be regarded as an additional fault reactance X_{N} according to Eq. (5.43) in the positive sequence system. For this reason, the VDE Specification up to 1971 [11] also had the axis designation for the line-to-line short circuit $\mu = f\left(I''_{\mathrm{k2GS}}/I_{\mathrm{rGS}}\right)$ in the representation of the factor $\mu = f\left(I''_{\mathrm{k3GS}}/I_{\mathrm{rGS}}\right)$, which was higher by a factor of $\sqrt{3}$ for the same μ values. This means that with a current ratio of $I''_{\mathrm{k3GS}}/I_{\mathrm{rGS}} = 5$ and

Table 5.3 Calculation of unbalanced short-circuit currents (peak, breaking and steady state short-circuit current)

Fault type	i_{p}	I_{b}	I_{k}
Line-to-earth	$\kappa\sqrt{2} \cdot I''_{\mathrm{k1}}$	I''_{k1}	I''_{k1}
Line-to-line	$\kappa\sqrt{2} \cdot I''_{\mathrm{k2}}$	I''_{k2}	I''_{k2}
Line-to-line (E)	$\kappa\sqrt{2} \cdot I''_{\mathrm{k2E}}$	I''_{k2E}	I''_{k2E}

i_{p} peak short-circuit current
I_{b} breaking current
I_{k} steady state short-circuit current
κ peak factor for the three phase short circuit
I''_{k1}, I''_{k2}, I''_{k2E} initial symmetrical short-circuit current according to Sect. 3.3

Table 5.4 Initial symmetrical short-circuit currents I''_k and breaking currents I_b in relation to the rated current I_{rGS}, (three phase, line-to-line) as a function of the minimum time delay t_{min}

Current ratio	Minimum time delay/s			
	0.02	0.05	0.10	0.25
i''_{k3}	5.00			
i_{b3}	4.55	4.12	3.83	3.50
i''_{k2}	4.33			
i_{b2} [a]	4.17	4.00	3.92	3.78
i_{b2} [b]	4.33			

[a]Calculation according to VDE 0102 (1971) [11]
[b]Calculation according to IEC 60909–0 (2016) [1]

a minimum time delay of $t_{min} = 0.1$ s, a value of $\mu = 0.765$ results for the three phase short circuit according to Eq. (5.49). This μ value corresponds to a ratio of $I''_{k2GS}/I_{rGS} = 8.66$. According to Sect. 3.3, the following applies to the line-to-line short-circuit current in relation to the three phase short-circuit current:

$$\frac{I''_{k2GS}}{I_{rGS}} = \frac{\sqrt{3}}{2} \cdot \frac{I''_{k3GS}}{I_{rGS}} \quad \Rightarrow \quad \frac{I''_{k2GS}}{I_{rGS}} = \frac{\sqrt{3}}{2} \cdot 5 = 4.33$$

According to [11], this results in a value for the line-to-line short-circuit of $\mu = 0.906$. The breaking currents are thus determined as a function of the fault types:

- Three phase: $\dfrac{I_{b3}}{I_{rGS}} = 0.765 \cdot \dfrac{I''_{k3GS}}{I_{rGS}}$

- Line-to-line: $\dfrac{I_{b2}}{I_{rGS}} = 0.906 \cdot \dfrac{\sqrt{3}}{2} \cdot \dfrac{I''_{k3GS}}{I_{rGS}} = 0.784 \cdot \dfrac{I''_{k3GS}}{I_{rGS}}$

The results show that the μ values for the three phase and line-to-line short circuit are almost identical if this minimum time delay is assumed. From the VDE 0102 Ed. (1990) [5] onwards, the Standard [1] that is still valid today was introduced in accordance with Table 5.3. If the result for the line-to-line breaking current is compared with the initial symmetrical short-circuit current, the following results:

VDE 0102 (1971) [11]: $\dfrac{I_{b2}}{I_{rGS}} = 0.784 \cdot \dfrac{I''_{k3GS}}{I_{rGS}}$

IEC 60909-0 (2016) [1]: $\dfrac{I_{b2}}{I_{rGS}} = \dfrac{\sqrt{3}}{2} \cdot \dfrac{I''_{k3GS}}{I_{rGS}} = 0.866 \cdot \dfrac{I''_{k3GS}}{I_{rGS}}$

The assumption that, according to Table 5.3, the breaking current is equal to the initial symmetrical short-circuit current is therefore on the safe side. However, this leads to a situation where the line-to-line short circuit leads to a larger current I_b than the three phase short circuit, which has also been determined for larger minimum time delays ($t_{min} > 0.1$ s) according to the measurements in [15]. The reason for this is that the line-to-line short-circuit currents are lower than for a three phase short-circuit due to the smaller armature reaction in synchronous machines. This means that the line-to-line short circuit is decisive for the dimensioning of the switching devices and the thermal stability of the systems.

Table 5.4 shows the breaking currents for different minimum time delays t_{min} for the three phase short-circuit current ratio i''_{k3} and the line-to-line short circuit derived from this as a function of the different calculation methods according to IEC.

If it is assumed that the calculation methods using the μ curves for the three phase and line-to-line short circuit correspond to the actual course, the following remarks can be derived:

- For minimum time delays $t_{min} \geq 0.1$ s, the line-to-line breaking current is greater than the three phase current [$i_{b2}{}^{a)} > i_{b3}$].
- The estimation $i_{b2}{}^{b)}$ is always on the safe side, in this case with approx. 10.5% at $t_{min} \geq 0.1$ s.

The behavior of asynchronous motors with unbalanced short circuits is dealt with separately in Sect. 12.4.

5.10 Influence of the Fictitious R_f/X Ratio

According to Sect. 5.4.1, a distinction is made between a real and a fictitious *R*/*X* ratio or resistance *R*, which is used differently during the short-circuit current calculation. The physical background is explained in more detail below. According to Eq. (5.7), the following relationship generally results for an *R*/*X* series circuit for the time course of the short-circuit current (short-circuit occurrence at voltage zero crossing):

$$i_k(t) = \frac{\sqrt{2} \cdot U}{X \cdot \sqrt{1 + (R/X)^2}} \cdot \left[\frac{e^{-t/T_{DC}}}{\sqrt{1 + (R/X)^2}} - \cos(\omega t + \alpha) \right] \tag{5.83}$$

With the DC component

$$T_{DC} = \frac{X}{\omega \cdot R} = \frac{X}{2\pi \cdot f \cdot R} \qquad \alpha = \arctan(R/X) \tag{5.84}$$

With

U active voltage (rms value),
R, *X* impedance of the short-circuit loop $\underline{Z}_k$,
f operating frequency (ω angular frequency).

The short-circuit current according to Eq. (5.83) consists of two time-dependent components.

- AC component:

$$i_{\mathrm{kAC}}(t) = -\frac{\sqrt{2} \cdot U}{X \cdot \sqrt{1 + (R/X)^2}} \cdot \cos(\omega t + \alpha) \tag{5.85}$$

- DC component:

$$i_{\mathrm{kDC}}(t) = \frac{\sqrt{2} \cdot U}{X \cdot \sqrt{1 + (R/X)^2}} \cdot \frac{\mathrm{e}^{-t/T_{\mathrm{DC}}}}{\sqrt{1 + (R/X)^2}} \tag{5.86}$$

A comparison of the two components shows that for constant values of the short-circuit impedance, the magnitude of the AC component is constant, while the DC component changes due to the DC time constant T_{DC}. In contrast, the components for a synchronous generator are determined according to Eqs. (4.39) and (4.40) for $X''_{\mathrm{d}} = X''_{\mathrm{q}}$:

- AC component:

$$i_{\mathrm{kAC}}(t) = \sqrt{\frac{2}{3}} \cdot U_{\mathrm{rGS}} \cdot \left\{ \left[\frac{1}{X''_{\mathrm{d}}} - \frac{1}{X'_{\mathrm{d}}}\right] \cdot \mathrm{e}^{-t/T''_{\mathrm{d}}} + \left[\frac{1}{X'_{\mathrm{d}}} - \frac{1}{X_{\mathrm{d}}}\right] \cdot \mathrm{e}^{-t/T'_{\mathrm{d}}} + \frac{1}{X_{\mathrm{d}}} \right\} \cdot \cos(\omega t) \tag{5.87}$$

- DC component:

$$i_{\mathrm{kDC}}(t) \approx \sqrt{\frac{2}{3}} \cdot U_{\mathrm{rGS}} \cdot \left(-\frac{1}{X''_{\mathrm{d}}} \cdot \mathrm{e}^{-t/T_{\mathrm{a}}}\right) \tag{5.88}$$

With

U_{rGS} voltage of the generator (rms value, line-to-line),
$X''_{\mathrm{d}}, T''_{\mathrm{d}}$ subtransient reactance/time constant,
$X'_{\mathrm{d}}, T'_{\mathrm{d}}$ transient reactance/ time constant,
X_{d} synchronous reactance,
T_{a} DC time constant according to Eq. (4.28).

While the DC component is derived from the constant values of the *R*/*X* ratio of the synchronous generator and is therefore comparable with the relationship according to Eq. (5.85), the AC component of the generator changes due to the different time constants.

According to Sect. 3.3, different short-circuit currents are determined as a function of time, which are generally made up of an AC and DC component. Since time constants are not taken into account when calculating the short-circuit current according to [1], e.g. for generators, but various factors (μ, q, λ) take into consideration the decay of the AC component, this process should also be considered when

Table 5.5 Consideration of the AC and DC components for different types of short circuit (short circuit near to the generator) as a function of the short-circuit duration

Type	t/ms	Factor	
		I_{AC}	I_{DC}
Initial symmetrical short-circuit current	0	–	–
Peak short-circuit current	≤10	κ	
Breaking current[a]	20–250	μ, q	–
Steady state short-circuit current	> 250	λ	–

I_{AC} AC component, I_{DC} DC component
[a]Symmetrical breaking current

calculating the peak short-circuit current. This means that both the decay of the AC and DC components must be taken into account when determining the peak short-circuit current. The DC component in the faulty conductor only occurs if the short-circuit occurs at a voltage that deviates from the maximum voltage.

If there is a short-circuit near to the generator, the individual components change as a function of the short-circuit time; the relationships are shown in Table 5.5. The factors μ, q and λ only have an influence on the AC component, as only this can be determined.

It makes sense to specify a fictitious R/X ratio, with $(R_f/X) > R/X$, since both the AC and DC components decay, when determining the peak short-circuit current in the event of a fault near to the generator. In this case, the decay of the DC component is artificially increased by increasing the effective resistance R of the short-circuit loop, with a constant AC component. This means that the factor κ must represent the reduction of both components in this case.

The consequence is that two different **Y** matrices must be set up if a peak short-circuit current i_p is calculated for short-circuits near to the generator, according to Eq. (5.2).

$$i_p = \sqrt{2} \cdot \kappa \cdot I''_k \tag{5.89}$$

With

κ: peak factor, calculation with the fictitious resistance R_f,
I''_k initial symmetrical short-circuit current, calculation with the real resistance R.

The following shows the error that can occur if the fictitious resistance R_f is also used to calculate the initial symmetrical short-circuit current for simplification purposes. The following applies in general, according to Eq. (5.1):

$$I''_k = \frac{c \cdot U_n/\sqrt{3}}{Z_k} = \frac{c \cdot U_n/\sqrt{3}}{X \cdot \sqrt{1 + (R/X)^2}} \tag{5.90}$$

Set for the real effective resistance R:

$$R = g \cdot R_{\mathrm{f}} \qquad g \leq 1 \tag{5.91}$$

Thus, the deviation (index A) of the initial symmetrical short-circuit current is determined by using the fictitious resistance (index f):

$$\frac{I''_{\mathrm{kf}}}{I''_{\mathrm{k}}} = i_{\mathrm{kA}} = \sqrt{\frac{1 + g^2 \cdot (R_{\mathrm{f}}/X)^2}{1 + (R_{\mathrm{f}}/X)^2}} \tag{5.92}$$

Figure 5.12 shows the error that occurs when the initial symmetrical short-circuit current is calculated with the fictitious effective resistance R_{f} and not with the real value R. This is illustrated by the ratio i_{kf} according to Eq. (5.92). If a maximum deviation to the unsafe side of 5% is assumed, the ratio R_{f}/X must not exceed the value of 0.33, if the following applies for the real effective resistance: $R \sim 0.0\ \Omega$.

According to Sect. 4.2.1.1, a fictitious resistance of $R_{\mathrm{GSf}} = 0.05 \cdot X''_{\mathrm{d}}$ can be assumed for a synchronous generator with a power of $S_{\mathrm{rGS}} \geq 100$ MVA and a voltage $U_{\mathrm{rGS}} > 1$ kV. If this value is also used to estimate the initial symmetrical short-circuit current, the maximum deviation is 0.13%. With a resistance ratio of 0.07 or 0.15, the incorrect calculation increases to 0.24 or 1.11% under these conditions. Since the stator resistance has a finite value in any case, the actual error will be lower in all cases. This is the reason why the IEC 909 (1988) Standard

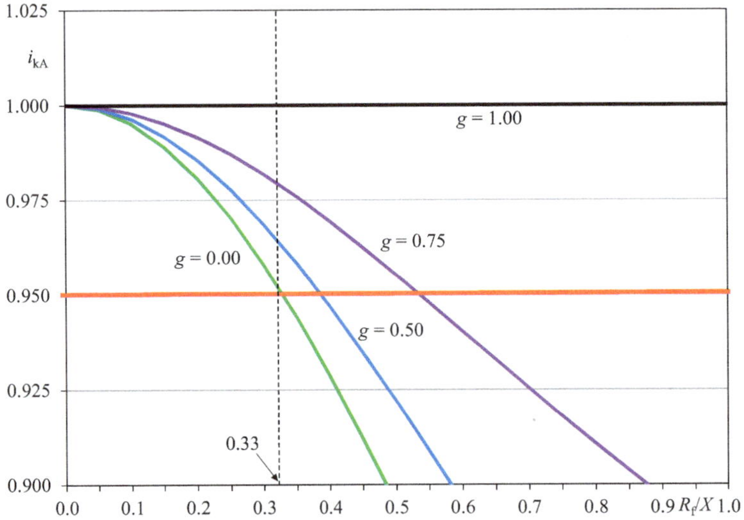

Fig. 5.12 Deviation i_{kA} in the calculation of the initial symmetrical short-circuit current g ratio of real to fictitious effective resistance Eq. (5.91)

refers to the difference between the real effective resistance (stator resistance) and the fictitious value, but all calculations could be carried out with the value R_f.

If, based on Eqs (5.89) and (5.92), the error is determined that occurs when using the real R/X ratio instead of the fictitious value when calculating the peak short-circuit current, the deviation (index A) is calculated according to Eq. (5.93). The index "f" makes it clear that the fictitious values are used in each case.

$$\frac{i_p}{i_{pf}} = i_{pA} = \frac{\kappa}{\kappa_f \cdot \left(I''_{kf}/I''_k\right)} = \frac{1.02 + 0.98 \cdot e^{-3 \cdot g \cdot R/X}}{\left(1.02 + 0.98 \cdot e^{-3 \cdot R/X}\right) \cdot i_{kA}} \tag{5.93}$$

Figure 5.13 shows the deviation i_{pA} according to Eq. (5.93) if the decay of the AC component is not taken into account when calculating the peak short-circuit current. With a fictitious ratio of $R_f/X = 0.15$ and a low stator resistance (corresponding to $g \sim 0$), the peak short-circuit current is determined to be approx. 23% too high, which could possibly lead to considerable oversizing of equipment.

This means that the fictitious ratio R_f/X should always be used when calculating the peak short-circuit current; when determining the initial symmetrical short-circuit current, the same ratio can be used for simplification in the case of synchronous generators.

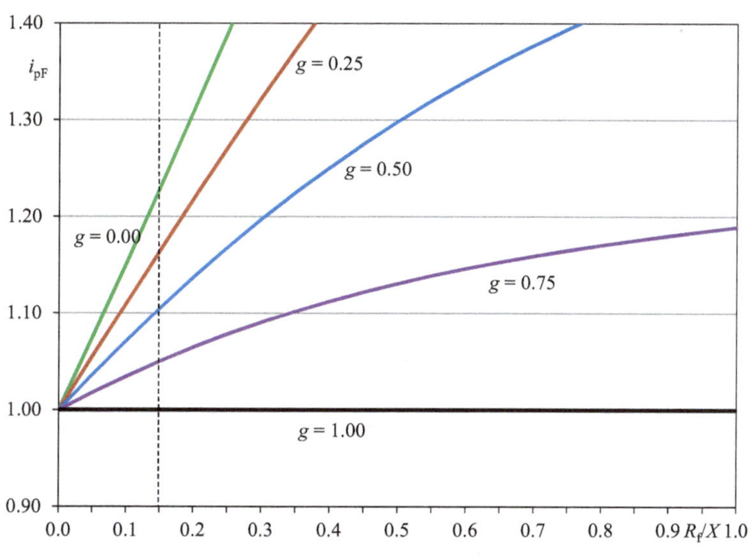

Fig. 5.13 Deviation i_{pA} in the calculation of the peak short-circuit current. g ratio of real to fictitious effective resistance Eq. (5.91)

5.11 Conclusions

This section lists the essential principles for calculating the characteristic short-circuit currents. In particular, the peak short-circuit current and the breaking current are considered, which are decisive for the mechanical strength of systems and the breaking capacity of the switching devices.

While the impedances of the short-circuit loop are responsible for calculating the peak short-circuit current, "maximum" envelopes from measurements and calculations are used to determine the breaking current. It is therefore likely that actual generator and motor data will deviate from these envelopes. As a rule, however, the μ and q curves should be an estimate that is on the safe side.

In addition, reference is made to the application and physical background of the fictitious ratio R_f/X in the calculation of peak short-circuit currents close to the generator.

References

1. IEC 60909-0:2016 Edition 2.0, 2016-01-28 (2016) Short-circuit current calculations in three-phase a.c. systems—Part 0: Calculation of currents. IEC, Geneva
2. IEC 60038: 2009-06-17 (2009) Standard voltages. IEC, Geneva
3. IEC 60865-1, Ed. 3.0, 2011-10 (2011) Short-circuit currents—Calculation of effects—Part 1: Definitions and calculation methods. IEC, Geneva
4. Koglin H-J (1971) Der abklingende Gleichstrom beim Kurzschluß in Energiever sorgungsnetzen. Dissertation TH Darmstadt, D-17
5. IEC 909, 1988 (1988) Short-circuit current calculations in three-phase a.c. systems—Part 0: Calculation of currents. IEC, Geneva
6. Balzer G, Deter O (1985) Berechnung der thermischen Kurzschlußbeanspruchung von Starkstromanlagen mithilfe der Faktoren *m* und *n* nach DIN VDE 0103/2.82. etz-Archiv 7, pp 287–290
7. Tsanakas D (1976) Beitrag zur Berechnung der elektromagnetischen Kurzschlusskräfte und der dynamischen Beanspruchung von Schaltanlagen. Dissertation TH Darmstadt, D 17
8. IEC TR 60909-1:2002 (2002) Short-circuit currents in three-phase a.c. systems, Factors for the calculation of short-circuit currents according to IEC 60909-0. IEC, Geneva
9. Ott G, Webs A (1971) Beitrag von Hochspannungs-Asynchronmotoren zum Kurzschlußstrom bei dreipoligem Kurzschluß. ETZ-Report 6. VDE, Berlin
10. VDE 0102, Teil 1/9.62 (1962) Leitsätze für die Berechnung der Kurzschlussströme. Teil 1 Drehstromanlagen mit Nennspannungen von 1 kV und darüber. VDE, Berlin
11. VDE 0102 Teil 1/11.71 (1971) Leitsätze für die Berechnung der Kurzschlussströme – Teil 1 Drehstromanlagen mit Nennspannungen über 1 kV. VDE, Berlin
12. Hosemann G, Balzer G (1984) Der Ausschaltwechselstrom bei dreipoligem Kurzschluß im vermaschten Netz. etzArchiv 6, pp 51–56
13. Balzer G, Nelles D, Tuttas C (2009) Kurzschlussstromberechnung nach IEC und DIN EN 60909–0 (VDE 0102):2002–07. VDE, Berlin
14. Funk G (1962) Der Kurzschluß im Drehstromnetz. R. Oldenbourg, München
15. Webs A (1972) Einfluss von Asynchronmotoren auf die Kurzschlussstromstärken in Drehstromanlagen. VDE Fachberichte 27:86–92

Chapter 6
Double Line-to-Earth Short Circuit

Until IEC Standard 909: 1988 [1], the double line-to-earth short circuit was part of the main standard. However, a revision of the entire text led to the content being considerably enlarged, so that it seemed advisable to provide a separate part for this. This was achieved by publishing the new standard [2].

6.1 General

The double line-to-earth short circuit (formerly: double earth fault) is a fault that occurs at the same time in two different lines of the three-phase system, Fig. 3.5. The renaming of the word earth fault → earth short circuit is because a short-circuit current flows although the system has an isolated or resonant earthed neutral point treatment. While an earth fault in an isolated system has a maximum current of 100 A at the fault location, values of >1 kA occur in case a short circuit.

The double line-to-earth short circuit often occurs in these systems (isolated, resonant earthed) because the second fault to earth is a consequence of the first. In the event of an earth fault, the conductors not affected by the fault take on $\sqrt{3}$ times the voltage equal to the line-to-earth voltage, so it is possible that a flashover occurs because of contamination of an insulator, which leads to a second earth fault. The result is that this fault scenario expands into a double line-to-earth short circuit and a short-circuit current starts to flow.

In low-impedance earthed systems, the double line-to-earth short circuit is not considered, as the first fault causes a high short-circuit current and is therefore detected and disconnected by the system protection. In addition, the voltage increase is lower so that the second fault location is less likely. In these cases, the line-to-line short circuit with earth contact is determined, whereby the fault is at the same location, Fig. 3.4c.

G. Balzer, *Short-Circuit Currents in Three-Phase Systems*,
https://doi.org/10.1007/978-3-658-50710-7_6

In principle, the double line-to-earth short-circuit currents can be determined with the aid of the symmetrical components, whereby the following examples are generally derived as a function of the feeders, in each case for the initial symmetrical short-circuit current:

- Single-fed line, Sect. 6.2,
- Two single-fed lines, Sect. 6.3,
- Double-fed line, Sect. 6.4.

6.2 Single-Fed Line

Figure 6.1 shows the network diagram for a double line-to-earth short circuit fed through a network. It is assumed that the neutral point of the transformer on the low-voltage side is not earthed, as the short-circuit current is determined when two earth faults occur in an isolated or resonant earthed system.

The short circuit is between the conductors R (fault location A) and S (fault location B) to earth. The initial symmetrical short-circuit current of the network is assumed to be infinite, so that the internal impedance of the network is zero. The currents I_{k1RA} and I_{k1SB} can be determined using the symmetrical components (Chap. 9).

The short-circuit conditions for fault location A result from a line-to-earth fault in accordance with Sect. 9.2.2.4 at component level [3]:

$$\underline{I}_{(0)\mathrm{A}} = \underline{I}_{(1)\mathrm{A}} = \underline{I}_{(2)\mathrm{A}} = \underline{I}_{\mathrm{RA}}/3 \tag{6.1}$$

$$\underline{U}_{(0)\mathrm{A}} + \underline{U}_{(1)\mathrm{A}} + \underline{U}_{(2)\mathrm{A}} = 0 \tag{6.2}$$

This means that the components at short-circuit location A are connected directly in series as shown in Fig. 9.10. For short-circuit location B, the fault conditions in the original system are as follows:

$$\underline{U}_{\mathrm{SB}} = 0 \quad \underline{I}_{\mathrm{RB}} = \underline{I}_{\mathrm{TB}} = 0 \tag{6.3}$$

Equations (9.25) and (9.42) apply to the transformation matrices for determining the component voltages and currents, resulting in the following:

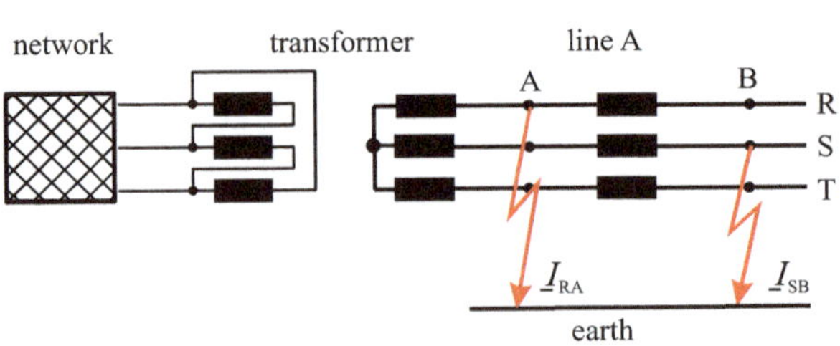

Fig. 6.1 Network circuit, single-fed line

$$\begin{pmatrix} \underline{U}_{(0)\mathrm{B}} \\ \underline{U}_{(1)\mathrm{B}} \\ \underline{U}_{(2)\mathrm{B}} \end{pmatrix} = \frac{1}{3} \cdot \begin{pmatrix} 1 & 1 & 1 \\ 1 & \underline{a} & \underline{a}^2 \\ 1 & \underline{a}^2 & \underline{a} \end{pmatrix} \cdot \begin{pmatrix} \underline{U}_{\mathrm{RB}} \\ \underline{U}_{\mathrm{SB}} \\ \underline{U}_{\mathrm{TB}} \end{pmatrix} \tag{6.4}$$

After inserting the following results:

$$\underline{U}_{(0)\mathrm{B}} + \underline{U}_{(1)\mathrm{B}} + \underline{U}_{(2)\mathrm{B}} = 0 \tag{6.5}$$

The following results from the current conditions according to Eq. (6.3):

$$\begin{pmatrix} \underline{I}_{(0)\mathrm{B}} \\ \underline{I}_{(1)\mathrm{B}} \\ \underline{I}_{(2)\mathrm{B}} \end{pmatrix} = \frac{1}{3} \cdot \begin{pmatrix} 1 & 1 & 1 \\ 1 & \underline{a} & \underline{a}^2 \\ 1 & \underline{a}^2 & \underline{a} \end{pmatrix} \cdot \begin{pmatrix} \underline{I}_{\mathrm{RB}} \\ \underline{I}_{\mathrm{SB}} \\ \underline{I}_{\mathrm{TB}} \end{pmatrix} = \frac{1}{3} \cdot \begin{pmatrix} 1 \\ \underline{a} \\ \underline{a}^2 \end{pmatrix} \cdot (\underline{I}_{\mathrm{SB}}) \tag{6.6}$$

Resp.

$$\underline{I}_{(1)\mathrm{B}} = \underline{a} \cdot \underline{I}_{(0)\mathrm{B}} \quad \underline{I}_{(2)\mathrm{B}} = \underline{a}^2 \cdot \underline{I}_{(0)\mathrm{B}} \tag{6.7}$$

From Eq. (6.7) it can be seen that with this fault, which is not symmetrical to the conductor R, the components are not connected directly, but through transformers with the transformation ratios a and $\underline{a}^2$, in contrast to the representation according to Eq. (6.1).

Figure 6.2 shows the equivalent circuit of the components for the considered short-circuit case of the double line-to-earth short circuit in the conductors R and S to earth. The following equations are introduced for simplification:

$$\begin{aligned} \underline{I}_{(1)\mathrm{A}} &= \underline{I}_{\mathrm{A}} + \underline{a} \cdot \underline{I}_{(0)\mathrm{B}} = (1 - \underline{a}) \cdot \underline{I}_{\mathrm{A}} \\ \underline{I}_{(2)\mathrm{A}} &= \underline{I}_{\mathrm{A}} + \underline{a}^2 \cdot \underline{I}_{(0)\mathrm{B}} = \left(1 - \underline{a}^2\right) \cdot \underline{I}_{\mathrm{A}} \\ 0 &= \underline{I}_{\mathrm{A}} + \underline{I}_{(0)\mathrm{B}} \end{aligned} \tag{6.8}$$

Based on the current conditions according to Eq. (6.8) and the equivalent circuit of the components according to Fig. 6.2, the following system of Eq. (6.9) can be set up. While the fault is symmetrical to the conductor R at fault location A, this is not the case at the fault location B, so that transformers must be used with the transformation ratios according to Eq. (6.7). For simplification, the current $\underline{I}_{(0)\mathrm{B}} = \underline{I}_{\mathrm{B}}$ is selected, which is therefore the current in the component system at the fault location B.

The zero sequence impedance of the transformer and the upstream network $\underline{Z}_{(0)\mathrm{A}}$ is not shown in Fig. 6.2, as it is ineffective due to the unearthed transformer neutral point.

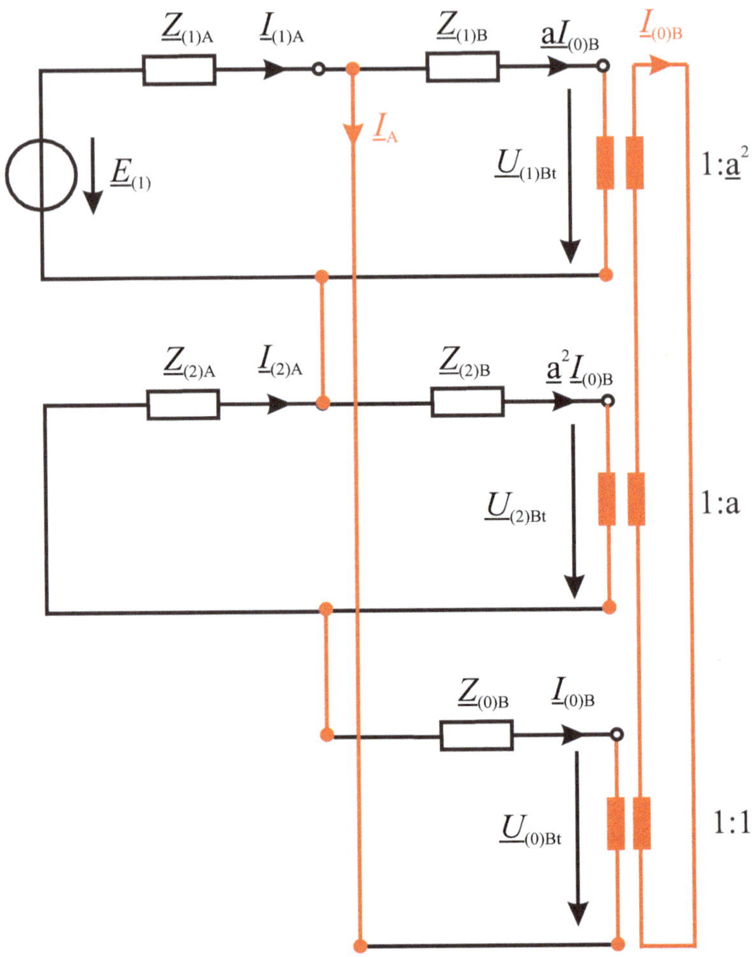

Fig. 6.2 Component equivalent circuit of a double line-to-earth short circuit, single-fed line

$$\begin{pmatrix} \underline{E}_{(1)} \\ \underline{E}_{(1)} \\ 0 \\ 0 \\ 0 \end{pmatrix} = \begin{pmatrix} \left(\underline{Z}_{(1)\mathrm{A}}+\underline{Z}_{(2)\mathrm{A}}\right) & \left(\underline{\mathrm{a}}\cdot\underline{Z}_{(1)\mathrm{A}}+\underline{\mathrm{a}}^2\cdot\left(\underline{Z}_{(2)\mathrm{A}}-\underline{Z}_{(0)\mathrm{B}}\right)\right) & 0 & 0 & -1 \\ \underline{Z}_{(1)\mathrm{A}} & \underline{\mathrm{a}}\cdot\left(\underline{Z}_{(1)\mathrm{A}}+\underline{Z}_{(1)\mathrm{B}}\right) & 1 & 0 & 0 \\ 1 & 1 & 0 & 0 & 0 \\ \underline{Z}_{(2)\mathrm{A}} & \underline{\mathrm{a}}^2\cdot\left(\underline{Z}_{(2)\mathrm{A}}+\underline{Z}_{(2)\mathrm{B}}\right) & 0 & 1 & 0 \\ 0 & 0 & \underline{\mathrm{a}}^2 & \underline{\mathrm{a}} & 1 \end{pmatrix} \cdot \begin{pmatrix} \underline{I}_{\mathrm{A}} \\ \underline{I}_{\mathrm{B}} \\ \underline{U}_{(1)\mathrm{Bt}} \\ \underline{U}_{(2)\mathrm{Bt}} \\ \underline{U}_{(0)\mathrm{Bt}} \end{pmatrix} \tag{6.9}$$

Since the positive sequence system $\underline{Z}_{(1)}$ is equal to the negative sequence system $\underline{Z}_{(2)}$, the simplification according to Eq. (6.10) results together with the 3rd equation.

$$\begin{pmatrix} \underline{E}_{(1)} \\ \underline{E}_{(1)} \\ 0 \\ 0 \end{pmatrix} = \begin{pmatrix} \left(3 \cdot \underline{Z}_{(1)\mathrm{A}} + \underline{Z}_{(0)\mathrm{B}}\right) & 0 & 0 & -1 \\ \left(\underline{Z}_{(1)\mathrm{A}} \cdot (1 - \underline{\mathrm{a}}) - \underline{Z}_{(1)\mathrm{B}} \cdot \underline{\mathrm{a}}\right) & 1 & 0 & 0 \\ \left(\underline{Z}_{(1)\mathrm{A}} \cdot (1 - \underline{\mathrm{a}}^2) - \underline{Z}_{(1)\mathrm{B}} \cdot \underline{\mathrm{a}}^2\right) & 0 & 1 & 0 \\ 0 & \underline{\mathrm{a}}^2 & \underline{\mathrm{a}} & 1 \end{pmatrix} \cdot \begin{pmatrix} \underline{I}_{\mathrm{A}} \\ \underline{U}_{(1)\mathrm{Bt}} \\ \underline{U}_{(2)\mathrm{Bt}} \\ \underline{U}_{(0)\mathrm{Bt}} \end{pmatrix} \tag{6.10}$$

By summarizing the equations, the component voltages can be eliminated so that the current $\underline{I}_{\mathrm{A}}$ can be determined.

$$\underline{I}_{\mathrm{A}} = \frac{\sqrt{3} \cdot \underline{E}_{(1)}}{6\underline{Z}_{(1)\mathrm{A}} + 2\underline{Z}_{(1)\mathrm{B}} + \underline{Z}_{(0)\mathrm{B}}} \tag{6.11}$$

According to Eq. (6.1), the short-circuit current in the original system results from three times the value of the component current, so that the following applies:

$$\underline{I}_{\mathrm{RA}} = \frac{3\sqrt{3} \cdot \underline{E}_{(1)}}{6\underline{Z}_{(1)\mathrm{A}} + \underline{Z}_{(0)\mathrm{B}} + 2\underline{Z}_{(1)\mathrm{B}}} \tag{6.12}$$

According to Eq. (6.8), the short-circuit current at fault location A also corresponds to the current at fault location B, taking into account the sign. Since the nominal system voltage U_{n} and the voltage factor c are also taken into account according to [2], the following applies to the double line-to-earth short-circuit current.

$$\underline{I}''_{\mathrm{kEE}} = \frac{3 \cdot cU_{\mathrm{n}}}{6\underline{Z}_{(1)\mathrm{A}} + \underline{Z}_{(0)\mathrm{B}} + 2\underline{Z}_{(1)\mathrm{B}}} \tag{6.13}$$

With

- $E_{(1)}$ voltage of the positive sequence system $(U_{\mathrm{n}}/\sqrt{3})$,
- $\underline{Z}_{(1)\mathrm{A}}$ positive sequence impedance of the network and transformer (up to the short-circuit location A),
- $\underline{Z}_{(1)\mathrm{B}}$ positive sequence impedance of the line between the short-circuit locations A and B,
- $\underline{Z}_{(0)\mathrm{B}}$ neutral sequence impedance of the line between the short-circuit locations A and B.

The short-circuit currents via the network feeder or the transformer correspond to the short-circuit currents in the faulty conductors to earth. If the two short-circuit locations in Fig. 6.1 coincide, the short-circuit current of a line-to-line short circuit without earth contact, Eq. (3.4), results according to Eq. (6.13).

In the following, the ratio of the short-circuit currents is calculated according to Eq. (3.4), line-to-line short circuit without earth contact and Eq. (6.13). The amounts according to Eq. (6.14) are used as a reference.

$$i_k = \frac{|\underline{I}''_{k2}|}{|\underline{I}''_{kEE}|} = \frac{|2\underline{Z}_{(1)A}|}{|2\underline{Z}_{(1)A} + 2\underline{Z}_{(1)B}/3 + \underline{Z}_{(0)B}/3|} \tag{6.14}$$

The following data are assumed for the calculation, the impedance $\underline{Z}_B$ corresponds to a 110 kV overhead line (5 km), the length of which is varied in 5 km steps, according to Table 4.2:

$$R'_{(1)} = 0.12\,\Omega/\text{km} \qquad X'_{(1)} = 0.393\,\Omega/\text{km}$$
$$R'_{(0)} = 0.34\,\Omega/\text{km} \qquad X'_{(0)} = 1.62\,\Omega/\text{km}$$

The value of the feeding impedance $\underline{Z}_A$ corresponds to the line length of $\ell = 5$ km (case 1) or $\ell = 10$ km (case 2) with a ratio of $R_{(1)A}/X_{(1)A} = 0.1$. In the first case, the impedance is equivalent to a three-phase initial symmetrical short-circuit current $I''_{k3} = 33,73$ kA of a 110 kV feeder.

According to Fig. 6.3, with a ratio $Z_{(1)B}/Z_{(1)A} = 0$, the double line-to-earth short-circuit currents are equal to the line-to-line short-circuit current with $i_k = 1$, since in

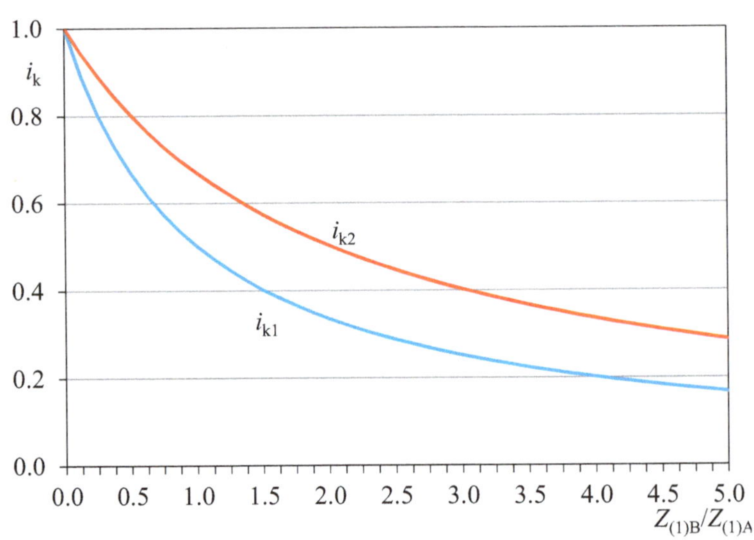

Fig. 6.3 Ratio i_k of the double line-to-earth short-circuit current in relation to the line-to-line short-circuit current as a function of the impedance ratio $Z_{(1)B}/Z_{(1)A}$. i_{k1}: $I''_{k3} = 33,73$ kA for $Z_{(1)B} = Z_{(0)B} = 0$, i_{k2}: $I''_{k3} = 16,87$ kA for $Z_{(1)B} = Z_{(0)B} = 0$

this case the two short-circuit nodes A and B according to Fig. 6.1 of the double line-to-earth short circuit are identical. With an additional impedance $Z_{(1)B}$, the short-circuit current becomes smaller, depending on the impedance of the feeder, so that after 10 km, for example, the short-circuit current is already reduced to approx. 50% of the current of a line-to-line short circuit on the busbar.

6.3 Two Single-Fed Lines

In contrast to the illustration in Fig. 6.1, in this case, the two short-circuit nodes are located on different lines that start from a common busbar as shown in Fig. 6.4. Since this example also involves a double line-to-earth short circuit, the same fault conditions apply in the original and component system, so that the component equivalent circuit can be derived as shown in Fig. 6.5.

Figure 6.5 shows the equivalent circuit of the components for the considered fault case of the double line-to-earth short circuit in the conductors R and S to earth. The following equations are introduced for simplification:

$$\begin{aligned} \underline{I}_{(1)T} &= \underline{I}_A + \underline{a} \cdot \underline{I}_{(0)B} \\ \underline{I}_{(2)T} &= \underline{I}_A + \underline{a}^2 \cdot \underline{I}_{(0)B} \\ 0 &= \underline{I}_A + \underline{I}_{(0)B} \end{aligned} \tag{6.15}$$

To simplify matters, the current $\underline{I}_{(0)B}$ is also referred to as $\underline{I}_B$ and $\underline{Z}_{(1)} = \underline{Z}_{(2)}$ in this example, so that the system of Eq. (6.16) can be derived.

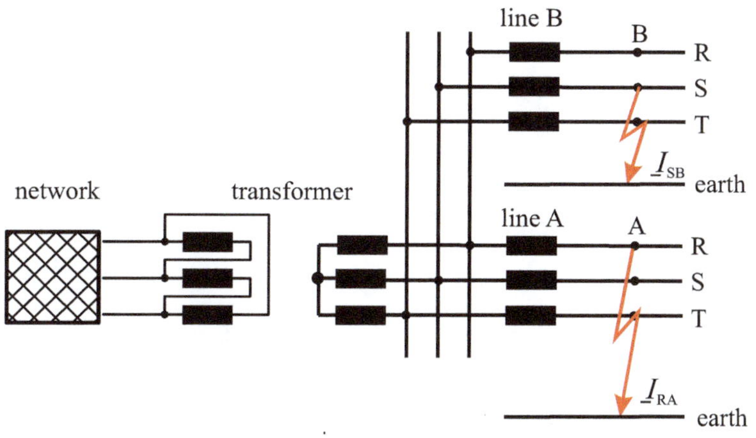

Fig. 6.4 Network circuit diagram, two single-fed lines

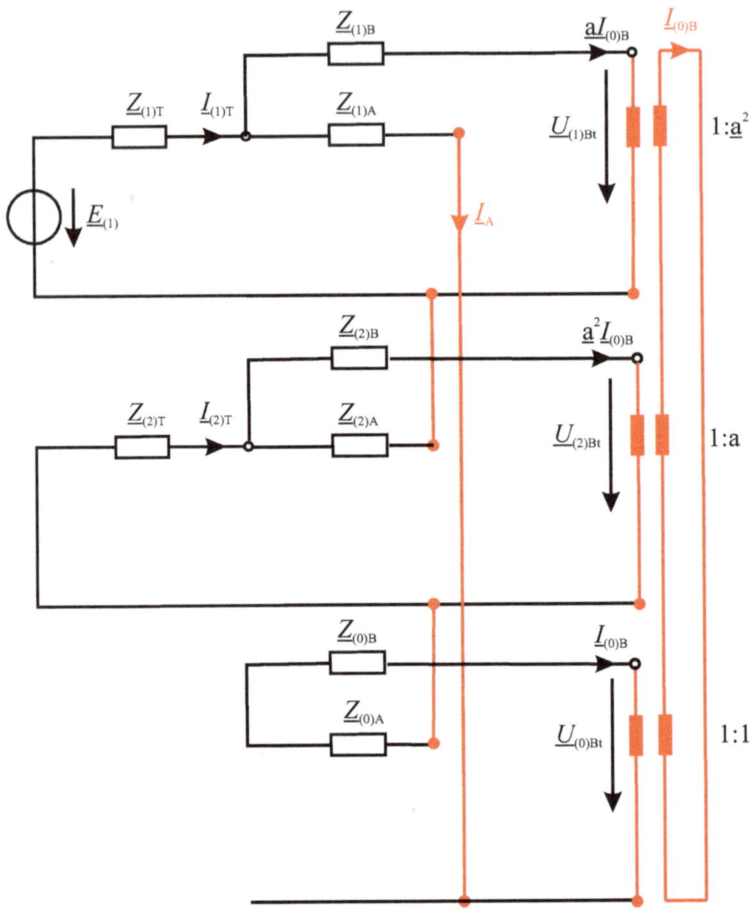

Fig. 6.5 Component equivalent circuit of a double line-to-earth short circuit, two single-fed lines

$$\begin{pmatrix} \underline{E}_{(1)} \\ \underline{E}_{(1)} \\ 0 \\ 0 \\ 0 \end{pmatrix} = \begin{pmatrix} \left(2\underline{Z}_{(1)T} + 2\underline{Z}_{(1)A} + \underline{Z}_{(0)A} + \underline{Z}_{(0)B}\right) & -\underline{Z}_{(1)T} & 0 & 0 & -1 \\ \underline{Z}_{(1)T} & \underline{a} \cdot \left(\underline{Z}_{(1)T} + \underline{Z}_{(1)B}\right) & 1 & 0 & 0 \\ 1 & 1 & 0 & 0 & 0 \\ \underline{Z}_{(1)T} & \underline{a}^2 \cdot \left(\underline{Z}_{(1)T} + \underline{Z}_{(1)B}\right) & 0 & 1 & 0 \\ 0 & 0 & \underline{a}^2 & \underline{a} & 1 \end{pmatrix} \cdot \begin{pmatrix} \underline{I}_A \\ \underline{I}_B \\ \underline{U}_{(1)Bt} \\ \underline{U}_{(2)Bt} \\ \underline{U}_{(0)Bt} \end{pmatrix} \tag{6.16}$$

After inserting the 3rd line of the system of Eq. (6.16), the calculation is simplified according to Eq. (6.17).

$$\begin{pmatrix} \underline{E}_{(1)} \\ \underline{E}_{(1)} \\ 0 \\ 0 \end{pmatrix} = \begin{pmatrix} \left(3\underline{Z}_{(1)\mathrm{T}}+2\underline{Z}_{(1)\mathrm{A}}+\underline{Z}_{(0)\mathrm{A}}+\underline{Z}_{(0)\mathrm{B}}\right) & 0 & 0 & -1 \\ \left(\underline{Z}_{(1)\mathrm{T}}-\underline{a}\cdot\left(\underline{Z}_{(1)\mathrm{T}}+\underline{Z}_{(1)\mathrm{B}}\right)\right) & 1 & 0 & 0 \\ \left(\underline{Z}_{(1)\mathrm{T}}-\underline{a}^2\cdot\left(\underline{Z}_{(1)\mathrm{T}}+\underline{Z}_{(1)\mathrm{B}}\right)\right) & 0 & 1 & 0 \\ 0 & \underline{a}^2 & \underline{a} & 1 \end{pmatrix} \cdot \begin{pmatrix} \underline{I}_{\mathrm{A}} \\ \underline{U}_{(1)\mathrm{Bt}} \\ \underline{U}_{(2)\mathrm{Bt}} \\ \underline{U}_{(0)\mathrm{Bt}} \end{pmatrix} \tag{6.17}$$

By replacing the transformer voltages at the fault location B, the component current $\underline{I}_{\mathrm{A}}$ can be determined so that the fault current to earth in the original system can be determined according to Eqs. (6.11) and (6.12).

$$\begin{aligned} \underline{I}_{\mathrm{A}} &= \frac{\sqrt{3}\cdot\underline{E}_{(1)}}{6\underline{Z}_{(1)\mathrm{T}}+2\cdot\left[\underline{Z}_{(1)\mathrm{B}}+\underline{Z}_{(1)\mathrm{B}}\right]+\underline{Z}_{(0)\mathrm{A}}+\underline{Z}_{(0)\mathrm{B}}} \\ \underline{I}''_{\mathrm{kEE}} &= \frac{3\cdot cU_{\mathrm{n}}}{6\underline{Z}_{(1)\mathrm{T}}+2\cdot\left[\underline{Z}_{(1)\mathrm{B}}+\underline{Z}_{(1)\mathrm{B}}\right]+\underline{Z}_{(0)\mathrm{A}}+\underline{Z}_{(0)\mathrm{B}}} \end{aligned} \tag{6.18}$$

With

- $\underline{E}_{(1)}$ voltage of the positive sequence system ($U_{\mathrm{n}}/\sqrt{3}$),
- $\underline{Z}_{(1)\mathrm{T}}$ positive sequence impedance of the network and the transformer (up to the busbar on the low-voltage side),
- $\underline{Z}_{(1,0)\mathrm{A}}$ positive sequence, zero sequence impedance of the line between busbar and fault location A,
- $\underline{Z}_{(1,0)\mathrm{B}}$ positive sequence, zero sequence impedance of the line between busbar and fault location B.

The short-circuit currents via the network feeder or the transformer correspond to the short-circuit currents in the faulty conductors to earth.

6.4 Double-Fed Line

In contrast to the representation in Sect. 6.2, in this case the single line is supplied from two sides. For simplification, it is assumed that the system voltages are the same so that they can be combined in the component representation. Figure 6.6 shows the network circuit, which is supplemented by an additional feeder based on Fig. 6.1.

Figure 6.7 shows the equivalent circuit of the components for the considered fault case of the double line-to-earth short circuit in the conductors R and S to earth.

The following current equations apply for the calculation, whereby the current relationship applies due to the fault: $\underline{I}_{\mathrm{A}} = -\underline{I}_{(0)\mathrm{B}}$.

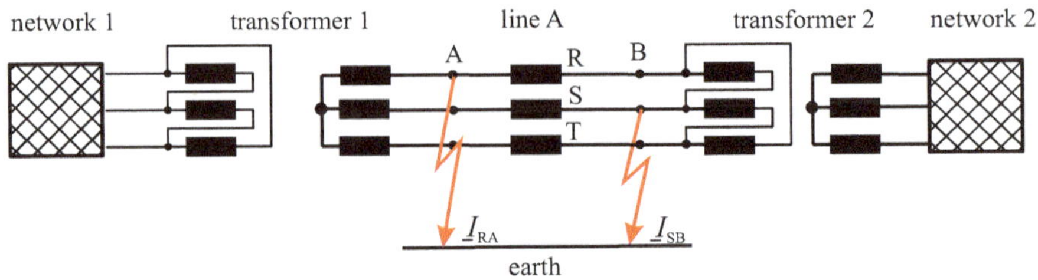

Fig. 6.6 Network circuit diagram, double-fed line

$$\begin{aligned} \underline{I}_{(1)T1} &= \underline{I}_A \cdot (1 - \underline{a}) - \underline{I}_{(1)T2} \\ \underline{I}_{(2)T1} &= \underline{I}_A \cdot \left(1 - \underline{a}^2\right) - \underline{I}_{(2)T2} \\ \underline{I}_{(1)B} &= -\underline{a} \cdot \underline{I}_A - \underline{I}_{(1)T2} \\ \underline{I}_{(2)B} &= -\underline{a}^2 \cdot \underline{I}_A - \underline{I}_{(2)T2} \end{aligned} \tag{6.19}$$

Taking into account the voltage cycles, the system of Eq. (6.20) can be set up, whereby the negative sequence impedance is set equal to the positive sequence impedance. This results in a 6 × 6 matrix with the following unknowns: $\underline{I}_A$, $\underline{I}_{(1)T2}$, $\underline{I}_{(2)T2}$, $\underline{U}_{(1)Bt}$, $\underline{U}_{(2)Bt}$, $\underline{U}_{(0)Bt}$.

$$\begin{pmatrix} \underline{E}_{(1)} \\ \underline{E}_{(1)} \\ 0 \\ \underline{E}_{(1)} \\ 0 \\ 0 \end{pmatrix} = \begin{pmatrix} \left(3\underline{Z}_{(1)T1} + \underline{Z}_{(0)B}\right) & -\underline{Z}_{(1)T1} & -\underline{Z}_{(1)T1} & 0 & 0 & -1 \\ \left(\underline{Z}_{(1)T1} \cdot (1-\underline{a}) - \underline{Z}_{(1)B} \cdot \underline{a}\right) & -\left(\underline{Z}_{(1)T1} + \underline{Z}_{(1)B}\right) & 0 & 1 & 0 & 0 \\ \left(\underline{Z}_{(1)T1} \cdot (1-\underline{a}^2) - \underline{Z}_{(1)B} \cdot \underline{a}^2\right) & 0 & -\left(\underline{Z}_{(1)T1} + \underline{Z}_{(1)B}\right) & 0 & 1 & 0 \\ 0 & \underline{Z}_{(1)T2} & 0 & 1 & 0 & 0 \\ 0 & 0 & \underline{Z}_{(1)T2} & 0 & 1 & 0 \\ 0 & 0 & 0 & \underline{a}^2 & \underline{a} & 1 \end{pmatrix} \cdot \begin{pmatrix} \underline{I}_A \\ \underline{I}_{(1)T2} \\ \underline{I}_{(2)T2} \\ \underline{U}_{(1)Bt} \\ \underline{U}_{(2)Bt} \\ \underline{U}_{(0)Bt} \end{pmatrix} \tag{6.20}$$

By eliminating the component voltages, the currents can be determined as a function of the voltage $\underline{E}_{(1)}$.

$$\begin{pmatrix} (1-\underline{a}^2) \cdot \underline{E}_{(1)} \\ 0 \\ 0 \end{pmatrix} = \begin{pmatrix} \left(3\underline{Z}_{(1)T1} + \underline{Z}_{(0)B}\right) & -\left(\underline{Z}_{(1)T1} + \underline{a}^2 \cdot \underline{Z}_{(1)T2}\right) & -\left(\underline{Z}_{(1)T1} + \underline{a} \cdot \underline{Z}_{(1)T2}\right) \\ \left(\underline{Z}_{(1)T1} \cdot (1-\underline{a}) - \underline{Z}_{(1)B} \cdot \underline{a}\right) & -\left(\underline{Z}_{(1)T1} + \underline{Z}_{(1)B} + \underline{Z}_{(1)T2}\right) & 0 \\ \left(\underline{Z}_{(1)T1} \cdot (1-\underline{a}^2) - \underline{Z}_{(1)B} \cdot \underline{a}^2\right) & 0 & -\left(\underline{Z}_{(1)T1} + \underline{Z}_{(1)B} + \underline{Z}_{(1)T2}\right) \end{pmatrix} \cdot \begin{pmatrix} \underline{I}_A \\ \underline{I}_{(1)T2} \\ \underline{I}_{(2)T2} \end{pmatrix} \tag{6.21}$$

The component currents $\underline{I}_{(1)T2}$ and $\underline{I}_{(2)T2}$ of the transformer feeder T2, Fig. 6.6, can be determined as a function of the fault current $\underline{I}_A$.

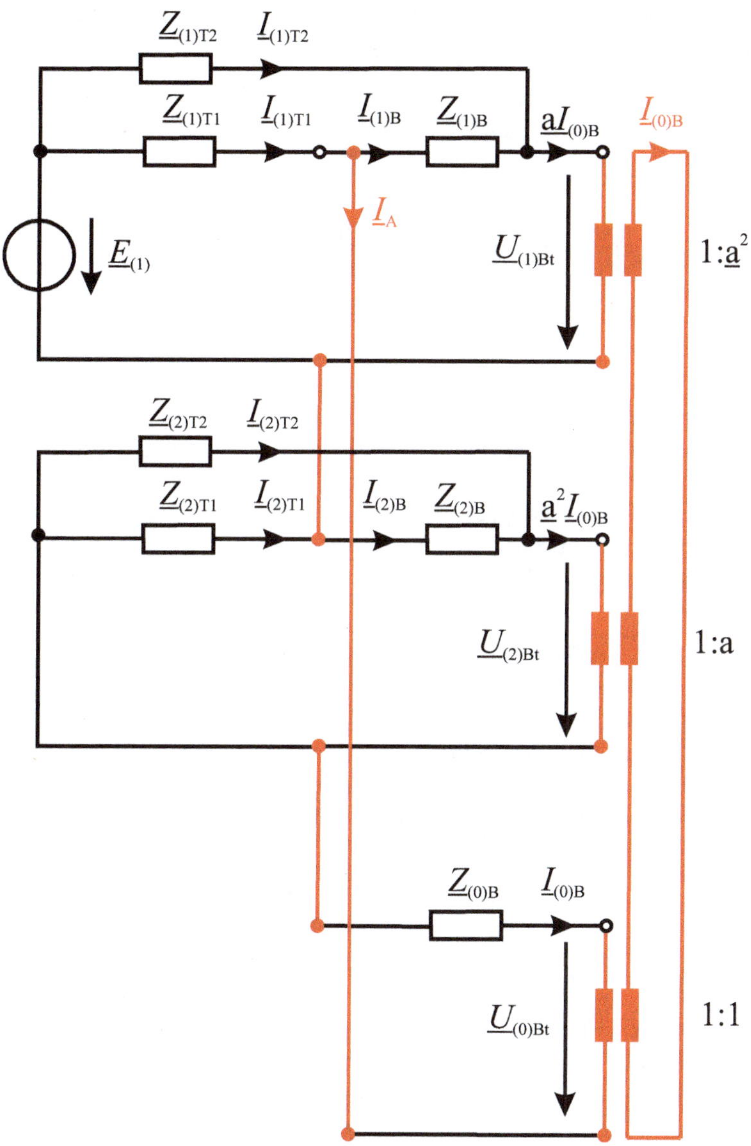

Fig. 6.7 Component equivalent circuit of a double line-to-earth short circuit, double-fed line

$$\underline{I}_{(1)T2} = \frac{\underline{Z}_{(1)T1} \cdot (1 - \underline{a}) - \underline{a} \cdot \underline{Z}_{(1)B}}{\underline{Z}_{(1)T1} + \underline{Z}_{(1)T2} + \underline{Z}_{(1)B}} \cdot \underline{I}_A$$
$$\underline{I}_{(2)T2} = \frac{\underline{Z}_{(1)T1} \cdot (1 - \underline{a}^2) - \underline{a}^2 \cdot \underline{Z}_{(1)B}}{\underline{Z}_{(1)T1} + \underline{Z}_{(1)T2} + \underline{Z}_{(1)B}} \cdot \underline{I}_A \tag{6.22}$$

The following applies to the fault current $\underline{I}_{\mathrm{A}}$ in the component system and to the short-circuit current in the original system:

$$\underline{I}_{\mathrm{A}} = \frac{(1-\underline{a}^2)\cdot E_{(1)}}{\dfrac{6Z_{(1)\mathrm{T1}}\cdot Z_{(1)\mathrm{T2}} + 2Z_{(1)\mathrm{B}}\cdot\left(Z_{(1)\mathrm{T1}} + Z_{(1)\mathrm{T2}}\right)}{Z_{(1)\mathrm{T1}} + Z_{(1)\mathrm{T2}} + Z_{(1)\mathrm{B}}} + Z_{(0)\mathrm{B}}}$$
$$\underline{I}''_{\mathrm{kEE}} = \frac{3\cdot cU_{\mathrm{n}}}{\dfrac{6Z_{(1)\mathrm{T1}}\cdot Z_{(1)\mathrm{T2}} + 2Z_{(1)\mathrm{B}}\cdot\left(Z_{(1)\mathrm{T1}} + Z_{(1)\mathrm{T2}}\right)}{Z_{(1)\mathrm{T1}} + Z_{(1)\mathrm{T2}} + Z_{(1)\mathrm{B}}} + Z_{(0)\mathrm{B}}} \quad (6.23)$$

With

$E_{(1)}$ voltage of the positive sequence system ($U_{\mathrm{n}}/\sqrt{3}$),
$\underline{Z}_{(1)\mathrm{T1}}$ positive sequence impedance of the network 1 and the transformer T1 (up to the fault location A on the low-voltage side),
$\underline{Z}_{(1)\mathrm{T2}}$ positive sequence impedance of the network 2 and the transformer T2 (up to the fault location B on the low-voltage side),
$\underline{Z}_{(1,0)\mathrm{B}}$ positive sequence, zero sequence impedance of the line between busbar and fault location B.

Since, according to Sects. 6.2 and 6.3, the short-circuit current injection during the short-circuit duration clearly results from the fault conditions, the short-circuit currents in this case depend on the impedances of the two network feeders according to Fig. 6.6. Using the transformation equations, Sect. 9.2.2.1, the short-circuit currents of the two feeders in the original system can be determined based on the currents according to Eqs. (6.22) and (6.19). The transformation equations apply:

$$\begin{pmatrix}\underline{I}_{\mathrm{R}}\\ \underline{I}_{\mathrm{S}}\\ \underline{I}_{\mathrm{T}}\end{pmatrix} = \begin{pmatrix}1 & 1 & 1\\ 1 & \underline{a}^2 & \underline{a}\\ 1 & \underline{a} & \underline{a}^2\end{pmatrix}\cdot\begin{pmatrix}\underline{I}_0\\ \underline{I}_1\\ \underline{I}_2\end{pmatrix} \quad (6.24)$$

Since transformers T1 and T2 are not earthed, $\underline{I}_{(0)\mathrm{T1}} = \underline{I}_{(0)\mathrm{T2}} = 0$ for the corresponding zero sequence currents, so that the short-circuit current contribution T2 is generally valid, taking into account Eq. (6.22):

$$\underline{I}_{\mathrm{RT2}} = \underline{I}_{(1)\mathrm{T2}} + \underline{I}_{(2)\mathrm{T2}} = \frac{3\underline{Z}_{(1)\mathrm{T1}} + \underline{Z}_{(1)\mathrm{B}}}{\underline{Z}_{(1)\mathrm{T1}} + \underline{Z}_{(1)\mathrm{T2}} + \underline{Z}_{(1)\mathrm{B}}}\cdot\underline{I}_{\mathrm{A}}$$
$$\underline{I}_{\mathrm{ST2}} = \underline{a}^2\cdot\underline{I}_{(1)\mathrm{T2}} + \underline{a}\cdot\underline{I}_{(2)\mathrm{T2}} = -\frac{3\underline{Z}_{(1)\mathrm{T1}} + 2\underline{Z}_{(1)\mathrm{B}}}{\underline{Z}_{(1)\mathrm{T1}} + \underline{Z}_{(1)\mathrm{T2}} + \underline{Z}_{(1)\mathrm{B}}}\cdot\underline{I}_{\mathrm{A}} \quad (6.25)$$
$$\underline{I}_{\mathrm{TT2}} = \underline{a}\cdot\underline{I}_{(1)\mathrm{T2}} + \underline{a}^2\cdot\underline{I}_{(2)\mathrm{T2}} = \frac{\underline{Z}_{(1)\mathrm{B}}}{\underline{Z}_{(1)\mathrm{T1}} + \underline{Z}_{(1)\mathrm{T2}} + \underline{Z}_{(1)\mathrm{B}}}\cdot\underline{I}_{\mathrm{A}}$$

By inserting the component current $\underline{I}_A$ at the fault location A according to Eq. (6.23), the conductor currents of transformer T2 can be determined.

6.5 Other Short-Circuit Currents

The κ factor for the three-phase short circuit is used to calculate the peak short-circuit current i_{pEE}, so that the following applies:

$$i_{pEE} = \kappa \cdot \sqrt{2} \cdot I''_{kEE} \tag{6.26}$$

The κ factor according to Eq. (6.26) is based on the larger value of the three-phase short-circuit current at fault locations A or B, so that the result is on the safe side.

In the case of a double line-to-earth short circuit, it can generally be assumed that it is a fault far from the generator, so that the following applies for the breaking current $\underline{I}_{bEE}$ and steady state short-circuit current $\underline{I}_{kEE}$:

$$\underline{I}_{kEE} = \underline{I}_{bEE} = \underline{I}''_{kEE} \tag{6.27}$$

6.6 Conclusion

For the frequently occurring network examples, the short-circuit currents are derived for a double line-to-earth short circuit. The initial symmetrical short-circuit current I''_{kEE} always leads to a smaller value than the corresponding line-to-line short-circuit with earth contact. Since edition [1], double line-to-earth short circuit has been published in a separate standard together with the calculation of partial short-circuit currents via earth.

References

1. IEC 909, 1988 (1988) Short-circuit current calculations in three-phase a.c. systems—Part 0: Calculation of currents. IEC, Geneva
2. IEC 60909-3:2009 Edition 3.0 (2009) Short-circuit currents in three-phase a.c. systems—Part 3: Currents during two separate simultaneous line-to-earth short circuits and partial short-circuit currents flowing through earth. IEC, Geneva
3. Funk G (1962) Der Kurzschluß im Drehstromnetz. R. Oldenbourg, München

Chapter 7
Calculation of the Voltage Factor *c*

When using the equivalent voltage source method at the short-circuit location, it is necessary to use a voltage that corresponds to the voltage at the short-circuit location before the short circuit occurred. A load flow calculation is therefore necessary in each case so that this voltage can then be used as a voltage source. The problem is that under these conditions, extensive calculations must first be carried out for each short-circuit case in order to calculate the desired maximum and minimum short-circuit currents. For this reason, considerations have been made in the past to generally estimate the value of this voltage source in order to calculate a maximum or minimum short-circuit current [1].

Not only the load, but also different components have an influence on the magnitude of the voltage drop. While separate impedance correction factors are used for generators, power station units and transformers in the short-circuit current calculation (Chap. 8), the effect of the line as a function of the load is also included in the voltage factor c, so that this influence is shown in detail below (Sect. 7.4).

7.1 Voltage Factor *c*

The use of the equivalent voltage source method at the short-circuit location in accordance with Sect. 3.6.6 always requires a voltage that is to be used depending on the load. As it is not possible to specify exact voltage values in a planning state, the IEC Standards specify corresponding values c for the voltage factor [2].

The voltage factor c is selected for the short-circuit current calculation according to Table 7.1, with c_{max} being the maximum and c_{min} the minimum short-circuit current. The voltage factors specified in Table 7.1 generally apply to all voltages, whereby the product $c_{max} \cdot U_n$ should not generally exceed the value of the highest voltage for equipment U_m. However, it must be pointed out that special considerations must be made for networks with voltages $U_m > 420$ kV and above, e.g. the

G. Balzer, *Short-Circuit Currents in Three-Phase Systems*,
https://doi.org/10.1007/978-3-658-50710-7_7

Table 7.1 Voltage factor c_{max} according to IEC 60909-0 [2] resp. the internal voltage of synchronous generators

Nominal voltage	Factor c
Synchronous generators	Eq. (8.10)
Nominal system voltage ($U_n \leq 1$ kV)	
– Networks with a voltage tolerance of +6%	1,05
– Networks with a voltage tolerance of +10%	1,10
High–/extra high voltage (>1 kV < $U_n \leq 230$ kV) and $U_m \leq 420$ kV	1,10

use of the superposition method, as the voltage factors specified in Table 7.1 should no longer be sufficient due to the long lines that are common there. This means that the user must consider his own values for calculations with these voltages, or the superposition method with selected load flows must be applied.

In summary, there are the following restrictions regarding the application of the voltage factors according to Table 7.1:

- The maximum voltage factor multiplied by the nominal system voltage should not exceed the corresponding highest voltage for equipment U_m,
- if no nominal system voltage is defined, e.g. for voltages $U_m > 300$ kV, the following voltage values are used:
  $c_{max} \cdot U_n = U_m$ resp. $c_{min} \cdot U_n = 0{,}90 \cdot U_m$,
- the voltage factors are not valid for voltages $U_m > 420$ kV.

The values of the voltage factors listed in Table 7.1 and the associated comments assume that at the high-voltage level ($U_n \geq 110$ kV) the steady-state value of the operating voltage U_b does not exceed the highest voltage for equipment U_m or does not exceed it for a longer period of time. To calculate the minimum short-circuit current, a reduced voltage factor c_{min} is taken into account, which is 0.1 p.u. lower than the relevant factor c_{max}. In this case too, the direct, metallic short-circuit at the short-circuit location is assumed, so that arcing resistances, for example, are not taken into account. The value of c_{min} therefore takes into account a lower operating voltage at the time the short circuit occurs.

According to the European Union regulation [EU] for power generation facilities [3]; voltages of e.g. $U_{max} = 1.1 \cdot 400$ kV are required for a minimum of 20 min and a maximum of 60 min, or according to the EU regulation for HVDC systems [4] voltages for a minimum period of 60 min. Since the EU regulations do not specify the frequency per year, the condition stated in the comments to Table 7.1 that the maximum voltage factor together with the nominal system voltage should not exceed the associated highest voltage for equipment U_m is no longer generally permissible. For this reason, a higher voltage for $1.1 \cdot U_n$ should be used, so that $c_{max} \cdot U_n = U_b$ with U_b highest operating voltage for $U_b > U_m$ applies if these voltage conditions occur in a network area over a longer period of time and frequency.

In general, it should be assumed that operational measures such as synchronous phase shifters, compensating reactors or STATCOM (Static Synchronous Compensator) would ensure that the voltage range $\leq U_m$ is not exceeded. However, the voltage range may be exceeded in exceptional network situations or if the measures listed above are not applied consistently.

7.2 Derivation of the Voltage Factor

In the following, a justification is given for the derivation of the voltage factor using a simple model network. Consisting of a voltage source without internal resistance and an overhead line, the factor c is determined as a function of the overhead line length. It is assumed that the line was unloaded before the short circuit occurred, i.e. the voltage factor c is determined solely on the basis of a comparison of the simulation of a line [5]. If an overhead line is represented by its quadripole equation (Sect. 4.7.1), the following results according to Eq. (7.1):

$$\begin{pmatrix} \underline{U}_1 \\ \underline{I}_1 \end{pmatrix} = \begin{pmatrix} \cosh\left(\underline{\gamma} \cdot \ell\right) & \underline{Z}_\mathrm{W} \cdot \sinh\left(\underline{\gamma} \cdot \ell\right) \\ \underline{Z}_\mathrm{W}^{-1} \cdot \sinh\left(\underline{\gamma} \cdot \ell\right) & \cosh\left(\underline{\gamma} \cdot \ell\right) \end{pmatrix} \cdot \begin{pmatrix} \underline{U}_2 \\ \underline{I}_2 \end{pmatrix} \tag{7.1}$$

With

Z_W surge impedance of the line,
γ propagation constant ($\alpha + j\beta$),
α damping constant,
β phase constant,
ℓ line length.

For a no-loss line, the matrix (7.1) can be simplified as follows:

$$\begin{aligned} \cosh(\alpha \cdot \ell + \mathrm{j}\beta \cdot \ell) &= \cosh(\alpha \cdot \ell) \cdot \cos(\beta \cdot \ell) + \mathrm{j} \sinh(\alpha \cdot \ell) \cdot \sin(\beta \cdot \ell) \\ \sinh(\alpha \cdot \ell + \mathrm{j}\beta \cdot \ell) &= \sinh(\alpha \cdot \ell) \cdot \cos(\beta \cdot \ell) + \mathrm{j} \cosh(\alpha \cdot \ell) \cdot \sin(\beta \cdot \ell) \end{aligned} \tag{7.2}$$

With $\alpha = 0$, Eq. (7.1) simplifies to.

$$\begin{pmatrix} \underline{U}_1 \\ \underline{I}_1 \end{pmatrix} = \begin{pmatrix} \cos(\beta \cdot \ell) & \mathrm{j}Z_\mathrm{W} \cdot \sin(\beta \cdot \ell) \\ \mathrm{j}Z_\mathrm{W}^{-1} \cdot \sin(\beta \cdot \ell) & \cos(\beta \cdot \ell) \end{pmatrix} \cdot \begin{pmatrix} \underline{U}_2 \\ \underline{I}_2 \end{pmatrix} \tag{7.3}$$

The relationship according to Eq. (7.3) results from the equivalent circuit according to Fig. 7.1, i.e. the variables at the beginning of the line are expressed by the variables at the end.

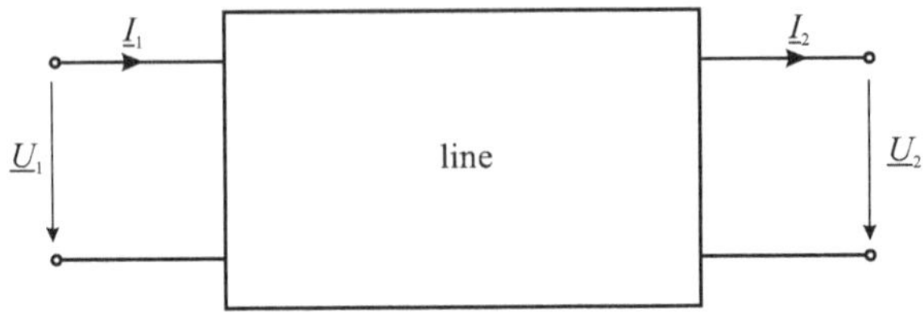

Fig. 7.1 General equivalent circuit of an overhead line

Assuming that there is a short circuit at the end of the line ($\underline{U}_2 = 0$), Eq. (7.3) results:

$$\underline{U}_1 = \mathrm{j}\underline{I}_2 \cdot Z_\mathrm{W} \cdot \sin(\beta \cdot \ell) \tag{7.4}$$

$$\underline{I}_1 = \underline{I}_2 \cdot \cos(\beta \cdot \ell) \tag{7.5}$$

From this follows for the short-circuit impedance $\underline{Z}_{\mathrm{kA}}$ of an overhead line, case A:

$$\frac{\underline{U}_1}{\underline{I}_1} = \mathrm{j}Z_\mathrm{W} \cdot \tan(\beta \cdot \ell) = \underline{Z}_{\mathrm{kA}} \tag{7.6}$$

Since shunt-admittances in the positive sequence system are not taken into account when using the method with the equivalent voltage source at the short-circuit location, case B applies to the short-circuit impedance $\underline{Z}_{\mathrm{kB}}$ (longitudinal impedance):

$$\underline{Z}_{\mathrm{kB}} = \mathrm{j}X' \cdot \ell \tag{7.7}$$

The calculation of the currents using Eq. (7.4) should provide the same result with an input voltage $\underline{U}_1 = \underline{E}$ as the calculation taking into account only the longitudinal reactance according to Eq. (7.7), but in this case with the voltage $c \cdot \underline{E}$. The voltage factor c can be determined from this condition. In this case, the line is unloaded and the voltage drop along the line is compared for the two different line simulations.

$$\frac{c \cdot \underline{E}}{\mathrm{j}X' \cdot \ell} = \frac{\underline{E}}{\mathrm{j}Z_\mathrm{W} \cdot \tan(\beta \cdot \ell)} \tag{7.8}$$

This results in the voltage factor c:

$$c = \frac{X' \cdot \ell}{Z_\mathrm{W} \cdot \tan(\beta \cdot \ell)} = \frac{\beta \cdot \ell}{\tan(\beta \cdot \ell)} \tag{7.9}$$

With

$$Z_\mathrm{W} = \sqrt{L'/C'} \qquad \beta = \omega \cdot \sqrt{L' \cdot C'} \tag{7.10}$$

The following values are assumed for two examples:

- 380 kV overhead line:

$$L'_{\mathrm{OHL}} = 0.799\ \mathrm{mH/km} \qquad C'_{\mathrm{OHL}} = 13.8\ \mathrm{nF/km}$$

- 110 kV cable:

$$L'_{\mathrm{C}} = 0.439\ \mathrm{mH/km} \qquad C'_{\mathrm{C}} = 160\ \mathrm{nF/km}$$

This results in the following values according to Eq. (7.10):

$\beta_{\mathrm{OHL}} = 59.770 \cdot 10^{-3}$ °/km resp. $\beta_{\mathrm{OHL}} = 1.0432 \cdot 10^{-3}$ rad/km
$\beta_{\mathrm{C}} = 150.857 \cdot 10^{-3}$ °/km resp. $\beta_{\mathrm{C}} = 2.6330 \cdot 10^{-3}$ rad/km

After inserting the overhead line and cable values, the following voltage factors c are calculated according to Eq. (7.9):

- Overhead line:

  $\ell = 200\ \mathrm{km} \rightarrow c = 0.985$
  $\ell = 400\ \mathrm{km} \rightarrow c = 0.941$

- Cable:

  $\ell = 50\ \mathrm{km} \rightarrow c = 0.994$

When calculating the short-circuit current, the factor $c = 1.0$ is not exceeded for a long overhead line that is unloaded if the nominal system voltage is not exceeded at the short-circuit location. This reduction in the factor c reflects the influence of the capacitance (Ferranti effect). As cables for three-phase current transmission can only be used up to a length of approx. 50 km without shunt compensation depending on the voltage level, c_{C} only changes in a small range depending on the cable length ($\geq$0.001 p.u.). For long, unloaded cable lengths, the calculation therefore leads to short-circuit currents that are too much on the "safe" side, so that the use of the superposition method is recommended in these cases.

If, on the other hand, the line is loaded with an impedance $\underline{Z}_{\mathrm{B}}$ at the end, the voltages $\underline{U}_1$ are derived from Eq. (7.3).

$$\underline{U}_1 = \underline{U}_2 \cdot \cos(\beta \cdot \ell) + \mathrm{j}Z_{\mathrm{W}} \cdot \underline{I}_2 \cdot \sin(\beta \cdot \ell) \tag{7.11}$$

With

$$\underline{Z}_{\mathrm{B}} = Z_{\mathrm{B}} \cdot (\cos\varphi + \mathrm{j}\sin\varphi) \qquad \underline{I}_2 = \frac{\underline{U}_2}{\underline{Z}_{\mathrm{B}}} \tag{7.12}$$

follows if the current $\underline{I}_2$, Eq. (7.1), is expressed by the voltage $\underline{U}_2$ and the impedance $\underline{Z}_B$ and the voltage factor c is denoted by the ratio U_1/U_2:

$$c = \frac{\underline{U}_1}{\underline{U}_2} = \cos(\beta \cdot \ell) + [(\sin\varphi + \mathrm{j}\cos\varphi) \cdot \sin(\beta \cdot \ell)] \cdot \frac{Z_W}{Z_B} \tag{7.13}$$

With a line length of $\ell = 300$ km, Eq. (7.13) with $\beta_{OHL} = 59.77 \cdot 10^{-3}\,°$ /km and $Z_W = 240.62\ \Omega$ (overhead line), is calculated as follows

$$c = \frac{\underline{U}_1}{\underline{U}_2} = 0.9514 + (\sin\varphi + \mathrm{j}\cos\varphi) \cdot \frac{74.08\,\Omega}{Z_B}$$

The value of the voltage factor c is shown in Fig. 7.2 as a function of the load impedances Z_B for different power factors. Figure 7.2 shows that the voltage increase of the voltage $\underline{U}_1$ depends on the load at the end of the line. With an inductive load, the current $\underline{I}_2$ along the line leads to a voltage drop $\Delta\underline{U}$, which increases the voltage $\underline{U}_1$ so that the voltage factor c assumes values that are higher than $c = 1.1$. In contrast, a capacitive load leads to a reduction in the voltage $\underline{U}_1$.

The diagram in Fig. 7.2 is in principle independent of the voltage, but a load impedance of $Z_B = 1000\ \Omega$ corresponds to a power of $S_B = 144.4$ MVA at a nominal system voltage of $U_n = 380$ kV.

The result is even clearer if the load S_B causes a voltage drop on the transformer with a large leakage reactance. The voltage factor $c = 1.1$ is then exceeded if there is a large inductive load. In order to be on the safe side when determining the maximum

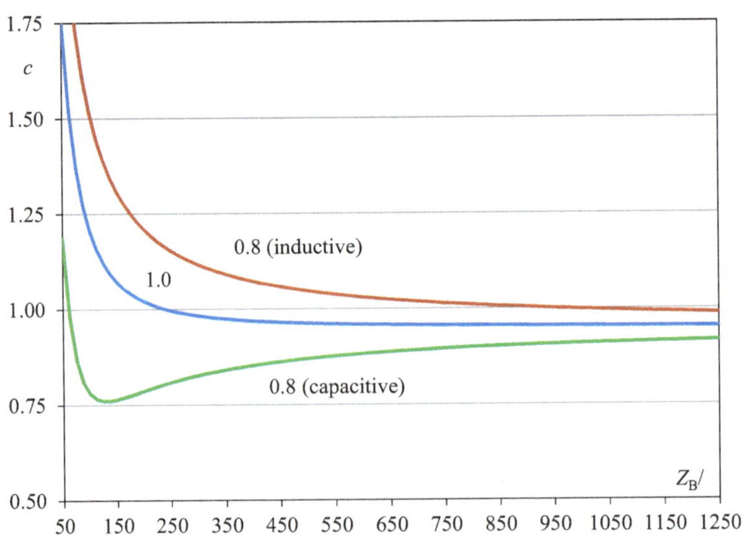

Fig. 7.2 Voltage factor c (p.u.) as a function of the load on an overhead line ($l = 300$ km); $Z_B = 1000\ \Omega$ corresponds to a load of $S_B = 144.4$ MVA at a voltage of $U_n = 380$ kV

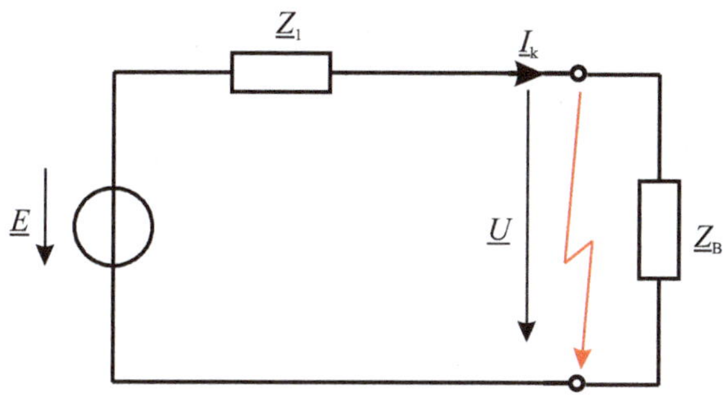

Fig. 7.3 Influence of the voltage factor c on the short-circuit current $\underline{I}_k = \underline{I}_k''$ as a function of the load before the short circuit occurs

short-circuit current, corrections are necessary. Instead of increasing the voltage factors for different source voltages, an impedance correction is introduced in the IEC Standard in order to keep the method with the equivalent voltage source at the short-circuit location with a constant voltage, as shown in Chap. 8.

Using Fig. 7.3, a relationship can be derived between the voltage factor c and the deviation in the simplified calculation of the short-circuit current, as shown in [6]. In this case, the line or the network feeder is modeled exclusively by the impedance $\underline{Z}_1$ in order to show only the influence of the different load on the voltage factor. The voltage $\underline{U}$ at the fault location is calculated as shown in Fig. 7.3:

$$\underline{U} = \frac{\underline{Z}_B}{\underline{Z}_B + \underline{Z}_1} \cdot \underline{E} \quad \underline{E} = \left(1 + \frac{\underline{Z}_1}{\underline{Z}_B}\right) \cdot \underline{U} \tag{7.14}$$

The following then applies to the relative voltage difference $\Delta \underline{u}$:

$$\Delta \underline{u} = \frac{\underline{E} - \underline{U}}{\underline{U}} = \frac{\underline{E}}{\underline{U}} - 1 = \left(1 + \frac{\underline{Z}_1}{\underline{Z}_B}\right) - 1 = \frac{\underline{Z}_1}{\underline{Z}_B} \tag{7.15}$$

The voltage difference is positive if the voltage $\underline{U}$ according to Fig. 7.3 is less than $\underline{E}$, i.e. an inductive load is connected at the network point of the voltage $\underline{U}$.

If the voltage $\underline{U}$ at the fault location is kept constant when calculating the short-circuit current, this results in the initial symmetrical short-circuit current $\underline{I}''_{k0}$ without load (Fig. 7.3, $\underline{Z}_B \rightarrow \infty$):

$$\underline{I}''_{k0} = \frac{\underline{U}}{\underline{Z}_1} \tag{7.16}$$

The following applies to the short-circuit current taking the load into account at constant voltage $\underline{E}$:

$$\underline{I}''_{k1} = \frac{1 + \underline{Z}_1/\underline{Z}_B}{\underline{Z}_1} \cdot \underline{U} = (1 + \underline{Z}_1/\underline{Z}_B) \cdot \underline{I}''_{k0} \tag{7.17}$$

The following applies to the relative short-circuit current deviation $\Delta \underline{i}''_k$:

$$\Delta \underline{i}''_k = \frac{\underline{I}''_{k0} - \underline{I}''_{k1}}{\underline{I}''_{k1}} = \frac{1}{1 + \underline{Z}_1/\underline{Z}_B} - 1 \tag{7.18}$$

With

$\underline{I}''_{k0}$ short-circuit current without load $\underline{Z}_B$ at the end of the line,
$\underline{I}''_{k1}$ short-circuit current with load $\underline{Z}_B$ at the end of the line.

After inserting Eq. (7.15) into (7.18), the following follows:

$$\Delta \underline{i}''_k = \frac{1}{1 + \Delta \underline{u}} - 1 = \frac{-\Delta \underline{u}}{1 + \Delta \underline{u}} \tag{7.19}$$

Equation (7.19) states that the value of the deviation in the calculation of the short-circuit current is always smaller than the relative voltage change at the faulty network node before the fault occurs [7]. This applies if the active voltage $\underline{E}$ of the network is kept constant and, for example, the voltage $\underline{U}$ at the connection point of an inductive load is reduced.

To ensure that the calculation without impedance $\underline{Z}_B$ also leads to the correct result in this simplified equivalent circuit according to Fig. 7.3, $c \cdot \underline{U}$ must be used at the fault location instead of $\underline{U}$, corresponding to the short-circuit current deviation. The following relationship is then obtained from Eqs. (7.16) and (7.17) using Eq. (7.12):

$$\frac{c \cdot \underline{U}}{\underline{Z}_1} = \left(1 + \frac{\underline{Z}_1}{\underline{Z}_B}\right) \cdot \frac{\underline{U}}{\underline{Z}_1} = \frac{\underline{E}}{\underline{Z}_1} \quad \text{resp.} \quad c = \left|\frac{\underline{E}}{\underline{U}}\right| \tag{7.20}$$

The voltage factor c is used in the calculation of the short-circuit current, taking into account the equivalent voltage source at the short-circuit location and thus takes into account the load of the network before the short-circuit occurs.

7.3 Voltage Stability in Case of Load

Based on the example in Sect. 3.6.1.1, the voltage at the load impedance is determined, Fig. 7.4, in order to determine the voltage factor c as a function of the impedance $\underline{Z}_B$ for two feeders. The load impedance $\underline{Z}_B$ not only changes the resulting short-circuit impedance, but also has a significant influence on the voltage stability in the network and thus in turn has an effect on the short-circuit current.

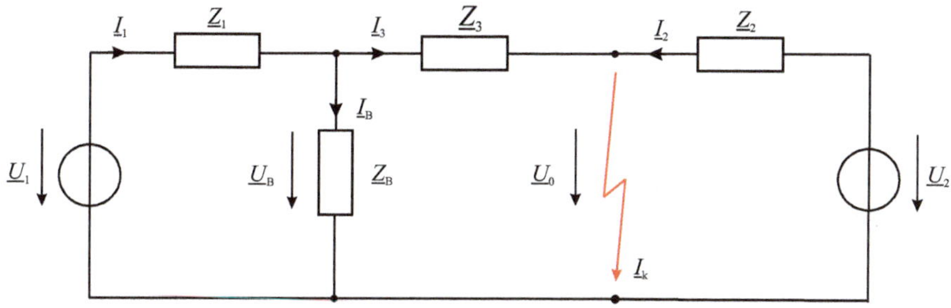

Fig. 7.4 Voltage stability in the network with a load $\underline{Z}_B$ (according to Fig. 3.9)

If the short circuit is not taken into account according to Fig. 7.4, the current and voltage distribution can be determined using the system of Eq. (7.21) (with $\underline{I}_3 = -\underline{I}_2$).

$$\begin{pmatrix} \underline{U}_1 \\ \underline{U}_2 \\ 0 \end{pmatrix} = \begin{pmatrix} \underline{Z}_1 & 0 & \underline{Z}_B \\ 0 & -(\underline{Z}_2 + \underline{Z}_3) & \underline{Z}_B \\ -1 & 1 & 1 \end{pmatrix} \cdot \begin{pmatrix} \underline{I}_1 \\ \underline{I}_3 \\ \underline{I}_B \end{pmatrix} \tag{7.21}$$

The current $\underline{I}_B$ is determined according to Eq. (3.18), so that the determinant **D** and the partial determinant $\mathbf{D}_B$ apply:

$$\mathbf{D} = -(\underline{Z}_2 + \underline{Z}_3) \cdot (\underline{Z}_1 + \underline{Z}_B) - \underline{Z}_1 \cdot \underline{Z}_B \tag{7.22}$$

$$\mathbf{D}_B = -(\underline{Z}_2 + \underline{Z}_3) \cdot \underline{U}_1 - \underline{Z}_1 \cdot \underline{U}_2 \tag{7.23}$$

Thus, the load current $\underline{I}_B$ follows from Eq. (7.24):

$$\underline{I}_B = \frac{(\underline{Z}_2 + \underline{Z}_3) \cdot \underline{U}_1 + \underline{Z}_1 \cdot \underline{U}_2}{(\underline{Z}_1 + \underline{Z}_B) \cdot (\underline{Z}_2 + \underline{Z}_3) + \underline{Z}_1 \cdot \underline{Z}_B} \tag{7.24}$$

Or for the voltage $\underline{U}_B$ across the load:

$$\underline{U}_B = \frac{(\underline{Z}_2 + \underline{Z}_3) \cdot \underline{U}_1 + \underline{Z}_1 \cdot \underline{U}_2}{\left(\frac{\underline{Z}_1}{\underline{Z}_B} + 1\right) \cdot (\underline{Z}_2 + \underline{Z}_3) + \underline{Z}_1} \tag{7.25}$$

If, for the purposes of simplification, it is assumed that the two voltages $\underline{U}_1$ and $\underline{U}_2$ are identical in magnitude and phase, then Eq. (7.25) follows:

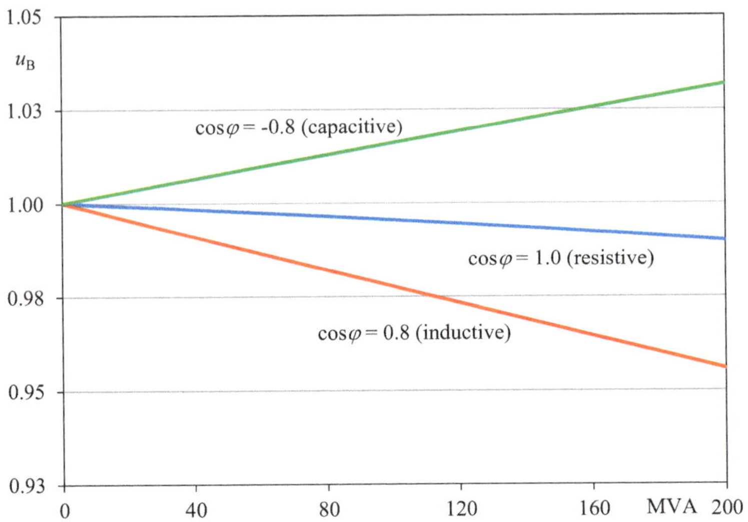

Fig. 7.5 Related load voltage $u_B = U_B/U_1$, Eq. (7.26) (p.u.) as a function of the load at a nominal system voltage of $U_n = 110$ kV

$$\underline{U}_B = \frac{\underline{Z}_1 + \underline{Z}_2 + \underline{Z}_3}{\frac{\underline{Z}_1}{\underline{Z}_B} \cdot (\underline{Z}_2 + \underline{Z}_3) + \underline{Z}_1 + \underline{Z}_2 + \underline{Z}_3} \cdot \underline{U}_1 \tag{7.26}$$

For the voltage $\underline{U}_0$ at the later short circuit location, taking into account Eqs. (7.21) and (7.24), the following results:

$$\underline{U}_0 = \underline{U}_1 \cdot \left\{ 1 - \frac{\underline{Z}_2}{\underline{Z}_2 + \underline{Z}_3} \cdot \left[1 - \frac{\underline{Z}_B \cdot (\underline{Z}_1 + \underline{Z}_2 + \underline{Z}_3)}{(\underline{Z}_1 + \underline{Z}_B) \cdot (\underline{Z}_2 + \underline{Z}_3) + \underline{Z}_1 \cdot \underline{Z}_B} \right] \right\} \tag{7.27}$$

Figure 7.5 shows the related voltage $u_B = U_B/U_1$ for different load impedances. The following values are assumed for the impedances, with a nominal system voltage of $U_n = 110$ kV:

$$\underline{Z}_1 = \underline{Z}_2 = 0.4634\,\Omega + \mathrm{j}4.6342\,\Omega \quad \text{because of} \quad I''_k = 15\,\mathrm{kA} \ \text{and} \ R/X = 0.1$$

$$c \cdot \underline{U}_1/\sqrt{3} = 1.1 \cdot 110\,\mathrm{kV}/\sqrt{3} \qquad \underline{Z}_3 = 6.00\,\Omega + \mathrm{j}19.65\,\Omega$$

The impedances $\underline{Z}_1$ and $\underline{Z}_2$ each correspond to a power station feeder with an initial symmetrical short-circuit current of 15 kA, while the impedance $\underline{Z}_3$ represents an overhead line length of 50 km, Table 4.2. The load impedance $\underline{Z}_B$ is changed as a function of the load between 0 and 200 MVA, whereby a power factor of $\cos\varphi = 0.8$ (inductive); 1.0 and −0.8 (capacitive) is assumed.

Figure 7.5 illustrates that the voltage in the network depends on the load, whereby an inductive load ($\cos\varphi = 0.8$) always leads to a reduction of the voltage, while a capacitive load leads to an increase. If the voltage in the network is required to be kept constant regardless of the load, this leads to an adjustment of the active voltages, e.g. $\underline{U}_1$, $\underline{U}_2$, in the network (Fig. 7.4). The consequence is that this leads to an increase in the short-circuit current when an inductive load is applied.

In this example, the voltage U_0 at the fault location decreases significantly less ($\Delta u < 1\%$), as this voltage is influenced by the proximity of the power station feeder, so that in this case the load has almost no influence on the voltage before the short circuit occurs.

The conclusion from this is that it makes sense to assume a constant driving voltage that results in the maximum short-circuit current so that the result is on the safe side. In the past, this has led to the introduction of a voltage factor c in the short-circuit current calculation in the planning phase, which reflects the most unfavorable load case (Table 7.1).

7.4 Calculation of the Voltage Factor for Different Configurations or Voltage Levels

In the following, voltage factors are derived based on lines with different boundary conditions (voltage level, line type, load, and line length) and their limits are shown, using the values used today according to Table 7.1. It should be noted that the voltage factor c refers to the nominal system voltage in which the short circuit occurs. If, in contrast to this, the usual operating voltage $U_b > U_n$, e.g. $U_b = 115$ kV, this has a corresponding effect on the voltage factor c, as shown in Sect. 7.4.2.

7.4.1 Basics

The equivalent circuit diagram shown in Fig. 7.6 is generally used for the further calculations. In a first step, only one line is assumed, which is loaded at the end by an impedance or a power is transmitted at a specified voltage. For simplification, the capacitances and impedances of the line are considered as concentrated elements. The transmitted power represents a load whose power factor is changed between $\cos\varphi = 1.0$ to 0.9 (capacitive/inductive). The voltage factor c describes the ratio of the voltages at the beginning of the line U_1 in relation to the voltage at the end U_2 according to Eq. (7.29) and is related to the nominal system voltage U_n:

$$c = \frac{\underline{U}_1 = f(\underline{U}_2)}{U_n} \tag{7.29}$$

U_2 represents the voltage at this location before a short circuit occurs. The maximum load on the line depends on the current-carrying capacity of the overhead line or cable used; the line length is also varied.

The currents and voltages in Fig. 7.6 are determined using Eq. (7.30), which describes the behavior of a quadripole.

$$\begin{pmatrix} \underline{U}_1 \\ \underline{I}_1 \end{pmatrix} = \begin{pmatrix} 1 + \underline{Z}_L \cdot \underline{Y}_q & \underline{Z}_L \\ \underline{Y}_q \cdot (2 + \underline{Z}_L \cdot \underline{Y}_q) & 1 + \underline{Z}_L \cdot \underline{Y}_q \end{pmatrix} \cdot \begin{pmatrix} \underline{U}_2 \\ \underline{I}_2 \end{pmatrix} \tag{7.30}$$

The admittance Y_q according to Eq. (7.30) represents the parallel circuit of the capacitance of the line and the impedance of the compensation $\underline{Z}_K$. The compensation is used when the voltage of the unloaded line is so high that a reduction is required. The degree of compensation of the line is assumed to be a maximum of 70%, which is only of interest for unloaded overhead lines at very high voltages and for cable networks.

In principle, inductive compensation is possible, e.g. up to 100%, but in these cases, missing current zero crossings occur during switch-on, which should be avoided, see Sect. 6.5.2.2 of [8].

The values in Table 7.2 are used for the calculations depending on the voltage level and the line (overhead line/cable). The cross-sections within a voltage level are selected so that the maximum transmissible continuous current I_d is almost identical or parallel systems are used if necessary. In addition, the characteristic values of the surge impedance Z_W of the cable and the natural power P_{nat} at the operating voltage U_b (line-to-line voltage) are determined in Table 7.2 according to Eq. (7.31).

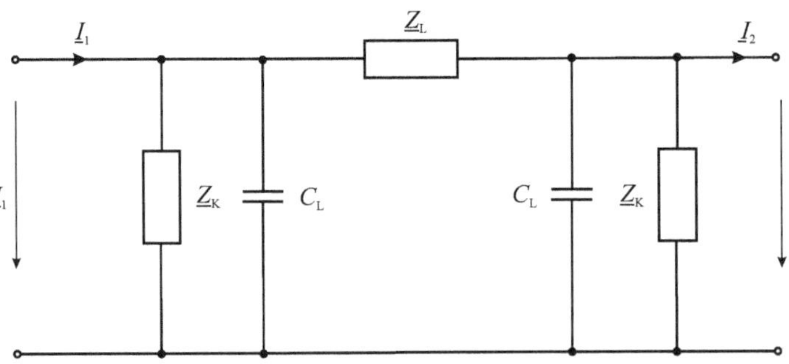

Fig. 7.6 Equivalent circuit diagram for calculating the voltage factor. $\underline{Z}_L$ Line impedance of the line, C_L Line capacitance, $\underline{Z}_K$ Impedance of the line compensation (if required)

Table 7.2 Data for the equivalent circuit (OHL overhead line, cable) according to Fig. 7.6 for different voltage levels (U_n nominal system voltage)

Term	Dim.	110 kV		380 kV		500 kV
		OHL	Cable	OHL	Cable	OHL
Type		a	b	c	d	e
R'_L	Ω/km	0.119	0.047[1)]	0.030	0.006[1)]	0.018
X'_L	Ω/km	0.401	0.123	0.251	0.108	0.26
C'_L	μF/km	0.0091	0.185	0.0134	0.48	0.0144
I_d	A	645	673	2580	2734	5320
Z	Ω	374.46	46.00	244.19	26.76	239.73
U_b	kV	115	115	410	410	530
I_{nat}	A	486	3954[2)]	969	8845[3)]	1276

a 240/40 mm², *b* A2XS(FL)2Y 630 mm², *c* 4 × 240/40 mm², *d* N2XS(FL)2Y 1 × 2500/250 mm² (two systems in parallel, *e* 4 × 795 kcmil, *1* 20 °C, DC resistance, *2* without compensation; with compensation of 70%, I_{nat} = 791 A, *3* without compensation; with compensation of 70%, I_{nat} = 4845 A

$$Z_W = \sqrt{\frac{L_L}{C_L}} \quad P_{nat} = \frac{U_b^2}{Z_W} \tag{7.31}$$

The value given in Figs. 7.7, 7.8, 7.9, 7.10 and 7.11 is assumed for the operating voltage U_b. From the natural power P_{nat}, the corresponding current I_{nat} can be determined according to Eq. (7.32), which is also listed in Table 7.2.

$$I_{nat} = \frac{P_{nat}}{\sqrt{3} \cdot U_b} = \frac{U_b/\sqrt{3}}{Z_W} \tag{7.32}$$

The current, which is determined from the natural power, is a measure of how the line works during a transmission. The following ranges are possible depending on the actual current I_2:

- $I_2 < I_{nat}$: The line has a capacitive effect,
- $I_2 = I_{nat}$: The line has a resistive effect,
- $I_2 > I_{nat}$: The line has an inductive effect.

The electrical values in Fig. 7.6 are determined according to the following steps:

- The voltage U_2 (real value) is set equal to the desired operating voltage at the fault location.
- Determination of the maximum current I_2 at the end of the line, which results from the maximum steady state current I_d, or half the value of the steady state current. The power factor cos φ is varied.
- The voltage U_1 at the beginning of the line and thus the voltage factor c is determined from the first line of Eq. (7.30), taking into account the respective nominal system voltage, Eq. (7.29).
- The second line of Eq. (7.30) is used to check whether the current I_1 at the beginning of the line exceeds the permissible value I_d.

The short-circuit current at the fault location (index 2 according to Fig. 7.6) is determined as a function of the voltage U_1 at the beginning of the line, so that only the influence of the line on the voltage drop is considered. From the various calculations, it can thus be derived under which boundary conditions the voltage factor $c = 1.1$ is justified for determining the maximum short-circuit current. For this reason, the voltage factor c is determined exclusively by the voltage ratio of $\underline{U}_2$ and $\underline{U}_1$ in relation to the nominal system voltage U_n according to Eq. (7.29).

The calculations assume that half the maximum steady state current is transmitted at the end of the line connection. This value is justified if the $(n-1)$ principle is applied in network planning.

7.4.2 110 kV Voltage Level

The information in Table 7.2 is used for the calculation, including the explanations. In addition, only an operating voltage of $U_b = \sqrt{3} \cdot U_2 = 115$ kV at the end of the line of is assumed.

7.4.2.1 Overhead Line

Figure 7.7 shows the voltage factor for the case where only 50% of the maximum steady state current I_d is transmitted at the end of the line. The overhead line is not inductively compensated due to the low capacitive charging current. As the transmitted current of $I_2 = 322.5$ A is less than the current calculated from the natural power of $I_{nat} = 486$ A, the line has a fundamentally capacitive effect under this condition, which is expressed by the line length, without exceeding the voltage factor of $c = 1.1$.

Based on the results, the following assessments can be derived according to Fig. 7.7:

- With a resistive load on the overhead line ($\cos \varphi = 1.0$), the specified voltage factor $c = 1.1$ is exceeded after a line length of $\ell \approx 81$ km. This means that in these cases the short-circuit results according to [2] are on the unsafe side.
- The voltage factor $c = 1.1$ is not exceeded up to a line length of $\ell \approx 250$ km with a load with a power factor of $\cos \varphi = 0.9$ (capacitive), i.e. the results are always on the safe side if practical transmission lengths are assumed.
- If, on the other hand, the line is loaded inductively ($\cos \varphi = 0.9$), the permissible line length is reduced to $\ell \approx 38$ km.
- The intersection of the curves shown with the y-axis results from the value $c = U_b/U_n = 115/110 = 1.045$.
- At higher operating voltages, for example $U_b = 123$ kV ($= U_m$, highest voltage for equipment), the curves consequently shift to higher c values.

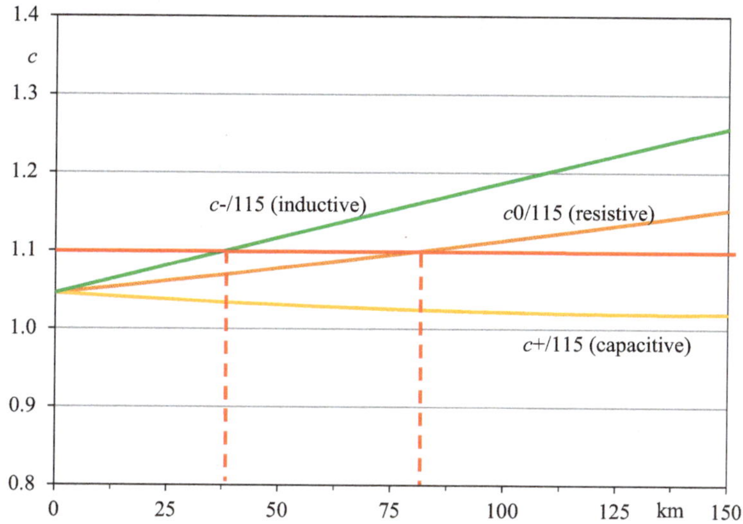

Fig. 7.7 Voltage factors c at an operating voltage of $U_b = 115$ kV as a function of the overhead line length (not compensated). $c0/115$ power factor $\cos \varphi = 1.0$, $c-/115$ power factor $\cos \varphi = 0.9$ (inductive), $c+/115$ power factor $\cos \varphi = 0.9$ (capacitive)

7.4.2.2 Cable

Due to the large capacitance of a 110 kV cable, it always behaves capacitively even under an inductive load. This means that the limitation in the short-circuit behavior does not occur as a result of exceeding the voltage factor $c = 1.1$, but rather the rated current of the cable is the dimensioning variable when determining the maximum cable length. If there is no compensation of the charging current, the calculations result in maximum cable lengths of $\ell = 60$ km (capacitive), 80 km (resistive) and 100 km (inductive) at an operating voltage of $U_b = 115$ kV, depending on the power factor $\cos \varphi$, if only 50% of the rated current is transmitted at the end of the cable.

For this reason, the 110 kV cables are compensated by 70% according to Fig. 7.8, which means that only 30% of the original cable capacity is effective. Even if the charging current is compensated, this results in a current of $I_{nat} = 791$ A, which is considerably higher than the operating current, so that the cable always has a capacitive effect in this case too.

Based on the results, the following assessments can be derived according to Fig. 7.8:

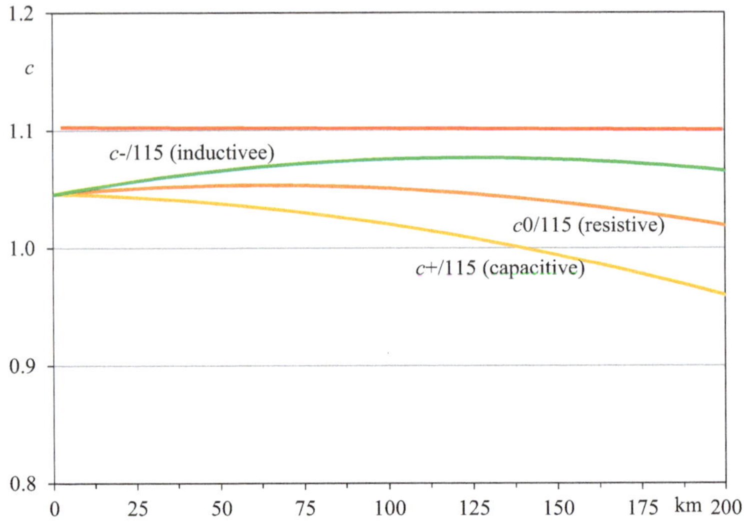

Fig. 7.8 Voltage factors c at an operating voltage of $U_b = 115$ kV as a function of the cable length (see Fig. 7.7 for legend)

- In the examples considered (115 kV), the voltage factor $c = 1.1$ is not exceeded in any case.
- Cable lengths of $\ell > 200$ km are possible without exceeding the rated current at the beginning of the cable (with an operating current of $I_b = I_d/2$).
- The voltage factor $c = 1.1$ is only exceeded at maximum operating voltages of $U_b = 123$ kV ($= U_m$, highest voltage for equipment), regardless of the power factor.

7.4.3 380 kV Voltage Level

The data of Table 7.2 is used for the calculation, whereby compensation of the charging current is also taken into account for the cable example. A value of $U_b = 410$ kV is selected as the operating voltage for the examples, which should correspond to the usual operating voltage at this voltage level. As a nominal system voltage according to [9] is not standardized at this voltage level, a value of $U_n = 380$ kV is assumed, with a maximum voltage for equipment of $U_m = 420$ kV.

7.4.3.1 Overhead Line

Figure 7.9 shows the results of the voltage calculations assuming that no compensation of the overhead line is taken into account. In this case, the load current

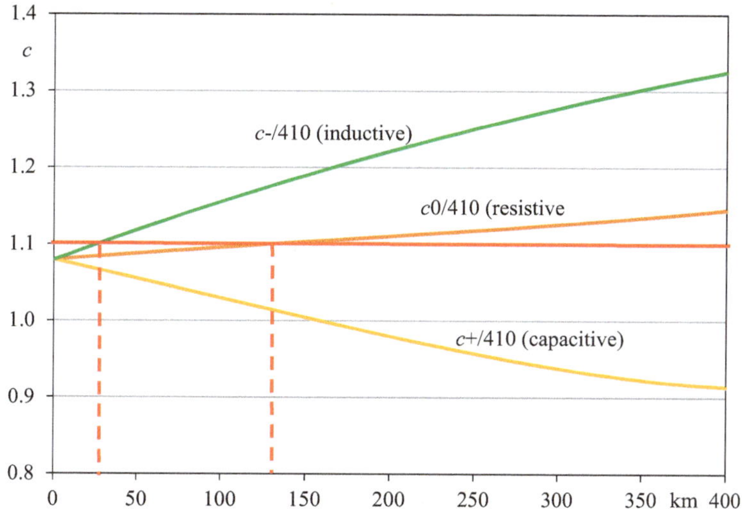

Fig. 7.9 Voltage factors c at an operating voltage of $U_b = 410\ \text{kV}/\sqrt{3}$ as a function of the overhead line length (without compensation). $c0/410$ power factor $\cos\varphi = 1.0$, $c-/410$ power factor $\cos\varphi = 0.9$ (inductive), $c+/410$ power factor $\cos\varphi = 0.9$ (capacitive)

(1290 A) is always above the current I_{nat} (969 A), so that the overhead line becomes inductive under resistive load, which influences the maximum line length.

Figure 7.9 shows the following evaluations:

- An inductive ($c-/410$) or resistive ($c0/410$) load results in maximum overhead line lengths of $\ell = 30$ km or 130 km without exceeding the voltage factor $c = 1.1$.
- In contrast, the voltage factor $c = 1.1$ is never exceeded with a capacitive load.
- With a reduced operating voltage of $U_b = 380$ kV, the permissible overhead line length increases to $\ell = 380$ km (resistive load, not shown in Fig. 7.9).
- Inductive compensation of the overhead line due to the Ferranti effect, e.g. with 50%, changes the possible overhead line lengths only insignificantly in some cases, with shorter lengths of $\ell = 97$ km occurring with a resistive load. However, it must be taken into account here that no inductive compensation should be made in the case of inductive transmission, but the operating voltage at the end of the line should be increased, which is possible with capacitive compensation (MSCDN, see Chap. 14).

7.4.3.2 Cable

Similar to 110 kV transmission, the two 380 kV cable systems used have a high charging current of $I_c = 36.57$ A/km due to the high charging capacity of $C = 0.48$ μF/km. This means that with a cable length of $\ell = 74.8$ km, the unloaded cable is already fully loaded due to the charging current. For this reason, an inductive compensation of 70% is provided for the calculation according to Fig. 7.10. Even in

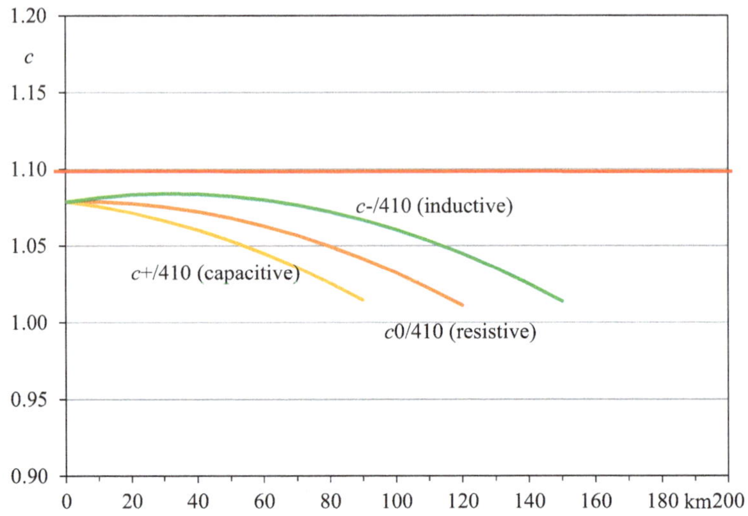

Fig. 7.10 Voltage factors c at an operating voltage of $U_b = 410$ kV as a function of the cable length (see Fig. 7.9 for legend)

this case, the cable is always operated below the natural power at the loads investigated.

Figure 7.10 shows the following evaluations:

- In the cases considered with an operating voltage of $U_b = 410$ kV, the voltage factor $c = 1.1$ is not exceeded in any operating case.
- With an inductive load with a power factor of $\cos \varphi = 0.9$, operation with a maximum cable length of $\ell \approx 150$ km is possible without exceeding the rated current at the beginning of the cable.
- With a maximum voltage at the end of the cable of $U_b = 420$ kV, the voltage factor also remains below $c < 1.1$, with the same maximum cable lengths.

7.4.4 500 kV Voltage Level, Overhead Line

The data in Table 7.2 is used for the calculation, whereby in this case only an overhead line is taken into account. A value of $U_b = 530$ kV is selected as the operating voltage for the examples, which should correspond to a normal operating voltage at this voltage level. As a nominal system voltage according to [6] is not standardized at this voltage level, a value of $U_n = 500$ kV is assumed, with a maximum voltage for equipment of $U_m = 550$ kV.

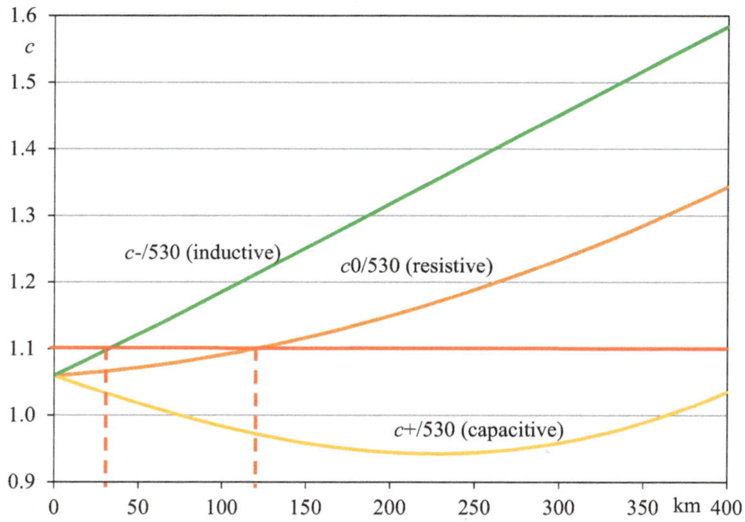

Fig. 7.11 Voltage factors c at an operating voltage of $U_b = 530$ kV as a function of the overhead line length (without compensation). $c0/530$ power factor $\cos\varphi = 1.0$, $c-/530$ power factor $\cos\varphi = 0.9$ (inductive), $c+/530$ power factor $\cos\varphi = 0.9$ (capacitive) $c0/530$

Figure 7.11 shows the results of the voltage calculations assuming that no compensation for the overhead line is taken into account. The voltage behavior is the same as when considering the 380 kV overhead line:

- An inductive ($c-/530$) or resistive ($c0/530$) load results in maximum overhead line lengths of $\ell = 35$ km or 120 km without exceeding the voltage factor $c = 1.1$.
- In contrast, the voltage factor $c = 1.1$ is never exceeded with a capacitive load (up to a line length of $\ell = 460$ km).
- With a reduced operating voltage of $U_b = 500$ kV, the permissible overhead line length increases to $\ell = 210$ km (resistive load, not included in Fig. 7.11).

7.4.5 *Summary of the Calculations*

The calculations show that the voltage factor c is a function of the voltage drop between the feeder and the short-circuit location, so that the relevant parameters are the inductance of the transmission line and the power factor of the load flow. It is therefore essential that the voltage factor $c = 1.1$ is exceeded for long lines, which are used particularly in high and extra-high voltage networks. As the reference voltage for the voltage factor is the nominal system voltage U_n, if the operating voltage U_b is close to the highest voltage for equipment U_m, the question arises as to whether the specified value according to [2] is sufficient, especially in systems with long overhead lines.

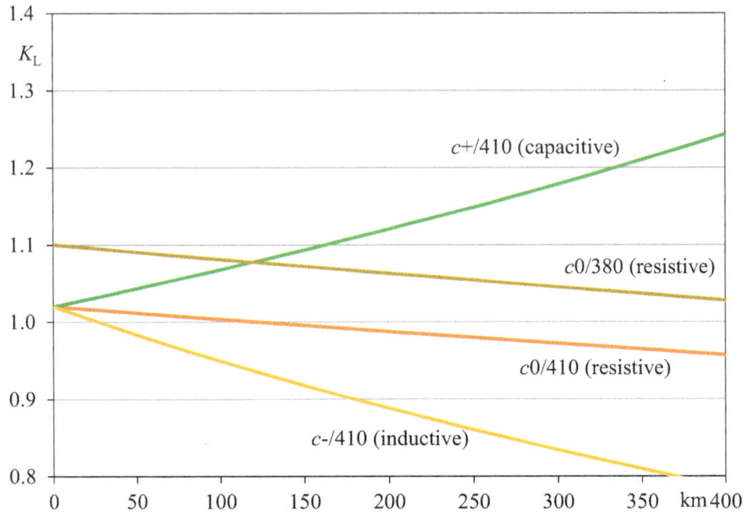

Fig. 7.12 Correction factor K_L, Eq. (7.41), for a 380 kV overhead line with $c_{max} = 1,1$

The statement that the calculation of the short-circuit currents in the aforementioned networks leads to results that may be on the unsafe side is consistent with the presentation in [1]. In principle, it is also possible to introduce a correction factor for long overhead lines, as shown as an example in Sect. 7.5.

In contrast, cable networks have a capacitive effect even with compensation (up to 70%), so that the voltage factor *c* is usually sufficient, even with an inductive load.

7.5 Correction Factor for Lines

The determination of the voltage factor $c = 1.1$ for the maximum short-circuit current is based, among other things, on the investigations presented in [7]. Here, the operating voltages in the various networks ($U_n = 0.4$ kV to 735 kV) are almost identical to the nominal system voltages ($\leq 2\%$). The main difference to the simplified calculations according to Sects. 7.4.2 to 7.4.4 is that the operating voltages before the fault occurs deviate from the nominal system voltages (115/110 = 1.045; 410/380 = 1.079; 530/500 = 1.06).

With high operating voltages, an attempt should be made to reduce the voltages and thus influence the power factor so that an active current is primarily transmitted. This will in any case have a reducing effect on the actual short-circuit current, or the voltage factor $c = 1.1$ should not be exceeded.

Since the evaluation of Sect. 7.4 shows that the voltage factor $c = 1.1$ is sufficient for a cable connection due to the capacitive operating behavior, even with an inductive load. The further explanations refer exclusively to the behavior with overhead lines, since in these cases the factor $c = 1.1$ can be exceeded even with relatively "short" line lengths and with a resistive/inductive load.

Since the voltage factor c takes into account the voltage difference between a future fault location and the active voltage of a network depending on the load flow, it is basically possible to express this voltage difference by a correction factor, which is taken into account when determining the line impedance.

The voltage difference can be determined using the quadripole Eq. (7.30), taking into account the phase voltages $\underline{U}_1$, $\underline{U}_2$:

$$\underline{U}_1 = \left(1 + \underline{Z}_{\mathrm{L}} \cdot \underline{Y}_{\mathrm{q}}\right) \cdot \underline{U}_2 + \underline{Z}_{\mathrm{L}} \cdot \underline{I}_2 \tag{7.33}$$

With

$$\text{Line impedance}: \underline{Z}_{\mathrm{L}} = R_{\mathrm{L}} + \mathrm{j}X_{\mathrm{L}} = R_{\mathrm{L}} + \mathrm{j}\omega L_{\mathrm{L}} \tag{7.34}$$

$$\text{Shunt admittance}: \underline{Y}_{\mathrm{q}} \approx \mathrm{j}\omega C_{\mathrm{L}} \tag{7.35}$$

$$\text{Load current (inductive)}: \underline{I}_2 = I_2 \cdot (\cos\varphi - j\sin\varphi) \tag{7.36}$$

Substituting Eq. (7.34) to (7.36) into (7.33) gives the following result:

$$\underline{U}_1 = \{1 + \mathrm{j}(R_{\mathrm{L}} + \mathrm{j}X_{\mathrm{L}}) \cdot \omega C\} \cdot U_2 + I_2 \cdot (\cos\varphi + \mathrm{j}\sin\varphi) \cdot (R_{\mathrm{L}} + \mathrm{j}X_{\mathrm{L}}) \tag{7.37}$$

$$\begin{aligned} \mathrm{Re}\{\underline{U}_1\} &= \left\{1 - \omega^2 \cdot L \cdot C\right\} \cdot U_2 + I_2 \cdot (R_{\mathrm{L}} \cdot \cos\varphi + X_{\mathrm{L}} \cdot \sin\varphi) \\ \mathrm{Im}\{\underline{U}_1\} &= \omega \cdot R \cdot C \cdot U_2 + I_2 \cdot (X_{\mathrm{L}} \cdot \cos\varphi - R_{\mathrm{L}} \cdot \sin\varphi) \end{aligned} \tag{7.38}$$

The voltage U_1 is determined with sufficient accuracy from the real part of Eq. (7.38), since the voltage U_2 is assumed to be real, whereby the "Ferranti effect" is also not taken into account, so that the following applies:

$$U_1 \approx U_2 + I_2 \cdot (R_{\mathrm{L}} \cdot \cos\varphi + X_{\mathrm{L}} \cdot \sin\varphi) \tag{7.39}$$

According to the calculations carried out, a larger voltage factor c should be selected for larger overhead line sections in order to calculate a short-circuit current that is on the safe side for these operating cases as well. However, if a uniform value c is to be used, a reduced impedance can also be used, as is the case with the impedance correction for transformers. For the exact short-circuit current at the fault location shown in Fig. 7.6, the following applies according to Eq. (7.30):

$$\underline{I}_{\mathrm{k}}'' = \underline{I}_{2\mathrm{k}} = \frac{U_1}{\underline{Z}_{\mathrm{L}}} \stackrel{!}{=} \frac{c_{\max} \cdot U_{\mathrm{n}}/\sqrt{3}}{K_{\mathrm{L}} \cdot \underline{Z}_{\mathrm{L}}} \tag{7.40}$$

With

Z_L line impedance,
K_L correction factor,
U_n nominal system voltage,
U_b operating voltage (line-to-line),
I_{2k} initial symmetrical short-circuit current at point 2 according to Fig. 7.6.

The correction factor K_L for an overhead line can be determined from Eqs. (7.39) and (7.40), where the index "b" denotes the operating conditions at the fault location before the short circuit occurred (Index 2).

$$K_L = \frac{c_{max} \cdot U_n/\sqrt{3}}{U_2 + I_2 \cdot (R_L \cdot \cos\varphi + X_L \cdot \sin\varphi)} = \frac{c_{max} \cdot U_n}{U_b + \sqrt{3} \cdot I_b \cdot (R_L \cdot \cos\varphi + X_L \cdot \sin\varphi)} \quad (7.41)$$

Figure 7.12 shows the correction factor K_L for a maximum voltage factor for $c_{max} = 1.1$. The factor is related to the 380 kV overhead line according to Fig. 7.9, as a function of the line length and the operating voltage U_b before the fault occurs at the short-circuit location, at an operating current equal to 50% of the maximum value according to Table 7.2.

According to Fig. 7.12, the following statements can be made in relation to the maximum three-phase initial symmetrical short-circuit current:

- $K_L \geq 1.0$: The calculation of the short-circuit current with $c_{max} = 1.1$ leads to a result that is on the safe side. This is particularly the case with a capacitive load (at $U_b = 410$ kV). In these cases, the correction factor will increase the overhead line impedance so that the calculated short-circuit current should correspond to the actual current.
- $K_L < 1.0$: Calculating the short-circuit current with $c_{max} = 1.1$ leads to a result that is on the unsafe side, e.g. with an inductive load and longer lines. In these cases, a larger voltage factor ($c > 1.1$) is therefore appropriate to calculate the maximum short-circuit current, so that the correction factor reduces the impedance of the overhead line accordingly.

7.6 Conclusions

To simplify the short-circuit current calculation, a single voltage $cU_n/\sqrt{3}$ is used as the equivalent voltage at the short-circuit location. The advantage of this is that no load flow variables have to be taken into account for a specific case. A value of $c = 1.1$ is used for the maximum short-circuit current in the high-voltage network. In general, this factor is sufficient in networks with short overhead lines and cable networks, even if an inductive load is assumed. This applies in particular if there are multiple feeders that keep the operating voltage constant.

In networks with a short circuit at the end of long overhead lines and high operating voltages that are close to the highest voltage for equipment, it is possible that this value ($c = 1.1$) is no longer sufficient. In these cases, the superposition method should be used.

References

1. Funk G (1967) Die Wirkungen von Belastungsimpedanzen und Leitungskapazitäten auf die Größe der Kurzschlussströme. Elektrizitatswirtschaft 66(15), pp. 437–440
2. IEC 60909–0:2016 Edition 2.0 (2016) Short-circuit current calculations in three-phase a.c. systems—Part 0: calculation of currents. IEC, Geneva
3. EU (2016) Commission Regulation (EU) 2016/1447 of 26 August 2016 establishing a network code on requirements for grid connection of high voltage direct current systems and direct current-connected power park modules
4. EU (2016) Commission Regulation (EU) 2016/631 of 14 April 2016 establishing a network code on requirements for grid connection of generators
5. Balzer G, Nelles D, Tuttas C (2009) Kurzschlussstromberechnung nach IEC und DIN EN 60909–0 (VDE 0102): 2002–07: Grundlagen, Anwendung der Normen, Auswirkungen der Kurzschlussströme (VDE-Schriftenreihe - Normen verständlich). VDE, Berlin
6. Taumberger H (1988) Ein Beitrag zur Neukonzeption der Vorschriften über Kurzschlussstromberechnung. Dissertation, TU Graz
7. Rittinghaus D (1983) Über den Einfluss der im Netzbetrieb veränderlichen Größen auf die Kurzschlussstromstärke. Dissertation, TU Erlangen-Nürnberg, Erlangen
8. Balzer G, Neumann C, Balzer C (2025) Schalt- und Ausgleichsvorgänge in elektrischen Netzen, 2nd Edn. Springer, Heidelberg
9. IEC 60038 Ed.7 (2009) IEC standard voltages. IEC, Geneva

Chapter 8
Correction Factors for Power Station Units and Power Transformers

In the IEC standard for short-circuit current calculation [1], the load of a network is taken into account by applying a voltage factor *c*. The voltage factor basically compensates the voltage drop between the active voltage and a possible short-circuit location in the network in the event of a load. This voltage drop has a particular effect with an inductive load, as the components of a network also act predominantly as inductive elements. In principle, it is possible to compensate the voltage drop using a general voltage factor *c* in the short-circuit current calculation, which has the same effect on all short-circuit locations in the network, or to make an individual impedance correction for a specific piece of equipment. Both options are used in the IEC Regulations.

While a correction factor for generators was already included in the first version of the VDE Regulation [2], the correction factors for transformers and power station units were added in later versions.

The causes for the derivation of the impedance correction factors are presented below. Additional information can be found in [3].

8.1 General

In general, both generators and transformers can be represented as an impedance $\underline{Z} = R + jX$ as shown in Fig. 8.1. Depending on the current, there is a voltage change at the terminals compared to the active voltage $\underline{U}_0$.

Equation (8.1) applies to the voltage $\underline{U}_0$:

$$\underline{U}_0 = \underline{U}_1 + I_1 \cdot (\cos\varphi - j\sin\varphi) \cdot (R + jX) \tag{8.1}$$

The change in voltage $\Delta\underline{U}$ is given by Eq. (8.1):

G. Balzer, *Short-Circuit Currents in Three-Phase Systems*,
https://doi.org/10.1007/978-3-658-50710-7_8

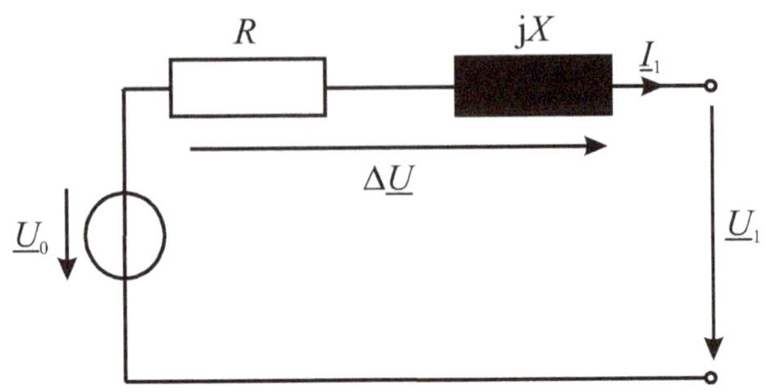

Fig. 8.1 Equivalent circuit diagram for determining the voltage change due to a load

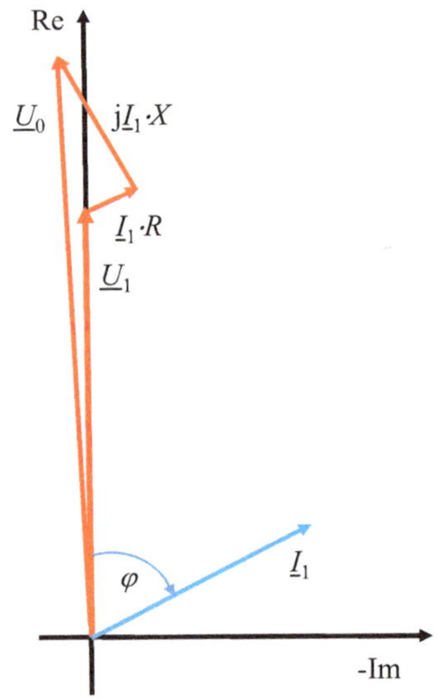

Fig. 8.2 Vector diagram of the equivalent circuit, Fig. 8.1

$$\begin{aligned}\Delta\underline{U} &= I_1 \cdot (\cos\varphi - \mathrm{j}\sin\varphi) \cdot (R + \mathrm{j}X) = \\ &I_1 \cdot \{R \cdot \cos\varphi + X \cdot \sin\varphi + \mathrm{j}(X \cdot \cos\varphi - R \cdot \sin\varphi)\}\end{aligned} \tag{8.2}$$

For the voltage difference, assuming that the voltage $\underline{U}_1$ is in the real plane, the following simplified equation applies:

$$\Delta U \approx I_1 \cdot (R \cdot \cos\varphi + X \cdot \sin\varphi) \tag{8.3}$$

Figure 8.2 shows the vector diagram of the voltages and the load current for the equivalent circuit according to Fig. 8.1.

If the voltage U_1 in the network is to be kept constant at a possible short-circuit location regardless of the load, it is necessary to increase the driving voltage U_0 by the value ΔU. In addition, for generators or transformers, $R << X$, so that the voltage U_0 applies:

$$U_0 = U_1 + I_1 \cdot X \cdot \sin\varphi = U_1 \cdot \left(1 + \frac{I_1}{U_1} \cdot X \cdot \sin\varphi\right) \stackrel{!}{=} c \cdot U_n/\sqrt{3} \tag{8.4}$$

If the active voltage U_0 is to be equal to the value $c \cdot U_n/\sqrt{3}$, the following applies to the voltage factor c:

$$c = \frac{U_1}{U_n/\sqrt{3}} \cdot \left(1 + \frac{I_1}{U_1} \cdot X \cdot \sin\varphi\right) \tag{8.5}$$

With

U_n nominal system voltage (line-to-line),
U_1 terminal voltage (line-earth),
I_1 load current before short-circuit initiation at the short-circuit location,
φ power factor, angle between $\underline{U}_1$ and $\underline{I}_1$,
X reactance of equipment.

In the derivation of Eq. (8.5), it is assumed that an inductive load current has the same direction as an expected short-circuit current before the short-circuit occurs. This leads to a voltage drop in such a way that the voltage U_1 at the terminals is reduced. This is based on the consideration that there is generally a load flow from the higher voltage levels to the lower ones, which has an inductive power factor and the generation is mainly in the upper voltage levels. If, on the other hand, there is a capacitive load flow before the short circuit occurs, this leads to an increase in the voltage U_1.

If, in contrast to Eq. (8.5), the voltage drop according to Eq. (8.3) is taken into account, the factor c becomes:

$$c = \frac{U_1}{U_n/\sqrt{3}} \cdot \left\{1 + \frac{I_1}{U_1} \cdot (R \cdot \cos\varphi + X \cdot \sin\varphi)\right\} \tag{8.6}$$

In the following, the general representation is applied to generators, transformers and power station units.

8.2 Synchronous Generator

The impedance correction according to the IEC Standard [1] for a generator is only applied to generators that feed directly into a substation, e.g. in a low-voltage network or industrial network. In contrast, generators that feed into the network via a step-up transformer are considered in connection with a power station unit, so that there is an impedance correction for a power station unit, Sect. 8.4.

For the calculation of the initial symmetrical short-circuit current $\underline{I}''_{kGS}$ at the terminals of the synchronous generator, Fig. 8.3 and Eq. (8.7) generally apply.

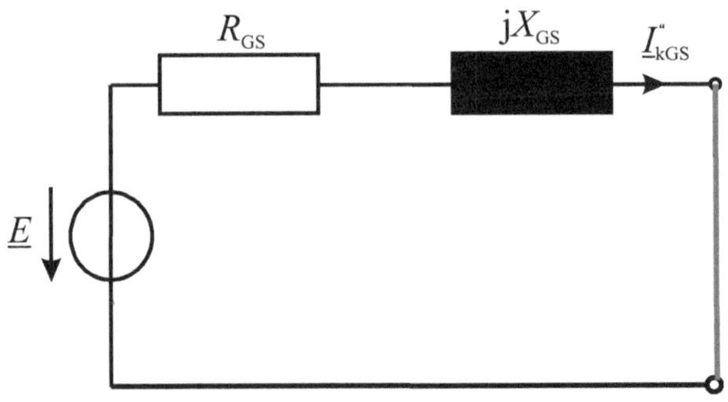

Fig. 8.3 Initial symmetrical short-circuit current at the generator terminals

$$I''_{kGS} = \frac{E}{R_{GS} + jX_{GS}} = \frac{E}{R_{GS} + jX''_{d}} \tag{8.7}$$

With

E synchronous internal voltage of the generator,
R_{GS} resistance of the generator,
X_{GS} subtransient reactance (direct-axis) of the generator ($= X''_{d}$).

According to Eq. (8.4), the following assignment results for the example synchronous generator:

$$E = \frac{U_{rGS}}{\sqrt{3}} \cdot \left(1 + \frac{I_{rGS} \cdot \sqrt{3}}{U_{rGS}} \cdot X''_{d} \cdot \sin\varphi\right) \tag{8.8}$$

This assumes that the voltage at the generator terminals is always kept constant with U_{rGS} and that the load current before the short-circuit occurs is equal to the rated current $\underline{I}_{rGS}$ of the generator. Because of

$$X''_{d} = x''_{d} \cdot \frac{U_{rGS}/\sqrt{3}}{I_{rGS}} \tag{8.9}$$

results for the synchronous internal voltage E of the generator

$$E = \frac{U_{rGS}}{\sqrt{3}} \cdot \left(1 + x''_{d} \cdot \sin\varphi\right) \tag{8.10}$$

With

U_{rGS} rated voltage of the generator,
I_{rGS} rated current of the generator,
φ power factor, angle between $\underline{U}_{rGS}$ and $\underline{I}_{rGS}$,
x''_{d} subtransient reactance related on X''_{d}

For the calculation of the maximum initial symmetrical short-circuit current, the following results according to Eq. (8.7):

$$\underline{I}''_{\mathrm{kGS}} = \frac{\frac{U_{\mathrm{rGS}}}{\sqrt{3}} \cdot \left(1 + x''_{\mathrm{d}} \cdot \sin\varphi\right)}{R_{\mathrm{GS}} + \mathrm{j}X''_{\mathrm{d}}} \tag{8.11}$$

The calculation of the initial symmetrical short-circuit current according to Eq. (8.11) assumes that only a single synchronous generator feeds the short-circuit location and this contribution is calculated. In general, a synchronous machine is installed in a network with other feeders, so that a total short-circuit current must be determined and the method of the equivalent voltage source at the short-circuit location is applied. Since the maximum voltage factor is set at $c = 1.1$, the voltage increase according to Eq. (8.11) must be reflected in a correction factor of the generator impedance. This results in the following equation for the correction factor K_{GS}:

$$\underline{I}''_{\mathrm{kGSmax}} = \frac{\frac{U_{\mathrm{rGS}}}{\sqrt{3}} \cdot \left(1 + x''_{\mathrm{d}} \cdot \sin\varphi\right)}{R_{\mathrm{GS}} + \mathrm{j}X''_{\mathrm{d}}} \stackrel{!}{=} \frac{c_{\max} \cdot U_{\mathrm{n}}/\sqrt{3}}{K_{\mathrm{GS}} \cdot \left(R_{\mathrm{GS}} + \mathrm{j}X''_{\mathrm{d}}\right)} \tag{8.12}$$

$$K_{\mathrm{GS}} = \frac{U_{\mathrm{n}}}{U_{\mathrm{rGS}}} \cdot \frac{c_{\max}}{1 + x''_{\mathrm{d}} \cdot \sin\varphi} \tag{8.13}$$

When using the equivalent voltage source at the short-circuit location, the generator impedance must be multiplied by the correction factor according to Eq. (8.13) in accordance with Eq. (8.12). The voltage factor $c_{\max}$ depends on the voltage level of the short-circuit location. Equation (8.8) assumes that the rated current of the generator I_{GS} flows before the short-circuit occurs. If this is taken into account, the result for the synchronous internal voltage E becomes:

$$E = \frac{U_{\mathrm{rGS}}}{\sqrt{3}} \cdot \left(1 + \frac{I_{\mathrm{GS}}}{U_{\mathrm{rGS}}/\sqrt{3}} \cdot X''_{\mathrm{d}} \cdot \sin\varphi\right) \tag{8.14}$$

Under these conditions, the correction factor is K_{GS}:

$$K_{\mathrm{GS}} = \frac{U_{\mathrm{n}}}{U_{\mathrm{rGS}}} \cdot \frac{c_{\max}}{1 + \frac{I_{\mathrm{GS}}}{I_{\mathrm{rGS}}} x''_{\mathrm{d}} \cdot \sin\varphi} \tag{8.15}$$

The correction factors for different operating points are determined for an exemplary synchronous generator and the resulting factors c for a short-circuit current calculation are also specified. The maximum active current is taken into account at different operating points and Eq. (8.15) is used. The following data is assumed:

Nominal system voltage $U_n = 10$ kV
Generator voltage $U_{rGS} = 10.5$ kV
Subtransient reactance $x''_d = 0.13$
Power factor $\cos\varphi_{rGS} = 0.8$
Voltage factor $c_{max} = 1.1$

- Rated operating point (inductive): $\cos\varphi = 0.8$; $I_{GS}/I_{rGS} = 1.0$

$$K_{GS} = \frac{10}{10.5} \cdot \frac{1.1}{1 + 1.0 \cdot 0.13 \cdot 0,6} = 0.9718$$

- Active power operating point: $\cos\varphi = 1.0$; $I_{GS}/I_{rGS} = 0.8$

$$K_{GS} = \frac{10}{10.5} \cdot \frac{1.1}{1 + 0.8 \cdot 0.13 \cdot 0.0} = 1.0476$$

- Capacitive operating point: $\cos\varphi = 0.9$; $I_{GS}/I_{rGS} = 0.8889$

$$K_{GS} = \frac{10}{10.5} \cdot \frac{1.1}{1 - (0.8889) \cdot 0.13 \cdot 0.4359} = 1.1032$$

A resulting voltage factor $c_{res} = c_{max}/K_G$ can be derived from Eq. (8.12). In this example, the following values result:

- $\cos\varphi = 0.8$ (inductive): $c_{res} = 1.1319$,
- $\cos\varphi = 1.0$: $c_{res} = 1.0500$,
- $\cos\varphi = 0.9$ (capacitive): $c_{res} = 0.9971$.

It can be seen that in the first case (inductive rated operating point) the resulting voltage factor c_{res} is greater than the voltage factor according to Table 7.1, while in the other cases the application of the correction factor leads to a lower short-circuit current. However, this assumes that the synchronous generator is operated at the terminals with a voltage of U_{rGS}. If this is not the case, the higher voltage must be used in Eq. (8.15).

The application of the correction factor always leads to larger short-circuit currents if the short-circuit is initiated from inductive operation, whereas smaller short-circuit currents result from capacitive operation. Since the maximum possible short-circuit current must always be taken into account when planning a system, it makes sense to take the overexcited range (rated operating point) of a synchronous generator as the basis for dimensioning. This means that Eq. (8.13) must be used with maximum operating current, with $K_{GS} = 0.9718$. This consideration also led to the factor K_{GS} being determined according to Eq. (8.16) in the Standard [1].

$$K_{GS} = \frac{c}{1 + x''_d \cdot \sqrt{1 - \cos^2\varphi_{GS}}} \cdot \frac{U_n}{U_{rGS}} \tag{8.16}$$

8.3 Power Transformers (Network)

Network transformers are referred to below as transformers that are connected to a distribution and transmission network on the high- and low-voltage side. In contrast, there are step-up transformers, which are installed in power station units and connect the generator to the power network.

The short-circuit impedance of transformers (two-winding transformers) is determined according to Eq. (8.17), in accordance with Sect. 4.8.1 or IEC regulations [4].

$$Z_{\mathrm{T}} = \frac{u_{\mathrm{kr}}}{100\%} \cdot \frac{U_{\mathrm{rT}}^2}{S_{\mathrm{rT}}} \tag{8.17}$$

With

$$R_{\mathrm{T}} = \frac{u_{\mathrm{Rr}}}{100\%} \cdot \frac{U_{\mathrm{rT}}^2}{S_{\mathrm{rT}}} = \frac{P_{\mathrm{krT}}}{S_{\mathrm{rT}}} \cdot \frac{U_{\mathrm{rT}}^2}{S_{\mathrm{rT}}} \tag{8.18}$$

and

$$X_{\mathrm{T}} = \sqrt{Z_{\mathrm{T}}^2 - R_{\mathrm{T}}^2} \tag{8.19}$$

It means:

U_{rT} rated voltage,
I_{rT} rated current,
S_{rT} rated apparent power,
P_{krT} winding losses at rated current,
u_{kr} short-circuit voltage at rated current in %,
u_{Rr} rated resistive component of the short-circuit voltage in %.

To calculate the short-circuit current, it is sufficient to determine the impedance Z_{T} for the main position of the tap-changer. The following note was given in the former Standard [4] in this context:

Special considerations are only necessary if:

- A single-fed short-circuit current is calculated and the short-circuit current has the same direction as the operating current before the short-circuit occurred (short-circuit on the LV side of a transformer or parallel transformers with tap changers,...),
- it is possible to change the transformer ratio within wide limits using a tap-changer, $U_{\mathrm{TOS}} = U_{\mathrm{rTOS}} \cdot (1 \pm p_{\mathrm{T}})$ with $p_{\mathrm{T}} > 0.05$,
- the minimum short-circuit voltage u_{kmin} is noticeably lower than the rated value of the short-circuit voltage at the main tap position ($u_{\mathrm{kmin}} < u_{\mathrm{kr}}$),
- the voltage during operation is noticeably higher than the rated system voltage ($U \geq 1.05 \cdot U_{\mathrm{n}}$).

Since the above considerations represent an "or" operation, it turned out that the first point corresponds to the standard application of a transformer in that the load flow is from a higher voltage level to a lower one and this load flow is usually inductive. The "special" considerations led to the introduction of a correction factor for network transformers [5]. The term "network transformer" was chosen here in contrast to transformers in power station units, as these have their own correction factor, Sect. 8.4.

The aim of this impedance correction factor for network transformers is therefore to compensate for the voltage drop, which is basically possible by increasing the voltage factor c, e.g. to $c = 1.15$, or by reducing the transformer impedance. The electrical conditions are therefore the same as those shown in Sect. 8.1.

The disadvantage of a general increase in the voltage factor c is that when using the method of the equivalent voltage source at the short-circuit location, all short-circuit currents in the network are increased accordingly, even if the voltage drop is low. In contrast, the correction factor in relation to a transformer is only noticeable if the transformer data has a high short-circuit reactance or the load flow is inductive. In this case, the effect on the short-circuit current should primarily be limited locally.

Since the voltage factor c according to Table 7.1 remains unchanged, the impedance correction K_T is only used to determine the deviation in relation to the factor $c_{max} = 1.1$. From Eqs. (8.4) and (8.5), the maximum short-circuit current, related to the LV side of the transformer, can be written according to Eqs. (8.20) to (8.23), with $U_1 = U_{bT}/\sqrt{3}$. In Eq. (8.20), operating variables before the short circuit occurs are only marked with the index "bT".

$$I''_{kTmax} = \frac{c_{max} \cdot U_n/\sqrt{3}}{K_T \cdot Z_T} = \frac{U_{bT}/\sqrt{3}}{U_n/\sqrt{3}} \cdot \left(1 + \frac{I_{bT}}{U_{bT}/\sqrt{3}} \cdot X_T \cdot \sin\varphi\right) \cdot \frac{U_n/\sqrt{3}}{Z_T} \tag{8.20}$$

$$\text{resp.with} \quad X_T = x_T \cdot \frac{U_{rT}/\sqrt{3}}{I_{rT}} \tag{8.21}$$

$$I''_{kTmax} = \frac{c_{max} \cdot U_n/\sqrt{3}}{K_T \cdot Z_T} = \frac{U_T/\sqrt{3}}{Z_T} \cdot \left(1 + \frac{I_{bT}}{I_{rT}} \cdot \frac{U_{rT}}{U_{bT}} \cdot x_T \cdot \sin\varphi\right) \tag{8.22}$$

This results in the following correction factor K_T:

$$K_T = \frac{U_n}{U_{bT}} \cdot \frac{c_{max}}{1 + \frac{I_{bT}}{I_{rT}} \cdot \frac{U_{rT}}{U_{bT}} \cdot x_T \cdot \sin\varphi} \tag{8.23}$$

With

c_{max} voltage factor c according to Table 7.1,
U_n nominal system voltage,
I_{rT} rated current of the transformer,

x_T relative transformer reactance,
U_{rT} rated voltage of the transformer,
U_{bT} operating voltage before short circuit (line-to-line),
I_{bT} operating current before short circuit,
φ angle of the power factor of the power flow before short circuit.

If it is assumed that the highest operating voltage is generally equal to the rated voltage of the transformer, then Eq. (8.23) is simplified to $U_{bT} = U_{rT}$:

$$K_T = \frac{U_n}{U_{bT}} \cdot \frac{c_{max}}{1 + \frac{I_{bT}}{I_{rT}} \cdot x_T \cdot \sin\varphi} \tag{8.24}$$

In Eqs. (8.23) and (8.24), it is assumed that the operating voltage refers to the low-voltage side of the transformer. This can be seen from the representation of the voltages in Fig. 8.1. Assuming that the following operating conditions are met, a further simplification can be selected:

- Maximum operating voltage: $U_{bT} < 1.05 \cdot U_n$,
- Maximum operating current: $I_{bT} < I_{rT}$,
- Power factor cos $\varphi > 0.8$ resp. sin $\varphi < 0.6$.

The product of the last two variables should not exceed a value of 0.6. Under these conditions, the correction factor can be written as a simplification:

$$K_T = 0.95 \cdot \frac{c_{max}}{1 + 0.6 \cdot x_T} \tag{8.25}$$

The advantage of using Eq. (8.25) is that no operating variables are used, which is the basic advantage of the method of the equivalent voltage source at the short-circuit location with the application of the voltage factor c. However, this means that the actual operating values before the short-circuit occurs should also be within the limits specified above.

According to Eq. (8.23), it is basically possible that a capacitive load current can also be taken into account, so that the correction factor increases. However, since an inductive current leads to larger short-circuit currents, which should therefore be assumed in the planning, this has been taken into account accordingly in the new version [1], with:

$$K_T = \frac{U_n}{U_{bT}} \cdot \frac{c_{max}}{1 + \frac{I_{bT}}{I_{rT}} \cdot x_T \cdot \sqrt{1 - \cos^2\varphi}} \tag{8.26}$$

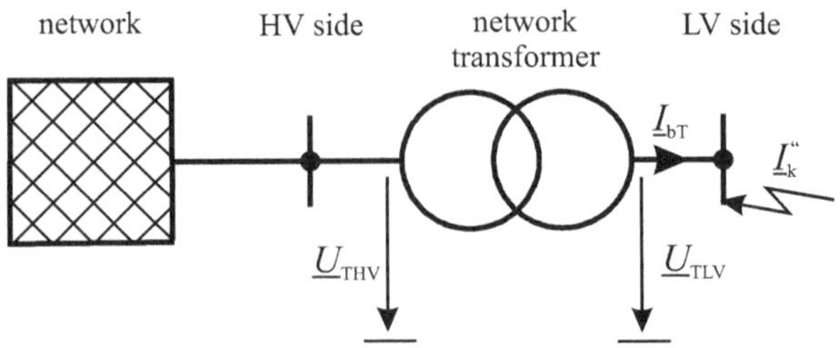

Fig. 8.4 Equivalent circuit for determining the short-circuit current downstream of a network transformer

For the three-phase initial symmetrical short-circuit current, taking into account the correction factor according to [1]:

$$\underline{I}''_{kT} = \frac{c_{max} \cdot U_n/\sqrt{3}}{K_T \cdot \underline{Z}_T} = \frac{U_{bT}/\sqrt{3}}{\underline{Z}_T} \cdot \left(1 + \frac{I_{bT}}{I_{rT}} \cdot x_T \cdot \sqrt{1 - \cos^2\varphi}\right) \tag{8.27}$$

or with the simplification according to Eq. (8.25):

$$\underline{I}''_{kT} = \frac{U_n/\sqrt{3}}{\underline{Z}_T} \cdot \frac{(1 + 0.6 \cdot x_T)}{0.95} \tag{8.28}$$

The use of the correction factors is shown below for four examples, whereby the difference is mainly illustrated in relation to the deviating operating conditions. The calculations for the following examples are each carried out using the equivalent circuit of the network according to Fig. 8.4. Further investigations are listed in [3, 6, 7].

Example 1

- **Transformer Data**

rated voltage	U_r	400 kV/120 kV/30 kV	
rated apparent power	S_{rHVMV}:	250 MVA	
rated apparent power	S_{rHVLV}:	50 MVA	
rated apparent power	S_{rMVHV}:	50 MVA	
rel. Short-circuit voltage	u_{krHVMV}:	20%	main position
rel. Short-circuit voltage	u_{krHVLV}:	10.5%	main position
rel. Short-circuit voltage	u_{krMVLV}:	5.7%	main position
winding losses	P_{krOSMS}:	693 kW	rated current
winding losses	P_{krOSUS}:	93 kW	rated current
winding losses	P_{krMSUS}:	85 kW	rated current

- **Network Data (380 kV)**
  nominal system voltage U_n: 380 kV
  short-circuit current (three phase) I''_{kN}: 60 kA
  ratio R_N/X_N: 0.1

According to Sect. 4.8.5.1, the short-circuit impedance can be determined for each winding from the winding losses and the power between the different sides, in each case in relation to the 110 kV network level, Eqs. (8.17) to (8.19).

$$\underline{Z}_{HV} = (0.103 + j12.673)\,\Omega \qquad \underline{Z}_{MV} = (0.957 - j1.154)\,\Omega$$
$$\underline{Z}_{LV} = (0.4333 + j17.563)\,\Omega$$

According to Eq. (4.212), the short-circuit impedance of the transformer is determined between the high and medium voltage sides:

$$\underline{Z}_T = \underline{Z}_{HV} + \underline{Z}_{MV} = (0.160 + j11.519)\,\Omega$$

According to Eqs. (4.1) and (4.2), the network impedance is calculated for the 110 kV level with $t_r = 400/120$:

$$Z_{Nt} = \frac{1.1 \cdot U_{nN}}{\sqrt{3} \cdot I''_{kN}} \cdot \frac{1}{t_r^2} = 0.362\,\Omega$$

$$X_{Nt} = \frac{Z_{Nt}}{\sqrt{1 - (R_N/X_N)^2}} = 0.360\,\Omega \qquad R_{Nt} = 0.036\,\Omega$$

The calculations of the three phase short-circuit current on the low voltage side of the transformer according to Fig. 8.4 are carried out under the following conditions:

- The operating current before the short circuit occurs is changed between $I_{bT} = (0.0\text{–}1.0) \cdot I_{rT}$.
- For the power factor, $\cos\varphi = 0.0$ and 0.8 (inductive) is applied.
- The transformer voltage of the low voltage side is $U_{bT} = U_{rT}$ and U_m.

The three phase short-circuit current is determined according to Eq. (8.29), whereby the voltage drop according to Eqs. (8.3) and (8.6) is considered as a reference current, which is a simplification compared to the actual current, since only the real part is taken into account when deriving the voltage drop. In addition, the voltage drop across the network feeder is not considered, so that the further representation only compares the behavior of the transformer as a function of the correction factor.

This reference current is compared with the short-circuit currents, which are determined using the impedance correction factors according to Eqs. (8.25) and (8.26).

$$\underline{I}''_{\mathrm{k}} = \frac{1.1 \cdot U_{\mathrm{n}}/\sqrt{3}}{\underline{Z}_{\mathrm{Nt}} + K_{\mathrm{T}} \cdot \underline{Z}_{\mathrm{T}}} \tag{8.29}$$

The following relationships therefore apply to the various correction factors K_{T}:

- Reference current $\underline{I}''_{\mathrm{kT1}}$

$$K_{\mathrm{T1}} = \frac{U_{\mathrm{n}}}{U_{\mathrm{bT}}} \cdot \frac{c_{\max}}{1 + \frac{I_{\mathrm{bT}}}{I_{\mathrm{rT}}} \cdot \frac{U_{\mathrm{rT}}}{U_{\mathrm{bT}}} \cdot (u_{\mathrm{Rr}} \cdot \cos\varphi + x_{\mathrm{T}} \cdot \sin\varphi)} \tag{8.30}$$

- Short-circuit current $\underline{I}''_{\mathrm{kT2}}$ according to [1] with simplified operating conditions

$$K_{\mathrm{T2}} = \frac{U_{\mathrm{n}}}{U_{\mathrm{bT}}} \cdot \frac{c_{\max}}{1 + \frac{I_{\mathrm{bT}}}{I_{\mathrm{rT}}} \cdot x_{\mathrm{T}} \cdot \sin\varphi} \tag{8.31}$$

- Short-circuit current $\underline{I}''_{\mathrm{kT3}}$ according to [1] without operating conditions

$$K_{\mathrm{T3}} = 0.95 \cdot \frac{c_{\max}}{1 + 0.6 \cdot x_{\mathrm{T}}} \tag{8.32}$$

With

U_{n} nominal system voltage (110 kV),
$c_{\max}$ voltage factor (= 1,1),
U_{rT} transformer rated voltage of the low voltage side,
U_{bT} transformer voltage of the low voltage side before short circuit,
I_{rT} transformer rated current of the low voltage side,
I_{bT} transformer current of the low voltage side before short circuit,
x_{T} related short-circuit reactance, p.u.,
u_{Rr} related short-circuit resistance, p.u., Eq. (8.18),
$\cos\varphi$ power factor before short circuit.

The deviation Δi in % as a function of the load current before the short circuit occurs and the power factor is determined according to Eq. (8.33).

$$\Delta i = \frac{I''_{\mathrm{kT2,T3}} - I''_{\mathrm{kT1}}}{I''_{\mathrm{kT1}}} \cdot 100\% \tag{8.33}$$

Figure 8.5 below shows the deviations Δi of the currents according to Eqs. (8.31) and (8.32) in relation to the reference current, Eq. (8.33). In addition, the transformer voltage U_{bT} is changed before the short circuit occurs.

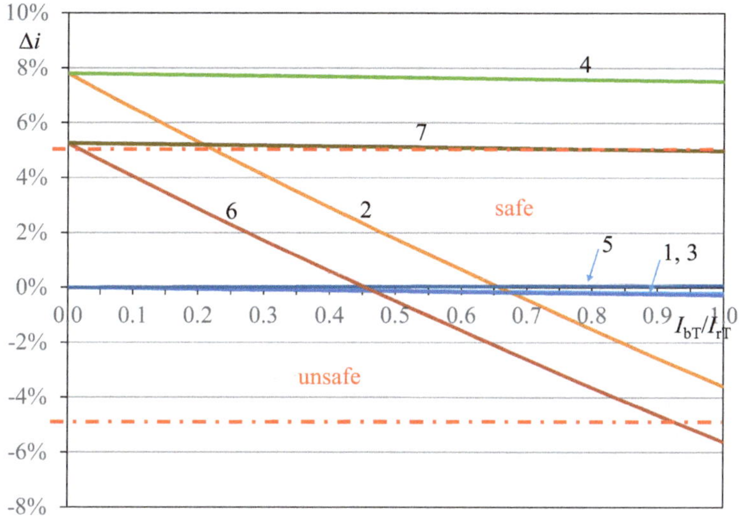

Fig. 8.5 Deviations in the calculation of the three phase short-circuit current according to Fig. 8.4 as a function of the transformer current before the short circuit occurs, short circuit: 110 kV level. (1) K_{T2}: cos $\varphi = 0.8$; $U_{bT} = U_{rT}$; (2) K_{T3}: cos $\varphi = 0.8$; $U_{bT} = U_{rT}$; (3) K_{T2}: cos $\varphi = 1.0$; $U_{bT} = U_{rT}$; (4) K_{T3}: cos $\varphi = 1.0$; $U_{bT} = U_{rT}$; (5) K_{T2}: cos $\varphi = 0.8$; $U_{bT} = U_m$; (6) K_{T3}: cos $\varphi = 0.8$; $U_{bT} = U_m$; (7) K_{T3}: cos $\varphi = 1.0$; $U_{bT} = U_m$

In general, a deviation of ±5% in relation to the exact calculation value is assumed to be acceptable when calculating the short-circuit current. For this reason, these two limit values are also shown in Fig. 8.5. The following results can be derived from the diagram as a function of the transformer current before the short-circuit occurs:

- The deviations in the calculation between the correction factors K_{T1} and K_{T2} remain below 0.1%, so that the use of the factor K_{T2} is justified, regardless of the load current, power factor and the transformer voltages assumed here (curves 1, 3, 5).
- If only active current (cos $\varphi = 1.0$) is transmitted, the simplified equation (K_{T3}, curve 4) leads to an average deviation of +7.67% (range: + 7.81% to +7.52%), which means that the short-circuit current is determined to be too high, with $U_{bT} = U_{rT}$.
- If the transformer voltage is higher than U_{rT}, this leads to a reduction of the correction factor K_{T1}, resulting in a higher current. This therefore leads to a lower deviation overall (K_{T3}, curve 7), resulting in an average value of +5.12%, with cos $\varphi = 1.0$.
- If the correction factor K_{T3} is used with a power factor of cos $\varphi = 0.8$ (inductive), this leads to short-circuit currents that are on the unsafe side if the ratio of the transformer currents $I_{bT}/I_{rT} > 0.65$ ($I_{bT} = U_{rT}$, curve 2) or $I_{bT}/I_{rT} > 0.45$ ($U_{bT} = U_m$, curve 6). If a deviation of ±5% is accepted, this only leads to a

violation of the tolerance limit with a voltage $U_{bT} = U_m$ and currents $I_{bT}/I_{rT} > 0.95$.
- With a power factor of $\cos\varphi = 0.9$ (inductive, not listed in Fig. 8.5), the limit shifts to $I_{bT}/I_{rT} > 0.90$ ($U_{bT} = U_{rT}$) if the factor K_{T3} is applied.

In summary, this example shows that the correction factor K_{T3} is always useful if the power factor of the load flow is $\cos\varphi > 0.9$ (inductive). With a value of $\cos\varphi = 0.8$ (inductive), the short-circuit current is calculated too small, but it remains within a tolerance limit of ±5%. If the load current I_{bT} before the short circuit occurs is an active current, the result is always on the safe side depending on the ratio U_{rT}/U_n, so that the short-circuit current is determined too large.

The use of K_{T2} always leads to a better result, but the load current must be clearly defined before the short circuit occurs, which should be problematic in the planning state of a network.

Example 2
In this example, the same transformer data as in Example 1 is selected, but the short circuit occurs on the high-voltage side of the transformer so that the supply comes from the 110 kV network. Due to the transformer impedance, the contribution in the event of a short circuit on the high-voltage side can only be $I''_k = 1.89$ kA, so that the short-circuit current contribution of the 110 kV network can be regarded as infinite for this example. The calculations are carried out under the following boundary conditions:

- The same load flow as in example 1 is assumed.
- The calculations take the correction factor K_{T3} into account.
- The correction factor K_{T3} is determined from the data of the low-voltage side of the transformer.
- The reference current is determined using the correction factor K_{T1}, but in relation to the high-voltage side of the transformer.

Since the correction factor is assigned directly to the transformer data set when calculating the short-circuit current during data input, the K_{T3} factor is determined exclusively for the low-voltage side of the transformer [1], regardless of the short-circuit location. In contrast, the calculation of the voltage drop according to Sect. 8.1 always refers to the short-circuit side of the transformer. Due to the changed load flow of the load current before the short circuit occurs, the following correction factors result:

$$K_{T1} = \frac{U_n}{U_T} \cdot \frac{c_{max}}{1 - \frac{I_T}{I_{rT}} \cdot \frac{U_{rT}}{U_T} \cdot (u_{Rr} \cdot \cos\varphi + x_T \cdot \sin\varphi)} \tag{8.34}$$

$$K_{T3} = 0.95 \cdot \frac{c_{max}}{1 + 0.6 \cdot x_T} \tag{8.35}$$

Figure 8.6 shows the deviations between the reference current and the short-circuit current calculation using the correction factor according to Eq. (8.35). It can be seen

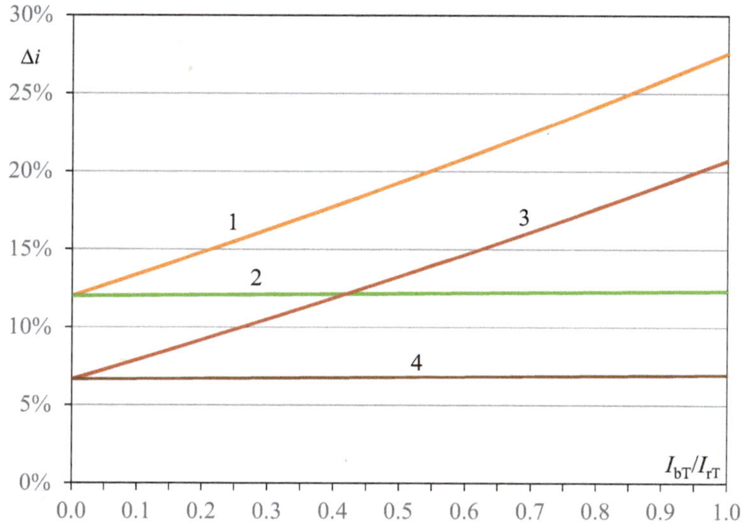

Fig. 8.6 Deviations in the calculation of the three phase short-circuit current according to Fig. 8.4 as a function of the transformer current before the short circuit occurs, short-circuit: 380 kV level. (1) K_{T3}: cos $\varphi = 0.8$ (cap.); $U_{bT} = U_{rT}$; (2) K_{T3}: cos $\varphi = 1.0$; $U_{bT} = U_{rT}$; (3) K_{T3}: cos $\varphi = 0.8$ (cap.); $U_{bT} = U_m$; (4) K_{T3}: cos $\varphi = 1.0$; $U_{bT} = U_m$

that the calculation with K_{T3} (according to [1]) is always on the safe side, i.e. too high currents are determined. Depending on the load and transformer voltage before the short circuit occurs, the deviation can be up to 27.6% ($I_{bT}/I_{rT} = 1.0$, $U_{bT} = U_{rT}$, curve 1). Since in this case the contribution to the total short-circuit current at the busbar (380 kV) is only 3.05% (1.89 kA of 61.89 kA), this deviation can be tolerated.

However, if these short-circuit current ratios do not apply and a "more precise" determination of the partial short-circuit current is necessary, the superposition method should be used. Or, if there is an inductive load flow to the downstream voltage level, the correction factor should not be applied in the event of a fault on the upstream voltage side ($K_T = 1.0$).

Exampel 3

- **Transformer Data**

rated voltage	U_r:	115 kV/22 kV
rated apparent power	S_r:	50 MVA
rel. short-circuit voltage	u_{krS}:	14% main position
winding losses	P_{kr}:	200 kW rated current

- **Network Data (110 kV)**

nominal system voltage	U_n:	110 kV
short-circuit current (three phase)	I''_{kN}:	35 kA
ratio	R_N/X_N:	0,1

A short circuit is assumed on the low-voltage side (20 kV side) according to Fig. 8.4. The following impedances can be determined from the specified data, referred to the 20 kV side:

$$\underline{Z}_T = \underline{Z}_{OS} + \underline{Z}_{MS} = (0.160 + j11.519)\,\Omega$$

According to Eqs. (4.1) and (4.2), the network impedance is calculated for the 110 kV level with $t_r = 400/120$:

$$Z_{Nt} = \frac{1,1 \cdot U_n}{\sqrt{3} \cdot I''_{kN}} \cdot \frac{1}{t_r^2} = 0.362\,\Omega$$

$$X_{Nt} = \frac{Z_{Nt}}{\sqrt{1 - (R_N/X_N)^2}} = 0.360\,\Omega \qquad R_{Nt} = 0.036\,\Omega$$

Figure 8.7 shows the deviations between the short-circuit currents Δi at the 20 kV level, taking into account the impedance correction factors K_{T1}, K_{T2} and K_{T3} according to Eqs. (8.31) to (8.33). In addition, the voltage before the short circuit occurs at the 20 kV level is changed between $U_{bT} = U_{rT}$ and $U_{bT} = U_m$.

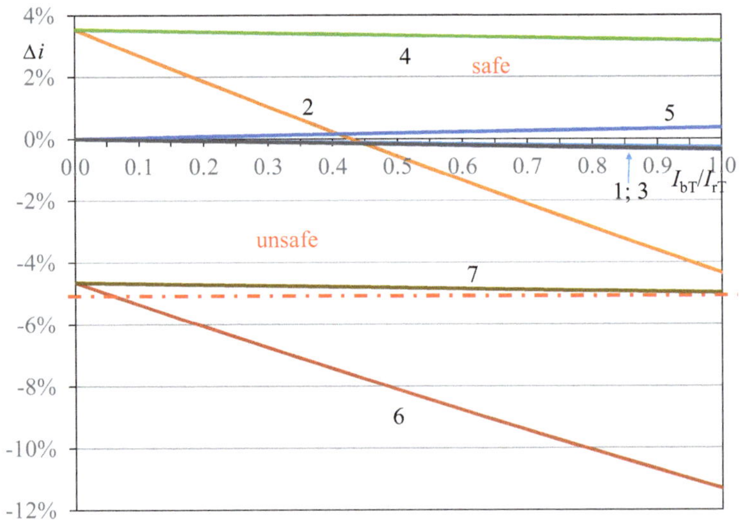

Fig. 8.7 Deviations in the calculation of the three phase short-circuit current according to Fig. 8.4 as a function of the transformer current before the short circuit occurs, short circuit: 20 kV level. (1) K_{T2}: $\cos\varphi = 0.8$; $U_{bT} = U_{rT}$; (2) K_{T3}: $\cos\varphi = 0.8$; $U_{bT} = U_{rT}$; (3) K_{T2}: $\cos\varphi = 1.0$; $U_{bT} = U_{rT}$; (4) K_{T3}: $\cos\varphi = 1.0$; $U_{bT} = U_{rT}$; (5) K_{T2}: $\cos\varphi = 0.8$; $U_{bT} = U_m$; (6) K_{T3}: $\cos\varphi = 0.8$; $U_{bT} = U_m$; (7) K_{T3}: $\cos\varphi = 1.0$; $U_{bT} = U_m$

The general curves are similar to Fig. 8.5, whereby the main deviations are caused by the ratio of the rated voltage to the voltage before the short circuit occurred (factor K_{T3}).

The following results can be derived from the diagram as a function of the transformer current before the short circuit occurs and the transformer voltage:

- The deviations in the calculation between the correction factors K_{T1} and K_{T2} remain below 0.4%, so that the use of the factor K_{T2} is justified, regardless of the load current, power factor and the transformer voltages assumed (curves 1, 3, 5).
- If only active current ($\cos \varphi = 1.0$) is transmitted, the simplified equation (K_{T3}, curve 4) leads to an average deviation of +3.34% (range: + 3.53% to +3.14%), which means that the short-circuit current is determined to be too high, with $U_{bT} = U_{rT}$. However, the deviations are within the tolerance limit of ±5% and are on the safe side.
- If the transformer voltage is higher than U_{rT}, this leads to a reduction of the correction factor K_{T1}, resulting in a higher current. This therefore leads to a lower deviation overall (K_{T3}, curve 7), resulting in an average value of—5.12%, with $\cos \varphi = 1.0$. The deviations are at the lower tolerance limit and are on the unsafe side.
- If the correction factor K_{T3} is used with a load factor of $\cos \varphi = 0.8$ (inductive), this leads to short-circuit currents that are on the unsafe side if the ratio of the transformer currents $I_{bT}/I_{rT} > 0.43$ ($U_{bT} = U_{rT}$, curve 2). A higher transformer voltage results exclusively in short-circuit currents that are on the unsafe side, with a maximum value of—11.3% at a value of $I_{bT}/I_{rT} = 1.0$ ($U_{bT} = U_m$, curve 6).
- With a power factor of $\cos \varphi = 0.9$ (inductive, not listed in Fig. 8.5), the limit shifts to $I_{bT}/I_{rT} > 0{,}58$ ($U_{bT} = U_{rT}$) if the factor K_{T3} is applied.

Example 4

In this example, the same transformer data is selected as in Example 3, but the short circuit takes place on the high-voltage side of the transformer (110 kV) so that the feeder comes from the 20 kV network. The basic procedure is therefore identical to Example 2. Due to the transformer impedance, the contribution in the event of a short circuit on the high-voltage side can only be $I''_k = 1.89$ kA, so that the short-circuit current contribution of the 20 kV network can be regarded as infinite for this example.

Under the same conditions as in Example 2, a power factor of $\cos \varphi = 0.8$ (cap.) results in the following deviation Δi, depending on the load and the transformer voltage:

- $U_{bT} = U_{rT}$: $\Delta i = 9.1\%$ to 19.6%,
- $U_{bT} = U_m$: $\Delta i = 2.0\%$ to 11.6%.

Since in this case the contribution to the total short-circuit current at the busbar (110 kV) is only 5.12% (1.89 kA of 36.89 kA), this deviation can be tolerated, as the deviation specified above only relates to the partial short-circuit current. With a maximum deviation of 19.6%, this example results in a total failure of approx. 1%.

8.4 Power Station Unit with Synchronous Generators

Similar to the load conditions in Sect. 8.1, impedance correction is also required for power station units, which has a particular effect at high reactances and inductive loads (operation in the overexcited range). The reason for different factors depending on the on-load tap-changer of the unit transformer is that the voltage regulation is different.

- Power station unit with on-load tap-changer:
  The generator voltage U_{rGS} is kept constant, while the regulation of the specified voltage U_Q or U_{nQ} on the HV side of the transformer (busbar Q) is carried out by the on-load tap-changer, Fig. 8.8.
- Power station unit without on-load tap-changer:
  The voltage U_Q or U_{nQ} on the OS side of the transformer is determined exclusively by the regulation p_{GS} of the generator voltage U_{rGS}, Fig. 8.9.

The short-circuit impedance of a power station unit is generally determined according to Eq. (8.36), taking into account the correction factor K_{PS}, Sect. 4.3, Eq. (4.49):

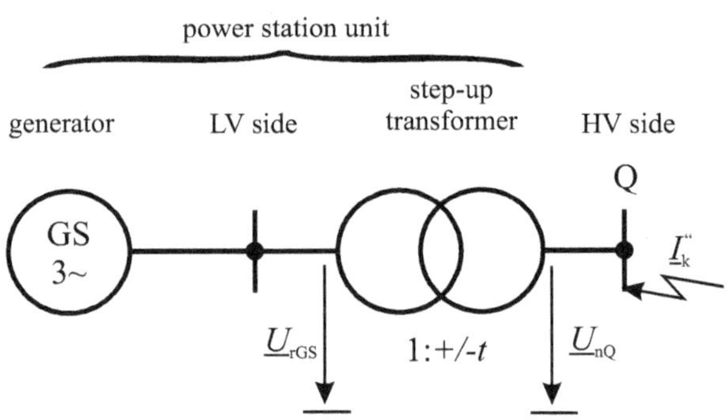

Fig. 8.8 Network diagram of power station unit with no-load tap-changer

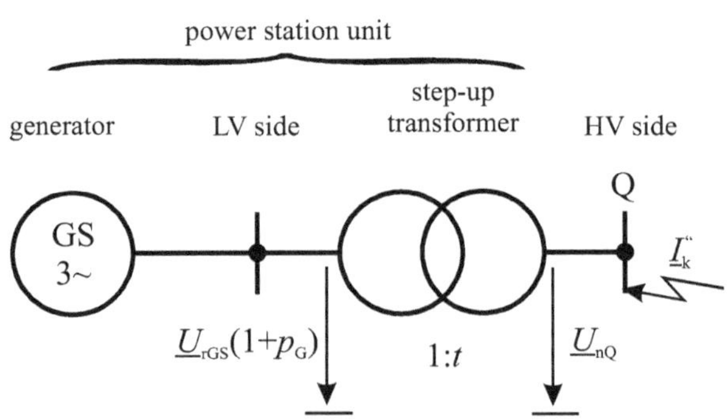

Fig. 8.9 Network diagram of power station unit without no-load tap-changer

$$\underline{Z}_{PSK} = K_{PS} \cdot \left(t_r^2 \cdot \underline{Z}_{GS} + \underline{Z}_{TOS}\right) \tag{8.36}$$

According to [1], the following correction factors are used for the impedances of the entire power station unit on the high-voltage side:

- Power station unit with on-load tap-changer:

$$K_{PS} = \frac{U_{nQ}^2}{U_{rGS}^2} \cdot \frac{1}{t_r^2} \cdot \frac{c_{max}}{1 + \left|x_d'' - x_T\right| \cdot \sqrt{1 - \cos^2\varphi_{rGS}}} \tag{8.37}$$

- Power station unit without on-load tap-changer:

$$K_{PS} = \frac{U_{nQ}}{U_{rGS} \cdot (1 + p_{GS})} \cdot \frac{1}{t_r} \cdot \frac{(1 \pm p_T) \cdot c_{max}}{1 + x_d'' \cdot \sqrt{1 - \cos^2\varphi_{rGS}}} \tag{8.38}$$

With:

$\underline{Z}_{PS}$	corrected impedance of the power station unit on the HV side,
c_{max}	voltage factor for the maximum short-circuit current acc. to Table 7.1,
$\underline{Z}_{GS}$	subtransient generator impedance,
$\underline{Z}_{TOS}$	transformer impedance on the HV side,
U_{nQ}	nominal system voltage of the HV side,
U_{rGS}	rated voltage of the generator,
$\cos\varphi_{rGS}$	power factor (rated value),
x_d''	saturated subtransient generator reactance (related value),
x_T	related reactance of the step-up transformer,
t_r	transformation ratio of the step-up transformer (rated value) ($t = U_{rTOS}/U_{rTUS}$),
$(1 \pm p_T)$	taps of the step-up transformer,
p_{GS}	voltage range of the generator.

In Eqs. (8.37) and (8.38), the voltage U_Q should be on the HV side of the transformer instead of U_{nQ}. However, calculations have shown that the maximum partial short-circuit currents due to regulation occur under the condition $U_Q = U_{nQ}$ [3, 8].

In the impedance correction method presented here, the power station is corrected as an entire unit. In practical application, this means that the "generator busbar" node does not exist. This means that network short circuits and a short circuit on the generator busbar cannot be determined at the same time. Since in practice the short-circuit current ratios at the generator busbar should only be of interest during the planning of a power station unit, there is therefore no need to repeat this calculation process for each network calculation.

If a generator supplies two decoupled networks (possibly with different voltage levels) via a three-winding transformer, these calculation examples are not covered by [1]. In these cases, suitable methods (e.g. the superposition method) should be used, as the power output to the individual networks is regulated. In addition, the load flow that leads to the maximum short-circuit current must be determined.

Special calculation rules are given in [1] for calculating the short-circuit currents at the generator busbar and in the auxiliaries service, with impedance correction methods adapted for these cases.

8.5 Conclusions

In general, the voltage factor c of the equivalent voltage source at the short-circuit location in the short-circuit current calculation compensates for the voltage drop in a network between the driving voltage and the later short-circuit location. This voltage drop has a particular effect with an inductive load flow. In some applications, however, it makes more sense to determine an "adjusted" voltage factor. This is particularly the case when generators or power station units feed in a short circuit or a short-circuit current flows via a network transformer. This "adjusted" voltage factor is used, for example, as an impedance correction factor for transformers and power station units.

References

1. IEC 60909-0:2016 Edition 2.0 (2016) Short-circuit current calculations in three-phase a.c. systems—Part 0: calculation of currents. IEC, Geneva
2. VDE 0102 Teil 1/9.62 (1962) Leitsätze für die Berechnung der Kurzschlußströme, Teil 1 Drehstromanlagen mit Nennspannungen von 1 kV und darüber. VDE, Berlin
3. IEC/TR 60909-1:2002 Edition 2.0 (2002): Short-circuit currents in three-phase a.c. systems—Part 1: factors for the calculation of short-circuit currents according to IEC 60909-0
4. IEC 909:1988 (1988) Short-circuit current calculations in three-phase a.c. systems. IEC, Geneva
5. IEC 60909-0:2001 (2001) Short-circuit current calculations in three-phase a.c. systems—Part 0: Calculation of currents. IEC, Geneva
6. Pitz V, Waider G (1993) Impedanzkorrekturfaktoren für Netztransformatoren bei der Kurzschlußstromberechnung mit der Ersatzspannungsquelle an der Kurzschlußstelle. Elektrie 47:301–304
7. Waider G (1992) Impedanzkorrekturfaktoren bei der Kurzschlußstromberechnung mit der Ersatzspannungsquelle an der Kurzschlußstelle. Dissertation TH Darmstadt, D17
8. Hunger T (1996) Beiträge zur Kurzschlussstromberechnung. Dissertation TH Darmstadt, D17

Chapter 9
Component Systems

Calculating the electrical variables of a three-phase system is relatively simple if it is a symmetrical system, which means that all voltages and currents are 120° out of phase and all components are identical and symmetrical. Determining the currents and voltages using a single-phase network then simultaneously simulates the three-phase behavior of the network. However, the analysis according to Kirchoff's laws is more complex if the network is not symmetrical, which is due, for example, to asymmetrical loads, unbalanced faults and equipment or short circuits that are not symmetrical in the three phases. In these cases, the symmetrical components introduced by Fortescue in 1918 are generally used [1].

To simplify the short-circuit current calculation, the symmetrical components are used if symmetrical equipment is assumed. The symmetrical components are derived below and the equivalent circuits for the most important fault cases are shown [2]. In addition to the symmetrical 0, 1, 2 components, 0, α, β components are also used in the networks if, for example, switching operations are to be simulated.

9.1 Three-Phase System

The main features of a three-phase system that are of interest for the simulation of networks and the derivation of component systems are described below.

9.1.1 Current and Voltage Relationships

In a three-phase system, three line-to-earth voltages U_R, U_S and U_T supply a network or a load. The voltages $u_R(t)$, $u_S(t)$ and $u_T(t)$ over time are defined by Eqs. (9.1)–(9.3), with the effective values U:

G. Balzer, *Short-Circuit Currents in Three-Phase Systems*,
https://doi.org/10.1007/978-3-658-50710-7_9

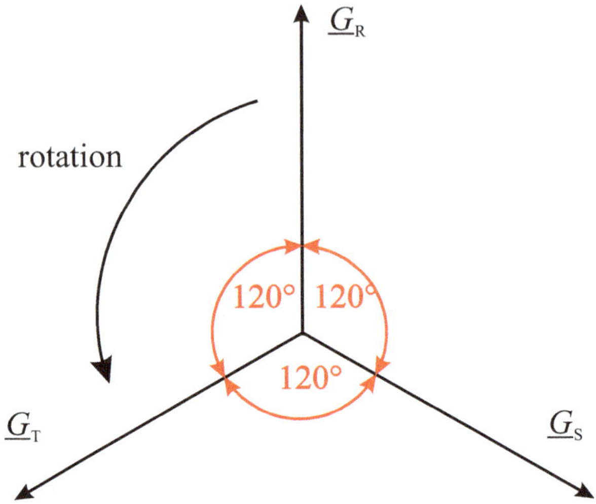

Fig. 9.1 Definition of a symmetrical current and voltage system (three-phase system) with the general variables G_R, G_S and G_T

$$u_R(t) = \sqrt{2} \cdot U \cdot \cos(\omega \cdot t) \tag{9.1}$$

$$u_S(t) = \sqrt{2} \cdot U \cdot \cos(\omega \cdot t - 120°) \tag{9.2}$$

$$u_T(t) = \sqrt{2} \cdot U \cdot \cos(\omega \cdot t - 240°) \tag{9.3}$$

The vector diagram in Fig. 9.1 can be used for better visualization and the following abbreviations a and a^2 are introduced for the complex operators to simplify the calculation, so that the following relationships apply:

$$\underline{a} = e^{j120°} = -\frac{1}{2} + j\frac{1}{2}\sqrt{3} \tag{9.4}$$

$$\underline{a}^2 = e^{j240°} = -\frac{1}{2} - j\frac{1}{2}\sqrt{3} \tag{9.5}$$

$$1 + \underline{a} + \underline{a}^2 = 0 \qquad \underline{a} = 1/\underline{a}^2 \tag{9.6}$$

$$\underline{a}^* = \underline{a}^2 \qquad \underline{a} - \underline{a}^2 = j\sqrt{3} \tag{9.7}$$

(* conjugate)

These abbreviations can be used to write the voltages and currents in a three-phase system (line–earth) if the values of line R are taken as a reference:

$$\underline{U}_S = \underline{a}^2 \cdot \underline{U}_R \quad \underline{I}_S = \underline{a}^2 \cdot \underline{I}_R \tag{9.8}$$

$$\underline{U}_T = \underline{a} \cdot \underline{U}_R \quad \underline{I}_T = \underline{a} \cdot \underline{I}_R \tag{9.9}$$

For the line-to-line voltages $\underline{U}_{RS}$, $\underline{U}_{ST}$ and $\underline{U}_{TR}$ according to Fig. 9.1, the following applies with the aid of the relationships according to Eqs. (9.8) and (9.9):

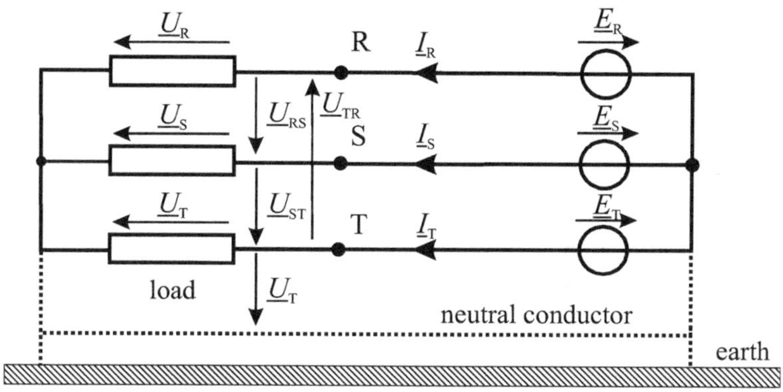

Fig. 9.2 Equivalent circuit of a three-phase system

$$\underline{U}_{RS} = \underline{U}_S - \underline{U}_R = \underline{a}^2 \cdot \underline{U}_R - \underline{U}_R = \left(\underline{a}^2 - 1\right) \cdot \underline{U}_R$$
$$= -\left[1.5 + j0.5 \cdot \sqrt{3}\right] \cdot \underline{U}_R \tag{9.10}$$

$$\underline{U}_{ST} = \underline{U}_T - \underline{U}_S = \underline{a} \cdot \underline{U}_R - \underline{a}^2 \cdot \underline{U}_R = \left(\underline{a} - \underline{a}^2\right) \cdot \underline{U}_R$$
$$= \left[0 + j1.0 \cdot \sqrt{3}\right] \cdot \underline{U}_R \tag{9.11}$$

$$\underline{U}_{TR} = \underline{U}_R - \underline{U}_T = \underline{U}_R - \underline{a} \cdot \underline{U}_R = (1 - \underline{a}) \cdot \underline{U}_R = \left[1.5 - j0.5 \cdot \sqrt{3}\right] \cdot \underline{U}_R \tag{9.12}$$

It can be seen that in a symmetrical three-phase system, both the line-to-line and the line-to-earth voltages can be represented by the relationships U_R and the complex operators $\underline{a}$ and $\underline{a}^2$. Equations (9.10)–(9.12) show that the following applies to the voltage magnitudes in each case:

$$U_{TR} = \sqrt{3} \cdot U_R \tag{9.13}$$

In electric power supply, the line-to-line voltage, e.g. U_{RS}, is used to describe a system. This value is referred to as the nominal system voltage of the network. In a 110 kV network, the line-to-earth voltage, U_R, therefore has a value of 110 kV/√3 = 63.51 kV.

Figure 9.2 describes a three-phase system consisting of a symmetrical voltage system $\underline{E}_R$, $\underline{E}_S$ and $\underline{E}_T$ and a load, which in this case is connected in a star. Assuming that the individual phase impedances of the load are equal, the amplitudes of the voltages $\underline{U}_R$, $\underline{U}_S$ and $\underline{U}_T$ and the currents $\underline{I}_R$, $\underline{I}_S$ and $\underline{I}_T$ are equal. The consequence is that in this case no current flows via the neutral conductor (return conductor) or earth. If, in contrast to this, AC consumers are also connected between a line and the neutral conductor, as is the case in a low-voltage network, an additional return conductor must be provided. Figure 9.2 shows that a three-phase system can also be seen as a superposition of three alternating current systems with different voltages and currents. This knowledge is also used when implementing a component system.

9.1.2 *Impedances of Three-Phase System*

Equipment of a three-phase system, e.g. a no-loss overhead line as shown in Fig. 9.3, not only has a conductor impedance (self-inductance), but also an inductive and capacitive coupling to the other conductors and to earth.

For an asset (e.g. overhead line) according to Fig. 9.3, this results in an impedance matrix $\underline{\mathbf{Z}}$ according to Eq. (9.14).

$$\underline{\mathbf{Z}} = \begin{pmatrix} \underline{Z}_A & \underline{Z}_B & \underline{Z}_C \\ \underline{Z}_C & \underline{Z}_A & \underline{Z}_B \\ \underline{Z}_B & \underline{Z}_C & \underline{Z}_A \end{pmatrix} \tag{9.14}$$

The general matrix according to Eq. (9.14) also applies to motors and generators, while the mutual impedances $\underline{Z}_B$, $\underline{Z}_C$ are identical for symmetrical equipment. The following generally applies to voltages and currents:

$$\underline{\mathbf{U}} = \underline{\mathbf{Z}} \cdot \underline{\mathbf{I}} \tag{9.15}$$

Taking into account the conditions according to Eqs. (9.8) and (9.9), the system of equations is as follows

$$\underline{U}_R = \underline{Z}_A \cdot \underline{I}_R + \underline{Z}_B \cdot \underline{a}^2 \cdot \underline{I}_R + \underline{Z}_C \cdot \underline{a} \cdot \underline{I}_R \tag{9.16}$$

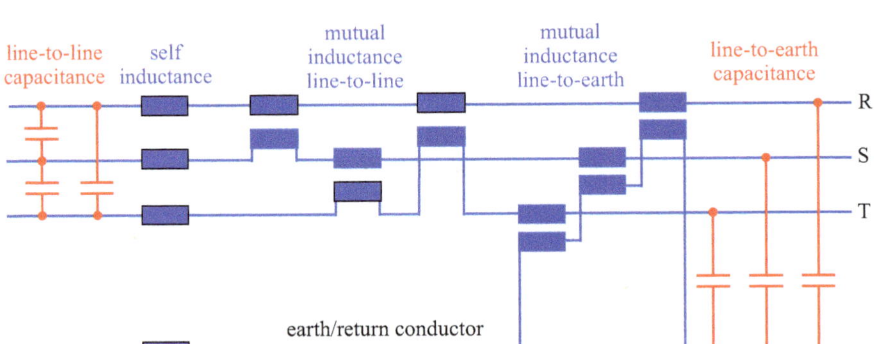

Fig. 9.3 Three-phase equivalent circuit of a no-loss overhead line

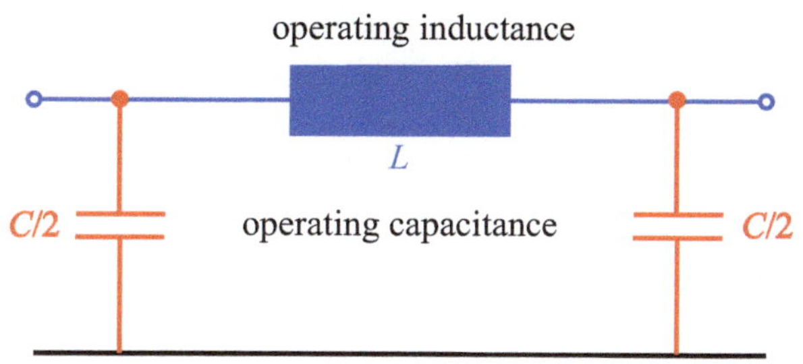

Fig. 9.4 Single-line equivalent circuit of a no-loss overhead line

$$\underline{U}_\mathrm{S} = \underline{Z}_\mathrm{C} \cdot \underline{a} \cdot \underline{I}_\mathrm{S} + \underline{Z}_\mathrm{A} \cdot \underline{I}_\mathrm{S} + \underline{Z}_\mathrm{B} \cdot \underline{a}^2 \cdot \underline{I}_\mathrm{S} \tag{9.17}$$

$$\underline{U}_\mathrm{T} = \underline{Z}_\mathrm{B} \cdot \underline{a}^2 \cdot \underline{I}_\mathrm{T} + \underline{Z}_\mathrm{C} \cdot \mathrm{a} \cdot \underline{I}_\mathrm{T} + \underline{Z}_\mathrm{A} \cdot \underline{I}_\mathrm{T} \tag{9.18}$$

or in matrix notation

$$\begin{pmatrix} \underline{U}_\mathrm{R} \\ \underline{U}_\mathrm{S} \\ \underline{U}_\mathrm{T} \end{pmatrix} = \begin{pmatrix} \underline{Z}_\mathrm{A} + \underline{a}^2 \cdot \underline{Z}_\mathrm{B} + \underline{a} \cdot \underline{Z}_\mathrm{C} & 0 & 0 \\ 0 & \underline{Z}_\mathrm{A} + \underline{a}^2 \cdot \underline{Z}_\mathrm{B} + \underline{a} \cdot \underline{Z}_\mathrm{C} & 0 \\ 0 & 0 & \underline{Z}_\mathrm{A} + \underline{a}^2 \cdot \underline{Z}_\mathrm{B} + \underline{a} \cdot \underline{Z}_\mathrm{C} \end{pmatrix} \cdot \begin{pmatrix} \underline{I}_\mathrm{R} \\ \underline{I}_\mathrm{S} \\ \underline{I}_\mathrm{T} \end{pmatrix} \tag{9.19}$$

Under the condition of a cyclically symmetrical component ($\underline{Z}_\mathrm{C} = \underline{Z}_\mathrm{B}$), matrix (9.19) is simplified to

$$\begin{pmatrix} \underline{U}_\mathrm{R} \\ \underline{U}_\mathrm{S} \\ \underline{U}_\mathrm{T} \end{pmatrix} = \begin{pmatrix} \underline{Z}_\mathrm{A} - \underline{Z}_\mathrm{B} & 0 & 0 \\ 0 & \underline{Z}_\mathrm{A} - \underline{Z}_\mathrm{B} & 0 \\ 0 & 0 & \underline{Z}_\mathrm{A} - \underline{Z}_\mathrm{B} \end{pmatrix} \cdot \begin{pmatrix} \underline{I}_\mathrm{R} \\ \underline{I}_\mathrm{S} \\ \underline{I}_\mathrm{T} \end{pmatrix} \tag{9.20}$$

Equations (9.19) and (9.20) show that in a symmetrical voltage and current system, the three systems R, S and T are decoupled and identical. In these cases, it is permissible to calculate with a single-line equivalent circuit according to Fig. 9.4 for a no-loss overhead line, which corresponds to the detailed equivalent circuit according to Fig. 9.3.

9.2 Component Systems for Network Calculations

Only in the case of a symmetrical voltage and current system, a decoupling of a cyclically symmetrical matrix takes place (Sect. 9.1.2), so that a calculation with a single-line model is possible. If, on the other hand, there is an asymmetry, the equation system must be solved with all couplings. In these cases, it has proven to be practical to carry out the calculation using a transformation into a component system [3].

9.2.1 General

Processing using a component system is carried out in the following steps according to Fig. 9.5:

1. Problem definition in the original system (R, S, T system),
2. Transformation to the component level,
3. Solution of the problem at the component level,
4. Transformation back into the original system (R, S, T-system),
5. Solving the problem in the R, S, T system.

In electrical power engineering, various component systems have been developed in the past [1, 4, 5], each of which is particularly suitable for certain processes, e.g.:

- Symmetrical components (Fortescue components): 0-1-2, calculation of stationary processes (Sect. 9.2.2),
- Diagonal components (Clarke components): 0-α-β, calculation of transient processes (Sect. 9.4),
- Two-axis components (Park components): 0-d-q, Calculation of transient processes in synchronous machines.

According to [6], the following requirements can be defined for a transformation:

- Same calculation rules for the conversion of currents and voltages.
- The impedance matrix must be converted to diagonal form.
- The transformation is identical for all equipment.
- A reverse transformation must be possible.

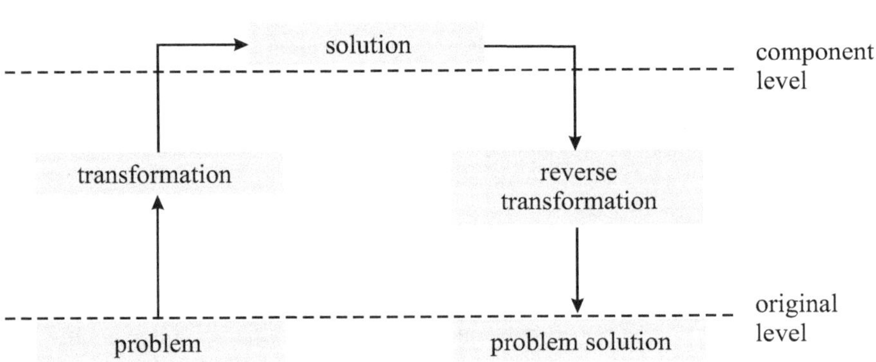

Fig. 9.5 Process flow for the solution using a component system

If the power in the original area and component area should not change, this transformation is referred to as orthonormal. Otherwise, if there is a scale factor between these two systems, it is an orthogonal transformation.

The application of symmetrical components for the short-circuit current calculation is presented below [7], which are mainly used in the calculation of short-circuit currents. For this reason, the corresponding equivalent circuit for typical short-circuit cases are derived in Sect. 9.2.2.3 using the symmetrical components. In principle, it is of course also possible to use the 0-α-β components for all short-circuit cases, as shown as an example in Sect. 9.4.

9.2.2 *Symmetrical Components (Fortescue Components)*

Symmetrical components are generally used for the calculation of steady-state processes (calculation of short-circuit currents in three-phase systems). They are characterized by a complex transformation matrix and for faults that are symmetrical to phase R, the connections between the component systems are implemented without additional ideal transformers. These components are also referred to as Fortescue components.

The basic idea is that the original system can be composed of three different systems (components), namely the

- Positive sequence system: Symmetrical three-phase system with natural phase sequence,
- Negative sequence system: Symmetrical three-phase system with opposite phase sequence,
- Zero sequence system: Currents/voltages with the same phase sequence.

Based on these three different component systems, the relevant current and voltage ratios are determined. The solution in the original system then results from the geometric addition of the component values.

9.2.2.1 Derivation

The essential basic requirement of a transformation matrix is that both the voltages and the currents are to be transformed with the same matrices and the impedance matrix is brought to diagonal form. The system of Eq. (9.15) can therefore be written for a three-phase system:

$$\underline{\mathbf{U}}_{\mathrm{RST}} = \underline{\mathbf{Z}}_{\mathrm{RST}} \cdot \underline{\mathbf{I}}_{\mathrm{RST}} \qquad \begin{pmatrix} \underline{U}_{\mathrm{R}} \\ \underline{U}_{\mathrm{S}} \\ \underline{U}_{\mathrm{T}} \end{pmatrix} = \begin{pmatrix} \underline{Z}_{\mathrm{A}} & \underline{Z}_{\mathrm{B}} & \underline{Z}_{\mathrm{C}} \\ \underline{Z}_{\mathrm{C}} & \underline{Z}_{\mathrm{A}} & \underline{Z}_{\mathrm{B}} \\ \underline{Z}_{\mathrm{B}} & \underline{Z}_{\mathrm{C}} & \underline{Z}_{\mathrm{A}} \end{pmatrix} \cdot \begin{pmatrix} \underline{I}_{\mathrm{R}} \\ \underline{I}_{\mathrm{S}} \\ \underline{I}_{\mathrm{T}} \end{pmatrix} \tag{9.21}$$

The question now arises whether a suitable transformation matrix can be used to convert the resulting matrix into diagonal form as follows, according to Eq. (9.22):

$$\underline{\mathbf{U}}_{(012)} = \underline{\mathbf{Z}}_{(012)} \cdot \underline{\mathbf{I}}_{(012)} \qquad \begin{pmatrix} \underline{U}_{(0)} \\ \underline{U}_{(1)} \\ \underline{U}_{(2)} \end{pmatrix} = \begin{pmatrix} \underline{Z}_{(0)} & 0 & 0 \\ 0 & \underline{Z}_{(1)} & 0 \\ 0 & 0 & \underline{Z}_{(2)} \end{pmatrix} \cdot \begin{pmatrix} \underline{I}_{(0)} \\ \underline{I}_{(1)} \\ \underline{I}_{(2)} \end{pmatrix} \tag{9.22}$$

This results in a decoupling of the three systems [here: (0), (1) and (2)] equivalent to Eq. (9.19) with the advantages mentioned above. If a regular matrix $\underline{\mathbf{T}}$ with the quantities according to Eq. (9.23) is selected for the transformation of the vectors $\underline{\mathbf{U}}$ and $\underline{\mathbf{I}}$,

$$\underline{\mathbf{T}} = \begin{pmatrix} \underline{t}_{11} & \underline{t}_{12} & \underline{t}_{13} \\ \underline{t}_{21} & \underline{t}_{22} & \underline{t}_{23} \\ \underline{t}_{31} & \underline{t}_{32} & \underline{t}_{33} \end{pmatrix} \tag{9.23}$$

using the following transformation equations results in

$$\underline{\mathbf{U}}_{\mathrm{RST}} = \underline{\mathbf{T}} \cdot \underline{\mathbf{U}}_{(012)} \qquad \underline{\mathbf{I}}_{\mathrm{RST}} = \underline{\mathbf{T}} \cdot \underline{\mathbf{I}}_{(012)} \tag{9.24}$$

and the inverse matrix $\mathbf{T}^{-1}$

$$\underline{\mathbf{U}}_{(012)} = \underline{\mathbf{T}}^{-1} \cdot \underline{\mathbf{U}}_{\mathrm{RST}} \qquad \underline{\mathbf{I}}_{(012)} = \underline{\mathbf{T}}^{-1} \cdot \underline{I}_{\mathrm{RST}} \tag{9.25}$$

by inserting Eqs. (9.24) and (9.25) into Eq. (9.26) to obtain the desired result.

$$\underline{\mathbf{U}}_{(012)} = \underline{\mathbf{T}}^{-1} \cdot \underline{\mathbf{Z}}_{\mathrm{RST}} \cdot \underline{\mathbf{T}} \cdot \underline{\mathbf{I}}_{(012)} \tag{9.26}$$

The comparison of Eqs. (9.22) and (9.26) then provides:

$$\underline{\mathbf{Z}}_{(012)} = \underline{\mathbf{T}}^{-1} \cdot \underline{\mathbf{Z}}_{\mathrm{RST}} \cdot \underline{\mathbf{T}} \tag{9.27}$$

If this problem is to be solvable, i.e. $\underline{\mathbf{Z}}_{(012)}$ is to have a diagonal structure, the transformation matrix must consist of the eigenvectors of $\underline{\mathbf{Z}}_{\mathrm{RST}}$. In this case, the matrix $\underline{\mathbf{T}}$ is composed column by column of the coordinates of the eigenvectors forming the new matrix. The main diagonal elements of $\underline{\mathbf{Z}}_{(012)}$ that are also being searched for result from the eigenvalues of $\underline{\mathbf{Z}}_{\mathrm{RST}}$. The non-trivial solutions of the system of equations are therefore to be determined

$$\underline{\mathbf{Z}}_{\mathrm{RST}} \cdot \underline{\mathbf{X}} = \underline{\lambda} \cdot \underline{\mathbf{X}} \qquad (\underline{\mathbf{Z}}_{\mathrm{RST}} - \underline{\lambda} \cdot \mathbf{E}) \cdot \underline{\mathbf{X}} = 0 \tag{9.28}$$

with any three-dimensional vector $\underline{\mathbf{X}}$ and a complex number λ. The eigenvalues λ_i of the matrix $\underline{\mathbf{Z}}_{\mathrm{RST}}$ are determined by the condition for the non-trivial solvability of the eigenvalue Eq. (9.29):

$$\det(\underline{\mathbf{Z}}_{\mathrm{RST}} - \underline{\lambda} \cdot \mathbf{E}) = \begin{vmatrix} \underline{Z}_{\mathrm{A}} - \underline{\lambda} & \underline{Z}_{\mathrm{B}} & \underline{Z}_{\mathrm{C}} \\ \underline{Z}_{\mathrm{C}} & \underline{Z}_{\mathrm{A}} - \underline{\lambda} & \underline{Z}_{\mathrm{B}} \\ \underline{Z}_{\mathrm{B}} & \underline{Z}_{\mathrm{C}} & \underline{Z}_{\mathrm{A}} - \underline{\lambda} \end{vmatrix} = 0 \tag{9.29}$$

This results in the characteristic equation:

$$\begin{aligned} &\underline{\lambda}^3 + \underline{\lambda}^2 \cdot (-3 \cdot \underline{Z}_{\mathrm{A}}) + \underline{\lambda} \cdot \left(3 \cdot \underline{Z}_{\mathrm{A}}^2 - 3 \cdot \underline{Z}_{\mathrm{B}} \cdot \underline{Z}_{\mathrm{C}}\right) + \\ &\left(-\underline{Z}_{\mathrm{A}}^3 - \underline{Z}_{\mathrm{B}}^3 - \underline{Z}_{\mathrm{C}}^3 + 3 \cdot \underline{Z}_{\mathrm{A}} \cdot \underline{Z}_{\mathrm{B}} \cdot \underline{Z}_{\mathrm{C}}\right) = 0 \end{aligned} \tag{9.30}$$

The solution can be determined using the Cardan solution formula for cubic equations.

$$\lambda^3 + r \cdot \lambda^2 + s \cdot \lambda + t = 0 \tag{9.31}$$

With

$$\sigma = \lambda \pm \frac{r}{3} \tag{9.32}$$

results

$$\begin{aligned} &\sigma^3 + p \cdot \sigma + q = 0 \qquad \underline{p} = \frac{3\underline{s} - \underline{r}^2}{3} \\ &\underline{q} = \frac{2\underline{r}^3}{27} - \frac{\underline{r} \cdot \underline{s}}{3} + \underline{t} \end{aligned} \tag{9.33}$$

and the solutions

$$\underline{\sigma}_1 = \underline{u} + \underline{v} \tag{9.34}$$

$$\underline{\sigma}_2 = -\frac{\underline{u} + \underline{v}}{2} + \frac{\underline{u} - \underline{v}}{2} \mathrm{j}\sqrt{3} = \underline{\mathrm{a}} \cdot \underline{u} + \underline{\mathrm{a}}^2 \cdot \underline{v} \tag{9.35}$$

$$\underline{\sigma}_3 = -\frac{\underline{u} + \underline{v}}{2} - \frac{\underline{u} - \underline{v}}{2} \mathrm{j}\sqrt{3} = \underline{\mathrm{a}}^2 \underline{u} + \underline{\mathrm{a}} \cdot \underline{v} \tag{9.36}$$

$$\underline{u} = \sqrt[3]{-\frac{\underline{q}}{2} + \sqrt{\underline{D}}} \qquad \underline{v} = -\frac{\underline{p}}{3 \cdot \underline{u}} \qquad \underline{D} = \frac{\underline{p}^3}{27} + \frac{\underline{p}^2}{4} \tag{9.37}$$

This means in the case under consideration

$$\underline{\lambda}_{(0)} = \underline{Z}_A + \underline{Z}_B + \underline{Z}_C \tag{9.38}$$

$$\underline{\lambda}_{(1)} = \underline{Z}_A + \underline{a}^2 \cdot \underline{Z}_B + \underline{a} \cdot \underline{Z}_C \tag{9.39}$$

$$\underline{\lambda}_{(2)} = \underline{Z}_A + \underline{a} \cdot \underline{Z}_B + \underline{a}^2 \cdot \underline{Z}_C \tag{9.40}$$

The eigenvectors are obtained by inserting Eqs. (9.38)–(9.40) in Eq. (9.28) and then solving the system of equations (e.g. using the Gaussian algorithm):

$$\underline{\mathbf{X}}_{(0)} = \mu_{(0)} \cdot \begin{pmatrix} 1 \\ 1 \\ 1 \end{pmatrix} \qquad \underline{\mathbf{X}}_{(1)} = \mu_{(1)} \cdot \begin{pmatrix} 1 \\ \underline{a}^2 \\ \underline{a} \end{pmatrix} \qquad \underline{\mathbf{X}}_{(2)} = \mu_{(2)} \cdot \begin{pmatrix} 1 \\ \underline{a} \\ \underline{a}^2 \end{pmatrix} \tag{9.41}$$

The required transformation matrix $\underline{\mathbf{T}}$ is formed column by column from the eigenvectors (9.41), whereby $\mu_0 = \mu_1 = \mu_2 = 1$ is selected:

$$\mathbf{T} = \begin{pmatrix} 1 & 1 & 1 \\ 1 & \underline{a}^2 & \underline{a} \\ 1 & \underline{a} & \underline{a}^2 \end{pmatrix} \qquad \mathbf{T}^{-1} = \frac{1}{3} \cdot \begin{pmatrix} 1 & 1 & 1 \\ 1 & \underline{a} & \underline{a}^2 \\ 1 & \underline{a}^2 & \underline{a} \end{pmatrix} \tag{9.42}$$

The eigenvalues, Eq. (9.41), form the required coefficients of $\underline{\mathbf{Z}}_{012}$.

$$\underline{\mathbf{Z}}(012) \cdot \begin{pmatrix} \underline{\lambda}_{(0)} & 0 & 0 \\ 0 & \underline{\lambda}_{(1)} & 0 \\ 0 & 0 & \underline{\lambda}_{(2)} \end{pmatrix} = \begin{pmatrix} \underline{Z}_A + \underline{Z}_B + \underline{Z}_C & 0 & 0 \\ 0 & \underline{Z}_A + \underline{Z}_B + \underline{Z}_C & 0 \\ 0 & 0 & \underline{Z}_A + \underline{Z}_B + \underline{Z}_C \end{pmatrix} \tag{9.43}$$

The transformation makes it possible to convert an unbalanced three-phase system into three single-phase systems. If the couplings between the individual conductors of a piece of equipment are identical, e.g. an overhead line (Fig. 9.3), the system of Eq. (9.43) is simplified to $\underline{Z}_B = \underline{Z}_C$:

$$\underline{Z}_{(0)} = \underline{Z}_A + 2 \cdot \underline{Z}_B \qquad \underline{Z}_{(1)} = \underline{Z}_{(2)} = \underline{Z}_A - \underline{Z}_B \tag{9.44}$$

In addition to calculating the impedance of a component system, it is also necessary to determine the active voltages. The following applies in this case

$$\begin{pmatrix} \underline{E}_{(0)} \\ \underline{E}_{(1)} \\ \underline{E}_{(2)} \end{pmatrix} = \frac{1}{3} \cdot \begin{pmatrix} 1 & 1 & 1 \\ 1 & \underline{a} & \underline{a}^2 \\ 1 & \underline{a}^2 & \underline{a} \end{pmatrix} \cdot \begin{pmatrix} \underline{U}_R \\ \underline{U}_S \\ \underline{U}_T \end{pmatrix} \tag{9.45}$$

Equation (9.45) shows that, based on the voltages $\underline{U}_R$, $\underline{U}_S$ and $\underline{U}_T$, the voltage in the zero sequence system $\underline{E}_{(0)}$ has the same phase position, while the voltage $\underline{E}_{(1)}$ is derived from a rotating field (equal to the original system RST) and $\underline{E}_{(2)}$ represents a

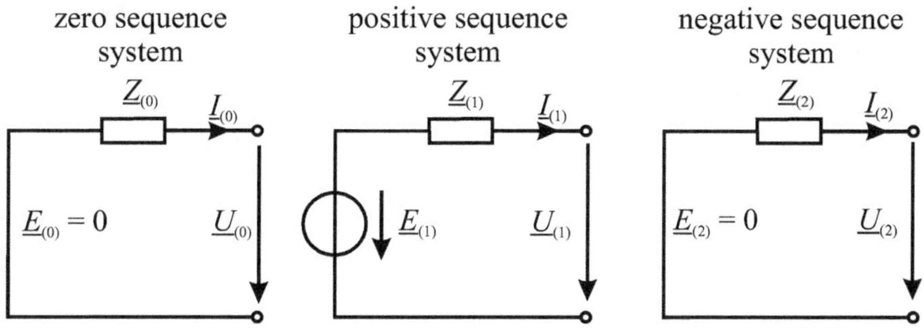

Fig. 9.6 Three single-phase equivalent circuits of the component systems to represent a balanced three-phase system

voltage system with the opposite phase sequence. If a symmetrical voltage system with the active voltages $\underline{\mathbf{E}}_{\text{RST}}$ is transformed according to Eq. (9.24), the following values result, (E_{R}: reference voltage)

$$\begin{pmatrix} \underline{E}_{(0)} \\ \underline{E}_{(1)} \\ \underline{E}_{(2)} \end{pmatrix} = \frac{1}{3} \cdot \begin{pmatrix} 1 & 1 & 1 \\ 1 & \underline{a} & \underline{a}^2 \\ 1 & \underline{a}^2 & \underline{a} \end{pmatrix} \cdot \begin{pmatrix} 1 \\ \underline{a}^2 \\ \underline{a} \end{pmatrix} \cdot E_{\text{R}} = \begin{pmatrix} 0 \\ E_{\text{R}} \\ 0 \end{pmatrix} \tag{9.46}$$

According to Eq. (9.46), in a symmetrical RST voltage system only one component system has an active voltage, the positive sequence system. In contrast to this, the zero sequence system and the negative sequence system have no voltages. Using the transformation described above, any circuit of the RST system can therefore be broken down into three identical equivalent circuits using the symmetrical components as shown in Fig. 9.6. If there is no symmetrical system, e.g. with unequal voltages, different loads on the individual conductors, short circuits or line disconnections, the individual component systems are interconnected and voltages exist in all component equivalent circuits. However, the calculation of the component currents is always simpler than the calculation in the original system.

The transformation according to Eq. (9.42) is a power-variant form of transformation, since the powers in the original system are greater by a factor of 3 than in the component system. The following applies to the power:

$$\underline{\mathbf{S}}_{\text{RST}} = \underline{U}_{\text{R}} \cdot \underline{I}_{\text{R}}^* + \underline{U}_{\text{S}} \cdot \underline{I}_{\text{S}}^* + \underline{U}_{\text{T}} \cdot \underline{I}_{\text{T}}^* = (\underline{\mathbf{U}}_{\text{RST}})^{\text{T}} \cdot (\underline{\mathbf{I}}_{\text{RST}})^* \tag{9.47}$$

$$\begin{aligned} \underline{\mathbf{S}}_{\text{RST}} &= \left(\underline{\mathbf{T}}^{-1} \cdot \underline{\mathbf{U}}_{(012)}\right)^{\mathbf{T}} \cdot \left(\underline{\mathbf{T}}^{-1} \cdot \underline{\mathbf{I}}_{(012)}\right)^* \\ &= \left(\underline{\mathbf{U}}_{(012)}\right)^{\mathbf{T}} \cdot \left(\underline{\mathbf{T}}^{-1}\right)^{\mathbf{T}} \cdot \left(\underline{\mathbf{T}}^{-1}\right)^* \cdot \left(\underline{\mathbf{I}}_{(012)}\right)^* \end{aligned} \tag{9.48}$$

$$\underline{\mathbf{S}}_{\mathrm{RST}} = \left(\underline{\mathbf{U}}_{(012)}\right)^{\mathbf{T}} \cdot \begin{pmatrix} 1 & 1 & 1 \\ 1 & \underline{a}^2 & \underline{a} \\ 1 & \underline{a} & \underline{a}^2 \end{pmatrix} \cdot \begin{pmatrix} 1 & 1 & 1 \\ 1 & \underline{a} & \underline{a}^2 \\ 1 & \underline{a}^2 & \underline{a} \end{pmatrix} \cdot \left(\underline{\mathbf{I}}_{(012)}\right)^{*}$$
$$= 3 \cdot \underline{\mathbf{S}}_{012} \tag{9.49}$$

If this disadvantage of the transformation is to be avoided, the power-invariant forms of the transformation matrices are to be used instead, according to Eq. (9.50):

$$\mathbf{T} = \frac{1}{\sqrt{3}} \cdot \begin{pmatrix} 1 & 1 & 1 \\ 1 & \underline{a}^2 & \underline{a} \\ 1 & \underline{a} & \underline{a}^2 \end{pmatrix} \qquad \mathbf{T}^{-1} = \frac{1}{\sqrt{3}} \cdot \begin{pmatrix} 1 & 1 & 1 \\ 1 & \underline{a} & \underline{a}^2 \\ 1 & \underline{a}^2 & \underline{a} \end{pmatrix} \tag{9.50}$$

The main advantages and disadvantages of symmetrical components are:

- In a cyclic, symmetrical network, the component systems are decoupled.
- In a cyclic, symmetrical network, the positive and negative sequence impedances of the component systems are equal.
- The fault circuits for unbalanced loads, short circuits and interruptions consist of galvanic couplings if these faults are each symmetrical to phase R. In the other case, complex transformers are required between the component systems.
- Phase-shifting transformers also require complex ideal transformers.
- Under certain conditions, the calculation of transient network processes is possible if the transformation does not cause a phase shift between the component and original system.

9.2.2.2 Transformation of Symmetrical Components via Power Transformers

It is possible to calculate the short-circuit currents using the symmetrical components within a voltage level. This is also considered in the corresponding standard [8], as only short-circuit currents or total currents in one voltage level are considered. If, in contrast to this, the current and voltage distribution is also to be determined on other voltage levels, the shifting of the phase position depending on the vector group or phase angle index of the transformers may have to be taken into account. The following conditions apply for the conversion from the primary side (HV) to the secondary side (LV) and vice versa.

Primary → secondary

$$\underline{\mathbf{U}}_{(012)}^{(\mathrm{s})} = \frac{1}{t} \cdot \underline{\ddot{\mathbf{U}}}^{(\mathrm{p})} \cdot \underline{\mathbf{U}}_{(012)}^{(\mathrm{p})} \qquad \underline{\mathbf{I}}_{(012)}^{(\mathrm{s})} = t \cdot \underline{\ddot{\mathbf{U}}}^{(\mathrm{p})} \cdot \underline{\mathbf{I}}_{(012)}^{(\mathrm{p})} \tag{9.51}$$

Secondary → primary

Table 9.1 Transfer matrices for taking phase shifting into account when transferring symmetrical components via transformers

No.	Vector group	$\underline{\ddot{\mathbf{U}}}^{(\mathrm{p})}$	$\underline{\ddot{\mathbf{U}}}^{(\mathrm{s})}$
1	Yy0	$\begin{pmatrix} 1^{\mathrm{a}} & 0 & 0 \\ 0 & 1 & 0 \\ 0 & 0 & 1 \end{pmatrix}$	$\begin{pmatrix} 1^{\mathrm{a}} & 0 & 0 \\ 0 & 1 & 0 \\ 0 & 0 & 1 \end{pmatrix}$
2	Dy5 Yd5 Yz5	$\begin{pmatrix} 0 & 0 & 0 \\ 0 & \mathrm{j}\underline{a} & 0 \\ 0 & 0 & -\mathrm{j}\underline{a}^2 \end{pmatrix}$	$\begin{pmatrix} 0 & 0 & 0 \\ 0 & -\mathrm{j}\underline{a}^2 & 0 \\ 0 & 0 & \mathrm{j}\underline{a} \end{pmatrix}$
3	Yd11	$\begin{pmatrix} 0 & 0 & 0 \\ 0 & -\mathrm{j}\underline{a} & 0 \\ 0 & 0 & -\mathrm{j}\underline{a}^2 \end{pmatrix}$	$\begin{pmatrix} 0 & 0 & 0 \\ 0 & \mathrm{j}\underline{a}^2 & 0 \\ 0 & 0 & -\mathrm{j}\underline{a} \end{pmatrix}$

[a] 1 only if both neutral points are earthed, otherwise 0

$$\underline{\mathbf{U}}_{(012)}^{(\mathrm{p})} = t \cdot \underline{\ddot{\mathbf{U}}}^{(\mathrm{s})} \cdot \underline{\mathbf{U}}_{(012)}^{(\mathrm{s})} \qquad \underline{\mathbf{I}}_{(012)}^{(\mathrm{p})} = \frac{1}{t} \cdot \underline{\ddot{\mathbf{U}}}^{(\mathrm{s})} \cdot \underline{\mathbf{I}}_{(012)}^{(\mathrm{s})} \tag{9.52}$$

The transformation matrices $\underline{\ddot{\mathbf{U}}}$ can be taken from Table 9.1 for the various vector groups. For line-to-earth short circuit on the high voltage side of a Yd5 transformer, the individual component currents are calculated according to Eqs. (9.53) and (9.71).

$$\underline{I}_{(1)} = \underline{I}_{(2)} = \underline{I}_{(0)} = \frac{c \cdot U_{\mathrm{n}}/\sqrt{3}}{2 \cdot \underline{Z}_{(1)} + \underline{Z}_{(0)}} \tag{9.53}$$

The conversion to the secondary side (s) results from Eq. (9.51) and No. 2 of Table 9.1 if the values of the primary side (p) are known.

$$\begin{pmatrix} \underline{I}_{(0)} \\ \underline{I}_{(1)} \\ \underline{I}_{(2)} \end{pmatrix}^{(\mathrm{s})} = t \cdot \begin{pmatrix} 0 & 0 & 0 \\ 0 & \mathrm{j}\underline{a} & 0 \\ 0 & 0 & -\mathrm{j}\underline{a}^2 \end{pmatrix} \cdot \begin{pmatrix} \underline{I}_{(0)} \\ \underline{I}_{(1)} \\ \underline{I}_{(2)} \end{pmatrix}^{(\mathrm{p})} \tag{9.54}$$

From this follows with $t = 1$:

$$\begin{aligned} \underline{I}_{(0)}^{(\mathrm{s})} &= 0 \\ \underline{I}_{(1)}^{(\mathrm{s})} &= \mathrm{j}\underline{a} \cdot \underline{I}_{(1)}^{(\mathrm{p})} \\ \underline{I}_{(2)}^{(\mathrm{s})} &= -\mathrm{j}\underline{a}^2 \cdot \underline{I}_{(2)}^{(\mathrm{p})} \end{aligned} \tag{9.55}$$

Using the transfer matrix (9.24), the following applies to the currents in the RST system.

$$\begin{pmatrix} \underline{I}_R \\ \underline{I}_S \\ \underline{I}_T \end{pmatrix}^{(s)} = \begin{pmatrix} 1 & 1 & 1 \\ 1 & \underline{a}^2 & \underline{a} \\ 1 & \underline{a} & \underline{a}^2 \end{pmatrix} \cdot \begin{pmatrix} 0 \\ j\underline{a} \cdot \underline{I}_{(1)} \\ -j\underline{a}^2 \cdot \underline{I}_{(2)} \end{pmatrix}^{(s)} \tag{9.56}$$

$$\begin{aligned} \underline{I}_R^{(s)} &= j \cdot (\underline{a} - \underline{a}^2) \cdot \underline{I}_{(1)}^{(p)} = -\sqrt{3} \cdot \underline{I}_{(1)}^{(p)} \\ \underline{I}_S^{(s)} &= 0 \\ \underline{I}_T^{(s)} &= j \cdot (\underline{a}^2 - \underline{a}) \cdot \underline{I}_{(1)}^{(p)} = \sqrt{3} \cdot \underline{I}_{(1)}^{(p)} \end{aligned} \tag{9.57}$$

A line-to-earth fault, for example on the high-voltage side of a step-up transformer, thus becomes a line-to-line short circuit due to the transformation or the vector group on the secondary side.

9.2.2.3 Measurement-Based Determination of the Symmetrical Components of Networks and Equipment

The impedances $\underline{Z}_0$, $\underline{Z}_1$ and $\underline{Z}_2$ in the equivalent circuit diagrams of the symmetrical components shown in Fig. 9.6 can be calculated using the system of Eq. (9.43). However, it is also possible to determine these impedances by measurement as shown in Fig. 9.7.

In the measuring circuit, it must be taken into account that, according to Eq. (9.45), the zero sequence voltage consists of an equal-direction voltage system, while a rotating field or a symmetrical rotating field with opposite direction is applied when determining the impedances $\underline{Z}_1$ or $\underline{Z}_2$. For this reason, the neutral point of equipment does not have to be connected to earth when measuring the positive and negative sequence impedances according to Fig. 9.6. In practice, the impedance $\underline{Z}_1$ is also referred to as the positive sequence impedance and $\underline{Z}_2$ as the negative sequence impedance, which in each case corresponds to the rotating field of the applied voltage system.

9.2.2.4 Application of the Symmetrical Components

The procedure for calculating the short-circuit currents with the aid of the symmetrical components is shown based on some examples.

In contrast to the calculation according to [8], the active voltages in these examples are carried out with the real voltage distribution (Sect. 3.6.1) and not with the method of the equivalent voltage source at the short-circuit location (Sect. 3.6.6). This has no influence on the magnitude of the short-circuit currents, but does have an influence on the voltages to earth that are calculated during the short circuit.

When simulating the different types of faults using the symmetrical components, it makes sense to consider the fault conditions “symmetrically” to the phase R. Only

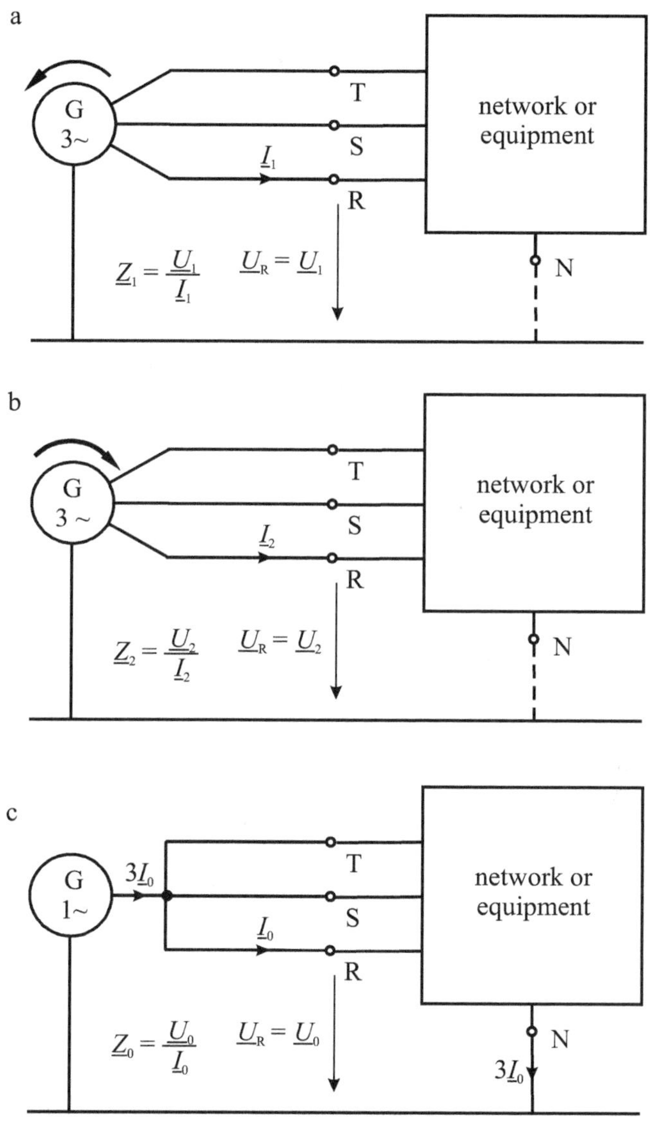

Fig. 9.7 Measuring circuit for determining the impedances of equipment and networks. (**a**) Positive sequence impedance, (**b**) negative sequence impedance, (**c**) zero sequence impedance

in these cases, there are real relationships between the component systems. This means:

- Line-to-line short circuit (with and without earth contact): Phases S–T,
- Line-to-earth short circuit: Phase R.

The reason for this is that line R counts as a reference value in the transformation matrix. In principle, the fault conditions can also be assumed in other phases, e.g. in the case of a line-to-earth short circuit phase S to earth, but in these cases complex ideal transformers must be used for the connections of the component systems.

Example 1: Three-Phase Short-Circuit Current

The calculation of the three-phase short-circuit current is essential for the mechanical stress (short-circuit forces) of equipment and systems. If the three-phase short-circuit current is determined at the end of an overhead line (Fig. 9.8), only the positive sequence system according to Fig. 9.6 must be considered, as this is a symmetrical equipment and a symmetrical process, since the short-circuit current flows in all three conductors. The short-circuit current calculation is divided into various steps as shown in Fig. 9.5.

- ***Step 1:*** Determination of the fault conditions in the original system, to earth
  The voltages $\underline{U}_{(012)}$ and the currents $\underline{I}_{(012)}$ of the component systems result from the fault conditions in the RST system:

$$\underline{U}_{\mathrm{R}} = \underline{U}_{\mathrm{S}} = \underline{U}_{\mathrm{T}} \quad \underline{I}_{\mathrm{R}} + \underline{I}_{\mathrm{S}} + \underline{I}_{\mathrm{T}} = 0 \tag{9.58}$$

- ***Step 2:*** Determining the fault conditions for the component system
  Taking into account the transformation matrix, Eq. (9.24), the fault conditions in the component system can be derived.

$$\begin{pmatrix} \underline{I}_{(0)} \\ \underline{I}_{(1)} \\ \underline{I}_{(2)} \end{pmatrix} = \frac{1}{3} \cdot \begin{pmatrix} 1 & 1 & 1 \\ 1 & \underline{a} & \underline{a}^2 \\ 1 & \underline{a}^2 & \underline{a} \end{pmatrix} \cdot \begin{pmatrix} \underline{I}_{\mathrm{R}} \\ \underline{I}_{\mathrm{S}} \\ \underline{I}_{\mathrm{T}} \end{pmatrix} \Rightarrow \underline{I}_{(0)} = 0 \tag{9.59}$$

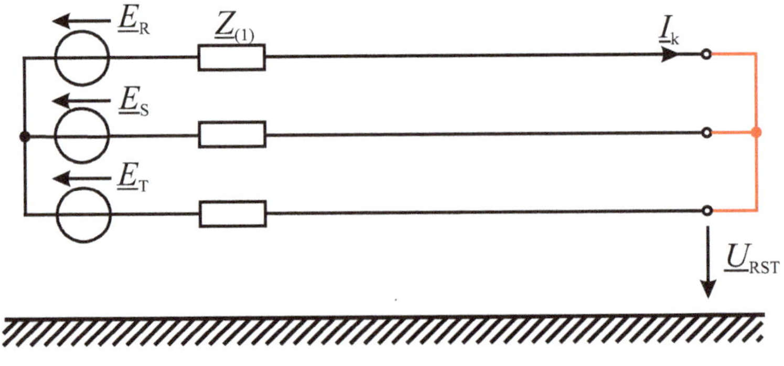

Fig. 9.8 Three-phase short circuit at the end of an overhead line with impedance $\underline{Z}_{(1)}$ in the positive sequence system

$$\begin{pmatrix} \underline{U}_{(0)} \\ \underline{U}_{(1)} \\ \underline{U}_{(2)} \end{pmatrix} = \frac{1}{3} \cdot \begin{pmatrix} 1 & 1 & 1 \\ 1 & \underline{a} & \underline{a}^2 \\ 1 & \underline{a}^2 & \underline{a} \end{pmatrix} \cdot \begin{pmatrix} \underline{U}_R \\ \underline{U}_S \\ \underline{U}_T \end{pmatrix} \Rightarrow \underline{U}_{(1)} = \underline{U}_{(2)} = 0 \qquad (9.60)$$

The fault conditions in the component system can be realized using the equivalent circuits shown in Fig. 9.6. Figure 9.9 shows the equivalent circuit in the positive sequence, negative sequence and zero sequence systems for this fault case. Since symmetrical voltages are assumed, an active voltage is only present in the positive sequence system and the other component systems are currentless.

- ***Step 3:*** Determine the current and voltage conditions in the component system: The following applies to the current $\underline{I}_{(1)}$, as there is only a voltage in the component system:

$$\underline{I}_{(1)} = \frac{\underline{E}_{(1)}}{\underline{Z}_{(1)}} = \frac{\underline{E}_R}{\underline{Z}_{(1)}} \qquad \underline{I}_{(2)} = \underline{I}_{(0)} = 0 \qquad (9.61)$$

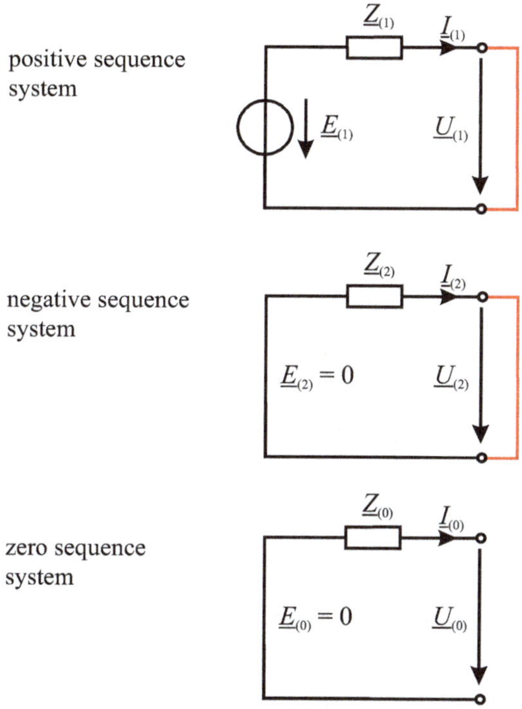

Fig. 9.9 Component equivalent circuits for the three-phase short-circuit current in the positive, negative and zero sequence system

- ***Step 4:*** Reverse Transformation of the current and voltage relationships into the original system:

$$\underline{\mathbf{I}}_{\mathrm{RST}} = \begin{pmatrix} 1 & 1 & 1 \\ 1 & \underline{a}^2 & \underline{a} \\ 1 & \underline{a} & \underline{a}^2 \end{pmatrix} \cdot \underline{\mathbf{I}}_{(012)} \tag{9.62}$$

$$\underline{I}_{\mathrm{R}} = \underline{I}_{(1)} \qquad \underline{I}_{\mathrm{S}} = \underline{a}^2 \cdot \underline{I}_{(1)} \qquad \underline{I}_{\mathrm{T}} = \underline{a} \cdot \underline{I}_{(1)} \tag{9.63}$$

The voltages at the short-circuit location to earth are zero, which corresponds to the fault conditions in the first step.

- ***Step 5***: Calculation of the currents at the fault location:
  The same amount of current flows in the individual conductors, so that the following applies for the three-phase initial symmetrical short-circuit current:

$$\underline{I}_{\mathrm{R}} = \underline{I}_{\mathrm{k}}'' = \frac{\underline{E}_{\mathrm{R}}}{\underline{Z}_{(1)}} \tag{9.64}$$

Due to the fact that the line-to-line voltage is selected to describe a voltage level, U_{n} (nominal system voltage) results:

$$E_{\mathrm{R}} = U_{\mathrm{n}}/\sqrt{3} \tag{9.65}$$

According to [9], a voltage factor $c = 1.1$ is selected to calculate the maximum short-circuit current in high-voltage systems, so that the following applies for the three-phase short-circuit current:

$$\underline{I}_{\mathrm{k}}'' = \frac{c \cdot U_{\mathrm{n}}/\sqrt{3}}{\underline{Z}_{(1)}} \tag{9.66}$$

- ***Step 6:*** Calculation of the voltage at the fault location (if necessary):
  The voltages of the individual conductors to earth can be determined using the current and voltage conditions. The component voltages are determined according to Fig. 9.8a–c:

$$U_{(1)} = U_{(2)} = U_{(0)} = 0 \tag{9.67}$$

This results in the voltages in the original system according to Eq. (9.62).

$$\begin{pmatrix} \underline{U}_{\mathrm{R}} \\ \underline{U}_{\mathrm{S}} \\ \underline{U}_{\mathrm{T}} \end{pmatrix} = \begin{pmatrix} 1 & 1 & 1 \\ 1 & \underline{a}^2 & \underline{a} \\ 1 & \underline{a} & \underline{a}^2 \end{pmatrix} \cdot \begin{pmatrix} \underline{U}_{(0)} \\ \underline{U}_{(1)} \\ \underline{U}_{(2)} \end{pmatrix} = \begin{pmatrix} 1 & 1 & 1 \\ 1 & \underline{a}^2 & \underline{a} \\ 1 & \underline{a} & \underline{a}^2 \end{pmatrix} \cdot \begin{pmatrix} 0 \\ 0 \\ 0 \end{pmatrix} \tag{9.68}$$

All voltages to earth have the value zero in the event of a three-phase short circuit.

The voltages at the fault location determined with Eq. (9.68) correspond to the actual voltages during a short-circuit according to Fig. 9.8. However, when using the equivalent voltage source method at the short-circuit location according to Sect. 3.6.6, these voltages represent the voltage changes during the short-circuit due to the changed voltage profile.

Example 2: Line-to-Earth Short Circuit

In the following, the short-circuit current is determined if a line-to-earth short circuit occurs according to Fig. 9.10. Since a current flows via earth due to the fault conditions, the zero sequence impedance of the overhead line is also included in the return line via earth. In this case, the following steps are required.

- ***Step 1:*** Determination of the fault conditions in the original system against earth:

$$\underline{U}_{\mathrm{R}} = 0 \quad \underline{I}_{\mathrm{S}} = \underline{I}_{\mathrm{T}} = 0 \tag{9.69}$$

The transformation provides the current and voltage conditions.

- ***Step 2:*** Determination of the fault conditions for the component system:

$$\begin{pmatrix} \underline{I}_{(0)} \\ \underline{I}_{(1)} \\ \underline{I}_{(2)} \end{pmatrix} = \frac{1}{3} \cdot \begin{pmatrix} 1 & 1 & 1 \\ 1 & \underline{a} & \underline{a}^2 \\ 1 & \underline{a}^2 & \underline{a} \end{pmatrix} \cdot \begin{pmatrix} \underline{I}_{\mathrm{R}} \\ 0 \\ 0 \end{pmatrix} \tag{9.70}$$

$$\underline{I}_{(0)} = \underline{I}_{(1)} = \underline{I}_{(2)} = \underline{I}_{\mathrm{R}}/3 \tag{9.71}$$

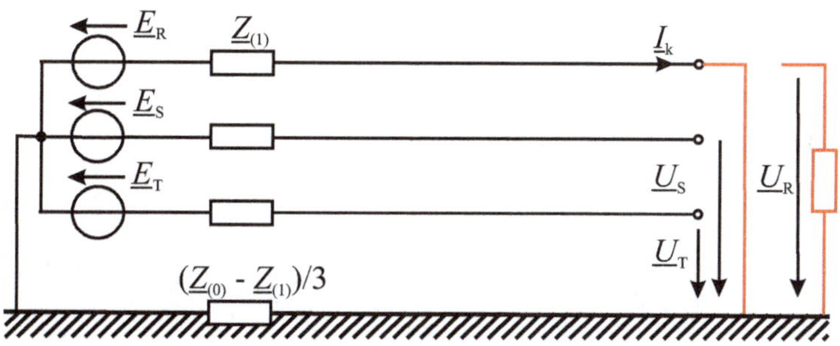

Fig. 9.10 Line-to-earth short-circuit at the end of an overhead line with positive sequence impedance $\underline{Z}_{(1)}$ and zero sequence impedance $\underline{Z}_{(0)}$, with and without fault resistance R

$$\begin{pmatrix} \underline{U}_{(0)} \\ \underline{U}_{(1)} \\ \underline{U}_{(2)} \end{pmatrix} = \frac{1}{3} \cdot \begin{pmatrix} 1 & 1 & 1 \\ 1 & \underline{a} & \underline{a}^2 \\ 1 & \underline{a}^2 & \underline{a} \end{pmatrix} \cdot \begin{pmatrix} 0 \\ \underline{U}_S \\ \underline{U}_T \end{pmatrix} \tag{9.72}$$

An addition of the equation system results:

$$\underline{U}_{(0)} + \underline{U}_{(1)} + \underline{U}_{(2)} = \frac{1}{3} \cdot \left[\underline{U}_S \cdot \left(1 + \underline{a} + \underline{a}^2\right) + \underline{U}_T \cdot \left(1 + \underline{a}^2 + \underline{a}\right)\right] = 0 \tag{9.73}$$

The results of the transformation lead to the equivalent circuit of the component systems as shown in Fig. 9.11 by connecting the three component systems in series. This allows the fault conditions to be realized.

- ***Step 3:*** Determination of the current and voltage conditions in the component system:

$$\underline{I}_{(1)} = \frac{\underline{E}_{(1)}}{\underline{Z}_{(1)} + \underline{Z}_{(2)} + \underline{Z}_{(0)}} = \frac{\underline{E}_{(1)}}{2 \cdot \underline{Z}_{(1)} + \underline{Z}_{(0)}} = \underline{I}_{(2)} = \underline{I}_{(0)} \tag{9.74}$$

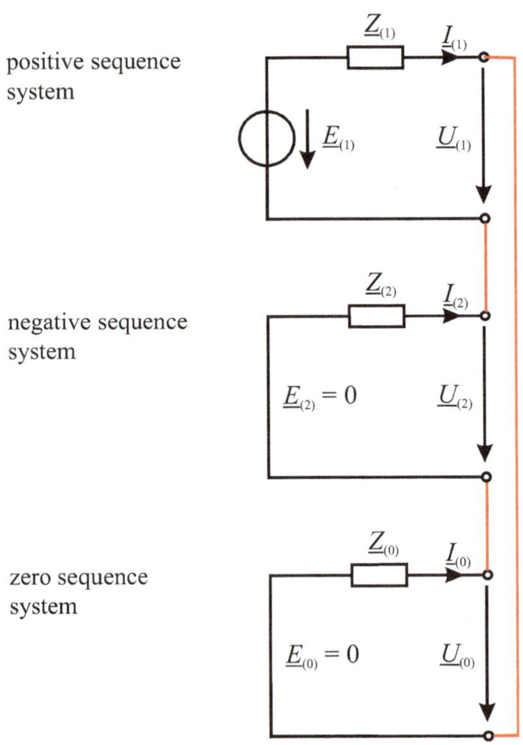

Fig. 9.11 Equivalent circuit of the symmetrical components in the event of line-to-earth short circuit

The calculated component currents can also be used to determine the component voltages.

$$\underline{U}_{(1)} = \underline{E}_{(1)} - \underline{Z}_{(1)} \cdot \underline{I}_{(1)} \quad \underline{U}_{(2)} = -\underline{Z}_{(2)} \cdot \underline{I}_{(2)}$$
$$\underline{U}_{(0)} = -\underline{Z}_{(0)} \cdot \underline{I}_{(0)} \tag{9.75}$$

- ***Step 4:*** Reverse Transformation of the current and voltage relationships into the original system:

  The line-to-earth short-circuit current is determined from the first line of the transformation Eq. (9.76), while the other two currents to earth have the value zero, which corresponds to the specified fault conditions (step 1).

$$\begin{pmatrix} \underline{I}_R \\ \underline{I}_S \\ \underline{I}_T \end{pmatrix} = \begin{pmatrix} 1 & 1 & 1 \\ 1 & \underline{a}^2 & \underline{a} \\ 1 & \underline{a} & \underline{a}^2 \end{pmatrix} \cdot \begin{pmatrix} \underline{I}_{(0)} \\ \underline{I}_{(1)} \\ \underline{I}_{(2)} \end{pmatrix} \tag{9.76}$$

$$\underline{I}_R = 3 \cdot \underline{I}_{(1)} = \frac{3 \cdot \underline{E}_{(1)}}{2 \cdot \underline{Z}_{(1)} + \underline{Z}_{(0)}} \qquad \underline{E}_{(1)} = \underline{E}_R \tag{9.77}$$

$$\underline{I}_S = \left(1 + \underline{a}^2 + \underline{a}\right) \cdot \underline{I}_{(1)} = 0 \quad \underline{I}_T = \left(1 + \underline{a} + \underline{a}^2\right) \cdot \underline{I}_{(1)} = 0 \tag{9.78}$$

The voltages result in:

$$\begin{aligned}
&\underline{U}_R = \underline{U}_{(0)} + \underline{U}_{(1)} + \underline{U}_{(2)} = 0 \\
&\underline{U}_S = \underline{U}_{(0)} + \underline{a}^2 \cdot \underline{U}_{(1)} + \underline{a} \cdot \underline{U}_{(2)} = \underline{E}_R \cdot \left(\underline{a}^2 - \frac{\underline{a}^2 \cdot \underline{Z}_{(1)} + \underline{a} \cdot \underline{Z}_{(1)} + \underline{Z}_{(0)}}{2 \cdot \underline{Z}_{(1)} + \underline{Z}_{(0)}}\right) \\
&= \underline{E}_R \cdot \left(\underline{a}^2 + \frac{\underline{Z}_{(1)}/\underline{Z}_{(0)} - 1}{2 \cdot \underline{Z}_{(1)}/\underline{Z}_{(0)} + 1}\right) \\
&\underline{U}_T = \underline{U}_{(0)} + \underline{a} \cdot \underline{U}_{(1)} + \underline{a}^2 \cdot \underline{U}_{(2)} = \underline{E}_R \cdot \left(\underline{a} + \frac{\underline{Z}_{(1)}/\underline{Z}_{(0)} - 1}{2 \cdot \underline{Z}_{(1)}/\underline{Z}_{(0)} + 1}\right)
\end{aligned} \tag{9.79}$$

- ***Step 5:*** Derivation of the final result:

  For the calculation of the initial symmetrical short-circuit current for a line-to-earth short-circuit, the following generally applies according to Eq. (9.80).

$$\underline{I}_R = \underline{I}''_{k1} = \underline{I}_{(1)} + \underline{I}_{(2)} + \underline{I}_{(0)} = \frac{3 \cdot \underline{E}_R}{2 \cdot \underline{Z}_{(1)} + \underline{Z}_{(0)}} \tag{9.80}$$

Taking into account the line-to-line voltage U_n and the voltage factor $c = 1.1$, the following applies to the line-to-earth short-circuit current:

$$\underline{I}''_{k1} = \frac{c \cdot \sqrt{3} \cdot U_n}{2 \cdot \underline{Z}_{(1)} + \underline{Z}_{(0)}} \tag{9.81}$$

- ***Step 6***: Calculation of the voltage at the fault location (if necessary):
  The voltages of the individual conductors to earth can be determined using the current and voltage conditions. The component voltages are determined according to Fig. 9.10:

$$\begin{aligned} \underline{U}_{(1)} &= \underline{E}_{(1)} - \underline{Z}_{(1)} \cdot \underline{I}_{(1)} \\ \underline{U}_{(2)} &= -\underline{Z}_{(1)} \cdot \underline{I}_{(1)} \\ \underline{U}_{(0)} &= -\underline{Z}_{(0)} \cdot \underline{I}_{(1)} \end{aligned} \tag{9.82}$$

With $\underline{I}_{(1)}$ according to Eq. (9.81).

$$\underline{I}_{(1)} = \frac{\underline{E}_{(1)}}{2 \cdot \underline{Z}_{(1)} + \underline{Z}_{(0)}} \tag{9.83}$$

This results in the voltages in the original system according to Eq. (9.67).

$$\underline{U}_R = \underline{E}_{(1)} - \left[2 \cdot \underline{Z}_{(1)} + \underline{Z}_{(0)}\right] \cdot \underline{I}_{(1)} = 0 \tag{9.84}$$

$$\begin{aligned} \underline{U}_S &= -\underline{Z}_{(0)} \cdot \underline{I}_{(1)} + \underline{a}^2 \cdot \left[\underline{E}_{(1)} - \underline{Z}_{(1)} \cdot \underline{I}_{(1)}\right] - \underline{a} \cdot \underline{Z}_{(1)} \cdot \underline{I}_{(1)} \\ &= \left[\underline{a}^2 + \frac{\underline{Z}_{(1)} - \underline{Z}_{(0)}}{2 \cdot \underline{Z}_{(1)} + \underline{Z}_{(0)}}\right] \cdot \underline{E}_{(1)} \end{aligned} \tag{9.85}$$

$$\underline{U}_T = \left[\underline{a} + \frac{\underline{Z}_{(1)} - \underline{Z}_{(0)}}{2 \cdot \underline{Z}_{(1)} + \underline{Z}_{(0)}}\right] \cdot \underline{E}_{(1)} \tag{9.86}$$

The characteristic values of the fault-free line $\underline{U}_S$ to earth result for the following values of the zero sequence impedance $\underline{Z}_{(0)}$:

- $\underline{Z}_{(0)} = \underline{Z}_{(1)}$; e.g. directly after a power transformer:

$$\underline{U}_S = \underline{a}^2 \cdot \underline{E}_{(1)} \tag{9.87}$$

  The voltage corresponds to the line-to-earth voltage of undisturbed operation.

- $\underline{Z}_{(0)} \to \infty$; e.g. in a network with isolated or compensated neutral points:

$$\underline{U}_S = (\underline{a}^2 - 1) \cdot \underline{E}_{(1)} \qquad |\underline{U}_S| = \sqrt{3} \cdot |\underline{E}_{(1)}| \tag{9.88}$$

The voltage corresponds to the line-to-line voltage of undisturbed operation.

Example 3: Line-to-Earth Short Circuit with a Fault Resistor

In the short-circuit current calculation according to [8], the metallic short-circuit is always assumed, i.e. no fault resistance is taken into account, e.g. between the conductor R and earth, Sect. 3.3. This fault resistance R can be, for example, an earth resistance or the resistance of an arc, Fig. 9.10. In the following, the changes compared to the derivation without fault resistance up to the configuration of the component equivalent circuit are shown.

- ***Step 1:*** Determination of the fault conditions in the original system against earth:

$$\underline{U}_R = R \cdot \underline{I}_R \tag{9.89}$$

- ***Step 2:*** Determination of the fault conditions for the component system:

$$\underline{I}_{(0)} = \underline{I}_{(1)} = \underline{I}_{(2)} = \underline{I}_R/3 \tag{9.90}$$

$$\begin{pmatrix} \underline{U}_{(0)} \\ \underline{U}_{(1)} \\ \underline{U}_{(2)} \end{pmatrix} = \frac{1}{3} \cdot \begin{pmatrix} 1 & 1 & 1 \\ 1 & \underline{a} & \underline{a}^2 \\ 1 & \underline{a}^2 & \underline{a} \end{pmatrix} \cdot \begin{pmatrix} \underline{U}_R \\ \underline{U}_S \\ \underline{U}_T \end{pmatrix} \tag{9.91}$$

Addition of the system of Eq. (9.91) results in:

$$\underline{U}_{(0)} + \underline{U}_{(1)} + \underline{U}_{(2)} = \frac{1}{3} \cdot \left[3 \cdot \underline{U}_R + \underline{U}_S \cdot \left(1 + \underline{a} + \underline{a}^2\right) + \underline{U}_T \cdot \left(1 + \underline{a}^2 + \underline{a}\right)\right] = \underline{U}_R = 3R \cdot \underline{I}_{(1)} \tag{9.92}$$

The results of the transformation lead to the equivalent circuit of the component systems as shown in Fig. 9.12 by connecting the three component systems in series. This allows the fault conditions to be realized.

Example 4: Simultaneous Faults

According to [10], simultaneous faults are short circuits that occur in the same phase at different locations in the network. As the component systems are directly galvanically connected when using symmetrical components for faults that are symmetrical

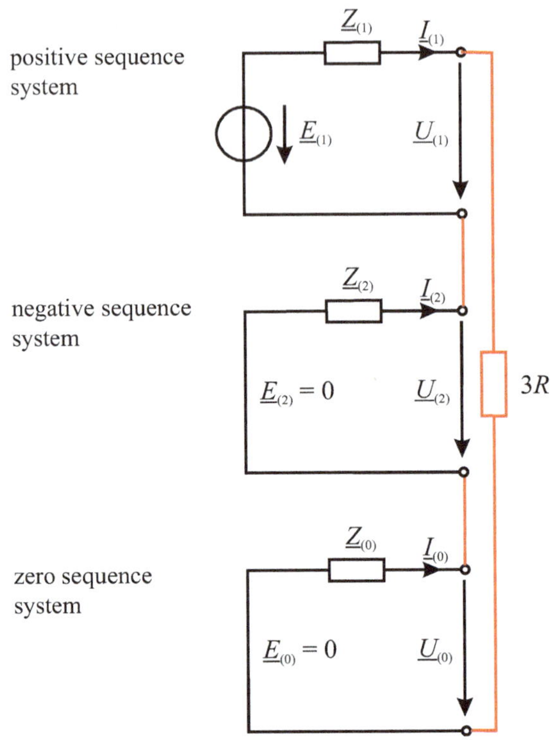

Fig. 9.12 Equivalent circuit of the symmetrical components for a line-to-earth short circuit with a fault resistance *R*

to phase R, ideal transformers must be used to determine simultaneous faults (with simultaneous faults at different locations) as shown in Fig. 9.13. In this case, line-to-earth short circuits occur in phase R to earth.

According to Fig. 9.13a, the direct connections introduce conditions that do not exist in reality. This applies to the voltages due to the continuous return line:

$$\underline{U}_{(1)\mathrm{B}} = \underline{U}_{(2)\mathrm{B}} = \underline{U}_{(0)\mathrm{B}} = 0 \tag{9.93}$$

This leads to incorrect current and voltage distribution in the network. This is avoided by using 1:1 ideal transformers as shown in Fig. 9.13b. This difficulty is avoided by switching to a representation using the 0, α, β components, as ideal transformers are always used to link the component systems in these cases.

Another example of a simultaneous fault or a multiple fault is shown in Chap. 6. This chapter shows the derivation of a double line-to-earth short circuit that occurs simultaneously in two phases at different locations.

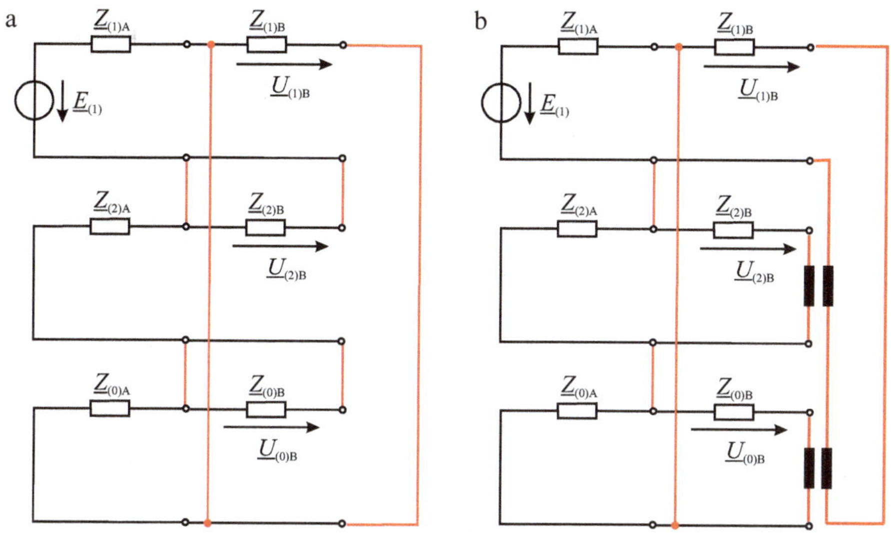

Fig. 9.13 Equivalent circuit of the component systems in the event of a multiple fault in phase R (line-to-earth short circuits) at different locations. (**a**) Incorrect, (**b**) correct

9.3 Fault Conditions, Fault Equations, Equivalent Circuits

Table 9.2 summarizes the fault conditions, currents and voltages for balanced and unbalanced short circuits.

To calculate the short-circuit currents using the symmetrical components in Table 9.2, the simplified equivalent circuit diagrams of frequently used equipment in Table 9.3 can be used. For other equivalent circuit diagrams, e.g. for transformers, use [11].

In general, generators are operated with an isolated neutral point, i.e. the generator neutral point is not connected to earth so that the earthing impedance $\underline{Z}_E$ has the value infinity. This may not be the case with low-voltage generators in order to avoid harmonics or to meet the protection requirements.

Based on the fault conditions according to Table 9.2 and the equivalent circuit diagrams of equipment, the links of the positive, negative and zero sequence systems can be formed for a three-phase, line-to-line and line-to-earth short circuit (without fault resistance at the short-circuit location), Fig. 9.14.

9.4 Comparison: 0, 1, 2 Components vs. 0, α, β Components

The steady state short-circuit currents are usually calculated using symmetrical components by applying the fault symmetrically to the conductor R in each case. In these cases, no ideal transformers are used when coupling the component

Table 9.2 Fault conditions, currents and voltages for balanced and unbalanced short circuits; at the first moment (subtransient) the following applies $\underline{Z}_{(2)} = \underline{Z}_{(1)}$

Short circuit	Fault conditions RST	Fault conditions (012)	Short-circuit currents and voltages (012) and RST
k3	$\underline{U}_R = \underline{U}_S = \underline{U}_T$ $\underline{I}_R + \underline{I}_S + \underline{I}_T = 0$	$\underline{I}_{(0)} = 0$ $\underline{U}_{(1)} = \underline{U}_{(2)} = 0$	$\underline{I}_{(1)} = \frac{\underline{E}_{(1)}}{\underline{Z}_{(1)}} \quad \underline{I}_{(2)} = \underline{I}_{(0)} = 0$ $\underline{I}_R = \frac{\underline{E}_{(1)}}{\underline{Z}_{(1)}} \quad \underline{I}_S = \underline{a}^2 \cdot \underline{I}_R \quad \underline{I}_T = \underline{a} \cdot \underline{I}_R$
k2	$\underline{U}_S = \underline{U}_T$ $\underline{I}_R = 0$ $\underline{I}_S + \underline{I}_T = 0$	$\underline{U}_{(1)} = \underline{U}_{(2)}$ $\underline{I}_{(1)} = -\underline{I}_{(2)}$ $\underline{I}_{(0)} = 0$	$\underline{I}_{(1)} = \frac{\underline{E}_{(1)}}{\underline{Z}_{(1)} + \underline{Z}_{(2)}} \quad \underline{I}_R = 0 \quad \underline{I}_S = -\underline{I}_T = \frac{-j\sqrt{3} \cdot \underline{E}_{(1)}}{\underline{Z}_{(1)} + \underline{Z}_{(2)}}$ $\underline{U}_R = \frac{2 \cdot \underline{E}_{(1)}}{1 + \underline{Z}_{(1)}/\underline{Z}_{(2)}} \quad \underline{U}_S = \underline{U}_T = -\frac{\underline{E}_{(1)}}{1 + \underline{Z}_{(1)}/\underline{Z}_{(2)}}$
k2E	$\underline{U}_S = \underline{U}_T = 0$ $\underline{I}_R = 0$	$\underline{U}_{(1)} = \underline{U}_{(2)} = \underline{U}_{(0)}$ $\underline{I}_{(1)} + \underline{I}_{(2)} + \underline{I}_{(0)} = 0$	$\underline{I}_{(1)} = \frac{\underline{E}_{(1)}}{\underline{Z}_{(1)} + \frac{\underline{Z}_{(2)} \cdot \underline{Z}_{(0)}}{\underline{Z}_{(2)} + \underline{Z}_{(0)}}}$ $\underline{I}_{(2)} = -\underline{I}_{(1)} \cdot \frac{\underline{Z}_{(0)}}{\underline{Z}_{(2)} + \underline{Z}_{(0)}} \quad \underline{I}_{(0)} = -\underline{I}_{(1)} \cdot \frac{\underline{Z}_{(2)}}{\underline{Z}_{(2)} + \underline{Z}_{(0)}}$ $\underline{I}_R = 0; \quad \underline{I}_S = \frac{-j\sqrt{3} \cdot \underline{E}_{(1)} \cdot \left(\underline{Z}_{(0)} - \underline{a} \cdot \underline{Z}_{(2)}\right)}{\underline{Z}_{(1)} \cdot \underline{Z}_{(2)} + \underline{Z}_{(1)} \cdot \underline{Z}_{(0)} + \underline{Z}_{(2)} \cdot \underline{Z}_{(0)}}$ $\underline{I}_T = \frac{-j\sqrt{3} \cdot \underline{E}_{(1)} \cdot \left(\underline{Z}_{(0)} - \underline{a}^2 \cdot \underline{Z}_{(2)}\right)}{\underline{Z}_{(1)} \cdot \underline{Z}_{(2)} + \underline{Z}_{(1)} \cdot \underline{Z}_{(0)} + \underline{Z}_{(2)} \cdot \underline{Z}_{(0)}}$ $\underline{I}_S + \underline{I}_T = 3 \cdot \underline{I}_{(0)} = \frac{-3 \cdot \underline{E}_{(1)}}{\underline{Z}_{(1)} + \underline{Z}_{(0)} \cdot \left(1 + \underline{Z}_{(1)}/\underline{Z}_{(2)}\right)}$ $\underline{U}_R = \frac{3 \cdot \underline{E}_{(1)} \cdot \underline{Z}_{(0)} \cdot \underline{Z}_{(2)}}{\underline{Z}_{(1)} \cdot \underline{Z}_{(2)} + \underline{Z}_{(1)} \cdot \underline{Z}_{(0)} + \underline{Z}_{(2)} \cdot \underline{Z}_{(0)}}$ $\underline{U}_S = 0; \quad \underline{U}_T = 0$

k1	$\underline{U}_{\mathrm{R}} = 0$ $\underline{I}_{\mathrm{S}} = \underline{I}_{\mathrm{T}} = 0$	$\underline{U}_{(1)} + \underline{U}_{(2)} + \underline{U}_{(0)} = 0$ $\underline{I}_{(1)} = \underline{I}_{(2)} = \underline{I}_{(0)}$	$\underline{I}_{(1)} = \underline{I}_{(2)} = \underline{I}_{(0)} = \frac{\underline{E}_{(1)}}{\underline{Z}_{(1)} + \underline{Z}_{(2)} + \underline{Z}_{(0)}}$ $\underline{I}_{\mathrm{R}} = 3 \cdot \underline{I}_{(1)} = 3 \cdot \underline{I}_{(2)} = 3 \cdot \underline{I}_{(0)}$ $\underline{U}_{\mathrm{R}} = 0 \quad \underline{U}_{\mathrm{S}} = \frac{-\mathrm{j}\sqrt{3} \cdot \underline{E}_{(1)} \cdot \left(\underline{Z}_{(2)} - \underline{\mathrm{a}} \cdot \underline{Z}_{(0)}\right)}{\underline{Z}_{(1)} + \underline{Z}_{(2)} + \underline{Z}_{(0)}}$ $\underline{U}_{\mathrm{T}} = \frac{\mathrm{j}\sqrt{3} \cdot \underline{E}_{(1)} \cdot \left(\underline{Z}_{(2)} - \underline{\mathrm{a}}^2 \cdot \underline{Z}_{(0)}\right)}{\underline{Z}_{(1)} + \underline{Z}_{(2)} + \underline{Z}_{(0)}}$

Table 9.3 Equivalent circuit diagrams of equipment for the positive and zero sequence system (extract)

Equipment	Positive sequence systems	Zero sequence systems
Line, (over-head line, cable)	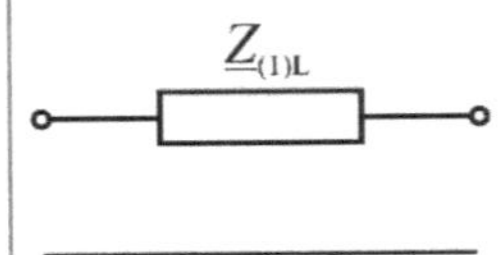	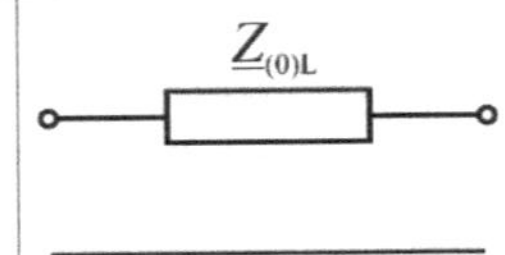
Transformer (Yy), HV side earthed	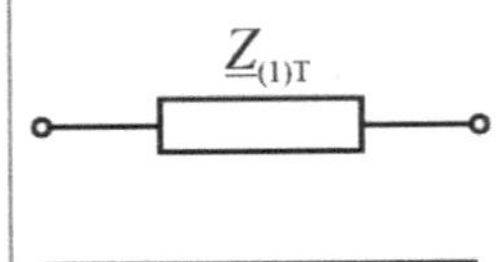	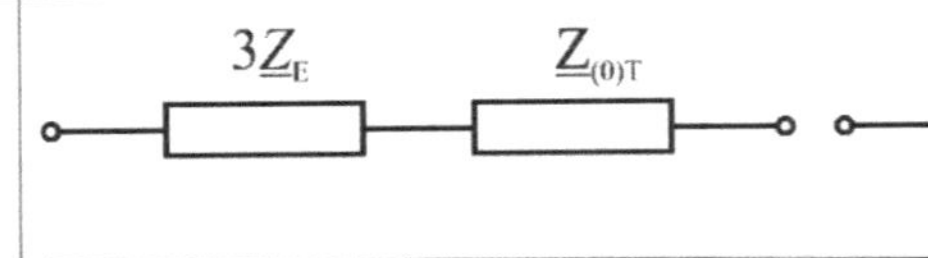
Transformer (Yd), HV side earthed	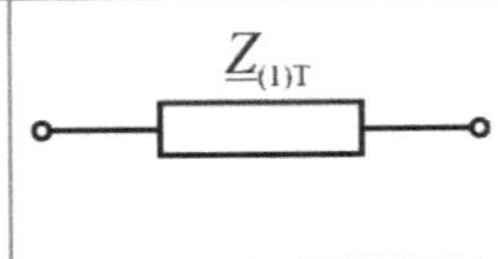	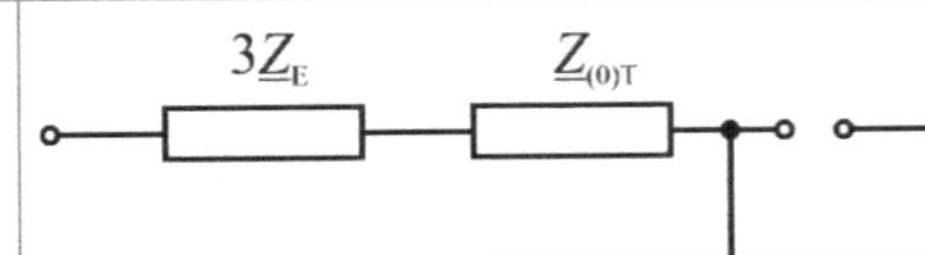
Generator[a]	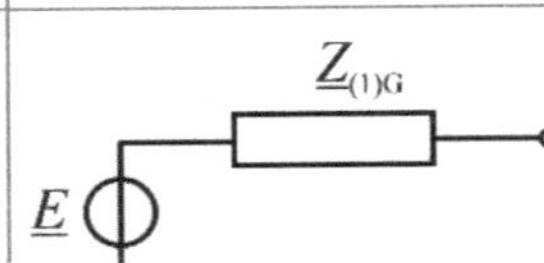	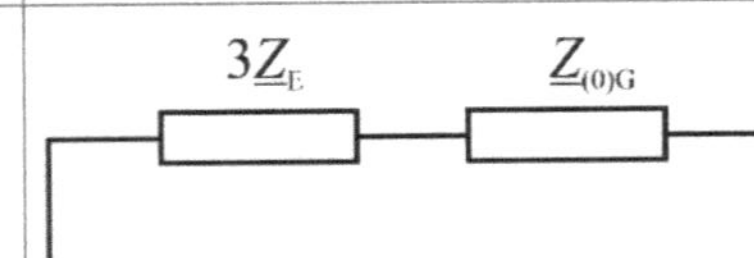
Network	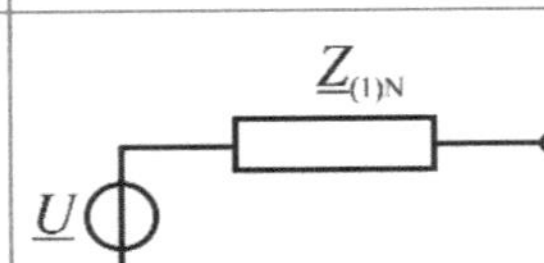	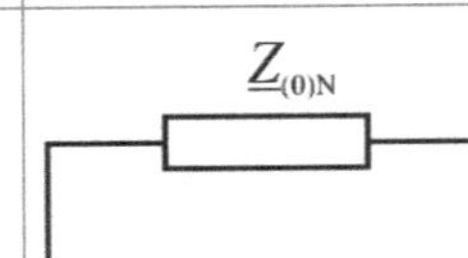

$\underline{Z}_E$ Earthing impedance in the original system
[a] On condition that the generator neutral point is earthed, otherwise $\underline{Z}_{(0)G} \rightarrow \infty$

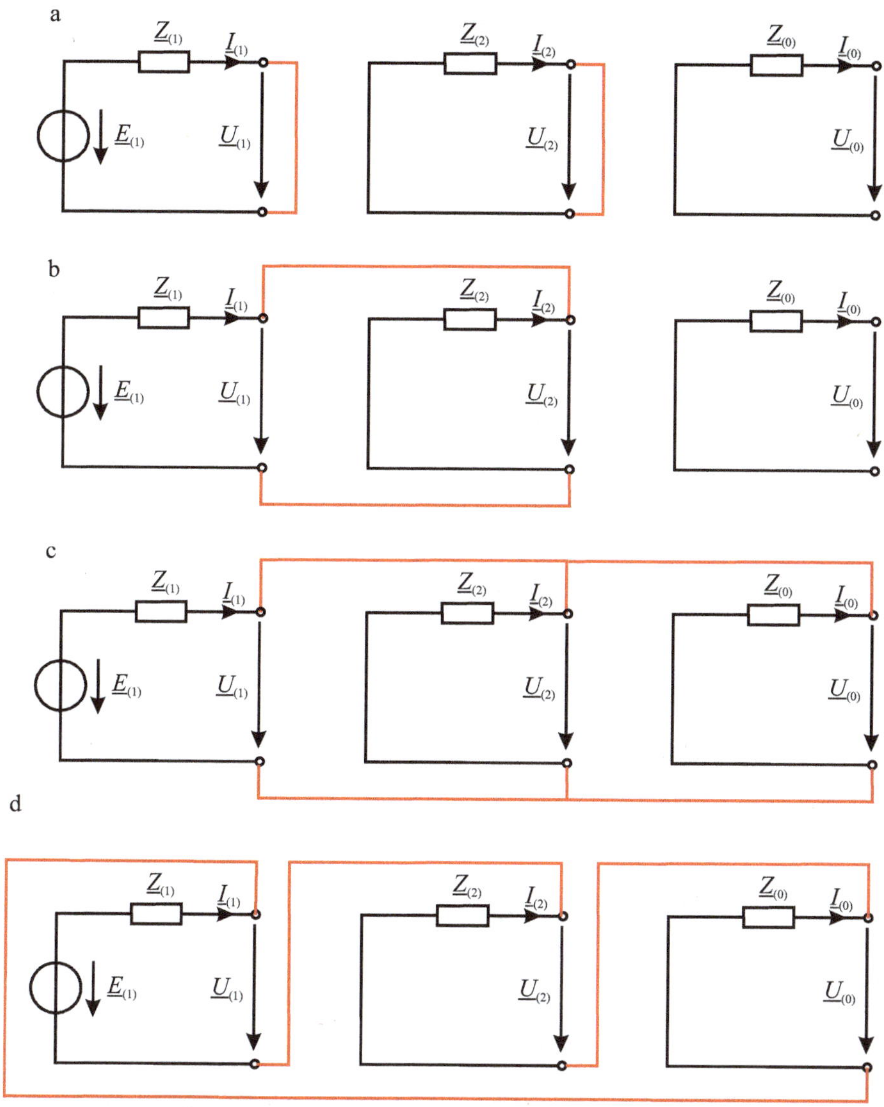

Fig. 9.14 Connections of the component systems depending on the short-circuit current. (**a**) three-phase, (**b**) line-to-line without earth connection, (**c**) line-to-line with earth connection, (**d**) line-to-earth

equivalent circuits, e.g. Fig. 9.11 for the line-to-line short circuit. In principle, it is also possible to use the 0, α, β components, which are generally used for transient processes. This section briefly describes the equivalent circuits for three-phase and line-to-line short circuits as shown in Figs. 9.8 and 9.10.

The following transformation equations according to [12] apply to the voltages and currents:

$$\mathbf{G}_{\mathrm{RST}} = \mathbf{C} \cdot \mathbf{G}_{0\alpha\beta} \qquad \mathbf{G}_{0\alpha\beta} = \mathbf{C}^{-1} \cdot \mathbf{G}_{\mathrm{RST}} \tag{9.94}$$

With the transformation matrices:

$$\mathbf{C} = \begin{pmatrix} 1 & 1 & 0 \\ 1 & -\frac{1}{2} & \frac{\sqrt{3}}{2} \\ 1 & -\frac{1}{2} & -\frac{\sqrt{3}}{2} \end{pmatrix} \qquad \mathbf{C}^{-1} = \frac{1}{3} \cdot \begin{pmatrix} 1 & 1 & 1 \\ 2 & -1 & -1 \\ 0 & \sqrt{3} & -\sqrt{3} \end{pmatrix} \tag{9.95}$$

For the impedance matrix, Eq. (9.96) results in:

$$\underline{\mathbf{Z}}_{(0\alpha\beta)} = \mathbf{C}^{-1} \cdot \underline{\mathbf{Z}}_{\mathrm{RST}} \cdot \mathbf{C} \tag{9.96}$$

After inserting the transformation matrices, a symmetrical impedance matrix can be converted into a diagonal matrix, so that three decoupled component systems exist in undisturbed operation.

$$\begin{aligned} \begin{pmatrix} \underline{Z}_{(0)} \\ \underline{Z}_{(\alpha)} \\ \underline{Z}_{(\beta)} \end{pmatrix} &= \frac{1}{3} \cdot \begin{pmatrix} 1 & 1 & 1 \\ 2 & -1 & -1 \\ 0 & \sqrt{3} & -\sqrt{3} \end{pmatrix} \cdot \begin{pmatrix} \underline{Z}_{\mathrm{A}} & \underline{Z}_{\mathrm{B}} & \underline{Z}_{\mathrm{B}} \\ \underline{Z}_{\mathrm{B}} & \underline{Z}_{\mathrm{A}} & \underline{Z}_{\mathrm{B}} \\ \underline{Z}_{\mathrm{B}} & \underline{Z}_{\mathrm{B}} & \underline{Z}_{\mathrm{A}} \end{pmatrix} \cdot \begin{pmatrix} 1 & 1 & 0 \\ 1 & -\frac{1}{2} & \frac{\sqrt{3}}{2} \\ 1 & -\frac{1}{2} & -\frac{\sqrt{3}}{2} \end{pmatrix} \\ &= \begin{pmatrix} \underline{Z}_{\mathrm{A}} + 2\underline{Z}_{\mathrm{B}} & 0 & 0 \\ 0 & \underline{Z}_{\mathrm{A}} - \underline{Z}_{\mathrm{B}} & 0 \\ 0 & 0 & \underline{Z}_{\mathrm{A}} - \underline{Z}_{\mathrm{B}} \end{pmatrix} \end{aligned} \tag{9.97}$$

The result of Eq. (9.97) shows that the impedances of the individual component systems are identical to the values of the symmetrical components, according to Eq. (9.44). Also in this case, the consideration is that the original system can be composed of three different systems (components), namely the

- α system: In this system, the current of the reference conductor (conductor R) flows back in the other two conductors,
- β system: In this system, only the currents of conductors S and T, which add up to zero, flow,
- zero system: Currents/voltages with the same phase position.

Using the conditions (9.94), the active voltages of the individual component systems can be determined if a symmetrical three-phase network is assumed according to Eqs. (9.8) and (9.9), $\underline{E}_{(1)} = \underline{U}_{\mathrm{R}}$.

$$\begin{pmatrix} \underline{U}_{(0)} \\ \underline{U}_{(\alpha)} \\ \underline{U}_{(\beta)} \end{pmatrix} = \frac{1}{3} \cdot \begin{pmatrix} 1 & 1 & 1 \\ 2 & -1 & -1 \\ 0 & \sqrt{3} & -\sqrt{3} \end{pmatrix} \cdot \begin{pmatrix} \underline{E}_{(1)} \\ \underline{\mathrm{a}}^2 \cdot \underline{E}_{(1)} \\ \underline{\mathrm{a}} \cdot \underline{E}_{(1)} \end{pmatrix} \tag{9.98}$$

$$\underline{U}_{(0)} = 0 \quad \underline{U}_{(\alpha)} = \underline{E}_{(1)} \quad \underline{U}_{(\beta)} = -\mathrm{j}\underline{E}_{(1)} \tag{9.99}$$

This means that, in contrast to the diagram in Fig. 9.6, there are active voltages in two component systems (α, β) under the condition of a symmetrical system. In the following, the currents for the three-phase and line-to-earth short circuit are determined using the 0, α, β components, similar to the illustration in Sect. 9.2.2.4.

Example 1: Three-Phase Short Circuit
From the fault conditions, Eq. (9.58) in the original system, the fault conditions of the voltages in the component system at the short-circuit location can be derived.

$$\begin{pmatrix} \underline{U}_{(0)} \\ \underline{U}_{(\alpha)} \\ \underline{U}_{(\beta)} \end{pmatrix} = \frac{1}{3} \cdot \begin{pmatrix} 1 & 1 & 1 \\ 2 & -1 & -1 \\ 0 & \sqrt{3} & -\sqrt{3} \end{pmatrix} \cdot \begin{pmatrix} \underline{U}_{\mathrm{R}} \\ \underline{U}_{\mathrm{R}} \\ \underline{U}_{\mathrm{R}} \end{pmatrix} \tag{9.100}$$

$$\underline{U}_{(\alpha)} = 0 \qquad \underline{U}_{(\beta)} = 0 \tag{9.101}$$

The addition of the three conductor currents provides the relationship:

$$\underline{I}_{\mathrm{R}} + \underline{I}_{\mathrm{S}} + \underline{I}_{\mathrm{T}} = 0 \qquad \underline{I}_{(0)} = 0 \tag{9.102}$$

The equivalent circuit of the component systems can be specified from this information, Fig. 9.15.

The component equivalent circuits can be used to determine the currents so that the short-circuit currents in the original system can be calculated under the condition $\underline{Z}_{(\alpha)} = \underline{Z}_{(\beta)}$.

$$\underline{I}_{(\alpha)} = \frac{\underline{E}_1}{\underline{Z}_{(\alpha)}} \qquad \underline{I}_{(\beta)} = -\mathrm{j}\frac{\underline{E}_1}{\underline{Z}_{(\beta)}} = -\mathrm{j}\underline{I}_{(\alpha)} \tag{9.103}$$

$$\begin{aligned} \underline{I}_{\mathrm{R}} &= \underline{I}_{(\alpha)} \\ \underline{I}_{\mathrm{S}} &= \underline{I}_{(\alpha)} \cdot \left(-\frac{1}{2} - \mathrm{j}\frac{\sqrt{3}}{2}\right) = \underline{\mathrm{a}}^2 \cdot \underline{I}_{(\alpha)} \\ \underline{I}_{\mathrm{T}} &= \underline{I}_{(\alpha)} \cdot \left(-\frac{1}{2} + \mathrm{j}\frac{\sqrt{3}}{2}\right) = \underline{\mathrm{a}} \cdot \underline{I}_{(\alpha)} \end{aligned} \tag{9.104}$$

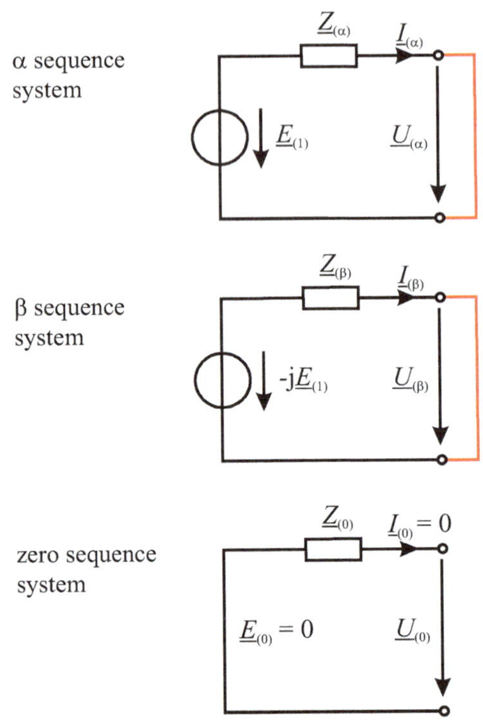

Fig. 9.15 Component equivalent circuits for the three-phase short-circuit current in the α-, β and zero sequence systems

Since the impedances and voltages of the component systems (0, 1, 2 and 0, α, β) are identical, the short-circuit current $\underline{I}_R$ according to Eq. (9.104) is equal to the calculated current according to Eq. (9.64).

Example 2: Line-to-Earth Short Circuit

As an example, the line-to-earth fault is assumed in conductor R to earth. In principle, it is also possible to assume the fault in other conductors, as ideal transformers are used in all cases, in contrast to the use of symmetrical components [13]. This results in the following fault conditions in the original and component system.

$$\underline{U}_R = 0 \qquad \underline{I}_S = \underline{I}_T = 0 \tag{9.105}$$

Current conditions:

$$\begin{pmatrix} \underline{I}_{(0)} \\ \underline{I}_{(\alpha)} \\ \underline{I}_{(\beta)} \end{pmatrix} = \frac{1}{3} \cdot \begin{pmatrix} 1 & 1 & 1 \\ 2 & -1 & -1 \\ 0 & \sqrt{3} & -\sqrt{3} \end{pmatrix} \cdot \begin{pmatrix} \underline{I}_{\mathrm{R}} \\ 0 \\ 0 \end{pmatrix} \tag{9.106}$$

$$\underline{I}_{(0)} = \frac{1}{2} \cdot \underline{I}_{(\alpha)} \quad \underline{I}_{(\beta)} = 0 \tag{9.107}$$

Voltage conditions:

$$\begin{pmatrix} \underline{U}_{(0)} \\ \underline{U}_{(\alpha)} \\ \underline{U}_{(\beta)} \end{pmatrix} = \frac{1}{3} \cdot \begin{pmatrix} 1 & 1 & 1 \\ 2 & -1 & -1 \\ 0 & \sqrt{3} & -\sqrt{3} \end{pmatrix} \cdot \begin{pmatrix} 0 \\ \underline{U}_{\mathrm{S}} \\ \underline{U}_{\mathrm{T}} \end{pmatrix} \tag{9.108}$$

$$\underline{U}_{(0)} + \underline{U}_{(\alpha)} = 0 \tag{9.109}$$

Taking into account the current and voltage conditions, the equivalent circuit can be determined as shown in Fig. 9.16. Since the zero sequence system is connected to an ideal transformer, a higher voltage ($2\underline{U}_{(0)}$) is present. For this reason, the zero sequence impedance $\underline{Z}_{(0)}$ must be multiplied by a factor of 2 to obtain the current $\underline{I}_{(0)}$. The component currents can be determined using the equivalent circuit in Fig. 9.16, so that the following applies:

$$\underline{I}_{(\alpha)} = \frac{\underline{E}_1}{\underline{Z}_{(\alpha)} + 2 \cdot \underline{Z}_{(0)}/t^2} = \frac{\underline{E}_1}{\underline{Z}_{(\alpha)} + \underline{Z}_{(0)}/2} \tag{9.110}$$

The currents in the original system are calculated using Eqs. (9.94) and (9.106):

$$\begin{aligned} \underline{I}_{\mathrm{R}} &= \underline{I}_{(0)} + \underline{I}_{(\alpha)} = \frac{3}{2} \cdot \underline{I}_{(\alpha)} = \frac{3 \cdot \underline{E}_1}{2 \cdot \underline{Z}_{(\alpha)} + \underline{Z}_{(0)}} \\ \underline{I}_{\mathrm{S}} &= \underline{I}_{(0)} - \frac{1}{2} \cdot \underline{I}_{(\alpha)} = 0 \\ \underline{I}_{\mathrm{S}} &= \underline{I}_{(0)} - \frac{1}{2} \cdot \underline{I}_{(\alpha)} = 0 \end{aligned} \tag{9.111}$$

In this case too, the determined line-to-earth short-circuit current $\underline{I}_{\mathrm{R}}$ corresponds to the calculation according to Eq. (9.81), taking into account the active voltages $\underline{E}_{(1)} = c \cdot U_{\mathrm{n}}/\sqrt{3}$.

Evaluation of Component Systems The fundamental advantage in the application of component systems is that in case of symmetrical equipment all unbalanced and balanced processes (e.g. short circuits) can be simulated in a simple way. According to [13], the advantage of using symmetrical components is that the component systems can also be given a physical assignment:

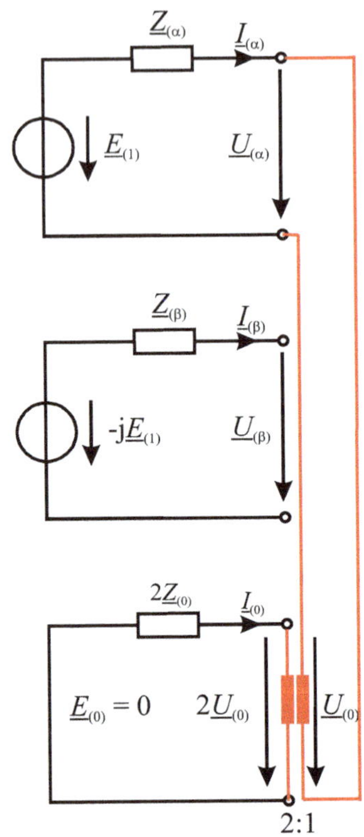

Fig. 9.16 Equivalent circuit of the 0, α, β components in the event of a line-to-earth short circuit

- The positive sequence system is decisive for the active power transmission in a network,
- the negative sequence system causes additional losses due to the opposing rotating field, and
- the zero sequence is a measure of the voltage displacement in a network.

All faults can be realized by simple interconnections of the component systems when using symmetrical components, if the fault is assumed to be symmetrical to the phase R. If this is not the case, complex ideal transformers must be used. In contrast, real ideal transformers are always used when applying the 0, α, β components.

In general, the symmetrical components are used for the steady state short-circuit current calculation, as all faults can be placed symmetrically to phase R according to Table 9.2.

The advantages of the 0, α, β components come into play when transient phenomenon are to be considered. This means, for example, starting from a line-to-line short-circuit with earth contact, the short-circuit currents in the individual conductors or to earth are to be interrupted at zero crossing, since in this case both the currents in the original system and in the component system must have the same phase position. This is possible by transformation with real values according to Eq. (9.95), in contrast to the complex transformation when using symmetrical components.

9.5 Conclusions

The calculation of processes in three-phase systems is fundamentally more extensive than in an AC system, as they are generally made up of three individual AC systems. Using the components, it is possible to break down a symmetrical three-phase system into three individual systems so that the calculations in this case can be simulated by a single AC system. When simulating steady-state electrical processes, the 0, 1, 2-component or symmetrical components have become established.

With the help of transformation matrices, a three-phase system can be broken down into three individual subsystems, which are referred to as 0, 1 and 2 systems, for example. While these three subsystems are decoupled with a symmetrical load, in the event of a fault (e.g. line-to-line short circuit, line-to-earth short circuit) these subsystems are connected to each other depending on the fault conditions, so that the fault currents and fault voltages in the component systems are calculated. The determined fault conditions are finally converted back into the original system.

While the symmetrical components are suitable for calculating steady-state faults if the system fault is assumed to be symmetrical to the conductor R, the 0, α, β components are particularly advantageous for simulating transient phenomenon.

References

1. Fortescue CL (1918) Method of symmetrical coordinates applied to the solution of polyphase networks. AIEE Trans 37(Part II):1027–1140
2. Balzer G, Neumann C, Balzer C (2025) Schalt- und Ausgleichsvorgänge in elektrischen Netzen, 2nd edn. Springer, Heidelberg
3. Funk G (1976) Symmetrische Komponenten. Elitera, Berlin
4. Clarke E (1950) Circuit-analysis of A-C power systems, vol 1 and 2. John Wiley and Sons, New York
5. Oswald BR (2009) Berechnung von Drehstromnetzen, 1st edn. Vieweg+Teubner, Wiesbaden
6. Nelles D, Tuttas C (1998) Elektrische Energietechnik. Vieweg+Teubner, Stuttgart
7. IEC 62428 2008-07-23 (2008) Electric power engineering—modal components in three-phase a.c. systems—quantities and transformations. IEC, Geneva
8. IEC 60909-0:2016 Edition 2.0 (2016) Short-circuit current calculations in three-phase a.c. systems—Part 0: calculation of currents. IEC, Geneva
9. IEC 60909-0:2002 (2002) Short-circuit current calculations in three-phase a.c. systems—Part 0: calculation of currents. IEC, Geneva
10. Edelmann H (1963) Berechnung elektrischer Verbundnetze. Springer, Berlin

11. IEC TR 60909-4:2021-06-24 (2021) Short-circuit currents in three-phase AC systems-Part 4: examples for the calculation of short-circuit currents. IEC, Geneva
12. Clarke E (1943) Circuit analysis of A-C-power systems, vol 1 and 2. John Wiley and Sons, New York
13. Funk G (1962) Der Kurzschluß im Drehstromnetz. R: Oldenbourg, München

Chapter 10
Breaking Current in Meshed Networks

In meshed electrical networks, deviations on the unsafe side with current values that are too small can occur when calculating the breaking current according to an old VDE Regulation 0102 from 1971 [1], as the network contribution was assumed to be constant regardless of the topology. The deviations are caused by generators and motors in the vicinity, which contribute significantly to the short-circuit current. Their influence can cause the contribution of the higher-level network to the short-circuit current to increase during the short-circuit duration, which cannot be neglected in all cases, as the contribution of a network is assumed to be constant during the period of the short-circuit current.

Equations are therefore given which considers these influences so that the network contribution increases during the short-circuit period. The following text largely corresponds to the reference [2] with a few amendments. In addition, the current curves over time during a short circuit are determined according to a combined impedance, taking into account a current converter (VSC technology), which is modeled as a constant current source.

The following considerations relate exclusively to synchronous machines, so the index “GS” only refers to this type of generator.

10.1 Introduction

If the three-phase short-circuit at the fault location Fl according to Fig. 10.1 is supplied by several voltage sources N (network), GS (synchronous generator) and M (motor) independently of each other, the breaking current $\underline{I}_b$ results from the sum of the partial currents, so that according to Eq. (10.1) follows:

G. Balzer, *Short-Circuit Currents in Three-Phase Systems*,
https://doi.org/10.1007/978-3-658-50710-7_10

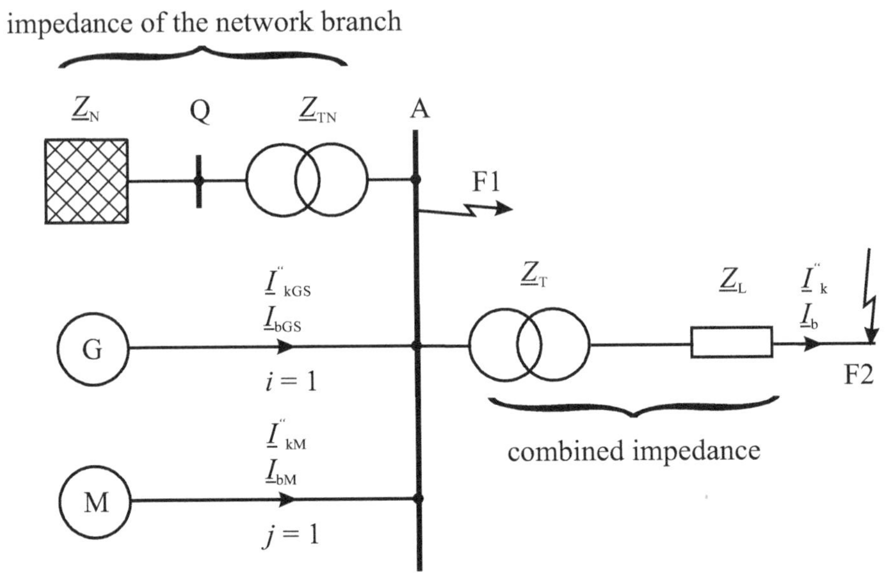

Fig. 10.1 Example of a single short circuit fed by several voltage sources in F1 and via a combined impedance in F2

$$\underline{I}_b = \underline{I}_{bN} + \underline{I}_{bGS} + \underline{I}_{bM} \tag{10.1}$$

With the individual contributions to the breaking current:

$$\text{Network (far} - \text{from} - \text{generator)}\ \underline{I}_{bN} = \underline{I}''_{kN} \tag{10.2}$$

$$\text{Generator}\ \underline{I}_{bGS} = \mu \cdot \underline{I}''_{kGS} \tag{10.3}$$

$$\text{Motor}\ \underline{I}_{bM} = \mu \cdot q \cdot \underline{I}''_{kM} \tag{10.4}$$

The factor μ takes into account the decrease of the short-circuit current from its initial value $\underline{I}''_k$ to the first contact separation (t_{min}) of a switching pole and is dependent on the ratio $\underline{I}''_{k3GS}/I_{rGS}$ or $\underline{I}''_{k3M}/I_{rM}$ and the minimum time delay t_{min}. Values for μ can be found in Eqs. (5.47)–(5.50) in Sect. 5.6.1. With a low-power asynchronous motor, the coil flux of the rotor decays faster than with a synchronous machine, as the excitation is taken from the motor terminals and is not externally excited, as with synchronous machines. This is taken into account by the additional factor q. Equations (5.52)–(5.55) in Sect. 5.6.1 represent the factor q and correspond to the model laws valid for motors. The factor q depends on the active power related to the pair of poles.

However, the calculation of the breaking current according to Eqs. (10.1)–(10.4) only applies if the individual partial short-circuit currents are decoupled during the entire short circuit duration, i.e. they do not influence each other. This is not the case for a short circuit at fault location F2 as shown in Fig. 10.1: The partial short-circuit currents of the three voltage sources N, GS and M flow through the combined impedance between busbar A and fault location F2. The consequence is that the calculation according to Eqs. (10.1)–(10.4) always leads to uncertain results [3].

The current component of the network branch does not remain constant during the short-circuit period, as assumed by Eq. (10.2). While the generator current and motor current components decrease, the potential at busbar A with respect to fault location F2 also decreases. If the network source voltage E is constant, the current flowing from the network must therefore increase in contradiction to Eq. (10.2).

If this is disregarded, the calculation of the breaking current with consideration of the motors may even result in a lower value than without motors, which can lead to considerable uncertainties, especially in industrial networks in which a large number of motors are installed, if switching devices with an insufficient switching capacity are used.

10.2 Factor μ

The magnitude of the short-circuit current I_{AC} on a single feeder at the fault location F according to Fig. 10.2 can be described by the time characteristic, Eq. (4.41), of the RMS value of the AC current component, Eq. (10.5).

$$\underline{I}_{AC} = \left(I_k'' - I_k'\right) \cdot e^{-t/T''} + \left(I_k' - I_k\right) \cdot e^{-t/T'} + I_k \tag{10.5}$$

With

I_k'' subtransient short-circuit current,
I_k' transient short-circuit current,
I_k steady-state short-circuit current,
T'', T' subtransient, transient time constant.

After the short-term subtransient component $e^{-t/T}$ of Eq. (10.5) has decayed, the approximation of the e-function is replaced by a power series.

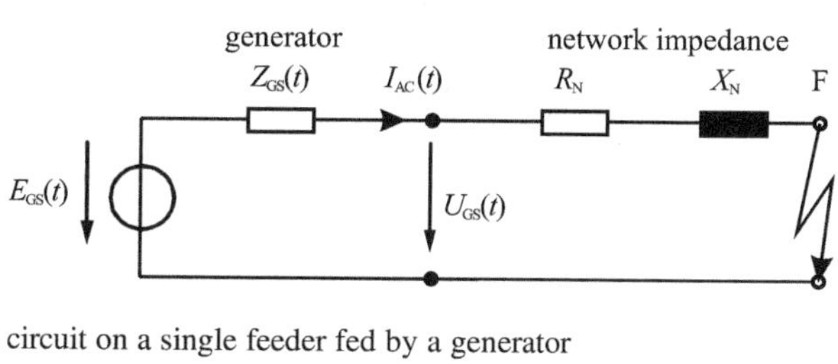

Fig. 10.2 Short circuit on a single feeder fed by a generator

$$\mathrm{e}^{-t/T'} \approx 1 + \frac{1}{1!} \cdot \left(-\frac{t}{T'}\right) + \frac{1}{2!} \cdot \left(-\frac{t}{T'}\right)^2 + \ldots \approx 1 - t/T' \tag{10.6}$$

$$I''_{\mathrm{k}} - I_{\mathrm{AC}} = \left(I''_{\mathrm{k}} - I'_{\mathrm{k}}\right) + \left(I'_{\mathrm{k}} - I_{\mathrm{k}}\right) \cdot t/T' \tag{10.7}$$

In addition to the internal generator data, this function also includes the external network impedance $R_{\mathrm{N}} + \mathrm{j}X_{\mathrm{N}}$. Of these, only the network reactance X_{N} is practically effective, because the network resistance R_{N} hardly changes the amounts I''_{k}, I'_{k}, I_{k}, of the short-circuit current as long as $R_{\mathrm{N}}/X_{\mathrm{N}} < 0.3$, nor does the transient time constant given in Eq. (10.7):

$$T' = \frac{\left(X'_{\mathrm{d}} + X_{\mathrm{N}}\right) \cdot \left(X_{\mathrm{q}} + X_{\mathrm{N}}\right) + R_{\mathrm{N}}^2}{\left(X_{\mathrm{d}} + X_{\mathrm{N}}\right) \cdot \left(X_{\mathrm{q}} + X_{\mathrm{N}}\right) + R_{\mathrm{N}}^2} \cdot T_{\mathrm{f}} \tag{10.8}$$

The indices d and q indicate the axes of the synchronous machine. T_{f} is the magnet wheel time constant. $R_{\mathrm{N}} = 0$ can therefore be set and the following follows:

$$T' = \frac{\left(X'_{\mathrm{d}} + X_{\mathrm{N}}\right)}{\left(X_{\mathrm{d}} + X_{\mathrm{N}}\right)} \cdot T_{\mathrm{f}} \tag{10.9}$$

The reactance X_{N} of the single feeder between the short-circuit location and the generator terminals can be simply substituted by the initial symmetrical short-circuit current I''_{k} as shown in Fig. 10.2:

$$X_{\mathrm{N}} = \frac{U_{\mathrm{GS}}/\sqrt{3}}{I''_{\mathrm{k}}} - X''_{\mathrm{d}} \tag{10.10}$$

The following applies to the currents according to Eq. (10.7):

$$I'_{\mathrm{k}} = \frac{U_{\mathrm{GS}}/\sqrt{3}}{X'_{\mathrm{d}} + X_{\mathrm{N}}} \qquad I_{\mathrm{k}} = \frac{U_{\mathrm{GS}}/\sqrt{3}}{X_{\mathrm{d}} + X_{\mathrm{N}}} \tag{10.11}$$

With

U_{GS} terminal voltage of generator, Fig. 10.2.

If X_{N} is replaced in the expressions for I'_{k}, I_{k} and T' according to Eq. (10.10), I_{AC} becomes dependent solely on the generator data, the time t and I''_{k}. It is advisable to refer $I''_{\mathrm{k}} - I_{\mathrm{AC}}$ to I''_{k} and use the abbreviation as follows.

$$\frac{I''_{\mathrm{k}} - I_{\approx}}{I''_{\mathrm{k}}} = 1 - \mu \tag{10.12}$$

In the short circuit on a single feeder considered in Fig. 10.2, the current is therefore only dependent on the generator data and the related variables $I''_{\mathrm{k}}/I_{\mathrm{r}}$ and t/T_{f}. The

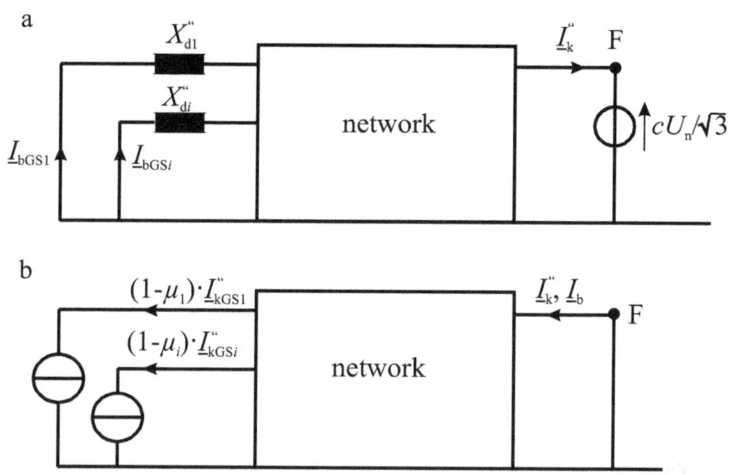

Fig. 10.3 Calculation of the breaking current in an arbitrarily meshed network by superposition (+). (**a**) Calculation of the generator components for the initial symmetrical short-circuit current according to the method with the equivalent voltage source at the short-circuit location, (**b**) calculation of the decay processes with ideal equivalent current sources at the generator terminals

expression $(1 - \mu) \cdot I''_k$ may therefore be replaced by an ideal current source without internal admittance.

If the short circuit is not on a single feeder as in Fig. 10.2, but in an arbitrarily meshed network that is also fed by other generators, the same factor μ remains valid. For the dynamic processes taking place in generator G_i, it is irrelevant how the currents are distributed in the network. The input reactance of the network represents the reactance X_N of a single feeder, this means that for each network topology, the terminals of generator G_i can be superimposed as shown in Fig. 10.3:

$$\mu = f\left(\frac{I''_{kGSi}}{I_{rGSi}}; t_{min}\right) \tag{10.13}$$

$$I_{bGSi} = I''_{kGSi} - (1 - \mu) \cdot I''_{kGSi} \tag{10.14}$$

With

I_{rGS} rated current of generator,
t_{min} minimum time delay.

In a first calculation step, the generator components I''_{kGSi} of the initial symmetrical short-circuit current are determined using the known method with the equivalent voltage source at the short-circuit location, Sect. 3.6.6, as shown in Fig. 10.3a. This is followed by the calculation of the decay processes according to Fig. 10.3b, whereby ideal equivalent current sources with the values $-(1 - \mu) \cdot I''_{kGi}$ at the generator terminals feed the short-circuited system (changed sign).

It is noticeable that the network in Fig. 10.3b has changed compared to Fig. 10.3a and does not contain any generator reactances $X''_{\mathrm{d}i}$. This is irrelevant in unmeshed networks, e.g. according to Sect. 10.4, but requires a change to the network matrix in meshed networks. If this circumstance is to be avoided, the generator reactance $X''_{\mathrm{d}i}$ can be left proximally with the equivalent current sources $(1 - \mu) \cdot I''_{\mathrm{kGS}i}$, thus producing results that are always on the safe side.

The factor μ is shown in Sect. 5.6.1 by Eqs. (5.47)–(5.50), which were derived from short-circuit tests and calculations [4]. It may also be used for asynchronous motors according to Eq. (10.4) together with the factor q. According to the theory of the rotating field, it also applies to the positive sequence current component for unbalanced short circuits.

10.3 Factor q

According to the model laws applicable to inductive machines, the time constant T_{f} in Eqs. (10.8) and (10.9) grows approximately with the square root of the rated power [5], which must be divided by the number of pair of pole. A time constant of 10 s at 100 MW therefore corresponds roughly to 0.1 s at 10 kW. The decay processes therefore take place considerably faster in machines with low power per number of pairs of poles than in the generators considered in the previous section.

The three-phase steady state short-circuit current of asynchronous motors is also zero. The factor q is shown in Sect. 5.6.1 as Eqs. (5.52)–(5.55) after short-circuit tests [5].

10.4 Single-Fed Short Circuit from Several Sources via a Combined Impedance

10.4.1 General

In the following, based on Fig. 10.1, the relationship for calculating the decaying short-circuit current is calculated when the current flows via a combined impedance. In contrast to the calculations in the case of a meshed network, Sect. 10.5.

10.4.2 Derivation of the General Equation Taking into Account Voltage Sources

The increase of the short-circuit current in the network branch N of Fig. 10.1 during the short-circuit duration can be derived with the aid of Fig. 10.4 by determining the current distribution over the short-circuit duration. To simplify matters, the motor branch is omitted and the network infeed is combined into an impedance $\underline{Z}_N$, and a coil $\underline{Z}_D$ represents the combined impedance, Fig. 10.4a. Here, $i_{kGS}(t)$ is the time characteristic of the generator current, $i_{kN}(t)$ is the network current transformed to the voltage level of busbar A and $i_k(t)$ is the total current at the fault location. The calculation is carried out using the equivalent voltage source at the short-circuit location.

In Fig. 10.4b, the equivalent current source represents the generator, which feeds a differential current into the network according to the short-circuit location, which leads to a reduction in the generator current over time. The current sharing results in differential currents that are superimposed on the initial symmetrical short-circuit currents.

The following applies to the initial symmetrical short-circuit currents of the individual branches and the total current.

Short-circuit location:

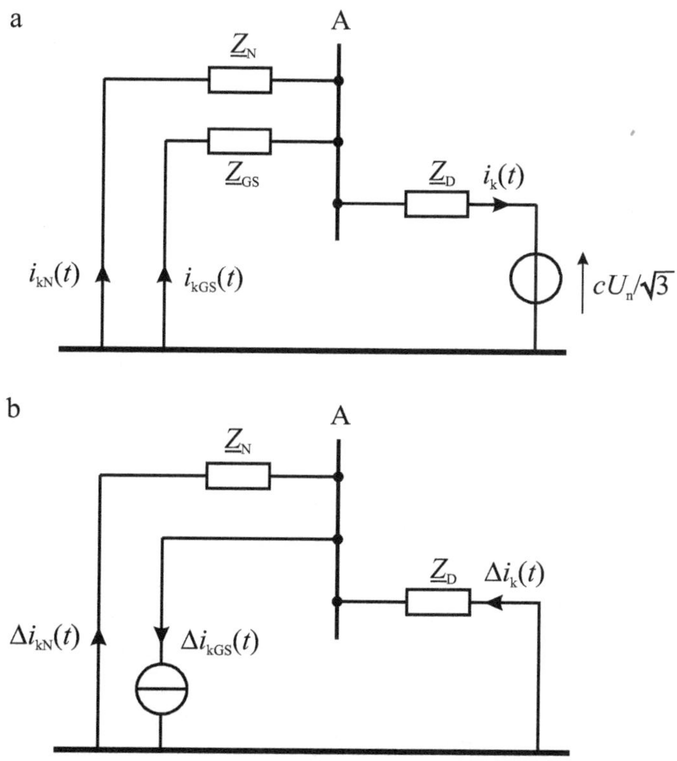

Fig. 10.4 Determination of the breaking current for a simple radial system. (**a**) Calculation of $i_{k(t)}$ at the short-circuit location, (**b**) calculation of the differential currents

$$\underline{I}''_{\mathrm{k}} = \frac{cU_{\mathrm{n}}}{\sqrt{3}} \cdot \frac{\underline{Z}_{\mathrm{N}} + \underline{Z}_{\mathrm{GS}}}{\underline{Z}_{\mathrm{D}} \cdot (\underline{Z}_{\mathrm{N}} + \underline{Z}_{\mathrm{G}}) + \underline{Z}_{\mathrm{N}} \cdot \underline{Z}_{\mathrm{GS}}} \tag{10.15}$$

Generator:

$$\underline{I}''_{\mathrm{kGS}} = \frac{cU_{\mathrm{n}}}{\sqrt{3}} \cdot \frac{\underline{Z}_{\mathrm{N}}}{\underline{Z}_{\mathrm{D}} \cdot (\underline{Z}_{\mathrm{N}} + \underline{Z}_{\mathrm{GS}}) + \underline{Z}_{\mathrm{N}} \cdot \underline{Z}_{\mathrm{GS}}} \tag{10.16}$$

Network (transformed to the voltage level of the fault):

$$\underline{I}''_{\mathrm{kN}} = \frac{cU_{\mathrm{n}}}{\sqrt{3}} \cdot \frac{\underline{Z}_{\mathrm{GS}}}{\underline{Z}_{\mathrm{D}} \cdot (\underline{Z}_{\mathrm{N}} + \underline{Z}_{\mathrm{GS}}) + \underline{Z}_{\mathrm{N}} \cdot \underline{Z}_{\mathrm{GS}}} \tag{10.17}$$

The superposition according to Fig. 10.4b applies to the breaking current $\underline{I}_{\mathrm{b}}$, as the decay of the generator current during the short-circuit time results in a reduced feed-in by the generator.

Breaking current at the short-circuit location:

$$\underline{I}_{\mathrm{b}} = \underline{I}''_{\mathrm{k}} - \frac{\underline{Z}_{\mathrm{N}}}{\underline{Z}_{\mathrm{D}} + \underline{Z}_{\mathrm{N}}} \cdot \Delta\underline{I}''_{\mathrm{kGS}} = \underline{I}''_{\mathrm{k}} - \frac{\underline{Z}_{\mathrm{N}}}{\underline{Z}_{\mathrm{D}} + \underline{Z}_{\mathrm{N}}} \cdot (1 - \mu) \cdot \underline{I}''_{\mathrm{kGS}} \tag{10.18}$$

Increase in the short-circuit current contribution of the network up to the time of disconnection because of the superposition according to Fig. 10.4:

$$\underline{I}_{\mathrm{bN}} = \underline{I}''_{\mathrm{kN}} + \frac{\underline{Z}_{\mathrm{D}}}{\underline{Z}_{\mathrm{D}} + \underline{Z}_{\mathrm{N}}} \cdot \Delta\underline{I}''_{\mathrm{kGS}} = \underline{I}''_{\mathrm{kN}} + \frac{\underline{Z}_{\mathrm{D}}}{\underline{Z}_{\mathrm{D}} + \underline{Z}_{\mathrm{N}}} \cdot (1 - \mu) \cdot \underline{I}''_{\mathrm{kGS}} \tag{10.19}$$

Equation (10.19) makes it clear that the network current for calculating the breaking current only remains constant if the impedance $\underline{Z}_{\mathrm{D}}$ between the short-circuit location and the connection point of the generator and the network feeder is zero. This means that the calculation with a constant power supply unit according to Eq. (10.2) is only applicable for those networks that are multiple single-fed. In these cases, the individual partial short-circuit currents are independent of each other and can therefore be added together. However, this does not apply to meshed networks or if the fault location is fed in via a combined impedance.

Based on Eqs. (10.18) and (10.19), the breaking current $\underline{I}_{\mathrm{b}}$ can be determined according to Fig. 10.4 (generator + motor; with the impedances according to Fig. 10.1) as follows:

$$\underline{I}_{\mathrm{b}} = \underline{I}''_{\mathrm{kN}} + \sum_{j} (\Delta\underline{I}_{\mathrm{N}j} + \underline{I}_{\mathrm{bM}j}) + \sum_{i} (\Delta\underline{I}_{\mathrm{N}i} + \underline{I}_{\mathrm{bGS}i}) \tag{10.20}$$

with the correction currents of the network $\Delta\underline{I}_{\mathrm{Nt}}$:

$$\begin{aligned}\Delta\underline{I}_{\mathrm{N}j} &= \frac{\underline{Z}_{\mathrm{T}}+\underline{Z}_{\mathrm{L}}}{\underline{Z}_{\mathrm{T}}+\underline{Z}_{\mathrm{L}}+\underline{Z}_{\mathrm{N}}+\underline{Z}_{\mathrm{TN}}}\cdot\left(1-\mu_j\cdot q_j\right)\cdot\underline{I}''_{\mathrm{kM}j}\\ \Delta\underline{I}_{\mathrm{N}i} &= \frac{\underline{Z}_{\mathrm{T}}+\underline{Z}_{\mathrm{L}}}{\underline{Z}_{\mathrm{T}}+\underline{Z}_{\mathrm{L}}+\underline{Z}_{\mathrm{N}}+\underline{Z}_{\mathrm{TN}}}\cdot\left(1-\mu_i\right)\cdot\underline{I}''_{\mathrm{kGS}i}\end{aligned} \tag{10.21}$$

These are:

$\underline{I}''_{\mathrm{kN}}, \underline{I}''_{\mathrm{kGS}i}, \underline{I}''_{\mathrm{kM}j}$	initial symmetrical short-circuit currents according to Fig. 10.4,
$\underline{I}_{\mathrm{bGS}i}$	breaking currents of the synchronous machine $i = 1 \ldots n$ according to Eq. (10.3),
$\underline{I}_{\mathrm{bM}j}$	breaking currents of the asynchronous machine $j = 1 \ldots m$ according to Eq. (10.4),
μ_i, μ_j, q_j	Factors for calculating the breaking alternating current components of synchronous machines $i = 1$... n and asynchronous motors $j = 1$ ... m according to Sect. 5.6.1.

In a first calculation step, the initial symmetrical short-circuit currents $\underline{I}''_{\mathrm{kN}}, \underline{I}''_{\mathrm{kGS}i}, \underline{I}''_{\mathrm{kM}j}$ and the factors μ_i, μ_j, q_j are determined. In a second calculation step, the breaking currents are calculated according to Eqs. (10.20) and (10.21) and, if necessary, the branch currents, e.g. the feeder according to Eq. (10.21).

If the current distribution in the entire network is to be determined during the breaking current, e.g. to determine the breaking capacity of circuit-breakers, the breaking current at the fault location, Eqs. (10.18) and (10.22) and the breaking currents of the generators and motors can be simulated as current sources. The result is the breaking current of the network feeder(s) as well as the current distribution in the entire network. This procedure also applies to the simulation of meshed networks, Sect. 10.5.

10.4.3 General Example (Voltage and Current Source)

The example in Fig. 10.4 takes into account the procedure and the influence of generators and motors in the breaking current calculation. By adding inverter feeders according to [6], the example can be extended according to Fig. 10.5. In addition to the generator feeder, a current source that feeds in a constant current during the short-circuit duration is taken into account in this case. According to the superposition principle (Sect. 3.6.7), the currents caused by the voltage sources (Fig. 10.5a, index: VS, voltage source) and then by the current source (Fig. 10.5b, index CS, current source) are calculated first. In this way, the current curve over time can be determined by superposition.

The calculation of the breaking alternating current is carried out in the following steps:

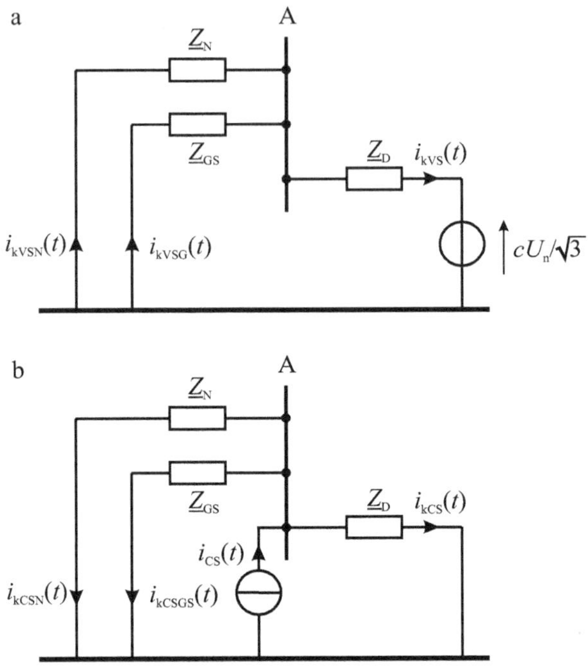

Fig. 10.5 Determination of the breaking current taking into account an additional inverter supply. (**a**) Influence of voltage source, (**b**) influence of current source

- *Voltage sources:* Determination of the initial symmetrical short-circuit current $\underline{I}''_{kVS}$ caused by the voltage sources in the network. The current sources are not taken into account.
- *Current sources:* Determination of the initial symmetrical short-circuit current $\underline{I}''_{kCS}$ caused by the current source in the network. The voltage sources are short-circuited.
- *Superposition:* Determination of the resulting initial symmetrical short-circuit currents in all network branches by superposition.
- *Correction of the network current:* Determination of the influence of the decaying currents of the voltage sources (generator) on the network current $\underline{I}''_{kVSN}$. It is assumed that the current source feeds in a constant current during the short-circuit period.

The calculation of the currents according to the steps listed above results in the following values:

Voltage Sources

The partial currents (network, generator) and the total current (short circuit) correspond to the values in Eqs. (10.15)–(10.17).

Current Source

The following equations result for the currents in the network (counting arrows according to Fig. 10.5b).

$$\text{Short-circuit current}: \underline{I}''_{\mathrm{kCS}} = \frac{\underline{Z}_{\mathrm{N}} \cdot \underline{Z}_{\mathrm{GS}}}{\underline{Z}_{\mathrm{N}} \cdot \underline{Z}_{\mathrm{GS}} + \underline{Z}_{\mathrm{D}} \cdot (\underline{Z}_{\mathrm{N}} + \underline{Z}_{\mathrm{GS}})} \cdot \underline{I}_{\mathrm{CS}} \tag{10.22}$$

$$\text{Network current}: \underline{I}''_{\mathrm{kCSN}} = -\frac{\underline{Z}_{\mathrm{D}} \cdot \underline{Z}_{\mathrm{GS}}}{\underline{Z}_{\mathrm{D}} \cdot \underline{Z}_{\mathrm{GS}} + \underline{Z}_{\mathrm{N}} \cdot (\underline{Z}_{\mathrm{D}} + \underline{Z}_{\mathrm{GS}})} \cdot \underline{I}_{\mathrm{CS}} \tag{10.23}$$

$$\text{Generator current}: \underline{I}''_{\mathrm{kCSG}} = -\frac{\underline{Z}_{\mathrm{D}} \cdot \underline{Z}_{\mathrm{N}}}{\underline{Z}_{\mathrm{D}} \cdot \underline{Z}_{\mathrm{N}} + \underline{Z}_{\mathrm{GS}} \cdot (\underline{Z}_{\mathrm{D}} + \underline{Z}_{\mathrm{N}})} \cdot \underline{I}_{\mathrm{CS}} \tag{10.24}$$

With

$\underline{I}_{\mathrm{CS}}$ Constant feeding of the current source during the short circuit.

Superposition

The total currents in the network branches result from the addition of the partial currents in Eqs. (10.15)–(10.17) and (10.22)–(10.24). While the current at the short-circuit location is increased by the additional current source, the initial symmetrical short-circuit currents of the network and generator are reduced in accordance with the impedance ratios.

$$\begin{aligned}\text{Short-circuit current}: \underline{I}''_{\mathrm{k}} &= \underline{I}''_{\mathrm{kVS}} + \underline{I}''_{\mathrm{kCS}} \\ &= \frac{(\underline{Z}_{\mathrm{N}} + \underline{Z}_{\mathrm{GS}}) \cdot cU_{\mathrm{n}}/\sqrt{3} + \underline{Z}_{\mathrm{N}} \cdot \underline{Z}_{\mathrm{GS}} \cdot \underline{I}_{\mathrm{CS}}}{\underline{Z}_{\mathrm{N}} \cdot \underline{Z}_{\mathrm{GS}} + \underline{Z}_{\mathrm{D}} \cdot (\underline{Z}_{\mathrm{N}} + \underline{Z}_{\mathrm{GS}})}\end{aligned} \tag{10.25}$$

$$\begin{aligned}\text{Network current}: \underline{I}''_{\mathrm{kN}} &= \underline{I}''_{\mathrm{kVSN}} + \underline{I}''_{\mathrm{kCSN}} \\ &= \frac{\underline{Z}_{\mathrm{GS}} \cdot cU_{\mathrm{n}}/\sqrt{3} - \underline{Z}_{\mathrm{D}} \cdot \underline{Z}_{\mathrm{GS}} \cdot \underline{I}_{\mathrm{CS}}}{\underline{Z}_{\mathrm{N}} \cdot \underline{Z}_{\mathrm{GS}} + \underline{Z}_{\mathrm{D}} \cdot (\underline{Z}_{\mathrm{N}} + \underline{Z}_{\mathrm{GS}})}\end{aligned} \tag{10.26}$$

$$\begin{aligned}\text{Generator current}: \underline{I}''_{\mathrm{kGS}} &= \underline{I}''_{\mathrm{kVSGS}} + \underline{I}''_{\mathrm{kCSGS}} \\ &= \frac{\underline{Z}_{\mathrm{N}} \cdot cU_{\mathrm{n}}/\sqrt{3} - \underline{Z}_{\mathrm{D}} \cdot \underline{Z}_{\mathrm{N}} \cdot \underline{I}_{\mathrm{CS}}}{\underline{Z}_{\mathrm{N}} \cdot \underline{Z}_{\mathrm{GS}} + \underline{Z}_{\mathrm{D}} \cdot (\underline{Z}_{\mathrm{N}} + \underline{Z}_{\mathrm{GS}})}\end{aligned} \tag{10.27}$$

Correction of the Network Current

The reduction of the generator current during the short circuit leads to an increase in the network current (the input of the current source is constant), so that the breaking current at the short-circuit location is calculated according to Eq. (10.28).

$$\underline{I}_{\mathrm{b}} = \underline{I}''_{\mathrm{k}} - \frac{\underline{Z}_{\mathrm{N}}}{\underline{Z}_{\mathrm{D}} + \underline{Z}_{\mathrm{N}}} \cdot \Delta \underline{I}''_{\mathrm{kGS}} = \underline{I}''_{\mathrm{k}} - \frac{\underline{Z}_{\mathrm{N}}}{\underline{Z}_{\mathrm{D}} + \underline{Z}_{\mathrm{N}}} \cdot (1 - \mu) \cdot \underline{I}''_{\mathrm{kGS}} \tag{10.28}$$

With

$\underline{I}''_{\mathrm{k}}$ resulting short-circuit current at the fault location, Eq. (10.25),
$\underline{I}''_{\mathrm{kGS}}$ resulting short-circuit current of the generator, Eq. (10.27),
μ factor for determining the breaking current.

As the generator current is lower than in the example in Fig. 10.4 due to the feeding of the current source (inverter), the factor μ will have a larger value. The influence of a current source that simulates an inverter during the short-circuit duration on the entire short-circuit current sequence is described in detail in Sect. 10.8 using a calculation program for simulating transient phenomena.

10.5 Short Circuit in the Meshed Network

The circuits of meshed networks cannot always be traced back to the simple structure shown in Fig. 10.1, e.g. through network conversions. In such cases, the breaking current I_b at the short-circuit location can be calculated with

$$\underline{I}_{\mathrm{b}} = \underline{I}''_{\mathrm{k}} - \sum_j \left(\frac{\sqrt{3} \cdot \Delta U''_{\mathrm{M}j}}{c \cdot U_{\mathrm{n}}} \right) \cdot (1 - \mu_j \cdot q_j) \cdot \underline{I}''_{\mathrm{kM}j} - \sum_i \left(\frac{\sqrt{3} \cdot \Delta U''_{\mathrm{G}i}}{c \cdot U_{\mathrm{n}}} \right) \cdot (1 - \mu_i) \cdot \underline{I}''_{\mathrm{kGS}i} \tag{10.29}$$

These are:

$c \cdot U_{\mathrm{n}}/\sqrt{3}$	equivalent voltage source at the short-circuit location,
$\underline{I}_{\mathrm{b}}$	breaking current at the short-circuit location, taking into account the network feeder, synchronous machines and asynchronous motors,
$\underline{I}''_{\mathrm{k}}$	initial symmetrical short-circuit current taking into account the network feeder, synchronous machines and asynchronous motors,
$\Delta U''_{\mathrm{M}j}, \Delta U''_{\mathrm{GS}i}$	difference between the voltage at the location of the asynchronous motor j or synchronous machine i at the occurrence of the short circuit and before the short circuit; the rotating machines, including their conductance values, are not taken into account,
$\underline{I}''_{\mathrm{kGS}i}, \underline{I}''_{\mathrm{kM}j}$	partial short-circuit currents of the synchronous machines i and the asynchronous motors j,
$\mu_{i,j}, q_j$	decay factors for calculating the breaking current of generators and asynchronous motors, Sect. 5.5.

If a calculation according to Eq. (10.29) is not possible or too time-consuming, the following applies to the breaking alternating current in meshed networks:

$$\underline{I}_{\mathrm{b}} = \underline{I}''_{\mathrm{k}} \tag{10.30}$$

This means that the result according to Eq. (10.30) is always on the safe side, since the decay behavior of the generators and motors is not taken into account.

The calculation according to Eqs. (10.20) and (10.29) leads to the same result for the equivalent circuit according to Fig. 10.1. In order to recognize the validity of

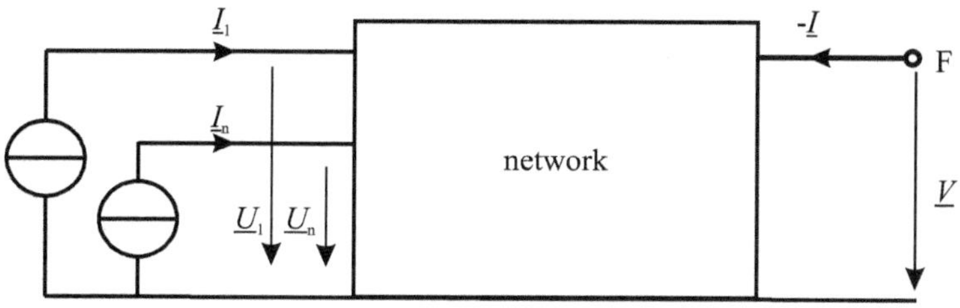

Fig. 10.6 (n + 1) Gate for the derivation of Eq. (10.29)

Eq. (10.29), the (n + 1) gate according to Fig. 10.6 is examined. Eq. (10.31) is obtained in matrix notation with the counting arrows inserted.

$$\begin{pmatrix} \underline{\mathbf{U}} \\ \underline{V} \end{pmatrix} = \begin{pmatrix} \underline{\mathbf{A}} & \underline{\mathbf{B}} \\ \underline{\mathbf{C}} & \underline{D} \end{pmatrix} \cdot \begin{pmatrix} \underline{\mathbf{J}} \\ -\underline{I} \end{pmatrix} \tag{10.31}$$

Here, $\underline{\mathbf{U}}$ is the column vector of the terminal voltages at the short-circuit current sources and V is the voltage at the future short-circuit location F. In the event of a short circuit, the following applies with $\underline{V} = 0$

$$\underline{D} \cdot \underline{I} = \underline{\mathbf{C}} \cdot \underline{\mathbf{J}} \tag{10.32}$$

The coefficients $\underline{D}$ and $\underline{\mathbf{C}}$ can be determined for the initial short-circuit $t = 0$. For this purpose, $\underline{\mathbf{U}}$, $\underline{V}$, $\underline{\mathbf{J}}$ and $\underline{I}$ in Eq. (10.31) are replaced by their change, and it is taken into account that the following applies for the rotating machines in the initial short circuit $\Delta\underline{\mathbf{J}}'' = 0$. The second line of Eq. (10.31) thus follows:

$$\Delta\underline{V}'' = -\underline{D} \cdot \Delta\underline{I}'' \tag{10.33}$$

$\underline{D} = \underline{Z}_{\mathrm{kF}}$ is obviously the short-circuit impedance at the fault location F. The same follows from the first line of Eq. (10.31):

$$\Delta\underline{\mathbf{U}}'' = -\underline{\mathbf{B}} \cdot \Delta\underline{\mathbf{I}}'' \tag{10.34}$$

For the hth line you get with $\Delta\underline{I}'' = \underline{I}''_{\mathrm{k}}$.

$$\underline{B}_h = \frac{-\underline{U}''_h}{\underline{I}''_{\mathrm{k}}}. \tag{10.35}$$

$\underline{U}''_h$ is the terminal voltage change at the generator h at the initial short circuit. Due to the diagonal symmetry of the impedance matrix in Eq. (10.31), the coefficients C:

$$\underline{\mathbf{C}} = \underline{\mathbf{B}}^{\mathrm{T}} \quad \text{or} \quad \underline{C}_h = \underline{B}_h \tag{10.36}$$

If the signs are selected so that they become positive when calculating with real numbers $\Delta\underline{I}$, $\Delta\underline{J}$, $\Delta\underline{U}''$, the following must be used.

$-\underline{J} = \left(\Delta\underline{J}_{\mathrm{M}j}, \Delta\underline{J}_{\mathrm{GS}i}\right)$ column vector of the source current change in M and G during the short-circuit duration, Eqs. (10.40) and (10.41),

$\underline{I} = \Delta\underline{I}$ requested short-circuit current change at the fault location according to Eq. (10.41),

$-\underline{U}'' = \left(\Delta\underline{U}''_{\mathrm{M}j}, \Delta\underline{U}''_{\mathrm{GS}i}\right)$ column vector of the terminal voltage change in M and G when a short circuit occurs.

This is obtained from Eq. (10.32) with $\underline{\mathrm{D}} = \underline{B}_{\mathrm{kF}}$ and C according to Eq. (10.36):

$$\underline{Z}_{\mathrm{kF}} \cdot \Delta\underline{I} = \sum_{j}\left(\frac{\Delta\underline{U}''_{\mathrm{M}j}}{\underline{I}''_{\mathrm{k}}}\right) \cdot \Delta\underline{J}_{\mathrm{M}j} + \sum_{i}\left(\frac{\Delta\underline{U}''_{\mathrm{GS}i}}{\underline{I}''_{\mathrm{k}}}\right) \cdot \Delta\underline{J}_{\mathrm{GS}i} \tag{10.37}$$

Because of

$$\underline{Z}_{\mathrm{kF}} \cdot \underline{I}''_{\mathrm{k}} = c \cdot U_{\mathrm{n}}/\sqrt{3} \tag{10.38}$$

results

$$\Delta\underline{I} = \sum_{j}\left(\frac{\Delta\underline{U}''_{\mathrm{M}j}}{c \cdot U_{\mathrm{n}}/\sqrt{3}}\right) \cdot \Delta\underline{J}_{\mathrm{M}j} + \sum_{i}\left(\frac{\Delta\underline{U}''_{\mathrm{GS}i}}{c \cdot U_{\mathrm{n}}/\sqrt{3}}\right) \cdot \Delta\underline{J}_{\mathrm{GS}i} \tag{10.39}$$

With the changes of the source current

$$\Delta\underline{J}_{\mathrm{M}j} = \left(1 - \mu_j \cdot q_j\right) \cdot \underline{I}''_{\mathrm{kM}j} \tag{10.40}$$

and

$$\Delta\underline{J}_{\mathrm{GS}i} = \left(1 - \mu_j\right) \cdot \underline{I}''_{\mathrm{kGS}i} \tag{10.41}$$

According to Eq. (10.39), the changes of the source current are transferred to the fault location F to the same extent as the voltage breakdown of the fault location to the position of the rotating machines. The breaking current $\underline{I}_{\mathrm{b}}$ at the fault location is therefore

$$\underline{I}_{\mathrm{b}} = \underline{I}''_{\mathrm{k}} - \Delta\underline{I}. \tag{10.42}$$

The procedure can also be applied to line-to-line short circuits, but these are generally not of interest.

The method described above does not require any additional assumptions or neglect other than the correctness of the μ-curves. However, it is a disadvantage that the quantities $\Delta U''_{\mathrm{M}j}$ and $\Delta U''_{\mathrm{GS}i}$ for Eq. (10.29) must be determined without the internal reactances of the motors and generators, i.e. the motors and generators are not present and the superimposed network alone feeds the short circuit. In many cases, it takes too much effort to calculate these variables $\Delta U''_{\mathrm{M}j}$ and $\Delta U''_{\mathrm{GS}i}$ in this way and one is content with

$$\Delta U''_{\mathrm{M}j} = \mathrm{j}X''_{\mathrm{M}j} \cdot \underline{I}''_{\mathrm{kM}j} \tag{10.43}$$

and

$$\Delta U''_{\mathrm{GS}i} = \mathrm{j}X''_{\mathrm{d}i} \cdot \underline{I}''_{\mathrm{kGS}i} \tag{10.44}$$

These results are then on the safe side, as can be seen from Table 10.2, Sect. 10.6.2.

By inserting the voltage values for the example in Fig. 10.4a, Eq. (10.29) for meshed networks can be converted into Eq. (10.18) for radial systems (combined impedance). The following applies to the voltage at the terminals of the generator $\underline{U}_{\mathrm{G}}$ if only the network is considered as an active element:

$$\underline{U}_{\mathrm{GS}} = \frac{\underline{Z}_{\mathrm{D}}}{\underline{Z}_{\mathrm{D}} + \underline{Z}_{\mathrm{N}}} \cdot \frac{c \cdot U_{\mathrm{n}}}{\sqrt{3}} \tag{10.45}$$

This can be used to determine the voltage change ΔU_{GS} caused by the short circuit.

$$\Delta \underline{U}_{\mathrm{GS}} = \left(1 - \frac{\underline{Z}_{\mathrm{D}}}{\underline{Z}_{\mathrm{D}} + \underline{Z}_{\mathrm{N}}}\right) \cdot \frac{c \cdot U_{\mathrm{n}}}{\sqrt{3}} = \frac{\underline{Z}_{\mathrm{N}}}{\underline{Z}_{\mathrm{D}} + \underline{Z}_{\mathrm{N}}} \cdot \frac{c \cdot U_{\mathrm{n}}}{\sqrt{3}} \tag{10.46}$$

By inserting $\Delta \underline{U}_{\mathrm{GS}}$ into Eq. (10.29), the breaking current $\underline{I}_{\mathrm{b}}$ can be derived for the example. The result obtained in this way agrees with Eq. (10.18).

If, instead, the result of the calculation of the initial symmetrical short-circuit current is used to determine the voltage change at the generator terminal, i.e. the generator is taken into account with its impedance in this example, the result is $\underline{U}_{\mathrm{GS}}$ for the voltage:

$$\underline{U}_{\mathrm{GS}} = \frac{\underline{Z}_{\mathrm{D}}}{\underline{Z}_{\mathrm{D}} + \dfrac{\underline{Z}_{\mathrm{N}} \cdot \underline{Z}_{\mathrm{GS}}}{\underline{Z}_{\mathrm{N}} + \underline{Z}_{\mathrm{GS}}}} \cdot \frac{c \cdot U_{\mathrm{n}}}{\sqrt{3}} \tag{10.47}$$

This results in the following for the voltage change $\Delta \underline{U}_{\mathrm{G}}$:

$$\Delta \underline{U}_{\mathrm{GS}} = \frac{\underline{Z}_{\mathrm{N}}}{\underline{Z}_{\mathrm{D}} + \underline{Z}_{\mathrm{N}} + \dfrac{\underline{Z}_{\mathrm{N}} \cdot \underline{Z}_{\mathrm{D}}}{\underline{Z}_{\mathrm{GS}}}} \cdot \frac{c \cdot U_{\mathrm{n}}}{\sqrt{3}} \tag{10.48}$$

A comparison of Eqs. (10.46) and (10.48) shows that the calculation of the terminal voltage according to Eq. (10.47) always provides a result that is on the safe side when calculating the breaking current. The calculated voltage $\underline{U}_{GS}$ is determined to be too high, which means that the current calculated in this way is greater than the current that actually occurs. Since, according to Eq. (10.49), the left-hand part of the inequality is always smaller than the right-hand part and therefore the voltage change is smaller, the reduction in the initial symmetrical short-circuit current is also smaller according to Eq. (10.29).

$$\frac{\underline{Z}_{N}}{\underline{Z}_{D}+\underline{Z}_{N}+\dfrac{\underline{Z}_{N}\cdot\underline{Z}_{D}}{\underline{Z}_{GS}}}\leq\frac{\underline{Z}_{N}}{\underline{Z}_{D}+\underline{Z}_{N}} \tag{10.49}$$

10.6 Examples

The calculation of the breaking current at the short-circuit location including the partial currents is shown below using two examples (short circuit after a combined impedance and in a meshed network). Only the influence of equipment that can be simulated as voltage sources is considered.

10.6.1 Example 1 (Combined Impedance)

In a first example, the current and voltage propagation for the network circuit as shown in Fig. 10.7 is calculated after a three-phase short circuit occurrence. Here, the short-circuit current is fed both by a generator, which feeds directly into a 6 kV busbar, and by a 110 kV network, which is connected to the busbar via a transformer. The short circuit occurs after a short-circuit current limiting coil.

The following values apply to equipment shown in Fig. 10.7:

Network: $U_{nN} = 110\text{ kV}$ $I''_{kN} = 1{,}574\text{kA}$ $R/X = 0.1$
Transformer: $S_{rT} = 25\text{ MVA}$ $t_r = 110\text{ kV}/6\text{ kV}$
$u_{kr} = 15\%$ $u_{kr} = 0.5\%$
Limiting coil: $U_{rD} = 6\text{ kV}$ $S_{rD} = 25\text{ MVA}$ $x_D = 6\%$
Generator: $U_{rGS} = 6.3\text{ kV}$ $S_{rGS} = 25\text{ MVA}$ $\sin\varphi_{rGS} = 0.8$
$x''_d = 11\%$ $x'_d = 16.5\%$ $x_d = 180\%$
$T''_d = 0.03\text{ s}$ $T'_{d0} = 7\text{ s}$ $T_a = 0.2\text{ s}$

Using the data of equipment given above, the impedances and currents with a minimum time delay of $t_{min} = 0.25$ s are calculated as follows.

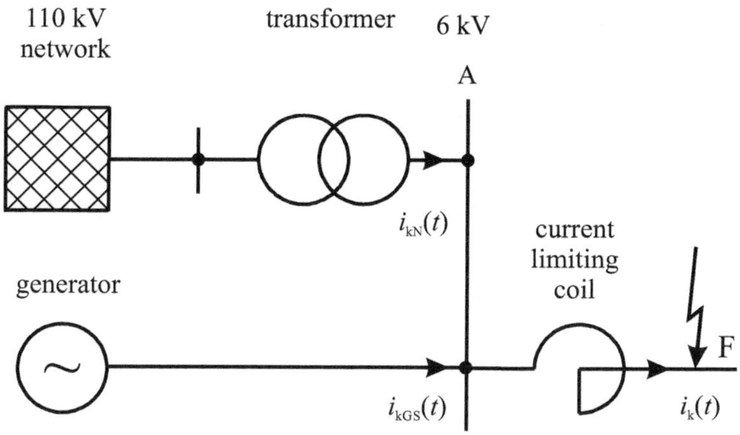

Fig. 10.7 Calculation of the breaking current I_b for a three-phase short circuit behind a combined impedance (limiting coil) $\underline{Z}_D$

Network

$$Z_N = \frac{c \cdot U_{nN}}{\sqrt{3} \cdot I''_{kN}} = \frac{1.1 \cdot 110 \text{ kV}}{\sqrt{3} \cdot 1.574 \text{ kA}} = 44.383 \ \Omega$$

$$\underline{Z}_N = (4.416 + \text{j}44.163) \ \Omega \quad 110 \text{ kV level}$$

$$\underline{Z}_{Nt} = (0.013 + \text{j}0.131) \ \Omega \quad 6 \text{ kV level}$$

Transformer

$$Z_T = \frac{u_{kr}}{100\%} \cdot \frac{U_{rT}^2}{S_{rT}} = 0.15 \cdot \frac{(6\,\text{kV})^2}{25 \text{ MVA}} = 0.216 \ \Omega$$

$$R_T = \frac{u_{Rr}}{100} \cdot \frac{U_{rT}^2}{S_{rT}} = 0.005 \cdot \frac{(6 \text{ kV})^2}{25 \text{ MVA}} = 0.007 \ \Omega \qquad X_T = \sqrt{Z_T^2 - R_T^2} = 0.216 \ \Omega$$

$$K_T = 0.95 \frac{c_{max}}{1 + 0.6 \cdot x_T} = 0.95 \cdot \frac{1,1}{1 + 0.6 \cdot 0.15} = 0.959$$

$$\underline{Z}_{TK} = K_T \cdot \underline{Z}_T = (0.007 + \text{j}0.207)\,\Omega$$

Current Limiting Coil

$$X_D \approx \frac{x_D}{100\%} \cdot \frac{U_{rD}^2}{S_{rD}} = 0.06 \cdot \frac{(6\,\text{kV})^2}{25\,\text{MVA}} = 0.086\,\Omega$$

Generator

$$X''_{\mathrm{d}} = \frac{x''_{\mathrm{d}}}{100\%} \cdot \frac{U^2_{\mathrm{rGS}}}{S_{\mathrm{rGS}}} = 0.11 \cdot \frac{(6.3\,\mathrm{kV})^2}{25\,\mathrm{MVA}} = 0.175\,\Omega$$

When calculating the initial symmetrical short-circuit current of the generator, the actual resistance of the generator is used, which is determined from the time constant T_{a}

$$R_{\mathrm{GS}} = \frac{X''_{\mathrm{d}}}{\omega \cdot T_{\mathrm{a}}} = \frac{0.175\ \Omega}{100 \cdot \pi \cdot 0.2} = 0.003\ \ \Omega$$

$$K_{\mathrm{GS}} = \frac{U_{\mathrm{n}}}{U_{\mathrm{rGS}}} \cdot \frac{c_{\mathrm{max}}}{1 + x''_{\mathrm{d}} \cdot \sin\varphi_{\mathrm{rGS}}} = \frac{6\,\mathrm{kV}}{6.3\,\mathrm{kV}} \cdot \frac{1.1}{1 + 0.11 \cdot 0.8} = 0.963$$

$$\underline{Z}_{\mathrm{GSK}} = K_{\mathrm{GS}} \cdot \underline{Z}_{\mathrm{GS}} = (0.003 + \mathrm{j}0.168)\,\Omega$$

The following applies to the rated current of the generator:

$$I_{\mathrm{rGS}} = \frac{S_{\mathrm{rGS}}}{\sqrt{3} \cdot U_{\mathrm{rGS}}} = \frac{25\ \ \mathrm{MVA}}{\sqrt{3} \cdot 6.3\ \ \mathrm{kV}} = 2.291\ \ \mathrm{kA}$$

The initial symmetrical short-circuit current is calculated according to Eqs. (10.15)–(10.17), all currents are related to the 6 kV level, so that the index t for the network feeder is omitted.

$$\underline{I}''_{\mathrm{k}} = \frac{1.1 \cdot 6\ \ \mathrm{kV}}{\sqrt{3}} \cdot (0.090 - \mathrm{j}5.042)\ \frac{1}{\Omega} = (0.343 - \mathrm{j}19.213)\ \ \mathrm{kA} \Rightarrow I''_{\mathrm{k}}$$
$$= 19.216\ \ \mathrm{kA}$$

$$\underline{I}''_{\mathrm{kN}} = \frac{1.1 \cdot 6\ \ \mathrm{kV}}{\sqrt{3}} \cdot (0.076 - \mathrm{j}1.671)\ \frac{1}{\Omega} = (0.289 - \mathrm{j}6.380)\ \ \mathrm{kA} \Rightarrow I''_{\mathrm{k}}$$
$$= 6.375\ \ \mathrm{kA}$$

$$\underline{I}''_{\mathrm{kGS}} = \frac{1.1 \cdot 6\ \ \mathrm{kV}}{\sqrt{3}} \cdot (0.014 - \mathrm{j}3.371)\ \frac{1}{\Omega} = (0.054 - \mathrm{j}12.845)\ \ \mathrm{kA} \Rightarrow I''_{\mathrm{k}}$$
$$= 12.845\ \ \mathrm{kA}$$

The ratio $I''_{\mathrm{kGS}}/I_{\mathrm{rGS}}$ is required to calculate the factor μ for determining the breaking current of the generator. This is calculated from the partial short-circuit current of the generator and its rated current I_{rGS}:

$$I''_{\mathrm{kGS}}/I_{\mathrm{rGS}} = \frac{12.845}{2.291} = 5.607$$

From this value, the factor $\mu = 0.679$ can be determined according to Eq. (5.51) for a minimum time delay of $t_{min} = 0.25$ s. This means that the short-circuit current contribution of the generator decreases to 67.9% of its original value after $t = 0.25$ s.

Equations (10.18) and (10.19) can be used to determine the breaking current at the fault location F. This results in the following values:

$$\underline{I}_b = (0.369 - j15.924)\ \text{kA} \Rightarrow I_b = 15.928\ \text{kA}$$

$$\underline{I}_{bN} = (0.332 - j7.215)\ \text{kA} \Rightarrow I_{bN} = 7.222\ \text{kA}$$

The breaking current $\underline{I}_{bGS}$ of the generator can be determined using the initial symmetrical short-circuit current and the factor μ, taking into account Eq. (5.52), or from the difference between the two breaking currents $\underline{I}_b$ and $\underline{I}_{bN}$ listed above.

$$\underline{I}_{bGS} = (0.037 - j8.710)\ \text{kA} \Rightarrow I_{bGS} = 8.710\ \text{kA}$$

A comparison of the network contributions I''_{kN} and I_{bQ} shows that in this example this part increases from 6.375 to 7.222 kA, i.e. by 13.2%, during the short-circuit duration. The short-circuit current calculation at fault location F therefore provides the following values:

$$\underline{I}''_k = \underline{I}''_{kN} + \underline{I}''_{kGS} = (0.343 - j19.213)\,\text{kA} \Rightarrow I''_k = 19.216\ \text{kA}$$

$$\underline{I}_b = (0.369 - j15.924)\ \text{kA} \Rightarrow I_b = 15.928\ \text{kA}$$

If, on the other hand, the short-circuit current part of the network is incorrectly kept constant, this results in a value of $I_b = 15.093$ kA, which is on the unsafe side, i.e. the breaking current is determined to be 4.8% too small.

$$\underline{I}_b = \underline{I}''_{kN} + \underline{I}_{bGS} = (0.326 - j15.090)\,\text{kA} \Rightarrow I_b = 15.093\ \text{kA}$$

Figure 10.8a–d show the corresponding results for a short-circuit at fault location F. The short-circuit occurs from the no-load operation of the generator at time $t = 15$ ms, the excitation voltage is considered constant during the short-circuit duration. The calculations are carried out with a digital calculation program for dynamic network processes.

The results of the calculation of the transient short-circuit current curves during the short-circuit duration up to $t = 0.5$ s show (Fig. 10.8) that, when the generator is taken into account, the breaking current I_{bN} of the network part is greater than the initial symmetrical short-circuit current I''_{kN} (Fig. 10.8c). This is because the decrease in the generator current during the short-circuit period results in a reduction in the voltage at the combined busbar. With a constant network source voltage, the current flowing from the network must therefore increase.

In order to estimate the accuracy of the method proposed in Sect. 10.4, Table 10.1 shows the increase of the network current ΔI_N according to Eq. (10.21)

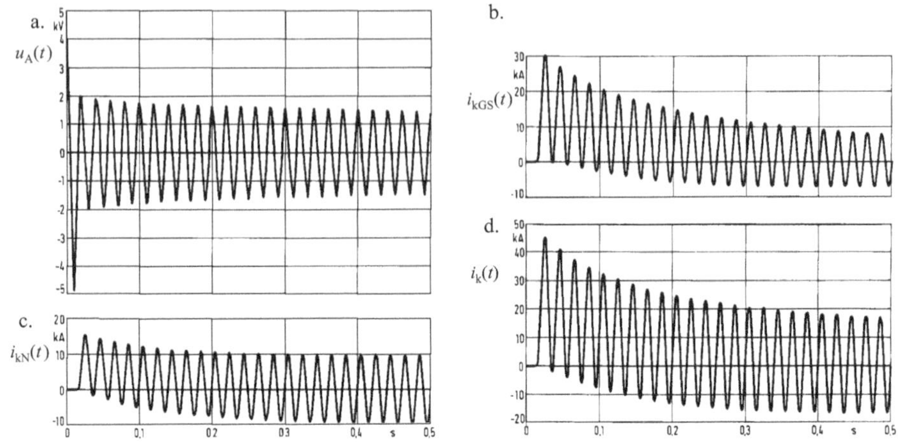

Fig. 10.8 Current and voltage curve for a three-phase short circuit across a combined impedance as shown in Fig. 10.6. (**a**) Voltage $u_A(t)$ at busbar A, (**b**) generator current $i_{kGS}(t)$, (**c**) network current $i_{kN}(t)$, (**d**) total current (short-circuit current at the fault location) $i_k(t)$

and the breaking currents according to Eq. (10.20). The factor μ in Eq. (10.21) was determined in two different ways: On the one hand with a decay curve $i_{k\approx}(t)/I''_k$ that only applies to the specific generator described in this example for the present case $I''_{kGS}/I_{rGS} = 5.6$ [Index b; Table 10.1] and on the other hand for the factor μ according to Eqs. (5.48)–(5.51), Sect. 5.6.1 [Index c; Table 10.1]. The values with index 1 are taken from the calculated oscillograms in Fig. 10.8. The values for the breaking current I_b (index a and b) agree well, so that the simulation using a calculation program for calculating transient phenomena and the calculation according to Eq. (10.18) are identical. In contrast, the deviations for the third method (Index 3) must be greater, as in this case the general curves μ are used, which do not have to correspond to the synchronous machine used.

The influence of a current source (inverter) on the short-circuit current curve is dealt with using this example in Sect. 10.8.

10.6.2 Example 2 (Meshed Network)

In the second example shown in Fig. 10.9, the combined impedance of the voltage sources cannot be specified directly, so that the breaking current $\underline{I}_b$ must be calculated according to Eq. (10.29). Equipment data are as follows:

Network: $U_{nN} = 110$ kV $I''_{kN} = 1.574$ kA $R/X = 0.1$

Transformer T1: $S_{rT} = 31.5$ MVA $t_r = 20$ kV/6 kV

$u_{kr} = 15.5\%$ $u_{kr} = 0.75\%$

Transformer T2: $S_{rT} = 1.6$ MVA $t_r = 6$ kV/0.4 kV

Table 10.1 Increases of the network current ΔI_{Nt} and breaking currents I_b, depending on the minimum time delay t_{min}

t_{min} (ms)	$I_{kAC}(t)/I''_k$	μ	ΔI_{Nt}[a] (kA)	ΔI_{Nt}[b] (kA)	ΔI_{Nt}[c] (kA)	I_b[a] (kA)	I_b[b] (kA)	I_b[c] (kA)
50	0.83	0.81	0.36	0.39	0.51	15.57	15.21	17.21
100	0.76	0.74	0.54	0.55	0.68	14.96	14.96	16.55
250	0.64	0.67	0.89	0.82	0.85	13.96	13.85	15.85

[a] Values according to digital calculation, Fig. 10.8b
[b] Values according to Eqs. (10.20) and (10.21) with for the generator described in this example
[c] Values according to Eqs. (10.20) and (10.21) with the factor μ according to Eqs. (5.48)–(5.51), Sect. 5.6.1

$u_{kr} = 5.0\%$ $u_{kr} = 1.25\%$
Cable L1: $R' = 0.059\ \Omega/\text{km}$ $X' = 0.153\ \Omega/\text{km}$ (three circuits)
Cable L2: $R' = 0.074\ \Omega/\text{km}$ $X' = 0.156\ \Omega/\text{km}$ (two circuits)
Cable L3: $R' = 0.15\ \Omega/\text{km}$ $X' = 0.168\ \Omega/\text{km}$ (two circuits)
Cable L4: $R' = 0.097\ \Omega/\text{km}$ $X' = 0.16\ \Omega/\text{km}$ (two circuits)
Current limiting coil D: $U_{rD} = 6$ kV $S_{rD} = 15$ MVA
$u_D = 3\%$ $u_r = 0.09\%$
Motors M1: $P_{rM} = 500$ kW $I_{an}/I_r = 5.5$
$\eta \cos\varphi = 0.86$ $p = 3$ $U_{rM} = 6$ kV
Motors M2: $P_{rM} = 250$ kW $I_{an}/I_r = 6.0$
$\eta \cos\varphi = 0.82$ $p = 3$ $U_{rM} = 380$ V

Table 10.2 shows the result of the calculation, assuming a short circuit at all network nodes. The value 100% corresponds to the calculation according to Eq. (10.29) with the voltage changes $\Delta \underline{U}''_{Mj}$ at the asynchronous motors for $1/X_{Mj} \rightarrow 0$, according to the derivation according to Eq. (10.29), Sect. 10.5. This requires the short-circuit current calculation to be carried out in a first calculation step *without* and in a second calculation step *with* the asynchronous motors. If this is to be avoided, $\Delta \underline{U}''_{Mj}$ must be calculated in a single calculation using Eq. (10.43). The result $I_b{}^b$ is shown in the right-hand columns and is always on the safe side when using the usual method of the equivalent voltage source at the short-circuit location. However, this means considerable deviations at node B, where the calculation with Eq. (10.1) already led to a correct result due to the decoupling of the currents.

If the calculation is carried out under the incorrect assumption that the network current remains constant during the entire short-circuit duration, the breaking currents are calculated too small, with the exception of node B, at which the partial short-circuit currents of the other voltage sources are decoupled from the network short-circuit current, Index 1, Table 10.2.

In addition, Table 10.3 shows that the calculations of the breaking current according to Eq. (10.1) with a constant current part of the network in nodes D and E, taking into account the motors, provides significantly smaller values for $t_{min} \geq 0.1$ s than without motors.

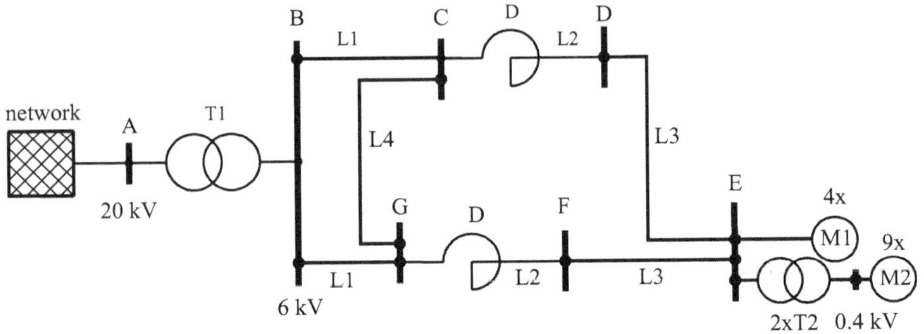

Fig. 10.9 Example for determining the breaking current according to Eq. (10.20); nodes B to G each have four high-voltage and nine low-voltage motors, as shown for node E

Table 10.2 Determination of the breaking current I_b and the initial symmetrical short-circuit current I''_k according to Fig. 10.9 as a function of the minimum time delay t_{min}

		$I_b{}^a/\%\ t_{min}$			$I_b{}^b/\%\ t_{min}$		
Nodes	I''_k/kA	0.05 s	0.1 s	0.25 s	0.05 s	0.1 s	0.25 s
B	26.7	100.0	100.0	100.0	116.1	118.2	121.0
C, G	24.5	97.7	96.9	96.3	110.8	111.9	113.9
D, F	18.7	87.8	85.4	80.6	107.1	107.7	108.4
E	18.3	85.9	83.2	79.5	103.3	103.9	104.3

[a] According to Eq. (10.1) with constant current of the network
[b] According to Eqs. (10.29) and (10.43), i.e. $X_{Mj} \neq \infty$

This incomprehensible result according to Table 10.3, that a calculation without motors results in a higher breaking current than with motors, led to a revision of the former VDE standard with regard to the breaking current. Further examples with a comparison of the calculation methods are shown in [7].

10.7 Application of the Result to the Calculation of the Steady-State Short-Circuit Current I_k

According to the result in Table 10.2, even greater errors should be expected when calculating the steady-state short-circuit current I_k, neglecting the coupling through a combined impedance, since according to Eq. (10.21) the network current corrections ΔI_{Nj} are maximum for asynchronous motors in the steady-state short circuit. On the other hand, the voltage regulator will intervene in the case of synchronous machines and try to maintain the terminal voltage, which prevents the short-circuit current from falling further.

It is therefore considered sufficient for the calculation of the steady-state short-circuit current to only take into account the increase in the network short-circuit

Table 10.3 Breaking current I_b according to Sect. 10.9

		$I_b{}^b$ (kA) t_{min}		
Node	$I_b{}^a$ (kA)	0.05 s	0.1 s	0.25 s
D	8.6	9.3	8.5	7.4
E	8.3	9.0	8.1	7.0

[a] Without motors
[b] With motors, but according to Eq. (10.1) with constant current part of the network

current due to the asynchronous motors. For this purpose, the steady-state short-circuit current is calculated in meshed networks according to [5] without the asynchronous motors present in the network.

10.8 Influence of a Current Source (Self-Commutated Inverter, IGBT Technology) on the Short-Circuit Current Curve

In the past, only voltage sources were assumed when deriving the calculation equation for the breaking current, and the equation for determining the breaking current [7, Eq. (77)] for meshed networks listed in the valid IEC standard was originally derived for this short-circuit case. This equation corresponds to Eq. (10.29) in this chapter.

Due to the use of converter systems with IGBT technology, which supply a constant current during the short circuit, the application of the calculation method must be checked for this process. In contrast to the presentation in Sect. 10.4, the calculation in this section is carried out using a calculation program to simulate transient processes in order to check and confirm the plausibility of the equations derived there.

10.8.1 Basis of Calculation

Based on the example in Fig. 10.7, with equipment data used in Sect. 10.6.1, the short-circuit current characteristic is determined by connecting a current source to busbar A. The current contribution of the converter during the short-circuit is assumed to be $I_{CS} = 4.0$ kA. The calculations are carried out using a calculation program for simulating transient processes (ATP). Since the voltage at the fault location is not regulated in the simulation and a preload is not assumed, the impedance correction factors (transformer, synchronous generator) are not applied when calculating the steady-state currents. For this reason, the impedances and short-circuit currents differ from those specified in Sect. 10.6.1; however, they can be compared with the values determined using the calculation program.

In the following, different examples are determined according to Fig. 10.7, in each case using the equations according to Sect. 10.4 and, in comparison, the calculations of the transient voltage curves. The results of the examples "Feeder: Synchronous Generator and Network" (Sect. 10.8.2) and "Feeder: Synchronous Generator, Network and Converter" (Sect. 10.8.3) are compared with each other.

10.8.2 Feeder: Synchronous Generator and Network

The breaking current at the fault location F (Fig. 10.7) is determined for this case in accordance with Sect. 10.4.2 (substitute voltage source at the short-circuit location) and the transient voltage curve is also calculated. The calculation in accordance with Sect. 10.4.2 is referred to below as the IEC calculation, as the increase in the network current is taken into account in these cases.

10.8.2.1 Calculation According to Sect. 10.4.2

The initial symmetrical short-circuit current is calculated using the equivalent voltage source method at the short-circuit location (Sect. 10.4.2). The following values result for the individual currents, corresponding to Eqs. (10.15)–(10.17):

Short-circuit current:

$$\underline{I}''_{\mathrm{k}} = cU_{\mathrm{n}}/\sqrt{3}\cdot\frac{(\underline{Z}_{\mathrm{N}}+\underline{Z}_{\mathrm{T}}+\underline{Z}_{\mathrm{GS}})}{\underline{Z}_{\mathrm{D}}\cdot(\underline{Z}_{\mathrm{N}}+\underline{Z}_{\mathrm{T}}+\underline{Z}_{\mathrm{GS}})+\underline{Z}_{\mathrm{GS}}\cdot(\underline{Z}_{\mathrm{N}}+\underline{Z}_{\mathrm{T}})} = (0.61-\mathrm{j}18.81)\,\mathrm{kA}$$

Generator current:

$$\underline{I}''_{\mathrm{kGS}} = cU_{\mathrm{n}}/\sqrt{3}\cdot\frac{(\underline{Z}_{\mathrm{N}}+\underline{Z}_{\mathrm{T}})}{\underline{Z}_{\mathrm{D}}\cdot(\underline{Z}_{\mathrm{N}}+\underline{Z}_{\mathrm{T}}+\underline{Z}_{\mathrm{GS}})+\underline{Z}_{\mathrm{GS}}\cdot(\underline{Z}_{\mathrm{N}}+\underline{Z}_{\mathrm{T}})} = (0.24-\mathrm{j}12.52)\,\mathrm{kA}$$

Network current:

$$\underline{I}''_{\mathrm{kN}} = cU_{\mathrm{n}}/\sqrt{3}\cdot\frac{\underline{Z}_{\mathrm{GS}}}{\underline{Z}_{\mathrm{D}}\cdot(\underline{Z}_{\mathrm{N}}+\underline{Z}_{\mathrm{T}}+\underline{Z}_{\mathrm{GS}})+\underline{Z}_{\mathrm{GS}}\cdot(\underline{Z}_{\mathrm{N}}+\underline{Z}_{\mathrm{T}})} = (0.37-\mathrm{j}6.29)\,\mathrm{kA}$$

With the impedances and the voltage factor $c = 1.1$:

$$\underline{Z}_{\rm N} = (0.013 + {\rm j}0.131)\,\Omega \qquad \underline{Z}_{\rm T} = (0.007 + {\rm j}0.216)\Omega$$
$$\underline{Z}_{\rm GS} = (0.003 + {\rm j}0.175)\Omega \qquad \underline{Z}_{\rm D} = (0.003 + {\rm j}0.086)\Omega$$

Since the rated current of the generator is $I_{\rm rGS} = 2.29$ kA, the short-circuit for the generator can be regarded as near to the generator due to $I''_{\rm kGS} > 2 \cdot I_{\rm rGS}$, so that a decay factor μ must be taken into account according to [7]. The breaking currents are calculated using Eqs. (10.18) and (10.19):
Short-circuit current:

$$\underline{I}_{\rm b} = \underline{I}''_{\rm k} - \frac{\underline{Z}_{\rm N} + \underline{Z}_{\rm T}}{\underline{Z}_{\rm D} + \underline{Z}_{\rm N} + \underline{Z}_{\rm T}} \cdot (1 - \mu) \cdot \underline{I}''_{\rm kGS} \tag{10.50}$$

Generator current:

$$\underline{I}_{\rm bGS} = \mu \cdot \underline{I}''_{\rm kGS} \tag{10.51}$$

Network current:

$$\underline{I}_{\rm bN} = \underline{I}''_{\rm kN} + \frac{\underline{Z}_{\rm D}}{\underline{Z}_{\rm D} + \underline{Z}_{\rm N} + \underline{Z}_{\rm T}} \cdot (1 - \mu) \cdot \underline{I}''_{\rm kGS} \tag{10.52}$$

The currents determined using Eqs. (10.50)–(10.52) are listed for various times in Table 10.4, column IEC.

10.8.2.2 Calculation Using a Calculation Program for Transient Processes (ATP)

Taking into account the data of equipment according to Sect. 10.6.1, a short circuit is simulated at the fault location F, Fig. 10.7. At the time of the short circuit occurrence ($t = 0.0$ s), the voltages are in phase, so that the current zero crossings of the various feeders are identical. The three-phase short circuit is initiated at the voltage maximum of conductor R, so that the DC component can be neglected. By starting the short circuit at the maximum of the conductor voltage R, only the change in the AC component is considered, which is also considered in Eq. (10.29). The current curve is shown in Fig. 10.10 for a time response up to $t = 0.1$ s.

A compensating curve can be formed from the current maxima and minima so that the size of the AC component can be determined at defined times. The results are contained in Table 10.4, column ATP, as well as the decay factors μ according to Eqs. (5.47)–(5.51) according to IEC as a function of the minimum time delay $t_{\rm min}$.

During the short-circuit time ($t = 0.5$ s; not included in Fig. 10.7), the generator current is reduced to the decay factor $\mu = 0.43$, whereas the network current increases by a factor of 1.22 over the entire time range.

Table 10.4 RMS values of the short-circuit currents for a three-phase short circuit according to Fig. 10.7 as a function of the time t_{min}, calculations according to IEC Sect. 10.4 and ATP

	μ		I_b (kA)		I_{bGS} (kA)		I_{bN} (kA)	
t_{min} (s)	IEC	ATP	IEC	ATP	IEC	ATP	IEC	ATP
0.00	1.00	1.00	18.82	18.82[a]	12.52	12.52[a]	6,.0	6.30[a]
0.02	0.90	0.85	17.84	17.33	11.30	10.67	6.55	6.67
0.05	0.81	0.75	16.90	16.31	10.13	9.40	6.78	6.91
0.10	0.75	0.67	16.26	15.54	9.33	8.44	6.94	7.10
0.25	0.68	0.57	15.59	14.51	8.48	7.18	7.10	7.34
0.50	–	0.43	–	13.08	–	5.38	–	7.71

μ decay factor
t_{min} minimum time delay
I_b breaking current, short-circuit location
I_{bGS} breaking current, generator
I_{bN} breaking current, network
[a] The value for $t_{min} = 0$ is taken from the IEC calculation

A comparison shows that the decay factors μ according to IEC always lead to larger values for the generator current than the results determined using the generator simulation (ATP). The determination of the generator current in accordance with Sect. 10.4.2 (IEC) is therefore on the safe side; in contrast, the contribution of the network is determined to be too small as a result. For the short-circuit current according to Sect. 10.4.2 at the short-circuit location F, values are also obtained that are greater than those determined from the transient curves.

10.8.3 Feeder: Synchronous Generator, Network and Converter

In the following, the influence of an additional current source (e.g. converter) on the transient short-circuit current curve is examined in comparison to the diagram in Sect. 10.8.2. The additional current source in Fig. 10.7 is connected to busbar A as shown in Fig. 10.5b. The injection of the current source is $\underline{I}_{CS} = -j4$ kA, the phase angle refers to the equivalent voltage source at the fault location.

In accordance with the illustration in Sect. 10.4.3, the currents caused by voltage sources are marked with the additional index “VS” (voltage source), while the currents of the current source have the index “CS” (current source).

10.8.3.1 Calculation According to Sect. 10.4.3

According to Sect. 10.4.3, the influence of the voltage sources is determined in a first step and the influence of the current source in a second step.

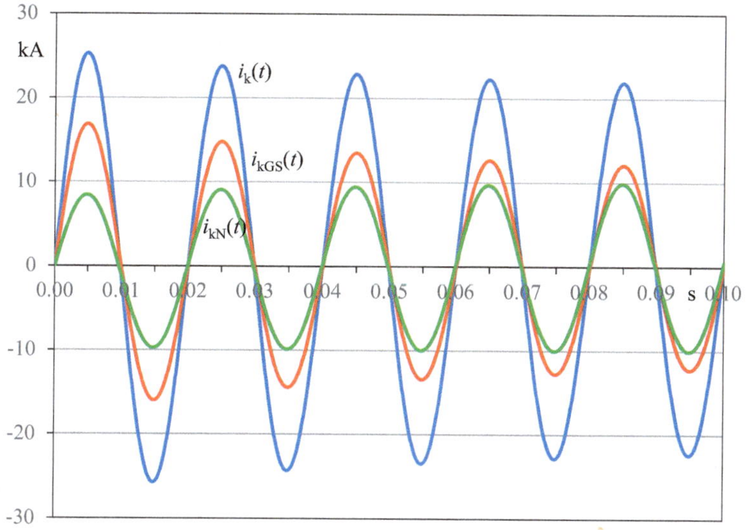

Fig. 10.10 Short-circuit current of conductor R for a three-phase short-circuit according to Fig. 10.7. $i_k(t)$ total short-circuit current at location F, $i_{kGS}(t)$ generator current, $i_{kN}(t)$ network current

- Initial symmetrical short-circuit current caused by the voltage sources, Fig. 10.5a.
  The partial currents (network, generator) and the residual current correspond to the values according to Sect. 10.8.2; calculation according to Sect. 10.4.2, index VS.
- Initial symmetrical short-circuit current caused by the current sources, Fig. 10.5b.
  According to Eqs. (10.22)–(10.24), the various currents are obtained taking into account the current source, with the impedances according to Sect. 10.8.2. The parts in the individual current branches are obtained with I_{CS} of the constant supply of the current source during the short-circuit duration; calculation according to Sect. 10.4.3, index CS.
  Short-circuit current:

$$\underline{I}''_{kCS} = \frac{\underline{Z}_{GS} \cdot (\underline{Z}_N + \underline{Z}_T)}{\underline{Z}_{GS} \cdot (\underline{Z}_N + \underline{Z}_T) + \underline{Z}_D \cdot (\underline{Z}_N + \underline{Z}_T + \underline{Z}_{GS})} \cdot \underline{I}_{CS} = (0.004 - \mathrm{j}2.300)\,\mathrm{kA}$$

Generator current:

$$\begin{aligned}\underline{I}''_{kCSGS} &= -\frac{\underline{Z}_D \cdot (\underline{Z}_N + \underline{Z}_T)}{\underline{Z}_{GS} \cdot (\underline{Z}_N + \underline{Z}_T) + \underline{Z}_D \cdot (\underline{Z}_N + \underline{Z}_T + \underline{Z}_{GS})} \cdot \underline{I}_{CS} \\ &= (-0.014 + \mathrm{j}0.569)\,\mathrm{kA}\end{aligned}$$

Network current:

$$\underline{I}''_{\mathrm{kCSN}} = -\frac{\underline{Z}_{\mathrm{D}} \cdot \underline{Z}_{\mathrm{GS}}}{\underline{Z}_{\mathrm{GS}} \cdot (\underline{Z}_{\mathrm{N}} + \underline{Z}_{\mathrm{T}}) + \underline{Z}_{\mathrm{D}} \cdot (\underline{Z}_{\mathrm{N}} + \underline{Z}_{\mathrm{T}} + \underline{Z}_{\mathrm{GS}})} \cdot \underline{I}_{\mathrm{CS}} = (0.018 + \mathrm{j}1.131)\,\mathrm{kA}$$

- Superimposition of the current values
  The superposition of the determined currents in the individual branches gives the resulting current distribution during the short circuit.

Short-circuit current:

$$\underline{I}''_{\mathrm{k}} = \underline{I}''_{\mathrm{kVS}} + \underline{I}''_{\mathrm{kCS}} = (0.615 - \mathrm{j}21.109)\,\mathrm{kA}$$

Generator current:

$$\underline{I}''_{\mathrm{kGS}} = \underline{I}''_{\mathrm{kVSGS}} + \underline{I}''_{\mathrm{kCSGS}} = (0.255 - \mathrm{j}11.386)\,\mathrm{kA}$$

Network current:

$$\underline{I}''_{\mathrm{kN}} = \underline{I}''_{\mathrm{kVSN}} + \underline{I}''_{\mathrm{kCSN}} = (0.360 - \mathrm{j}5.724)\,\mathrm{kA}$$

This is also a short circuit near to the generator, so that the decay factor μ must be applied when determining the generator current. The same Eqs. (5.48)–(5.51) are decisive for the calculation of the breaking current, in accordance with Sect. 10.8.2. The currents determined are entered for various times in Table 10.5, column IEC (calculation in accordance with Sect. 10.4.3). The proportion of the current source is not listed, as the amount is constant with $I_{\mathrm{CS}} = 4.0$ kA.

10.8.3.2 Calculation Using a Calculation Program for Transient Phenomena (ATP)

In accordance with Sect. 10.8.2, the curves of the currents during the three-phase short circuit are shown in Fig. 10.11 (up to $t = 0.1$ s). The three-phase short-circuit is initiated at the voltage maximum of the conductor R, so that the DC component can be neglected. The phase position of the current of the current source is 90° lagging behind the corresponding line voltage before the short circuit occurs and therefore almost in phase with the short-circuit current of the voltage sources (network, generator). The generator and the network are unloaded before the short circuit occurs.

The results are included in Table 10.5, column ATP, as well as the decay factors μ. The contribution of the current source is constant $I_{\mathrm{CS}} = 4$ kA and is not listed in the table.

Table 10.5 RMS values of the short-circuit currents for a three-phase short-circuit according to Fig. 10.7 (additional current source) as a function of the time t_{min}, calculations according to Sect. 10.4.3 (IEC) and ATP

	μ		I_b (kA)		I_{bGS} (kA)		I_{bN} (kA)	
t_{min} (s)	IEC	ATP	IEC	ATP	IEC	ATP	IEC	ATP
0.00	1.00	1.00	21.12	21.12[a]	11.39	11.39[a]	5.74	5.74[a]
0.02	0.91	0.85	20.31	19.76	10.38	9.70	5.94	6.06
0.05	0.82	0.75	19.52	18.83	9.39	8.55	6.13	6.29
0.10	0.77	0.67	18.99	18.13	8.73	7.68	6.26	6.45
0.25	0.70	0.57	18.40	17.20	8.00	6.53	6.41	6.68
0.50	–	0.43	–	15.89	–	4.89	–	7.01

μ decay factor
t_{min} minimum time delay
I_b breaking current, short-circuit location
I_{bGS} breaking current, generator
I_{bN} breaking current, network
[a] The value for $t_{min} = 0$ is taken from the IEC calculation

In this case, too, the calculation of the decay factors μ according to IEC always leads to smaller values than those determined using the transient current curves, so that the generator currents are determined too large. For this reason, the network currents according to Sect. 10.4 are also calculated too small.

10.8.4 Assessment of the Influence of the Converter on Grid and Generator Current

Table 10.6 compares the results of the transient calculations resulting from the current curves according to Sect. 10.8.2 (Table 10.4, ATP) with the values determined taking into account a current source or converter according to Sect. 10.8.3 (Table 10.5, ATP). The values are listed as differences in Table 10.6.

The contribution of the current source (converter) is distributed differently over the individual current branches, is not constant, and leads to an increase in the short-circuit current at the fault location, while the currents of the voltage sources (generator, network) are reduced. The influence of the current source on the generator current is caused by the time-dependent course of the generator current according to Sect. 4.2.1.2, Eq. (4.38), so that the generator branch can be simulated at a constant voltage by a fictitious increase in the generator impedance.

While Table 10.6 shows the influence of a current source on the short-circuit current, which is determined using the different impedances and time constants, Table 10.7 shows the same network conditions, but in this case the current values are determined according to the equations in Sect. 10.8.2 and 10.8.3 (calculations according to Sect. 10.4). The current values shown correspond to Tables 10.4 and 10.5, column IEC.

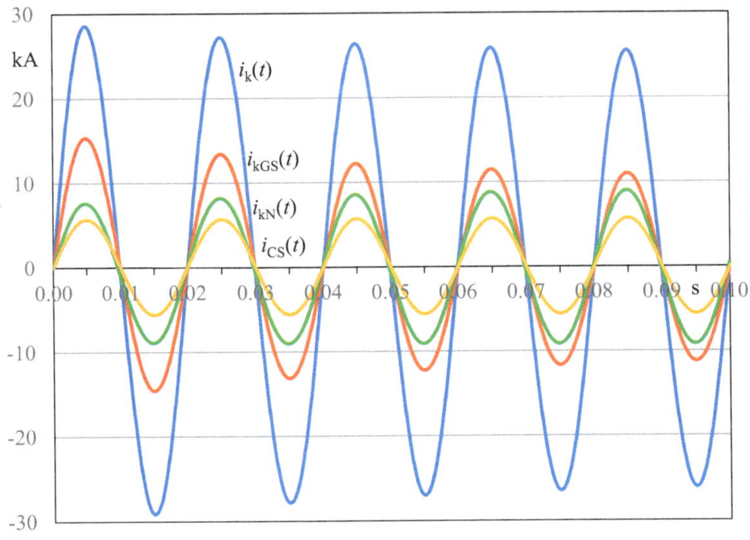

Fig. 10.11 Short-circuit current of conductor R for a three-phase short circuit according to Fig. 10.7 with additional current source on busbar A. $i_k(t)$ short-circuit current at the short-circuit location, $i_{kGS}(t)$ generator current, $i_{kN}(t)$ network current, $i_{CS}(t)$ converter current (current source)

A comparison of the changes in the calculation of the breaking currents taking into account a converter feeder (comparison of Tables 10.6 and 10.7) shows that the calculation according to IEC reproduces the current conditions correctly. The result is on the safe side compared to the calculation according to Table 10.6, whereby the difference is a consequence of the different decay factors μ.

10.8.5 Determination of the Breaking Current in Meshed Networks According to Sect. 10.5

While in Sect. 10.8.2 and 10.8.3 the calculation of the breaking current is carried out according to Eqs. (10.18) and (10.28) by determining the calculation of the currents from the superposition of the currents, the equivalent circuit diagram can also be regarded as a meshed network, so that Eq. (10.29) must be applied.

Due to the derivation of the breaking alternating current according to Eq. (10.29), these results can be regarded as an approximation of the exact solution, as already shown in Sect. 10.5.

Table 10.6 RMS values of the short-circuit currents for a three-phase short circuit according to Fig. 10.7 (with and without additional current source) as a function of the time t_{min}, transient results (ATP)

t_{min} (s)	I_b (kA)	ΔI_b (kA)	I_{bGS} (kA)	ΔI_{bGS} (kA)	I_{bN} (kA)	ΔI_{bN} (kA)
0.00	18.82[a]	2.30	12.52[a]	−1.13	6.30[a]	−0.56
0.02	17.33	2.43	10.67	−0.97	6.67	−0.61
0.05	16.31	2.52	9.40	−0.85	6.91	−0.62
0.10	15.54	2.59	8.44	−0.76	7.10	−0.65
0.25	14.51	2.69	7.18	−0.65	7.34	−0.66
0.50	13.08	2.81	5.38	−0.49	7.71	−0.70

μ decay factor
t_{min} minimum time delay,
I_b breaking current, short-circuit location (without current source)
I_{bGS} breaking current, generator (without current source)
I_{bN} breaking current, network (without current source)
Δ-values differential values of the breaking AC currents each with current source
[a] The value for $t_{min} = 0$ is taken from the IEC calculation

Table 10.7 RMS values of the short-circuit currents for a three-phase short circuit according to Fig. 10.7 (with and without additional current source) as a function of time t_{min}, steady-state results (IEC)

t_{min} (s)	I_b (kA)	ΔI_b (kA)	I_{bG} (kA)	ΔI_{bG} (kA)	I_{bN} (kA)	ΔI_{bN} (kA)
0.00	18.82	2.30	12.52	−1.13	6.30	−0.56
0.02	17.84	2.47	11.30	−0.92	6.55	−0.61
0.05	16.90	2.62	10.13	−0.74	6.78	−0.65
0.10	16.26	2.73	9.33	−0.60	6.94	−0.68
0.25	15.59	2.81	8.48	−0.48	7.10	−0.69

Legend see Table 10.6

10.8.5.1 Breaking Current without Converter

In a first step, the influence of the converter is not considered, so that the current I_b at the short-circuit location according to [7, Eq. (77)] or Eq. (10.29) applies, whereby only the contribution of a generator is taken into account.

$$\underline{I}_b = \underline{I}''_k - \frac{\underline{Z}_{GS} \cdot \underline{I}''_{kG}}{cU_n/\sqrt{3}} \cdot (1-\mu) \cdot \underline{I}''_{kGS} = \underline{I}''_k - \frac{\underline{U}''_{GS}}{cU_n/\sqrt{3}} \cdot (1-\mu) \cdot \underline{I}''_{kGS} \tag{10.53}$$

With

I''_k initial symmetrical short-circuit current at the short-circuit location,
U_n nominal system voltage,
$\underline{Z}_{GS}$ generator impedance,
I''_{kGS} initial symmetrical short-circuit current (generator),
U''_{GS} initial voltage drop at the terminals of the generator.

With the initial symmetrical short-circuit currents according to the equations (Sect. 10.8.2), the following results for the breaking current after insertion into Eq. (10.53)

$$\underline{I}_{\mathrm{b}} = \underline{I}''_{\mathrm{k}} - \frac{\underline{Z}_{\mathrm{N}} + \underline{Z}_{\mathrm{T}}}{\underline{Z}_{\mathrm{D}} + \underline{Z}_{\mathrm{N}} + \underline{Z}_{\mathrm{T}} + \dfrac{\underline{Z}_{\mathrm{D}} \cdot (\underline{Z}_{\mathrm{N}} + \underline{Z}_{\mathrm{T}})}{\underline{Z}_{\mathrm{GS}}}} \cdot (1 - \mu) \cdot \underline{I}''_{\mathrm{kGS}} \tag{10.54}$$

The result of Eq. (10.54) is to be compared with Eq. (10.50), which gives the deviation in the calculation methods. This deviation corresponds to the representation according to Eq. (10.50). In this case, the estimation also leads to a breaking current at the short-circuit location that is determined too small and is therefore on the safe side. In the event of a short circuit on the busbar ($\underline{Z}_{\mathrm{D}} = 0\ \Omega$), both calculations lead to an identical result.

Table 10.8 shows the deviations F in % for the breaking current at the short-circuit location as a function of the minimum time delay t_{min} in relation to the calculation according to Eq. (10.50). The values in Table 10.4 (column IEC) are used for the decay factors μ.

This example results in a maximum deviation of +4.12% with a minimum time delay of $t_{\mathrm{min}} = 0.25$ s when applying the equation for meshed networks. The breaking currents, calculated with Eq. (10.54), therefore provide deviations that are always on the safe side and are therefore determined to be too large.

10.8.5.2 Breaking Current with Converter

In this case, the initial symmetrical short-circuit currents are considered taking into account the short-circuit current feeding of a current source (converter). Equation (10.54) applies to the breaking current $\underline{I}_{\mathrm{b}}$ with the initial symmetrical short-circuit currents according to Sect. 10.8.3.

The results determined in this way must be compared with the values according to Sect. 10.8.3 (Table 10.4) in order to determine the difference between the calculation methods if a current source is also acting. Table 10.9 shows the deviations F in % for the breaking current at the short-circuit location as a function of the minimum time delay t_{min} in relation to the calculation according to Sect. 10.8.3 (Table 10.5, column IEC). The values in Table 10.5 are used for the decay factors μ.

According to Table 10.9, the breaking current I_{b} at the short-circuit location is only insignificantly changed by the supply of a constant current source compared to the results according to Table 10.8 (without converter) due to the low voltage drop of the current source across the combined impedance $\underline{Z}_{\mathrm{D}}$. In this example, there is a maximum deviation of +4.21% with a minimum time delay of $t_{\mathrm{min}} = 0.25$ s when applying the equation for meshed networks, Eq. (10.53), so that the result is also on the safe side in this case, based on the total short-circuit current.

Table 10.8 Deviations F in % for the breaking current at the short-circuit location according to Eq. (10.54) compared to the calculation according to Eq. (10.50), Table 10.4 (column 10.4), without converter

t_{min} (s)	μ	I_b (kA)	F (%)
0.00	1.00	18.82	0.00
0.02	0.90	17.84	+1.43
0.05	0.81	16.90	+2.60
0.10	0.75	16.26	+3.33
0.25	0.68	15.59	+4.12

μ decay factor
t_{min} minimum time delay
I_b breaking current according to Eq. (10.50), Table 10.4
F deviations in %, Eq. (10.54) related to Eq. (10.50)

Table 10.9 Deviations F in % for the breaking current at the short-circuit location according to Eq. (10.54) compared to the calculation according to Sect. 10.8.3 (Table 10.5, column IEC)

t_{min} (s)	μ	I_b (kA)	F (%)
0.00	1.00	21.12	0.00
0.02	0.91	20.31	+1.14
0.05	0.82	19.52	+2.39
0.10	0.77	18.99	+3.12
0.25	0.70	18.40	+4.21

μ decay factor
t_{min} minimum time delay
I_b breaking current, short-circuit location according to Table 10.5 (IEC)
F deviations in %, Eq. (10.54) related to Eq. (10.50)

10.9 Conclusions

If a short circuit is supplied by network feeders and additional voltage sources (asynchronous motors and synchronous machines) via a combined impedance or via an arbitrarily meshed network, the current components of the network increase during the short-circuit duration and do not remain constant, as was assumed in the short circuit current determination of 1971 [1].

If this is neglected, the determination of the breaking alternating current I_b is on the uncertain side. Equations (10.18) and (10.29), on the other hand, lead to sufficiently accurate results. This also applies in the event that additional current sources are injected into the fault location during a short circuit.

References

1. VDE 0102 Teil 1/11.71 (1971) Leitsätze für die Berechnung der Kurzschlussströme, Drehstromanlagen mit Nennspannungen über 1 kV. VDE, Berlin
2. Hosemann G, Balzer G (1984) Der Ausschaltwechselstrom beim dreipoligen Kurzschluss im vermaschten Netz. etzArchiv Bd 6(2):51–56
3. Neuendorf H (1974) Untersuchungen über die Bestimmung von Ausschaltwechselströmen bei dreipoligem Kurzschluss. Diss. TU Berlin
4. Webs A (1972) Einfluß von Asynchronmotoren auf die Kurzschlußstromstärken in Drehstromanlagen. VDE-Fachbericht 27, Berlin & Offenbach, VDE, pp 86–92
5. Hosemann G, Boeck W (1983) Grundlagen der elektrischen Energietechnik, 2nd Edn. Springer, Berlin
6. IEC 60909-0:2016 Edition 2.0 (2016) Short-circuit current calculations in three-phase a.c. systems—Part 0: calculation of currents. IEC, Geneva
7. Oswald RB (2018) Berechnung der Kurzschlussströme in Elektroenergieversorgungsnetzen mit Bezug auf die Normen IEC 60909-0 und DIN 60909-0 (VDE0102). Leipziger Universitätsverlag, Leipzig

Chapter 11
Calculation of the Thermal Equivalent Short-Circuit Current

The mechanical and thermal short-circuit strength of electrical equipment was previously determined in accordance with the former German VDE Standard 0103 [1] using the factors *m* and *n*. These factors take into account the decay of the direct and alternating current components during the short-circuit duration. In the course of the continuous revision of the two IEC Standards 60909 and 60865, the calculation of the thermally equivalent short-circuit current has been presented in the IEC 60909 edition since 2002, as this regulation deals exclusively with the calculation of short-circuit currents, while IEC 60865 deals with the mechanical and thermal effects of short-circuit currents.

While the factor *m* can be clearly determined from the ratio *R*/*X* of the short-circuit loop and the short-circuit duration, the factor *n* is different for each synchronous machine in the case of short-circuits close to the generator. In this section, the equations for a generator pattern are given, which are used to determine the *m* and *n* factors and are identical to the specifications according to IEC 60909-0 (Figs. 18 and 19) [2].

This section essentially reproduces the content of a paper published in 1985 [3], but corrections, additions and adaptations have been made. The aim of the study was to reproduce the *m* and *n* factors shown exclusively in images using equations in order to implement them in computer programs. For this reason, a general generator pattern is used that meets these requirements, i.e. using the data from this generator results in values that correspond to the *m* and *n* curves.

The *m*/*n* curves shown in [2] were originally derived in [4] using the required equations for a range of values for different generators (turbo-generators and salient-pole generators). To simplify the calculation of the thermally equivalent short-circuit current, the results of these calculations were represented by suitable curves in figures. This avoided the detailed determination depending on the parameters that characterize the short-circuit current curve. In contrast, these *m*/*n* curves were used in [3] to determine a generator model in reverse in order to derive these *m* and *n* factors mathematically using the appropriate equations.

G. Balzer, *Short-Circuit Currents in Three-Phase Systems*,
https://doi.org/10.1007/978-3-658-50710-7_11

In principle, it is possible to use special generator data for an application so that customized n-values are taken into account. In this case, however, the user is responsible for creating the equations themselves.

11.1 Introduction

Equipment of power installations, e.g. conductors, circuit-breakers and transformers, are subjected to mechanical and thermal stress in the event of a short circuit. The former German VDE Standard 0103/2.82 [1] contains the necessary calculation methods to determine the mechanical and thermal stresses on equipment for the most severe short-circuit case in advance. If the stresses are below the permissible strength values, the system is short-circuit-proof, as required.

In addition to the short-circuit current, the aperiodically decaying DC component of the short-circuit current must also be taken into account for the thermal short-circuit stress, for which the factor m is used in [1]. In the case of short-circuits far from the generator, the operating-frequency short-circuit current remains constant, whereas in the case of short-circuits near the generator, this component decreases with the duration of the short circuit, which is expressed in [1] by the factor n.

11.2 General

The thermally equivalent short-time current is calculated in accordance with IEC 60865 or IEC 60909 (from 2002), Eq. (11.1):

$$I_{\mathrm{th}} = I_{\mathrm{k}}'' \cdot \sqrt{m+n} \tag{11.1}$$

With

I_{k}'' initial symmetrical short-circuit current (subtransient),
m factor for the heat effect of the DC component,
n factor for the heat effect of the AC component.

According to these definitions, I_{th} is the effective value of an operating-frequency current of constant amplitude, which generates the same amount of heat with a current flow duration of the size of the short-circuit duration T_{k} as the short-circuit current, which is variable in its DC and AC components, so that the following applies:

$$I_{\mathrm{th}}^2 = \frac{1}{T_{\mathrm{k}}} \cdot \int_0^{T_{\mathrm{k}}} i^2(t)\,\mathrm{d}t = I_{\mathrm{k}}''^2 \cdot (m+n) \tag{11.2}$$

If the decay processes occurring in the quadrature-axis are neglected, then the time characteristic of the short-circuit current $i(t)$ can be set with sufficient accuracy in the event of a terminal short circuit of a synchronous generator:

$$i(t) = \sqrt{2} \cdot \begin{bmatrix} \left(I''_{\mathrm{k}} - I'_{\mathrm{k}}\right) \cdot \mathrm{e}^{-t/T''_{\mathrm{d}}} \cdot \cos(\omega t + \vartheta) + \left(I'_{\mathrm{k}} - I_{\mathrm{k}}\right) \cdot \mathrm{e}^{-t/T'_{\mathrm{d}}} \cdot \cos(\omega t + \vartheta) + \\ I_{\mathrm{k}} \cdot \cos(\omega t + \vartheta) - I''_{\mathrm{k}} \cdot \mathrm{e}^{-t/T_{\mathrm{a}}} \cdot \cos\vartheta \end{bmatrix} \tag{11.3}$$

Here is:

I''_{k} initial symmetrical short-circuit current,
I'_{k} transient short-circuit current,
I_{k} steady state short-circuit current,
T''_{d} subtransient time constant of the axis d,
T'_{d} transient time constant of the axis d,
T_{a} time constant of the DC component,
ϑ zero phase angle,
ω angular frequency.

For the calculation of the maximum direct current element, the zero phase angle is calculated according to Eq. (11.4):

$$\vartheta = 0 \tag{11.4}$$

If the root mean square value from Eq. (11.3) is formed according to Eq. (11.2), the following constant and aperiodic summands result for the maximum DC component [4]:

$$\begin{aligned} \int_0^{T_{\mathrm{k}}} i^2(t)\,\mathrm{d}t = {} & \left(I''_{\mathrm{k}} - I'_{\mathrm{k}}\right)^2 \cdot \frac{T''_{\mathrm{d}}}{2} \cdot \left(1 - \mathrm{e}^{-2 \cdot T_{\mathrm{k}}/T''_{\mathrm{d}}}\right) + \\ & \left(I'_{\mathrm{k}} - I_{\mathrm{k}}\right)^2 \cdot \frac{T'_{\mathrm{d}}}{2} \cdot \left(1 - \mathrm{e}^{-2 \cdot T_{\mathrm{k}}/T'_{\mathrm{d}}}\right) + I''^2_{\mathrm{k}} \cdot T_{\mathrm{a}} \cdot \left(1 - \mathrm{e}^{-2 \cdot T_{\mathrm{k}}/T_{\mathrm{a}}}\right) + \\ & I_{\mathrm{k}}^2 \cdot T_{\mathrm{k}} + \left(I''_{\mathrm{k}} - I'_{\mathrm{k}}\right) \cdot I_{\mathrm{k}} \cdot 2 \cdot T''_{\mathrm{d}} \cdot \left(1 - \mathrm{e}^{-T_{\mathrm{k}}/T''_{\mathrm{d}}}\right) + \\ & \left(I'_{\mathrm{k}} - I_{\mathrm{k}}\right) \cdot I_{\mathrm{k}} \cdot 2 \cdot T'_{\mathrm{d}} \cdot \left(1 - \mathrm{e}^{-T_{\mathrm{k}}/T'_{\mathrm{d}}}\right) + \\ & \left(I''_{\mathrm{k}} - I'_{\mathrm{k}}\right) \cdot \left(I'_{\mathrm{k}} - I_{\mathrm{k}}\right) \cdot 2 \cdot T_{\mathrm{SS}} \cdot \left(1 - \mathrm{e}^{-T_{\mathrm{k}}/T_{\mathrm{SS}}}\right) \end{aligned} \tag{11.5}$$

With

$$\frac{1}{T_{\mathrm{SS}}} = \frac{1}{T''_{\mathrm{d}}} + \frac{1}{T'_{\mathrm{d}}} \tag{11.6}$$

The components oscillating at 50 and 100 Hz are negligibly small and are therefore not included in Eq. (11.5). To determine the thermally equivalent short-time current according to Eq. (11.5), first calculate I_k'' and I_k according to [2]. In addition, the short-circuit duration T_k is assumed to be known. The transient short-circuit current and the effective time constants are determined in Sects. 11.3 and 11.4.

11.3 Transient Short-Circuit Current I_k'

The short-circuit currents I_k'', I_k' and I_k can be set by neglecting the ohmic components of the short-circuit loop, assuming the network circuit shown in Fig. 11.1. The reactances X_d'', X_d' und X_d represent the time-dependent behavior of a synchronous generator, while the reactance X_N represents the network between the terminals of the generator and the short-circuit location.

In general, the internal voltage $\underline{U}_0$ for the terminal voltage $\underline{U}_1$ under load conditions before the short-circuit occurs as shown in Fig. 8.1 can be written according to Eq. (8.1):

$$\underline{U}_0 = \underline{U}_1 + I_1 \cdot (\cos\varphi - \mathrm{j}\sin\varphi) \cdot (R + \mathrm{j}X) \tag{11.7}$$

or neglecting the resistance R.

$$\begin{aligned} &\underline{U}_0 \approx \underline{U}_1 + \mathrm{j}X \cdot I_1 \cdot (\cos\varphi - \mathrm{j}\sin\varphi) = \\ &\underline{U}_1 + X \cdot I_1 \cdot \sin\varphi + \mathrm{j}X \cdot I_1 \cdot \cos\varphi \end{aligned} \tag{11.8}$$

If the terminal voltage $\underline{U}_1$ is assumed real, the magnitude of the internal voltage U_0 is generally calculated according to Eq. (11.9):

$$\begin{aligned} &U_0 \approx \sqrt{(U_1 + X \cdot I_1 \cdot \sin\varphi)^2 + (X \cdot I_1 \cdot \cos\varphi)^2} = \\ &\sqrt{U_1^2 + 2 \cdot X \cdot U_1 \cdot I_1 \cdot \sin\varphi + (X \cdot I_1)^2} \end{aligned} \tag{11.9}$$

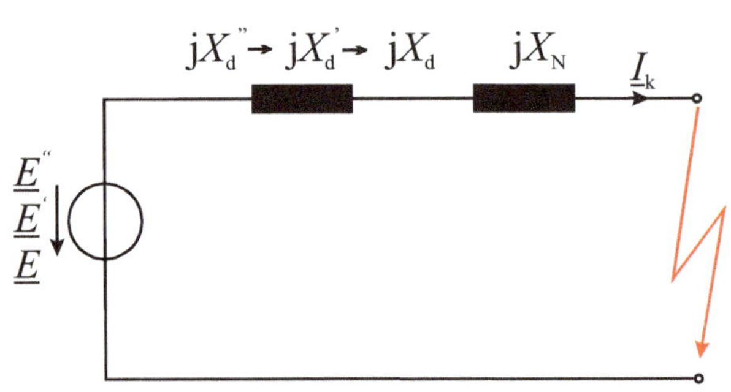

Fig. 11.1 Network circuit diagram for calculating the different short-circuit currents of a synchronous generator

Taking into account the rated values of a synchronous generator, the values according to Eq. (11.9) apply:

$$U_1 = U_{rGS}/\sqrt{3} \qquad I_1 = I_{rGS} = \frac{U_{rGS}/\sqrt{3}}{Z_{rGS}} \tag{11.10}$$

With the generator data:

U_{rGS} rated voltage, line-to-line,
I_{rGS} rated current of the generator,
Z_{rGS} impedance of the generator under rated conditions.

After insertion into Eq. (11.10), the following results:

$$U_0 \approx \frac{U_{rGS}}{\sqrt{3}} \cdot \sqrt{1 + 2 \cdot \left(\frac{X}{Z_{rGS}}\right) \cdot \sin\varphi_{GS} + \left(\frac{X}{Z_{rGS}}\right)^2} \tag{11.11}$$

Using the general relationship according to Eq. (11.11), the different short-circuit currents can be determined as a function of the different internal voltages according to Fig. 11.1, so that the following applies to the amounts of the time-dependent short-circuit currents as a function of the load current during rated operation:

$$I_k'' = \frac{\sqrt{1 + 2 \cdot \left(\frac{X_d''}{Z_{rGS}}\right) \cdot \sin(\varphi_{rGS}) + \left(\frac{X_d''}{Z_{rGS}}\right)^2}}{(X_d'' + X_N)/Z_{rGS}} \cdot I_{rGS} = \frac{E''}{X_d'' + X_N} \tag{11.12}$$

$$I_k' = \frac{\sqrt{1 + 2 \cdot \left(\frac{X_d'}{Z_{rGS}}\right) \cdot \sin(\varphi_{rGS}) + \left(\frac{X_d'}{Z_{rGS}}\right)^2}}{(X_d' + X_N)/Z_{rGS}} \cdot I_{rGS} = \frac{E'}{X_d' + X_N} \tag{11.13}$$

$$I_k = \frac{p \cdot \sqrt{1 + 2 \cdot \left(\frac{X_d}{Z_{rGS}}\right) \cdot \sin(\varphi_{rGS}) + \left(\frac{X_d}{Z_{rGS}}\right)^2}}{(X_d + X_N)/Z_{rGS}} \cdot I_{rGS} = \frac{E}{X_d + X_N} \tag{11.14}$$

With (saturated reactances of the generator):

X_d'' subtransient reactance axis d,
X_d' transient reactance axis d,
X_d synchronous reactance axis d,
$\cos(\varphi_{rGS})$ power factor
p ratio of the maximum excitation voltage to the rated excitation voltage of the generator; according to [2], the following can be set: $p = 1.3$ for turbo generators, $p = 1.6$ for salient pole generators,

X_N fault reactance of the network between the short-circuit location and the generator terminal,

E'', E', E internal voltages for calculating the short-circuit currents I''_k, I'_k, I_k.

From Eqs. (11.12), and (11.14) it follows for the external network reactance:

$$\frac{I''_k}{I_k} = \frac{(X_d + X_N) \cdot E''}{\left(X''_d + X_N\right) \cdot E} \tag{11.15}$$

$$X_N = \frac{I_k \cdot X_d \cdot E'' - I''_k \cdot X''_d \cdot E}{I''_k \cdot E - I_k \cdot E''} \tag{11.16}$$

The short-circuit current ratio can be determined from Eqs. (11.13) and (11.14):

$$\frac{I'_k}{I_k} = \frac{E' \cdot (X_d + X_N)}{E \cdot \left(X'_d + X_N\right)} \tag{11.17}$$

After inserting the network reactance X_N according to Eq. (11.16), the short-circuit current ratio follows:

$$\frac{I'_k}{I_k} = \frac{\left(I''_k / I_k\right)}{\frac{E''}{E'} \cdot \frac{X_d - X'_d}{X_d - X''_d} + \left(\frac{I''_k}{I_k}\right) \cdot \frac{E}{E'} \cdot \frac{X'_d - X''_d}{X_d - X''_d}} \tag{11.18}$$

The internal voltages E'', E', E in Eq. (11.18) are determined from Eqs. (11.12)–(11.14).

11.4 Time Constants

The subtransient time constant T''_d can be set according to [5], taking into account the eddy currents in the solid rotor iron:

$$T''_d \approx 0.1\, T'_d \tag{11.19}$$

The transient time constant T'_d is hardly influenced by the network resistance, so that [6], Sect. 4.2.1.2 applies with sufficient accuracy:

$$\frac{T'_d}{T'_{d0}} = \frac{X'_d + X_N}{X_d + X_N} \tag{11.20}$$

After inserting Eq. (11.17), the ratio of the transient time constants follows:

$$\frac{T'_d}{T'_{d0}} = \left(\frac{I_k}{I'_k}\right) \cdot \left(\frac{E'}{E}\right) \tag{11.21}$$

with the values according to Eqs. (11.18), (11.13) and (11.14). Here, T'_{d0} represents the no-load time constant of the magnet wheel. The time constant T_a of the aperiodically decaying DC component depends essentially on the resistance R_N of the network and is also included in the factor κ, with which the peak short-circuit current i_p can be determined:

$$\kappa = \frac{i_p}{\sqrt{2} \cdot I''_k} \tag{11.22}$$

The peak short-circuit current i_p reaches the highest possible value after approximately half a period 1/(2*f*). A comparison of the functions for κ, Eqs. (11.3) and (11.16), and the DC component shows the following with sufficient accuracy:

$$\kappa = 1 + e^{-1/(2 \cdot f \cdot T_a)} \tag{11.23}$$

This results in:

$$T_a = -\frac{1}{2 \cdot f \cdot \ln(\kappa - 1)} \tag{11.24}$$

11.5 Determining the Factors *m* and *n*

By simplifying Eq. (11.3), the short circuit far from the generator results in:

$$i(t) = \sqrt{2} \cdot I''_k \cdot \left[\cos(\omega \cdot t) - e^{-t/T_a}\right] \tag{11.25}$$

or

$$i^2(t) = \sqrt{2} \cdot I''^2_k \cdot \left\{1 + \cos(2\omega \cdot t) - 4 \cdot \cos(\omega \cdot t) \cdot e^{-t/T_a} + 2 \cdot e^{-2t/T_a}\right\} \tag{11.26}$$

If the summands oscillating at 50 and 100 Hz are neglected, the result for the factor m is with $n = 1$:

$$\begin{aligned}(m+1) \cdot I_k^2 \cdot T_k &= \int_0^{T_k} i^2(t)\mathrm{d}t = I''^2_k \cdot \left[t - T_a \cdot e^{-2t/T_a}\right]_0^{T_k} \\ m &= \frac{T_a}{T_k} \cdot \left[1 - e^{-2T_k/T_a}\right]\end{aligned} \tag{11.27}$$

The time constant T_a is conveniently determined using Eq. (11.24) from the factor κ of Eq. (11.22), which must be determined for every short-circuit current calculation anyway. After inserting Eq. (11.24) into Eq. (11.27), the factor m becomes:

$$m = \frac{1}{2 \cdot f \cdot T_k \ln(\kappa - 1)} \cdot \left[e^{4T_k \cdot f \cdot \ln(\kappa - 1)} - 1 \right] \tag{11.28}$$

Figure 11.2 shows the course of the curves m as a function of the factor κ according to Eq. (11.28). In the currently valid version of the IEC Standard [2], the abscissa value is not the frequency, but the product $T_k{\cdot}f$, so that the representation there is frequency-independent.

The influence of the AC component on the heat effect is then described solely by the factor n according to Eq. (11.29):

$$n = \frac{\int_0^{T_k} i^2(t) \mathrm{d}t}{I_k''^2 \cdot T_k} - m \tag{11.29}$$

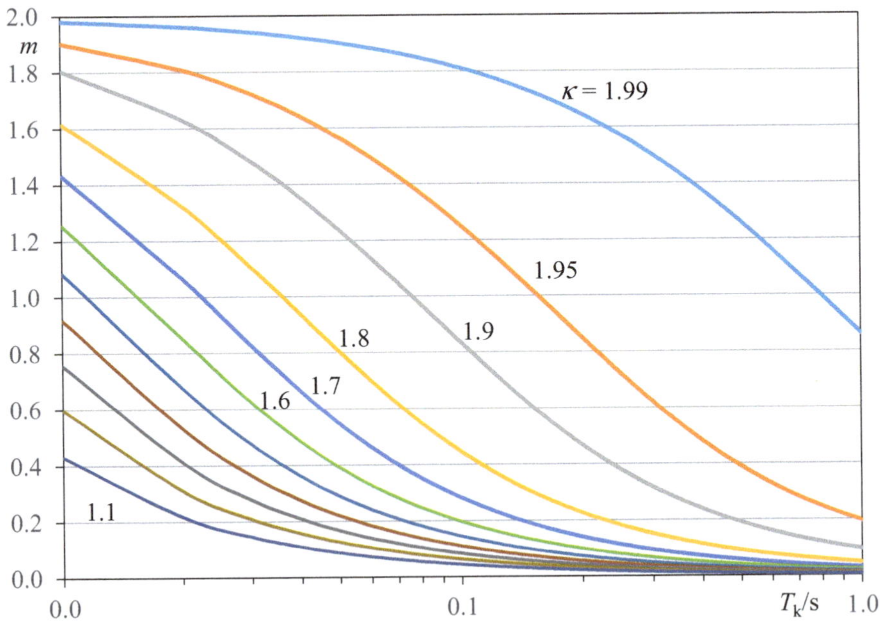

Fig. 11.2 Curves m of the DC component, representing the thermal effect, as a function of the short-circuit duration T_k (in s) for different values of the factor κ at $f = 50$ Hz

Taking Eqs. (11.1) and (11.27) into account, the expression for the factor n can be determined.

$$n = \frac{1}{I_k''^2} \times \left\{ \begin{array}{l} \left(I_k'' - I_k'\right)^2 \cdot \frac{T_d''}{2 \cdot T_k} \cdot \left(1 - e^{-2 \cdot T_k / T_d''}\right) + \left(I_k' - I_k\right)^2 \cdot \frac{T_d'}{2 \cdot T_k} \cdot \left(1 - e^{-2 \cdot T_k / T_d'}\right) \\ + I_k''^2 \cdot \frac{T_a}{T_k} \cdot \left(1 - e^{-2 \cdot T_k / T_a}\right) + I_k^2 + \left(I_k'' - I_k'\right) \cdot I_k \cdot 2 \cdot \frac{T_d''}{T_k} \cdot \left(1 - e^{-T_k / T_d''}\right) \\ + \left(I_k' - I_k\right) \cdot I_k \cdot 2 \cdot \frac{T_d'}{T_k} \cdot \left(1 - e^{-T_k / T_d'}\right) \\ + \left(I_k'' - I_k'\right) \cdot \left(I_k' - I_k\right) \cdot 2 \cdot \frac{T_{SS}}{T_k} \cdot \left(1 - e^{-T_k / T_{SS}}\right) \end{array} \right\} - \frac{T_a}{T_k} \cdot \left(1 - e^{-2 \cdot T_k / T_a}\right) \tag{11.30}$$

$$n = \frac{1}{\left(I_k''/I_k\right)^2} \cdot \left\{ \begin{array}{l} 1 + \left(\frac{I_k''}{I_k} - \frac{I_k'}{I_k}\right)^2 \cdot \frac{T_d'}{20 \cdot T_k} \cdot \left(1 - e^{-20 \cdot T_k / T_d'}\right) \\ + \left(\frac{I_k'}{I_k} - 1\right)^2 \cdot \frac{T_d'}{2 \cdot T_k} \cdot \left(1 - e^{-2 \cdot T_k / T_d'}\right) \\ + \left(\frac{I_k''}{I_k} - \frac{I_k'}{I_k}\right) \cdot \frac{T_d'}{5 \cdot T_k} \cdot \left(1 - e^{-10 \cdot T_k / T_d'}\right) \\ + \left(\frac{I_k'}{I_k} - 1\right) \cdot 2 \cdot \frac{T_d'}{T_k} \cdot \left(1 - e^{-T_k / T_d'}\right) \\ + \left(\frac{I_k''}{I_k} - \frac{I_k'}{I_k}\right) \cdot \left(\frac{I_k'}{I_k} - 1\right) \cdot \frac{T_d'}{5.5 \cdot T_k} \cdot \left(1 - e^{-11 \cdot T_k / T_d'}\right) \end{array} \right\} \tag{11.31}$$

After inserting the generator pattern data from Table 11.1 into Eq. (11.20), the time constant T_d', can be set at a value of $p = 1.3$ (turbo-generator):

$$T_d' \approx \frac{3.1\,\mathrm{s}}{I_k'/I_k} \tag{11.32}$$

The factors n are displayed as a function of the short-circuit duration T_k and the current ratio I_k''/I_k, so that the ratio must be determined on the basis of the generator pattern. After inserting the data of the generator pattern, the following results:

$$\frac{I_k'}{I_k} = \frac{I_k''/I_k}{0.88 + 0.17 \cdot I_k''/I_k} \tag{11.33}$$

The representation of the curves n is shown in Fig. 11.3, which is based on the generator pattern in Table 11.1. This generator pattern has been selected so that the

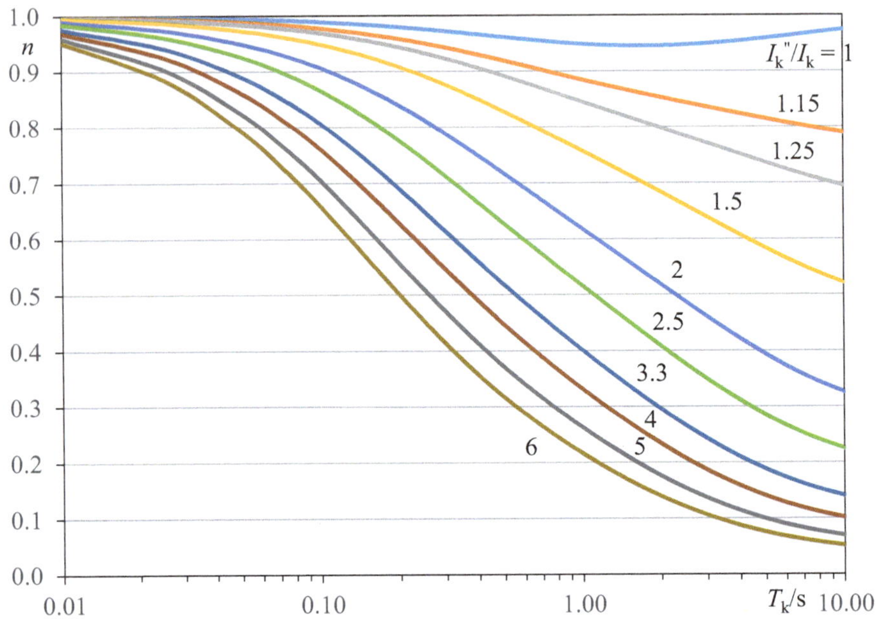

Fig. 11.3 Curves n of the generator pattern as a function of the short-circuit duration T_k (in s) for different values of the short-circuit current ratio I''_k/I_k

Table 11.1 Data of the generator pattern

Description	Variables	Value
Subtransient reactance axis d	x''_d	0.225 p.u.
Transient reactance axis d	x'_d	0.340 p.u.
Synchronous reactance axis d	x_d	2.350 p.u.
No-load time constant	T''_{d0}	10 s
Power factor	$\cos\varphi$	0.8

curves n shown in the former VDE Standard [1] could be reproduced with the least possible deviations.

The set of curves n (T_k, I''_k/I_k) corresponds to Fig. 19 in IEC 60909-0 [2] for $I''_k/I_k \geq 1.25$. For $I''_k/I_k \leq 1.15$, the value $n = 1$ is used for simplification, as in this case the calculated curve n does not result in a value $n = 1$.

The derivation of the curves m and n refers to the isolated operation of a single generator. By inserting the values of the generator pattern according to Table 11.1, only the variables I''_k, κ, T_k and I''_k/I_k are necessary for the calculation of the thermal short-circuit stress. The factors m and n can therefore also be used approximately for arbitrarily meshed grids.

In addition, further 21 generators were examined using Eqs. (11.18)–(11.21), inserted in Eq. (11.31), in accordance with Table 11.2. The last two columns show the deviations F in % between the values of the investigated generators calculated

Table 11.2 Synchronous generators designed and percentage deviations F of the factor n; stress of a short-circuit duration of $T_k = 1$ s, compared with the factor n of the generator pattern; designations according to Table 11.1

No.	S_r/MVA	X_d/Z_{GS}	x'_d/Z_{GS}	x''_d/Z_{GS}	T''_{d0}/s	F/% for I''_k/I_k	
						1.25	3.30
1	27.7	1.80	0.217	0.181	6.5	+3.95	+4.43
2	27.7	2.16	0.180	0.130	8.3	+3.38	+5.63
3S	60.0	1.31	0.310	0.240	5.9	−4.81	−15.45
4	68.8	1.45	0.182	0.123	5.9	+2.38	+3.57
5	93.3	2.16	0.286	0.201	7.5	+0.05	−4.93
6	100.0	1.90	0.187	0.145	8.1	+4.65	+10.98
7	166.7	1.97	0.270	0.172	8.4	−0.05	−1.91
8	248.3	2.12	0.377	0.336	5.9	+2.98	−1.57
9	312.5	1.54	0.318	0.230	5.0	−2.31	−14.66
10	400.0	2.56	0.300	0.220	11.1	+3.00	+9.38
11S	615.0	0.90	0.270	0.230	7.4	+1.73	+9.92
12	675.0	1.79	0.320	0.245	3.4	−4.68	−28.94
13S	737.0	0.90	0.300	0.200	8.7	−6.72	−7.22

S_r rated power
S salient pole generator

with Eq. (11.31) and the factor n of the generator pattern. Negative deviations mean that the results according to IEC 600909-0 are on the safe side.

A short circuit $I''_k/I_k = 1.25$ far from the generator results in an average deviation of −0.3% with a bandwidth of −6.7% to +4.7%, while a short circuit near to the generator ($I''_k/I_k = 3.3$) results in a deviation of −4.6%, with an increased bandwidth of −28.9% to +4.7% (without outliers of 11%). The stated positive deviations can be regarded as acceptable.

The studies presented confirm that the procedure in accordance with IEC 60865/60909-0 is practical and technically justifiable.

11.6 Conclusions

The factors m and n, which are shown in diagrams in IEC 60909: 2016-01-28 [2] in Figs. 18 and 19, are used to calculate the thermal short-circuit stresses. The considerations and assumptions on which the diagrams are based are summarized.

The factor m is clearly described by Eq. (11.28) and can be used in this form in computer programs.

The factor n is clearly described by Eqs. (11.12)–(11.14), (11.18), (11.21), and (11.31) and the factor m. If the data of the generator pattern according to Table 11.1 is used, the diagram according to Fig. 19 of IEC 60909: 2016-01-28 [2] . is obtained.

References

1. DIN VDE·0103/2.82 (1982) Bemessung von Starkstromanlagen auf mechanische und thermische Kurzschlussfestigkeit. VDE, Berlin
2. IEC 60909-0:2016 Edition 2.0 (2016) Short-circuit current calculations in three-phase a.c. systems—Part 0: calculation of currents. IEC, Geneva
3. Balzer G, Deter O (1985) Berechnung der thermischen Kurzschlußbeanspruchung von Starkstromanlagen mit Hilfe der Faktoren *m* und *n* nach DIN VDE 0103/2.82. ETZ Arch, pp 287–290
4. Roeper R (1949) Ermittlung der thermischen Beanspruchung bei nichtstationären Kurzschlussströmen. ETZ Elektrotechnik Zeitung 70, H. 4, pp 131–135
5. Tittel J (1942) Ausgleichsvorgänge in Synchronmaschinen bei plötzlichen Blindlaständerungen. Wiss Veröffentlichungen Siemenswerk 21, pp 38–74
6. Jacottet P (1932) Dämpfung und Wärmewirkung des Stoßkurzschlussstroms bei einfach gespeistem Netzkurzschluss. Arch Elektrotechnik 26, pp 679–692

Chapter 12
Contribution of Asynchronous Motors to the Short-Circuit Current

In principle, asynchronous motors must be taken into account when calculating the maximum short-circuit currents, if they are not switched on at the same time due to the operating mode (interlocking or reserve operation). To simplify the network calculation, low-voltage motors can be neglected in the case of short-circuits in the public power supply, in contrast to short circuits in auxiliary networks of power plants or similar industrial plants (e.g. chemical and steel industry).

While Sects. 12.2 and 12.3 show the estimation of the short-circuit contribution for a three-phase short circuit, Sect. 12.4 deals with unbalanced short circuits. The calculation of the short-circuit current for a three-phase fault is shown in Chap. 5.

12.1 General

The contribution of motors must be taken into account when calculating the following currents if there is a short circuit:

- Peak short-circuit current i_p,
- breaking current I_b,
- steady state short-circuit current I_k (line-to-line and line-to-earth only),
- thermal equivalent short-circuit current I_{th} (line-to-line and line-to-earth only).

Due to the type of excitation, asynchronous motors do not contribute to the three-phase steady state short-circuit current I_k, as the excitation voltage is taken from the terminals. However, this does not apply to unbalanced short circuits (e.g. line-to-line), as shown at the end of this section using the example of the steady state short-circuit current. Motors also contribute to the line-to-earth short-circuit current, even if the neutral points of the motors are not earthed, provided that a neutral point of the network is earthed.

G. Balzer, *Short-Circuit Currents in Three-Phase Systems*,
https://doi.org/10.1007/978-3-658-50710-7_12

The motor impedance Z_M is determined according to Eq. (4.52), Sect. 4.6.1, as well as the representation of several low-voltage motors to an equivalent motor.

12.2 Estimation of the Short-Circuit Current Contribution

The older German VDE Standard [1] included an estimate that is no longer included in the IEC Revision [2]. The reason for this is that manual calculations are no longer used in the majority of cases and it then makes sense to use all the data from high-voltage motors in a computer program. This estimate stated that, in general, the contribution of motors to the initial symmetrical short-circuit current I''_k need not be taken into account if the following relationship is observed:

$$\Sigma I_{rM} \leq 0.01 \cdot I''_k \tag{12.1}$$

With

ΣI_{rM} total value of the rated currents of all motors,
I''_k initial symmetrical short-circuit current without motors.

Since a deviation of 5% between the actual and calculated result is tolerated in accordance with Sect. 1.1 if the method of the equivalent voltage source at the short-circuit location is used, Eq. (12.1) considers this value if the ratio of the locked-rotor current of $I_{LR}/I_{rM} = 5$ is assumed. In addition, it is assumed that the motor voltage is $U_{rM} = 1.1\ U_n$.

When calculating networks into which motors feed, it always makes sense to consider all motors if the calculations are carried out using network computer programs. If, in contrast, short-circuit currents are determined using a manual calculation, a further estimation of the motor influence can lead to a simplification of the calculation under certain circumstances. This is particularly the case if the networks are meshed. The following estimation formula applies, according to [1]:

$$\frac{\sum P_{rM}}{\sum S_{rT}} \leq \frac{0.8}{\left| \dfrac{c \cdot 100 \cdot \sum S_{rT}}{\sqrt{3} \cdot U_n \cdot I''_k} - 0.3 \right|} \tag{12.2}$$

$\sum P_{rM}$ sum of the active power of the high- and low-voltage motors,
$\sum S_{rT}$ sum of the rated power of the transformers feeding these motors,
I''_k initial symmetrical short-circuit current at connection node Q of the motor group without the motor contribution,
U_n nominal system voltage at the connection node Q.

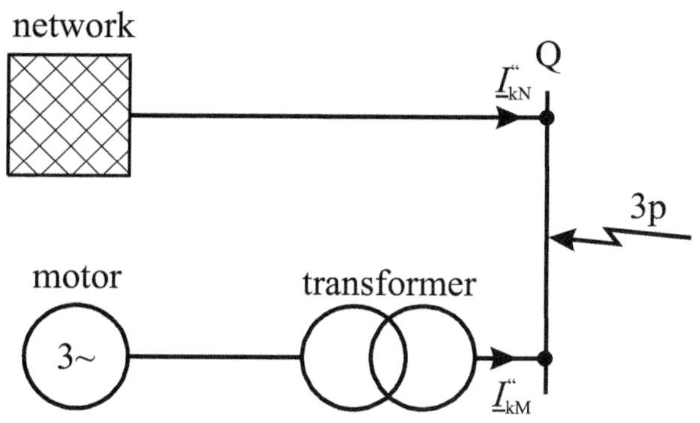

Fig. 12.1 Network circuit to determine the contribution of motors, Eq. (12.2)

Equation (12.2) applies under the condition that the contribution of the motors in relation to the initial symmetrical short-circuit current without motors according to Fig. 12.1 should not exceed the value of 5%, so that the following applies:

$$I''_{kM} \leq 0.05 \cdot I''_{k} \tag{12.3}$$

In the further derivation of the estimation formula, the amounts are used as the resistances can be neglected, so that the result is on the safe side.

The contribution of the motors to the short-circuit current at the short-circuit location is determined by the motor impedance (Z_M) and transformer impedance (Z_T):

$$I''_{kM} = \frac{c \cdot U_n/\sqrt{3}}{Z_M + Z_T} \tag{12.4}$$

With

U_n nominal system voltage, short-circuit location Q,
c voltage factor Table 7.1.

According to Eqs. (4.52) and (4.189), the following relationships apply:

$$Z_M = \frac{1}{I_{LR}/I_{rM}} \cdot \frac{U_{rM}^2}{S_{rM}} = \frac{1}{I_{LR}/I_{rM}} \cdot \frac{U_{rM}^2}{P_{rM}} \cdot \eta \cdot \cos\varphi \tag{12.5}$$

$$P_{rM} = \eta \cdot \cos(\varphi) \cdot S_{rM} \tag{12.6}$$

$$Z_T = \frac{u_k}{100\%} \cdot \frac{U_{rT}^2}{S_{rT}} \tag{12.7}$$

With

U_{rM} rated voltage of the motor, converted to the voltage level of the short circuit with the transformation ratio t of the transformer,

I_{LR}/I_{rM} symmetrical locked-rotor current, related to the rated current,
S_{rM} rated apparent power of the motor,
P_{rM} rated active power of the motor,
η efficiency factor,
$\cos\varphi$ power factor,
U_{rT} rated voltage of the transformer, HV side,
S_{rM} rated apparent power of the transformer,
u_k rated short-circuit voltage.

The following applies to the contribution of the motors:

$$I''_{kM} = \frac{c \cdot U_n/\sqrt{3}}{\frac{1}{I_{LR}/I_{rM}} \cdot \frac{U_{rM}^2}{P_{rM}} \cdot \eta \cdot \cos\varphi + \frac{u_k}{100\%} \cdot \frac{U_{rT}^2}{S_{rT}}} \quad (12.8)$$

Taking into account the condition, Eq. (12.3), the following results:

$$\frac{c \cdot U_n/\sqrt{3}}{I''_k} \leq 0.05 \cdot \left\{ \frac{1}{I_{LR}/I_{rM}} \cdot \frac{U_{rM}^2}{P_{rM}} \cdot \eta \cdot \cos\varphi + \frac{u_k}{100\%} \cdot \frac{U_{rT}^2}{S_{rT}} \right\}$$
$$= 0.05 \cdot \frac{1}{S_{rT}} \cdot \left\{ \frac{1}{I_{LR}/I_{rM}} \cdot \frac{S_{rT}}{P_{rM}} \cdot U_{rM}^2 \cdot \eta \cdot \cos\varphi + \frac{u_k}{100\%} \cdot U_{rT}^2 \right\} \quad (12.9)$$

After transformation, the result is:

$$\frac{P_{rM}}{S_{rT}} \leq \frac{\frac{1}{I_{an}/I_{rM}} \cdot U_{rM}^2 \cdot \eta \cdot \cos\varphi}{\frac{c \cdot U_n \cdot S_{rT}/\sqrt{3}}{0.05 \cdot I''_k} - \frac{u_k}{100\%} \cdot U_{rT}^2} \quad (12.10)$$

The following simplifications or specifications are made for the derivation of the general Eq. (12.2):

- voltages: $U_n = U_{rM} = U_{rT}$
- efficiency factor, power factor motor: $\eta \cdot \cos\varphi = 0.8$
- locked-rotor current: $I_{LR}/I_{rM} = 5$
- short-circuit voltage, transformer: $u_k = 6\%$

If these values are used in Eq. (12.10), the result is:

$$\frac{P_{rM}}{S_{rT}} \leq \frac{0.01 \cdot 0.8}{\frac{c \cdot S_{rT}}{\sqrt{3} \cdot U_n \cdot I''_k} - 0.003} = \frac{0.8}{\frac{c \cdot 100 \cdot S_{rT}}{\sqrt{3} \cdot U_n \cdot I''_k} - 0.3} \quad (12.11)$$

This means that under the above conditions, the estimation formula corresponds to the exact calculation according to Eq. (12.10). The consequence is, for example, a lower short-circuit voltage of the transformer or a higher ratio of the locked-rotor current can lead to a higher current contribution compared to the result of the estimation formula.

As this estimation is quite practical for manual calculation, its application is shown in an example, Sect. 12.3.

12.3 Example of the Contribution of Motors

In the following, the validity of the estimation formula, Eq. (12.2), is checked using an example according to Fig. 12.2.

In the example shown in Fig. 12.2, five HV motors (6.9 kV, $P_{rM} = 400$ kW each) feed into a 10 kV busbar via transformers. In addition, a LV motor group (0.69 kV, $P_{rM} = 500$ kW) is supplied from this busbar. The initial symmetrical short-circuit current of the network without the influence of the motors is $I''_{kN} = 25$ kA at the 10 kV busbar. The following values are used for the exact calculation of the partial short-circuit currents of the motors (resistances are neglected):

- Transformers (number: $n_T = 2$):
  10 kV/6.6 kV: $S_{rT} = 4$ MVA; $u_k = 10\%$; $U_{rT} = 11/6.6$ kV $= t$
  10 kV/0.69 kV: $S_{rT} = 1$ MVA; $u_k = 6\%$; $U_{rT} = 11/0.69$ kV $= t$
- Motors (number: $n_M = 5$):
  6.9 kV: $P_{rM} = 400$ kW; $\eta \cos \varphi = 0.87$; $U_{rM} = 6$ kV $I_{LR}/I_r = 6$
  0.69 kV: $P_{rM} = 90$ kW; $\eta \cos \varphi = 0.81$; $U_{rM} = 0.69$ kV $I_{LR}/I_r = 5$

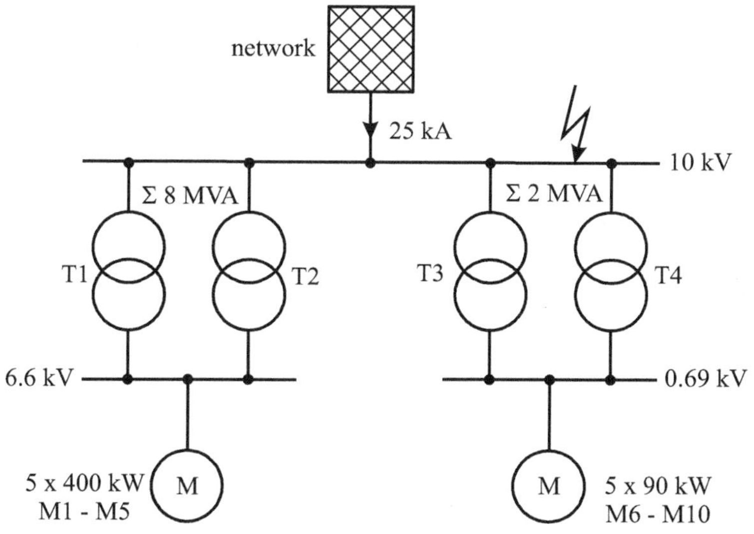

Fig. 12.2 Network diagram for determining the short-circuit current contribution of asynchronous motors in the event of a three-phase short-circuit on the 10 kV busbar

- Network:
  Short circuit (3p): $I''_{\text{kN}} = 25\ \text{kA}$

When applying Eq. (12.2) in relation to Fig. 12.2, the following allocations of transformer and motor power apply:

$$\sum_{i=1}^{4} S_{\text{rT}i} = S_{\text{rT1}} + S_{\text{rT2}} + S_{\text{rT3}} + S_{\text{rT4}} = 10\ \text{MVA}$$

$$\sum_{i=1}^{10} P_{\text{rM}i} = \sum_{i=1}^{5} P_{\text{rM}i} + \sum_{i=6}^{10} P_{\text{rM}i} = 2.45\ \text{MW}$$

After inserting the following results:

$$\frac{P_{\text{rM}}}{S_{\text{rT}}} = \frac{2.45}{10} \leq \frac{0.8}{\dfrac{c \cdot 100 \cdot S_{\text{rT}}}{\sqrt{3} \cdot U_{\text{nN}} \cdot I''_{\text{k}}} - 0.3} = \frac{0.8}{\dfrac{1,1 \cdot 100 \cdot 10}{\sqrt{3} \cdot 10 \cdot 25} - 0.3} = 0.3571$$

The estimation results in ($0.245 \leq 0.3571$), so that all motors can be disregarded for this example if a three-phase short circuit is considered at the overlaid 10 kV level.

The same estimation result is also obtained if the low-voltage level feeds into the 10 kV level via the 6.6 kV level, as the sum of the transformer power is always used, regardless of whether they are connected in parallel or in series. With a series connection, however, the short-circuit current contribution is lower, so that the estimation formula is on the safe side.

The resulting three-phase short-circuit current contribution of the asynchronous motors is determined by neglecting the resistances:

$$I''_{\text{kM}} = \frac{1.1 \cdot U_{\text{n}}/\sqrt{3}}{X_{\text{k}}} \tag{12.12}$$

with the short-circuit reactance X_{k}, transformed to the short-circuit level:

$$\frac{1}{X_{\text{k}}} = \frac{1}{X_{\text{k1}}} + \frac{1}{X_{\text{k2}}} \tag{12.13}$$

X_{k1} short-circuit reactance of HV motors (6.9 kV), including transformers,
X_{k2} short-circuit reactance of LV motors (0.69 kV), including transformers.

The reactances are calculated using the following equations, taking into account the transformation ratios t of the transformers:

- HV motors: Eq. (4.52)
- LV group of motors: Sect. 4.6.1

- transformers: Eq. (4.189)
- impedance correction (transformer): Eq. (4.190)

$$X_{k1} = \frac{0.95 \cdot c_{max}}{1 + 0.6 \cdot x_T} \cdot \frac{u_{kr}}{100\%} \cdot \frac{(U_{rT})^2}{S_{rT}} \cdot \frac{1}{n_T} + \frac{1}{I_{LR}/I_{rM}} \cdot \frac{(U_{rM})^2 \cdot \eta \cos\varphi}{P_{rM}} \cdot \frac{1}{n_M} \cdot (t)^2$$
$$= \frac{0.95 \cdot 1.1}{1 + 0.6 \cdot 0,10} \cdot \frac{10\%}{100\%} \cdot \frac{(11\,\text{kV})^2}{4\ \text{MVA}} \cdot \frac{1}{2} + \frac{1}{6} \cdot \frac{(6\ \text{kV})^2 \cdot 0.87}{0.4\ \text{MW}} \cdot \frac{1}{5} \cdot \left(\frac{11\,\text{kV}}{6.6\,\text{kV}}\right)^2$$
$$= 1.491\ \Omega + 7.250\ \Omega = 8.741\ \Omega$$

$$X_{k2} = \frac{0.95 \cdot 1.05}{1 + 0.6 \cdot 0,06} \cdot \frac{6\%}{100\%} \cdot \frac{(11\ \text{kV})^2}{1\ \text{MVA}} \cdot \frac{1}{2} + \frac{1}{5} \cdot \frac{(0.69\ \text{kV})^2 \cdot 0.81}{0.09\ \text{MW}} \cdot \frac{1}{5} \cdot \left(\frac{11}{0.69}\right)^2$$
$$= 3.495\ \Omega + 43.560\ \Omega = 47.055\ \Omega$$

The following applies to the total short-circuit reactance $X_K = 7.372\ \Omega$, so that the short-circuit contribution of the motors is calculated as follows:

$$I''_{kM} = \frac{1.1 \cdot 10\ \text{kV}/\sqrt{3}}{7.385\ \Omega} = 861.53\ \text{A}$$

Compared with the three-phase short-circuit current at the fault location without a motor contribution of $I''_{kN} = 25$ kA, the value of $I''_{kM} = 0.860$ kA can be neglected, as this contribution is <5% of 25 kA (corresponding to 1.25 kA), which shows that the estimation formula is correct.

In this case, the calculation of the three-phase short-circuit current is carried out taking into account the impedance correction factors for network transformers. Section 8.3 shows that these factors must be taken into account if the inductive short-circuit current has the same current direction as the inductive load current before the short-circuit occurs. The consequence would be that in the event of a short circuit on the 10 kV busbar, as shown in Fig. 12.2, it is not necessary to use the impedance correction factors for the transformers. If these factors are not taken into account in this case, the result is a value of $I''_{kM} = 859.36$ A, which is slightly lower than determined above. As the transformer impedance is considerably lower than the motor impedance, the impedance correction of the transformers has only a minor effect.

12.4 Contribution of Asynchronous Motors in the Event of Unbalanced Short Circuits

The calculation of the short-circuit current for a line-to-line and line-to-earth short circuit in asynchronous motors is different to the calculation for a three-phase short-circuit, so that these differences are shown separately below.

12.4.1 Line-to-Line Short Circuit

According to Sect. 4.6.2, the negative sequence impedance $\underline{Z}_2$ of asynchronous motors is equal to the positive sequence impedance $\underline{Z}_1$, so that the following applies to the line-to-line initial symmetrical short-circuit current I''_{k2}:

$$\underline{I}''_{k2} = \frac{c \cdot U_n}{\underline{Z}_1 + \underline{Z}_2} = \frac{c \cdot U_n}{2 \cdot \underline{Z}_M} = \frac{\sqrt{3}}{2} \cdot \underline{I}''_{k3} \tag{12.14}$$

With

$\underline{Z}_M$ impedance of the motor according Eq. (4.52),
$\underline{I}''_{k3}$ three-phase initial symmetrical short-circuit current of the motor.

To simplify the calculation of the peak short-circuit current, the factor κ is determined from the calculation of the positive sequence system. For a line-to-line short-circuit without earth contact between conductors S and T, the terminal voltages are determined according to Table 9.2:

$$\underline{U}_R = \frac{2 \cdot \underline{E}_{(1)}}{1 + \underline{Z}_{(1)}/\underline{Z}_{(2)}} \qquad \underline{U}_S = \underline{U}_T = -\frac{\underline{E}_{(1)}}{1 + \underline{Z}_{(1)}/\underline{Z}_{(2)}} \tag{12.15}$$

With

$$\underline{E}_{(1)} = c \cdot U_n/\sqrt{3} \qquad \underline{Z}_{(1)} \approx \underline{Z}_{(2)} \tag{12.16}$$

follows:

$$\underline{U}_R \approx c \cdot U_n/\sqrt{3} \qquad \underline{U}_S = \underline{U}_T \approx -0.5 \cdot c \cdot U_n/\sqrt{3} \tag{12.17}$$

This means that, in contrast to the three-phase short circuit, there is always a voltage at the terminals in this case, so that the excitation can remain partial. The consequence is that, according to [2], no decay of the line-to-line short-circuit current is assumed. This is always set:

$$I_{b2M} = \frac{\sqrt{3}}{2} \cdot I''_{k3M} \qquad I_{k2M} \approx \frac{\sqrt{3}}{2} \cdot I''_{k3M} \tag{12.18}$$

The assumptions are essentially based on tests, the results of which are presented in [3, 4]. Since the three-phase steady state short-circuit current $I_k = 0$ (Sect. 5.7.4), this means that when designing equipment with a high proportion of asynchronous motors, the line-to-line short circuit should be considered and not the three-phase short circuit if the breaking current or the steady state short-circuit current is decisive for the dimensioning.

12.4.2 Line-to-Earth Short Circuit

Although the neutral points of motors are generally not earthed, they contribute to the initial symmetrical short-circuit current in the event of a line-to-earth short circuit, as shown in the following example and specified in [2]. According to Fig. 12.3, a low-voltage motor group (AM) is supplied via a 10 kV transformer. Due to the protection system, the low-voltage network is a low-impedance earthed network, meaning that the neutral point of the transformer is directly earthed. In contrast, the motors are not earthed, which is standard practice. The line-to-earth short-circuit current is compared with the magnitude of the three-phase current. For this reason, the three-phase short-circuit current is calculated in a first calculation step.

The following data is assumed for the calculation using the equivalent voltage source at the short-circuit location:

- Network: $U_{\text{nN}} = 10\text{ kV}$ $I''_{\text{kN}} = 10\text{ kA}$ $R_{\text{N}}/X_{\text{N}} = 0.1$
- Transformer: $S_{\text{rT}} = 2\text{ MVA}$ $u_{\text{k}} = 6\%$ $u_{\text{Rr}} = 0.8\%$
  $t_{\text{r}} = 11\text{ kV}/0.69\text{ kV}$ $\underline{Z}_{(0)\text{T}}/\underline{Z}_{(1)\text{T}} = 1$ $\underline{Z}_{(1)\text{T}} = \underline{Z}_{(2)\text{T}}$
- Group of motors: $S_{\text{rM}} = 1.5\text{ MVA}$ $I_{\text{LR}}/I_{\text{r}} = 5$
  $U_{\text{rM}} = 690\text{ V}$ $R_{\text{M}} = 0.42\,X_{\text{M}}$
  $\underline{Z}_{(1)\text{M}} = \underline{Z}_{(2)\text{M}}$ $\underline{Z}_{(0)\text{M}} \to \infty$

The impedances are calculated from this data according to Eqs. (4.1), (4.52), (4.189), and (4.190) in relation to $U_{\text{n}} = 0.69\text{ kV}$:

Network:

$$Z_{\text{N}} = \frac{cU_{\text{nN}}/\sqrt{3}}{I''_{\text{kN}}} \cdot \frac{1}{t_{\text{r}}^2} = \frac{1.1 \cdot 10\,\text{kV}}{\sqrt{3} \cdot 10\,\text{kA}} \cdot \left(\frac{0.69}{11}\right)^2 \Omega = 2.5\text{ m}\Omega$$

$$X_{\text{N}} = \frac{Z_{\text{N}}}{\sqrt{1 + (R_{\text{N}}/X_{\text{N}})^2}} = \frac{2.5\,\text{m}\Omega}{\sqrt{1 + (0.1)^2}} \approx 2.5\,\text{m}\Omega \qquad R_{\text{N}} = 0.1 \cdot X_{\text{N}} \approx 0.2\,\text{m}\Omega$$

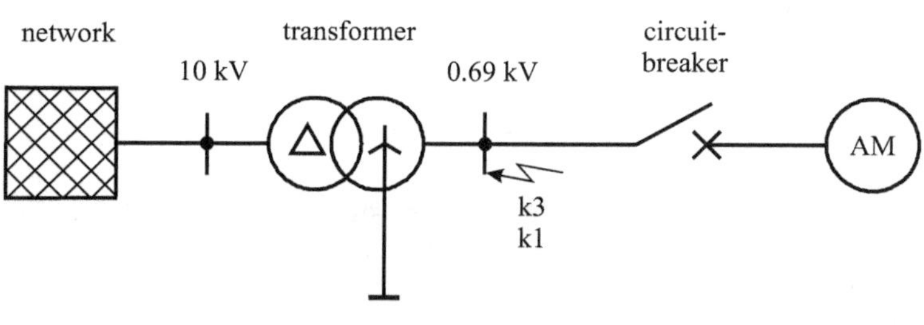

Fig. 12.3 Network circuit for determining the contribution of a LV motor group (AM asynchronous motors) in the event of a line-to-earth and three-phase short circuit

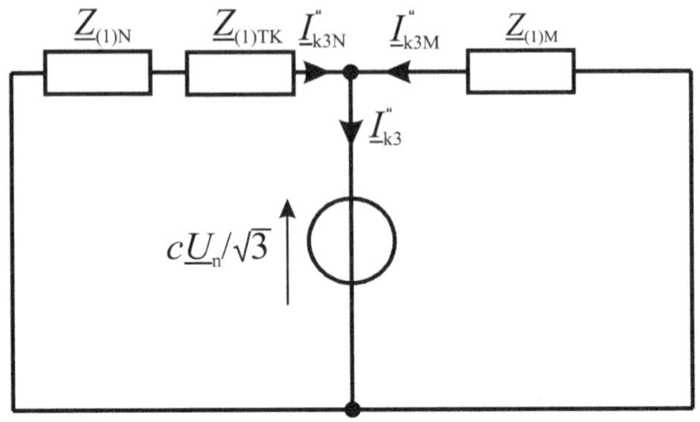

Fig. 12.4 Equivalent circuit for determining the three-phase short-circuit current I''_{k3}

$$\underline{Z}_{(1)N} = (0.2 + j2.5)\ \text{m}\Omega$$

Transformer, including impedance correction:

$$Z_T = \frac{u_{kr}}{100\%} \cdot \frac{U_{rT}^2}{S_{rT}} = 0.06 \cdot \frac{(0.69\,\text{kV})^2}{2\ \text{MVA}} = 14.3\ \text{m}\Omega$$

$$R_T = \frac{u_{Rr}}{100} \cdot \frac{U_{rT}^2}{S_{rT}} = 0.008 \cdot \frac{(0.69\ \text{kV})^2}{2\ \text{MVA}} = 1.9\ \text{m}\Omega \qquad X_T = \sqrt{Z_T^2 - R_T^2} = 14.2\ \text{m}\Omega$$

$$K_T = 0.95 \cdot \frac{c_{max}}{1 + 0.6 \cdot x_T} = 0.95 \cdot \frac{1.05}{1 + 0.6 \cdot 0.06} = 0.9628$$

$$\underline{Z}_{(1)TK} = K_T \cdot \underline{Z}_T = (1.8 + j13.6)\,\text{m}\Omega = \underline{Z}_{(0)TK}$$

Group of motors:

$$Z_M = \frac{1}{I_{LR}/I_{rM}} \cdot \frac{U_{rM}^2}{S_{rM}} = \frac{1}{5} \cdot \frac{(0.69\ \text{kV})^2}{1.5\ \text{MVA}} = 63.5\ \text{m}\Omega$$

$$X_M = \frac{Z_M}{\sqrt{1 + (R_M/X_M)^2}} = \frac{63.5\,\text{m}\Omega}{\sqrt{1 + (0.42)^2}} = 58.5\,\text{m}\Omega \qquad R_M = 0.42 \cdot X_M = 24.6\,\text{m}\Omega$$

$$\underline{Z}_{(1)M} = (24.6 + j58.5)\,\text{m}\Omega$$

Figure 12.4 applies to the calculation of the three-phase short-circuit current I''_{k3} at the fault location. It is made up of the short-circuit current components of the feeder and the motor group. Since the short-circuit location shown in Fig. 12.3 is a multiple single fed short circuit, the partial short-circuit currents can be determined separately and added together.

The following applies to the feeder contribution:

$$\underline{I}''_{\mathrm{k3N}} = \frac{cU_{\mathrm{n}}/\sqrt{3}}{\underline{Z}_{(1)\mathrm{N}} + \underline{Z}_{(1)\mathrm{TK}}} = \frac{1.05 \cdot 0.690\ \mathrm{kV}/\sqrt{3}}{(2.0 + \mathrm{j}16.1)\ \mathrm{m}\Omega} = (3.18 - \mathrm{j}25.59)\ \mathrm{kA}$$

$$I''_{\mathrm{k3N}} = 25.78\ \mathrm{kA}$$

Short-circuit current contribution of the motor group:

$$\underline{I}''_{\mathrm{k3M}} = \frac{cU_{\mathrm{n}}/\sqrt{3}}{\underline{Z}_{(1)\mathrm{M}}} = \frac{1.05 \cdot 0.690\ \mathrm{kV}/\sqrt{3}}{(24.6 + \mathrm{j}58.5)\ \mathrm{m}\Omega} = (2.55 - \mathrm{j}6.08)\ \mathrm{kA}$$

$$I''_{\mathrm{k3M}} = 6.59\ \mathrm{kA}$$

The following results for the total current:

$$\underline{I}''_{\mathrm{k3}} = \underline{I}''_{\mathrm{k3N}} + \underline{I}''_{\mathrm{k3M}} = (5.73 - \mathrm{j}31.67)\ \mathrm{kA} \qquad I''_{\mathrm{k3}} = 32.18\ \mathrm{kA}$$

If, on the other hand, the resulting initial symmetrical short-circuit current is determined from the parallel connection of the impedances according to Eq. (12.21), the result is given by:

$$\underline{Z}_{(2)\mathrm{N}} + \underline{Z}_{(2)\mathrm{TK}} = \underline{Z}_{(1)\mathrm{N}} + \underline{Z}_{(1)\mathrm{TK}} \qquad \underline{Z}_{(2)\mathrm{M}} = \underline{Z}_{(1)\mathrm{M}} \tag{12.19}$$

$$\underline{Z}_{(0)\mathrm{TK}} = \underline{Z}_{(1)\mathrm{TK}} \qquad \underline{Z}_{(0)\mathrm{M}} = \infty \tag{12.20}$$

$$\underline{I}''_{\mathrm{k3}} = \frac{cU_{\mathrm{n}}/\sqrt{3}}{\underline{Z}_{\mathrm{k}}} = cU_{\mathrm{n}}/\sqrt{3} \cdot \frac{\underline{Z}_{(1)\mathrm{N}} + \underline{Z}_{(1)\mathrm{TK}} + \underline{Z}_{(1)\mathrm{M}}}{\left(\underline{Z}_{(1)\mathrm{N}} + \underline{Z}_{(1)\mathrm{TK}}\right) \cdot \underline{Z}_{(1)\mathrm{M}}} \tag{12.21}$$

$$\underline{I}''_{\mathrm{k3}} = \frac{1.05 \cdot 0.69}{\sqrt{3}} \cdot \frac{26.6 + \mathrm{j}74.6}{(2.0 + \mathrm{j}16.1) \cdot (24.6 + \mathrm{j}58.5)} = (5.73 - \mathrm{j}31.66)\,\mathrm{kA}$$

The result is identical compared to the addition of the individual initial symmetrical short-circuit currents, but this does not apply to the calculation of the resulting peak short-circuit current.

The line-to-earth short-circuit current is calculated using the equivalent circuit of the component systems according to Fig. 12.5, so that the individual component currents are determined first. In this case, the negative sequence and zero sequence systems must also be taken into account in order to calculate the different currents in the various conductors. As the neutral point of the motor group is not earthed, the motor impedance in the neutral system is infinite. The following therefore applies to the component current $\underline{I}$, taking the impedances into account:

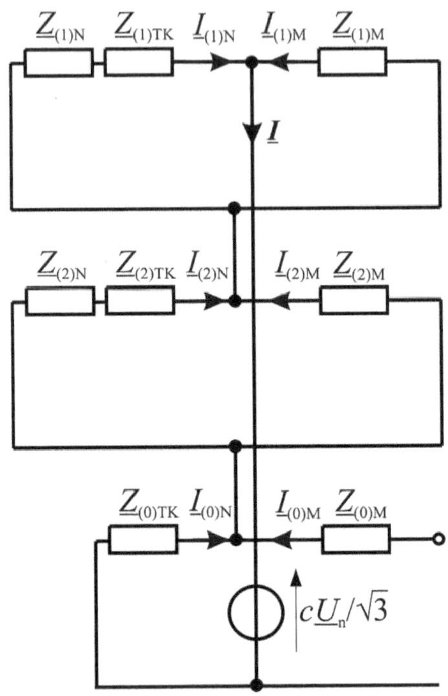

Fig. 12.5 Equivalent circuit for determining the line-to-earth short-circuit current I''_{k1}

$$\underline{I} = \frac{cU_n/\sqrt{3}}{2 \cdot \frac{\left(\underline{Z}_{(1)N} + \underline{Z}_{(1)TK}\right) \cdot \underline{Z}_{(1)M}}{\underline{Z}_{(1)N} + \underline{Z}_{(1)TK} + \underline{Z}_{(1)M}} + \underline{Z}_{(1)TK}} \tag{12.22}$$

The partial currents of the component systems are calculated as follows:

$$\begin{aligned} \underline{I}_{(1)N} = \underline{I}_{(2)N} &= \frac{\underline{Z}_{(1)M}}{\underline{Z}_{(1)N} + \underline{Z}_{(1)TK} + \underline{Z}_{(1)M}} \cdot \underline{I} \\ &= \frac{\underline{Z}_{(1)M} \cdot cU_n/\sqrt{3}}{2 \cdot \left(\underline{Z}_{(1)N} + \underline{Z}_{(1)TK}\right) \cdot \underline{Z}_{(1)M} + \underline{Z}_{(1)TK} \cdot \left(\underline{Z}_{(1)N} + \underline{Z}_{(1)TK} + \underline{Z}_{(1)M}\right)} \end{aligned} \tag{12.23}$$

$$\underline{I}_{(0)N} = \underline{I} = \frac{cU_n/\sqrt{3}}{2 \cdot \frac{\left(\underline{Z}_{(1)N} + \underline{Z}_{(1)TK}\right) \cdot \underline{Z}_{(1)M}}{\underline{Z}_{(1)N} + \underline{Z}_{(1)TK} + \underline{Z}_{(1)M}} + \underline{Z}_{(1)TK}} \tag{12.24}$$

$$\underline{I}_{(1)\mathrm{M}} = \underline{I}_{(2)\mathrm{M}} = \frac{\left(\underline{Z}_{(1)\mathrm{N}} + \underline{Z}_{(1)\mathrm{TK}}\right) \cdot cU_{\mathrm{n}}/\sqrt{3}}{2 \cdot \left(\underline{Z}_{(1)\mathrm{N}} + \underline{Z}_{(1)\mathrm{TK}}\right) \cdot \underline{Z}_{(1)\mathrm{M}} + \underline{Z}_{(1)\mathrm{TK}} \cdot \left(\underline{Z}_{(1)\mathrm{N}} + \underline{Z}_{(1)\mathrm{TK}} + \underline{Z}_{(1)\mathrm{M}}\right)} \tag{12.25}$$

$$\underline{I}_{(0)\mathrm{M}} = 0 \tag{12.26}$$

Taking into account Eq. (9.76) for the reverse transformation, the currents in the original range (R, S, T) can be determined, Eqs. (12.27)–(12.29).

$$\underline{I}_{\mathrm{R}} = \underline{I}_{(0)} + \underline{I}_{(1)} + \underline{I}_{(2)} = \underline{I}_{(0)} + 2 \cdot \underline{I}_{(1)} \tag{12.27}$$

$$\underline{I}_{\mathrm{S}} = \underline{I}_{(0)} + \underline{a}^2 \cdot \underline{I}_{(1)} + \underline{a} \cdot \underline{I}_{(2)} = \underline{I}_{(0)} - \underline{I}_{(1)} \tag{12.28}$$

$$\underline{I}_{\mathrm{T}} = \underline{I}_{(0)} + \underline{a} \cdot \underline{I}_{(1)} + \underline{a}^2 \cdot \underline{I}_{(2)} = \underline{I}_{(0)} - \underline{I}_{(1)} \tag{12.29}$$

The short-circuit current contribution in the individual phases of the feeder is determined as follows:

$$\underline{I}_{\mathrm{RN}} = \underline{I}_{(0)\mathrm{N}} + 2 \cdot \underline{I}_{(1)\mathrm{N}} = \frac{\left(\underline{Z}_{(1)\mathrm{N}} + \underline{Z}_{(1)\mathrm{TK}} + 3 \cdot \underline{Z}_{(1)\mathrm{M}}\right) \cdot cU_{\mathrm{n}}/\sqrt{3}}{2 \cdot \left(\underline{Z}_{(1)\mathrm{N}} + \underline{Z}_{(1)\mathrm{TK}}\right) \cdot \underline{Z}_{(1)\mathrm{M}} + \underline{Z}_{(1)\mathrm{TK}} \cdot \left(\underline{Z}_{(1)\mathrm{N}} + \underline{Z}_{(1)\mathrm{TK}} + \underline{Z}_{(1)\mathrm{M}}\right)} \tag{12.30}$$

$$\underline{I}_{\mathrm{SN}} = \underline{I}_{(0)\mathrm{N}} - \underline{I}_{(1)\mathrm{N}} = \underline{I}_{\mathrm{TN}} = \frac{\left(\underline{Z}_{(1)\mathrm{N}} + \underline{Z}_{(1)\mathrm{TK}}\right) \cdot cU_{\mathrm{n}}/\sqrt{3}}{2 \cdot \left(\underline{Z}_{(1)\mathrm{N}} + \underline{Z}_{(1)\mathrm{TK}}\right) \cdot \underline{Z}_{(1)\mathrm{M}} + \underline{Z}_{(1)\mathrm{TK}} \cdot \left(\underline{Z}_{(1)\mathrm{N}} + \underline{Z}_{(1)\mathrm{TK}} + \underline{Z}_{(1)\mathrm{M}}\right)} \tag{12.31}$$

The motor contribution in the individual phases can be derived accordingly, with $\underline{I}_{(0)\mathrm{M}} = 0$.

$$\underline{I}_{\mathrm{RM}} = 2 \cdot \underline{I}_{(1)\mathrm{M}} = \frac{2 \cdot \left(\underline{Z}_{(1)\mathrm{N}} + \underline{Z}_{(1)\mathrm{TK}}\right) \cdot cU_{\mathrm{n}}/\sqrt{3}}{2 \cdot \left(\underline{Z}_{(1)\mathrm{N}} + \underline{Z}_{(1)\mathrm{TK}}\right) \cdot \underline{Z}_{(1)\mathrm{M}} + \underline{Z}_{(1)\mathrm{TK}} \cdot \left(\underline{Z}_{(1)\mathrm{N}} + \underline{Z}_{(1)\mathrm{TK}} + \underline{Z}_{(1)\mathrm{M}}\right)} \tag{12.32}$$

$$\underline{I}_{\mathrm{SM}} = \underline{I}_{\mathrm{TM}} = -\underline{I}_{(1)\mathrm{M}} = \frac{-\left(\underline{Z}_{(1)\mathrm{N}} + \underline{Z}_{(1)\mathrm{TK}}\right) \cdot cU_{\mathrm{n}}/\sqrt{3}}{2 \cdot \left(\underline{Z}_{(1)\mathrm{N}} + \underline{Z}_{(1)\mathrm{TK}}\right) \cdot \underline{Z}_{(1)\mathrm{M}} + \underline{Z}_{(1)\mathrm{TK}} \cdot \left(\underline{Z}_{(1)\mathrm{N}} + \underline{Z}_{(1)\mathrm{TK}} + \underline{Z}_{(1)\mathrm{M}}\right)} \tag{12.33}$$

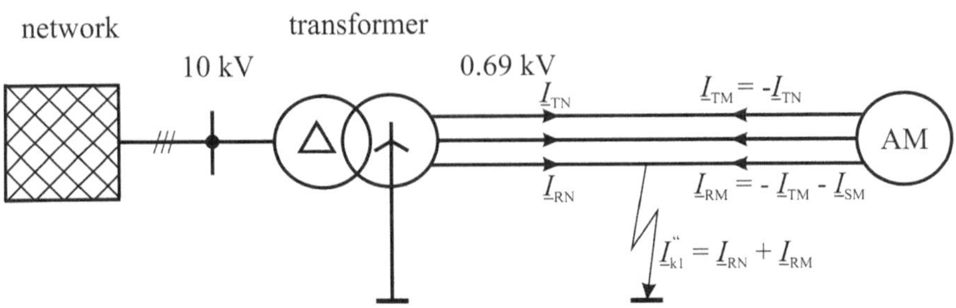

Fig. 12.6 Representation of the currents in the event of a line-to-earth fault, according to Fig. 12.3

The short-circuit currents to earth can then be calculated from the partial short-circuit currents determined above.

$$\underline{I}''_{k1R} = \underline{I}_{RN} + \underline{I}_{RM} = \frac{3 \cdot \left(\underline{Z}_{(1)N} + \underline{Z}_{(1)TK} + \underline{Z}_{(1)M}\right) \cdot cU_n/\sqrt{3}}{2 \cdot \left(\underline{Z}_{(1)N} + \underline{Z}_{(1)TK}\right) \cdot \underline{Z}_{(1)M} + \underline{Z}_{(1)TK} \cdot \left(\underline{Z}_{(1)N} + \underline{Z}_{(1)TK} + \underline{Z}_{(1)M}\right)} \tag{12.34}$$

$$\underline{I}''_{k1S} = \underline{I}''_{k1T} = \underline{I}_{SN} + \underline{I}_{SM} = 0 \tag{12.35}$$

The currents given in Eqs. (12.30)–(12.33) can be represented in the original network as shown in Fig. 12.6.

The resulting line-to-earth short-circuit current at the fault location can also be determined directly using Eq. (12.36)

$$\underline{I}''_{k1} = \frac{c \cdot \sqrt{3} \cdot U_n}{2 \cdot \underline{Z}_{(1)} + \underline{Z}_{(0)}} \tag{12.36}$$

With the impedances of the components according to Fig. 12.5.

$$\underline{Z}_{(1)} = \frac{\left(\underline{Z}_{(1)N} + \underline{Z}_{(1)TK}\right) \cdot \underline{Z}_{(1)M}}{\underline{Z}_{(1)N} + \underline{Z}_{(1)TK} + \underline{Z}_{(1)M}} \tag{12.37}$$

$$\underline{Z}_{(0)} = \underline{Z}_{(1)TK} \tag{12.38}$$

After inserting the impedances, the result is the same as shown in Eq. (12.34). Considering the impedances, the currents for a line-to-earth short circuit can be specified according to Table 12.1.

The example shows that the contribution of the motor group cannot be neglected even in the event of a line-to-earth short circuit, although the neutral points of the

Table 12.1 Initial symmetrical short-circuit current in kA for a line-to-earth short-circuit according to Fig. 12.6

Conductor	Feeder	Motor	Short-circuit location
R	27.41	4.32	31.60[a]
S	2.16	2.16	–
T	2.16	2.16	–

[a] Calculation according to Eq. (12.36)

motor group itself are not earthed. Under these conditions, the conductor current in the faulty phase is related to the three-phase short-circuit current:

$$I''_{\mathrm{k1M}}/I''_{\mathrm{k3M}} = I_{\mathrm{RM}}/I''_{\mathrm{k3M}} = \frac{4.32\ \mathrm{kA}}{6.59\ \mathrm{kA}} = 0.656$$

In addition, the non-faulty conductors also carry a short-circuit current.

If, in contrast to this, the impedance $\underline{Z}_{\mathrm{N}}$ of the feeding network is assumed to be zero, the maximum ratio of the conductor currents of the motor for a line-to-earth short-circuit in relation to the three-phase short-circuit can be determined according to Eq. (12.38). This means that the network-side contribution to the three-phase short-circuit current is only determined by the impedance of the transformer.

$$\begin{aligned} I''_{\mathrm{k1M}}/I''_{\mathrm{k3M}} &= \frac{2 \cdot \underline{Z}_{(1)\mathrm{TK}} \cdot cU_{\mathrm{n}}/\sqrt{3}}{2 \cdot \underline{Z}_{(1)\mathrm{TK}} \cdot \underline{Z}_{(1)\mathrm{M}} + \underline{Z}_{(1)\mathrm{TK}} \cdot \left(\underline{Z}_{(1)\mathrm{TK}} + \underline{Z}_{(1)\mathrm{M}}\right)} \cdot \frac{\underline{Z}_{(1)\mathrm{M}}}{cU_{\mathrm{n}}/\sqrt{3}} \\ &= \frac{2 \cdot \underline{Z}_{(1)\mathrm{M}}}{3 \cdot \underline{Z}_{(1)\mathrm{M}} + \underline{Z}_{(1)\mathrm{TK}}} = \frac{2}{3 + \underline{Z}_{(1)\mathrm{TK}}/\underline{Z}_{(1)\mathrm{M}}} \end{aligned} \tag{12.39}$$

Under these conditions, the maximum line-to-earth short-circuit current contribution of a motor can be a maximum of 66.7% of the three-phase current contribution if a terminal short-circuit is assumed, whereby the impedance ratio is $\underline{Z}_{(1)\mathrm{TK}}/\underline{Z}_{(1)\mathrm{M}} = 0$.

If the circuit-breaker is open as shown in Fig. 12.3, the following short-circuit currents result at the fault location for the three-phase short circuit:

$$\underline{I}''_{\mathrm{k3N}} = \frac{cU_{\mathrm{n}}/\sqrt{3}}{\underline{Z}_{(1)\mathrm{N}} + \underline{Z}_{(1)\mathrm{TK}}} = (3.18 - \mathrm{j}25.59)\ \mathrm{kA} \qquad I''_{\mathrm{k3N}} = 25.78\ \mathrm{kA}$$

For the line-to-earth short-circuit current, the following applies according to Eq. (12.34) with $\underline{Z}_{(1)\mathrm{M}} \to \infty$:

$$\underline{I}''_{\mathrm{k1R}} = \frac{\sqrt{3} \cdot cU_{\mathrm{n}}}{2 \cdot \underline{Z}_{(1)\mathrm{N}} + 3 \cdot \underline{Z}_{(1)\mathrm{TK}}} = (3.41 - \mathrm{j}26.37)\ \mathrm{kA} \qquad I''_{\mathrm{k1N}} = 27.18\ \mathrm{kA}$$

In this case, the line-to-earth short-circuit current is greater than the three-phase current due to the lower zero sequence impedance compared to the positive sequence impedance. The conductor current R differs only slightly from the network contribution according to Table 12.1, but the resulting short-circuit current is significantly lower due to the motor contribution, 27.18 kA compared to 31.60 kA.

12.5 Conclusions

Asynchronous motors contribute to the short-circuit current, whereby the amount depends on the type of short-circuit. While in the case of a three-phase short-circuit, the missing terminal voltage leads to a decay process so that the time-dependent short-circuit current contribution is reduced. The short-circuit current contribution is constant as a function of time in the case of unbalanced faults.

Motors also contribute to the line-to-earth short-circuit current, although the neutral point of motors is generally not earthed. However, this only applies if a neutral point of the supplying network is earthed.

References

1. IEC 60909-0:2002 (2002) Short-circuit current calculations in three-phase a.c. systems—Part 0: calculation of currents. IEC, Geneva
2. IEC 60909-0:2016 Edition 2.0 (2016) Short-circuit current calculations in three-phase a.c. systems—Part 0: calculation of currents. IEC, Geneva
3. Ott G, Webs A (1971) Beitrag von Hochspannungs-Asynchronmotoren zum Kurzschlußstrom bei dreipoligem Kurzschluß. ETZ-Report 6. VDE, Berlin
4. Webs A (1972) Einfluss von Asynchronmotoren auf die Kurzschlussstromstärken in Drehstromanlagen. VDE-Fachberichte 27, Berlin, pp 86–92

Chapter 13
Short-Circuit Current Calculation with Converter Systems

Due to the increasing use of renewable energies, it is necessary to transport electrical energy over longer distances, which means that HVDC transmissions are increasingly being used as a result of the low transmission losses. In addition, converters are often used in the generation of electrical energy from renewable energy sources (PV feeder, offshore wind farms). In principle, two different technologies can be distinguished here:

- IGBT Technology, full-size converter (VSC Voltage Source Converter),
- Thyristor Technology (LCC Line Commutated Converter).

The following shows how to calculate a short-circuit current at the fault location, in these cases in accordance with IEC standard 60909-0 [1], and provides information on how this equipment can be used. In addition, the application rules VDE-AR-N 4120 [2] and VDE-AR-N4131 [3] published by the German Standardization Group FNN are discussed. In principle, converters are also used in drive technology (Sect. 4.6.3), but these are not part of the description in this chapter.

13.1 Network Configuration

The simulation of a converter system in the short-circuit current calculation is generally based on the following network configurations:

- Connection of two three-phase systems as shown in Fig. 13.1a. With LCC technology, three-phase filters must also be installed on the high-voltage side for reactive current compensation and harmonic compensation, which are not shown in Fig. 13.1a.
- Connection of an offshore wind farm as a submarine cable connection, Fig. 13.1b.
- Connection of a PV installation to the three-phase network, Fig. 13.1c.

G. Balzer, *Short-Circuit Currents in Three-Phase Systems*,
https://doi.org/10.1007/978-3-658-50710-7_13

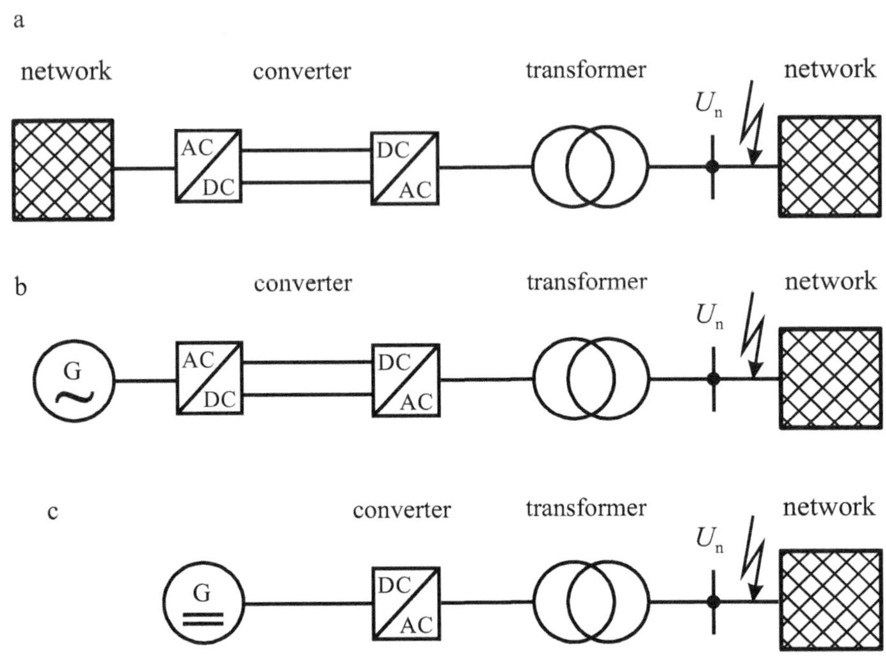

Fig. 13.1 Network connection of a converter system. (**a**) HVDC transmission, (**b**) wind farm with full-size converter, (**c**) PV system with full-size converter

The converter installation consists of the converter and the associated transformer as shown in Fig. 13.1. Similar to synchronous machines, the following currents must also be calculated when feeding converters into the three-phase network:

- Initial symmetrical short-circuit current,
- peak short-circuit current,
- breaking current,
- steady state short-circuit current, and
- thermal equivalent short-circuit current.

The currents that should be taken into account when designing equipment depend on the technology (VSC, LCC).

13.2 VSC Technology (Full-Size Converter)

In the following sections, the equations for the three-phase, line-to-line and line-to-earth short circuits are derived in general and the equations based on IEC Standard 60909-0 [1] for the different characteristic current values are listed first. In general, systems using VSC technology contribute to the short-circuit current during the entire short-circuit duration [4]. In addition, examples are calculated for the three-phase short-circuit.

13.2.1 Equations for Calculating the Three-Phase Short-Circuit Current

According to IEC Standard 60909-0 [1], the short-circuit current calculation has also taken full-size converters into account since 2016, such as those used when connecting wind and PV installations. While a voltage source is used to simulate the consideration of networks, generators and motors, a full-size converter is a current source that may feed in an (inductive) current under voltage control in the event of a fault, depending on the system operator's specifications.

The three-phase short-circuit current at any network node is determined according to Eq. (13.1), in that the total short-circuit current consists of two parts, namely the contribution consisting of network, synchronous generators and motors, and the contribution caused by the feed-in from full-size converter installation. The indexing is modified compared to the representation in the IEC Standard [1] and is based on the notation according to Eq. (3.71). In the following, the index "VS" is therefore used for a short-circuit current contribution that has a voltage source as its cause, as opposed to a contribution that is the result of a current source (CS). Furthermore, the complex notation is preferred.

$$\underline{I}''_{\mathrm{k}} = \frac{1}{\underline{Z}_{ii}} \frac{c \cdot U_{\mathrm{n}}}{\sqrt{3}} + \frac{1}{\underline{Z}_{ii}} \sum_{j=1}^{n} \underline{Z}_{ij} \cdot \underline{I}_{\mathrm{CS}j} = \underline{I}''_{\mathrm{kVS}} + \underline{I}''_{\mathrm{kCS}} \tag{13.1}$$

With

$\underline{I}_{\mathrm{CS}j}$ rms value of the current source (full-size converter, positive sequence system) in the event of a three-phase short circuit on the HV side of the step-up transformer. This value is specified by the manufacturer,

$\underline{Z}_{ii}$, $\underline{Z}_{ij}$ absolute values of the node impedance matrix in the positive sequence system, where *i* is the short-circuit node and *j* are the nodes to which feeders with full-size converters are connected,

$\underline{I}''_{\mathrm{kVS}}$ maximum initial symmetrical short-circuit current without the influence of the full-size converter,

$\underline{I}''_{\mathrm{kCS}}$ sum of the currents for the initial symmetrical short-circuit current of the full-size converters.

In addition to the representation of the IEC Standard [1], a feeder with full-size converters is simulated as follows:

- Feeders with a full-size converter (e.g. wind, photovoltaic, HVDC converter) are to be simulated in the positive sequence system using a current source. The

current source depends on the short-circuit fault and the level of the current feeder is provided by the manufacturer. The angle of the converter current $\underline{I}_{CS}$ is assumed to be 90° lagging in relation to the equivalent voltage source at the short-circuit location (inductive). The shunt impedance of the current source $\underline{Z}_{(1)CS}$ is assumed to be infinite.
- In the case of an unbalanced short circuit, the shunt impedance in the negative sequence system $\underline{Z}_{(2)CS}$ depends on the type and the control strategy. Values for the shunt impedance are specified by the manufacturer or are assumed to be infinite. The current in the negative sequence system is a consequence of the system operator's control strategy in the event of a short circuit and depends on the current injection in the positive sequence system, as the permissible conductor current must not be exceeded.
- The impedance in the zero sequence system $\underline{Z}_{(0)CS}$ is infinite. However, this value only refers to the connected converter and not to the associated transformer, which should be taken into account according to its earthing and vector group. However, this is not taken into account in the actual Standard [1].
- Feeders with full-size converters can be neglected if their contribution to the short-circuit current is not higher than 5% of the initial symmetrical short-circuit current without these feeders.

According to the description of Eq. (13.1), the current injection is 90° lagging in relation to the equivalent voltage at the short-circuit location. This assumption is independent of the current phase position of the voltage at the connection node of the converter and the *R*/*X* ratios of the impedances $\underline{Z}_{ij}$ of the network. These influences in the event of deviating ratios are examined in Sects. 13.2.5.3 and 13.2.5.4.

The current $\underline{I}''_{kVS}$ corresponds to the result of the first step $\underline{I}''_{kIi}$, Eq. (13.12), while $\underline{I}''_{kCS}$, the result of the second step $\underline{I}''_{kIIi}$, Eq. (13.17), represents Sect. 13.2.2. According to [1], calculations are not performed with complex variables, but exclusively with amounts. For this reason, the result will be on the safe side. According to Eq. (13.1), the entire current I_{CSj} of the full-size converter is always injected at the terminals, regardless of the voltage difference before and during the short circuit. It is therefore assumed that the reactive current feeding has a constant value during the short circuit, regardless of the short-circuit location.

The peak short-circuit current is determined from the two components (without and with full-size converter) and superimposed on the resulting short-circuit current. The peak short-circuit current of the full-size converter is calculated for a feeder according to Eq. (13.2):

$$i_p = \sqrt{2} \cdot I''_{kCS} \tag{13.2}$$

With

I''_{kCS} short-circuit contribution of a full-size converter on the HV side of a transformer, this value is specified by the manufacturer, Eq. (13.2).

In a radial network, the resulting peak short-circuit current results from the addition of the partial currents (network and converter). If, in contrast to Eq. (13.2), a meshed network is assumed, the resulting peak short-circuit current at any network node is calculated according to Eq. (13.3):

$$i_{\mathrm{p}} = \kappa \cdot \sqrt{2} \cdot I''_{\mathrm{kVS}} + \sqrt{2} \cdot I''_{\mathrm{kCS}} \tag{13.3}$$

With

I''_{kVS} three-phase initial symmetrical short-circuit current at the fault location without the contribution of the full-size converter according to Eq. (13.1),

I''_{kCS} contribution of the full-size converter to the three-phase short-circuit current at the fault location, Eq. (13.1).

When calculating the peak short-circuit current according to Eq. (13.3), the DC current displacement is taken into account when determining the part without converter (factor κ), but not when determining the contribution of the converter. The example in Sect. 13.2.5.5 shows the simulation of a transient short-circuit current curve during a network short circuit in the 380 kV network of a high-voltage direct current transmission (HVDC) using IGBT technology. Here, a three-phase short circuit is assumed at the end of a short overhead line, which is initiated at the voltage zero crossing of phase R. A shift of the converter contribution is not recognizable, Fig. 13.14b, so that the assumption according to Eq. (13.3) is justified.

The breaking current of a full-size converter is based on the magnitude of the steady state short-circuit current, as the current is kept at a constant value due to the regulation. The three-phase breaking current I_{b} is therefore equal to the steady state short-circuit current I_{kCS}, which is specified by the manufacturer, so that the following generally applies:

$$I_{\mathrm{b}} = I_{\mathrm{kCS}} \tag{13.4}$$

Since according to Eq. (13.4) the breaking current is equal to the steady state short-circuit current and according to Eq. (13.1) not necessarily equal to the initial symmetrical short-circuit current, a correction according to Eq. (5.59) should be applied under these conditions if the breaking current is determined in meshed networks. This depends on the relationship between $I''_{\mathrm{kCS}} \approx I_{\mathrm{kCS}}$, so that a fictitious value for the factor μ must be determined from this current ratio. If these two currents are identical, there is no need for a correction as this contribution is already included in the initial symmetrical short-circuit current.

In the case of unbalanced short-circuits, the following conditions apply for the breaking currents (I_b) and steady state short-circuit currents (I_k, max/min):

$$I_{b2} = I''_{k2} \qquad I_{k2} = I''_{k2} \tag{13.5}$$

$$I_{b2E} = I''_{k2E} \qquad I_{k2E} = I''_{k2E} \tag{13.6}$$

$$I_{b1} = I''_{k1} \qquad I_{k1} = I''_{k1} \tag{13.7}$$

13.2.2 Calculation of the Three-Phase Initial Symmetrical Short-Circuit Current

The basic procedure for calculating the short-circuit current when full-size converters are to be taken into account is shown using the three-phase short-circuit in a 110 kV example network, which also has three feeders from full converters in addition to a network feeder.

13.2.2.1 Procedure

The network connection shown in Fig. 13.2 is used as an example [5]. A total of three wind farms are connected to a 110 kV network, which is supplied from an overlaid network. The short circuit is located on busbar 2, so that one converter (U2) feeds directly and the other two converters (U3, U4) feed via an impedance to the fault location. The simulation assumes that the full-size converter and the associated step-up transformer form a single generation unit. All specifications always refer to the HV side of the transformer.

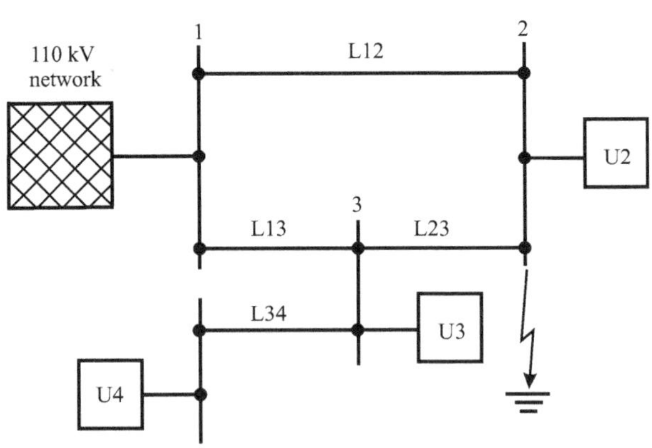

Fig. 13.2 Network circuit for determining the three-phase short-circuit current (busbar 2) in a 110 kV network, taking into account full-size converters (U)

The illustration in Fig. 13.2 is based on an overhead line network. The following component data is assumed for the calculation:

- 110 kV network:
  nominal system voltage $U_n = 110$ kV
  initial symmetrical short-circuit current $\underline{I}''_{kN} = 20\,\text{kA}$
  R/*X* ratio $R/X = 0.1$
- Overhead lines:
  impedance $\underline{Z}'_L = (0.120 + j0.393)\,\Omega/\text{km}$
  length $\ell_{12} = 100\,\text{km}$ $\ell_{13} = 50\,\text{km}$
  $\ell_{23} = 50\,\text{km}$ $\ell_{34} = 25\,\text{km}$
- Full-size converter unit (wind farm):
  rated voltage (HV side) $U_{rU} = 110\,\text{kV}$
  rated power $P_{rU2} = 100\,\text{MW}$ $P_{rU3} = 50\,\text{MW}$
  $P_{rU4} = 50\,\text{MW}$
  short-circuit current contribution $I_{CS} = k_U \cdot I_{rU}$ $k_U = 1.0$

In the case of full-size converters, it is assumed that the converter supply corresponds to the rated current in the event of a short circuit. This is an inductive current, as required by Sect. 13.4 for voltage support during the short circuit. The feeder of the network according to Fig. 13.2 is assumed to be a voltage source, while the full-size converter systems each represent a constant current source. For this reason, a total of three work steps are defined for calculating the short-circuit currents, namely:

- Calculation of the short-circuit current on busbar 2, caused by the network supply and without the contribution of the full-size converters, fault current $\underline{I}''_{kVS}$ (index: VS, Voltage Source).
- Calculation of the short-circuit current at busbar 2, caused by the feeder of the full-size converter installations and without contribution from the network, fault current $\underline{I}''_{kCS}$ (Index: CS, Current Source). Only an inductive current is assumed here. The phase angle of the converter currents is based on the equivalent voltage source at the short-circuit location, which lies in the real axis and is therefore independent of the phase angle of the terminal voltage before the fault occurs. In addition, the shunt impedances of the current sources in the positive sequence system are assumed to be infinite in accordance with Sect. 13.2.1.
- The two fault currents are then superimposed to determine the total short-circuit current $\underline{I}''_k$.

Figure 13.3 shows the equivalent circuit in the positive sequence system for the three-phase short circuit; both the current sources of the full-size converters and the equivalent voltage source at short-circuit location 2 are shown.

To simplify matters, Fig. 13.3 can be converted by combining the impedances and current sources as an example (Fig. 13.4).

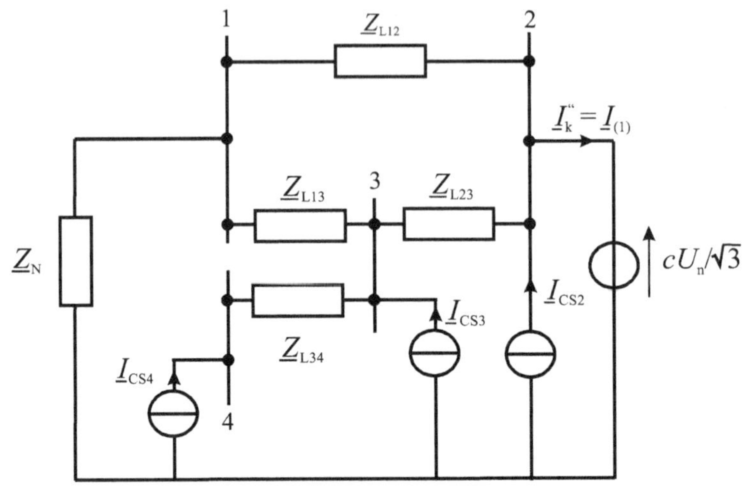

Fig. 13.3 Equivalent circuit diagram in the positive sequence system for the three-phase short circuit on busbar 2 according to Fig. 13.2 (equivalent voltage source at the short-circuit location, current source)

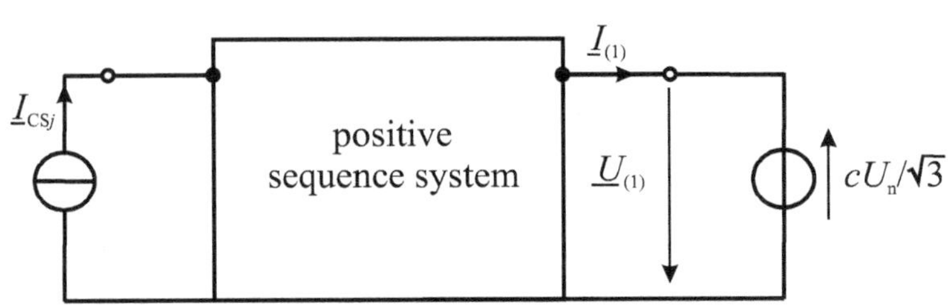

Fig. 13.4 Simplified equivalent circuit diagram of the positive sequence system

13.2.2.2 Work Step 1: Equivalent Voltage Source at the Short-Circuit Location

In the first step, the current sources are not taken into account, so that according to Fig. 13.5 they represent an interruption at which a voltage to earth arises due to the network feeder. This step only determines the short-circuit current contribution resulting from the voltage source (network).

Basically, the admittance matrix $\underline{\mathbf{Y}}_N$ of the network is set up for the calculation in order to determine the currents at the network nodes $\underline{\mathbf{I}}_i$ using the impedance matrix $\underline{\mathbf{Z}}_N$ and the voltage distribution $\underline{\mathbf{U}}_i$ (see Sect. 3.6.6), so that the following generally applies:

$$\underline{\mathbf{I}}_i = \underline{\mathbf{Y}}_N \cdot \underline{\mathbf{U}}_i \qquad \underline{\mathbf{U}}_i = \underline{\mathbf{Z}}_N \cdot \underline{\mathbf{I}}_i \tag{13.8}$$

The following generally applies to the admittance or impedance matrix of a network:

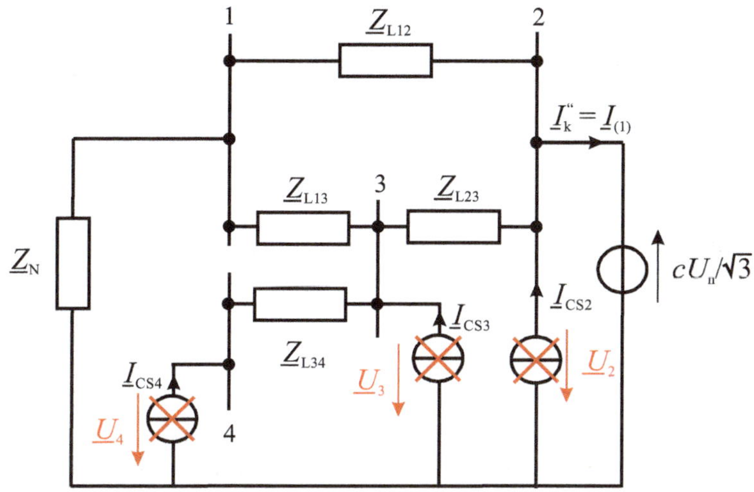

Fig. 13.5 Equivalent circuit for determining the network contribution to the short-circuit current (positive sequence system)

$$\underline{\mathbf{Y}}_N = \begin{pmatrix} \underline{Y}_{11} & \cdots & \underline{Y}_{1i} & \cdots & \underline{Y}_{1m} \\ \vdots & \ddots & \vdots & \ddots & \vdots \\ \underline{Y}_{i1} & \cdots & \underline{Y}_{ii} & \cdots & \underline{Y}_{im} \\ \vdots & \ddots & \vdots & \ddots & \vdots \\ \underline{Y}_{m1} & \cdots & \underline{Y}_{mi} & \ddots & \underline{Y}_{mm} \end{pmatrix} \quad \underline{\mathbf{Z}}_N$$
$$= \begin{pmatrix} \underline{Z}_{11} & \cdots & \underline{Z}_{1i} & \cdots & \underline{Z}_{1m} \\ \vdots & \ddots & \vdots & \ddots & \vdots \\ \underline{Z}_{i1} & \cdots & \underline{Z}_{ii} & \cdots & \underline{Z}_{im} \\ \vdots & \ddots & \vdots & \ddots & \vdots \\ \underline{Z}_{m1} & \cdots & \underline{Z}_{mi} & \ddots & \underline{Z}_{mm} \end{pmatrix} \tag{13.9}$$

To calculate the short-circuit current using the equivalent voltage source method at the short-circuit location F (busbar 2), Fig. 13.5 applies, with the voltage according to Eq. (13.10) as the only voltage in the system. The current sources according to Fig. 13.5 are not taken into account. The node voltages are marked with an asterisk (*) in accordance with Sect. 3.6.6.

$$\underline{U}_i^* = -c \cdot U_n/\sqrt{3} \tag{13.10}$$

Using the impedance matrix, it is possible to calculate the short-circuit current at the fault location. Here, the equivalent voltage at node i is used according to Eq. (13.10), while all node currents have the value $\underline{I}_i = 0$, except for the fault current $\underline{I}''_{kVSi}$. Equations (3.57) and (13.11) then apply in general:

$$\begin{pmatrix} \underline{U}_1^* \\ \vdots \\ -c \cdot U_\mathrm{n}/\sqrt{3} \\ \vdots \\ \underline{U}_\mathrm{m}^* \end{pmatrix} = \begin{pmatrix} \underline{Z}_{11} & \cdots & \underline{Z}_{1i} & \cdots & \underline{Z}_{1\mathrm{m}} \\ \vdots & \ddots & \vdots & \ddots & \vdots \\ \underline{Z}_{i1} & \cdots & \underline{Z}_{ii} & \cdots & \underline{Z}_{i\mathrm{m}} \\ \vdots & \ddots & \vdots & \ddots & \vdots \\ \underline{Z}_{\mathrm{m}1} & \cdots & \underline{Z}_{\mathrm{m}i} & \ddots & \underline{Z}_{\mathrm{mm}} \end{pmatrix} \cdot \begin{pmatrix} 0 \\ \vdots \\ -\underline{I}''_{\mathrm{kVS}i} \\ \vdots \\ 0 \end{pmatrix} \tag{13.11}$$

Equation (13.11) gives the short-circuit current $\underline{I}''_{\mathrm{kVS}i}$ without the influence of the full-size converter systems:

$$c \cdot U_\mathrm{n}/\sqrt{3} = \underline{Z}_{ii} \cdot \underline{I}''_{\mathrm{kVS}i} \qquad \underline{I}''_{\mathrm{kVS}i} = \frac{c \cdot U_\mathrm{n}/\sqrt{3}}{\underline{Z}_{ii}} \tag{13.12}$$

Subsequently, all node voltages against reference earth (see remark after Sect. 3.6.6) can be derived from the system of Eq. (13.11), so that, for example, the following applies to the voltage $\underline{U}_1^*$ at network node 1 for a short circuit at node i:

$$\underline{U}_1^* = \underline{Z}_{1i} \cdot \underline{I}''_{\mathrm{kVS}i} = -\frac{\underline{Z}_{1i}}{\underline{Z}_{ii}} \cdot c \cdot U_\mathrm{n}/\sqrt{3} = -\Delta \underline{U}_1 \tag{13.13}$$

It should be noted that the voltage according to Eq. (13.13) is the result of the changed voltage profile caused by the equivalent voltage source at the short-circuit location. The voltage according to Eq. (13.13) also corresponds to the voltage change caused by the short circuit as a result of the equivalent voltage source. The actual voltage $\underline{U}_1$ at node 1 results in:

$$\underline{U}_1 = c \cdot U_\mathrm{n}/\sqrt{3} - \frac{\underline{Z}_{1i}}{\underline{Z}_{ii}} \cdot c \cdot U_\mathrm{n}/\sqrt{3} = c \cdot U_\mathrm{n}/\sqrt{3} \cdot \left(1 - \frac{\underline{Z}_{1i}}{\underline{Z}_{ii}}\right) \tag{13.14}$$

This voltage should be used if, as shown later, the current contribution of the full-size converter is dependent on the voltage reduction at the connection node (Sect. 13.4).

13.2.2.3 Work Step 2: Constant Current Source

The simulation of the full-size converter installations is carried out in the positive sequence system by a current source with the internal resistance $\underline{Z}_{\mathrm{CS}i} \rightarrow \infty$, so that the equivalent circuit is carried out according to Fig. 13.6.

Similarly, according to the method of the equivalent voltage source at the short-circuit location, Eq. (13.11), the current distribution is calculated exclusively by the influence of the full-size converter installations, so that the system of Eq. (13.15) applies. At the short-circuit location, the voltage $\underline{U}_{\mathrm{CS}i}^*$ is short-circuited, while the other voltages are determined from the current conditions. The feeders of the full-size converter installations are represented by a current vector on the right-hand side.

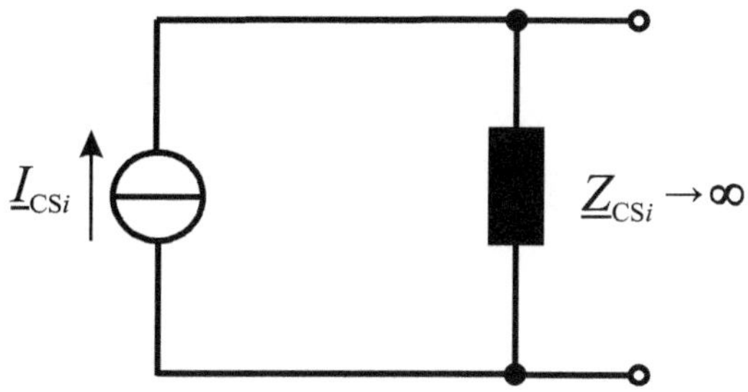

Fig. 13.6 Simulation of the full-size converter installation as a current source (positive sequence system)

$$\begin{pmatrix} \underline{U}^*_{\mathrm{CS1}} \\ \vdots \\ 0 \\ \vdots \\ \underline{U}^*_{\mathrm{CSm}} \end{pmatrix} = \begin{pmatrix} \underline{Z}_{11} & \cdots & \underline{Z}_{1i} & \cdots & \underline{Z}_{1\mathrm{m}} \\ \vdots & \ddots & \vdots & \ddots & \vdots \\ \underline{Z}_{i1} & \cdots & \underline{Z}_{ii} & \cdots & \underline{Z}_{i\mathrm{m}} \\ \vdots & \ddots & \vdots & \ddots & \vdots \\ \underline{Z}_{\mathrm{m1}} & \cdots & \underline{Z}_{\mathrm{m}i} & \ddots & \underline{Z}_{\mathrm{mm}} \end{pmatrix} \cdot \left[\begin{pmatrix} 0 \\ \vdots \\ -\underline{I}''_{\mathrm{kCS}i} \\ \vdots \\ 0 \end{pmatrix} + \begin{pmatrix} \underline{I}_{\mathrm{CS1}} \\ \vdots \\ \underline{I}_{\mathrm{CS}i} \\ \vdots \\ \underline{I}_{\mathrm{CSm}} \end{pmatrix} \right] \tag{13.15}$$

This results in the short-circuit current contribution $\underline{I}''_{\mathrm{kCS}i}$ at fault location i as a result of the feeder of the full-size converter installations by multiplying out the ith line:

$$\begin{aligned} 0 &= -\underline{Z}_{ii} \cdot \underline{I}''_{\mathrm{kCS}i} + (\underline{Z}_{i1} \cdot \underline{I}_{\mathrm{CS1}} + \cdots + \underline{Z}_{i\mathrm{m}} \cdot \underline{I}_{\mathrm{CSm}}) \\ &= -\underline{Z}_{ii} \cdot \underline{I}''_{\mathrm{kCS}i} + \sum_{j}^{\mathrm{m}} \underline{Z}_{ij} \cdot \underline{I}_{\mathrm{CS}j} \end{aligned} \tag{13.16}$$

By transformation follows:

$$\underline{I}''_{\mathrm{kCS}i} = \frac{1}{\underline{Z}_{ii}} \cdot \sum_{j}^{\mathrm{m}} \underline{Z}_{ij} \cdot \underline{I}_{\mathrm{CS}j} \tag{13.17}$$

The current $\underline{I}''_{\mathrm{kCS}i}$ represents the short-circuit current contribution at the fault location, which is caused by all feeders of the various converters, taking into account the network impedance. If there is no network impedance in the theoretical case, the short-circuit current contribution corresponds to the sum of all converter feeds in the network.

13.2.2.4 Work Step 3: Superposition

The two current components according to Eqs. (13.12) and (13.17) must be added in order to calculate the total short-circuit current at the fault location, so that the following applies.

$$\underline{I}''_{\mathrm{k}i} = \underline{I}''_{\mathrm{kVS}i} + \underline{I}''_{\mathrm{kCS}i} = \frac{c \cdot U_{\mathrm{n}}/\sqrt{3}}{\underline{Z}_{ii}} + \frac{1}{\underline{Z}_{ii}} \cdot \sum_{j}^{\mathrm{m}} \underline{Z}_{ij} \cdot \underline{I}_{\mathrm{CS}j} \tag{13.18}$$

The values $\underline{Z}_{ii}$ and $\underline{Z}_{ij}$ result from the impedance matrix according to Eq. (13.9), where the index i (in this example 2) represents the short-circuit location and j represents the nodes to which full-size converter installations are connected (in this example 2, 3 and 4, according to Fig. 13.2). The value $\underline{Z}_{ii}$ corresponds to the short-circuit impedance $\underline{Z}_{\mathrm{k}}$.

13.2.2.5 Work Step 4: Calculation of the Voltage Distribution

The voltage distribution at the network nodes can be determined based on the short-circuit currents according to Eq. (13.18). The system of Eq. (13.15) is used here by determining the total short-circuit current $\underline{I}''_{\mathrm{k}i}$ according to Sect. 13.2.2.4, Eq. (13.19). With $\underline{U}_i$ of the equivalent voltage source at the short-circuit location according to Eq. (13.10). In this case, too, the voltages actually represent the voltage change due to the short circuit. The real voltages $\underline{U}_i$ at the network nodes are determined according to Eq. (13.14).

$$\begin{pmatrix} \underline{U}_1^* \\ \vdots \\ \underline{U}_i^* \\ \vdots \\ \underline{U}_{\mathrm{m}}^* \end{pmatrix} = \begin{pmatrix} \underline{Z}_{11} & \cdots & \underline{Z}_{1i} & \cdots & \underline{Z}_{1\mathrm{m}} \\ \vdots & \ddots & \vdots & \ddots & \vdots \\ \underline{Z}_{i1} & \cdots & \underline{Z}_{ii} & \cdots & \underline{Z}_{i\mathrm{m}} \\ \vdots & \ddots & \vdots & \ddots & \vdots \\ \underline{Z}_{\mathrm{m}1} & \cdots & \underline{Z}_{\mathrm{m}i} & \ddots & \underline{Z}_{\mathrm{mm}} \end{pmatrix} \cdot \left[\begin{pmatrix} 0 \\ \vdots \\ -\underline{I}''_{\mathrm{k}i} \\ \vdots \\ 0 \end{pmatrix} + \begin{pmatrix} \underline{I}_{\mathrm{CS}1} \\ \vdots \\ \underline{I}_{\mathrm{CS}i} \\ \vdots \\ \underline{I}_{\mathrm{CSm}} \end{pmatrix} \right] \tag{13.19}$$

13.2.3 *Determining the Short-Circuit Current for a Line-to-Line Short Circuit*

The equivalent circuits in the positive and negative sequence system must be used to derive the short-circuit current at the fault location in the event of a line-to-line short circuit. While the positive sequence system is shown in Fig. 13.5, Fig. 13.7 shows the corresponding negative sequence system. This system is generally characterized by two significant differences to the positive sequence system with regard to the full-size converters to be considered:

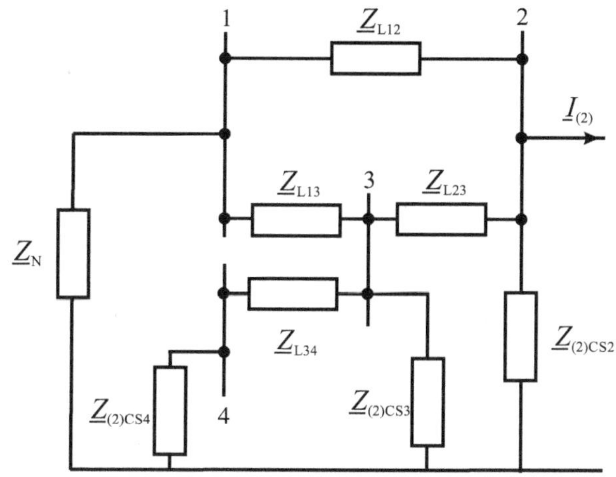

Fig. 13.7 Equivalent circuit diagram of the negative sequence system of the example shown in Fig. 13.2

- The current in the negative sequence system is basically a consequence of the control strategy in the event of a short circuit and depends on the current source in the positive sequence system, as the permissible rated current must not be exceeded. According to [1], however, it is assumed that only one current source acts in the positive sequence system.
- The negative sequence impedance of the full-size converter has a value that must be specified by the manufacturer and depends on the control strategy, or is assumed to be infinite

The negative sequence impedances of the full-size converters, if they have a defined value, are labeled $\underline{Z}_{(2)CSi}$, while the impedances of the overhead lines and the network are equal to the positive sequence impedances. This means that when calculating the line-to-line short-circuit current, a repeated calculation of the impedance matrix for the negative sequence system is required. In this example, it is assumed that the current in the negative sequence system has a value of zero, as the current in the positive sequence system has the maximum rated value.

Taking into account the fault conditions for the line-to-line short-circuit current without earth contact, the following conditions can be specified for fault node i using the symmetrical components:

$$\begin{aligned} \underline{I}_{(1)i} &= -\underline{I}_{(2)i} \\ \underline{U}_{(1)i} &= \underline{U}_{(2)i} - c \cdot U_n/\sqrt{3} \end{aligned} \tag{13.20}$$

When deriving the fault conditions, it is taken into account that this voltage source is not in the positive sequence system due to the method of the equivalent voltage source at the short-circuit location. This results in the equivalent circuit for the line-to-line short circuit as shown in Fig. 13.8.

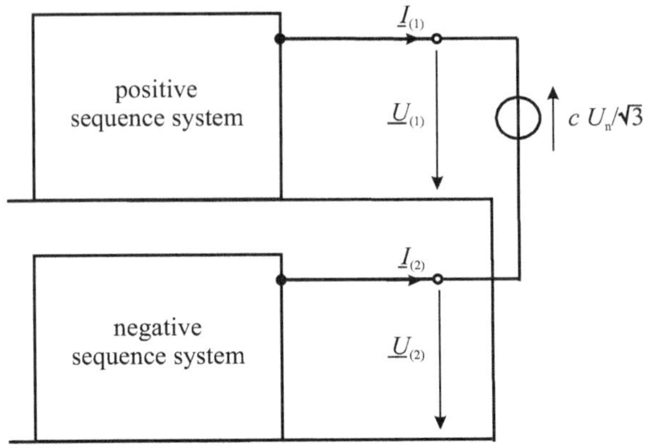

Fig. 13.8 Equivalent circuit of the symmetrical components in the event of a line-to-line short circuit; calculation method: equivalent voltage source at the short-circuit location

The current in the positive sequence system is determined according to Eq. (13.18) by replacing the voltage $c \cdot U_{\mathrm{n}}/\sqrt{3}$ with the voltage at the terminals of the positive sequence system $\underline{U}_{(1)i}$ (Fig. 13.4). Since the influence of the full-size converter (current source) in this example only affects the positive sequence system, these current contributions are only taken into account in Eq. (13.21). When setting up the equation, the voltage and current counter arrows are taken into account as shown in Fig. 13.8, whereby the following applies to the voltage at the short-circuit location: $U_{(1)i} = -c \cdot U_{\mathrm{n}}/\sqrt{3}$.

$$\underline{I}_{(1)i} = \frac{-\underline{U}_{(1)i}}{\underline{Z}_{(1)ii}} + \frac{1}{\underline{Z}_{(1)ii}} \cdot \sum_{j}^{\mathrm{m}} \underline{Z}_{(1)ij} \cdot \underline{I}_{(1)\mathrm{CS}j} \tag{13.21}$$

$$\underline{I}_{(2)i} = -\frac{\underline{U}_{(2)i}}{\underline{Z}_{(2)ii}} \tag{13.22}$$

Taking into account the component voltages according to Eq. (13.20), the following results:

$$\underline{I}_{(1)i} = -\frac{U_{(2)i} - c \cdot U_{\mathrm{n}}/\sqrt{3}}{\underline{Z}_{(1)ii}} + \frac{1}{\underline{Z}_{(1)ii}} \cdot \sum_{j}^{\mathrm{m}} \underline{Z}_{(1)ij} \cdot \underline{I}_{(1)\mathrm{CS}j} \tag{13.23}$$

Substituting Eqs. (13.20) and (13.22) gives:

$$\underline{I}_{(1)i} = \frac{\underline{Z}_{(2)ii}}{\underline{Z}_{(1)ii}} \cdot \underline{I}_{(2)i} + \frac{c \cdot U_{\mathrm{n}}/\sqrt{3}}{\underline{Z}_{(1)ii}} + \frac{1}{\underline{Z}_{(1)ii}} \cdot \sum_{j}^{\mathrm{m}} \underline{Z}_{(1)ij} \cdot \underline{I}_{(1)\mathrm{CS}j} \tag{13.24}$$

$$\underline{I}_{(1)i} = \frac{1}{\underline{Z}_{(1)ii} + \underline{Z}_{(2)ii}} \cdot \left(c \cdot U_{\mathrm{n}}/\sqrt{3} + \sum_{j}^{\mathrm{m}} \underline{Z}_{(1)ij} \cdot \underline{I}_{(1)\mathrm{CS}j} \right) = -\underline{I}_{(2)i} \tag{13.25}$$

The transformation matrix can be used to determine the initial symmetrical short-circuit current in the original system, $\underline{I}_{(0)i} = 0$

$$\begin{pmatrix} \underline{I}''_{\mathrm{R}i} \\ \underline{I}''_{\mathrm{S}i} \\ \underline{I}''_{\mathrm{T}i} \end{pmatrix} = \begin{pmatrix} 1 & 1 & 1 \\ 1 & \underline{\mathrm{a}}^2 & \underline{\mathrm{a}} \\ 1 & \underline{\mathrm{a}} & \underline{\mathrm{a}}^2 \end{pmatrix} \cdot \begin{pmatrix} \underline{I}_{(0)i} \\ \underline{I}_{(1)i} \\ \underline{I}_{(2)i} \end{pmatrix} \tag{13.26}$$

$$\underline{I}''_{\mathrm{R}i} = 0 \quad \underline{I}''_{\mathrm{S}i} = -\underline{I}''_{\mathrm{T}i} = -\mathrm{j}\sqrt{3} \cdot \underline{I}_{(1)i} \tag{13.27}$$

This results in complex values for the line-to-line initial symmetrical short-circuit current (without earth contact):

$$\underline{I}''_{\mathrm{k}2i} = \underline{I}''_{\mathrm{T}i} = \frac{\mathrm{j}\sqrt{3}}{\underline{Z}_{(1)ii} + \underline{Z}_{(2)ii}} \cdot \left(c \cdot U_{\mathrm{n}}/\sqrt{3} + \sum_{j}^{\mathrm{m}} \underline{Z}_{(1)ij} \cdot \underline{I}_{(1)\mathrm{CS}j} \right) \tag{13.28}$$

As the short-circuit impedance $\underline{Z}_{(1)ij} = \mathrm{j}X_{(1)ij}$ and the currents of the full-size converters $\underline{I}''_{(1)\mathrm{CS}j} = -\mathrm{j}I_{(1)\mathrm{CS}j}$ are usually also inductive, the two parts of the bracket have the same sign.

13.2.4 Determining the Short-Circuit Current for a Line-to-Earth Short Circuit

Analogous to the line-to-line short circuit, the contribution in the event of a line-to-earth short circuit can also be determined by applying the component equivalent circuit as shown in Fig. 13.9. The simulation of the line-to-earth short circuit results in the series connection of the component systems. In this case, it is also assumed that only the infeed of the current source is present in the positive sequence system. When simulating the zero sequence system, the note in Sect. 13.2.1 must be observed that the zero sequence impedance of the converter installation is characterized by the behavior of the associated transformer (vector group, earthing), as shown in Fig. 13.10.

In addition to Eqs. (13.21) and (13.22), the zero sequence current $\underline{I}_{(0)}$ is determined as follows:

$$\underline{I}_{(0)i} = -\frac{\underline{U}_{(0)i}}{\underline{Z}_{(0)ii}} \tag{13.29}$$

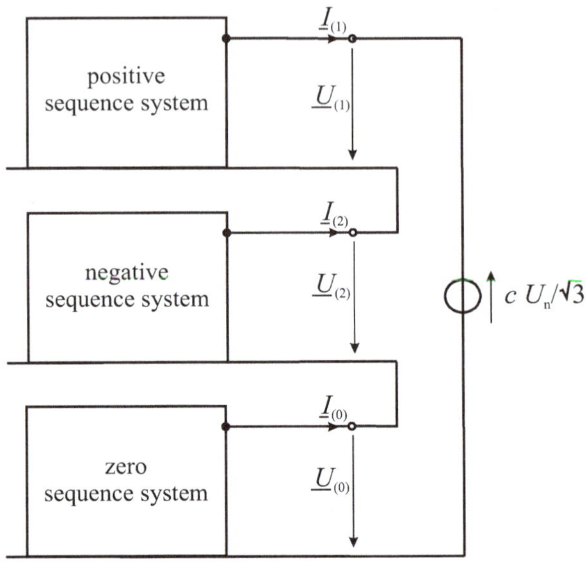

Fig. 13.9 Equivalent circuit of the symmetrical components in the event of a line-to-earth short circuit; calculation method: equivalent voltage source at the short-circuit location

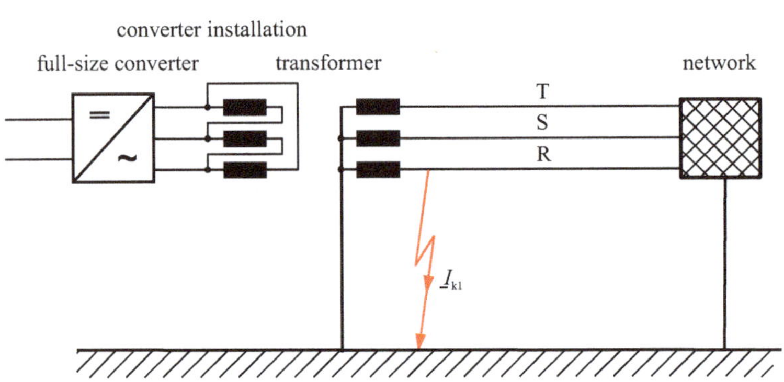

Fig. 13.10 Equivalent circuit for a line-to-earth short circuit

This gives the component current $\underline{I}_{(1)i}$ according to Eq. (13.30):

$$\underline{I}_{(1)i} = \frac{U_{(2)i} + U_{(0)i} + c \cdot U_{\mathrm{n}}/\sqrt{3}}{\underline{Z}_{(1)ii}} + \frac{1}{\underline{Z}_{(1)ii}} \cdot \sum_{j}^{m} \underline{Z}_{(1)ij} \cdot \underline{I}_{(1)\mathrm{CS}j} \tag{13.30}$$

According to the specifications of IEC Standard 2016 [1], the zero sequence system of the full-size converter installation $\underline{Z}_{(0)\mathrm{PF}}$ must be set to infinity. However, this formulation refers exclusively to the converter, as it is assumed that the associated transformer is not earthed on both sides (network and converter side). According to

Fig. 13.10, the network contribution of the system to the short-circuit current depends on the neutral point treatment of the transformer on the HV side.

The delta winding of the transformer shown in Fig. 13.10 represents a short circuit in the zero sequence system, with a line-to-earth short circuit on the HV side, so that the zero sequence impedance of the transformer is parallel to the zero sequence impedance of the network.

13.2.5 Examples for the Calculation of the Three-Phase Short-Circuit Current

In the following, the initial symmetrical short-circuit current and the peak short-circuit current are determined for the example in Sect. 13.2.2.1 (Fig. 13.2), in each case for the three-phase short circuit. A distinction is made here between two different procedures. First, the current injection of the converter is independent of the voltage change at the terminals during the short circuit, Sect. 13.2.5.1. In the second case, the injection is corrected as a result of the voltage-dependent current injection depending on the fault location, Sect. 13.2.5.2.

According to Sect. 13.2.1, the converter current is fed lagging ($-90°$) in relation to the voltage of the equivalent voltage source at the short-circuit location. For this reason, Sect. 13.2.5.3 also considers the influence on the short-circuit current if the injected current deviates from this specification.

In addition, Sect. 13.2.5.4 examines, in deviation from the specification in Sect. 13.2.1 [1], whether the specification of an exclusively inductive current injection leads to the largest short-circuit current at the fault location.

13.2.5.1 Initial Symmetrical Short-Circuit Current without Correction of the Full-Size Converter Injection

In this section, it is assumed that the connected full-size converters supply an inductive current contribution during the short-circuit duration that is independent of the voltage reduction at the terminals of the converter installation. The short-circuit current contribution therefore corresponds to the maximum required current regardless of the short-circuit location. This procedure corresponds to the representation according to the IEC Standard 2016 [1].

Based on the impedances of equipment, the corresponding matrices are set up to determine the initial symmetrical short-circuit current according to Eq. 13.1 or 13.18. This results in the following impedances or admittances, according to the specifications in Sect. 13.2.2.1:

- Network:

$$Z_{\mathrm{N}} = \frac{c \cdot U_{\mathrm{n}}}{\sqrt{3} \cdot I''_{\mathrm{k}}} = \frac{1.1 \cdot 110}{\sqrt{3} \cdot 20}\,\Omega = 3.4930\Omega \qquad \underline{Z}_{\mathrm{N}} = (0.3476 + \mathrm{j}3.4756)\Omega$$

$$\underline{Y}_{\mathrm{N}} = (0.02849 - \mathrm{j}0.28487) \cdot 1/\Omega$$

- Overhead lines:

$$\underline{Z}_{\mathrm{L12}} = (12{,}000 + \mathrm{j}39{,}300)\,\Omega; \qquad \underline{Y}_{\mathrm{L12}} = (0{,}00711 - \mathrm{j}0{,}02328) \cdot 1/\Omega$$
$$\underline{Z}_{\mathrm{L13}} = (6{,}000 + \mathrm{j}19{,}650)\,\Omega = \underline{Z}_{\mathrm{L12}}/2; \qquad \underline{Y}_{\mathrm{L13}} = (0{,}01421 - \mathrm{j}0{,}04655) \cdot 1/\Omega = 2 \cdot \underline{Y}_{\mathrm{L12}}$$
$$\underline{Z}_{\mathrm{L23}} = (6{,}000 + \mathrm{j}19{,}650)\,\Omega = \underline{Z}_{\mathrm{L12}}/2; \qquad \underline{Y}_{\mathrm{L23}} = (0{,}01421 - \mathrm{j}0{,}04655) \cdot 1/\Omega = 2 \cdot \underline{Y}_{\mathrm{L12}}$$
$$\underline{Z}_{\mathrm{L34}} = (3{,}000 + \mathrm{j}9{,}825)\,\Omega = \underline{Z}_{\mathrm{L12}}/4; \qquad \underline{Y}_{\mathrm{L34}} = (0{,}02843 - \mathrm{j}0{,}09310) \cdot 1/\Omega = 4 \cdot \underline{Y}_{\mathrm{L12}}$$

Full-size converter installation (wind farms):

$$I_{\mathrm{rU2}} = \frac{P_{\mathrm{rU2}}}{\sqrt{3} \cdot U_{\mathrm{n}}} = \frac{100}{\sqrt{3} \cdot 110}\,\mathrm{kA} = 0.5249\,\mathrm{kA} \qquad I_{\mathrm{rU3}} = 0.2624\,\mathrm{kA} = I_{\mathrm{rU4}}$$

As the current, fed in by the full-size converter installations, is assumed to be inductive, the respective currents are calculated taking into account a current contribution in the event of a short circuit of $I_{\mathrm{CS}} = 1.0 \cdot I_{\mathrm{rU}}$:

$$\underline{I}_{\mathrm{CS2}} = -\mathrm{j}0.5249\,\mathrm{kA} \qquad \underline{I}_{\mathrm{CS3}} = -\mathrm{j}0.2624\,\mathrm{kA} = \underline{I}_{\mathrm{CS4}}$$

For the example in Fig. 13.2, the admittance matrix according to Eq. (13.9) can generally be set up for a system.

$$\underline{\mathbf{Y}}_{\mathrm{S}} = \begin{pmatrix} \underline{Y}_{\mathrm{N}} + \underline{Y}_{\mathrm{L12}} + \underline{Y}_{\mathrm{L13}} & -\underline{Y}_{\mathrm{L12}} & -\underline{Y}_{\mathrm{L13}} & 0 \\ -\underline{Y}_{\mathrm{L12}} & \underline{Y}_{\mathrm{L12}} + \underline{Y}_{\mathrm{L23}} & -\underline{Y}_{\mathrm{L23}} & 0 \\ -\underline{Y}_{\mathrm{L13}} & -\underline{Y}_{\mathrm{L23}} & \underline{Y}_{\mathrm{L13}} + \underline{Y}_{\mathrm{L23}} + \underline{Y}_{\mathrm{L34}} & -\underline{Y}_{\mathrm{L34}} \\ 0 & 0 & -\underline{Y}_{\mathrm{L34}} & \underline{Y}_{\mathrm{L34}} \end{pmatrix} \tag{13.31}$$

Taking into account the component data, the following results:

$$\underline{\mathbf{Y}}_{\mathrm{S}} = \begin{pmatrix} 0.04981 - \mathrm{j}0.35470 & -0.00711 + \mathrm{j}0.02328 & -0.01421 + \mathrm{j}0.04655 & 0 \\ -0.00711 + \mathrm{j}0.02328 & 0.02132 - \mathrm{j}0.06983 & -0.01421 + \mathrm{j}0.04655 & 0 \\ -0.01421 + \mathrm{j}0.04655 & -0.01421 + \mathrm{j}0.04655 & 0.05685 - \mathrm{j}0.18620 & -0.02843 + \mathrm{j}0.09310 \\ 0 & 0 & -0.02843 + \mathrm{j}0.09310 & 0.02843 - \mathrm{j}0.09310 \end{pmatrix}$$

Inverting the admittance matrix provides the impedance matrix for calculating the short-circuit currents.

$$\underline{\mathbf{Z}}_\mathrm{S} = \begin{pmatrix} 0.3476+\mathrm{j}3.4756 & 0.3476+\mathrm{j}3.4756 & 0.3476+\mathrm{j}3.4756 & 0.3476+\mathrm{j}3.4756 \\ 0.3476+\mathrm{j}3.4756 & 6.3469+\mathrm{j}23,1238 & 3.3473+\mathrm{j}13.2997 & 3.3473+\mathrm{j}13.2997 \\ 0.3476+\mathrm{j}3.4756 & 3.3473+\mathrm{j}13.2997 & 4.8468+\mathrm{j}18.2132 & 4.8468+\mathrm{j}18.2132 \\ 0.3476+\mathrm{j}3.4756 & 3.3473+\mathrm{j}13.2997 & 4.8468+\mathrm{j}18.2132 & 7.8471+\mathrm{j}28.0381 \end{pmatrix}$$

The example is first calculated using complex variables according to Eq. (13.18), in contrast to the IEC Standard 2016 [1]. The impedances are determined from the corresponding rows and columns of the impedance matrix $\underline{\mathbf{Z}}_\mathrm{S}$ specified above. For the total short-circuit current at busbar 2 (marked: fault location F2), the following results for this example:

$$\underline{I}''_\mathrm{kF2} = \frac{c \cdot U_\mathrm{n}/\sqrt{3}}{\underline{Z}_{22}} + \frac{1}{\underline{Z}_{22}} \cdot [\underline{Z}_{21} \cdot \underline{I}_\mathrm{CS1} + \underline{Z}_{22} \cdot \underline{I}_\mathrm{CS2} + \underline{Z}_{23} \cdot \underline{I}_\mathrm{CS3} + \underline{Z}_{24} \cdot \underline{I}_\mathrm{CS4}] \tag{13.32}$$

Since no full-size converter feeds in at busbar 1 ($\underline{I}_\mathrm{kCS1} = 0$) and the current contribution is directly effective at the faulty busbar 2, the following simplifies the short-circuit current:

$$\underline{I}''_\mathrm{kF2} = \frac{c \cdot U_\mathrm{n}/\sqrt{3}}{\underline{Z}_{22}} + \frac{1}{\underline{Z}_{22}} \cdot [\underline{Z}_{23} \cdot \underline{I}_\mathrm{CS3} + \underline{Z}_{24} \cdot \underline{I}_\mathrm{CS4}] + \underline{I}_\mathrm{CS2} \tag{13.33}$$

After inserting the numerical values, the following results from Eq. (13.33) with $\underline{Z}_{23} = \underline{Z}_{24}$ and $\underline{I}_\mathrm{CS3} = \underline{I}_\mathrm{CS4}$:

$$\underline{I}''_\mathrm{kF2} = \frac{1.1 \cdot 110/\sqrt{3}}{6.3469 + \mathrm{j}23.1238}\,\mathrm{kA} - \mathrm{j}\frac{3.3473 + \mathrm{j}13.2997}{6.3469 + \mathrm{j}23.1238} \cdot 2 \cdot 0.2625\,\mathrm{kA} - \mathrm{j}0.5249\,\mathrm{kA}$$
$$= (0.7711 - \mathrm{j}2.8094)\,\mathrm{kA} + (0.0064 - \mathrm{j}0.8250)\,\mathrm{kA} = 0.7775 - \mathrm{j}3.6345\,\mathrm{kA}$$

The first part refers to the contribution caused by the network feeder ($I''_\mathrm{kVS} = 2.9134\,\mathrm{kA}$), while the second is caused exclusively by the feeder of the full-size converters ($I''_\mathrm{kCS} = 0.8251\,\mathrm{kA}$). The following applies to the absolute value of the total short-circuit current:

$$I''_\mathrm{kF2} = 3.7167\,\mathrm{kA}$$

If only absolute values are used in the calculation according to Eq. (13.1), the following results:

$$I''_{\text{kF2}} = \frac{1.1 \cdot 110/\sqrt{3}}{23.9790}\,\text{kA} + \frac{13.7144}{23.9790} \cdot 0.5249\,\text{kA} + 0.5249\,\text{kA} =$$
$$2.9134\,\text{kA} + 0.8251\,\text{kA} = 3.7385\,\text{kA}$$

In this case, the calculation with absolute values provides a result that is 0.59% greater than the value with complex variables. When calculating the largest short-circuit current, the result is therefore on the safe side.

The voltage distribution at the network nodes, taking into account the network feeder and the full-size converters, can be determined using the system of Eq. (13.19).

$$\begin{pmatrix} \underline{U}_1^* \\ \underline{U}_2^* \\ \underline{U}_3^* \\ \underline{U}_4^* \end{pmatrix} = \begin{pmatrix} \underline{Z}_{11} & \underline{Z}_{12} & \underline{Z}_{13} & \underline{Z}_{14} \\ \underline{Z}_{21} & \underline{Z}_{22} & \underline{Z}_{23} & \underline{Z}_{24} \\ \underline{Z}_{31} & \underline{Z}_{32} & \underline{Z}_{33} & \underline{Z}_{34} \\ \underline{Z}_{41} & \underline{Z}_{42} & \underline{Z}_{43} & \underline{Z}_{44} \end{pmatrix} \cdot \left[\begin{pmatrix} 0 \\ -\underline{I}''_{\text{kF2}} \\ 0 \\ 0 \end{pmatrix} + \begin{pmatrix} 0 \\ \underline{I}_{\text{CS2}} \\ \underline{I}_{\text{CS3}} \\ \underline{I}_{\text{CS4}} \end{pmatrix} \right] \tag{13.34}$$

The following then results for the individual node voltage levels:

$$\begin{aligned} \underline{U}_1^* &= -\underline{Z}_{12} \cdot \underline{I}''_{\text{kF2}} + (\underline{Z}_{12} \cdot \underline{I}_{\text{CS2}} + \underline{Z}_{13} \cdot \underline{I}_{\text{CS3}} + \underline{Z}_{14} \cdot \underline{I}_{\text{CS4}}) \\ \underline{U}_2^* &= -\underline{Z}_{22} \cdot \underline{I}''_{\text{kF2}} + (\underline{Z}_{22} \cdot \underline{I}_{\text{CS2}} + \underline{Z}_{23} \cdot \underline{I}_{\text{CS3}} + \underline{Z}_{24} \cdot \underline{I}_{\text{CS4}}) \\ \underline{U}_3^* &= -\underline{Z}_{32} \cdot \underline{I}''_{\text{kF2}} + (\underline{Z}_{32} \cdot \underline{I}_{\text{CS2}} + \underline{Z}_{33} \cdot \underline{I}_{\text{CS3}} + \underline{Z}_{34} \cdot \underline{I}_{\text{CS4}}) \\ \underline{U}_4^* &= -\underline{Z}_{42} \cdot \underline{I}''_{\text{kF2}} + (\underline{Z}_{42} \cdot \underline{I}_{\text{CS2}} + \underline{Z}_{43} \cdot \underline{I}_{\text{CS3}} + \underline{Z}_{44} \cdot \underline{I}_{\text{CS4}}) \end{aligned} \tag{13.35}$$

Taking into account the values of the impedance matrix and the current relationships, the simplified result is:

$$\underline{Z}_{12} = \underline{Z}_{13} = \underline{Z}_{14} \quad \underline{Z}_{23} = \underline{Z}_{24} = \underline{Z}_{32} = \underline{Z}_{42} \tag{13.36}$$

$$\underline{Z}_{33} = \underline{Z}_{34} = \underline{Z}_{43} \quad \underline{I}_{\text{CS3}} = \underline{I}_{\text{CS4}} \tag{13.37}$$

With

$$\underline{I}''_{\text{kF2}} = \frac{c \cdot U_{\text{n}}/\sqrt{3}}{\underline{Z}_{22}} + 2 \cdot \frac{\underline{Z}_{23}}{\underline{Z}_{22}} \cdot \underline{I}_{\text{CS3}} + \underline{I}_{\text{CS2}} \tag{13.38}$$

follows according to Eq. (13.19):

$$\begin{aligned} \underline{U}_1^* &= -\frac{\underline{Z}_{12}}{\underline{Z}_{22}} \cdot \left[c \cdot U_{\text{n}}/\sqrt{3} + 2 \cdot (\underline{Z}_{23} - \underline{Z}_{22}) \cdot \underline{I}_{\text{CS3}} \right] \\ \underline{U}_2^* &= -c \cdot U_{\text{n}}/\sqrt{3} \\ \underline{U}_3^* &= -\frac{\underline{Z}_{23}}{\underline{Z}_{22}} \cdot \left[c \cdot U_{\text{n}}/\sqrt{3} + 2 \cdot \left(\underline{Z}_{23} - \frac{\underline{Z}_{22} \cdot \underline{Z}_{33}}{\underline{Z}_{23}} \right) \cdot \underline{I}_{\text{CS3}} \right] \\ \underline{U}_4^* &= -\frac{\underline{Z}_{23}}{\underline{Z}_{22}} \cdot \left[c \cdot U_{\text{n}}/\sqrt{3} + \left(2 \cdot \underline{Z}_{23} - \frac{\underline{Z}_{22} \cdot (\underline{Z}_{33} + \underline{Z}_{44})}{\underline{Z}_{23}} \right) \cdot \underline{I}_{\text{CS3}} \right] \end{aligned} \tag{13.39}$$

Taking into account the values of the impedance matrix, the real voltage distribution U_i can be calculated according to Eq. (13.14) when the full-size converters are feeding in:

$$U_1 = 0.8679 \cdot c \cdot U_\mathrm{n}/\sqrt{3} \qquad U_2 = 0.0$$
$$U_3 = 0.5088 \cdot c \cdot U_\mathrm{n}/\sqrt{3} \qquad U_4 = 0.5465 \cdot c \cdot U_\mathrm{n}/\sqrt{3}$$

If the full converters are not taken into account ($I_{\mathrm{CS}i} = 0$), the following relationships result for the node voltages if the absolute value of the full-size converters is set to zero in Eq. (13.39):

$$\underline{U}_1^* = -\frac{\underline{Z}_{12}}{\underline{Z}_{22}} \cdot c \cdot U_\mathrm{n}/\sqrt{3} \qquad \underline{U}_2^* = -c \cdot U_\mathrm{n}/\sqrt{3}$$
$$\underline{U}_3^* = -\frac{\underline{Z}_{23}}{\underline{Z}_{22}} \cdot c \cdot U_\mathrm{n}/\sqrt{3} \qquad \underline{U}_2^* = -\frac{\underline{Z}_{23}}{\underline{Z}_{22}} \cdot c \cdot U_\mathrm{n}/\sqrt{3}$$

After inserting the values, the following results for the node voltages U_i during the short-circuit duration under these conditions:

$$U_1 = 0.8567 \cdot c \cdot U_\mathrm{n}/\sqrt{3} \quad U_2 = 0.0$$
$$U_3 = U_4 = 0.4284 \cdot c \cdot U_\mathrm{n}/\sqrt{3}$$

The inductive supply of the full-size converters during the short-circuit phase raises the node voltages, which of course does not apply to the short-circuit location.

13.2.5.2 Initial Symmetrical Short-Circuit Current with Correction of the Full-Size Converter Feeder

In contrast to the illustration in Sect. 13.2.5.1, in this case the short-circuit current contribution of the full-size converters is a result of the voltage reduction during the short-circuit, so that a distinction can be made between a short-circuit near to the generator and a short-circuit far from the generator in relation to the full-size converters. The maximum current contribution should be equal to the rated current of the full-size converter, $I_\mathrm{CS} = 1.0 \cdot I_\mathrm{rU}$.

According to Sect. 13.4, the feed-in current of a full-size converter during the short circuit can depend on the voltage reduction, as shown in Fig. 13.33, so that the feeder is variable depending on the "electrical" distance. If an amplification factor $k = 2$ is assumed, the supply voltage Δi_CS depends on the voltage change Δu according to Eq. (13.40). In this case, the referenced voltage is given $c = 1.1$, as this is intended for the maximum short-circuit current according to the calculation.

Table 13.1 Data for determining the short-circuit current contribution of the full-size converters as a function of the node (example, Fig. 13.2), first iteration: without full-size converter

Node	U_k (kV)	$\|\Delta u\|$	I_{rU} (kA)	I_{CS} (kA)
2	0.000	1.000	0.5249	–
3	29.925	0.572	0.2624	–
4	29.925	0.572	0.2624	–

U_k voltage during the short circuit
$|\Delta u|$ related voltage change during the short circuit
I_{rU} rated current of the full-size converter in kA

$$\frac{I_{CS}}{I_{rU}} = \Delta i_{CS} = k \cdot \Delta u = 2 \cdot \left| \frac{U_k - U_{ref}}{c \cdot U_n/\sqrt{3}} \right| \tag{13.40}$$

With

I_{CS} short-circuit contribution of the full-size converter,
I_{rU} rated current,
Δi_{CS} related current feeder,
Δu related voltage change,
U_k voltage at the transformer on the HV voltage side during the short-circuit (line-to-earth),
U_{ref} reference voltage at the transformer (HV side) before the short-circuit occurs $= c \cdot U_n/\sqrt{3}$ with $c = 1.1$,
U_n nominal system voltage.

Taking into account the voltages specified above according to Sect. 13.2.5.1 during the short-circuit time (without the contribution of the converters), the short-circuit current contributions of the converters can be determined in a first step if only the influence of the equivalent voltage source at the short-circuit location is considered, Table 13.1. In addition, the rated current of the full-size converters is listed in the table. Only nodes 2–4 are considered, as a converter only feeds in at these nodes. The values listed correspond to the results according to Sect. 13.2.5.1.

Due to the network structure (Fig. 13.5), both network nodes 3 and 4 have the same voltage reduction if only the equivalent voltage source method is used at the short-circuit location (without the contribution of the full converters).

An amplification factor of $k = 2$ results in a relative current feed of $\Delta i_{CS} = 2.0$ (node 2) or $\Delta i_{CS} = 1.14$ (nodes 3 and 4) according to Eq. (13.42). As both exceed the specified value of $\Delta i_{CS} = 1.0$, the converter feeder remains limited to this value ($I_{CS} = 1.0 \cdot I_{rU}$). According to the definition in Fig. 13.33, this fault location is a short circuit near to generator in relation to the use of the full-size converter if the first iteration step is taken as the basis. The consequence is that the short-circuit current at node 2 is equal to $I''_{kF2} = 3.72\text{kA}$, which corresponds to the result of the

Table 13.2 Data for determining the short-circuit current contribution of the full converter as a function of the node (example, Fig. 13.2), second iteration with full-size converter

Node	U_k (kV)	$\|\Delta u\|$	I_{rU} (kA)	I_{CS} (kA)
2	0.00	1.00	0.5249	–j0.5249
3	35.546	0.491	0.2624	–j0.2578
4	38.177	0.454	0.2624	–j0.2380

Legend see Table 13.1

short-circuit current calculation according to Sect. 13.2.5.1 and is therefore not listed again here.

If, in a second iteration step, the voltage at the terminals of the converter installation is recalculated based on the inductive feed-in of the full-size converter according to Table 13.1, the changed values result according to Table 13.2.

As a result of the changed voltages at nodes 3 and 4, the converters have different feeders, taking into account Eq. (13.40), resulting in the following values: $\Delta i_{CS} = 0.98$ (node 3) and $\Delta i_{CS} = 0.91$ (node 4). This results in modified current values for these converters with $I_{CS3} = 0.98{\cdot}0.2624$ A $= 0.2578$ A and $I_{CS4} = 0.91{\cdot}0.2624$ A $= 0.2380$ A. Due to the lower supply of the converters, a new short-circuit current results at the fault location, according to Eq. (13.35), so that the following applies:

$$\underline{I}''_{kF2} = \frac{c \cdot U_n/\sqrt{3}}{\underline{Z}_{22}} + \frac{\underline{Z}_{23}}{\underline{Z}_{22}} \cdot [\underline{I}_{CS3} + \underline{I}_{CS4}] + \underline{I}_{CS2} \tag{13.41}$$

$$\underline{I}''_{kF2} = \frac{1.1 \cdot 110/\sqrt{3}}{6.3469 + j23.1238}\,\text{kA} - j\frac{3.3473 + j13.2997}{6.3469 + j23.1238} \cdot 0.4958\,\text{kA} - j0.5249\,\text{kA}$$

$$\underline{I}''_{kF2} = (0.7711 - j2.8094)\,\text{kA} + (0.0060 - j0.8083)\,\text{kA} = 0.7772 - j3.6177\,\text{kA}$$

This results in the new short-circuit current at node 2 after correction to $I''_{kF2} = 3.7003$kA, which represents a reduction of 0.44% compared to the original value of $I''_{kF2} = 3.7003$kA. As a result of the lower voltage reduction Δu according to Table 13.2, the fault location for the converters at nodes 3 and 4 now represents a short circuit remote from the generator, in accordance with the definition in Fig. 13.33.

13.2.5.3 Short-Circuit Current Contribution as a Function of the Network Voltage at the Node of the Converter

According to [1], the angle of the converter current $\underline{I}_{CS}$ is assumed to be 90° lagging in relation to the equivalent voltage source at the short-circuit location (inductive). In principle, the phase angle at which the converter current feeds in during the short-circuit period must be determined so that the maximum current is expected at the short-circuit location or the current feeding in has a voltage-supporting effect at the connection location. In addition, in practice the converter can only take the phase position of the voltage at the installation location as a benchmark, but this is fundamentally dependent on the load and is not taken into account in the calculation according to the method of the equivalent voltage source at the short-circuit location [1]. The influence of a changed voltage at the installation location and thus a different current injection on the resulting short-circuit current is therefore investigated below. The simplified example in Sect. 3.6.7 is used for the illustration, so that Fig. 13.11 applies

The following equipment data is assumed for the calculations, whereby two different cables are taken into account for the line connection:

- 110 kV network: $U_{\mathrm{n}} = 110\ \mathrm{kV}\ c = 1.1$
  $I''_{\mathrm{k}} = 25\ \mathrm{kA}\ \ R/X = 0.1$
- Transformer: $S_{\mathrm{rT}} = 20.0\ \mathrm{MVA}\ \ u_{\mathrm{kr}} = 14\%$
  $u_{\mathrm{Rr}} = 0.5\%\ t_{\mathrm{r}} = 110\,\mathrm{kV}/10\,\mathrm{kV}$
- 10 kV cable (2x): $\ell = 5$ km
  150 mm^2 (319 A) $R' = 0.206\ \Omega/\mathrm{km}\ \ L' = 0.34\ \mathrm{mH/km}$
  240 mm^2 (417 A) $R' = 0.125\ \Omega/\mathrm{km}\ \ L' = 0.32\ \mathrm{mH/km}$
- Converter: $\underline{I}_{\mathrm{CS}} = -\mathrm{j}1.0\ \mathrm{kA}$

The impedances can be determined from these values, in each case in relation to 10 kV.

$$\begin{aligned}
&\underline{Z}_{\mathrm{Nt}} = 0.002\ \Omega + \mathrm{j}0.023\ \Omega \qquad \underline{Z}_{\mathrm{T}} = 0.025\ \Omega + \mathrm{j}0.700\ \Omega \\
&\underline{Z}_{\mathrm{L}} = 0.515\ \Omega + \mathrm{j}0.267\ \Omega \quad (150\ \mathrm{mm}^2) \\
&\underline{Z}_{\mathrm{L}} = 0.313\ \Omega + \mathrm{j}0.251\ \Omega \quad (240\ \mathrm{mm}^2)
\end{aligned}$$

The total short-circuit current at the fault location F (Fig. 13.11) is calculated according to Eq. (13.18), so that:

$$\underline{I}''_{\mathrm{k}i} = \underline{I}''_{\mathrm{kVS}} + \underline{I}''_{\mathrm{kCS}} = \frac{1}{\underline{Z}_{\mathrm{Nt}} + \underline{Z}_{\mathrm{T}} + \underline{Z}_{\mathrm{L}}} \cdot \left[c \cdot U_{\mathrm{n}}/\sqrt{3} + (\underline{Z}_{\mathrm{Nt}} + \underline{Z}_{\mathrm{T}}) \cdot \underline{I}_{\mathrm{CS}} \right] \tag{13.42}$$

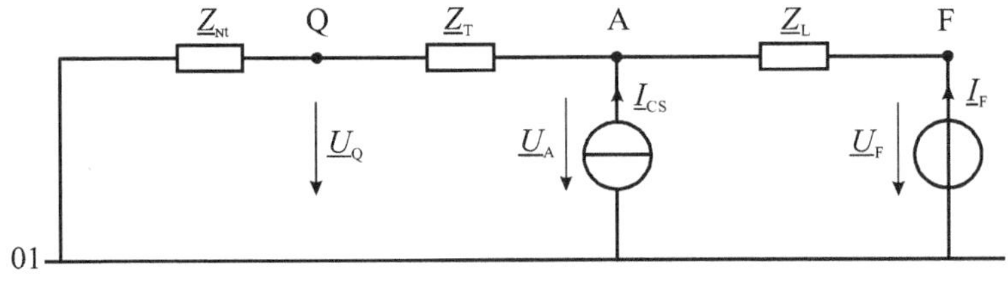

Fig. 13.11 Equivalent circuit for the network circuit according to Fig. 3.18

In a first step, the short-circuit current at the fault location is determined without the contribution of the current source.

$$\underline{I}''_{\mathrm{k}} = \underline{I}''_{\mathrm{kVS}} = (2.705 - \mathrm{j}4.936)\ \mathrm{kA} \qquad I''_{\mathrm{k}} = 5.628\ \mathrm{kA}\ (150\ \mathrm{mm}^2)$$
$$\underline{I}''_{\mathrm{k}} = \underline{I}''_{\mathrm{kVS}} = (2.028 - \mathrm{j}5.814)\ \mathrm{kA} \qquad I''_{\mathrm{k}} = 6.157\ \mathrm{kA}\ (240\ \mathrm{mm}^2)$$

The different cable connections result in different R/X ratios, which will have an influence on the short-circuit currents. The values at the fault location are: $R/X = 0.548$ (150 mm^2) and $R/X = 0.349$ (240 mm^2), this results in a phase angle (between $\underline{U}$ and $\underline{I}$) of $\varphi = -61.28°$ ($\cos\varphi = 0.48$) or $\varphi = -70.77°$ ($\cos\varphi = 0.33$).

If the current source is taken into account, the calculation of the short-circuit current at the fault location is extended according to Eq. (13.18). In this example, an inductive current of $\underline{I}_{\mathrm{CS}}$ is fed in. The additional contribution at the short-circuit location results in:

$$\underline{I}''_{\mathrm{kCS}} = (0.287 - \mathrm{j}0.573)\mathrm{kA} \quad I''_{\mathrm{kCS}} = 0.641\mathrm{kA} \quad (150\ \mathrm{mm}^2)$$
$$\underline{I}''_{\mathrm{kCS}} = (0.206 - \mathrm{j}0.670)\mathrm{kA} \quad I''_{\mathrm{kCS}} = 0.701\mathrm{kA} \quad (240\ \mathrm{mm}^2)$$

This means that the total current at the short-circuit location results from the addition of the two partial currents:

$$\underline{I}''_{\mathrm{k}} = (2.991 - \mathrm{j}5.509)\mathrm{kA} \quad I''_{\mathrm{k}} = 6.268\mathrm{kA} \quad (150\ \mathrm{mm}^2)$$
$$\underline{I}''_{\mathrm{k}} = (2.234 - \mathrm{j}6.484)\mathrm{kA} \quad I''_{\mathrm{k}} = 6.858\mathrm{kA} \quad (240\ \mathrm{mm}^2)$$

The phase angle can be calculated from this, resulting in the following values: $\varphi = -61.50°$ ($\cos\varphi = 0.48$) or $\varphi = -70.99°$ ($\cos\varphi = 0.33$).

In these calculations, it is assumed that the infeed of the current source should be out of phase (−90°) in relation to the equivalent voltage source at the short-circuit location. In the following, a fictitious load flow is assumed before the short circuit occurs, in each case with different power factors. The following load flows are assumed for the network simulation according to Fig. 13.11 (or Fig. 3.18):

- Network: $S_{\mathrm{N}} = 100$ or 200 MVA
- Transformer: $S_{\mathrm{T}} = 10$ or 20 MVA

A power factor of $\cos\varphi = 1.0$; $=0.95$; $=0.9$ is taken into account in each case. The voltages during the load flow can be determined at the node of the converter as shown in Fig. 13.12.

The voltage at node A of the converter can be determined using Eq. (13.45), assuming that the network voltage U_{Nt} is in the real axis. The currents are determined on the basis of the power specifications given above.

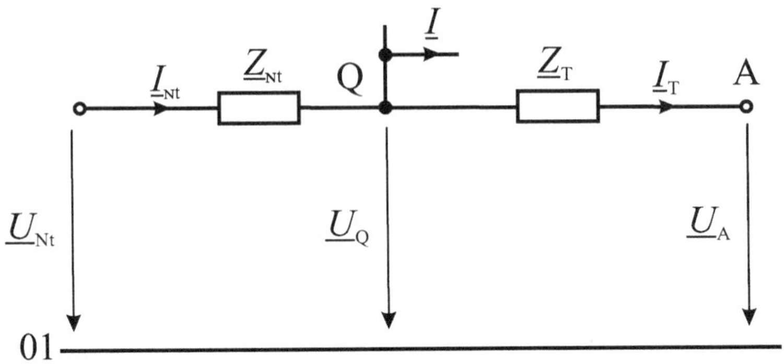

Fig. 13.12 Simplified equivalent circuit for determining the voltages as a function of the load

Table 13.3 Voltage U_A and voltage angle δ at the node of the converter as a function of the load flow (transformers)

	100 MVA/10 MVA		200 MVA/20 MVA	
cos φ	U_A (kV)	δ (°)	U_A (kV)	δ (°)
1.00	6.35	−4.85	6.39	−9.67
0.95	6.18	−4.65	6.05	−9.54
0.90	6.11	−4.42	5.91	−9.17

$$\underline{U}_A = U_{Nt} - \underline{Z}_{Nt} \cdot \underline{I}_{Nt} - \underline{Z}_T \cdot \underline{I}_T \tag{13.43}$$

The following generally applies to the currents as a function of the load factor, with an inductive load flow:

$$\underline{I} = I \cdot \cos(\varphi) - jI \cdot \sin(\varphi) \tag{13.44}$$

Taking into account the impedances and the power factors, the following voltages $\underline{U}_A$ result, depending on the power according to Table 13.3:

If a maximum voltage angle at node A of $\delta_{max} = -10°$ is assumed under these conditions, this must be taken into account when feeding in the current source in the event of a short circuit, as the converter can only utilize the current feed as a function of the voltage angle at its node. If this additional angle is taken into account in relation to the voltage source at the short-circuit location, the following current results for this example:

$$\underline{I}_{CS} = I_{CS} \cdot e^{(-j100°)} = I_{CS} \cdot [\cos(100°) - j\sin(100°)]$$

Under these conditions, the following current values at the short-circuit location result according to Table 13.4 or the deviations in percent.

This example shows that the changes in the maximum short-circuit current due to a different voltage angle can be neglected, whereby the total current is

Table 13.4 Deviations as a function of a changed voltage angle depending on the cable cross-section

Cross-section	I''_{kCS} (kA)	F (%)	I''_{k} (kA)	F (%)
150^2	0.641	0.0	6.256	-0.2
240^2	0.701	0.0	6.844	-0.2

I''_{kCS} short-circuit current contribution of the converter in relation to the fault location
I''_{k} total current at the short-circuit location

slightly lower as a result of the different *R*/*X* ratio (approx. -0.2%). This means that, due to the small difference in the voltage angles (between the connection node and the equivalent voltage source), for simplicity, $\varphi = -90°$ can be used between the equivalent voltage and current of the current source when determining the maximum short-circuit current. Even if the converter injection is doubled to $I_{CS} = 2$ kA, the total short-circuit current is only reduced by approx. -0.3%. The simplified result is therefore on the safe side when determining the maximum short-circuit current. The amount of current in relation to the short-circuit location is always independent of the angle of the current. When calculating the minimum initial short-circuit current, the calculated current is too small compared to the actual current under these conditions. In these considerations, the current before the short-circuit occurs is not taken into account, in contrast to [6], which should always be represented by the application of the voltage factor *c*.

According to [1], however, when calculating the minimum initial symmetrical short-circuit current in a network under the condition of the supply of a full-size converter, it is assumed that in this case there is no current from the generation unit (e.g. wind, solar), which means that the following applies to the current contribution: $I_{CSmin} = 0$.

13.2.5.4 Short-Circuit Current Contribution as a Function of the Network Impedances

According to [1], the current source is injected according to Eq. (13.1) with an angle of 90° lagging behind the equivalent voltage source at the short-circuit location. However, this value of the source current is a simplification and does not generally lead to the maximum short-circuit current contribution for a three-phase fault. It is assumed that the *R*/*X* ratio of the network impedances is almost zero. In the following, the converter $\underline{I}_{CS}$ is determined which leads to the largest contribution depending on the impedances. For this reason, Eq. (13.1) is simplified for a current source so that the following applies:

$$\underline{I}_\mathrm{k}'' = \frac{1}{\underline{Z}_{ii}} \cdot \left(\frac{c \cdot U_\mathrm{n}}{\sqrt{3}} + \underline{Z}_{ij} \cdot \underline{I}_{\mathrm{CS}j} \right) \tag{13.45}$$

With:

$$\underline{Z}_{ij} = \underline{Z} = R + \mathrm{j}X \quad \underline{I}_{\mathrm{CS}j} = \underline{I}_\mathrm{CS} = I_\mathrm{Re} - \mathrm{j}I_\mathrm{Im} \tag{13.46}$$

To calculate the maximum short-circuit current, the phase angle of the current $\underline{I}_\mathrm{CS}$ must be determined as a function of the *R*/*X* ratio of the impedances, subject to the secondary condition:

$$I_\mathrm{CS} = \sqrt{I_\mathrm{Re}^2 + I_\mathrm{Im}^2} \tag{13.47}$$

The condition according to Eq. (13.47) means that the source current should not exceed a specified current of the converter. In general, the maximum of the active voltage at the short-circuit location according to Eq. (13.48) applies:

$$\mathrm{Max} \stackrel{!}{=} c \cdot U_\mathrm{n}/\sqrt{3} + \underline{Z} \cdot \underline{I}_\mathrm{CS} = U + (R + \mathrm{j}X) \cdot (I_\mathrm{Re} - \mathrm{j}I_\mathrm{Im}) \tag{13.48}$$

The following applies to the absolute value:

$$\mathrm{Max} \stackrel{!}{=} \sqrt{(U + R \cdot I_\mathrm{Re} + X \cdot I_\mathrm{Im})^2 + (X \cdot I_\mathrm{Re} - R \cdot I_\mathrm{Im})^2} \tag{13.49}$$

After using Eq. (13.47), the following results after transformation:

$$\mathrm{Max} \stackrel{!}{=} U^2 + 2 \cdot U \cdot R \cdot \sqrt{I_\mathrm{CS}^2 - I_\mathrm{Im}^2} + 2 \cdot U \cdot X \cdot I_\mathrm{Im} + I^2 \cdot \left(R^2 + X^2\right) \tag{13.50}$$

From Eq. (13.50), the condition for the maximum value of the short-circuit current contribution of I_Im or I_Re can be determined by calculating the derivative

$$\frac{\mathrm{d}}{\mathrm{d}I_\mathrm{b}} \mathrm{Max} \stackrel{!}{=} 2 \cdot U \cdot R \cdot \frac{1}{2} \cdot \frac{-2I_\mathrm{Im}}{\sqrt{I_\mathrm{CS}^2 - I_\mathrm{Im}^2}} + 2 \cdot U \cdot X = 0 \tag{13.51}$$

From this follows:

$$I_\mathrm{Im} = \frac{I_\mathrm{CS}}{\sqrt{1 + (R/X)^2}} \quad I_\mathrm{Re} = \sqrt{I_\mathrm{CS}^2 - I_\mathrm{Im}^2} = \frac{I_\mathrm{CS} \cdot (R/X)}{\sqrt{1 + (R/X)^2}} \tag{13.52}$$

$$\underline{I}_{\mathrm{CS}} = \frac{I_{\mathrm{CS}}}{\sqrt{1 + (R/X)^2}} \cdot \{R/X - \mathrm{j}\} \tag{13.53}$$

The inductive part of the current contribution, which leads to the maximum short-circuit current I''_{kCS}, is therefore dependent on the *R*/*X* ratio of the impedance $\underline{Z}_{ij}$ according to Eq. (13.46), and the real part changes accordingly. The same current conditions also result if, based on Eq. (13.48), it is required that the imaginary part of this equation assumes the value zero, since the voltage of the equivalent voltage source is placed in the real axis and thus leads to the maximum voltage value. The following applies to the impedance according to Eq. (13.46) as a function of the *R*/*X* ratio:

$$\underline{Z}_{ij} = \frac{Z_{ij}}{\sqrt{1 + (R/X)^2}} \cdot \{R/X + \mathrm{j}\} \tag{13.54}$$

If the impedance $\underline{Z}_{ij}$ of the network is assumed to be inductive, the additional current that increases the equivalent voltage at the short-circuit location as a result of the converter's feeder increases:

$$\underline{Z}_{ij} \cdot \underline{I}_{\mathrm{CS}} = \frac{Z_{ij} \cdot I_{\mathrm{CS}}}{\sqrt{1 + (R/X)^2}} \cdot (R/X + \mathrm{j}) \cdot (R/X - \mathrm{j}) = Z_{ij} \cdot I_{\mathrm{CS}} \tag{13.55}$$

If the maximum short-circuit current contribution of a current source (converter) is to be determined so that the result is on the safe side, the following can be written for Eq. (13.45):

$$\underline{I}''_{\mathrm{k}} = \frac{1}{\underline{Z}_{ii}} \cdot \left(\frac{c \cdot U_{\mathrm{n}}}{\sqrt{3}} + Z_{ij} \cdot I_{\mathrm{CS}j} \right) \tag{13.56}$$

The second term in Eq. (13.56) therefore increases the equivalent voltage source by the real absolute value $Z_{ij} \cdot I_{\mathrm{CSj}}$. The deviation between the specification that only an inductive current is fed in and the value that the contribution leads to the maximum short-circuit current can be estimated according to Eq. (13.56) by considering the two variants listed below:

- Variant A: Short-circuit current contribution according to [1] with an angle of 90°, related to the equivalent voltage source at the short-circuit location and an inductive current, index A.
- Variant B: The short-circuit current contribution is calculated according to Eq. (13.55) as a function of the *R*/*X* ratio of the impedance $\underline{Z}_{ij}$ according to Eq. (13.45), index B.

For the error $\Delta \underline{I}_{kCS}$ the following applies according to Eq. (13.57), after inserting the initial symmetrical short-circuit current

$$\Delta \underline{I}_{kCS} = \underline{I}''_{kCSA} - \underline{I}''_{kCSB} = \underline{Z}_{ij} \cdot (\underline{I}_{CSA} - \underline{I}_{CSB}) \tag{13.57}$$

The following applies to the feeder of the current sources depending on the variant:

$$\underline{I}_{CSA} = -jI_{CS} \quad \underline{I}_{CSB} = \frac{I_{CS} \cdot (R/X - j)}{\sqrt{1 + (R/X)^2}} \tag{13.58}$$

Substituting Eqs. (13.54) and (13.58) gives the deviation $\Delta \underline{I}_{CS}$::

$$\Delta \underline{I}_{CS} = -Z_{ij} \cdot I_{CS} \cdot \left\{ \frac{R/X + j}{\sqrt{1 + (R/X)^2}} \right\} \cdot \left\{ j + \frac{R/X - j}{\sqrt{1 + (R/X)^2}} \right\} \tag{13.59}$$

With the information given in Sect. 13.2.5.3, the impedance $\underline{Z}_{ij}$ is determined according to Eq. (13.42):

$$\underline{Z}_{ij} = \underline{Z}_{Nt} + \underline{Z}_{T} = (0.027 + j0.723)\Omega \quad Z_{ij} = 0.723\,\Omega \quad R/X = 0.0378$$

With a current contribution of the converter of $I_{CS} = 1.0$ kA in the short-circuit case, the deviation according to Eq. (13.59) is:

$$\Delta \underline{I}_{kCS} = -0.723 \cdot 1.0 \cdot \left\{ \frac{0.0378 + j}{\sqrt{1 + (0.0378)^2}} \right\} \cdot \left\{ j + \frac{0.0378 - j}{\sqrt{1 + (0.0378)^2}} \right\} \text{ kA}$$

$$\Delta I_{kCS} = -0.0273 \quad \text{kA}$$

This means that the influence of the converter contribution is less than $\Delta \underline{I}_{CS} = 0.03$ kA. The maximum total current at the short-circuit location (including the network component) can be calculated for this case according to Eq. (13.56), with the short-circuit impedance Z_{ii} for a cable cross-section of 150 mm^2:

$$\underline{Z}_{ii} = \underline{Z}_{Nt} + \underline{Z}_{T} + \underline{Z}_{L} = (0.542 + j0.990)\Omega$$

$$\underline{I}''_{k} = \frac{1.1 \cdot 10/\sqrt{3} + 0.723 \cdot 1,0}{0.542 + j0.990} \text{ kA} = (3.013 - j5.497) \text{ kA} \quad I''_{k} = 6.269\text{kA}$$

Compared to the calculation according to Sect. 13.2.5.3, the deviation in this example is approx. 0.001 kA, which can be neglected. In principle, the specification according to [1] that the contribution of the current source is applied lagging behind the voltage source leads to a result that may be on the uncertain side, but it should be noted that the R/X ratio of the infeed according to Fig. 13.11 ($\underline{Z}_{Nt} + \underline{Z}_T$) is decisive. In addition, the maximum contribution of the current source is used in the short-circuit current calculation, so that a reduction, e.g. according to Sect. 13.4, is not taken into account.

13.2.5.5 Peak Short-Circuit Current

To calculate the peak short-circuit current taking into account a feed-in from a high-voltage direct current transmission (VSC, HVDC), the initial symmetrical short-circuit current is first determined and then the peak short-circuit current is derived from this according to [1]. In addition, the calculations are compared with transient current curves [5].

According to Fig. 13.13, a three-phase short circuit is fed in from the connected 380 kV network and an HVDC system. The combined overhead line has a length of $\ell = 1$ km, so that the short circuit can almost be regarded as a terminal short circuit in relation to the HVDC system.

The following data is used for the simulation, whereby the inductive current of the transformer corresponds to the short-circuit current contribution of the converter:

- Network: $U_n = 380\ \text{kV}$ $\quad I''_k = 20\ \text{kA}$ $\quad R/X = 0.1$
  $Z_N = 12.0666\ \Omega$ $\quad \underline{Z}_N = 1.2007\ \Omega + \text{j}12.0067\ \Omega$
- HVDC Installation: $t_{rT} = 380\,\text{kV}/370\,\text{kV}$ $\quad u_{kr} = 15\%$
  $S_{rT} = 1000\ \text{MVA}$ $\quad \underline{I}_{rT} = -\text{j}1.519\,\text{kA}$
- Overhead line: $\underline{Z}_L = 0.030\,\Omega + \text{j}0.251\,\Omega$

The following admittance or impedance matrix for determining the short-circuit current contribution of the network can be derived from the data provided:

$$\underline{\mathbf{Y}}_S = \begin{pmatrix} \underline{Y}_N + \underline{Y}_L & -\underline{Y}_L \\ -\underline{Y}_L & \underline{Y}_L \end{pmatrix} \qquad \underline{\mathbf{Z}}_S = \begin{pmatrix} \underline{Z}_N & \underline{Z}_N \\ \underline{Z}_N & \underline{Z}_N + \underline{Z}_L \end{pmatrix} \tag{13.60}$$

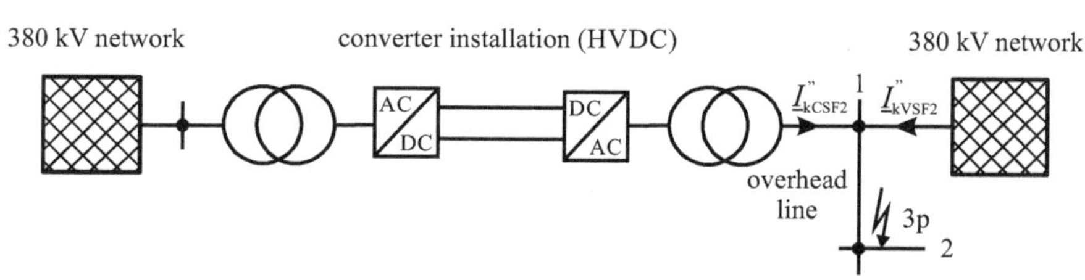

Fig. 13.13 Equivalent circuit for determining the contribution of a high-voltage direct current (HVDC) transmission in VSC technology to the short-circuit current

Here, node 2 represents the fault node, while node 1 represents the node of the HVDC feeder and the network feeder. Taking into account Eq. (13.18), the resulting short-circuit current $\underline{I}''_{\mathrm{kF2}}$ at fault node 2 is as follows:

$$\underline{I}''_{\mathrm{kF2}} = \underline{I}''_{\mathrm{kVSF2}} + \underline{I}''_{\mathrm{kCSF2}} = \frac{c \cdot U_{\mathrm{n}}/\sqrt{3}}{\underline{Z}_{\mathrm{N}} + \underline{Z}_{\mathrm{L}}} + \frac{\underline{Z}_{\mathrm{N}}}{\underline{Z}_{\mathrm{N}} + \underline{Z}_{\mathrm{L}}} \cdot \underline{I}_{\mathrm{rT}} \tag{13.61}$$

With

$\underline{I}''_{\mathrm{kVSF2}}$ short-circuit current contribution of network,
$\underline{I}''_{\mathrm{kCSF2}}$ short-circuit current of the HVDC installation.

In the simulation, it is assumed that the short-circuit current contribution of the HVDC installation is equal to the rated current of the transformer (inductive). Taking into account the equipment data and Eq. (13.61), the following short-circuit currents result:

$$\underline{I}''_{\mathrm{kF2}} = \underline{I}''_{\mathrm{kVSF2}} + \underline{I}''_{\mathrm{kCSF2}} = (1.9570 - \mathrm{j}19.4917)\,\mathrm{kA} + (0.0006 - \mathrm{j}1.4878)\,\mathrm{kA}$$
$$= (1.9560 - \mathrm{j}20.9796)\,\mathrm{kA}$$

or for the absolute value: $I''_{\mathrm{kF2}} = 21.0707\,\mathrm{kA}$

The following applies to the individual contributions of the partial short-circuit currents:

Network part: $I''_{\mathrm{kVSF2}} = 19.5897\,\mathrm{kA}$
HVDC part: $I''_{\mathrm{kCSF2}} = 1.4878\,\mathrm{kA}$

Due to the short overhead line, the HVDC component does not correspond exactly to the rated current of the transformer, as the network impedance is parallel to the overhead line impedance. The following applies to the peak short-circuit current according to Eq. (13.3)

$$i_{\mathrm{pF2}} = \kappa \cdot \sqrt{2} \cdot I''_{\mathrm{kVSF2}} + \sqrt{2} \cdot I''_{\mathrm{kCSF2}} \tag{13.62}$$

From the *R*/*X* ratio, a value $\kappa = 1.7451$ can be determined for the peak short-circuit current, resulting in:

$$i_{\mathrm{pF2}} = 1.7451 \cdot \sqrt{2} \cdot 19.5897\,\mathrm{kA} + \sqrt{2} \cdot 1.4878\,\mathrm{kA} =$$
$$48.3470\,\mathrm{kA} + 2.1041\,\mathrm{kA} = 50.4511\,\mathrm{kA}$$

In relation to the total current of $i_{\mathrm{pF2}} = 50.4511$ kA, the HVDC part makes a contribution of 4.17%. Figure 13.14 shows the current curve $i_{\mathrm{kVSF2}}(t)$ on conductor R for the network part (Fig. 13.14a) and HVDC part $i_{\mathrm{kCSF2}}(t)$ (Fig. 13.14b). Due to the decaying current curve in Fig. 13.14b, this part will be on the safe side. The effective value calculated from the three conductor voltages is also shown. The short

a

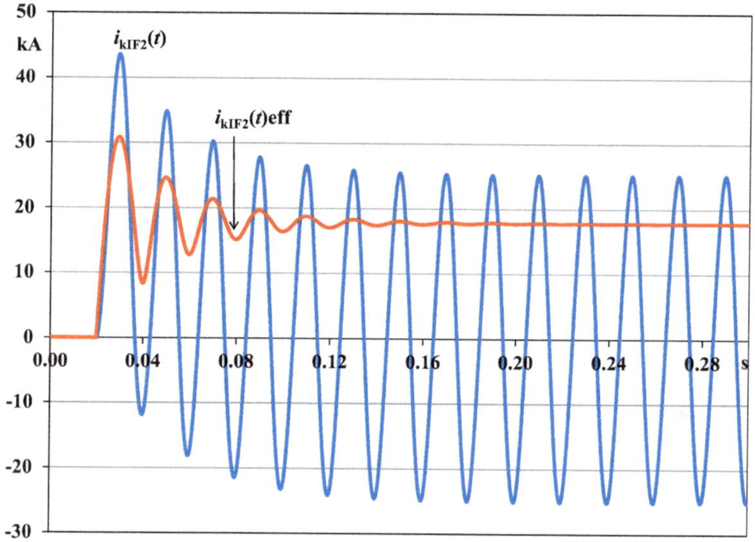

b

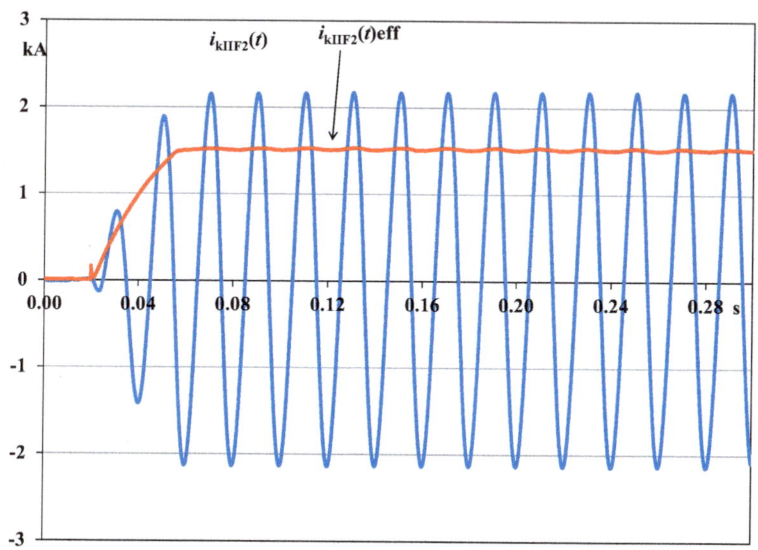

Fig. 13.14 Current curve of phase R for a three-phase short circuit at fault location 2 according to Fig. 13.13 [5]. (**a**) Network part (VS), (**b**) HVDC part (CS). $i_{kVSF2}(t)$ current contribution of the network (transient), $i_{kVSF2}(t)$eff current curve of the network part (rms value), $i_{kCSF2}(t)$ current curve of the HVDC (transient), $i_{kCSF2}(t)$eff current curve of the HVDC (rms value)

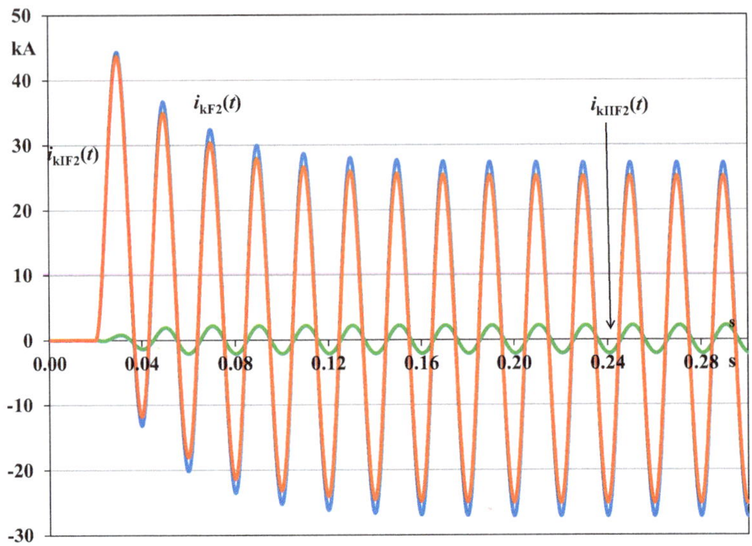

Fig. 13.15 Total and partial currents at fault location F2, conductor R in each case for a fault occurrence at voltage zero crossing [5]. $i_{kIF2}(t)$ network part, $i_{kIIF2}(t)$ HVDC (converter) part, $i_{kF2}(t)$ total current

circuit is always initiated at the voltage zero crossing of phase R, as this leads to the maximum DC displacement and thus to the maximum peak short-circuit current, fault occurrence at $t = 20$ ms.

After the initiation of the short circuit, the current contribution of the HVDC system is regulated so that the rated current is fed in as inductive current. The current reaches its steady-state value after approx. two periods. The short-circuit current contribution of the HVDC installation is independent of the short-circuit occurrence (voltage zero crossing, maximum). Figure 13.15 shows the superposition of the current curves in conductor R. This inductive current of the HVDC installation is superimposed on the network part at the fault location, which is predominantly inductive due to the *R*/*X* ratio and therefore has a DC current displacement.

The following results can be obtained from Figs. 13.14 and 13.15:

- The short-circuit current $i_{kF2}(t)$ at the fault location F2 is mainly characterized by the network part, which results from the ratio of the currents I''_{kVSF2}/I_{rT}, Fig. 13.15.
- The converter part (HVDC) does not have a DC component, so that Eqs. (13.2) and (13.3) are justified, Fig. 13.14b.
- The maximum value of the fault current is only increased from $i_{pVSF2} = 44.574$ kA to $i_{pF2} = 45.129$ kA, which corresponds to an increase of 1.25%.

The deviating value of the network short-circuit results from the fact that a voltage of $c \cdot U_n/\sqrt{3}$ is effective at the short-circuit location in the calculation using the equivalent voltage source, compared to the voltage of $U_1 = 385$ kV.

13.3 LCC Technology

LCC converters (line commutated converters) are equipped with thyristors instead of IGBTs (VSC) and therefore cannot make a controlled contribution to the short-circuit current compared to VSC technology. For this reason, the behavior is fundamentally different in the event of a network short circuit. The contribution of these HVDC installations to a short-circuit in the three-phase system is not currently addressed in the applicable Standard [1].

13.3.1 Current Profile with a Three-Phase Short Circuit

The short-circuit current contribution of an installation with LCC technology depends on the following parameters [7]:

- Operating mode (DC or AC),
- load flow before the short-circuit occurs,
- commutation behavior of the thyristors,
- type and extent of the DC network.

In contrast to full-size converter installations (VSC), LCC applications require additional filters on the three-phase side to compensate for reactive power and to absorb harmonic currents. The filters from the HV side are simulated by the *RLC* elements, Fig. 13.16 shows the corresponding equivalent circuit diagram.

According to Fig. 13.16, a distinction is made between the following parameters, each of which relates to the time-varying currents:

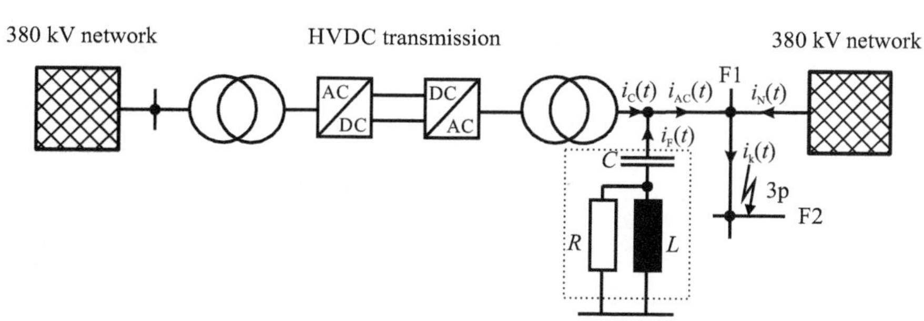

Fig. 13.16 Equivalent circuit of an HVDC connection using LCC technology (the filters of the remote station are not shown)

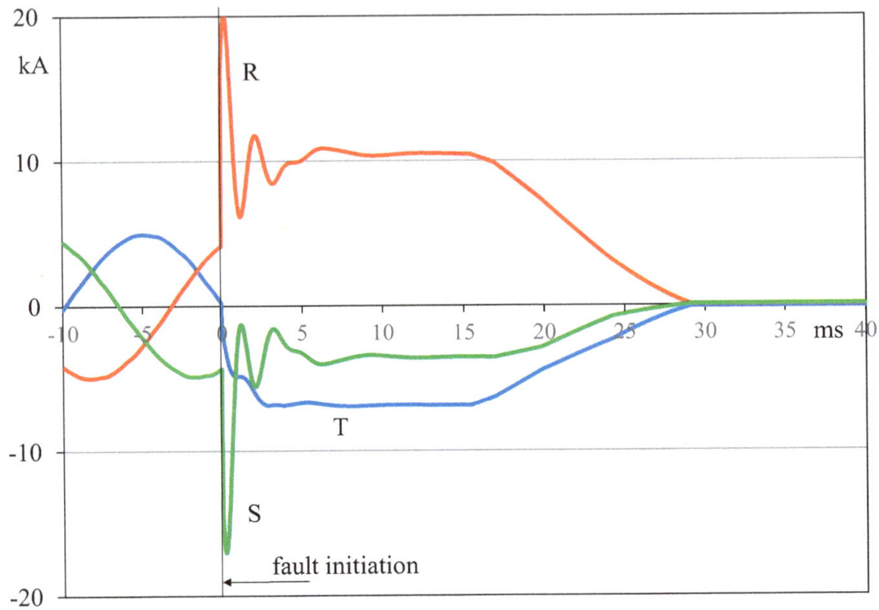

Fig. 13.17 Current profile $i_{AC}(t)$ for a three-phase fault (without network part) [2, 10], voltage zero crossing conductor T

- $i_k(t)$ resulting short-circuit current at the fault location F2,
- $i_C(t)$ contribution of the converter installation on the HV side of the transformer,
- $i_F(t)$ contribution of the filter unit,
- $i_{AC}(t)$ contribution of the entire installation (converter + filter) on the three-phase side,
- $i_N(t)$ contribution of the 380 kV network.

Due to the use of filters on the three-phase side, the short-circuit current of the entire HVDC installation consists of two components, as shown in Fig. 13.17. A short circuit at node F2 as shown in Fig. 13.16 is assumed here, for simplification purposes, the connection between F1 and F2 is assumed to be impedance-free for this example. The short-circuit initiation of the three-phase fault is initiated at the voltage zero crossing of phase T. The maximum current values are reached in the two remaining phases (R, S) due to the simulation (short-circuit in the voltage zero crossing of phase T).

In principle, the Cigre benchmark model according to [8, 9] is used with some deviations. In the next example, the following data from the HVDC system and the 380 kV network are used:

- HVDC installation
  Configuration: bipolar,
  DC voltage: $U_{DC} = \pm 600$ kV,

Rated active power: $P_{\mathrm{rDC}} = 2500$ MW,
- Transformer (4×)
  Rated power: $S_{\mathrm{rT}} = 590$ MVA,
  Short-circuit impedance: $u_{\mathrm{kr}} = 18\%$,
  Zero sequence impedance: $Z_{(0)\mathrm{T}}/Z_{(1)\mathrm{T}} = 1$
- Filter
  Reactive power: $Q_{\mathrm{F}} = 600$ Mvar,
- Network
  Nominal system voltage: $U_{\mathrm{n}} = 380$ kV,
  Short circuit current: $I''_{\mathrm{kN}} = 50$ kA,
  R/*X* ratio: $R/X = 0.1$,
  Zero sequence impedance: $Z_{(0)\mathrm{N}}/Z_{(1)\mathrm{N}} = 1$.

As Fig. 13.17 shows, the current contribution $i_{\mathrm{AC}}(t)$ consists firstly of the contribution of the converter $i_{\mathrm{C}}(t)$, Fig. 13.18a, and secondly of the discharge current of the filter $i_{\mathrm{F}}(t)$, Fig. 13.18b. The magnitude of the discharge current of the filter depends on the voltage level in the three-phase circuit at the time the fault occurs. The maximum value occurs after $t_{\mathrm{pC}} = 0.2$ ms and has already decayed after 10 ms. In contrast to this, the converter current makes a contribution up to approx. 30 ms. This means that the discharge current and the converter current do not contribute to the breaking current and the steady state short-circuit current.

Due to the current profile, the converter current therefore contributes to the peak short-circuit current in the three-phase system, but not the discharge current of the filter. This is also because the peak short-circuit current is initiated at the voltage zero crossing of a phase, while the maximum peak value of the discharge current occurs at the voltage maximum.

The amplitudes of the individual currents before ($t < 0$ ms) and during ($t \geq 0$ ms) the short circuit are shown in Table 13.5. The maximum currents occur in conductors R and S due to the voltage distribution at the occurrence of the short circuit.

According to Fig. 13.18a, three different ranges can be distinguished during the short-circuit current contribution in the converter current, namely:

- Zone I: Discharge of the DC side (capacity), whereby the discharge current is superimposed on the DC load current in this range; the HVDC control is not active in this time range.
- Zone II: Constant short-circuit current contribution on the three-phase side, as a result of commutation errors on the DC side.
- Zone III: The DC side (capacitance) is discharged, the HVDC control system is activated.

The discharge current on the DC side, which leads to a contribution on the three-phase side, can be derived using the equivalent circuit diagram Fig. 13.19 [2, 10]. In this case, three thyristors are active, namely thyristors 1, 2 and 3.

To calculate the discharge current, Fig. 13.19 can be converted into a simplified equivalent circuit, Fig. 13.20.

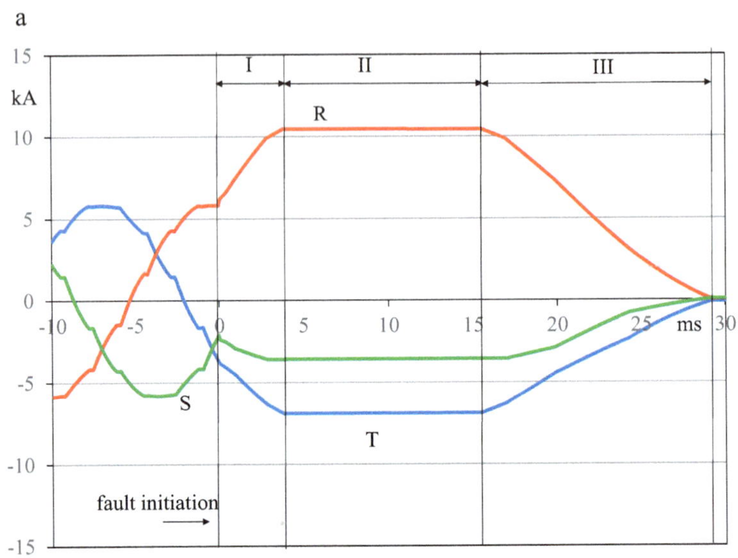

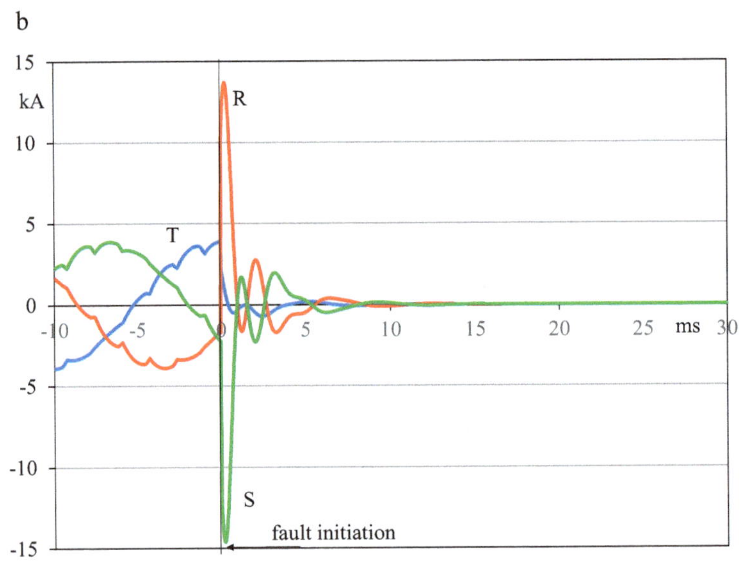

Fig. 13.18 Current contribution of an HVDC installation in LCC technology with a three-phase short circuit [10]. (**a**) Contribution of the converter $i_C(t)$, (**b**) current of the filter $i_F(t)$

Based on the equivalent circuit Fig. 13.20, the discharge current of the capacitance (overhead line, cable) on the DC side can be determined. The following applies to the current $i_C(t)$:

Table 13.5 Maximum values of the individual currents in p.u.; voltage zero crossing of the conductor T at short-circuit occurrence for a three-phase short-circuit on the three-phase side; 1.0 p.u. = 4.94 kA (according to Fig. 13.17)

Conductor	i_{Cmax} (p.u.)	i_{Fmax} (p.u.)	i_{ACmax} (p.u.)
$t < 0$			
R, S, T	1.18	0.79	1.00
$t \geq 0$			
R	2.12	2.77	4.07
S	0.72	2.96	3.45
T	1.39	0.44	1.41

i_{C} converter current
i_{F} filter current
i_{AC} total current of the HVDC installation

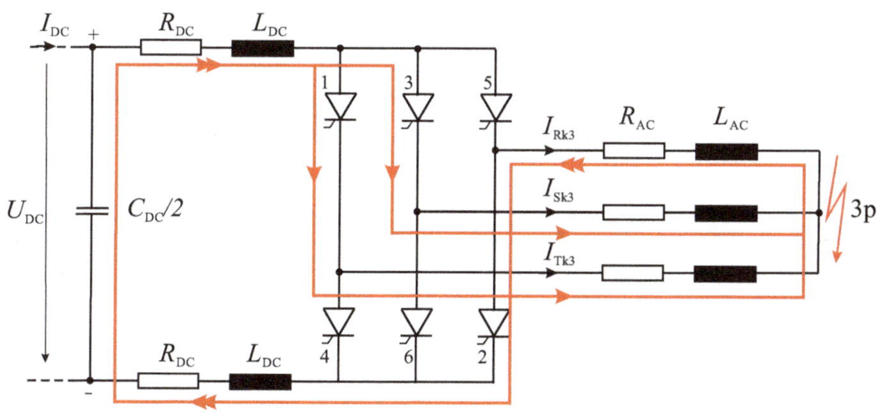

Fig. 13.19 Equivalent circuit diagram for calculating the short-circuit current contribution on the three-phase side

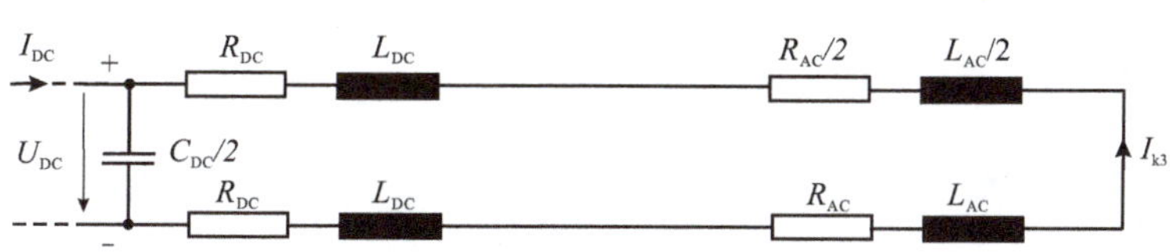

Fig. 13.20 Simplified equivalent circuit [10]

$$i_{\text{C}}(t) = \frac{U_{\text{DC}}}{\omega \cdot L} \cdot \mathrm{e}^{-\delta \cdot t} \cdot \sin(\omega \cdot t) \tag{13.63}$$

$$\delta = \frac{3 \cdot R_{\text{AC}}/2 + 2 \cdot R_{\text{DC}}}{2 \cdot (3 \cdot L_{\text{AC}}/2 + 2 \cdot L_{\text{DC}})} \qquad \omega = \sqrt{\omega_0^2 - \delta^2} \tag{13.64}$$

$$\omega_0^2 - \delta^2 > 0 \qquad \omega_0^2 = \frac{2}{(3 \cdot L_{\text{AC}}/2 + 2 \cdot L_{\text{DC}}) \cdot C_{\text{DC}}} \tag{13.65}$$

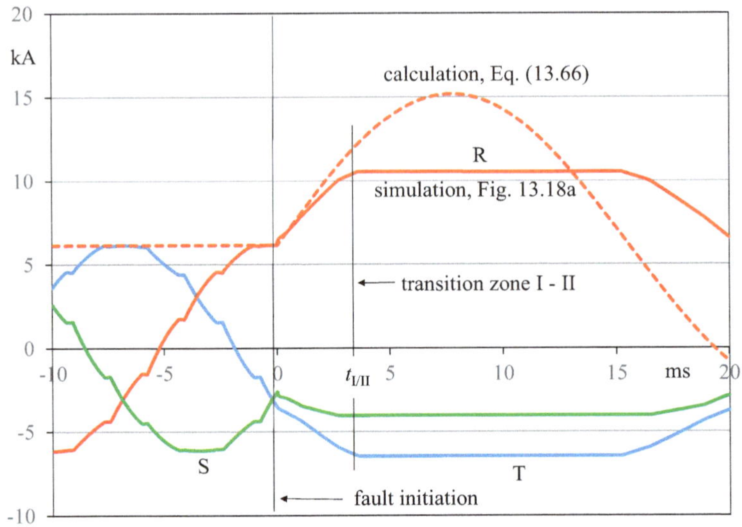

Fig. 13.21 Comparison of the short-circuit current contribution according to Eq. (13.66) calculation, and Fig. 13.18a (simulation) [10]

With

U_{DC} DC voltage,
R_{AC}, L_{AC} impedance of the AC side,
R_{DC}, L_{DC} impedance of the DC side.

The total contribution to the three-phase short-circuit current on the three-phase side $i_{AC}(t)$ results from the superposition of the discharge current $i_C(t)$, Eq. (13.63), with the load current I_{DC} before the short-circuit occurs, taking into account the transformer's transformation ratio for a 12-pulse bridge circuit [11], whereby this ratio is selected as a conservative approximation.

$$i_{AC}(t) = 2 \cdot t_T \cdot \left(1 + \frac{2}{\sqrt{3}}\right) \cdot [I_{DC} + i_C(t)] \qquad (13.66)$$

With

t_T transformation ration of the transformer

Figure 13.21 shows the superposition of the short-circuit current contribution according to Eq. (13.66) with the simulation according to Fig. 13.18a. To determine the maximum short-circuit current contribution of an LCC station, it is sufficient to determine the current at the time of the transition from zone I to II, as the current value then remains constant.

The comparison according to Fig. 13.21 shows that the estimation of the maximum short-circuit current of the converter according to Eq. (13.68) at the time $t = t_{I/II} = 3.33$ ms leads to a result that is on the safe side. In this case, the result is a value of $\Delta i_{Cmax} = 1.43$ kA, which corresponds to a deviation of 13.7%. The deviation is partly due to the assumption of the number of conductive thyristors and the simplification of the commutation process. In relation to the steady state current before the short circuit occurs, this results in a short-circuit current due to the LCC converter of:

Conductor R: $i_{pCR} = 1.71 \cdot \sqrt{2} \cdot I_r = 10.53$ kA
Conductor T: $i_{pCT} = 1.05 \cdot \sqrt{2} \cdot I_r = 6.47$ kA

This means that the largest contribution to the short-circuit current occurs in the conductor (R) that is not considered for the calculation of the peak short-circuit current in a three-phase network, as the maximum peak short-circuit current of the 380 kV network in this example occurs during the voltage zero crossing of phase T (short-circuit initiation). If the same HV network $U_n = 380$ kV as in Sect. 13.2.5.5 is assumed, the above-mentioned contribution of the HVDC system must be related to the network part of $i_{pN} = 48.35$ kA.

13.3.2 Simulation of Short Circuits with Changed Boundary Conditions

In [12], various simulations are shown for a network example according to Fig. 13.16, depending on the type of fault in the three-phase network, the load and the type of line in the DC network (cable $\ell = 100$ km, overhead line $\ell = 500$ km). The results are listed in Table 13.6. Here, not only terminal short-circuits are considered (fault location F1, Fig. 13.10), but the length of the combined overhead line is changed (fault location F2 with 50 and 100 km), in contrast to the presentation in Sect. 13.3.1. The data for the three-phase network correspond to the values in Sect. 13.3.1, where the short-circuit impedance of the network is determined using the voltage factor $c = 1.1$, but the simulations are carried out with the operating voltage $U_b = 380$ kV.

According to Table 13.6, the contribution to the peak short-circuit current at a power of $P_{DC} = 2.5$ GW leads to a slightly lower contribution compared to a load of $P_{DC} = 1.1$ GW. This is due to the fact that the voltage on the DC side decreases with a higher power in these simulations, which has a particular effect on a cable DC network. If, on the other hand, the voltage on the DC side is kept constant regardless of the load, this leads to a higher short-circuit current contribution on the three-phase side.

The contribution to the peak short-circuit current is significantly higher if a cable connection is assumed on the HVDC side, which is a consequence of the higher discharge current due to the higher cable capacity or energy. In addition, the current contribution is reduced in the event of a short circuit that is not initiated directly at

Table 13.6 Maximum peak short-circuit current ($t \leq 10$ ms), $i_{pC}(t)$, of a converter installation in the event of a fault on the three-phase side as a function of the fault location in kA, peak value, I_{rAC} at $P_{DC} = 2.5$ GW or $P_{DC} = 1.1$ GW and $U_{bAC} = 380$ kV [12]

	Cable				Overhead line			
Fault	1.1 GW F1	2.5 GW F1	2.5 GW F2 50 km	2.5 GW F2 100 km	1.1 GW F1	2.5 GW F1	2.5 GW F2 50 km	2.5 GW F2 100 km
k3	11.4	10.5	2.4	1.3	4.6	5.6	1.6	0.8
k2	20.1	18.0	1.8	0.5	8.3	8.1	0.9	0.3
kE2E	22.0	20.7	3.7	1.8	11.7	11.6	2.7	0.9
k1	26.4	24.0	3.3	1.3	17.1	16.3	2.0	1.0

k3 three-phase short circuit
k2 line-to-line short circuit
kE2E line-to-line short circuit with earth contact
k1 line-to-earth short circuit
F1, F2 fault locations according to Fig. 13.16

the terminals of the HVDC installation, fault location F2. It should be noted in the diagram that the maximum peak short-circuit current of the converter installation is not congruent with the operating frequency short-circuit current of the network, except for the line-to-earth fault. The short-circuit initiation on the three-phase side occurs at the voltage zero crossing of the faulty conductor (three-phase, line-to-earth, conductor R: line-to-earth; line-to-line/with earth, conductors S-T). This is in contrast to the representation of the simulation according to Sect. 13.3.1.

The values in Table 13.6 are given in kA; the rated current of the converter installation is $I_{rAC} \approx 3.8$ kA, based on the nominal system voltage of the three-phase side of $U_n = 380$ kV. Table 13.6 lists the maximum current values that occur up to a short-circuit duration of $t \leq 10$ ms. The reason for this is that the maximum value of the system peak short-circuit current occurs in the same time range due to the R_N/X_N ratio. As a result of the required short-circuit current ratio when connecting an LCC installation, the network component should be considerably larger, so that a superposition under these conditions leads to a resulting peak short-circuit current that should be a sufficient estimate on the safe side.

According to [12], the following basic remarks can be derived from Table 13.6, based on the three-pole short-circuit, if only a short circuit at fault location F1 is considered. In addition, Fig. 13.22a–d show the short-circuit current contributions of the converter installation $i_C(t)$.

- Three-phase short circuit (k3): Due to the discharging process of the DC side and the commutation failures, there is a contribution from the converter installation which has decayed after approx. 35 ms, Fig. 13.22a.
- Line-to-line short circuit (k2): The line-to-line short-circuit is larger than the three-phase short circuit and results from different commutation failures, as the voltage breakdown only occurs in two phases. There is only a contribution to the peak short-circuit current, as the process has decayed after approx. 35 ms, Fig. 13.22b.

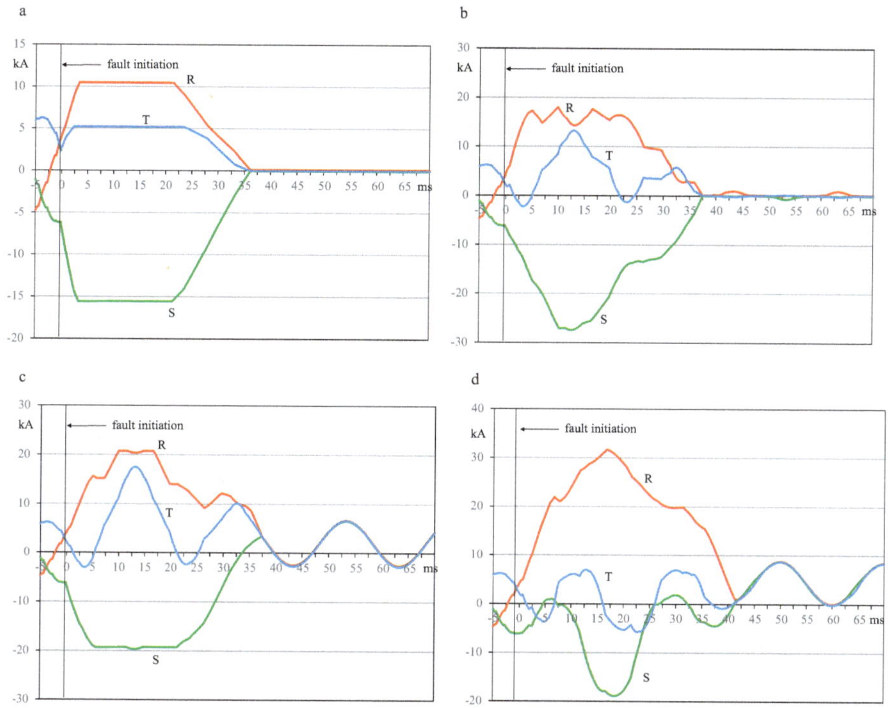

Fig. 13.22 Short-circuit current contribution $i_C(t)$ of the converter installation in the event of a terminal short-circuit on the three-phase side as a function of the fault type [12]. (**a**) Three-phase; (**b**) line-to-line; (**c**) line-to-line with earth contact; (**d**) line-to-earth

- Line-to-line short circuit with earth contact (kE2E): In contrast to the three-phase and line-to-line short circuits, the HVDC installation or the transformer makes a contribution to the breaking current and the steady state short-circuit current, as a constant zero sequence current is fed in due to the transformer neutral point treatment (earthing of the transformer neutral point of the HVDC transformer). If the zero sequence current contribution is deducted, the current curve of the line-to-line short-circuit with earth contact corresponds to the curve without earth contact, Fig. 13.22c.
- Line-to-earth short circuit (k1): In this case, the maximum short-circuit current occurs in the faulty conductor (R), in contrast to the other short-circuit types. In this case, a constant short-circuit current (constant zero sequence current) also occurs due to the neutral point treatment of the transformer. As the voltage breakdown only occurs in one conductor in the case of a line-to-earth short circuit, the commutation faults and therefore the short circuit in the DC system occur later, so that the discharge of the DC side is longer and more pronounced. From this curve it can be seen that after a discharge process, the impedances of the positive and negative sequence systems of the entire converter installation have the value infinity and only the zero sequence system (transformer) is effective,

comparable to the line-to-line short circuit with earth contact, Sect. 13.3.3. According to [12], the same behavior is also shown for a line-to-line short circuit with earth contact, Fig. 13.22d.

While according to Fig. 13.22a–c the maximum values of the discharge current occur in the time range $t \leq 10$ ms and thus correspond to the currents according to Table 13.6, this is not the case with a line-to-earth short circuit. In contrast to the illustration in Fig. 13.22d, Fig. 13.23a shows the current curve $i_N(t)$ of the network. After the initiation of the short-circuit in the voltage zero crossing of phase R, conductor R provides the displaced short-circuit current, while the other conductors first carry the discharge current of the converter. After the decay process ($t > 40$ ms), the conductors not affected by the fault not only carry the zero sequence current of the transformer, but also the charging current of the three-phase filters (Fig. 13.16). Figure 13.23b shows both the partial [$i_C(t)$, $i_N(t)$] and the total short-circuit current [$i_{F1}(t)$] in the event of a line-to-earth fault of conductor R. The following peak short-circuit currents result according to Fig. 13.23b or calculation according to Sect. 13.3.3:

- $i_{pC} = 31.65$ kA, $t_p = 16.80$ ms (contribution of the converter installation),
- $i_{pN} = 111.59$ kA, $t_p = 9.45$ ms (contribution of the HV network, with converter installation),
- $i_{pF1} = 134.81$ kA, $t_p = 9.75$ ms (peak short-circuit current at fault location F1, total current, with converter installation),
- $i_{pF1} = i_{pN} = 123.46$ kA (contribution of the high-voltage network if no converter installation is available, Sect. 13.3.3).

In this case, the contribution of the converter installation leads to a resulting peak short-circuit current at fault location F1, which is increased by approx. 9.2%, so that this proportion must be taken into account when designing the installations if the line-to-earth short-circuit current is relevant to the design.

The results of the simulations can be summarized as follows:

- The converter installation only contributes to the short-circuit current on the three-phase side when the converter is in inverter mode.
- In the case of a balanced (three-phase) or unbalanced fault without earth contact (or in the case of a transformer with an isolated neutral point on the HV side), the converter installation only contributes to the peak short-circuit current, although the maximum value of the current due to the discharge does not occur at the same time as the operating frequency short-circuit current.
- The contribution to the peak short-circuit current results from the discharge of the DC side and is influenced by the time of the commutation errors.
- In this case, the maximum short-circuit current contribution occurs with a line-to-earth short circuit.
- In the case of unbalanced short circuits with earth contact and an earthed transformer neutral point, the zero sequence impedance of the converter transformer, which is parallel to the network impedance, has an effect. This causes a stationary zero sequence current to flow, which is fed from the network, so that a

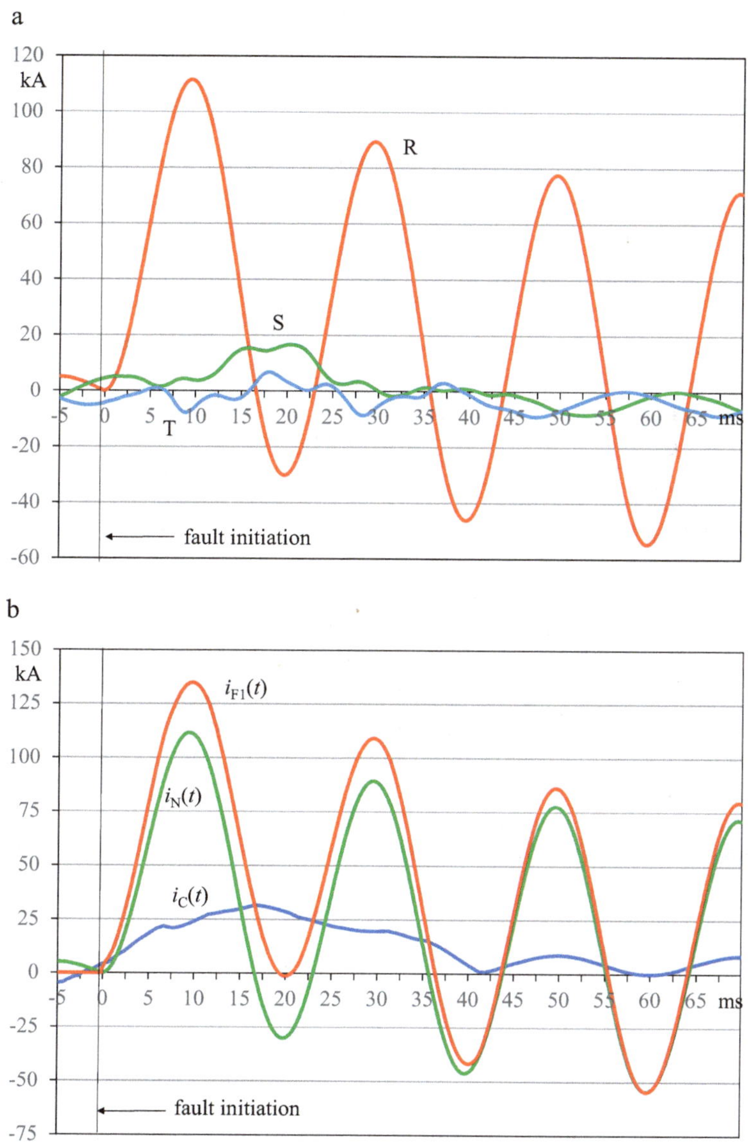

Fig. 13.23 Short-circuit current curve for a line-to-earth short-circuit at fault location F1 according to Fig. 13.16 [12]. (**a**) Contribution of the network $i_N(t)$, (**b**) partial and total currents of the faulty conductor R ($i_{F1}(t)$, $i_N(t)$, $i_C(t)$)

contribution is also made to the breaking current and the steady state short-circuit current.

13.3.3 Determining the Operating Frequency Component of the Network in the Event of a Line-to-Earth Short Circuit

The short-circuit current curve after the discharging process (Fig. 13.22d) can be explained using the component equivalent circuit in Fig. 13.24 after the discharging process has ended and the impedance of the converter installation in the positive and negative sequence system can therefore be assumed to be infinite. This procedure is used to determine the current ratios after the equalization processes have subsided, which means that the steady state short-circuit current and the breaking current are determined. Since the zero sequence systems are decoupled as a result of the neutral point treatment of the converter transformers, only the zero sequence impedance of the transformer acts in parallel with the zero sequence impedance of the network. The three-phase filters are not taken into account as they have no influence on the operating frequency component of the short-circuit current. In this analysis, only the influence of the zero sequence impedance of the converter transformer on the operating frequency current is shown. In contrast to this, Sect. 13.3.5.3 provides a possible simulation of the overall system in order to estimate the peak short-circuit current at the short-circuit location.

The following component currents according to Eqs. (13.67)–(13.69) apply to the steady state short-circuit currents if $\underline{Z}_{(2)\mathrm{N}} = \underline{Z}_{(1)\mathrm{N}}$ is assumed, as shown in Fig. 13.24:

Component currents AC network:

$$\underline{I}_{(1)\mathrm{N}} = \underline{I}_{(2)\mathrm{N}} = \underline{I} \qquad \underline{I}_{(0)\mathrm{N}} = \frac{\underline{Z}_{(0)\mathrm{T}}}{\underline{Z}_{(0)\mathrm{T}} + \underline{Z}_{(0)\mathrm{N}}} \cdot \underline{I} \tag{13.67}$$

Component currents converter installation:

$$\underline{I}_{(1)\mathrm{C}} = \underline{I}_{(2)\mathrm{C}} = 0 \qquad \underline{I}_{(0)\mathrm{C}} = \frac{\underline{Z}_{(0)\mathrm{N}}}{\underline{Z}_{(0)\mathrm{T}} + \underline{Z}_{(0)\mathrm{N}}} \cdot \underline{I} \tag{13.68}$$

Component currents $\underline{I}$:

$$\underline{I} = \frac{c \cdot U_{\mathrm{n}}/\sqrt{3}}{2 \cdot \underline{Z}_{(1)\mathrm{N}} + \dfrac{\underline{Z}_{(0)\mathrm{N}} \cdot \underline{Z}_{(0)\mathrm{T}}}{\underline{Z}_{(0)\mathrm{N}} + \underline{Z}_{(0)\mathrm{T}}}} \tag{13.69}$$

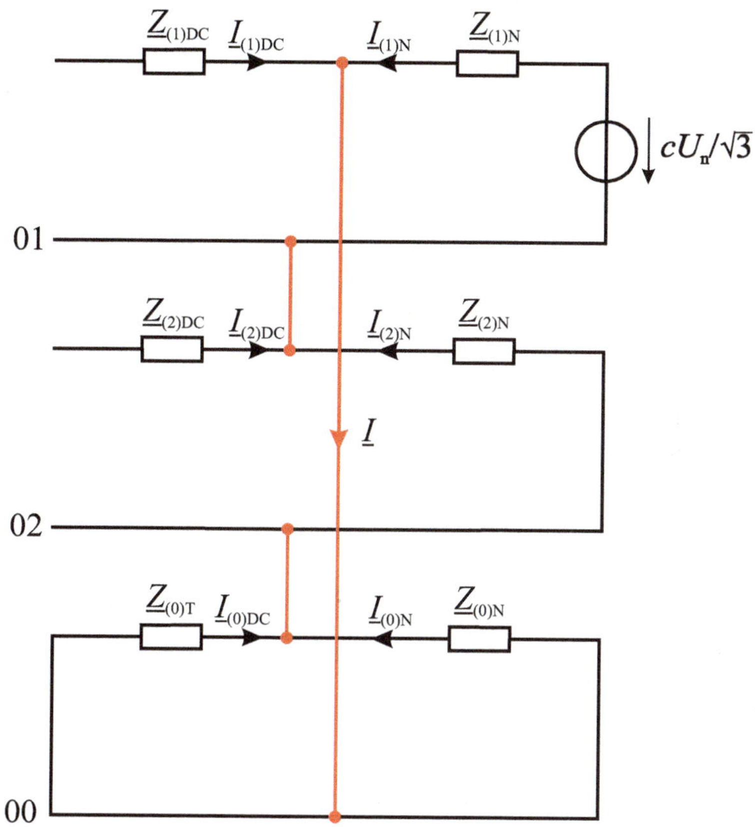

Fig. 13.24 Component equivalent circuit for a line-to-earth short-circuit for the stationary short-circuit current curve

With

U_n nominal system voltage at the short-circuit location,
c voltage factor,
$\underline{Z}_{(1)N}$ positive sequence system, network,
$\underline{Z}_{(0)N}$ zero sequence system, network,
$\underline{Z}_{(0)T}$ zero sequence system, converter transformer.

Equation (9.24) can be used to determine the conductor currents in the original system, so that the system of Eq. (13.70) can be applied.

$$\begin{pmatrix} \underline{I}_R \\ \underline{I}_S \\ \underline{I}_T \end{pmatrix} = \begin{pmatrix} 1 & 1 & 1 \\ 1 & \underline{a}^2 & \underline{a} \\ 1 & \underline{a} & \underline{a}^2 \end{pmatrix} \cdot \begin{pmatrix} \underline{I}_{(0)} \\ \underline{I}_{(1)} \\ \underline{I}_{(2)} \end{pmatrix} \tag{13.70}$$

Short-circuit current contribution of the network (conductor currents):

$$\underline{I}_{RN} = \left(2 + \frac{\underline{Z}_{(0)T}}{\underline{Z}_{(0)N} + \underline{Z}_{(0)T}}\right) \cdot \underline{I} = \frac{\left(2 \cdot \frac{\underline{Z}_{(0)N}}{\underline{Z}_{(0)T}} + 3\right) \cdot c \cdot U_n/\sqrt{3}}{2 \cdot \underline{Z}_{(1)N} \cdot \left(\frac{\underline{Z}_{(0)N}}{\underline{Z}_{(0)T}} + 1\right) + \underline{Z}_{(0)N}} \tag{13.71}$$

$$\begin{aligned} \underline{I}_{SN} = \underline{I}_{TN} &= \left(\frac{\underline{Z}_{(0)T}}{\underline{Z}_{(0)N} + \underline{Z}_{(0)T}} - 1\right) \cdot \underline{I} \\ &= \frac{-c \cdot U_n/\sqrt{3}}{2 \cdot \frac{\underline{Z}_{(1)N}}{\underline{Z}_{(0)N}} \cdot \left(\underline{Z}_{(0)N} + \underline{Z}_{(0)T}\right) + \underline{Z}_{(0)T}} \end{aligned} \tag{13.72}$$

Short-circuit current contribution of the converter (conductor currents):

$$\underline{I}_{RC} = \underline{I}_{SC} = \underline{I}_{TC} = \frac{\underline{Z}_{(0)N}}{\underline{Z}_{(0)T} + \underline{Z}_{(0)N}} \cdot \underline{I} = \frac{c \cdot U_n/\sqrt{3}}{2 \cdot \frac{\underline{Z}_{(1)N}}{\underline{Z}_{(0)N}} \cdot \left(\underline{Z}_{(0)T} + \underline{Z}_{(0)N}\right) + \underline{Z}_{(0)T}} \tag{13.73}$$

Line-to-earth short-circuit current at the fault location:

$$\underline{I}''_{k1} = 3 \cdot \underline{I} = \frac{\sqrt{3} \cdot c \cdot U_n}{2 \cdot \underline{Z}_{(1)N} + \frac{\underline{Z}_{(0)N} \cdot \underline{Z}_{(0)T}}{\underline{Z}_{(0)N} + \underline{Z}_{(0)T}}} \tag{13.74}$$

According to Eq. (13.73), a rectified zero sequence current flows in all conductors of the converter transformer, which is clearly recognizable (Fig. 13.22d), so that the converter itself represents an open end at this point. At a transformer zero sequence impedance value of $\underline{Z}_{(0)T} \rightarrow \infty$, which means that the transformer neutral point is isolated, these currents disappear, which also applies to the current distribution in the conductors S and T of the network. In principle, the network currents of conductors S and T correspond to the current contribution of the converter transformer, Eq. (13.72), if the three-phase filters are not taken into account.

If the equipment data for the network and the transformers of the converter installation (two transformers in parallel that are earthed on the HV voltage side) used in accordance with Sect. 13.3.1, neglecting the resistances, the amounts can be used for simplification. In addition, the voltage factor c and the transformer corrections are not taken into account, as this corresponds to the boundary conditions of the simulation. The following values result if the voltage factor $c = 1.0$ is set:

$$Z_{(1)\mathrm{N}} = Z_{(0)\mathrm{N}} = 4.388\,\Omega \qquad Z_{(1)\mathrm{T}} = Z_{(0)\mathrm{T}} = 11.014\,\Omega$$

According to Eqs. (13.73)–(13.75), the conductor currents of the network or the converter transformers are determined as follows:

$$I_{\mathrm{RN}} = \frac{\left(2\cdot\frac{Z_{(0)\mathrm{N}}}{Z_{(0)\mathrm{T}}}+3\right)\cdot U_{\mathrm{n}}/\sqrt{3}}{2\cdot Z_{(1)\mathrm{N}}\cdot\left(\frac{Z_{(0)\mathrm{N}}}{Z_{(0)\mathrm{T}}}+1\right)+Z_{(0)\mathrm{N}}} = \frac{U_{\mathrm{n}}/\sqrt{3}}{Z_{(1)\mathrm{N}}} = \frac{380\,\mathrm{kV}/\sqrt{3}}{4.388\,\Omega} = 50.0\,\mathrm{kA}$$

$$I_{\mathrm{SN}} = I_{\mathrm{TN}} = I_{\mathrm{RC}} = I_{\mathrm{SC}} = I_{\mathrm{TC}} = \frac{Z_{(0)\mathrm{N}}\cdot U_{\mathrm{n}}/\sqrt{3}}{2\cdot Z_{(1)\mathrm{N}}\cdot\left(Z_{(0)\mathrm{N}}+Z_{(0)\mathrm{T}}\right)+Z_{(0)\mathrm{N}}\cdot Z_{(0)\mathrm{T}}}$$

$$= \frac{U_{\mathrm{n}}/\sqrt{3}}{2\cdot Z_{(1)\mathrm{N}}+3\cdot Z_{(1)\mathrm{T}}} = \frac{380\,\mathrm{kV}/\sqrt{3}}{41.82\,\Omega} = 5.25\,\mathrm{kA}$$

For the line-to-earth short-circuit current at the fault location after discharge of the DC system ($t > 40$ ms), the following results according to Eq. (13.74):

$$I''_{\mathrm{k1}} = \frac{\sqrt{3}\cdot U_{\mathrm{n}}}{2\cdot \underline{Z}_{(1)\mathrm{N}}+\frac{\underline{Z}_{(0)\mathrm{N}}\cdot\underline{Z}_{(0)\mathrm{T}}}{\underline{Z}_{(0)\mathrm{N}}+\underline{Z}_{(0)\mathrm{T}}}} = \frac{\sqrt{3}\cdot U_{\mathrm{n}}}{\underline{Z}_{(1)\mathrm{N}}\cdot\left(2+\frac{\underline{Z}_{(0)\mathrm{T}}}{\underline{Z}_{(0)\mathrm{N}}+\underline{Z}_{(0)\mathrm{T}}}\right)}$$

$$= \frac{\sqrt{3}\cdot 380\,\mathrm{kV}}{11.91\,\Omega} = 55.25\,\mathrm{kA}$$

The short-circuit current at the fault location therefore consists of the network component (50 kA) and the zero sequence current of the converter transformers, which in this case increases the line-to-earth short-circuit current by 5.25 kA. This corresponds to an increase of approx. 10.5%. Under the network conditions ($\underline{Z}_{(0)\mathrm{N}} = \underline{Z}_{(1)\mathrm{N}}$, $R_{\mathrm{N}}/X_{\mathrm{N}} = 0.1$), this results in a peak short-circuit current for a line-to-earth short circuit without the contribution of the converter installation of $i_{\mathrm{p1}} = \sqrt{2}\cdot 1.75\cdot 50\,\mathrm{kA} = 123.46\,\mathrm{kA}$, which is to be compared with the peak short-circuit current taking into account the discharge process of the converter installation, Fig. 13.23b.

13.3.4 *Simulation of the Discharge Process*

The contribution of an LCC installation to the short circuit on the three-phase side is influenced by the discharge process of the line on the DC side. For this reason, an equivalent circuit diagram is presented to simplify matters compared to Fig. 13.19, which is intended to represent the discharge process for a six-pulse converter bridge, Fig. 13.25. The discharge current on the HV side of the transformer is labeled $i_T(t)$ and should represent the contribution of the LCC installation to a three-phase short circuit.

While the DC line is modeled by the impedance $\underline{Z}_{(1)LDC}$, the associated transformer is indicated by $\underline{Z}_{(1)T}$ or $\underline{Z}_{(0)T}$ depending on the component systems. The zero sequence impedance of the transformer is determined as a result of the vector group, Y(N)y or Y(N)d, with a transformation ratio of $t = 1$, depending on the neutral point earthing. The impedances of a connected three-phase system are also listed.

It should always be noted that the unaffected discharge currents are simulated in this section so that the commutation errors are not taken into account. The following equipment data is used to simulate the transient processes:

- Three-phase system
  system voltage: $U_N = 420$ kV
  Impedance ratio (network): $R_{(1)}/X_{(1)} = 0.1$; $X_{(0)}/X_{(1)} = 3.0$
  $R_{(0)}/R_{(1)} = 4.0$
  Initial symmetrical short-circuit current: $I''_{kN} = 50$ kA (three phase)
- Transformer
  rated voltage: $U_{rT} = 420$ kV/420 kV
  short-circuit voltage per cent: $u_k = 18\%$
  rated restive component per cent: $u_{Rr} = 0.3\%$
  rated power: $S_{rT} = 600$ MVA
  zero sequence impedance (transformer): $\underline{Z}_{(0)}/\underline{Z}_{(1)} = \infty$ (Yy)
  $\underline{Z}_{(0)}/\underline{Z}_{(1)} = 1$ (Yd)

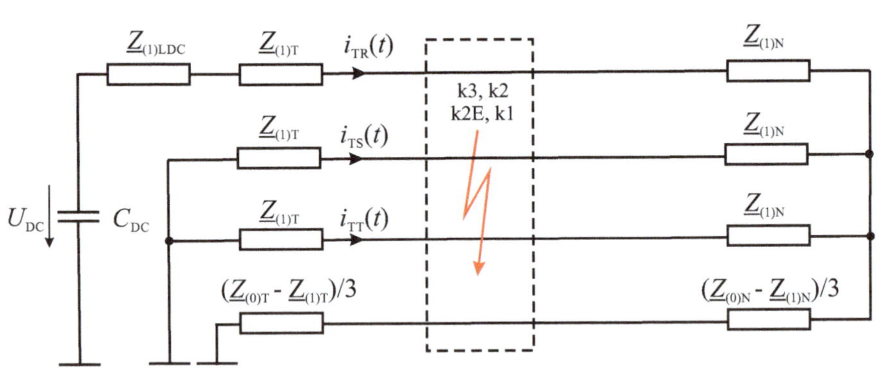

Fig. 13.25 General equivalent circuit diagram for simulating the discharge process of the DC side with different short circuits on the three-phase side. U_{DC} voltage of the DC side, C_{DC} capacitance of the DC side, $\underline{Z}_{(1)LDC}$ impedance of the DC line, $\underline{Z}_{(1)T}$, $\underline{Z}_{(0)T}$ positive and zero sequence impedance of the transformer, $\underline{Z}_{(1)N}$, $\underline{Z}_{(0)N}$ positive and zero sequence impedance of the three-phase network

Table 13.7 Maximum discharge currents depending on the type of fault and the influence of the connected network according to Fig. 13.25

		Without network		With network	
Fault	Vector group	i_{max} (kA)	F	i_{max} (kA)	F
k3	Yy	4.45	1.00	4.31	1.00
	Yd	3.85	0.87	3.73	0.87
k2	Yy	3.96	0.89	4.30	1.00
	Yd	4.23	0.95	3.74	0.87
kE2E	Yy	3.96	0.89	4.30	1.00
	Yd	5.78	1.30	3.74	0.87
k1	Yy	–	0.00	4.28	0.99
	Yd	5.78	1.30	3.74	0.87

k3 three phase short circuit
k2 line-to-line short circuit
kE2E line-to-line short circuit with earth contact
k1 line-to-earth short circuit

- DC side
  voltage: $U_{DC} = 300$ kV
  cable (overhead line): $R' = 122.6$ mΩ/km (61.7 mΩ/km)
  $L' = 0.241$ mH/km (1.9 mH/km
  $C' = 0.217$ μF/km (9.35 nF/km)

For the line parameters, a frequency of $f_0 = 50$ Hz is assumed as a result of the discharge that can occur during the oscillation processes.

Table 13.7 shows the maximum currents on the HV side of the converter transformer as a function of the short-circuit type (k3, k2, kE2E, k1) and the transformer's vector group. A cable connection of $\ell = 400$ km is taken into account as the direct current line.

The current contribution of the HVDC installation for the different fault types is based on the value for a three-phase short circuit with the transformer vector group Yy and is designated with the factor F. When taking the three-phase network into account, only the influence on the discharge process is considered and not the transient phenomena resulting from the contribution of the 50 Hz network.

The results in Table 13.7 show that the discharge current is approx. 30% higher for a line-to-earth fault than for a three-phase fault, without taking the three-phase system into account. In addition, the values depend on the transformer's vector group, as the zero sequence impedance is different. If, in contrast, the three-phase system is taken into account, the maximum discharge currents are equalized, so that the deviation is not more than 13%.

In the following, further simulations are presented for a three-phase short-circuit on the three-phase side by changing additional parameters, but without taking into

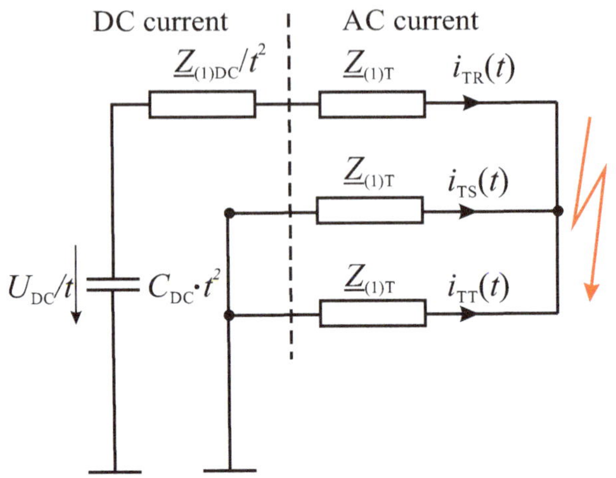

Fig. 13.26 Simplified equivalent circuit diagram to determine the influence of different parameters on the short-circuit current contribution of an HVDC system. *t* transformer ratio of the transformer

account the impedances of the three-phase system (AC) as shown in Fig. 13.25. The parameters include:

- Length of the DC cable connection.
- Changing the transformer power with the same cable connection:
  Two transformers are connected in parallel with the vector group Y(n)y and Y (n)d.
- Deviating DC voltage:
  In this case, the DC voltage and the transformation ratio are changed, with constant transformer power, Y(n)y.

Figure 13.25 can be simplified to show the influence of the various parameters on the three-phase short-circuit current, resulting in Fig. 13.26. Here, the discharging process of a DC line with return line via earth is considered. In addition, the capacitance of the DC line is modeled as a concentrated capacitance C_{DC} at the end of the line. Since different DC voltage levels are taken into account, the transformation ratios must be used when converting the characteristic values of the DC side. The nominal system voltage of the three-phase system is $U_n = 380$ kV in each case.

The dependence of the discharge current (current of the conductor R) on various parameters can be estimated on the basis of Eqs. (13.63)–(13.65), neglecting the resistances according to Eq. (13.75).

$$\begin{aligned} i_{TR}(t) &\approx \frac{U}{\omega_0 \cdot (3 \cdot L_T/2 + L_{DC}/t^2)} \cdot \sin(\omega_0 \cdot t) \\ &= \frac{U_{DC} \cdot t}{\sqrt{(3 \cdot L_T \cdot t^2/2 + L_{DC})/C_{DC}}} \cdot \sin(\omega_0 \cdot t) \end{aligned} \tag{13.75}$$

Table 13.8 Maximum discharge currents i_{max} as a function of DC voltage, cable length and transformer power (Fig. 13.26)

Power (MW)	Voltage (kV)	Length (km)	Vector group	Transformation (kV)	i_{max} (kA)	F
600	300	400	Y(n)y	420/420	4.45	1.00
600	300	300	Y(n)y	420/420	4.07	0.91
600	300	200	Y(n)y	420/420	3.56	0.80
600	300	100	Y(n)y	420/420	2.03	0.60
600	300	50	Y(n)y	420/420	1.93	0.43
600	300	25	Y(n)y	420/420	1.37	0.31
600	200	400	Y(n)y	420/280	2.78	0.63
600	400	400	Y(n)y	420/560	6.25	1.40
600	500	400	Y(n)y	420/700	8.03	1.80
1200	300	400	Y(n)y//Y(n)d	420/420	5.55	1.25
2400	300	400	Y(n)y//Y(n)d	420/420	7.58	1.70

With

$$\omega_0^2 = \frac{1}{(3 \cdot L_T/2 + L_{DC}/t^2) \cdot C_{DC} \cdot t^2} \tag{13.76}$$

The results of the discharge as a function of the parameters can be roughly estimated from Eq. (13.77), whereby, in contrast to this, the transient simulations of the discharge current simulate the line connection as π element, Table 13.8. The deviations of the currents due to deviating parameters are each related to the standard case ($i_{max} = 4.45$ kA) and are characterized by the factor F.

In principle, the discharge currents increase at higher voltages on the DC side, as the energy is dependent on the level of the DC voltage being discharged. This means that no discharge current flows when the length of the DC line approaches zero. As the transformer power increases, the discharge current increases because the transformer reactance decreases.

If the cable connection on the DC side is replaced by an overhead line of the same length (example 1 in Table 13.8), the discharge current is reduced from $i_{TRmax} = 4.45$ kA to $i_{TRmax} = 0.97$ kA due to the changed line parameters.

In the results of the simulations that simulate the discharging process, it should be noted that, when simulating the entire process, the commutation errors cause a limitation on the one hand and the non-consideration of the load current according to Eq. (13.68) causes an increase in the current on the other. The results shown in Table 13.8 use a simple, linear example to show the dependence of various variables on the three-phase short-circuit current contribution in the event of a fault in the three-phase system. However, it does not include the parameters that are affected by the use of a converter bridge, e.g. commutation failures. For this reason, Sect. 13.3.5.1 presents simulation results that arise from the results of using the HVDC Cigre model for transient calculations.

13.3.5 Transient Calculation of the Resulting Contribution to the Peak Short-Circuit Current in the Event of a Short-Circuit in Three-Phase Systems

Sections 13.3.2–13.3.4 estimate the contribution of a single converter installation using thyristor technology in the event of a fault in the three-phase network. This shows that there is only a contribution to the peak short-circuit current. However, when calculating the short-circuit currents in a network, it is useful if the short-circuit current at a fault location can be determined together with the other current and voltage sources. For this reason, the following subsections present a method for performing a joint calculation.

13.3.5.1 Evaluation of the Various Parameters for Determining the Peak Short-Circuit Current

Based on the illustrations in Sect. 13.3.2, the influence of various parameters on the contribution to the short-circuit current has been investigated according to [13]. In the following, these results are briefly presented in order to subsequently implement them in a network calculation by appropriate simulation. The basic equivalent circuit is shown in Fig. 13.16 and a bipolar 12-pulse converter bridge is considered for the simulations, which essentially corresponds to the Cigre model [8, 9]. The essential data for the basic version are listed in Table 13.9.

Two different simulations are examined, one calculation with and one without the converter installation. The difference in the result when determining the peak short-circuit current (maximum value) is defined as the contribution i_{pDC} of the converter installation. This involves not only the influence of the loading and discharged current of the system, but also the effect of the zero impedance of the transformer on the operating frequency short-circuit current, Sect. 13.3.3. In addition, the influence of commutation failures on the short-circuit current behavior is taken into account due to the complete transient simulation.

With regard to the contribution to the short-circuit current at the terminals of a converter installation (point F1, Fig. 13.6), different parameters can be identified according to [13], which are listed in Table 13.10.

Table 13.9 Base system data for the simulations [13]

Equipment	Size	Dimension	Value
DC network	Voltage U_{DC} Power P_{rDC} Inductivity (coil) L_{DC}	kV MW mH	±600 2500 250/80[a]
Three-winding transformer	Throughput power S_{rT} Transformation U_{OS}/U_{US} Short-circuit voltage u_{kr} Resistive part s.c. voltage u_{Rr}	MVA kV/kV % %	625.5 380/240 14/18[a] 0.3
AC network[b]	Voltage U_n s.c. current I_{k3} $Z_{(0)N}/Z_{(1)N}$ $R_{(1)}/X_{(1)}$	kV kA – –	380 50 1 0.1

[a] Long-distance transmission/back-to-back
[b] The short-circuit impedance is calculated using the voltage factor $c = 1.1$, while the simulations are carried out with the operating voltage $U_b = 380$ kV

Table 13.10 Parameters that influence the short-circuit current contribution of a converter installation [13]

Parameter	Dimension	Range
Operating point before fault initiation	p.u.	0...1
DC type	–	Cable, overhead line, back-to-back
Length of DC line	km	50...500
Type of faults	–	k3, k2, kE2E, k1
Power/voltage	MW/kV	2500, ±600 1000, ±500 1000, ±400 1000, ±320 700, ±450

k3 three phase
k2 line-to-line
kE2E line-to-line with earth contact
k1 line-to-earth

The following figures show the additional contribution to the peak short-circuit current, as defined above, depending on the parameters listed.

Operating Point Before Fault Initiation

Figure 13.27 shows the electricity contribution for the converter technologies under consideration. A cable length of $\ell = 300$ km is taken into account for long-distance transmission. Due to the higher capacity of a cable, this current contribution is significantly greater than for the other variants. It can be seen that the largest contribution is made at an output of 80% of the rated output. This operating point is therefore also used as the basis for the other simulations.

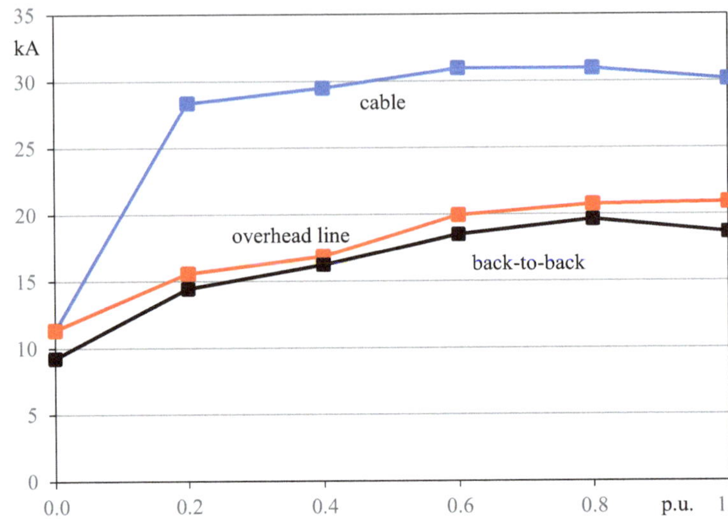

Fig. 13.27 Contribution to the line-to-earth peak short-circuit current as a function of the technology and the load (in p.u.) for a line-to-earth short circuit [13]

Length of DC Line

The influence of the cable length on the DC side is shown in Fig. 13.28. It can be seen that the contribution is relatively constant, so that the cable length can be omitted in the further analysis with regard to the estimation of the maximum current.

Type of Faults

Figure 13.29 shows the dependence of the current contribution as a function of the fault type for a converter installation as an example. It can be seen that the values are almost constant from a power of the converter installation $P_{DC} > 0.2$ p.u. The contribution is greater for a fault with earth contact due to the involvement of the zero sequence system. Furthermore, the additional, operational frequency component of the connected three-phase system also has an effect.

Assessment As a result of Figs. 13.27, 13.28 and 13.29, it can be determined that special consideration of the line length of the DC transmission and the operating point before the short circuit occurs can be dispensed with. However, this does not apply to the technology (cable, overhead line, back-to-back) and type of fault on the AC side. Depending on the network and installation parameters, the maximum basic contribution Δi_p of a converter installation with LCC technology to the peak short-circuit current in the event of a short-circuit at the terminals of the transformer on the HV side can generally be estimated according to [13], Eq. (13.77).

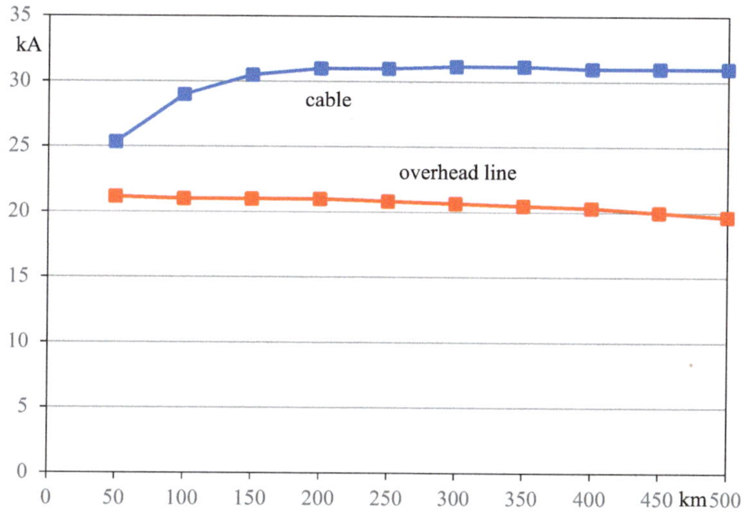

Fig. 13.28 Contribution to the line-to-earth peak short-circuit current as a function of the transmission line [13]

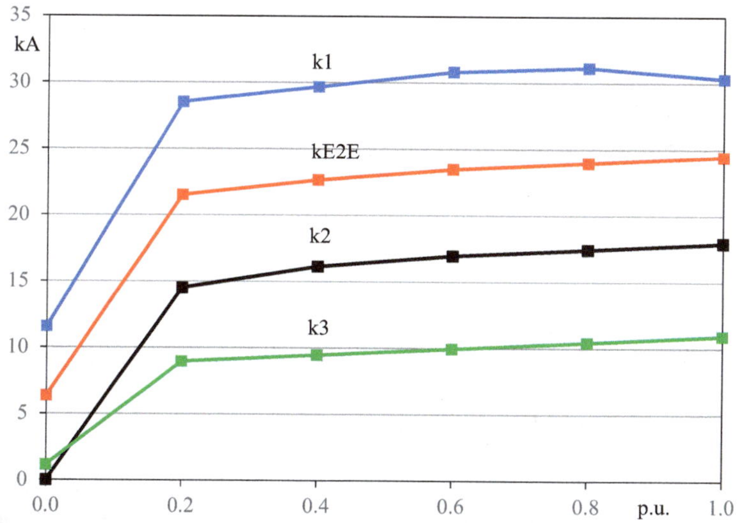

Fig. 13.29 Contribution to the peak short-circuit current depending on the type of fault as a function of the load (DC cable), example [13]. k3 three-phase, k2 line-to-line, kE2E line-to-line with earth contact, k1 line-to-earth

Table 13.11 Factor β, type of fault

Type of fault	Faktor β
Line-to-earth, k1	1.00
Line-to-line with earth, k2E2	0.85
Line-to-line, k2	0.65
Three-phase, k3	0.50

Table 13.12 Factor α, technology

Technology	Faktor α
Cable	1.00
Overhead line	0.70
Back-to-back	0.65

$$\Delta i_{\mathrm{p}} = \alpha \cdot \beta \cdot i_{\mathrm{p0}} \tag{13.77}$$

and

$$i_{\mathrm{p0}} \approx \left[0.01 \cdot \frac{P_{\mathrm{rDC}}}{\mathrm{MW}} + 16 \cdot 10^{-6} \left(\frac{U_{\mathrm{DC}}}{\mathrm{kV}}\right)^2\right] \mathrm{kA} \tag{13.78}$$

With

i_{p0} base current for a line-to-earth short circuit (cable connection),
P_{rDC} rated active power of the inverter system, MW,
U_{DC} DC voltage of the converter installation, kV,
α correction factor for fault type, Table 13.11,
β correction factor for technology, Table 13.12.

The base current according to Eq. (13.78) has been optimized for the determination of the line-to-earth short-circuit current in such a way that it provides the maximum contribution to the peak short-circuit current for this type of fault, taking into account the various boundary conditions according to Table 13.10. In Eq. (13.78), the discharge currents of the connected filters on the AC side are not taken into account, as they occur at a time that deviates from the operating frequency current. The correction factors α, β, which are an estimation on the safe side, describe the reduction of the short-circuit current if deviating boundary conditions (cable, line-to-earth fault) are used. The values of the correction factors are listed in Tables 13.11 and 13.12.

To determine the short-circuit currents in accordance with the IEC Standard on "Calculation of short-circuit currents in three-phase systems" as per Sect. 13.3.5.2, the three-phase peak short-circuit current is used as the output current value. In this case the short-circuit current contribution of the converter installation is unaffected by the parameters of the AC system and thus the currents are decoupled due to the topology, in contrast to the estimation according to Eq. (13.78). According to [14], the contribution of the converter to the peak short-circuit current i_{pk3C} is a value depending on the power and voltage according to Eq. (13.79) for a DC cable connection.

$$i_{pk3C} \approx \left[0.002 \cdot \frac{P_{rDC}}{MW} + 0.01 \cdot \frac{U_{DC}}{kV}\right] kA \tag{13.79}$$

The reduction of the current in the case of a DC overhead line or short-circuit is determined according to the factors in Table 13.12. With the peak short-circuit current determined according to Eq. (13.79), the short-circuit current calculation according to Sect. 13.3.5.2 can be carried out as an estimate.

If, for example, the contribution in the case of a three-phase fault is determined for an LCC installation with a DC cable, a voltage $U_{DC} = \pm 600$ kV and power $P_{DC} = 2500$ MW, the following results are obtained using Eq. (13.79):

$$i_{pk3C} \approx [0.002 \cdot 2500 + 0.01 \cdot 600] \cdot kA = 11\ kA$$

In this case, the estimate with $i_{pk3C} \approx 11$ kA is to be compared with the result of the simulation according to Fig. 13.22a or the table value according to Table 13.6, with a value of $i_{pk3C} \approx 10.4$ kA.

13.3.5.2 Proposal for a Simulation of the Converter Installation (LCC Technology) for Short-Circuit Current Calculation in Three-Phase Networks

Converter installations can be neglected when calculating the initial symmetrical short-circuit current and the peak short-circuit current if their contribution is not higher than 5% of the short-circuit current without these installations. Even if a converter installation is not taken into account, the zero sequence impedance of the transformer should still be taken into account in order to obtain a more accurate result in the event of a short circuit with earth contact.

Based on the contributions of an LCC installation to the peak short-circuit current, a suitable simulation is proposed below in order to include the contribution in the context of a comprehensive short-circuit current calculation in the three-phase system, in accordance with [14]. While the short-circuit current contribution of a VSC installation according to Sect. 13.2 is simulated using a current source, it makes sense to consider the discharge process of the DC side as a voltage source for an LCC installation. The simulation is carried out in the following steps, using the network circuit shown in Fig. 13.30:

- Determination of the three-phase peak short-circuit current based on Sect. 13.3.5.1, Eq. 13.79,
- Determination of the initial symmetrical short-circuit current to simulate a fictitious impedance $\underline{Z}_{DC}$.
- Calculation of the transformer impedance ($\underline{Z}_{TDC}$) if a short circuit with earth contact is determined.
- Calculation of the short-circuit contribution in the event of a short circuit using the equivalent voltage source at the short-circuit location.

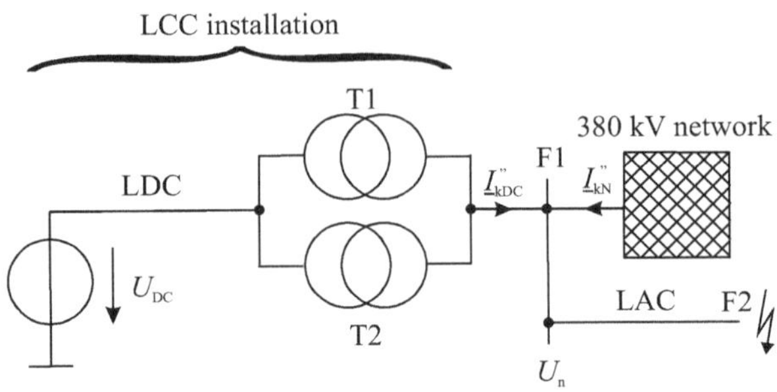

Fig. 13.30 Network circuit for determining the three-phase short-circuit current at network nodes F1 and F2 close to a converter installation (example with two parallel transformers T1, T2). LDC DC line, LAC AC line

Table 13.13 Positive, negative and zero sequence impedances of a converter installation (LCC) as a function of the short-circuit current

Short-circuit current	Symbol	Positive, negative sequence impedance	Zero sequence impedance
Initial symmetrical short circuit current	I''_k	$\underline{Z}_{(1)DC}$	$\underline{Z}_{(0)T}$
Peak short circuit current	i_p	$\underline{Z}_{(1)DC}$	$\underline{Z}_{(0)T}$
Breaking current	I_b	∞	$\underline{Z}_{(0)T}$
Steady state short-circuit current	I_k	∞	$\underline{Z}_{(0)T}$

$\underline{Z}_{(1)DC}$ impedance according Eq. (13.81)
$\underline{Z}_{(0)T}$ impedance according Eqs. (13.83a), (13.83b) and (13.83c)

The impedances are only simulated to calculate the initial symmetrical short-circuit current and the peak short-circuit current. To determine the breaking and the steady state short-circuit current, it should be noted that the positive and negative sequence impedances of the converter installation are infinite, but this does not apply to the zero sequence impedance, which results from the transformer impedances, the vector group and the earthing, so that the data in Table 13.13 should be used if all short-circuit currents are determined.

According to Fig. 13.30, the entire LCC installation consists of two parallel three-winding transformers T1 and T2 (e.g. in Y(N)yd configuration) as a function of the DC power, the DC line LDC and the fictitious voltage U_{DC} of the DC side. The task is therefore to convert the network circuit according to Fig. 13.30 into a simplified equivalent circuit (Fig. 13.31) so that the initial symmetrical short-circuit current can then be determined using the equivalent voltage source at the short-circuit location.

Figure 13.31a shows the equivalent circuit with the two short-circuit locations F1 and F2, while Fig. 13.31b shows the application of the equivalent voltage source method at the short-circuit location. The original voltage sources U_N and U_{DC} are short-circuited and only the equivalent voltage $c \cdot U_n/\sqrt{3}$ acts at the short-circuit locations.

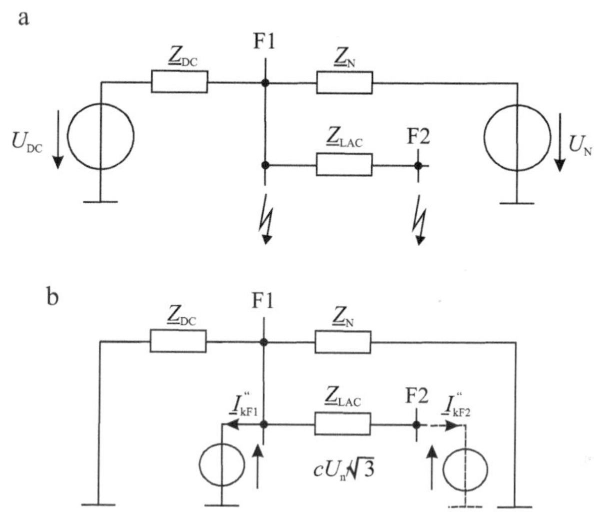

Fig. 13.31 Equivalent circuit for determining the three-phase short-circuit currents, taking into account an LCC installation (line-to-line voltages), short-circuit location F1 or F2. (**a**) Equivalent circuit, (**b**) equivalent voltage source at the short-circuit location

In the following, the individual work steps are described in general terms as listed above and illustrated in Sect. 13.3.5.3 using an example.

Determination of the Three-Phase Initial Symmetrical Short-Circuit Current

According to Sect. 13.3.5.1, for an LCC installation with a power rating of P_{DC} and a voltage of U_{DC}, a peak short-circuit current of $\Delta i_p = i_{pDC}$ can be determined as an additional short-circuit current contribution if a three-phase terminal short circuit is assumed on the AC side. All impedances refer to the HV side of the power converter transformer.

Determination of the Fictitious Initial Symmetrical Short-Circuit Current of the Converter Installation

The *R*/*X* ratio of the converter installation (power converter transformer and DC line) can be estimated using the factor $\kappa = 1.7$, so that the maximum peak short-circuit current occurs at approximately the same time as the network component, thus resulting in the maximum value. This determines the fictitious initial symmetrical short-circuit current as:

$$I''_{kDC} = \frac{i_{pDC}}{\kappa_{DC} \cdot \sqrt{2}} \tag{13.80}$$

With

i_{pDC} peak short-circuit current of the converter system in the event of a three-phase short circuit,
I''_{kDC} fictitious initial symmetrical short-circuit current of a converter system,
κ_{DC} factor $\kappa = 1.7$.

The peak short-circuit current i_{pDC} can either be specified by the manufacturer (simulations) or determined according to Eq. (13.79).

Determination of the Impedances of the Converter Installation

The fictitious impedance $\underline{Z}_{(1)\text{DC}}$ of the converter installation is determined from the initial symmetrical short-circuit current, Eq. (13.80), as:

$$\begin{aligned} Z_{(1)\text{DC}} &= \frac{1.1 \cdot U_{\text{n}}/\sqrt{3}}{I''_{\text{kDC}}} = \frac{1,1 \cdot U_{\text{n}}}{\sqrt{3}} \cdot \frac{\sqrt{2} \cdot \kappa}{i_{\text{pDC}}} \\ \underline{Z}_{(1)\text{DC}} &= R_{\text{DC}} + \text{j}X_{\text{DC}} = \frac{Z_{(1)\text{DC}} \cdot (R_{\text{DC}}/X_{\text{DC}} + \text{j})}{\sqrt{1 + (R_{\text{DC}}/X_{\text{DC}})^2}} \end{aligned} \tag{13.81}$$

With

U_{n} nominal system voltage of the AC network, Fig. 13.30,
$R_{\text{DC}}/X_{\text{DC}}$ fictitious *R*/*X* ratio of the converter installation [$= f(\kappa_{\text{DC}})$].

The following positive and zero sequence impedances can be used to calculate the currents for balanced and unbalanced short-circuit cases.

$$\underline{Z}_{(2)\text{DC}} = \underline{Z}_{(1)\text{DC}} \quad \underline{Z}_{(0)\text{DC}} = \underline{Z}_{(0)\text{T}} \tag{13.82}$$

Calculation of the Zero Sequence Impedance of Transformers

According to Fig. 13.30, the entire converter installation consists of the corresponding transformers (T1, T2) and the line (LDC) including the converter. To simplify the determination of the zero sequence impedance, these transformers can also be divided into two parallel two-winding transformers (Yy, Yd) [15] to take into account the neutral point treatment. For the zero sequence impedance $\underline{Z}_{(0)\text{T}}$, the following simplified equation can be used, depending on the transformer vector group (Table 4.7):

$$\text{Y(N)d} \quad \underline{Z}_{(0)\text{T}} \approx 0.9 \cdot \underline{Z}_{(1)\text{T}} \tag{13.83a}$$

1

$$\underline{Z}_{(0)\text{T}} \approx (7...10) \cdot \underline{Z}_{(1)\text{T}} \tag{13.83b}$$

$$\text{Y(N)dy} \quad \underline{Z}_{(0)\text{T}} \approx 2.4 \cdot \underline{Z}_{(1)\text{T}} \tag{13.83c}$$

[1] three-limb≈ transformer, in case of three single-phase transformer: $Z_{(0)T} \approx 100 \cdot Z_{(1)T}$ The positive sequence impedances of transformers T1 and T2 are determined in accordance with Sect. 4.8.1 using the manufacturer's specifications for:

$$Z_{(1)\text{T}} = \frac{u_{\text{kr}}}{100\%} \cdot \frac{U_{\text{rT}}^2}{S_{\text{rT}}} \tag{13.84}$$

$$R_{(1)\text{T}} = \frac{u_{\text{Rr}}}{100} \cdot \frac{U_{\text{rT}}^2}{S_{\text{rT}}} \qquad X_{(1)\text{T}} = \sqrt{Z_{(1)\text{T}}^2 - R_{(1)\text{T}}^2} \tag{13.85}$$

With

- U_{rT} rated voltage of the transformer (HV),
- S_{rT} rated apparent power of the transformer,
- u_{kr} rated short-circuit voltage (%),
- u_{Rr} rated resistive part of the short-circuit voltage,

The correction factor K_{T} has the value 1.

Determination of the Impedances of the AC Network

The positive, negative, and zero sequence impedance of the AC network is determined in accordance with Sect. 4.1, Eq. (4.1), taking into account the ratios $R_{(1)}/X_{(1)}$ and $Z_{(0)}/Z_{(1)}$.

13.3.5.3 Component Equivalent Circuit Diagrams

In the following, the equivalent circuit diagrams in the positive, negative, and zero sequence systems are shown for two different fault cases according to Fig. 13.31 in order to determine the initial symmetrical short-circuit current and the peak short-circuit current using the equivalent voltage source method at the short-circuit location in a first step.

---

[1] : $\underline{Z}_{(0)\text{T}} \approx 100 \cdot_{(1)\text{T}}$

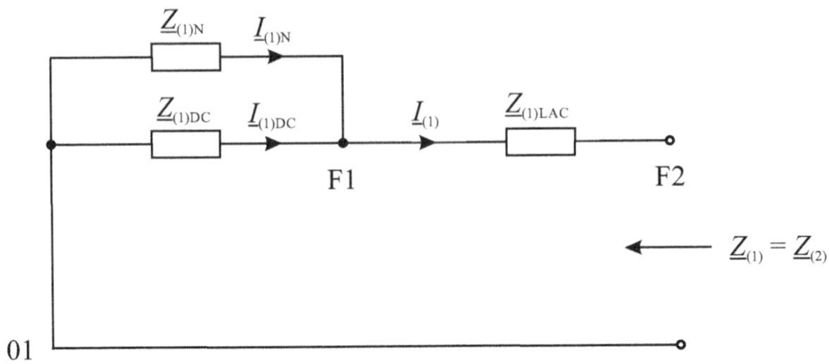

Fig. 13.32 Positive sequence system (= negative sequence system) for calculating the three-phase short-circuit current at fault locations F1 and F2

- Three-phase short circuit, fault location F1 and F2

The three-phase short-circuit current is calculated according to Fig. 13.32 with fault locations F1 and F2.

For the resulting three-phase initial symmetrical short-circuit currents $\underline{I}''_{k3}$ at fault locations F1 and F2, the following equations result with $c = 1.1$: Fault location F1:

$$\underline{I}''_{k3F1} = \frac{c \cdot U_n/\sqrt{3}}{\underline{Z}_{(1)F1}} \tag{13.86}$$

$$\underline{Z}_{(1)F1} = \frac{\underline{Z}_{(1)N} \cdot \underline{Z}_{(1)DC}}{\underline{Z}_{(1)N} + \underline{Z}_{(1)DC}} \tag{13.87}$$

Fault location F1:

$$\underline{I}''_{k3F2} = \frac{c \cdot U_n/\sqrt{3}}{\underline{Z}_{(1)F2}} \tag{13.88}$$

Fault location F2:

$$\underline{Z}_{(1)F2} = \underline{Z}_{(1)LAC} + \frac{\underline{Z}_{(1)N} \cdot \underline{Z}_{(1)DC}}{\underline{Z}_{(1)N} + \underline{Z}_{(1)DC}} \tag{13.89}$$

With

U_n nominal system voltage,
c voltage factor,
$\underline{Z}_{(1)N}$ positive sequence impedance of the 380 kV network,
$\underline{Z}_{(1)DC}$ fictitious positive sequence impedance of the converter installation according to Eq. (13.81),
$\underline{Z}_{(1)LAC}$ positive sequence impedance of the DC line according to Fig. 13.31.

Under these conditions, the contribution of the converter system is determined in each case according to Eqs. (13.90) and (13.91).
Fault location F1:

$$\underline{I}''_{\mathrm{k3F1DC}} = \frac{c \cdot U_{\mathrm{n}}/\sqrt{3}}{\underline{Z}_{(1)\mathrm{DC}}} \tag{13.90}$$

Fault location F2:

$$\underline{I}''_{\mathrm{k3F2DC}} = \frac{\underline{Z}_{(1)\mathrm{N}} \cdot \underline{I}''_{\mathrm{k3F2}}}{\underline{Z}_{(1)\mathrm{N}} + \underline{Z}_{(1)\mathrm{DC}}} = \frac{c \cdot U_{\mathrm{n}}/\sqrt{3}}{\underline{Z}_{(1)\mathrm{DC}} + \underline{Z}_{(1)\mathrm{LAC}} \cdot \left[1 + \underline{Z}_{(1)\mathrm{DC}}/\underline{Z}_{(1)\mathrm{N}}\right]} \tag{13.91}$$

After the discharge process on the DC side, or if the discharge process does not need to be taken into account due to the low short-circuit current contribution, the current contribution of the converter installation is determined with $\underline{Z}_{(1)\mathrm{DC}} \rightarrow \infty$ (Table 13.13) for a three-phase short circuit as follows:

$$\underline{I}''_{\mathrm{k3F1DC}} = \underline{I}''_{\mathrm{k3F2DC}} = 0 \tag{13.92}$$

This condition therefore also applies to the determination of the breaking current and the steady state short-circuit current.

- Line-to-earth short circuit, fault location F1 and F2

The line-to-earth short-circuit current is calculated according to Fig. 13.32 (positive sequence system) and 13.33 (zero sequence system) with fault locations F1 and F2. For the zero sequence system, only the transformers are taken into account due to the vector group and the neutral point earthing (Fig. 13.33).

For the line-to-earth initial symmetrical short-circuit current, follows with $c = 1.1$ for fault locations F1 and F2 ($\underline{Z}_{(2)} = \underline{Z}_{(1)}$): Fault location F1:

$$\underline{I}''_{\mathrm{k1F1}} = \frac{c \cdot \sqrt{3} \cdot U_{\mathrm{n}}}{2\underline{Z}_{(1)\mathrm{F1}} + \underline{Z}_{(0)\mathrm{F1}}} \tag{13.93}$$

$$\underline{Z}_{(0)\mathrm{F1}} = \frac{\underline{Z}_{(0)\mathrm{N}} \cdot \underline{Z}_{(0)\mathrm{T}}}{\underline{Z}_{(0)\mathrm{N}} + \underline{Z}_{(0)\mathrm{T}}} \tag{13.94}$$

Fault location F1:

$$\underline{I}''_{\mathrm{k1F2}} = \frac{c \cdot \sqrt{3} \cdot U_{\mathrm{n}}}{2 \cdot \underline{Z}_{(1)\mathrm{F2}} + \underline{Z}_{(0)\mathrm{F2}}} \tag{13.95}$$

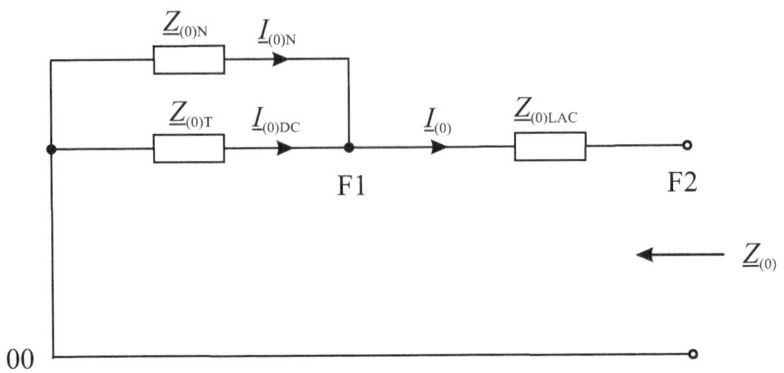

Fig. 13.33 Zero sequence system for calculating the line-to-earth short-circuit current at fault locations F1 and F2

Fault location F2:

$$\underline{Z}_{(0)F2} = \underline{Z}_{(0)LAC} + \frac{\underline{Z}_{0N} \cdot \underline{Z}_{(0)T}}{\underline{Z}_{(0)N} + \underline{Z}_{(0)T}} \tag{13.96}$$

With

U_n	nominal system voltage,
c	voltage factor,
$\underline{Z}_{(1)}$	positive sequence impedances according to Eqs. (13.87) and (13.89),
$\underline{Z}_{(0)N}$	zero sequence impedance of the 380 kV network,
$\underline{Z}_{(0)T}$	zero sequence impedance of transformers according to Eqs. (13.83a), (13.83b) and (13.83c),
$\underline{Z}_{(0)LAC}$	positive sequence impedances of the DC line according to Fig. 13.31.

The conductor currents of the converter installation can be calculated during discharge in accordance with the diagram in Sect. 9.2.2.4 (line-to-earth short-circuit current, Fig. 9.11) as a function of the component current $\underline{I}$.

$$\underline{I}''_{Rk1DC} = \left\{ \frac{\underline{Z}_{(0)N}}{\underline{Z}_{(0)N} + \underline{Z}_{(0)T}} + 2 \cdot \frac{\underline{Z}_{(1)N}}{\underline{Z}_{(1)N} + \underline{Z}_{(1)DC}} \right\} \cdot \underline{I} \tag{13.97}$$

$$\underline{I}''_{Sk1DC} = \underline{I}''_{Tk1DC} = \left\{ \frac{\underline{Z}_{(0)N}}{\underline{Z}_{(0)N} + \underline{Z}_{(0)T}} - \frac{\underline{Z}_{(1)N}}{\underline{Z}_{(1)N} + \underline{Z}_{(1)DC}} \right\} \cdot \underline{I} \tag{13.98}$$

Fault location F1:

$$\underline{I} = \frac{1.1 \cdot U_n / \sqrt{3}}{2 \cdot \underline{Z}_{(1)F1} + \underline{Z}_{(0)F1}} \tag{13.99}$$

Fault location F2:

$$\underline{I} = \frac{1.1 \cdot U_{\mathrm{n}}/\sqrt{3}}{2 \cdot \underline{Z}_{(1)\mathrm{F2}} + \underline{Z}_{(0)\mathrm{F2}}} \tag{13.100}$$

After the discharge process on the DC side (breaking current and steady state short-circuit current), or if the discharge process does not need to be taken into account due to the low short-circuit current contribution, the current contribution of the converter installation in the event of a line-to-earth short circuit to the breaking current is determined as follows:

$$\underline{I}_{\mathrm{Rb1DC}} = \underline{I}_{\mathrm{Sb1DC}} = \underline{I}_{\mathrm{Tb1DC}} = \left\{\frac{\underline{Z}_{(0)\mathrm{N}}}{\underline{Z}_{(0)\mathrm{N}} + \underline{Z}_{(0)\mathrm{T}}}\right\} \cdot \underline{I} \tag{13.101}$$

Fault location F1:

$$\underline{I} = \frac{1.1 \cdot U_{\mathrm{n}}/\sqrt{3}}{2 \cdot \underline{Z}_{(1)\mathrm{N}} + \underline{Z}_{(0)\mathrm{F1}}} \tag{13.102}$$

Fault location F2:

$$\underline{I} = \frac{1.1 \cdot U_{\mathrm{n}}/\sqrt{3}}{2 \cdot (\underline{Z}_{(1)\mathrm{N}} + \underline{Z}_{(1)\mathrm{LAC}}) + \underline{Z}_{(0)\mathrm{F2}}} \tag{13.103}$$

According to Eq. (13.101), after the discharge process, an identical current will flow in the conductors on the HV side, which is caused by the zero sequence impedance of the transformer. Here, Eq. (13.102) corresponds to the calculation according to Sect. 13.3.3, Eq. (13.73).

- Line-to-line short circuit with and without earth contact, fault location F1 and F2

The line-to-line short-circuit currents can be determined in the same way for all short-circuit current types in accordance with the specifications for the component systems.

13.4 Voltage-Dependent Short-Circuit Current Feeder for Converter Installations

According to [3], HVDC installations must be capable of dynamically stabilizing the network voltage so that even in the event of a short circuit in the network, a contribution in the form of an inductive current can be made. The consequence is that, in the event of a short circuit, the inductive feeder of a converter installation is voltage-dependent and therefore, under certain circumstances, not all installations contribute to the maximum short-circuit current at the fault location. In contrast, [1] takes into account in the design of equipment that all feeders contribute the maximum possible short-circuit current. A calculation according to [1] should therefore be on the safe side.

According to [3], the relative voltage change $\Delta u_{1,2}$ in the positive and negative sequence systems according to Eq. (13.104) is decisive for the consideration of the current contribution:

$$\Delta u_{(1,2)} = \frac{\Delta U_{(1,2)}}{U_{(1,2)\text{ref}}} = \frac{U_{(1,2)\text{ref}} - U_{(1,2)}}{U_{(1,2)\text{ref}}} \tag{13.104}$$

With

$U_{(1,2)}$ absolute voltage in the positive and negative sequence system,
$U_{(1,2)\text{ref}}$ reference voltage in the positive and negative sequence systems.

The reference voltage in the positive sequence system is the voltage before the short circuit occurs, while in the negative sequence system this value is zero, as symmetrical operation is assumed. With a constant additional reactive current feeder $\Delta I_{\text{CS}1,2}$, the feeder is proportional to the relative voltage change according to Eq. (13.105), so that the following applies:

$$\Delta i_{\text{CS}} = k \cdot \Delta u \tag{13.105}$$

With

Δi_{CS} relative, additional reactive current injection ($= \Delta I_{\text{CS}1,2}/I_{\text{rU}}$),
I_{rU} rated current of the converter,
k amplification factor for the positive and negative sequence systems ($k = 2$–6),
Δu related voltage reduction according to Eq. (13.104).

The maximum reactive current $I_{\text{CS}1,2}$ during the short circuit must be specified here and should correspond to at least the rated current I_{rU}. Since the short-circuit current calculation using the equivalent voltage source at the short-circuit location does not take into account any load flow before the short circuit occurred, the additional reactive current feeder ΔI_{rU} corresponds to the actual feed I_{CS} in this case. Figure 13.34 shows an example of reactive current feeder depending on the voltage change for the positive and negative sequence systems. The diagram shows the difference between a short circuit near to the generator and one far from it. In the case of a small voltage change ($\Delta u < \Delta u_{\text{k}}$), which corresponds to a short circuit far from generator, the maximum reactive current (i_{kCSmax}) is not fed in, but rather a reduced value according to Eq. (13.105).

In principle, the reactive current feeding can be constant in this process, based on the voltage deviation according to Eq. (13.104), so that no change occurs during the short-circuit duration. In contrast to this, it is possible that the feeding reactive

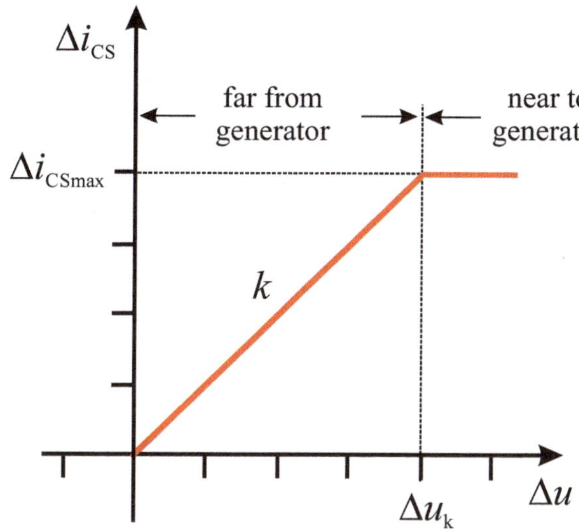

Fig. 13.34 Relative reactive current injection Δi_{CS} as a function of the relative voltage change Δu (absolute value representation)

current is adjusted as a result of the resulting voltage change at the terminals of the converter transformer. In any case, additional inductive feeder current must not lead to a violation of the permissible voltage band.

According to application rule [2], a converter installation in the positive and negative sequence system can be simulated by a source internal reactance. This satisfies the requirement that reactive current feed-in should be proportional to the voltage change at the terminals. In the zero sequence system, it is assumed that the zero sequence system of the converter side has no effect due to the neutral point treatment of the transformer. Figure 13.35 shows the corresponding equivalent circuit diagrams (current sources) in the positive and negative sequence systems, whereby the designations of the variables are taken from Sect. 13.2.2.

According to [2, 7], an equivalent voltage source can be used instead of the current source shown in Fig. 13.35 to calculate the short-circuit current. In this case, the following values are used as reference values:

- Voltage source: $u_0 = \mathrm{j}k$
- Internal impedance: $\underline{Z}_\mathrm{i} = +\mathrm{j}\ 1/k$

The internal reactance of the current source according to Fig. 13.34 is given in relative or real values [2]:

$$x_{CS} = \frac{\Delta u}{i_{CS}} = \frac{\Delta u}{k \cdot \Delta u} = \frac{1}{k} \qquad X_U = \frac{1}{k} \cdot \frac{U_{rU}/\sqrt{3}}{I_{rU}} \tag{13.106}$$

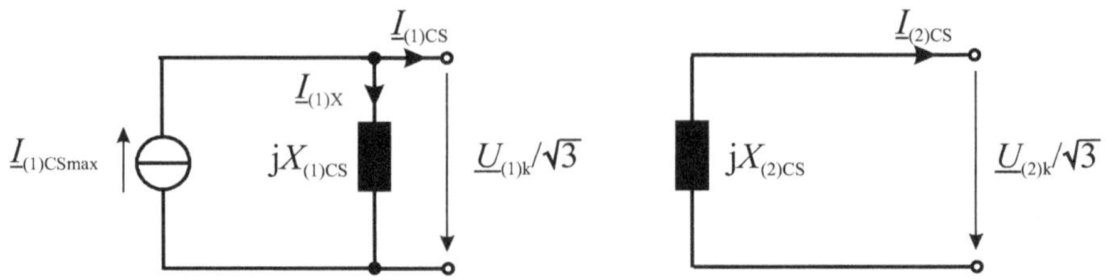

Fig. 13.35 Simulation of the converter installation in the positive and negative sequence system during a short circuit. $\underline{U}_{(1,2)}$ voltage on the HV side of the transformer, $\underline{I}_{(1,2)CS}$ feeder current of the converter installation, $\underline{I}_{(1)CSmax}$ maximum source current in the positive sequence system, $X_{(1,2)CS}$ reactance of the converter installation

With

U_{rU} rated voltage of the converter installation (line-to-line).

The current of source $\underline{I}_{(1)CSmax}$ is determined according to Eq. (13.107).

$$\underline{I}_{(1)kCSmax} = \underline{I}_{x} + \underline{I}_{(1)CS} \tag{13.107}$$

Taking into account the reactance X and the relationship according to Eq. (13.105), if $U_{ref} = U_{rU}$ is set, the following applies:

$$I_{(1)CS} = k \cdot I_{rU} \cdot \Delta u_{(1)} = k \cdot I_{rU} \cdot \frac{U_{rU} - U_{(1)}}{U_{rU}} \tag{13.108}$$

$$\begin{aligned} I_{(1)CSmax} &= \frac{U_{(1)}/\sqrt{3}}{X_{(1)CS}} + k \cdot I_{rU} \cdot \left(1 - \frac{U_{(1)}}{U_{rU}}\right) \\ &= k \cdot I_{rU} \cdot \frac{U_{(1)}/\sqrt{3}}{U_{rU}/\sqrt{3}} + k \cdot I_{rU} \cdot \left(1 - \frac{U_{(1)}}{U_{rU}}\right) = k \cdot I_{rU} \end{aligned} \tag{13.109}$$

According to Eq. (13.109), this means, for example, that with a permissible current feeding of I_r during the short circuit, twice the rated current is reached if the factor $k = 2$ is applied. The consequence is that, in the case of short circuits near to generator, the current feeding is corrected in accordance with Fig. 13.34.

13.5 Conclusions

Depending on their technology, converter installations can contribute to the short-circuit current in the AC network. While LCC installations with thyristor technology involve a discharge process on the DC side and the three-phase filters, the short-circuit current contribution in VSC installations with IGBT technology depends on the control strategy and the requirements of the network operator during the short circuit.

According to the IEC Standard [1], information about the simulation as a current source is provided when VSC installations are considered in the short-circuit current calculation. In contrast, this is not easily possible with LCC installations, but depends on various boundary conditions (DC circuit, fault location, fault type, etc.), so that special considerations must be made in each individual case. For this reason, contrary to this representation, a proposal is made under which conditions the contribution of an LCC system to the short-circuit current for different fault types and time ranges can be simulated as a voltage source, so that this influence can be estimated within a short-circuit current calculation [14].

References

1. IEC 60909-0:2016 Edition 2.0 (2016) Short-circuit current calculations in three-phase a.c. systems—Part 0: calculation of currents. IEC, Geneva
2. VDE-AR-N 4120:2018-11 (2018) Technische Regeln für den Anschluss von Kundenanlagen an das Hochspannungsnetz und deren Betrieb (TAR Hochspannung). VDE, Berlin
3. VDE-AR-N 4131:2019-3 (2019) Technische Regeln für den Anschluss von HGÜ-Systemen und über HGÜ-Systeme angeschlossene Erzeugungsanlagen (TAR HGÜ). VDE, Berlin
4. Jiang-Häfner Y, Hyttinen M, Pääjärvi B (2002) On the short circuit current contribution of HVDC light. In: IEEE/PES transmission and distribution conference and exhibition, Yokohama, Japan, vol 3, pp 1926–1932
5. Balzer G (2016) Short-circuit calculation with fullsize converters according to IEC 60909. In: 21st conference of the electric power supply industry (CEPSI), 23.–27. October 2016, Bangkok, Thailand
6. Schäkel N, Hofmann L (2021) Exakte und genäherte Kurzschlussstromberechnung in Netzen mit Vollumrichteranlagen. Automatisierungstechnik (DE GRUYTER Oldenbourg), vol 69(5), pp 389–400
7. Wassserrab A, Balzer G, Müller H, Krontiris T (2017) Short-circuit calculation in AC networks in case of HVDC stations. In: CIGRE Winnipeg Colloquium, SC A3, B4 & D1, September 30.–October 6., rep. A3-034
8. Cigre Working Group (WG) 14.02 (1991) First benchmark model for HVDC control studies. Electra 135: pp 55–73
9. Cigre Working Group (WG) 14.02 (1994) The Cigre benchmark model—a new proposal with revised parameters. Electra 157: pp 61–66
10. Balzer G, Saciak A, Wasserrab A, Hanson J (2018) Short-circuit calculation in AC networks in case of HVDC stations with line-commutated converters (LCC). In: 20th European conference on power electronics and applications (EPE'18 ECCE EUROPE), September 17-21, 2018, Riga, Latvia, rep. 0261
11. Mohan N, Undeland T, Robbins W (2003) Power electronics—converters, applications and design, 3rd edn. Wiley, New York
12. Saciak A, Balzer G, Hanson J (2018) Short-circuit current contribution in AC networks in case of HVDC stations with line-commutated converters. In: CEPSI 2018, Kuala Lumpur, Malaysia, September, C1
13. Saciak A, Mora U, Balzer G, Hanson J (2021) Calculation method for the contribution of HVDC stations with line-commutated converters (LCC) to the AC short-circuit current. In: 17th

international conference on AC and DC power transmission, December 7–8, Glasgow, Session 3, rep. 002
14. Balzer G, Saciak A, Hanson J (2023) Calculation of short-circuit currents of HVDC installations for the planning and operation of secure AC systems. Speedam ISCHIA 2024, June 19.–21, rep. PEC5962
15. IEC 60099-9:2014-06-26 (2014) Surge arresters—Part 9: metal-oxide surge arresters without gaps for HVDC converter stations. IEC, Geneva

Chapter 14
Influence of Capacitors or Capacitive Compensation Units on the Short-Circuit Current

In principle, capacitors are used in two different applications in electrical power supply networks, firstly to compensate for inductive reactive power as shunt capacitance between the conductor and earth and secondly as series capacitors to compensate for line inductance. Both installations have an influence on the short-circuit current, which are considered in more detail in the following sections.

14.1 Capacitors as Shunt Compensation for Inductive Reactive Power

In the event of a short circuit near to a capacitor, a discharge current will flow which varies depending on the voltage at the time the short circuit occurs. This process therefore always leads to an increase in the short-circuit current. Section 14.1 largely corresponds to the description in [1] with a few additions concerning the reduction of the discharge current of capacitors.

14.1.1 Introduction

Due to the transmission of electrical energy over long distances, which is, for example, a consequence of the increased wind feed-in in the north of Germany and the shutdown of nuclear power plants in the south, voltage problems may occur, so that additional technical measures are required to keep the voltage stable. If overhead lines are operated above their natural capacity, it is generally advisable to use compensation systems, which can be either series or parallel capacitors. As capacitors store energy, a short circuit in the nearby such a system can be expected to contribute to the short-circuit current. In the following, the contribution of parallel

G. Balzer, *Short-Circuit Currents in Three-Phase Systems*,
https://doi.org/10.1007/978-3-658-50710-7_14

capacitors (line-to-earth) is determined and related to the operating frequency short-circuit current, whereby only the three-phase short-circuit current is considered, as this leads to the highest stress on the systems [1, 2].

In addition to the use of compensation installations to provide capacitive reactive power, these systems are also used as filters to reduce harmonics, as this is the case with HVDC systems using thyristor technology (LCC).

14.1.2 Short-Circuit Current Calculation According to IEC 60909-0

The calculation of the short-circuit current in accordance with the applicable Standard IEC 60909-0 "Short-circuit currents in three-phase a.c. systems, Part 0: Calculation of currents, 2016-01" [3] is based on the equivalent voltage source at the short-circuit location. The scope section of the standard states: "However, this does not exclude the use of special calculation methods, e.g. the superposition method, which are adapted to special conditions if they have at least the same accuracy". In addition, Sect. 6.12 of this IEC Standard "Capacitors and non-rotating loads" states: "The discharge current of the shunt capacitors may be neglected for the calculation of the peak short-circuit current". Furthermore, with regard to capacitors of HVDC transmissions, it states that in these cases special considerations are necessary when calculating the short-circuit current.

When formulating the above sentence, the IEC commission TC73, MT1 ("Short-Circuit Calculation") primarily considered installations in MV networks for compensating inductive loads (motors). Because of the increased use in EHV networks (e.g. $U_{\mathrm{n}} = 380$ kV) and high rated power, the question arises as to whether this statement will still be valid in the future.

In addition, it is essential that shunt admittances must not be taken into account when using the equivalent voltage source method at the short-circuit location, which would mean that this practical method for calculating the short-circuit currents could no longer be used, but would always have to be switched to the superposition method.

14.1.3 Shunt Compensation Design (Example)

The compensation installation for the generation of capacitive reactive power as shown in Fig. 14.1 is also referred to as a "Mechanical Switched Capacitor with Damping Network" (MSCDN) and is installed, for example, in the 380 kV network in Germany for voltage stability. In addition to providing capacitive reactive power, harmonics can also be reduced due to the filter properties of the installation. The following values of the installation are used as an example for further consideration:

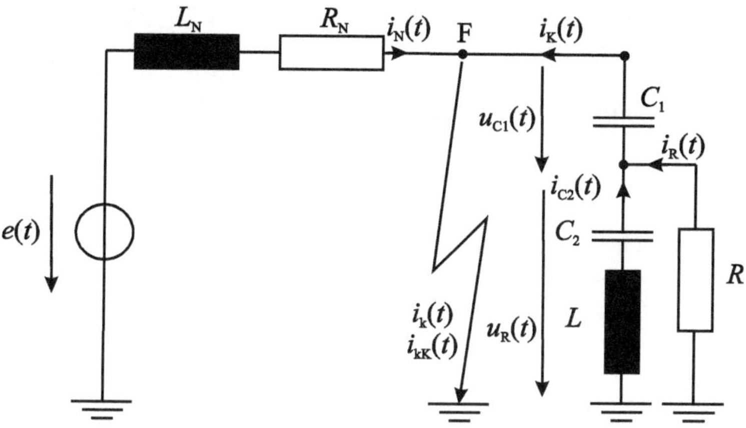

Fig. 14.1 Network equivalent circuit of a compensation installation (MSCDN) for determining the three-phase short-circuit current at the fault location F

- Main capacitor $C_1 = 5.51\ \mu F$,
- auxiliary capacitor $C_2 = 44.08\ \mu F$,
- coil $L = 230.05$ mH,
- damping resistor $R = 1000\ \Omega$.

The electrical data listed above correspond to the values of an installation with a rated power of $S_{rK} = 250$ Mvar and a rated voltage of $U_{rK} = 380$ kV, under these conditions, the rated current is $I_{rK} = 379.8$ A. The values of the individual components are determined from the following boundary conditions:

- The main capacitor C_1 is determined from the rated power and the steady-state voltage increase as result of the switching on of the compensation installation, Eqs. (14.1) and (14.2).
- The auxiliary capacitor C_2 and the coil L are tuned to the operating frequency so that the damping resistor R is currentless during operation and thus no losses occur during symmetrical operation, Eq. (14.3).
- The series capacitors ($C_1 + C_2$) and the coil L are tuned to the third harmonic as a filter unit in this example, Eq. (14.3).

The following variables are shown in Fig. 14.1:

e active voltage,
u_{C1} voltage of main capacitor C_1,
u_R voltage of damping resistor R,
i_k resulting short-circuit current without compensation,
i_{kK} resulting short-circuit current with compensation,
i_N short-circuit current contribution of the network,
i_K short-circuit current contribution of the compensation,
i_{C2} short-circuit current of the auxiliary capacitor C_2,
i_R short-circuit current of the damping resistor R.

An additional condition is that only a steady-state voltage change of e.g. $\Delta u = 2\%$ is permitted when the compensation installation is switched on. From this, the maximum power can be calculated as a function of the initial symmetrical short-circuit current and the nominal system voltage U_n according to Eq. (14.1).

$$S_{rK} = \Delta u \cdot \sqrt{3} \cdot U_n \cdot I''_{kN} \tag{14.1}$$

Taking into account a rated power of $S_{rK} = 250$ MVA and a ratio $R_N/X_N = 0.1$ of the network, the network reactance at the specified rated power is determined as: $X_N = 12.64\ \Omega$ ($L_N = = 40.25$ mH), which corresponds to a smallest possible initial symmetrical short-circuit current of $I''_{kN} = 19$ kA due to the rated power S_{rK}. With this information, the contributions of the compensation to the three-phase short-circuit current are calculated below. Equation (14.2) can be used to determine the main capacitor C_1.

$$S_{rK} = \frac{U_r^2}{X_{C1}} \qquad C_1 = \frac{S_{rK}}{\omega \cdot U_{rK}^2} = \frac{\Delta u \cdot \sqrt{3} \cdot U_n \cdot I''_{kN}}{\omega \cdot U_{rK}^2} \tag{14.2}$$

The other components C_2 and L of the compensation are determined according to the boundary conditions specified above:

$$C_2 = \left[\left(\frac{\omega_0}{\omega}\right)^2 - 1\right] \cdot C_1 \quad L = \frac{1}{\omega^2 \cdot C_2} \tag{14.3}$$

With

ω angular frequency in relation to the operating frequency f,
ω_0 angular frequency in relation to the resonance frequency f_0 of the filter.

Equations (14.2) and (14.3) can be used to determine the specified compensation components, taking into account the network specifications (voltage change, initial symmetrical short-circuit current).

14.1.4 Influence on the Three-Phase Short-Circuit Current

In principle, capacitive compensation affects the short-circuit current in two ways, firstly because the operating conditions (voltages) and the short-circuit impedance are changed and secondly because it itself contributes to the short-circuit current as an active voltage source. These two influences are considered separately in the following sections.

14.1.4.1 Change in the Operating Frequency Voltage at the Short-Circuit Location

The maximum three-phase short-circuit current I_k'' can generally be determined according to Eq. (14.4), taking into account the voltage factor c according to [3]:

$$I_k'' = \frac{c \cdot U_n/\sqrt{3}}{\underline{Z}_k} \tag{14.4}$$

With

c voltage factor with $c = 1.1$ for the maximum short-circuit current,
U_n nominal system voltage,
$\underline{Z}_k$ short-circuit impedance.

The voltage or voltage factor c or the short-circuit impedance can generally be influenced by using a shunt capacitor. The operating voltage U_b before the short-circuit occurs at the fault location F is generally dependent on the load in the network ($\underline{Z}_V$), so that under the condition that this operating voltage is kept constant, the required voltage factor c can be determined as a function of the load and the network impedance $\underline{Z}_N$ according to Fig. 14.2.

Equation (14.5) applies to the factor c_1 (without shunt capacitance) if the operating voltage $\underline{U}_b$ is to be kept constant at the subsequent fault location F, with a real, active voltage U_0:

$$\frac{\underline{U}_b}{\sqrt{3}} = \frac{\underline{Z}_V}{\underline{Z}_V + \underline{Z}_N} \cdot \frac{\underline{U}_0}{\sqrt{3}} \tag{14.5}$$

According to Eq. (14.5), the operating voltage U_b in relation to the driving voltage U_0 is always lower if an inductive load $\underline{Z}_V$ is assumed. Since this voltage drop can be compensated for in the short-circuit current calculation, the voltage factor c_1 results from the reciprocal of the ratio of the impedances of Eq. (14.5) to

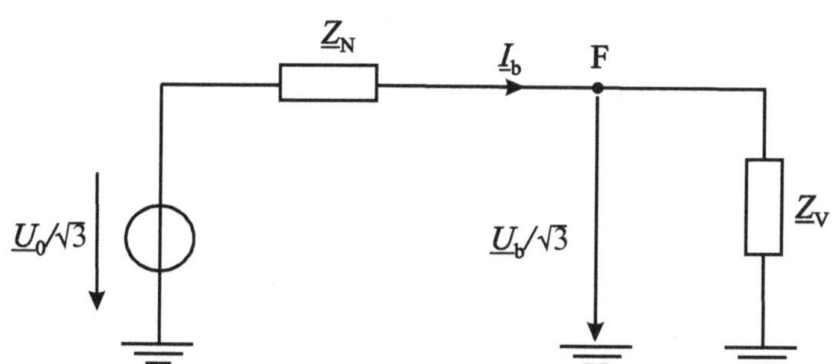

Fig. 14.2 Voltage factor c as a function of load and overhead line length

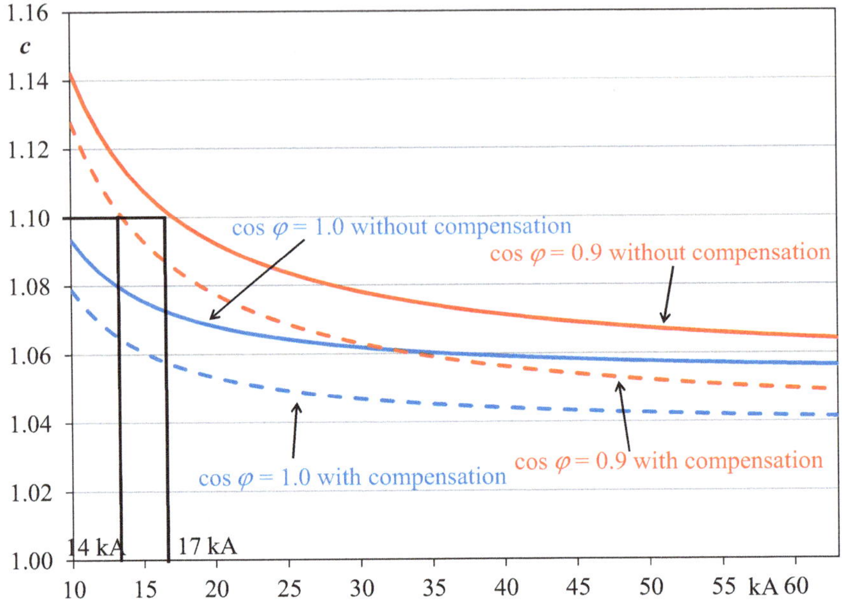

Fig. 14.3 Influence of compensation on the voltage factor c as a function of load and overhead line length depending on the initial symmetrical short-circuit current of the network _______ without compensation, -------- with compensation

$$c_1 = \left|1 + \frac{\underline{Z}_\mathrm{N}}{\underline{Z}_\mathrm{V}}\right| \tag{14.6}$$

Since an inductive load is usually the case in practice, a value of $c = 1.1$ is also generally used for high-voltage networks according to [3]. Taking into account the impedance of a shunt capacitance ($\underline{Z}_\mathrm{C}$) installed parallel to the load, the voltage factor c_2 is given by Eq. (14.6):

$$c_2 = \left|1 + \frac{\underline{Z}_\mathrm{N} \cdot (\underline{Z}_\mathrm{V} + \underline{Z}_\mathrm{C})}{\underline{Z}_\mathrm{V} \cdot \underline{Z}_\mathrm{C}}\right| \tag{14.7}$$

The range of voltage factors c_1 and c_2 is shown below using an example, depending on the network short-circuit current, the load and the compensation.

In this case, the network impedance (short-circuit impedance) at a nominal system voltage of $U_\mathrm{n} = 380$ kV is determined from the three-phase short-circuit current and a ratio $R/X = 0.1$. The capacitance of the compensation is determined according to Eq. (14.1) from the permissible voltage change, e.g. $\Delta u = 2.0\%$ after the compensation is switched on. The results are plotted in Fig. 14.3 for different power factors of the load. The operating voltage at point F in Fig. 14.2 is assumed to be $U_\mathrm{b} = 400$ kV.

It can be seen that the required voltage factor according to Eq. (14.7) with $c_2 = 1.1$ is only permissible for short-circuit currents $\underline{I}''_{\mathrm{k}} \geq 17$ kA with a transmitted power of $S_{\mathrm{V}} = 2000$ MVA, a power factor of $\cos \varphi = 0.9$, and an operating voltage of $U_{\mathrm{b}} = 400$ kV, if no capacitive compensation is assumed. For lower values of the network short-circuit current, a higher value for c is generally required, or the use of a capacitor changes the load flow in such a way that the inductive current is partially compensated. With capacitive compensation, the short-circuit current is reduced to $\underline{I}''_{\mathrm{k}} \geq 14$ kA without exceeding the voltage factor $c_1 = 1.1$. Under these conditions, voltage factors c occur during the transmission of pure active power, which always remain below a value of 1.1. If, in contrast, an operating voltage of $U_{\mathrm{b}} = 410$ kV is assumed, only short-circuit currents of $\underline{I}''_{\mathrm{k}} \geq 36.5$ kA are permissible in order to comply with the voltage factor $c = 1.1$ ($\cos \varphi = 0.9$; without compensation).

The use of cross compensation always reduces the voltage factor c, so that the results of the short-circuit current calculation according to Eq. (14.4) with a factor of $c = 1.1$ will be more on the safe side than would be the case without compensation. For network loads of $S_{\mathrm{V}} > 2000$ MVA and values of $\Delta u > 2.0\%$, the difference is even more pronounced.

14.1.4.2 Change in the Operating Frequency Voltage at the Short-Circuit Location

The resulting short-circuit impedance $\underline{Z}_{\mathrm{kN+}}$ under consideration of compensation, is calculated according to Eq. (14.8), based on Fig. 14.4. In this case, the short-circuit location moves from the voltage source to the overhead line end Z_{L}, which means that the factor α changes from $\alpha = 0 \rightarrow 1$. For simplification, the resistance is neglected. In this general example, it is assumed that the compensation is located on the line between the network feeder and the short-circuit location, so that the network contribution to the short-circuit current changes and the feeder is therefore not independent, which results in a different indexing of the network current.

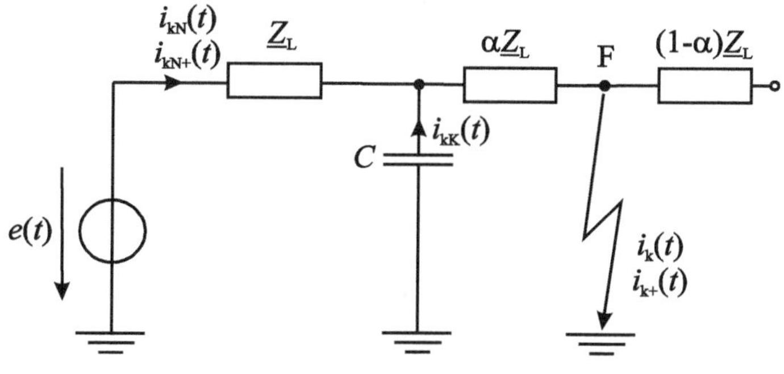

Fig. 14.4 Influence of compensation on the resulting short-circuit impedance, fault location F. Index: kN+ short-circuit with compensation, Index: kN short-circuit without compensation

$$\underline{Z}_{\text{kN+}} = \underline{Z}_{\text{L}} + \frac{\alpha \cdot \underline{Z}_{\text{L}} \cdot \underline{Z}_{\text{C}}}{\alpha \cdot \underline{Z}_{\text{L}} + \underline{Z}_{\text{C}}} \approx \text{j}X_{\text{L}} + \text{j}\frac{\alpha \cdot X_{\text{L}} \cdot X_{\text{C}}}{X_{\text{C}} - \alpha \cdot X_{\text{L}}}$$
$$= \text{j}X_{\text{L}} \cdot \left[1 + \frac{\alpha \cdot X_{\text{C}}}{X_{\text{C}} - \alpha \cdot X_{\text{L}}}\right] \quad (14.8)$$

With

$$\underline{Z}_{\text{L}} \approx \text{j}\omega L = \text{j}X_{\text{L}} \qquad \underline{Z}_{\text{C}} \approx \frac{1}{\text{j}\omega C} = -\text{j}X_{\text{C}} \quad (14.9)$$

If the short-circuit impedance according to Eq. (14.8) is related to the impedance without compensation $\underline{Z}_{\text{kN}}$, the result is according to Eq. (14.10):

$$\frac{\underline{Z}_{\text{kN+}}}{\underline{Z}_{\text{kN}}} \approx \frac{\text{j}X_{\text{L}} \cdot \left[1 + \frac{\alpha \cdot X_{\text{C}}}{X_{\text{C}} - \alpha \cdot X_{\text{L}}}\right]}{\text{j}X_{\text{L}} \cdot [1 + \alpha]} = \frac{1 + \frac{\alpha \cdot X_{\text{C}}}{X_{\text{C}} - \alpha \cdot X_{\text{L}}}}{1 + \alpha} \geq 1 \quad (14.10)$$

With the short-circuit impedance $\underline{Z}_{\text{kN}}$ without compensation depending on the fault location:

$$\underline{Z}_{\text{kN}} \approx \text{j}X_{\text{L}} \cdot [1 + \alpha] \quad (14.11)$$

According to Eq. (14.10), it can be seen that the short-circuit impedance increases in the event of a fault behind the location of the capacitor, which according to Eq. (14.4) leads to a reduction in the short-circuit current contribution $\underline{I}''_{\text{kN+}}$ according to Fig. 14.4. The calculation of the short-circuit current is on the safe side, if the shunt impedance of the compensation is neglected when the maximum value is considered. The short-circuit current contribution of the feeding voltage source is referred to as $\underline{I}''_{\text{kN+}}$. This only applies, if the short-circuit current is calculated using the superposition method, for example. If, in contrast, the calculation is carried out using the method of the equivalent voltage source at the short-circuit location, the capacitor has no influence on the level of the short-circuit current, as shunt impedances must not be taken into account in this case, in accordance with the short-circuit standard [3].

If the short-circuit current contribution of the network is calculated considering the compensation, the network contribution is calculated using the impedance according to Eq. (14.8) with $E = 1.1 \cdot U_{\text{n}}/\sqrt{3}$:

$$\underline{I}''_{\text{kN+}} = \frac{1{,}1 \cdot U_{\text{n}}/\sqrt{3}}{\text{j}X_{\text{L}} \cdot \left[1 + \frac{\alpha \cdot X_{\text{C}}}{X_{\text{C}} - \alpha \cdot X_{\text{L}}}\right]} \quad (14.12)$$

For the current $\underline{I}''_{\text{k+}}$ at the short-circuit location F (with compensation) this results in:

$$\underline{I}''_{k+} = \frac{-X_C}{\alpha \cdot X_L - X_C} \cdot \frac{1,1 \cdot U_n/\sqrt{3}}{jX_L \cdot \left[1 + \frac{\alpha \cdot X_C}{X_C - \alpha \cdot X_L}\right]} = \frac{1,1 \cdot U_n/\sqrt{3}}{jX_L \cdot [(1+\alpha) - \alpha \cdot X_L/X_C]} \tag{14.13}$$

The short-circuit current $\underline{I}''_k$ at the short-circuit location F without compensation is determined according to Eq. (14.14), so that the current ratio can be determined in order to evaluate the influence of the compensation.

$$\underline{I}''_k = \frac{1,1 \cdot U_n/\sqrt{3}}{jX_L \cdot (1+\alpha)} \tag{14.14}$$

$$\frac{\underline{I}''_{k+}}{\underline{I}''_k} = \frac{1}{1 - \frac{\alpha}{1+\alpha} \cdot \frac{X_L}{X_C}} \geq 1 \tag{14.15}$$

This means that by using capacitive compensation, the current $\underline{I}''_{k+}$ at the short-circuit location is always greater than without compensation ($\underline{I}''_k$). If the following values are assumed in accordance with Fig. 14.4:

$jX_L = j25.1\ \Omega$ (380 kV overhead line, 100 km, reactance according to Table 4.2),
$-jX_C = -j577.69\ \Omega$ (main capacitor C_1 according to Sect. 14.1.3),

The maximum deviation in the event of a short circuit at the end of the line ($\alpha = 1$) is calculated as 2.2%; this means that the actual short-circuit current in this example is underestimated by 2.2% if the capacitor is not taken into account for compensation.

14.1.4.3 Short-Circuit Current Contribution of Compensation in the Event of a Network Fault (Example, Transient Calculations)

While Sect. 14.1.4.2 examines the influence of capacitive compensation on the operating frequency short-circuit current, this section estimates the discharge current of a capacitor in the event of a terminal short circuit. This assumes that a three-phase short circuit occurs in the immediate vicinity of the compensation. The calculations refer to both the total current and the short-circuit current contribution of the compensation and the network contribution. The system data corresponds to the values in Sect. 14.1.3. The equivalent circuit for simulating the network and compensation corresponds to Fig. 14.1, so that the two feeders (network and capacitor) are decoupled. As a result, the current curves $i_{kN}(t)$ and $i_{kN+}(t)$ are identical.

Figure 14.5 shows the voltage and current curves when a fault occurs after $t = 2.5$ ms at the maximum operating voltage ($U_b = 408.8$ kV) at the terminals of the compensation unit. The network impedance is set to $L_N = 40.25$ mH and $R_N = 1.264\ \Omega$ in accordance with Sect. 14.1.3. The voltages $u_{C1}(t)$ and $u_R(t)$

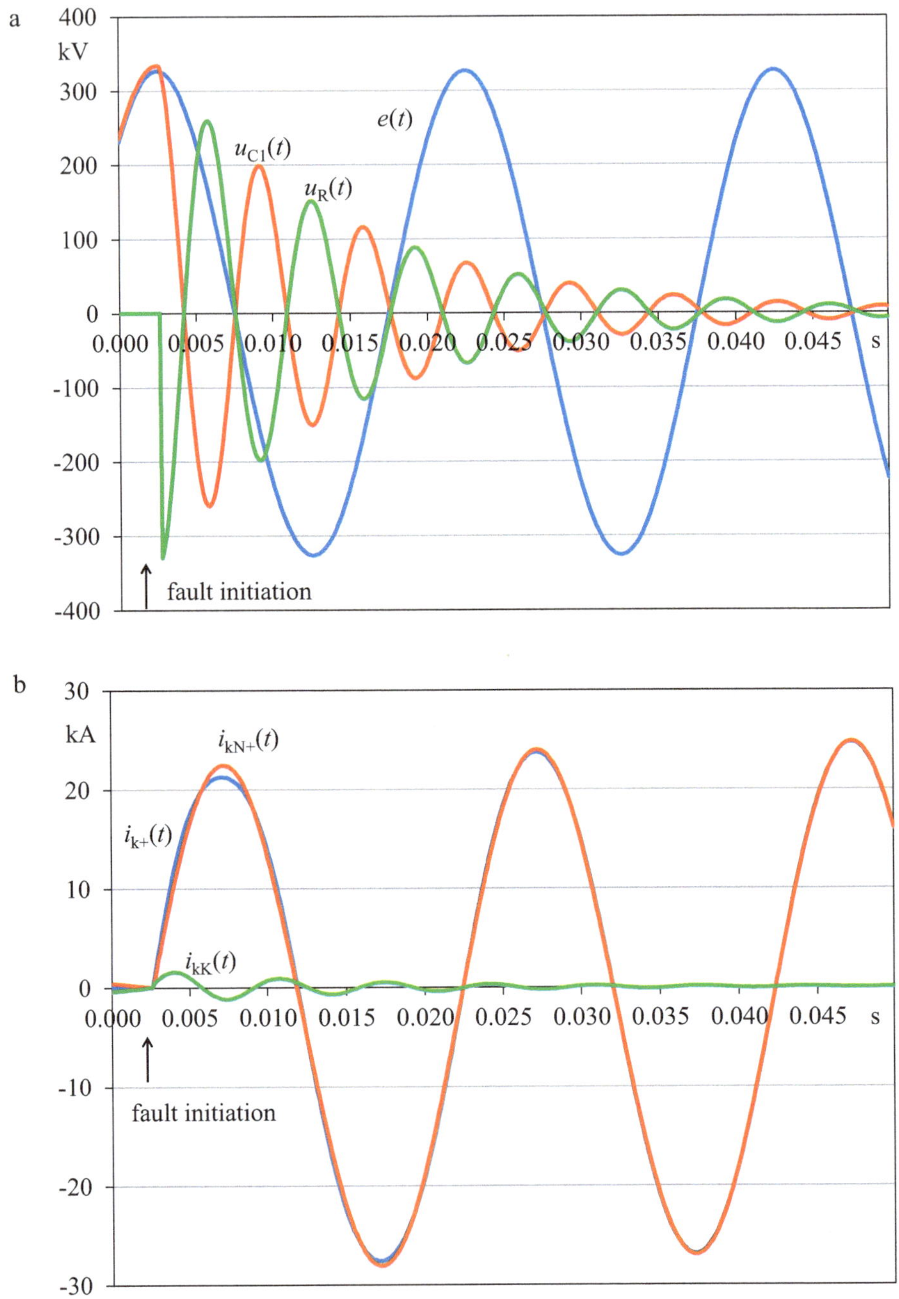

Fig. 14.5 Electrical quantities in a three-phase terminal short circuit, short circuit occurring at maximum voltage, fault occurring after $t = 2.5$ ms, Fig. 14.1 [1]. (**a**) Voltages, (**b**) currents, e active voltage, u_{C1} voltage of main capacitor C_1, u_R voltage of damping resistor R, i_{k+} resulting short-circuit current, i_{kN+} short-circuit current contribution of the network, i_{kK} short-circuit current contribution of the compensation

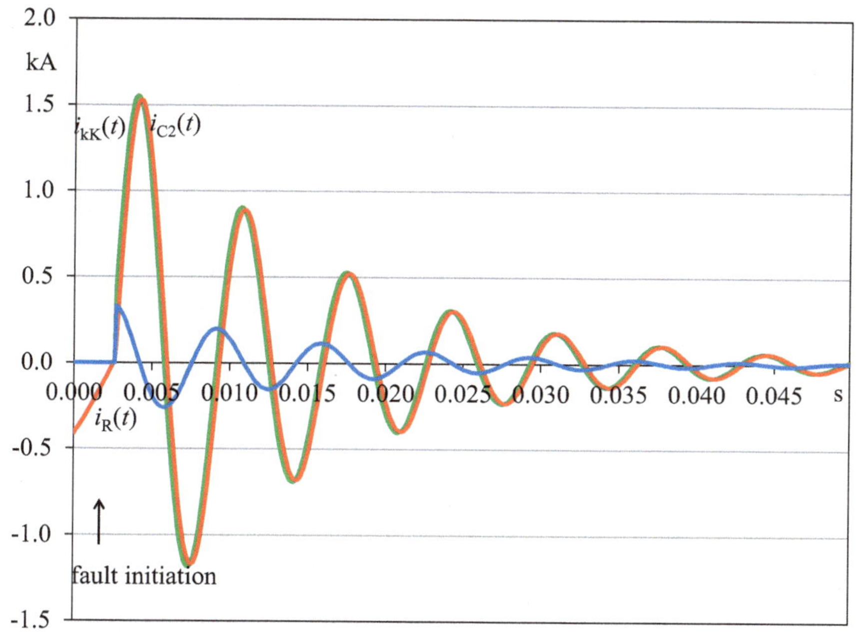

Fig. 14.6 Partial short-circuit currents of the compensation unit in the event of a three-phase terminal short circuit, short circuit occurring at maximum voltage, Fig. 14.1 [1]. i_{kK} Short-circuit current contribution of the compensation, i_{C2} Short-circuit current of the auxiliary capacitor C_2, i_R Short-circuit current of the damping resistor R

according to Fig. 14.1 decay after the short circuit occurs at the natural frequency of the compensation ($f_0 = 150$ Hz). Due to the design of the compensation unit, the voltage $u_R(t)$ has a value of zero before the fault occurs, as the series resonant circuit is tuned to the operating frequency.

An evaluation of the first maxima of the current curves $i_{k+}(t)$ und $i_{kN+}(t)$ yields the following initial symmetrical short-circuit currents (the network short-circuit current is determined from the minimum and maximum half-waves):

- Total short-circuit current at the short-circuit location: $I''_{k+} = 17.3\,\text{kA}$,
- short-circuit current contribution of the network: $I''_{kN+} = 18.1\,\text{kA}$.

The total short-circuit current is reduced at the peak value due to the contribution of the compensation unit because of the 150 Hz component. The partial currents within the compensation unit are shown in Fig. 14.6. The total short-circuit current contribution corresponds primarily to the partial current of capacitance C_2.

Using the short-circuit current curve $i_{kK}(t)$, the time constant of the discharge process can be determined as $T = 12.3$ ms, and the decaying short-circuit current contribution is then determined as:

$$i_{kK}(t) = \widehat{i}_{kK} \cdot e^{-t/T} \cdot \sin(\omega_0 \cdot t) \tag{14.16}$$

Taking into account the first peak value, the contribution of the initial symmetrical short-circuit current of the compensation unit to $I''_{kK} = 1.25\,\text{kA}$, so that this contribution is significantly lower than the total short-circuit current at the fault location with $I''_{k+} = 17.3\,\text{kA}$. Since the maximum operating frequency peak short-circuit current occurs when the fault is initiated at zero voltage crossing, Fig. 14.7a, b show the current curves under this boundary condition.

In this case, the partial short-circuit current of the compensation unit results in an initial symmetrical short-circuit current of $I''_{K} = 0.41\,\text{kA}$ and a peak short-circuit current of $i_{pK} = 0.45$ kA, with a total peak short-circuit current of $i_{p+} = 45.3$ kA.

When dimensioning equipment and systems, different currents must be taken into account (peak short-circuit current, breaking current, steady state, and thermally equivalent short-circuit current). However, due to the low time constant of the short-circuit current contribution of a compensation unit, only the peak current is of interest. For this reason, Table 14.1 shows the evaluation for the considered network arrangement under various voltage conditions before the short circuit occurs.

It can be seen that the maximum peak short-circuit current in this example is determined exclusively by the network contribution, while the contribution of the compensation unit can be neglected (<3.5%). Furthermore, the maximum values occur at different time t_{max}, so that no superposition occurs. In addition, the voltage conditions at the occurrence of a short circuit are different (maximum, zero crossing).

14.1.4.4 Estimating the Contribution of the Compensation Unit

The following section determines the short-circuit current contribution that can generally be expected as a function of the network short-circuit current without calculating the transient processes. To simplify the calculation of the discharge current in a three-phase short circuit, the compensation unit can be modeled at the resonance frequency as a series resonant circuit consisting of the resulting *RLC* elements. The current $i_{kK}(t)$ is determined when the short circuit occurs at zero voltage crossing as follows:

$$\begin{aligned} i_{kK}(t) &= \frac{\widehat{E}}{\omega_1 \cdot L_{res}} \cdot e^{-\delta \cdot t} \cdot \sin(\omega_1 \cdot t) \\ &= \frac{\widehat{E}}{\sqrt{\frac{L_{res}}{C_1} - \left(\frac{R_{res}}{2}\right)^2}} \cdot e^{-t/T} \cdot \sin(\omega_1 \cdot t) \end{aligned} \tag{14.17}$$

With:

$$2 \cdot \delta = \frac{R_{res}}{L_{res}} = \frac{2}{T} \qquad \omega_{01}^2 = \frac{1}{C_1 \cdot L_{res}} \qquad \omega_{01}^2 - \delta^2 = \omega_1^2 \tag{14.18}$$

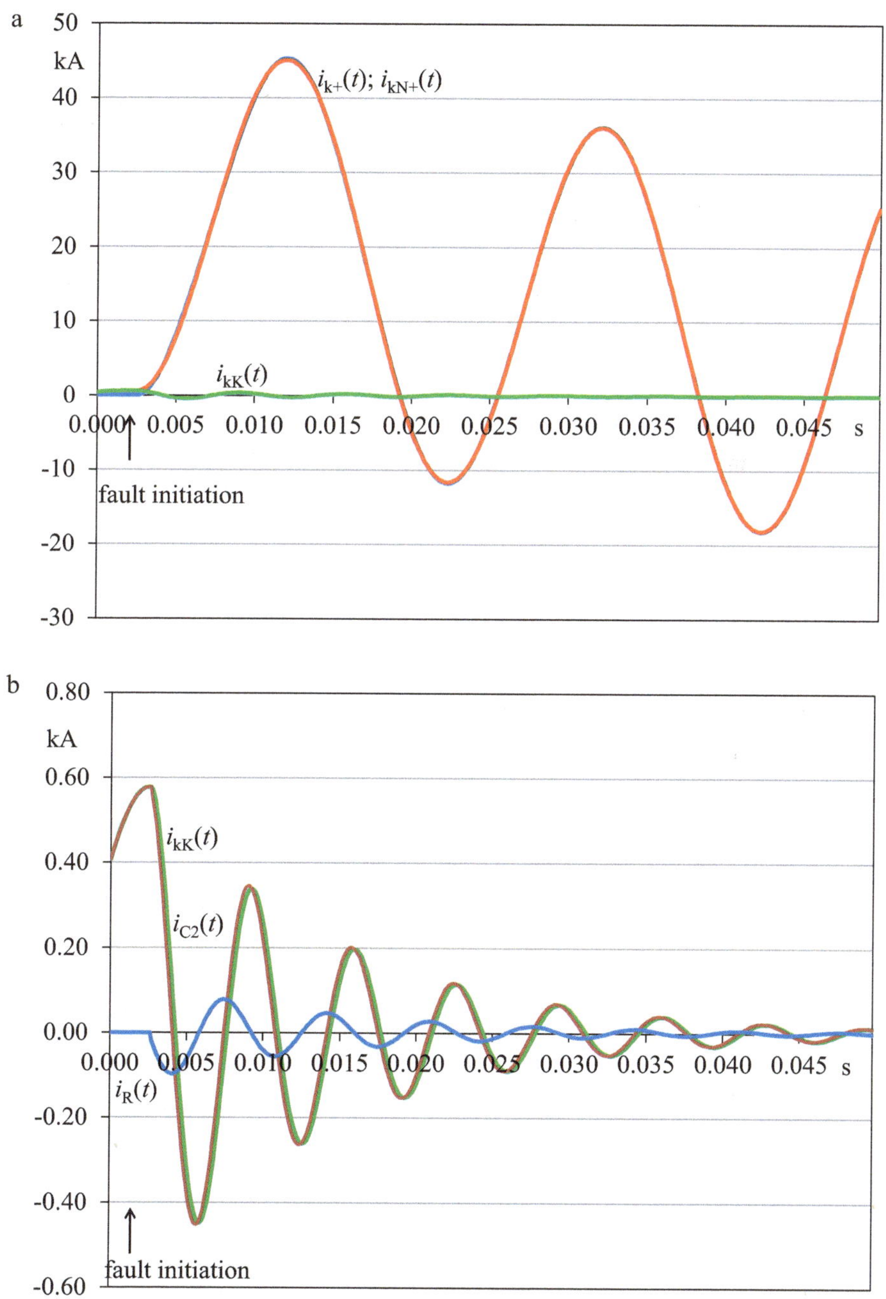

Fig. 14.7 Currents in a three-phase terminal short circuit, short circuit occurring at zero voltage crossing, fault occurring after $t = 2.5$ ms, Fig. 14.1 [1]. (**a**) Total current, (**b**) partial currents of the compensation unit. i_{k+} resulting short-circuit current, i_{kN+} short-circuit current contribution of the network, i_{kK} short-circuit current contribution of the compensation, i_{C2} short-circuit current of the auxiliary capacitor C_2, i_R short-circuit current of the damping resistor R

Table 14.1 Peak short-circuit current taking into account a compensation unit, example according to Fig. 14.1, in each case the first current maximum after the short circuit occurs

Voltage	Current	i_p (kA)	t_{max} (ms)
Maximum	Total	21.25	4.6
	Network	22.42	4.7
	Compensation	1.55	1.5
Zero crossing	Total	45.29	9.5
	Network	45.04	9.4
	Compensation	0.45	3.3

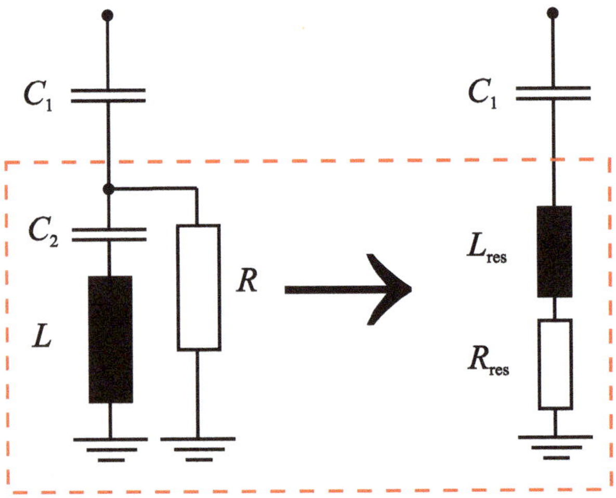

Fig. 14.8 Conversion of the compensation unit into a series resonant circuit

The general variables R_{res}, L_{res} and C_1 for calculating the short-circuit current are to be determined by converting the equivalent circuit of the compensation unit according to Fig. 14.8 at the resonance frequency ω_0. The resulting resistance R_{res} and inductance L_{res} are determined according to Eqs. (14.19) and (14.20).

$$R_{res} = \frac{R \cdot \left(\omega_0 L - \frac{1}{\omega_0 C_2}\right)^2}{R^2 + \left(\omega_0 L - \frac{1}{\omega_0 C_2}\right)^2} = \frac{R \cdot (\omega_0 L)^2 \cdot \left[1 - \left(\frac{\omega}{\omega_0}\right)^2\right]^2}{R^2 + (\omega_0 L)^2 \cdot \left[1 - \left(\frac{\omega}{\omega_0}\right)^2\right]^2} \tag{14.19}$$

$$\omega_0 \cdot L_{res} = \frac{R^2 \cdot \left(\omega_0 L - \frac{1}{\omega_0 C_2}\right)}{R^2 + \left(\omega_0 L - \frac{1}{\omega_0 C_2}\right)^2} = \frac{R^2 \cdot (\omega_0 L) \cdot \left[1 - \left(\frac{\omega}{\omega_0}\right)\right]}{R^2 + (\omega_0 L)^2 \cdot \left[1 - \left(\frac{\omega}{\omega_0}\right)^2\right]^2} \tag{14.20}$$

With the operating frequency ω, the following applies according to Sect. 14.1.3:

$$\omega = \frac{1}{L \cdot C_2} \tag{14.21}$$

Taking into account the values for a compensation unit according to Sect. 14.1.3, the following resulting values are determined according to Eqs. (14.19) and (14.20), with a resonance frequency of $\omega_0 = 2\pi \cdot 150$ 1/s:

$$L_{res} = 197.2\ \text{mH} \qquad R_{res} = 35.81\ \Omega$$

If the values of the equivalent circuit C_1, L_{res} und R_{res} are inserted into Eq. (14.17), the result is $L_{res}/C_1 \gg (R_{res}/2)^2$ or $\delta = 0$:

$$i_{kK}(t) \approx \frac{E}{\sqrt{L_{res}/C_1}} \cdot e^{-t/T} \cdot \sin(\omega_1 \cdot t) \tag{14.22}$$

To calculate the peak short-circuit current, the time of the peak value is determined from the derivative of Eq. (14.22):

$$0 \stackrel{!}{=} \frac{di_{kK}}{dt} = -\frac{1}{T} \cdot e^{-t_{max}/T} \cdot \sin(\omega_1 \cdot t_{max}) + \omega_1 \cdot e^{-t_{max}/T} \cdot \sin(\omega_1 \cdot t_{max}) \tag{14.23}$$

$$\tan(\omega_1 \cdot t_{max}) = T \cdot \omega_1 \quad \omega_1 \cdot t_{max} = \arctan(T \cdot \omega_1) \tag{14.24}$$

For the angular frequency ω_1 and the time constant T, the following simplified formula results according to Eq. (14.18):

$$\omega_1^2 \approx \frac{1}{C_1 \cdot L_{res}} = \frac{10^6}{0.1972 \cdot 5.51} \cdot \frac{1}{s^2} \qquad \omega_1 \approx 959.42 \cdot 1/s$$

$$T = \frac{2 \cdot L_{res}}{R_{res}} = \frac{2 \cdot 0.1972}{35.81} \cdot s = 11.01\,\text{ms}$$

The timing of the peak short-circuit current is thus determined as follows:

$$t_{max} = \frac{\arctan(T \cdot \omega_1)}{\omega_1} = \frac{\arctan(0.001101 \cdot 959.42)}{959.42} \cdot s = 1.5389\,\text{ms}$$

From this, the characteristic values for determining the short-circuit contribution can be determined, provided that the voltage according to Eq. (14.22) $E = 1.1 \cdot 380\ \text{kV}/\sqrt{3}$ applies:

$$\begin{aligned} i_{kK}(t) &= \frac{\sqrt{2} \cdot E}{\omega_1 \cdot L_{res}} \cdot e^{-t/T} \cdot \sin(\omega_1 \cdot t) \\ &= \frac{1,1 \cdot \sqrt{2} \cdot 380\text{kA}}{\sqrt{3} \cdot 0.1972 \cdot 959.42} \cdot e^{-t/0,01101} \cdot \sin\left(\frac{959.42}{s} \cdot t\right) \end{aligned}$$

Table 14.2 Influence of compensation in a three-phase terminal short circuit depending on the voltage change Δu and the order number n of the harmonic filter

Value	I''_{kK}/I''_{kN}	i_{pK}/i_{pN}
$n = 3$		
$\Delta u = 0.025$	0.084	0.041
$\Delta u = 0.050$	0.168	0.083
$n = 5$		
$\Delta u = 0.025$	0.140	0.069
$\Delta u = 0.050$	0.281	0.138

I''_{kK} initial symmetrical short-circuit current, contribution compensation unit
I''_{kN} initial symmetrical short-circuit current, contribution network
i_{pK} peak short-circuit current, contribution compensation unit
i_{pN} peak short-circuit current, contribution network
n order number of the harmonic
Δu voltage increase

$$I''_{kK} = \frac{1.1 \cdot 380\,\text{kA}}{\sqrt{3} \cdot 0.1972 \cdot 959.42} = 1.276\,\text{kA}$$

$$i_p = \sqrt{2} \cdot 1.276 \cdot e^{-1.5389/11.01} \cdot \sin(0.95942 \cdot 1.5389) = 1.562\,\text{kA}$$

While the result of the initial symmetrical short-circuit current according to Eq. (14.22) shows a deviation of less than 3.5%, the difference in determining the peak short-circuit current is even smaller (less than 1%). In addition, the voltage before the short circuit occurs according to Fig. 14.5 is approx. 2.2% lower than the estimate according to Eq. (14.22).

According to Eq. (14.1), the capacitive reactive power and thus the main capacitor C_1 are determined from the voltage increase Δu at the connection point. Assuming that the quality factor G of the parallel circuit shown in Fig. 14.8 always has the value $G = 0.2$ at the resonance frequency, the current contribution of a compensation in a three-phase terminal short circuit can be determined as a function of the network short-circuit current and the tuning to the harmonic $n = \omega_0/\omega_{50}$. The results are listed in Table 14.2. The quality of the series resonant circuit is determined according to Eq. (14.25), from which the resistance R_{ers} can be calculated.

$$G = \frac{1}{R_{res}} \cdot \sqrt{\frac{L_{res}}{C_1}} \tag{14.25}$$

The initial symmetrical short-circuit current of the network in this comparison is $I''_{kN} = 19\,\text{kA}$ with a peak short-circuit current $i_{pN} = 46.915$ kA ($R/X = 0.1$).

It can be seen that, in principle, the contribution of compensation increases with higher order number n of the harmonic and the voltage increase Δu. However, it turns out that the contribution to the peak short-circuit current can be a maximum of 14%. In addition, it should be noted that the currents (network, compensation) must not be added together, as the voltage conditions are different when a short circuit occurs, namely:

- Peak short-circuit current of the network in the event of a fault occurring at zero voltage crossing,
- peak short-circuit current of the compensation unit in the event of a fault occurring at maximum voltage.

14.1.5 Evaluation of the Contribution of Shunt Compensation

The following statements therefore apply when assessing the influence of a capacitive compensation unit in the event of a three-phase terminal short circuit:

- The compensation unit only contributes to the initial symmetrical short-circuit current and to the peak short-circuit current.
- The maximum peak short-circuit current is initiated at the voltage maximum (the compensation is fully charged), whereas according to [3] the calculation assumes that the short circuit occurs at the voltage zero crossing, meaning that the values cannot be added together.
- Due to the design of the compensation unit, which also acts as a harmonic filter, the maximum values do not occur at the same time.
- According to Table 14.2, the short-circuit currents of the compensation are significantly lower than the network contribution, so that the peak short-circuit current is not significantly increased by a compensation unit, provided that the voltage increase Δu occurs.

14.1.6 Contribution of a Capacitor to a Three-Phase Terminal Short Circuit

According to Table 14.2, the contribution of capacitor is determined by the order number of the harmonic if the compensation is additionally designed as a harmonic filter. Here, the short-circuit current contribution $i_{\mathrm{kK}}(t)$ represents the discharge of the capacitor. If a capacitor is not designed as a filter, its contribution to the initial symmetrical short-circuit current or peak short-circuit current depends on the inductance of the feeder or an equivalent inductance for reducing the current contribution. In the following, this inductance L is determined so that the short-circuit current via the circuit-breaker S according to Fig. 14.9 does not exceed the permissible current $I''_{\mathrm{kSmax}} = 19\,\mathrm{kA}$ or i_{pSmax}.

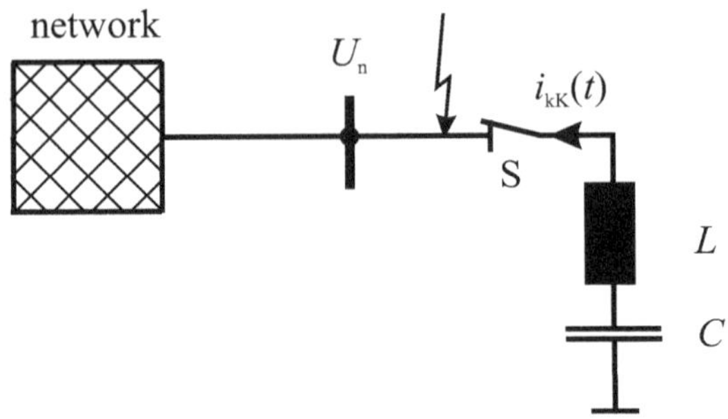

Fig. 14.9 Equivalent circuit for determining the maximum short-circuit current via the circuit-breaker S

According to Eq. (14.17), the following applies to the current contribution of the capacitor when the resistance is neglected:

$$i_{kK}(t) \approx \frac{\widehat{E}}{\sqrt{L/C}} \cdot \sin(\omega_1 \cdot t) = \frac{1.1 \cdot \sqrt{2} \cdot U_n/\sqrt{3}}{\sqrt{L/C}} \cdot \sin(\omega_1 \cdot t) \tag{14.26}$$

A comparison of the current curve with the permissible initial symmetrical short-circuit current of the switchgear (S) provides the necessary inductance for reducing the current contribution:

$$I''_{kSmax} = \frac{1.1 \cdot U_n/\sqrt{3}}{\sqrt{L/C}} \qquad L = \left[\frac{1.1 \cdot U_n/\sqrt{3}}{I''_{kSmax}}\right]^2 \cdot C \tag{14.27}$$

If the peak short-circuit current is considered, the result is as follows:

$$L = \left[\frac{1.1 \cdot \sqrt{2} \cdot U_n/\sqrt{3}}{i_{pSmax}}\right]^2 \cdot C \tag{14.28}$$

For the natural frequency ω_1 of the discharge process, this results in.

$$\omega_1 = \frac{I''_{kSmax}}{1.1 \cdot U_n/\sqrt{3}} \cdot \frac{1}{C} \tag{14.29}$$

With

I''_{kSmax} permissible initial symmetrical short-circuit current of the circuit breaker or peak short-circuit current i_{pSmax},
C capacitance of the capacitor,
U_n nominal system voltage.

Taking into account a maximum initial symmetrical short-circuit current of $I''_{\text{kSmax}} = 40\,\text{kA}$ and a capacitance of $C = 5.51\,\mu\text{F}$, corresponding to the dimensioning according to Sect. 14.1.3, the required inductance is determined to be:

$$L = \left[\frac{1.1 \cdot 380/\sqrt{3}}{40} \cdot \Omega\right]^2 \cdot 5.51\,\mu\text{F} = 0.2\,\text{mH}$$

This inductance must be installed at least in series with the capacitance according to Fig. 14.9 in order not to exceed the permissible peak short-circuit current, e.g. of the circuit-breaker, with the frequency ω_1 or f_1:

$$\omega_1 = \frac{40}{1.1 \cdot 380/\sqrt{3}} \cdot \frac{10^6}{5.51} \cdot \frac{1}{\text{s}} = 30.081 \cdot 10^3 \cdot \frac{1}{\text{s}} \qquad f_1 = 4.788\,\text{kHz}$$

Because of the higher frequency, the ever-present resistance will cause the short-circuit current to decay rapidly. In this case, too, the maximum current contribution only occurs during a short circuit at the voltage maximum, so that superimposition with the network component does not make physical sense.

14.2 Series Capacitor

Series capacitors are used to improve power transmission in HV grids with an extensive network of lines. These installations consist of capacitor banks connected in series with the transmission lines, which reduces the inductive impedance of the lines [4]. Series capacitors increase transmission capacity and dynamic system stability [5].

According to [3]:“*The effect of series capacitors can be neglected in the calculation of short-circuit currents, if they are equipped with voltage-limiting devices in parallel, acting if a short circuit occurs*”. For this reason, the influence of these arrangements on the short-circuit current is described below. In this context, the line is fed from a network and not from individual generators, as no processes relating to sub-synchronous resonances (SSR) according to [6] are investigated. However, when installing series capacitors, it is important to note that there is a risk of sub-synchronous resonances, so these processes should be taken into account.

14.2.1 Introduction

The power transferred between two networks according to Fig. 14.10 can be determined using Eq. (14.30). For simplicity, the impedance of the line between these two networks is represented exclusively by the reactance.

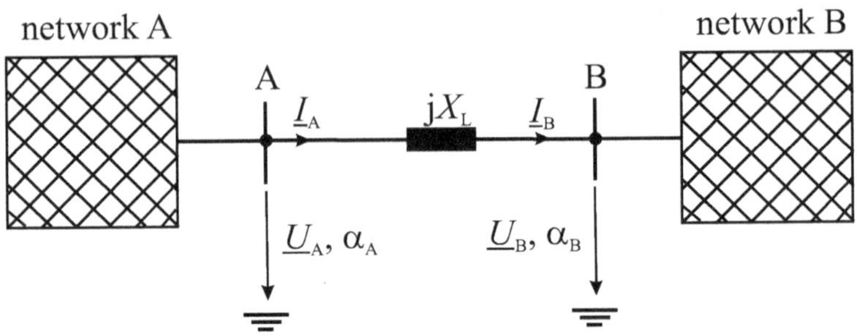

Fig. 14.10 Example of power transfer between two networks A and B

$$\begin{pmatrix} \underline{U}_A \\ \underline{I}_A \end{pmatrix} = \begin{pmatrix} 1 & jX_L \\ 0 & 1 \end{pmatrix} \cdot \begin{pmatrix} \underline{U}_B \\ \underline{I}_B \end{pmatrix} \tag{14.30}$$

The transmitted power $\underline{S}_B$ at the end of the line is determined taking into account Eq. (14.30) to (* conjugate complex):

$$\underline{S}_B = 3 \cdot \underline{U}_B \cdot \underline{I}_B^* = \frac{3 \cdot \underline{U}_B \cdot \left(\underline{U}_A^* - \underline{U}_B^*\right)}{(jX_L)^*} \tag{14.31}$$

If the voltages are expressed in polar coordinates, the resulting apparent power transmitted is:

$$\underline{S}_B = j\frac{3}{X_L} \cdot \left[U_A \cdot U_B \cdot e^{j(\alpha_B - \alpha_A)} - U_B^2\right] \tag{14.32}$$

The following therefore applies to the active power (P_B) and reactive power (Q_B) that can be transmitted at the end of a line:

$$P_B = \frac{3 \cdot U_A \cdot U_B}{X_L} \cdot \sin(\alpha_A - \alpha_B) \tag{14.33}$$

$$Q_B = \frac{3}{X_L} \cdot \left[U_A \cdot U_B \cdot \sin(\alpha_A - \alpha_B) - U_B^2\right] \tag{14.34}$$

With

U_A voltage at busbar A at the beginning of the line (line-earth),
U_B voltage at busbar B at the end of the line (line-earth),
X reactance of the line,
P_B active power transmitted at the end of the line,
Q_B reactive power transmitted at the end of the line,
α_A, α_B Voltage angle of the voltages at busbars A and B.

If, for clarity, the voltage angle $\alpha_B = 0$ is set, Eqs. (14.33) and (14.34) are simplified to:

$$P_B = \frac{3 \cdot U_A \cdot U_B}{X_L} \cdot \sin(\alpha_A) \qquad Q_B = \frac{3}{X_L} \cdot \left[U_A \cdot U_B \cdot \sin(\alpha_A) - U_B^2\right] \quad (14.35)$$

According to Eq. (14.35), it can be seen that short lines have a high transmission capacity (active power, *P*), which naturally depends on the thermal design. In contrast, the transmission capacity of longer lines decreases due to higher reactance, and the voltage drop along the line. The voltage angle increase as a result of the load flow. If the transmission capacity is to be kept constant even for a line with a long distance, it is possible to correct the parameters of Eq. (14.35) accordingly.

14.2.2 Series Compensation

By using a series compensator or a capacitance X_C in an overhead line, the reactance X_L of the line can be continuously or constantly changed in accordance with Eq. (14.33). This means that the reactance is reduced and thus the line is electrically reduced. The transmission capacity is thus increased, resulting in the network connection shown in Fig. 14.11. In this case, the capacitance is concentrated near the busbar B for illustrative purposes. In practice, the capacitance should be installed in the middle of the line, for example.

By using a capacitor, the effective reactance between network nodes A and B can be reduced as shown in Fig. 14.11, allowing a compensation factor to be defined according to Eq. (14.36).

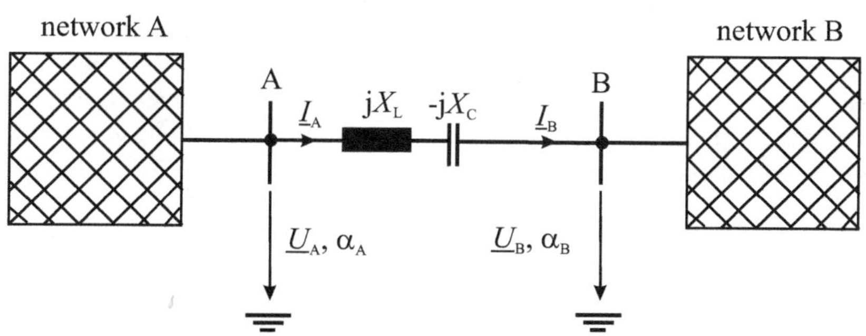

Fig. 14.11 Example of power transfer between two networks taking into account series compensation

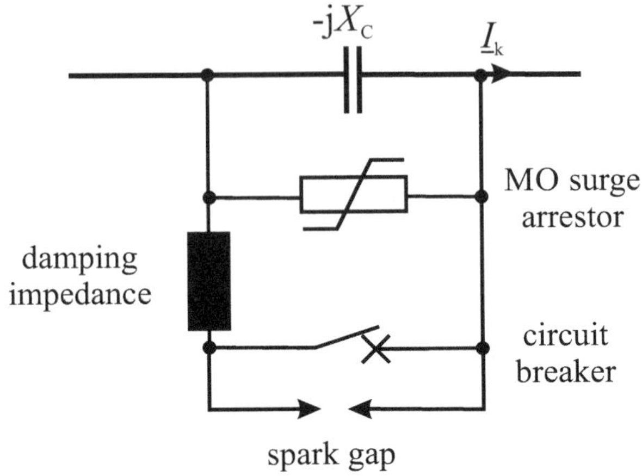

Fig. 14.12 Example of a series capacitor with additional protective device

$$k = \left| \frac{X_C}{X_L} \right| \tag{14.36}$$

In transmission networks, the compensation factor is selected between $0.3 \leq k \leq 0.7$ according to [7]. After substitution, this results in the following for the transmissible active power according to Eq. (14.35):

$$P_B = \frac{3 \cdot U_A \cdot U_B}{X_L \cdot (1 - k)} \cdot \sin(\alpha_A) \tag{14.37}$$

In practice, not only is a capacitor connected in series with a line, usually in the middle of the line, but protective devices are also required to protect the capacitor against overloading, resulting in the arrangement shown in Fig. 14.12 [8, 9].

The protective device basically has the task of protecting the capacitor from overvoltages and then disconnecting it from the network. Due to its limited energy absorption, the surge arrester is discharged by a triggered spark gap, which in turn is short-circuited by the circuit-breaker, so that the series capacitor is bypassed. According to [7], switch-off times of less than $t = 5$ ms are possible. In addition, a damping unit is required to protect the circuit-breaker in order to reduce the discharge current and its steepness. These processes, which occur during a short circuit, have different effects on the short-circuit current over time.

Examples

According to [9], the maximum compensation level should be selected so that a capacitive short-circuit current is avoided in order to prevent self-excitation of the connected generators. This prevents the occurrence of sub-synchronous resonances (SSR) [6]. In this section, both steady-state and transient calculations are performed with and without series compensation.

Data of Equipment

In the following, the data shown in Fig. 14.11 is assumed for the network connection in order to illustrate the entire process. For the purpose of simplification, it is assumed that only network A feeds in a short-circuit current.

- Network A
  nominal system voltage: $U_n = 380$ kV
  short-circuit current: $I''_k = 63\,\text{kA}$
  R/X ratio: $R/X = 0.1$
- Overhead line
  type: 4 × 240/40 mm^2
  resistance: $R'_{(1)} = 0.03\ \Omega/\text{km}$ $R'_{(0)} = 0.24\ \Omega/\text{km}$
  reactance: $X'_{(1)} = 0.251\ \Omega/\text{km}$ $X'_{(0)} = 1.4\ \Omega/\text{km}$
  length of line: $\ell = 300$ km
  steady state current: $I_d = 4 \times 645$ A
- Load
  operating voltage: $U_b = 400$ kV
  operating current: $I_{bmax} = 1290$ A
  power factor: cos $\varphi = 1.0;\ 0.8$

For the design of the surge arrester shown in Fig. 14.12, the maximum voltage along the capacitance is decisive, as the arrester must not be overloaded at the maximum operating current I_d of the line. For the continuous operating voltage U_c, the following applies, taking into account the reactance X_C of the capacitance:

$$U_c > X_C \cdot I_d = 41.415\ \Omega \cdot 2580\,\text{A} = 106.85\ \text{kV (rms value)}$$

According to [8], the protection level for compensation should be between 2.0 and 2.5, based on the longitudinal voltage drop under the condition of the rated current of the line.

14.2.2.1 Stationary Calculation

Taking into account the initial symmetrical short-circuit current of the network at busbar A, a maximum compensation factor of $k = 0.50$ is reasonable if the series compensation is located in the middle of the overhead line, in contrast to the representation in Fig. 14.11. The initial symmetrical short-circuit current in a three-phase short circuit is thus determined depending on the location of the fault:

$$\text{Upto } \ell < 150\ \text{km} \qquad \underline{I}''_k = \frac{1.1 \cdot U_n/\sqrt{3}}{\underline{Z}_N + \underline{Z}'_L \cdot \ell} \tag{14.38}$$

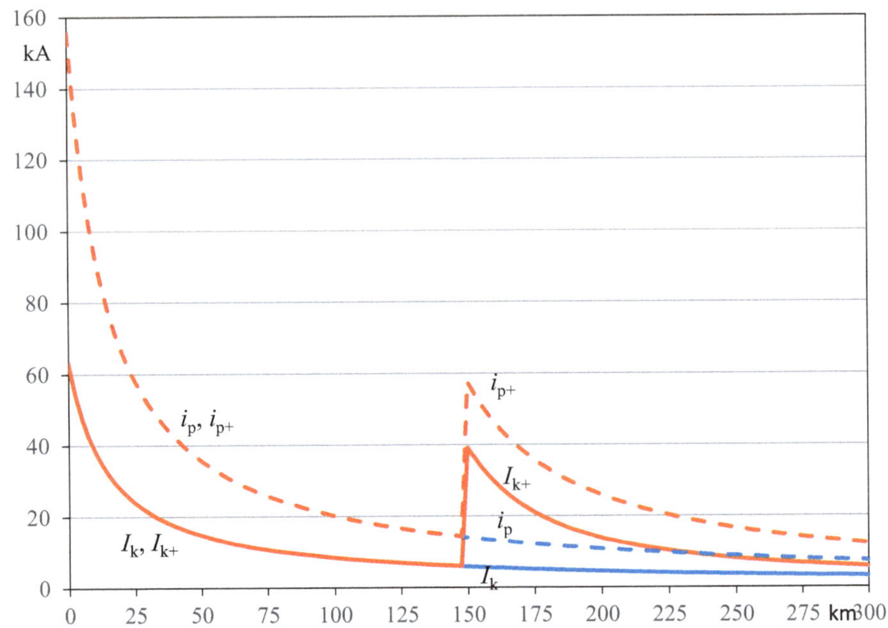

Fig. 14.13 Short-circuit currents according to Fig. 14.11 at a compensation factor of $k = 0.50$. I_k Initial symmetrical short-circuit current without compensation (rms values), I_{k+} initial symmetrical short-circuit current with compensation (rms values), i_p peak short-circuit current without compensation (peak values), i_{p+} peak short-circuit current with compensation (peak values)

$$\text{From } \ell \geq 150\text{ km} \qquad \underline{I}''_k = \frac{1.1 \cdot U_n/\sqrt{3}}{\underline{Z}_N + \underline{Z}'_L \cdot \ell - X_C} \tag{14.39}$$

With

U_n nominal system voltage,
c voltage factor (= 1.1),
$\underline{Z}_N$ impedance of the network,
$\underline{Z}'_L$ length-related overhead line impedance,
ℓ fault location,
X_C reactance of the compensation, $C = 84.544$ μF.

Under these conditions, Fig. 14.13 shows the initial symmetrical short-circuit current (rms value) as a function of the cable length ℓ at a distance from busbar A. In addition, the peak short-circuit current (peak value) based on the *R*/*X* ratio and the initial symmetrical short-circuit current is also shown. Figure 14.13 shows that immediately behind the compensation (at ℓ = 150 km), the short-circuit current increases again, but not to the initial value at busbar A, because although

Table 14.3 Calculation of the stationary initial symmetrical short-circuit currents $\underline{I}''_{k}$ and peak short-circuit currents i_p at locations A, M, and B in kA

Current	Location A	Location M		Location B	
	a, b	a	b	a	b
$\underline{I}''_{k}$	63.00	5.78	38.97	3.03	5.68
i_p	155.56	13.57	57.37	7.31	12.18

A busbar A
M immediately behind the series compensation
B busbar B
[a] Without series compensation
[b] With series compensation

compensation eliminates the inductance of the overhead line, it does not eliminate the existing resistance of the line. This has a particular effect on the *R*/*X* ratio when determining the peak short-circuit current.

For the specific short-circuit locations A ($\ell = 0$ km), M ($\ell = 150$ km) and B ($\ell = 300$ km), the currents determined are shown in Table 14.3. These values should be compared with the results of the transient calculations, Sect. 14.2.3.3.

In principle, the use of series compensation leads to a higher short-circuit current in the middle and in installation B (Fig. 14.11), which is relevant for the design of the switchgear and installations. However, it should be noted that compensation is generally equipped with a voltage limiting device. For this reason, the entire transient process is considered in the further description. In a first step, the short-circuit current is determined without this voltage limiter.

14.2.2.2 Transient Calculation without Voltage Limitation

The various simulations are performed using Fig. 14.14, whereby short circuits in the middle (M) and at the end (B) are considered. For simplicity, only one voltage source (network A) is assumed.

While the stationary calculations according to Sect. 14.2.3.2 are performed on the basis of the existing standard [3] (without load, short circuit occurring at zero voltage), the influence of these parameters is also examined below. Taking into account the load specifications according to Sect. 14.2.3.1, the following values of the load impedance $\underline{Z}_V$ result when the load current I_{bmax} according to Sect. 14.2.3.1 is assumed.

$$\underline{Z}_V = \frac{U_b/\sqrt{3}}{I_{bmax} \cdot (\cos\varphi - j\sin\varphi)} = \frac{400/\sqrt{3}}{1290} \cdot (\cos\varphi - j\sin\varphi)\Omega$$

Depending on the power factor, the following values result for the load impedance $\underline{Z}_V$, which are used in the simulations according to Fig. 14.14:

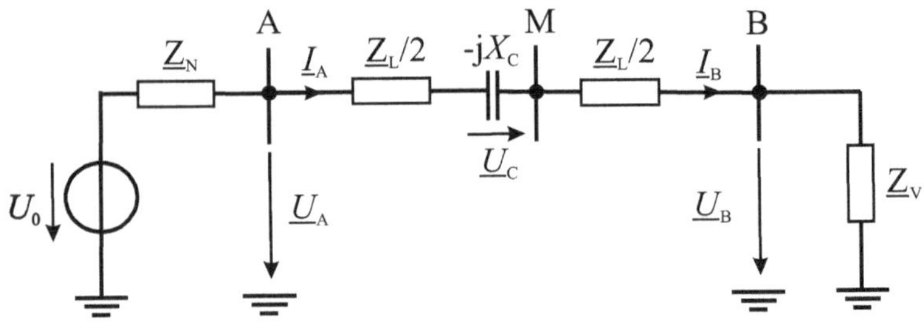

Fig. 14.14 Equivalent circuit diagram for determining the short circuit in series compensation

- $\cos \varphi = 1.0$: $\underline{Z}_{\mathrm{V}} = 179.02\ \Omega$ $L_{\mathrm{V}} = 0.00$ mH
- $\cos \varphi = 0.8$: $\underline{Z}_{\mathrm{V}} = 68.03 + \mathrm{j}51.02\ \Omega$ $L_{\mathrm{V}} = 341.91$ mH

In addition, the load condition "no-load" is also considered in the further simulations.

Due to the three-phase short circuit, e.g., station B, an oscillation process is excited, consisting of an LC series circuit, so that the short-circuit current can be determined in a simplified manner for the unloaded case and without resistance according to Eqs. (14.43) and (14.45) [6]. Here, the result of the temporal short-circuit current curve depends on the condition at the time of the fault occurrence (voltage maximum, voltage zero crossing). In the Laplace domain, the following generally applies, when the variables L and C are used for a series resonant circuit:

$$I_{\mathrm{A}}(p) = \frac{1}{L} \cdot \frac{p}{p^2 + \omega_0^2} \cdot E(p) \tag{14.40}$$

With the natural frequency:

$$\omega_0^2 = \frac{1}{L \cdot C} \tag{14.41}$$

- Voltage maximum:

$$E(p) = \frac{1.1 \cdot \sqrt{2} \cdot U_{\mathrm{n}}}{\sqrt{3}} \cdot \frac{p}{p^2 + \omega^2} \tag{14.42}$$

$$i_{\mathrm{A}}(t) = \frac{1.1 \cdot \sqrt{2} \cdot U_{\mathrm{n}}/\sqrt{3} \cdot \omega \cdot C}{1 - (\omega/\omega_0)^2} \cdot \left\{ \frac{\omega_0}{\omega} \cdot \sin(\omega_0 \cdot t) - \sin(\omega \cdot t) \right\} \tag{14.43}$$

- Voltage zero:

$$E(p) = \frac{1.1 \cdot \sqrt{2} \cdot U_{\mathrm{n}}}{\sqrt{3}} \cdot \frac{\omega}{p^2 + \omega^2} \tag{14.44}$$

$$i_{\mathrm{A}}(t) = \frac{1.1 \cdot \sqrt{2} \cdot U_{\mathrm{n}}/\sqrt{3} \cdot \omega \cdot C}{1 - (\omega/\omega_0)^2} \cdot \{\cos(\omega \cdot t) - \cos(\omega_0 \cdot t)\} \tag{14.45}$$

The equations listed above show that, in addition to the operating frequency part of the short-circuit current, an additional current is formed as a result of the resonant circuit. Furthermore, the operating frequency current part can assume high values depending on the degree of compensation, since a resonant circuit is excited whose natural frequency can be close to the operating frequency. Referring to network location A in Fig. 14.14, the natural frequency according to Eq. (14.41) for a short circuit directly behind the compensation (short-circuit location M) is obtained when the compensation compensates for the first half of the entire line:

$$\omega_0^2 = \frac{1}{L \cdot C} = \frac{X_{\mathrm{C}}}{X_{\mathrm{L}}/2} \cdot \omega^2 \stackrel{!}{=} \omega^2 \tag{14.46}$$

Taking into account the network inductance, the following frequencies result from Eq. (14.46) depending on the short-circuit location.

- Short-circuit location M: $\omega_0 = 299.37$ 1/s $f_0 = 47.65$ Hz
- Short-circuit location B: $\omega_0 = 216.73$ 1/s $f_0 = 34.49$ Hz

Based on the frequency ratio ω/ω_0, it can be seen that a short circuit occurring at zero voltage results in greater oscillation than a short circuit occurring at maximum voltage. This also applies to a short circuit directly behind the compensation, compared to short circuit location B.

The current profile according to Eq. (14.45) can generally be converted using trigonometric functions, so that the following applies:

$$i_{\mathrm{A}}(t) = \frac{1.1 \cdot \sqrt{2} \cdot U_{\mathrm{n}}/\sqrt{3} \cdot \omega \cdot C}{1 - (\omega/\omega_0)^2} \cdot 2 \cdot \sin\frac{(\omega + \omega_0) \cdot t}{2} \cdot \cos\frac{(\omega - \omega_0) \cdot t}{2} \tag{14.47}$$

$$i_{\mathrm{A}}(t) = \frac{1.1 \cdot \sqrt{2} \cdot U_{\mathrm{n}}/\sqrt{3} \cdot \omega \cdot C}{1 - (\omega/\omega_0)^2} \cdot 2 \cdot \sin(\omega_{\mathrm{R}} \cdot t) \cdot \cos(\omega_{\mathrm{S}} \cdot t) \tag{14.48}$$

With

ω_{R} resulting frequency as the average of the two output frequencies $(\omega + \omega_0)/2$,
ω_{S} beat frequency $(\omega - \omega_0)/2$.

The condition according to Eq. (14.47) applies in the case where the amplitudes of the two oscillation components are identical, which is true according to Eq. (14.45). Equation (14.48) shows that the resulting amplitude is twice the value of the individual oscillations, with the main frequency being the average of the two frequencies. In addition, a superimposed beat appears as a function of the difference between these frequencies.

If the short-circuit current is determined according to Eq. (14.39), e.g., at short-circuit location B, the calculation for the peak value of the steady-state, operating-frequency current i_{pmax}, neglecting the resistances, yields:

$$i_{\mathrm{pmax}} = \frac{1.1 \cdot \sqrt{2} \cdot U_{\mathrm{n}}/\sqrt{3}}{X_{\mathrm{N}} + X_{\mathrm{L}} - X_{\mathrm{C}}} = \frac{1.1 \cdot \sqrt{2} \cdot U_{\mathrm{n}}/\sqrt{3}}{\omega L - \dfrac{1}{\omega C}} = \frac{1.1 \cdot \sqrt{2} \cdot U_{\mathrm{n}}/\sqrt{3} \cdot \omega C}{1 - (\omega/\omega_0)^2} \quad (14.49)$$

A comparison with Eq. (14.48) shows that the calculation according to Eq. (14.49) only takes into account the operating frequency current component and is therefore lower than the actual current. When calculating the maximum peak value (peak short-circuit current) according to [3], the shift in current due to the short circuit occurring at zero voltage is taken into account by introducing the κ factor. If the resistance of the short-circuit path is not taken into account in this case either, the factor κ assumes the value 2. Under these conditions, the maximum value is identical, but the physical background is completely different, meaning that these processes are not comparable.

The transient current curves for the different load conditions are plotted for a short circuit occurring at zero voltage crossing at short-circuit locations B and M in Fig. 14.15.

In the simulations shown in Fig. 14.15, traveling wave processes excited by overhead lines are not considered, as the line is modeled exclusively by inductance and resistance. Since the calculations show that the highest current values occur at the voltage zero-crossing during fault initiation, neglecting the line capacitances has no significant effect on the result, as these are uncharged at the time of the fault.

The results show that the maximum short-circuit currents at locations B and M are almost independent of the load specified above if the short circuit occurs during the zero crossing of the voltage at the fault location. The maximum peak short-circuit currents do not occur in the range $t \leq 10$ ms, as is the case with an *LR* resonant circuit (Sect. 3.4) or [3], but the maximum peak value occurs at a fault in station B after $t = 33.5$ ms with a value of $i_{\mathrm{pBmax}} = 13.3$ kA. In the event of a fault immediately behind the capacitor (M), the maximum current is not yet reached in the time range of $t < 100$ ms. The characteristic maximum current values for assessing the peak short-circuit current are listed in Table 14.4 for the range $t \leq 10$ ms and $t \leq 100$ ms.

A comparison of the values for peak short-circuit currents between the results in Table 14.3 (i_{p}, column 2) and Table 14.4 (column 2) shows only a slight deviation with regard to the unloaded load case (location B: 12.18–13.27 kA or location M: 57.37–57.86 kA). However, the peak values occur at different times due to the

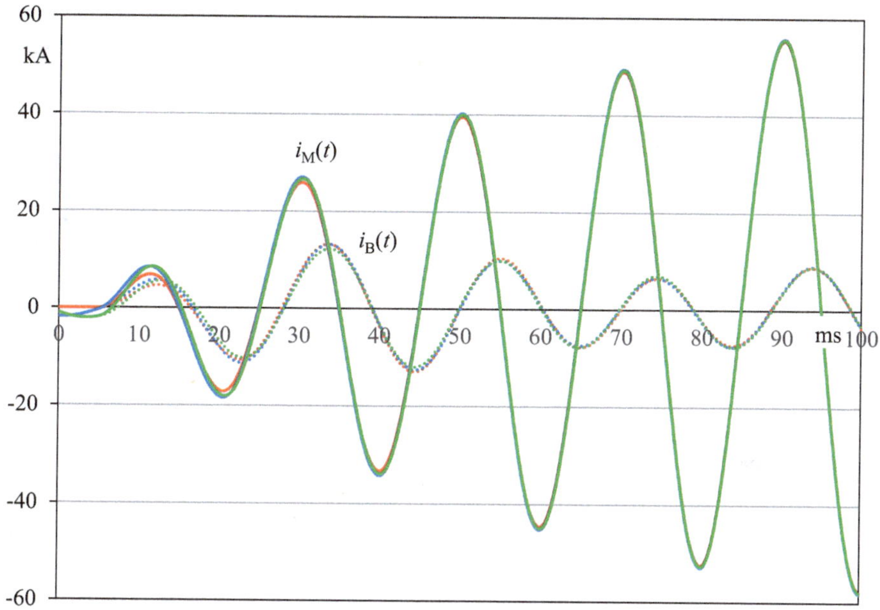

Fig. 14.15 Short-circuit currents in a three-phase short circuit (fault occurrence at voltage zero crossing at $t = 5$ ms) in the middle of the line connection $i_M(t)$ and at the end $i_B(t)$ (station B) at different loads. Solid line: short-circuit location M, dotted line: short-circuit location B

Table 14.4 Maximum currents i_p at short-circuit locations M and B, depending on load I_b (with series compensation)

Load	Location M		Location B	
	a	b	a	b
$I_b = 0.0$ kA	6.88	57.86	4.66	13.27
I_{bmax}; $\cos\varphi = 1.0$	8.45	57.16	5.81	13.27
I_{bmax}; $\cos\varphi = 0.8$	8.46	58.03	5.62	12.33

M immediately behind the series compensation
B busbar B
I_b operating current before short-circuit initiation
$\cos\varphi$ power factor before short-circuit initiation
[a] Time range $t \leq 10$ ms after short-circuit initiation
[b] Time range $t \leq 100$ ms after short-circuit initiation

different current curves, which cannot be compared with each other, as shown above. The consequence is that a calculation of the short-circuit current (peak short-circuit current, breaking current) using standard [3] is not permissible if the series capacity is taken into account as an operating element, since the short-circuit current model under consideration and the quantities derived from it are not comparable.

A rough estimate of the short-circuit current can be derived from the amplitude of the current waveform for the initial symmetrical short-circuit current according to Eq. (14.47) (without resistance), so that the following applies:

$$I_k'' \approx 2 \cdot \frac{1.1 \cdot U_n/\sqrt{3} \cdot \omega \cdot C}{1 - (\omega/\omega_0)^2} = 2 \cdot \frac{1.1 \cdot U_n/\sqrt{3}}{X_k} = 2 \cdot \frac{1.1 \cdot U_n/\sqrt{3}}{(X_N + X_L/2) - X_C} \quad (14.50)$$

With

U_n	nominal system voltage,
c	voltage factor,
ω	operating frequency (1/s),
ω_0	natural frequency of the resonant circuit (1/s), Eq. (14.46),
C	capacitance of the general resonant circuit,
X_k	short-circuit reactance,
$X_{N,L,C}$	values of the network according to Fig. 14.14.

The following relationships apply to the remaining currents:

$$\text{Max.peak short-circuit current}: i_{pmax} \approx \sqrt{2} \cdot I_k'' \quad (14.51)$$

$$\text{Breaking current}: I_b \approx I_k'' \quad (14.52)$$

$$\text{Steady state short-circuit current}: I_k \approx I_k'' \quad (14.53)$$

Equations (14.50)–(14.53) take into account the superposition of the oscillation process in the event of a short circuit occurring at zero voltage, as is also applied in the calculation of the steady state short-circuit current of generators in meshed networks according to Sect. 5.7.3.

14.2.2.3 Transient Calculation with Voltage Limitation

To limit the capacitor voltage during the short circuit, a protective circuit is introduced in accordance with Fig. 14.12, which consists, among other things, of a surge arrester installed in parallel with the capacitor. Under this condition, the currents and voltages in the event of a fault on busbar B are simulated and listed in Fig. 14.16 for different boundary conditions. No load according to Table 14.4 is taken into account so that the results are comparable with the calculation according to [3].

The maximum peak value i_p of the current is given in a time range of $t \leq 10$ ms after the short circuit occurs, by Fig. 14.16 as:

- $i_p = 7.27$ kA without series compensation (as a result of the direct current displacement when a short circuit occurs at zero voltage), and
- $i_{p+A} = 4.66$ kA with series compensation and surge arrester (corresponds to the curve without surge arrester in this time range).

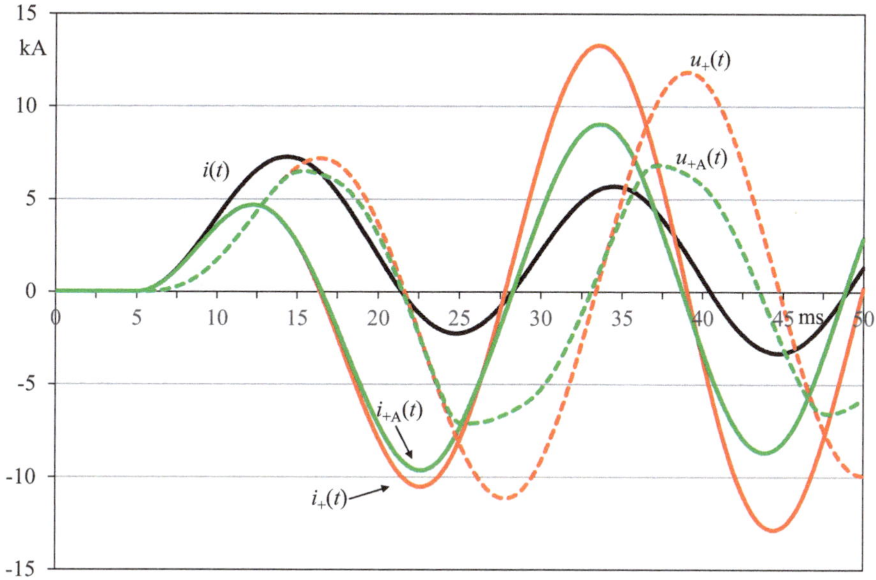

Fig. 14.16 Transient currents $i(t)$ and voltages $u(t)$ during a three-phase short circuit (short circuit location B) depending on compensation. $i(t)$ current without series compensation, $i_{+}(t)$ current with series compensation, $i_{+A}(t)$ current with series compensation and surge arrester, $u_{+}(t)$ capacitor voltage with series compensation, $u_{+A}(t)$ capacitor voltage with series compensation and surge arrester

Since the voltage limiter should not be overloaded because of the energy absorption of the surge arrester, the parallel spark gap responds in the time range of 5–10 ms. This short-circuits the series capacitor, making it ineffective. Assuming a short circuit in the middle of the line (short circuit location M), as shown in Fig. 14.16, Fig. 14.17 shows the results.

In this case, the following currents result for $t \leq 10$ ms after the short circuit occurs:

- $i_{p} = 13.89$ kA without series compensation (as a result of the direct current displacement when a short circuit occurs at zero voltage), and
- $i_{p+A} = 6.88$ kA with series compensation and surge arrester (corresponds to the curve without surge arrester in this time range).

In this case, too, it is necessary to bypass the arrester and thus also the capacitor in order to avoid overloading. Figure 14.18 shows the current curve for a short circuit at fault locations M and B for both variants:

- $i_{M}(t)$, $i_{B}(t)$: without series compensation,
- $i_{+M}(t)$, $i_{+B}(t)$: with series compensation and short circuit of the capacitor after $t = 5$ ms of the short circuit occurs.

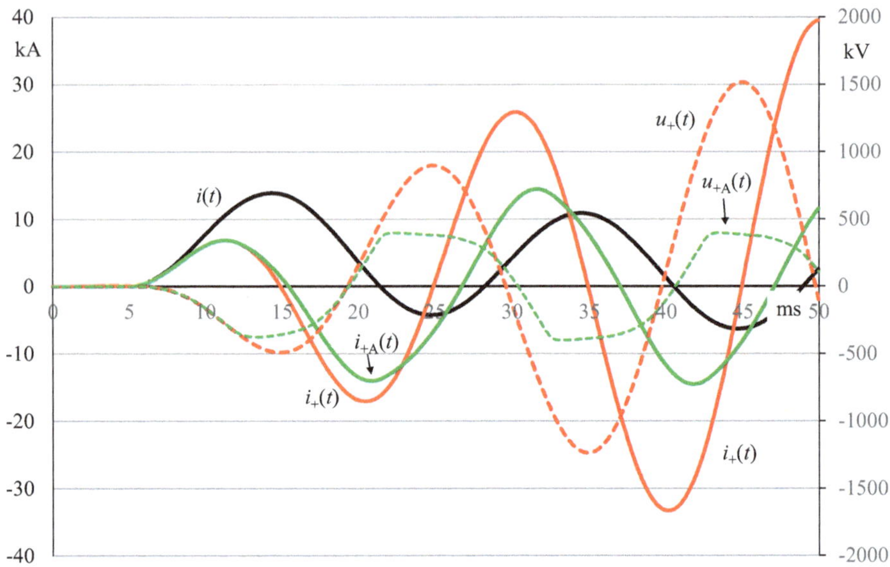

Fig. 14.17 Transient currents $i(t)$ and voltages $u(t)$ in the case of a three-phase short circuit (short-circuit location M) as a function of compensation (legend Fig. 14.16)

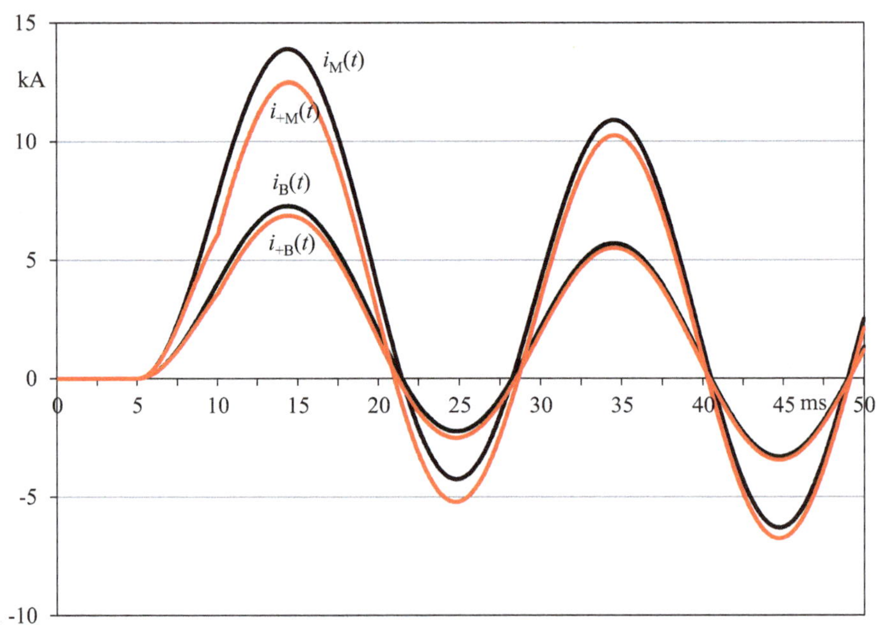

Fig. 14.18 Short-circuit current curve in the event of a fault at locations M and B. i_M, i_B without series compensation, i_{+M}, i_{+B} with series compensation and short circuit after 5 ms

A short circuit in the series capacitor discharges or bypasses it, with the amplitude and frequency of the discharge current being influenced by the damping impedance (Fig. 14.12). Figure 14.18 shows that the assumption according to [3] is justified, namely that the series capacitor should not be taken into account in the short-circuit current calculation if the voltage limitation device is activated.

14.3 Conclusions

According to IEC 600909-0 [3], it is not necessary to take shunt capacitors into account when calculating short-circuit current. In principle, capacitors contribute to the initial symmetrical short-circuit current and to the peak short-circuit current. However, due to the different boundary conditions (voltage maximum, voltage zero crossing) and the design criteria (maximum voltage increase during connection), it is not necessary to consider them.

When using large capacitors in networks with a high short-circuit current or when connecting several capacitors in parallel, it may be necessary to install inductances in series with the capacitance so that the peak short-circuit current of the component is not exceeded.

Series capacitors are used in electrical power transmission to compensate for reactance of long lines, thereby reducing the voltage drop depending on the load and thus increasing power transmission. In principle, series compensation reduces short-circuit reactance, which leads to an increase in short-circuit current. According to [3], a series capacitor does not need to be taken into account in the short-circuit current calculation if, due to an overvoltage protection device, this capacitor is short-circuited in a short time, e.g., between 5 and 10 ms, and thus becomes ineffective.

If this is not the case, the maximum short-circuit current can be roughly estimated from the superposition of the currents with the operating frequency and natural frequency of the oscillating circuit. In principle, the short-circuit current should then be simulated using computer programs to simulate transient processes in this network state, especially since there is also the possibility that sub-synchronous resonances may be excited [5].

References

1. Balzer G (2013) Einfluss von Kondensatoren auf den dreipoligen Kurzschlussstrom. ew Jg 112 (9), pp 48–53
2. Balzer G, Petranovic R (2014) Impact of H.V. capacitors for reactive power compensation in transmission systems on the short-circuit current. In: 20th conference of the electric power supply industry (CEPSI), 2014 October 26-30, Jeju, South Korea, rep. D1
3. IEC 60909-0:2016 Edition 2.0 (2016) Short-circuit current calculations in three-phase a.c. systems—Part 0: calculation of currents. IEC, Geneva
4. Gruenbaum R, Redlund J, Rollin LP (2007) Safety in speed. ABB Rev 3, pp 46–49

5. McTaggart C, Adam D, Brozio C, Stokoe J, Adobes R (2018) The application of series compensation to the existing Scottish 400 kV transmission system. CIGRE 2018, Paris, C4-125
6. Balzer G, Neumann C, Balzer C (2025) Schalt- und Ausgleichsvorgänge in elektrischen Netzen, 2nd edn. Springer, Heidelberg
7. Gruenbaum R, Rasmussen J, Li C (2012) Series capacitors for increased power transmission capability of a 500 kV grid intertie. In: IEEE electrical and energy conference October 10–12, 2012, London, Ontario, Canada
8. Cigre Brochure 411 (2010) Protection, control and monitoring of series compensated networks. WG B5.10, April
9. Cigre Brochure 829 (2021) Challenges with series compensation applications in power systems when overcompensating lines. JWG C4/B5.41

Chapter 15
Calculation and Operation with Parallel Transformers

In the case of parallel transformers, different results may be obtained when determining short-circuit currents, depending on the calculation method used and taking into account different transformation ratios. The reasons for this are explained below; along with how this should be taken into account when calculating short-circuit currents in accordance with IEC 60909. Parts of this chapter have been published in [1, 2].

In addition, the second part of the chapter (Sect. 15.2) lists various options for grounding transformers that have an impact on short-circuit current calculations when faults involving earth contact are considered [4–6].

15.1 Calculation with Parallel Transformers

Short-circuit current calculations in electrical networks are generally performed using IEC Standard 60909-0 [3], which applies the equivalent voltage source method at the short-circuit location as the calculation method. This method has the significant advantage that only one voltage source is taken into account, even in an extensive network with many feeders, and that preload before the short circuit occurs does not need to be considered. Here, the load on the network is simulated by applying a voltage factor c and, if necessary, by applying impedance correction. When transformers with unequal transformation ratios are connected in parallel, an error may occur in the short-circuit current calculation, taking into account the transformer equivalent circuit diagram used, so that a current is calculated in the event of a short circuit even though there is no feed from an active network.

The reason for this investigation was that during the standard check of the influence of short-circuit current injection from a lower voltage level on the high-voltage side of parallel transformers, it was found that short-circuit current injection, e.g., at the 110 kV level, occurred even though there was no active feed at the 10 kV

G. Balzer, *Short-Circuit Currents in Three-Phase Systems*,
https://doi.org/10.1007/978-3-658-50710-7_15

level. The calculation was performed using the equivalent voltage source method at the short-circuit location for different transformation ratios of the parallel transformers. Contrary to the instructions in the existing standard [3], which requires the arithmetic mean to be taken in such cases, the different transformation ratios of the parallel transformers were used in this case.

The following sections show the different options for taking parallel transformers with unequal transformation ratios into account in short-circuit current calculations when using the equivalent voltage source method at the short-circuit location.

15.1.1 Transformer Equivalent Circuit

In the short-circuit current calculation, the transformer is modeled using the short-circuit impedance (longitudinal impedance) and the transformation ratio. No-load losses and the magnetizing reactance are not taken into account (Sect. 4.8).

15.1.1.1 General Equivalent Circuit

The equivalent circuit shown in Fig. 15.1 is used to calculate the short-circuit currents, so that only the short-circuit impedance $\underline{Z}_T$ and the transformation ratio t are considered. The advantage of using this equivalent circuit diagram is that calculations can be made using the actual voltages and currents, and there is no need to convert impedances to the respective voltage level.

In principle, the transformer's transformation ratio can be complex, but in the following, only a real ratio is assumed.

The equivalent circuit diagram shown in Fig. 15.1 can be represented by two chain matrices: a longitudinal branch with impedance $\underline{Z}_T$ and a shunt branch with transformation ratio t, resulting in the following individual matrices:

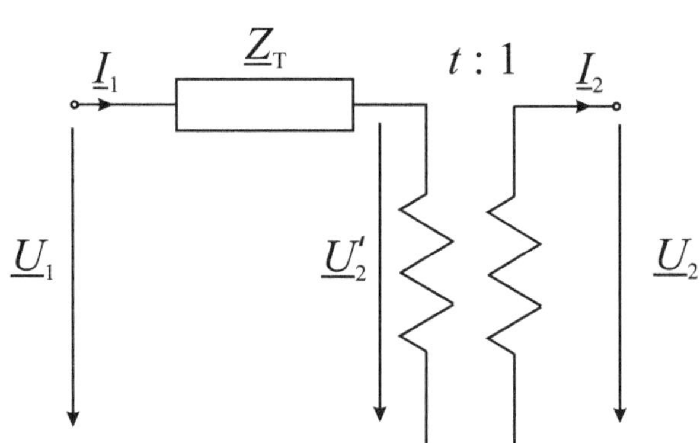

Fig. 15.1 General equivalent circuit of a power transformer

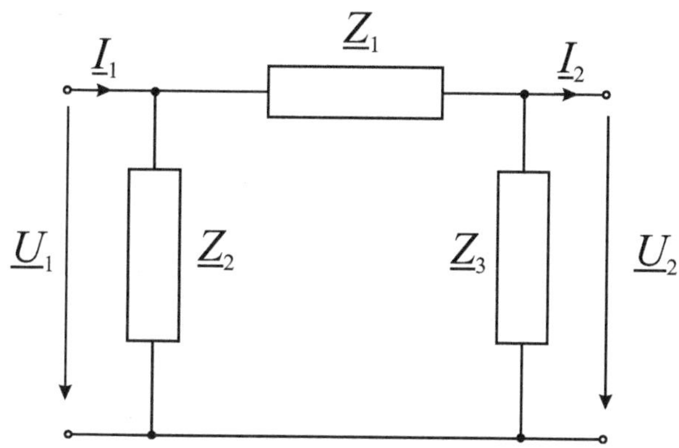

Fig. 15.2 Equivalent π circuit of a transformer

$$\begin{pmatrix} \underline{U}_1 \\ \underline{I}_1 \end{pmatrix} = \begin{pmatrix} 1 & \underline{Z}_\mathrm{T} \\ 0 & 1 \end{pmatrix} \cdot \begin{pmatrix} \underline{U}'_2 \\ \underline{I}'_2 \end{pmatrix} \qquad \begin{pmatrix} \underline{U}'_2 \\ \underline{I}'_2 \end{pmatrix} = \begin{pmatrix} t & 0 \\ 0 & 1/t \end{pmatrix} \cdot \begin{pmatrix} \underline{U}_2 \\ \underline{I}_2 \end{pmatrix} \tag{15.1}$$

Multiplying the matrices will result in the chain matrix, Eq. (15.2), and the corresponding admittance matrix, Eq. (15.3), where the transformer impedance $\underline{Z}_\mathrm{T}$ is related to the high-voltage side.

$$\begin{pmatrix} \underline{U}_1 \\ \underline{I}_1 \end{pmatrix} = \begin{pmatrix} t & \underline{Z}_\mathrm{T}/t \\ 0 & 1/t \end{pmatrix} \cdot \begin{pmatrix} \underline{U}_2 \\ \underline{I}_2 \end{pmatrix} \tag{15.2}$$

$$\begin{pmatrix} \underline{I}_1 \\ \underline{I}_2 \end{pmatrix} = \begin{pmatrix} \underline{Y}_\mathrm{T} & -t \cdot \underline{Y}_\mathrm{T} \\ t \cdot \underline{Y}_\mathrm{T} & -t^2 \cdot \underline{Y}_\mathrm{T} \end{pmatrix} \cdot \begin{pmatrix} \underline{U}_1 \\ \underline{U}_2 \end{pmatrix} \tag{15.3}$$

Based on the chain matrix, an equivalent π circuit can be derived (Fig. 15.2) with the following elements:

$$\underline{Z}_1 = \frac{\underline{Z}_\mathrm{T}}{t} \qquad \underline{Z}_2 = \frac{\underline{Z}_\mathrm{T}}{1-t} \qquad \underline{Z}_3 = \frac{\underline{Z}_\mathrm{T}}{t^2 - t} \tag{15.4}$$

In contrast to the equivalent circuit diagram shown in Fig. 15.2, according to [3] the transformer is considered exclusively by its short-circuit impedance and converted to the voltage level under consideration using the transformation ratio.

If two transformers with different transformation ratios t_1 and t_2 are assumed to be in parallel, the admittance matrix for the two transformers is determined according to Eq. (15.5).

$$\begin{pmatrix} \underline{I}_1 \\ \underline{I}_2 \end{pmatrix} = \begin{pmatrix} \underline{Y}_\mathrm{T1} + \underline{Y}_\mathrm{T2} & -(t_1 \cdot \underline{Y}_\mathrm{T1} + t_2 \cdot \underline{Y}_\mathrm{T2}) \\ t_1 \cdot \underline{Y}_\mathrm{T1} + t_2 \cdot \underline{Y}_\mathrm{T2} & -(t_1^2 \cdot \underline{Y}_\mathrm{T1} + t_2^2 \cdot \underline{Y}_\mathrm{T2}) \end{pmatrix} \cdot \begin{pmatrix} \underline{U}_1 \\ \underline{U}_2 \end{pmatrix} \tag{15.5}$$

From the admittance matrix according to Eq. (15.5), an equivalent chain matrix A can be formed according to Eq. (15.6).

$$\begin{pmatrix} \underline{U}_1 \\ \underline{I}_1 \end{pmatrix} = \begin{pmatrix} \underline{A}_{11} & \underline{A}_{12} \\ \underline{A}_{21} & \underline{A}_{22} \end{pmatrix} \cdot \begin{pmatrix} \underline{U}_2 \\ \underline{I}_2 \end{pmatrix} = \begin{pmatrix} -\dfrac{\underline{Y}_{22}}{\underline{Y}_{21}} & \dfrac{1}{\underline{Y}_{21}} \\ \underline{Y}_{12} - \dfrac{\underline{Y}_{11} \cdot \underline{Y}_{22}}{\underline{Y}_{21}} & \dfrac{\underline{Y}_{11}}{\underline{Y}_{21}} \end{pmatrix} \cdot \begin{pmatrix} \underline{U}_2 \\ \underline{I}_2 \end{pmatrix} \tag{15.6}$$

The elements $\underline{A}_{ij}$ of the chain matrix can be expressed by the elements of the admittance matrix, Eq. (15.5).

$$\underline{A}_{11} = \frac{\underline{Y}_{T1} \cdot t_1^2 + \underline{Y}_{T2} \cdot t_2^2}{\underline{Y}_{T1} \cdot t_1 + \underline{Y}_{T2} \cdot t_2} \qquad \underline{A}_{12} = \frac{1}{\underline{Y}_{T1} \cdot t_1 + \underline{Y}_{T2} \cdot t_2} \tag{15.7}$$

$$\underline{A}_{21} = -(\underline{Y}_{T1} \cdot t_1 + \underline{Y}_{T2} \cdot t_2) + \frac{(\underline{Y}_{T1} + \underline{Y}_{T2}) \cdot (\underline{Y}_{T1} \cdot t_1^2 + \underline{Y}_{T2} \cdot t_2^2)}{\underline{Y}_{T1} \cdot t_1 + \underline{Y}_{T2} \cdot t_2} \tag{15.8}$$

$$\underline{A}_{22} = \frac{\underline{Y}_{T1} + \underline{Y}_{T2}}{\underline{Y}_{T1} \cdot t_1 + \underline{Y}_{T2} \cdot t_2} \tag{15.9}$$

If the admittances $\underline{Y}_T = \underline{Y}_{T1} = \underline{Y}_{T2}$ are identical, this can be simplified from Eq. (15.5).

$$\begin{pmatrix} \underline{I}_1 \\ \underline{I}_2 \end{pmatrix} = \begin{pmatrix} 2 \cdot \underline{Y}_T & -2 \cdot \underline{Y}_T \cdot \dfrac{t_1 + t_2}{2} \\ 2 \cdot \underline{Y}_T \cdot \dfrac{t_1 + t_2}{2} & -2 \cdot \underline{Y}_T \cdot \dfrac{t_1^2 + t_2^2}{2} \end{pmatrix} \cdot \begin{pmatrix} \underline{U}_1 \\ \underline{U}_2 \end{pmatrix} \tag{15.10}$$

A comparison of matrices (15.3) and (15.10) yields two different approximations with regard to a resulting transformation ratio, namely:

$$t_{res1} = \frac{t_1 + t_2}{2} \text{ from elements } \underline{A}_{12} \text{ and } \underline{A}_{21}\text{, respectively} \tag{15.11}$$

$$t_{res2} = \sqrt{\frac{t_1^2 + t_2^2}{2}} \text{ from the element } \underline{A}_{22} \tag{15.12}$$

The deviations between the two translation ratios are small (a difference of Δt of 25% results in a difference of 0.61%), so that the values for t_{res1} and t_{res2} can be considered equivalent.

For different transformers, all impedances $\underline{Z}_1$, $\underline{Z}_2$ und $\underline{Z}_3$ of the individual transformers are connected in parallel (Fig. 15.2), so that the elements of the equivalent π circuit can be calculated according to Eq. (15.13).

$$\begin{aligned} \underline{Z}_1 &= \frac{\underline{Z}_{T1} \cdot \underline{Z}_{T2}}{\underline{Z}_{T1} \cdot t_2 + \underline{Z}_{T2} \cdot t_1} \\ \underline{Z}_2 &= \frac{\underline{Z}_{T1} \cdot \underline{Z}_{T2}}{\underline{Z}_{T1} \cdot (1 - t_2) + \underline{Z}_{T2} \cdot (1 - t_1)} \\ \underline{Z}_3 &= \frac{\underline{Z}_{T1} \cdot \underline{Z}_{T2}}{\underline{Z}_{T1} \cdot (t_2^2 - t_2) + \underline{Z}_{T2} \cdot (t_1^2 - t_1)} \end{aligned} \tag{15.13}$$

15.1.1.2 Input Impedance

Section 15.1.2 considers the example of a three-phase short circuit whose feeder is exclusively from the LV side (10 kV) of the transformer (secondary side, Fig. 15.4), while the short circuit is located on the HV side (110 kV). Taking into account the two calculation methods (equivalent voltage source at the short-circuit location and actual voltage distribution), different input impedances can be determined, whereby the short-circuit current contribution as a result of the network feeder on the secondary side can also be varied. The determination of the various input impedances is shown in Fig. 15.3. The transformers are always modeled using the π equivalent circuit diagram shown in Fig. 15.2.

- Variant 1: Equivalent voltage source at the short-circuit location
  Voltage source $\underline{U}_1$ on the primary side (110 kV) and secondary side (10 kV) is either short-circuited ($\underline{Z}_N = 0$, large short-circuit current contribution on the secondary side, variant 1a) or open ($\underline{Z}_N \to \infty$, small short-circuit current contribution on the secondary side, variant 1b), Fig. 15.3a.
- Variant 2: Actual voltage distribution
  Voltage source $\underline{U}_2$ on the secondary side (10 kV), primary side (110 kV) is short-circuited, which corresponds to the actual boundary conditions in this fault case, Fig. 15.3b.

The short-circuit current can then be determined using the respective input impedance, taking into account the actual voltage. Various conditions are possible when determining the input impedance, which are discussed below. The transformer impedance $\underline{Z}_T$ corresponds to the longitudinal impedance according to Fig. 15.1 and is related to the HV side.

- Variant 1a: Impedance $\underline{Z}_{k1}$, seen from the 110 kV side, when the impedance $\underline{Z}_N$ on the 10 kV side is zero. The input impedance $\underline{Z}_{k1}$ is calculated according to Eq. (15.14). This corresponds to the network example in which the active network feeds an infinite short-circuit current into the LV side (the impedance of the feeding network is zero).

$$\underline{Z}_{k1} = \frac{\underline{Z}_1 \cdot \underline{Z}_2}{\underline{Z}_1 + \underline{Z}_2} \tag{15.14}$$

  - The impedance of the transformers and the transformation ratio are identical. Under these conditions, the individual impedances are as follows:

$$\underline{Z}_1 = \frac{\underline{Z}_T/2}{t} \qquad \underline{Z}_2 = \frac{\underline{Z}_T/2}{1-t} \tag{15.15}$$

It follows that:

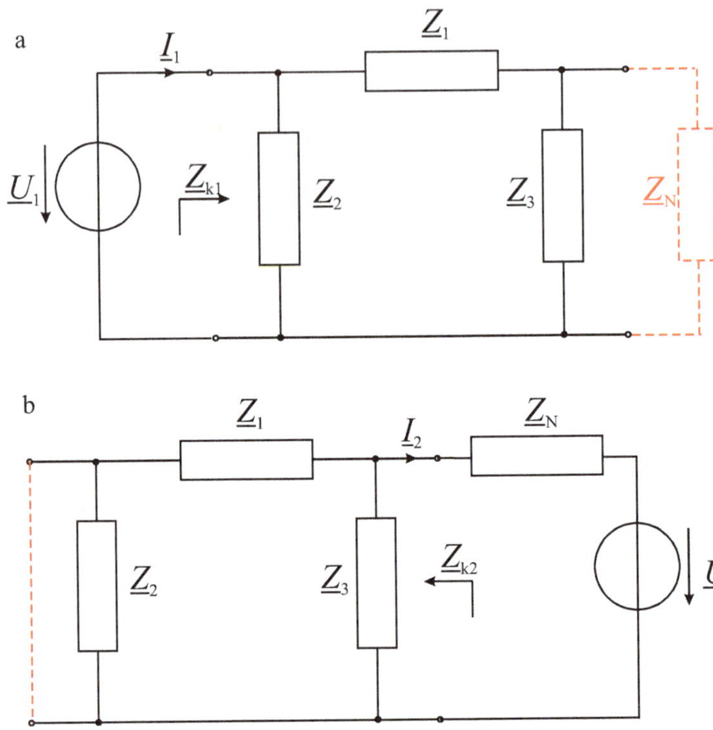

Fig. 15.3 Representation of the input impedance $\underline{Z}_k$, short circuit or open circuit on the secondary side. (**a**) Method of the equivalent voltage source at the short circuit location, (**b**) with actual voltage distribution

$$\underline{Z}_{k1} = \frac{\frac{\underline{Z}_T/2}{1-t} \cdot \frac{\underline{Z}_T/2}{t}}{\frac{\underline{Z}_T/2}{1-t} + \frac{\underline{Z}_T/2}{t}} = \frac{\underline{Z}_T}{2} \cdot \frac{1}{1-t+t} = \underline{Z}_T/2 \tag{15.16}$$

In this case, the resulting input impedance $\underline{Z}_{k1}$ is equal to the parallel connection of the short-circuit impedances $\underline{Z}_T$ of the transformers.

– The impedances of the transformers are different, but the transformation ratio is the same. Under these conditions, the individual impedances are as follows:

$$\begin{aligned} \underline{Z}_1 &= \frac{\underline{Z}_{T1} \cdot \underline{Z}_{T2}}{t \cdot (\underline{Z}_{T1} + \underline{Z}_{T2})} \\ \underline{Z}_2 &= \frac{\underline{Z}_{T1} \cdot \underline{Z}_{T2}}{(1-t) \cdot (\underline{Z}_{T1} + \underline{Z}_{T2})} \end{aligned} \tag{15.17}$$

$$\underline{Z}_{k1} = \frac{\underline{Z}_{T1} \cdot \underline{Z}_{T2}}{\underline{Z}_{T1} + \underline{Z}_{T2}} \cdot \frac{\frac{1}{(1-t) \cdot t}}{\frac{1}{(1-t)} + \frac{1}{t}} = \frac{\underline{Z}_T}{2} \cdot \frac{\underline{Z}_{T1} \cdot \underline{Z}_{T2}}{\underline{Z}_{T1} + \underline{Z}_{T2}} \cdot \frac{1}{1-t+t} = \frac{\underline{Z}_{T1} \cdot \underline{Z}_{T2}}{\underline{Z}_{T1} + \underline{Z}_{T2}} \tag{15.18}$$

In this case, the resulting input impedance $\underline{Z}_{k1}$ is equal to the parallel connection of the short-circuit impedances $\underline{Z}_T$ of the transformers.

– The impedances of the transformers are the same, with different transformation ratios. Under these conditions, the individual impedances are as follows:

$$\underline{Z}_1 = \frac{\underline{Z}_T}{t_1 + t_2} \qquad \underline{Z}_2 = \frac{\underline{Z}_T}{(1-t_1)+(1-t_2)} \tag{15.19}$$

$$\underline{Z}_{k1} = \underline{Z}_T \cdot \frac{\frac{1}{(1-t_1)+(1-t_2)} \cdot \frac{1}{t_1+t_2}}{\frac{1}{(1-t_1)+(1-t_2)} + \frac{1}{t_1+t_2}} = \frac{1}{(1-t_1)+(1-t_2)+t_1+t_2} = \frac{\underline{Z}_T}{2} \tag{15.20}$$

In this case, the resulting input impedance $\underline{Z}_{k1}$ is equal to the parallel connection of the short-circuit impedances $\underline{Z}_T$ of the transformers.

- Variant 1b: Impedance $\underline{Z}_{k1}$, seen from the 110 kV side, when the impedance $\underline{Z}_N$ on the 10 kV side is infinite ($\underline{I}_2 = 0$). The impedance $\underline{Z}_{k1}$ is calculated according to Eq. (15.21). This corresponds to the network example in which no active network is connected to the LV side.

$$\underline{Z}_{k1} = \frac{(\underline{Z}_1 + \underline{Z}_3) \cdot \underline{Z}_2}{\underline{Z}_1 + \underline{Z}_2 + \underline{Z}_3} \tag{15.21}$$

– The impedance of the transformers and the transformation ratio are identical. Under these conditions, the individual impedances are as follows:

$$\underline{Z}_1 = \frac{\underline{Z}_T/2}{t} \qquad \underline{Z}_2 = \frac{\underline{Z}_T/2}{1-t} \qquad \underline{Z}_3 = \frac{\underline{Z}_T/2}{t^2-t} \tag{15.22}$$

It follows that:

$$\underline{Z}_1 + \underline{Z}_3 = \frac{\underline{Z}_T/2}{t-1} \quad \rightarrow \quad \underline{Z}_1 + \underline{Z}_2 + \underline{Z}_3 = 0 \tag{15.23}$$

In this case, the input impedance $\underline{Z}_{k1} = \infty$ and therefore the short-circuit current $I_k = 0$.

– The impedances of the transformers are different, but the transformation ratio is the same. Under these conditions, the individual impedances are as follows:

$$\underline{Z}_1 = \frac{\underline{Z}_{T1} \cdot \underline{Z}_{T2}}{t \cdot (\underline{Z}_{T1} + \underline{Z}_{T2})} \qquad \underline{Z}_2 = \frac{\underline{Z}_{T1} \cdot \underline{Z}_{T2}}{(1-t) \cdot (\underline{Z}_{T1} + \underline{Z}_{T2})}$$
$$\underline{Z}_3 = \frac{\underline{Z}_{T1} \cdot \underline{Z}_{T2}}{(t^2-t) \cdot (\underline{Z}_{T1} + \underline{Z}_{T2})} \tag{15.24}$$

$$\underline{Z}_1 + \underline{Z}_3 = \frac{\underline{Z}_{T1} \cdot \underline{Z}_{T2}}{\underline{Z}_{T1} + \underline{Z}_{T2}} \cdot \frac{1}{t-1} \quad \rightarrow \quad \underline{Z}_1 + \underline{Z}_2 + \underline{Z}_3 = 0 \tag{15.25}$$

In this case, the input impedance $\underline{Z}_{k1} = \infty$ and therefore the short-circuit current $I_k = 0$.

– The impedances of the transformers are the same, with different transformation ratios. Under these conditions, the individual impedances are as follows:

$$\begin{aligned} \underline{Z}_1 &= \frac{\underline{Z}_T}{t_1 + t_2} \qquad \underline{Z}_2 = \frac{\underline{Z}_T}{(1-t_1) + (1-t_2)} \\ \underline{Z}_3 &= \frac{\underline{Z}_T}{\left(t_1^2 - t_1\right) + \left(t_2^2 - t_2\right)} \end{aligned} \tag{15.26}$$

$$\underline{Z}_1 + \underline{Z}_3 = \underline{Z}_T \cdot \frac{t_1^2 + t_2^2}{(t_1 + t_2) \cdot [t_2 \cdot (t_2 - 1) + t_1 \cdot (t_1 - 1)]} \tag{15.27}$$

$$\underline{Z}_1 + \underline{Z}_2 + \underline{Z}_3 = \underline{Z}_T \cdot \frac{(t_1 - t_2)^2}{(t_1 + t_2) \cdot [t_2 \cdot (t_2 - 1) + t_1 \cdot (t_1 - 1)] \cdot [(1 - t_1) + (1 - t_2)]} \tag{15.28}$$

$$\underline{Z}_{k1} = \underline{Z}_T \cdot \frac{t_1^2 + t_2^2}{(t_1 - t_2)^2} = \frac{\underline{Z}_T}{2} \cdot \frac{2 \cdot \left(t_1^2 + t_2^2\right)}{(t_1 - t_2)^2} \tag{15.29}$$

In this case, the input impedance $\underline{Z}_{k1}$ has a finite value, so that the current $\underline{I}_k$ also has a finite value.

- Variant 2: Impedance $\underline{Z}_{k2}$, seen from the 10 kV side, when there is a short circuit on the HV side. The short-circuit impedance is calculated as follows:

$$\underline{Z}_{k2} = \frac{\underline{Z}_1 \cdot \underline{Z}_3}{\underline{Z}_1 + \underline{Z}_3} \tag{15.30}$$

– The impedance of the transformers and the transformation ratio are identical. Under these conditions, the individual impedances are as follows:

$$\underline{Z}_1 = \frac{\underline{Z}_T/2}{t} \qquad \underline{Z}_3 = \frac{\underline{Z}_T/2}{t^2 - t} \tag{15.31}$$

It follows that:

$$\underline{Z}_{k2} = \frac{\frac{\underline{Z}_T/2}{t} \cdot \frac{\underline{Z}_T/2}{t^2 - t}}{\frac{\underline{Z}_T/2}{t} + \frac{\underline{Z}_T/2}{t^2 - t}} = \frac{\underline{Z}_T}{2} \cdot \frac{1}{t^2} \tag{15.32}$$

In this case, the short-circuit impedance $\underline{Z}_{k2}$ is calculated from half the value of the transformer impedance and the square of the same transformation ratio.

– The impedances of the transformers are different, but the transformation ratio is the same. Under these conditions, the individual impedances are as follows:

$$\underline{Z}_1 = \frac{\underline{Z}_{T1} \cdot \underline{Z}_{T2}}{t \cdot (\underline{Z}_{T1} + \underline{Z}_{T2})} \qquad \underline{Z}_3 = \frac{\underline{Z}_{T1} \cdot \underline{Z}_{T2}}{(t^2 - t) \cdot (\underline{Z}_{T1} + \underline{Z}_{T2})} \tag{15.33}$$

$$\underline{Z}_{k2} = \frac{\frac{\underline{Z}_{T1} \cdot \underline{Z}_{T2}}{t \cdot (\underline{Z}_{T1} + \underline{Z}_{T2})} \cdot \frac{\underline{Z}_{T1} \cdot \underline{Z}_{T2}}{(t^2 - t) \cdot (\underline{Z}_{T1} + \underline{Z}_{T2})}}{\frac{\underline{Z}_{T1} \cdot \underline{Z}_{T2}}{t \cdot (\underline{Z}_{T1} + \underline{Z}_{T2})} + \frac{\underline{Z}_{T1} \cdot \underline{Z}_{T2}}{(t^2 - t) \cdot (\underline{Z}_{T1} + \underline{Z}_{T2})}} = \frac{\underline{Z}_{T1} \cdot \underline{Z}_{T2}}{\underline{Z}_{T1} + \underline{Z}_{T2}} \cdot \frac{1}{t^2} \tag{15.34}$$

In this case, too, the short-circuit impedance $\underline{Z}_{k2}$ is determined by the parallel connection of the individual transformer impedances and the square of the same transformation ratio.

– The impedances of the transformers are the same, with different transformation ratios. Under these conditions, the individual impedances are as follows:

$$\underline{Z}_1 = \frac{\underline{Z}_T}{t_1 + t_2} \qquad \underline{Z}_3 = \frac{\underline{Z}_T}{(t_1^2 - t_1) + (t_2^2 - t_2)} \tag{15.35}$$

$$\underline{Z}_{k2} = \frac{\frac{\underline{Z}_T}{t_1 + t_2} \cdot \frac{\underline{Z}_T}{(t_1^2 - t_1) + (t_2^2 - t_2)}}{\frac{\underline{Z}_T}{t_1 + t_2} + \frac{\underline{Z}_T}{(t_1^2 - t_1) + (t_2^2 - t_2)}} = \frac{\underline{Z}_T}{2} \cdot \frac{1}{(t_1^2 + t_2^2)/2} \tag{15.36}$$

Assessment Depending on the different simulations, the various results can be summarized as follows:

- Variant 1a: When the secondary side is short-circuited (infinite short-circuit current feeding), the resulting input impedance depends on the parallel connection of the transformer impedances, regardless of the transformation ratios.
- Variant 1b: If the short-circuit current contribution on the secondary side is zero, with identical transformation ratios, the input impedance is infinite. With different transformation ratios, the result is a finite input impedance.
- Variant 2: In this case, the input impedance depends on the transformer impedances and the transformation ratios.

It can be seen that in the case of variant 1b, the result depends on the transformation ratios, which will have an effect on the short-circuit current calculation according to Sect. 15.1.2. The results are summarized in Table 15.1 for the variants described, with the same longitudinal impedances of the transformers $\underline{Z}_T = \underline{Z}_{T1} = \underline{Z}_{T2}$.

Table 15.1 Input impedances of the arrangement according to Fig. 15.3

Variant	$t_1 \neq t_2$	$t = t_1 = t_2$
1a	$\underline{Z}_{k1} = \frac{\underline{Z}_T}{2}$	$\underline{Z}_{k1} = \frac{\underline{Z}_T}{2}$
1b	$\underline{Z}_{k1} = \frac{2\cdot(t_1^2+t_2^2)}{(t_1-t_2)^2} \cdot \frac{\underline{Z}_T}{2}$	$\underline{Z}_{k1} \rightarrow \infty$
2	$\underline{Z}_{k2} = \frac{2}{(t_1^2+t_2^2)} \cdot \frac{\underline{Z}_T}{2}$	$\underline{Z}_{k2} = \frac{1}{t^2} \cdot \frac{\underline{Z}_T}{2}$

15.1.2 Determination of the Three-Phase Short-Circuit Current

For simplicity, the influence of the transformer ratio is illustrated using the three-phase short circuit, since only the parallel system is effective in these cases.

15.1.2.1 Network Circuit and Calculation Method

Since the computer programs used generally transformer equivalent circuits shown in Fig. 15.2 (π equivalent circuit diagram), the question arises as to whether the consideration of shunt admittances contradicts the requirement that shunt admittances must not be taken into account when applying the equivalent voltage source method at the short-circuit location.

As an example, the short-circuit current contribution of a lower voltage level (active network, i.e., with its own power plant feeder), e.g., 10 kV, to the higher voltage level (110 kV) is determined according to Fig. 15.4. Here, two transformers with different transformation ratios are connected in parallel, and the short-circuit current calculation is performed using different methods for comparison. This example is possible when the short-circuit current contribution of a lower voltage level to the higher voltage level is to be investigated. Here, the network impedance $\underline{Z}_N$ is related to 10 kV and the transformer impedances $\underline{Z}_{T1} = \underline{Z}_{T2}$ to 110 kV. Four different methods (A–D) are investigated to compare the calculation options.

- Method A: Calculation according to IEC 60909-0 [3] using the arithmetic mean of the transformer ratios, application of the equivalent voltage source method at the short-circuit location and transformer simulation with the π equivalent circuit diagram.
- Method B: Simulation of transformers using only the longitudinal impedances and the actual voltage distribution (10 kV network).
- Method C: Simulation of transformers with π equivalent circuit and actual voltage distribution (10 kV network).

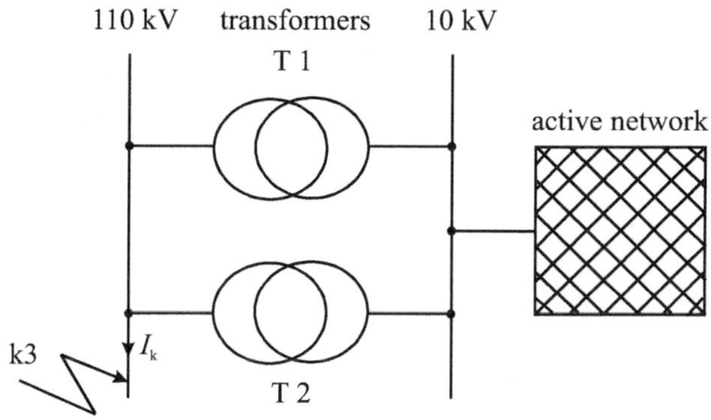

Fig. 15.4 Network circuit for determining the three-phase short-circuit current contribution in the 110 kV network

- Method D: Simulation of transformers with equivalent π equivalent circuit diagrams and consideration of the equivalent voltage source at the short-circuit location (110 kV network).

A general equivalent circuit for calculating the short-circuit current can be derived from the network circuit shown in Fig. 15.4, depending on the different feeders according to the various calculation methods.

The voltages $\underline{U}_{1\mathrm{A,D}}$ and $\underline{U}_{3\mathrm{B,C}}$ according to Fig. 15.5 are applied in accordance with methods (A–D), and the respective opposite voltage is short-circuited in these cases. The impedances are converted according to the voltage level.

Taking into account the chain matrix, Eq. (15.6), a total chain matrix can be derived for the equivalent circuit shown in Fig. 15.5, in which the values of the HV side (1) are linked to the values of the network feeder (3).

$$\begin{pmatrix}\underline{U}_1\\ \underline{I}_1\end{pmatrix}=\begin{pmatrix}\underline{A}_{11} & \underline{A}_{12}\\ \underline{A}_{21} & \underline{A}_{22}\end{pmatrix}\cdot\begin{pmatrix}\underline{U}_3\\ \underline{I}_3\end{pmatrix}=\begin{pmatrix}-\dfrac{\underline{Y}_{22}}{\underline{Y}_{21}} & \dfrac{1}{\underline{Y}_{21}}\\ \underline{Y}_{12}-\dfrac{\underline{Y}_{11}\cdot\underline{Y}_{22}}{\underline{Y}_{21}} & \dfrac{\underline{Y}_{11}}{\underline{Y}_{21}}\end{pmatrix}\cdot\begin{pmatrix}1 & \underline{Z}_{\mathrm{N}}\\ 0 & 1\end{pmatrix}\cdot\begin{pmatrix}\underline{U}_3\\ \underline{I}_3\end{pmatrix}$$

$$\begin{pmatrix}\underline{U}_1\\ \underline{I}_1\end{pmatrix}=\begin{pmatrix}-\dfrac{\underline{Y}_{22}}{\underline{Y}_{21}} & -\dfrac{\underline{Y}_{22}\cdot\underline{Z}_{\mathrm{N}}}{\underline{Y}_{21}}+\dfrac{1}{\underline{Y}_{21}}\\ \underline{Y}_{12}-\dfrac{\underline{Y}_{11}\cdot\underline{Y}_{22}}{\underline{Y}_{21}} & \left(\underline{Y}_{12}-\dfrac{\underline{Y}_{11}\cdot\underline{Y}_{22}}{\underline{Y}_{21}}\right)\cdot\underline{Z}_{\mathrm{N}}+\dfrac{\underline{Y}_{11}}{\underline{Y}_{21}}\end{pmatrix}\cdot\begin{pmatrix}\underline{U}_3\\ \underline{I}_3\end{pmatrix}\tag{15.37}$$

Depending on the calculation method (A–D), the various variables according to Eq. (15.5) must be entered into matrix (15.37). The same applies to the active voltages, so that the short-circuit current at the fault location can be determined. Using the $\underline{\mathbf{Y}}$ matrices of the two parallel transformers, the current distribution can be determined according to Fig. 15.6 if the total current $\underline{I}_1$ on the HV side is known.

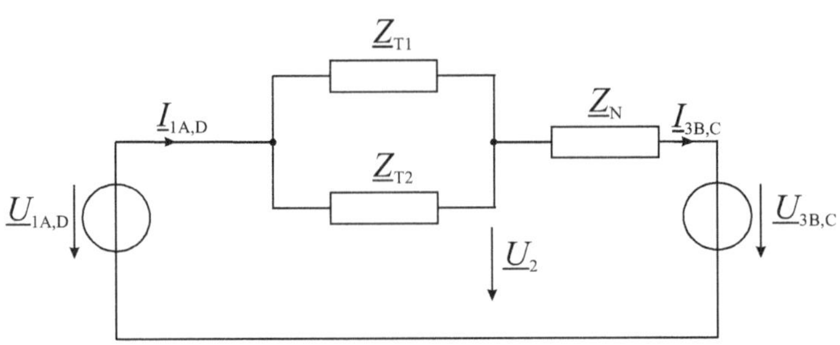

Fig. 15.5 Equivalent circuit for determining the short-circuit current (Fig. 15.4). $\underline{Z}_{T1}$, $\underline{Z}_{T2}$ impedance of transformers T1 and T2 (based on 110 kV), $\underline{Z}_N$ impedance of the 10 kV network (based on 10 kV), $\underline{U}_{1,3}$ active voltage depending on calculation method A–D, $\underline{U}_2$ voltage at the transformer terminals, $\underline{I}_{1,3}$ short-circuit currents depending on calculation method A–D

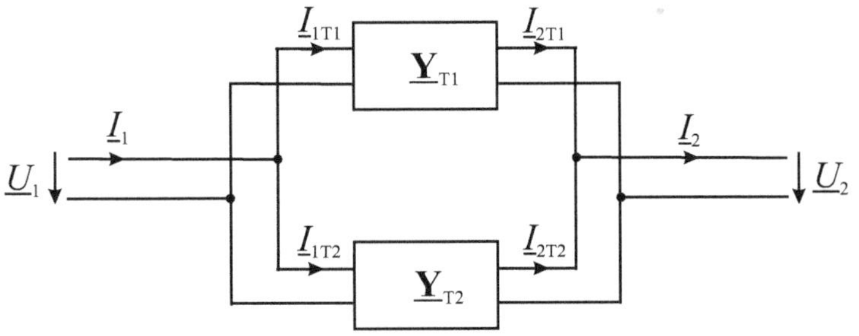

Fig. 15.6 Determining the current distribution of transformers

The system of Eq. (15.38) applies to the $\mathbf{\underline{Y}}$ matrices of the two transformers.

$$\begin{pmatrix} \underline{I}_{1T1} \\ \underline{I}_{2T1} \end{pmatrix} = \begin{pmatrix} \underline{Y}_{T1} & -t_1 \cdot \underline{Y}_{T1} \\ t_1 \cdot \underline{Y}_{T1} & -t_1^2 \cdot \underline{Y}_{T1} \end{pmatrix} \cdot \begin{pmatrix} \underline{U}_{1T1} \\ \underline{U}_{2T1} \end{pmatrix}$$
$$\begin{pmatrix} \underline{I}_{1T2} \\ \underline{I}_{2T2} \end{pmatrix} = \begin{pmatrix} \underline{Y}_{T2} & -t_2 \cdot \underline{Y}_{T2} \\ t_2 \cdot \underline{Y}_{T2} & -t_2^2 \cdot \underline{Y}_{T2} \end{pmatrix} \cdot \begin{pmatrix} \underline{U}_{1T2} \\ \underline{U}_{2T2} \end{pmatrix} \tag{15.38}$$

Since the respective voltages ($\underline{U}_1$, $\underline{U}_2$) are identical due to the parallel connection, the following simplified equation can be written for the first line:

$$\begin{aligned} \underline{I}_{1T1} &= \underline{Y}_{T1} \cdot \underline{U}_1 - t_1 \cdot \underline{Y}_{T1} \cdot \underline{U}_2 \\ \underline{I}_{1T2} &= \underline{Y}_{T2} \cdot \underline{U}_1 - t_2 \cdot \underline{Y}_{T2} \cdot \underline{U}_2 \end{aligned} \tag{15.39}$$

In addition, the current sum applies:

$$\underline{I}_1 = \underline{I}_{1T1} + \underline{I}_{1T2} \tag{15.40}$$

A distinction can be made between the two cases: equivalent voltage source at the short-circuit location and the actual voltage distribution. From this, the corresponding voltage conditions for solving the system of Eq. (15.39) can be derived. The unknown voltage $\underline{U}_2$ can be eliminated from Eq. (15.39) if the further derivation for the method of the equivalent voltage source at the short-circuit location is assumed. This results in:

$$\underline{I}_{1\mathrm{T2}} = \underline{Y}_{\mathrm{T2}} \cdot \left(\frac{t_1 - t_2}{t_1}\right) \cdot \underline{U}_1 + \frac{t_2}{t_1} \cdot \frac{\underline{Y}_{\mathrm{T2}}}{\underline{Y}_{\mathrm{T1}}} \cdot \underline{I}_{1\mathrm{T1}} \tag{15.41}$$

Taking into account the current condition according to Eq. (15.40), the following applies to the partial current $\underline{I}_{1\mathrm{T2}}$:

$$\underline{I}_{1\mathrm{T2}} = \frac{t_1 \cdot \underline{Y}_{\mathrm{T1}}}{t_1 \cdot \underline{Y}_{\mathrm{T1}} + t_2 \cdot \underline{Y}_{\mathrm{T2}}} \cdot \left\{ \underline{Y}_{\mathrm{T2}} \cdot \left(\frac{t_1 - t_2}{t_1}\right) \cdot \underline{U}_1 + \frac{t_2}{t_1} \cdot \frac{\underline{Y}_{\mathrm{T2}}}{\underline{Y}_{\mathrm{T1}}} \cdot \underline{I}_1 \right\} \tag{15.42}$$

$$\underline{I}_{1\mathrm{T2}} = \frac{(t_1 - t_2) \cdot \underline{Y}_{\mathrm{T1}} \cdot \underline{Y}_{\mathrm{T2}}}{t_1 \cdot \underline{Y}_{\mathrm{T1}} + t_2 \cdot \underline{Y}_{\mathrm{T2}}} \cdot \underline{U}_1 + \frac{t_2 \cdot \underline{Y}_{\mathrm{T2}}}{t_1 \cdot \underline{Y}_{\mathrm{T1}} + t_2 \cdot \underline{Y}_{\mathrm{T2}}} \cdot \underline{I}_1 \tag{15.43}$$

The partial current $\underline{I}_{1\mathrm{T1}}$ can then be determined using Eq. (15.40). For simplicity, it is assumed that the impedances of the transformers are identical on the HV side, which allows Eq. (15.43) to be simplified.

$$\begin{aligned} \underline{I}_{1\mathrm{T1}} &= \frac{1}{t_1 + t_2} \cdot [t_1 \cdot \underline{I}_1 - (t_1 - t_2) \cdot \underline{Y}_{\mathrm{T}} \cdot \underline{U}_1] \\ \underline{I}_{1\mathrm{T2}} &= \frac{1}{t_1 + t_2} \cdot [t_2 \cdot \underline{I}_1 + (t_1 - t_2) \cdot \underline{Y}_{\mathrm{T}} \cdot \underline{U}_1] \end{aligned} \tag{15.44}$$

In Eq. (15.44), the variable $\underline{I}_1$ represents the short-circuit current at the fault location, while $\underline{U}_1$ is the equivalent voltage source, so that the individual partial currents can then be determined.

The indexing of the electrical quantities (U and I) in the following sections is based on the representation in Fig. 15.5. The current $-I_1$ is the initial symmetrical short-circuit current at the fault location I''_{k}.

15.1.2.2 Method A

According to [3], the arithmetic mean of the two different transformation ratios t_{resA} may be used for parallel transformers. This corresponds to the value t_{res1} according to Eq. (15.11). The calculation is performed using the equivalent voltage source method at the short-circuit location and the transformer impedances $\underline{Z}_{\mathrm{T}}$. The active voltage of the downstream network is short-circuited. Taking the boundary conditions into account, Eq. (15.37) yields the system of Eq. (15.45):

$$\begin{pmatrix} \underline{U}_1 \\ \underline{I}_1 \end{pmatrix} = \begin{pmatrix} \underline{A}_{11} & \underline{A}_{12} \\ \underline{A}_{21} & \underline{A}_{22} \end{pmatrix} \cdot \begin{pmatrix} 0 \\ \underline{I}_3 \end{pmatrix} \tag{15.45}$$

With

$\underline{U}_1$ equivalent voltage at the short-circuit location $-c \cdot U_{n1}/\sqrt{3}$,
c voltage factor $c = 1{,}1$,
$\underline{I}_1$ short-circuit current at the fault location $-I''_k$,
$\underline{I}_3$ short-circuit current of the feeding network (LV side).

This results in:

$$\underline{U}_1 = \underline{A}_{12} \cdot \underline{I}_3 \qquad \underline{I}_1 = \underline{A}_{22} \cdot \underline{I}_3 \qquad \underline{I}_1 = \frac{\underline{A}_{22}}{\underline{A}_{12}} \cdot U_1 \tag{15.46}$$

Considering Eq. (15.38), the following applies to the short-circuit current:

$$\begin{aligned} \underline{I}_1 &= \frac{\left(\underline{Y}_{12} - \frac{\underline{Y}_{11} \cdot \underline{Y}_{22}}{\underline{Y}_{21}}\right) \cdot \underline{Z}_N + \frac{\underline{Y}_{11}}{\underline{Y}_{21}}}{-\frac{\underline{Y}_{22} \cdot \underline{Z}_N}{\underline{Y}_{21}} + \frac{1}{\underline{Y}_{21}}} \cdot \underline{U}_1 \\ &= \frac{(\underline{Y}_{12} \cdot \underline{Y}_{21} - \underline{Y}_{11} \cdot \underline{Y}_{22}) \cdot \underline{Z}_N + \underline{Y}_{11}}{-\underline{Y}_{22} \cdot \underline{Z}_N + 1} \cdot \underline{U}_1 \end{aligned} \tag{15.47}$$

Using the relationships in Eq. (15.5), if the transformer impedances and transformation ratios are identical, it follows that:

$$\begin{aligned} \underline{Y}_{11} &= 2 \cdot \underline{Y}_T & \underline{Y}_{12} &= -2 \cdot t_{res1} \cdot \underline{Y}_T \\ \underline{Y}_{21} &= 2 \cdot t_{res1} \cdot \underline{Y}_T & \underline{Y}_{22} &= -2 \cdot t_{res1}^2 \cdot \underline{Y}_T \end{aligned} \tag{15.48}$$

$$\underline{I}_1 = \frac{\underline{U}_1}{\underline{Z}_Q \cdot t_{res1}^2 + Z_T/2} \tag{15.49}$$

The short-circuit current according to Eq. (15.49) is calculated from the parallel connection of the transformers and the network impedance, which is converted to the voltage of side 1 using the square of the transformation ratio, so that the following applies:

$$\underline{I}_1 = \frac{\underline{U}_1}{\frac{\underline{Z}_T}{2} + \underline{Z}_N \cdot \left(\frac{t_1 + t_2}{2}\right)^2} \qquad \underline{I}''_k = \frac{c \cdot U_{n1}/\sqrt{3}}{\frac{\underline{Z}_T}{2} + \underline{Z}_N \cdot \left(\frac{t_1 + t_2}{2}\right)^2} \tag{15.50}$$

The following short-circuit current values result for the two borderline cases of network impedance

- $\underline{Z}_N = 0$: The short-circuit current is determined exclusively by half the transformer impedance. The partial short-circuit currents on the HV side each correspond to half the total current.
- $\underline{Z}_N \to \infty$: The short-circuit current is zero, i.e., the short-circuit current contribution from the downstream network is also zero.

With the same transformation ratio for the transformers, the current is calculated from the parallel connection of the transformers and the series connection of the network impedance, which is converted to the HV side using the square of the transformation ratio. If only the longitudinal impedances of the transformers are taken into account in the calculation, and not the π equivalent circuit diagram, the same results are obtained, as shown in Eq. (15.50).

15.1.2.3 Method B

In this case, the short-circuit current is calculated according to Fig. 15.7, assuming the actual voltage ratios, i.e., the active voltage is set in the 10 kV network. The transformers are considered using only their longitudinal impedances, the impedances are converted to the LV side (10 kV) using the square of the transformation ratios and connected in parallel, Eq. (15.51).

$$\underline{Z}_{T1} = \frac{\underline{Z}_T}{t_1^2} \qquad \underline{Z}_{T2} = \frac{\underline{Z}_T}{t_2^2} \tag{15.51}$$

This results in the short-circuit current, relative to the LV side, being:

$$\underline{I}_3 = \frac{-\underline{U}_3}{\underline{Z}_N + \frac{\underline{Z}_T}{2} \cdot \frac{2}{t_1^2 + t_2^2}} \tag{15.52}$$

The partial short-circuit currents on the LV and HV sides are determined by the impedance ratios $\underline{Z}_T/t_1^2$ and $\underline{Z}_T/t_2^2$ of the transformers. The total current at the fault location is calculated by adding the partial currents, taking into account the transformation ratios:

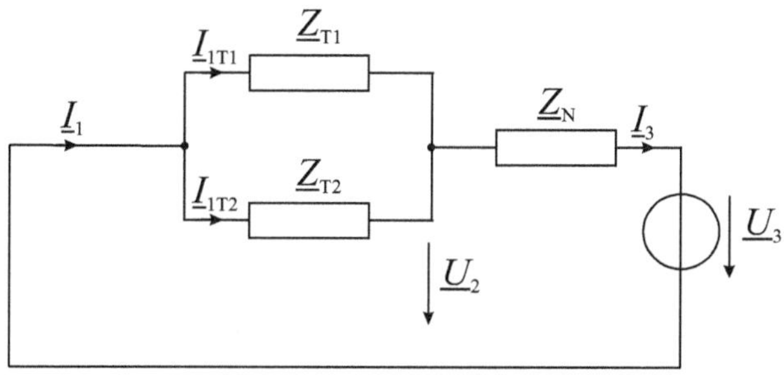

Fig. 15.7 Calculation using the actual voltage distribution (10 kV network)

$$\begin{aligned} \underline{I}_1 &= \underline{I}_{1\mathrm{T}1} + \underline{I}_{1\mathrm{T}2} = -\frac{t_1 + t_2}{2} \cdot \frac{\underline{U}_3}{\frac{\underline{Z}_\mathrm{T}}{2} + \underline{Z}_\mathrm{N} \cdot \frac{t_1^2 + t_2^2}{2}} \\ \underline{I}_\mathrm{k}'' &= \frac{t_1 + t_2}{2} \cdot \frac{c \cdot U_{\mathrm{n}3}/\sqrt{3}}{\frac{\underline{Z}_\mathrm{T}}{2} + \underline{Z}_\mathrm{N} \cdot \frac{t_1^2 + t_2^2}{2}} \end{aligned} \tag{15.53}$$

According to Eq. (15.53), the voltage is determined using the arithmetic mean t_{res1} according to Eq. (15.11) of the transformation ratios t_{resA}, while the conversion of the network impedance is performed using t_{res2}, Eq. (15.12). In this case, too, the two boundary cases lead to similar results as in method A:

- $\underline{\mathbf{Z}}_\mathbf{N} = \mathbf{0}$: The short-circuit current is determined exclusively from half the transformer impedance, taking into account the arithmetic mean of the transformation ratios.
- $\underline{\mathbf{Z}}_\mathbf{N} \to \infty$: The short-circuit current is zero, i.e., the short-circuit current contribution from the downstream network is also zero.

15.1.2.4 Method C

In contrast to method B, the transformers are modeled using the equivalent circuit shown in Fig. 15.2 (π equivalent circuit). The short-circuit current is therefore calculated using Eq. (15.37), where the voltage $\underline{U}_1$ is short-circuited and the current $\underline{I}_1$ corresponds to the short-circuit current at the fault location. It follows that:

$$\begin{pmatrix} 0 \\ \underline{I}_1 \end{pmatrix} = \begin{pmatrix} \underline{A}_{11} & \underline{A}_{12} \\ \underline{A}_{21} & \underline{A}_{22} \end{pmatrix} \cdot \begin{pmatrix} \underline{U}_3 \\ \underline{I}_3 \end{pmatrix} \tag{15.54}$$

With

$\underline{U}_3$ voltage of the feeding network $-c \cdot U_{\mathrm{n}3}/\sqrt{3}$,
c voltage factor $c = 1{,}1$,
$\underline{I}_1$ short-circuit current at the fault location $-\underline{I}_\mathrm{k}''$,
$\underline{I}_3$ short-circuit current of the feeding network.

The short-circuit current at the fault location is obtained by solving the system of Eq. (15.52):

$$0 = \underline{A}_{11} \cdot \underline{U}_3 + \underline{A}_{12} \cdot \underline{I}_3 \qquad \underline{I}_1 = \underline{A}_{21} \cdot \underline{U}_3 + \underline{A}_{22} \cdot \underline{I}_3 \tag{15.55}$$

$$\underline{I}_1 = -\underline{U}_3 \cdot \left(\underline{A}_{21} - \underline{A}_{22} \cdot \frac{\underline{A}_{11}}{\underline{A}_{12}} \right) \tag{15.56}$$

The elements of the chain matrix are determined from Eqs. (15.10) and (15.37), resulting in:

$$
\begin{aligned}
\underline{A}_{11} &= \frac{t_1^2 + t_2^2}{t_1 + t_2} & \underline{A}_{12} &= \frac{1}{t_1 + t_2} \cdot [(t_1^2 + t_2^2) \cdot \underline{Z}_{\mathrm{N}} + \frac{1}{\underline{Y}_{\mathrm{T}}}] \\
\underline{A}_{21} &= \underline{Y}_{\mathrm{T}} \cdot \frac{(t_1 - t_2)^2}{t_1 + t_2} & \underline{A}_{22} &= \frac{1}{t_1 + t_2} \cdot [(t_1 - t_2)^2 \cdot \underline{Z}_{\mathrm{N}} + \frac{1}{\underline{Y}_{\mathrm{T}}}]
\end{aligned}
\tag{15.57}
$$

After substitution into Eq. (15.56), it follows that:

$$
\begin{aligned}
\underline{I}_3 &= -\underline{U}_3 \cdot \frac{\underline{Y}_{\mathrm{T}}}{t_1 + t_2} \left\{ (t_1 - t_2)^2 - \frac{\left[\underline{Y}_{\mathrm{T}} \cdot \underline{Z}_{\mathrm{N}} \cdot (t_1 - t_2)^2 + 2\right] \cdot (t_1^2 + t_2^2)}{(t_1^2 + t_2^2) \cdot \underline{Y}_{\mathrm{T}} \cdot \underline{Z}_{\mathrm{N}} + 1} \right\} \\
&= -\underline{U}_3 \cdot \frac{\underline{Y}_{\mathrm{T}}}{t_1 + t_2} \left\{ (t_1 - t_2)^2 - \frac{\left[\underline{Y}_{\mathrm{T}} \cdot \underline{Z}_{\mathrm{N}} \cdot (t_1 - t_2)^2 + 2\right] \cdot (t_1^2 + t_2^2)}{(t_1^2 + t_2^2) \cdot \underline{Y}_{\mathrm{T}} \cdot \underline{Z}_{\mathrm{N}} + 1} \right\}
\end{aligned}
\tag{15.58}
$$

After transformation, the three-phase short-circuit current can be derived according to Eq. (15.58).

$$
\underline{I}_1 = \frac{t_1 + t_2}{2} \cdot \frac{-\underline{U}_3}{\frac{\underline{Z}_{\mathrm{T}}}{2} + \underline{Z}_{\mathrm{N}} \cdot \frac{t_1^2 + t_2^2}{2}} \qquad \underline{I}_{\mathrm{k}}'' = \frac{t_1 + t_2}{2} \cdot \frac{c \cdot U_{\mathrm{n3}}/\sqrt{3}}{\frac{\underline{Z}_{\mathrm{T}}}{2} + \underline{Z}_{\mathrm{N}} \cdot \frac{t_1^2 + t_2^2}{2}}
\tag{15.59}
$$

According to Eq. (15.59), the voltage is transformed using the arithmetic mean, Eq. (15.11), while the network impedance is converted using the quadratic values according to Eq. (15.12). The result thus corresponds to the result according to method B, Eq. (15.57), so that these two methods are equivalent. The partial short-circuit currents of transformers T1 and T2 on the HV side can be determined from Eq. (15.44).

$$
\underline{I}_{\mathrm{1T1}} = \frac{c \cdot t_1 \cdot U_{\mathrm{n2}}/\sqrt{3}}{\underline{Z}_{\mathrm{T}} + 2 \cdot \underline{Z}_{\mathrm{N}} \cdot \frac{t_1^2 + t_2^2}{2}} \qquad I_{\mathrm{1T2}} = \frac{c \cdot t_2 \cdot U_{\mathrm{n2}}/\sqrt{3}}{\underline{Z}_{\mathrm{T}} + 2 \cdot \underline{Z}_{\mathrm{N}} \cdot \frac{t_1^2 + t_2^2}{2}}
\tag{15.60}
$$

In this case, too, the two borderline cases lead to the same results as method B:

- $\underline{Z}_{\mathrm{N}} = 0$: The short-circuit current is determined exclusively by half the transformer impedance, taking into account the arithmetic mean of the transformation ratios.
- $\underline{Z}_{\mathrm{N}} \to \infty$: The short-circuit current is zero, i.e., the short-circuit current contribution from the downstream network is also zero.

15.1.2.5 Method D

In contrast to method A, the calculation is performed using the π equivalent circuit diagram of the transformer. Equations (15.45)–(15.47) apply to the calculation of the short-circuit current. Taking into account the different transformation ratios, the following values of the admittance matrix are obtained according to Eq. (15.5):

$$\begin{aligned} \underline{Y}_{11} &= 2 \cdot \underline{Y}_{\mathrm{T}} & \underline{Y}_{12} &= -2 \cdot \underline{Y}_{\mathrm{T}} \cdot \frac{t_1 + t_2}{2} \\ \underline{Y}_{21} &= 2 \cdot \underline{Y}_{\mathrm{T}} \cdot \frac{t_1 + t_2}{2} & \underline{Y}_{22} &= -2 \cdot \underline{Y}_{\mathrm{T}} \cdot \frac{t_1^2 + t_2^2}{2} \end{aligned} \tag{15.61}$$

The values according to Eq. (15.63) are used in Eq. (15.64) to calculate the short-circuit current.

$$\underline{I}_1 = \frac{(\underline{Y}_{12} \cdot \underline{Y}_{21} - \underline{Y}_{11} \cdot \underline{Y}_{22}) \cdot \underline{Z}_{\mathrm{N}} + \underline{Y}_{11}}{-\underline{Y}_{22} \cdot \underline{Z}_{\mathrm{N}} + 1} \cdot \underline{U}_1 \tag{15.62}$$

This results in the short-circuit current at the fault location with $\underline{I}''_{\mathrm{k}} = -\underline{I}_1$ and $\underline{U}_1 = -c \cdot U_{\mathrm{n1}}/\sqrt{3}$:

$$\begin{aligned} \underline{I}_1 &= \frac{\frac{(t_1 - t_2)^2}{2} \cdot \frac{\underline{Z}_{\mathrm{N}}}{\underline{Z}_{\mathrm{T}}} + 1}{\frac{\underline{Z}_{\mathrm{T}}}{2} + \frac{t_1^2 + t_2^2}{2} \cdot \underline{Z}_{\mathrm{N}}} \cdot U_1 \\ I''_{\mathrm{k}} &= \frac{\frac{(t_1 - t_2)^2}{2} \cdot \frac{\underline{Z}_{\mathrm{N}}}{\underline{Z}_{\mathrm{T}}} + 1}{\frac{\underline{Z}_{\mathrm{T}}}{2} + \frac{t_1^2 + t_2^2}{2} \cdot \underline{Z}_{\mathrm{N}}} \cdot c \cdot U_{\mathrm{n1}}/\sqrt{3} \end{aligned} \tag{15.63}$$

The following section examines the two borderline cases for network impedance on the LV side.

- $\underline{Z}_{\mathbf{N}} = \mathbf{0}$**:** The short-circuit current at the fault location is determined exclusively by half the transformer impedance, Eq. (15.64).

$$\underline{I}''_{\mathrm{k}} = \frac{c \cdot U_{\mathrm{n1}}/\sqrt{3}}{\underline{Z}_{\mathrm{T}}/2} \tag{15.64}$$

- $\underline{Z}_{\mathbf{N}} \to \infty$**:** The short-circuit current has a finite value, Eq. (15.65).

$$I''_{\mathrm{k}} = \frac{c \cdot U_{\mathrm{n1}}/\sqrt{3}}{\underline{Z}_{\mathrm{T}}} \cdot \frac{(t_2 - t_1)^2}{t_1^2 + t_2^2} \tag{15.65}$$

With the partial short-circuit currents on the HV side according to Eq. (15.44):

$$\underline{I}_{1\mathrm{T1}} = \frac{c \cdot U_{\mathrm{n1}}/\sqrt{3}}{\underline{Z}_{\mathrm{T}}} \cdot t_2 \cdot \frac{t_2 - t_1}{t_1^2 + t_2^2} \qquad \underline{I}_{1\mathrm{T2}} = -\frac{c \cdot U_{\mathrm{n1}}/\sqrt{3}}{\underline{Z}_{\mathrm{T}}} \cdot t_1 \cdot \frac{t_2 - t_1}{t_1^2 + t_2^2} \tag{15.66}$$

This does not correspond to the result of the other methods, as in this case no active network is connected on the LV side and yet a short-circuit current flows on the HV side. With the same transformation ratio of the parallel transformers, this short-circuit current becomes zero.

15.1.2.6 Summary

In principle, the correct short-circuit current must be calculated using the superposition method, which takes into account the load on a network before the short circuit occurs. The short-circuit current can also be determined using the actual voltage distribution, which is represented by methods B and C, so that these calculation methods represent the "correct" value. The simulation of the transformers (exclusively longitudinal impedance or π equivalent circuit) has no influence on this.

The general results of the various methods A–D are summarized in Table 15.2 as a function of the network impedance $\underline{Z}_N$ on the LV side, with the same transformer impedance $\underline{Z}_T$, relative to the HV side.

In the case of an infinite short-circuit current on the LV side ($\underline{Z}_N = 0\ \Omega$), the results of the different methods are identical if the voltage of the LV side $U_{n2}/\sqrt{3}$ is converted to the HV side using the arithmetic mean of the transformation ratios. In contrast, with a network impedance ($\underline{Z}_N \rightarrow \infty$), calculation according to method D results in a short-circuit current on the HV side, even though in this case there is no active, downstream network and therefore no short-circuit current contribution is possible. It can be seen that this short-circuit current contribution depends on the difference in the transformation ratios.

The consequence of the information in Table 15.2 is that when applying the method with the equivalent voltage source at the short-circuit location and the equivalent π circuit of the transformers; it must always be assumed that the parallel transformers have the same transformation ratio.

Table 15.3 lists the general equations for calculating the short-circuit current on the HV side as a function of the transformation ratios and the condition $t_1 = t_2$, with the same transformer impedance $\underline{Z}_T$, relative to the HV side.

Table 15.2 Short-circuit current at the fault location (HV) depending on the network impedance $\underline{Z}_N$

Method	$\underline{Z}_N = 0\ \Omega$	$\underline{Z}_N \rightarrow \infty$
A	$\underline{I}''_k = \dfrac{c \cdot U_{n1}/\sqrt{3}}{\underline{Z}_T/2}$	0
B	$\underline{I}''_k = \dfrac{t_1 + t_2}{2} \cdot \dfrac{c \cdot U_{n2}/\sqrt{3}}{\underline{Z}_T/2}$	0
C	$\underline{I}''_k = \dfrac{t_1 + t_2}{2} \cdot \dfrac{c \cdot U_{n2}/\sqrt{3}}{\underline{Z}_T/2}$	0
D	$\underline{I}''_k = \frac{c \cdot U_{n1}/\sqrt{3}}{\underline{Z}_T/2}$	$\underline{I}''_k = \frac{c \cdot U_{n1}/\sqrt{3}}{\underline{Z}_T} \cdot \frac{(t_2 - t_1)^2}{t_1^2 + t_2^2}$

Table 15.3 General equations for calculating the short-circuit current as a function of the transformation ratios

Method	$t_1 \neq t_2$	$t = t_1 = t_2$
A	$\underline{I}''_{k} = \dfrac{c \cdot U_{n1}/\sqrt{3}}{\dfrac{\underline{Z}_T}{2} + \underline{Z}_N \cdot \left(\dfrac{t_1 + t_2}{2}\right)^2}$	$\underline{I}''_{k} = \dfrac{c \cdot U_{n1}/\sqrt{3}}{\underline{Z}_T/2 + \underline{Z}_N \cdot t^2}$
B	$\underline{I}''_{k} = \dfrac{t_1 + t_2}{2} \cdot \dfrac{c \cdot U_{n2}/\sqrt{3}}{\dfrac{\underline{Z}_T}{2} + \underline{Z}_N \cdot \dfrac{t_1^2 + t_2^2}{2}}$	$\underline{I}''_{k} = t \cdot \dfrac{c \cdot U_{n2}/\sqrt{3}}{\underline{Z}_T/2 + \underline{Z}_N \cdot t^2}$
C	$\underline{I}''_{k} = \dfrac{t_1 + t_2}{2} \cdot \dfrac{c \cdot U_{n2}/\sqrt{3}}{\dfrac{\underline{Z}_T}{2} + \underline{Z}_N \cdot \dfrac{t_1^2 + t_2^2}{2}}$	$\underline{I}''_{k} = t \cdot \dfrac{c \cdot U_{n2}/\sqrt{3}}{\underline{Z}_T/2 + \underline{Z}_N \cdot t^2}$
D	$\underline{I}''_{k} = \dfrac{\dfrac{(t_1 - t_2)^2}{2} \cdot \dfrac{\underline{Z}_N}{\underline{Z}_T} + 1}{\dfrac{\underline{Z}_T}{2} + \underline{Z}_N \cdot \dfrac{t_1^2 + t_2^2}{2}} \cdot c \cdot U_{n1}/\sqrt{3}$	$\underline{I}''_{k} = \dfrac{c \cdot U_{n1}/\sqrt{3}}{\underline{Z}_T/2 + \underline{Z}_N \cdot t^2}$

Table 15.3 shows that, with the same transformer ratio, the calculated short-circuit currents are independent of the calculation method, taking into account the different voltages. When the transformation ratio is different, methods A and D (equivalent voltage source at the short-circuit location) differ, although the deviation according to A is negligible due to the representation according to Eqs. (15.11) and (15.12). This means that when applying the equivalent voltage source at the short-circuit location; the arithmetic mean value must always be used.

15.1.3 Examples

Two examples from the distribution and transmission network show the different results depending on the method used to calculate the three-phase short-circuit current.

15.1.3.1 Distribution Network

The following example shows the results of the calculation according to method D (equivalent voltage source at the short-circuit location and π equivalent circuit of the transformers) for two transformers connected in parallel. The short-circuit current contribution of the LV side is zero, which means that there is no additional feed into this voltage level; the following transformer data is assumed, neglecting the resistances:

$$U_{rT1} = 110\ \text{kV}/10\ \text{kV}, \quad u_{Xr} = 10\%, \quad S_{rT1} = 100\ \text{MVA}$$
$$U_{rT2} = 110\ \text{kV}/12\ \text{kV}, \quad u_{Xr} = 10\%, \quad S_{rT2} = 100\ \text{MVA}$$

Taking into account Eqs. (15.65) and (15.66), the following currents result in the event of a three-phase short circuit on the 110 kV side:

Short-circuit current: $I''_{\mathrm{k}} = 95\,\mathrm{A}$
Partial currents: $I''_{\mathrm{kT1}} = -473\,\mathrm{A}$ $I''_{\mathrm{kT2}} = 568\,\mathrm{A}$

Figure 15.8 illustrates that partial short-circuit currents are calculated in the transformer branches that cannot flow under the given network conditions because there is no feeder at the lower voltage level. Figure 15.9 shows that the cause is an additional voltage source resulting from the different transformation ratio, which causes a loop current. This circulating current is an operating current that is not taken into account in the equivalent voltage source method at the short-circuit location due to the lack of superposition of the currents before the short circuit occurs. In these

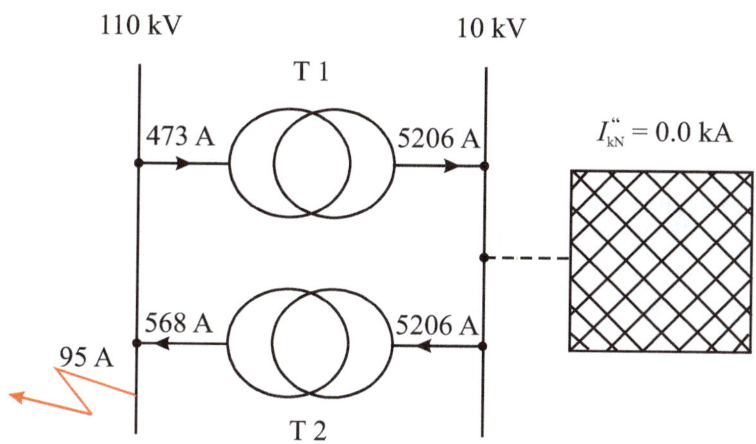

Fig. 15.8 Short-circuit current calculation using the equivalent voltage source at the short-circuit location for transformers with different transformation ratios

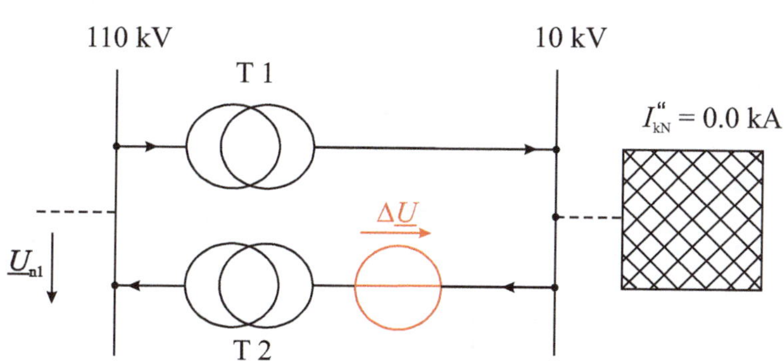

Fig. 15.9 Influence of different transformation ratios

cases, the shunt admittances affect the transformer simulation in such a way that the result is incorrect.

If, instead, the arithmetic mean of the two transformation ratios is taken, Eq. (15.53), method A, with $\underline{Z}_N \rightarrow \infty$, yields a short-circuit current of $I''_k = 0\,\text{A}$, which is physically correct.

15.1.3.2 Transmission Network

A network calculation program is used to calculate the short-circuit currents at the 380 kV voltage level according to Fig. 15.10, using transformers with different transformation ratios:

$U_{rT1} = 400\ \text{kV}/220\ \text{kV}, \quad u_{kr} = 14\%, \quad u_{Rr} = 0.15\%, S_{rT} = 400\ \text{MVA}$
$U_{rT2} = 400\ \text{kV}/240\ \text{kV}, \quad u_{kr} = 14\%, \quad u_{Rr} = 0.15\%, \quad S_{rT} = 400\ \text{MVA}$

Initial symmetrical short-circuit currents of the two networks for determining the network impedances $\underline{Z}_{N1}$ (380 kV) and $\underline{Z}_{N2}$ (220 kV) at a ratio $R/X = 0.1$ are:

380 kV network: $I''_{kN1} = 10\ \text{kA}$
220 kV network: $I''_{kN2} = 2\ \text{kA}$

The following different calculation methods are used, whereby the transformers are always modeled using the π equivalent circuit diagram:

- Figure 15.10a Calculation according to method D, Sect. 15.1.2.5 (different transformation ratios, equivalent voltage source at the short-circuit location),
- Figure 15.10b Calculation according to method A, Sect. 15.1.2.2 (arithmetic mean of the transformation ratios, equivalent voltage source at the short-circuit location),
- Figure 15.10c Calculation according to method C, Sect. 15.1.2.4 (with different transformation ratios, actual voltages). This calculation method uses the superposition method, whereby no load flow is assumed before the short circuit occurs.

Due to the location of the fault, the two networks are decoupled (380 and 220 kV) and the short-circuit current contributions are therefore independent of each other. The total short-circuit currents are approximately 11.0 kA for all methods and differ only minimally from each other ($\Delta I = 0.028$ kA), which is due to the strong feeder from the 380 kV grid. The partial short-circuit currents of the transformers deviate by up to 41.1% in the case of uncorrected transformation ratios (Fig. 15.10a) when the result of the calculation with actual voltages (Fig. 15.10c) is taken as a reference. The deviation with the arithmetic mean of the transformation ratios (Fig. 15.10b) results in a maximum deviation of 7.6%. The deviations become greater when the feeder from the subordinate 220 kV network is reduced.

It can be seen that, in this case too, using the arithmetic mean rather than the exact calculation produces a result that is sufficiently accurate.

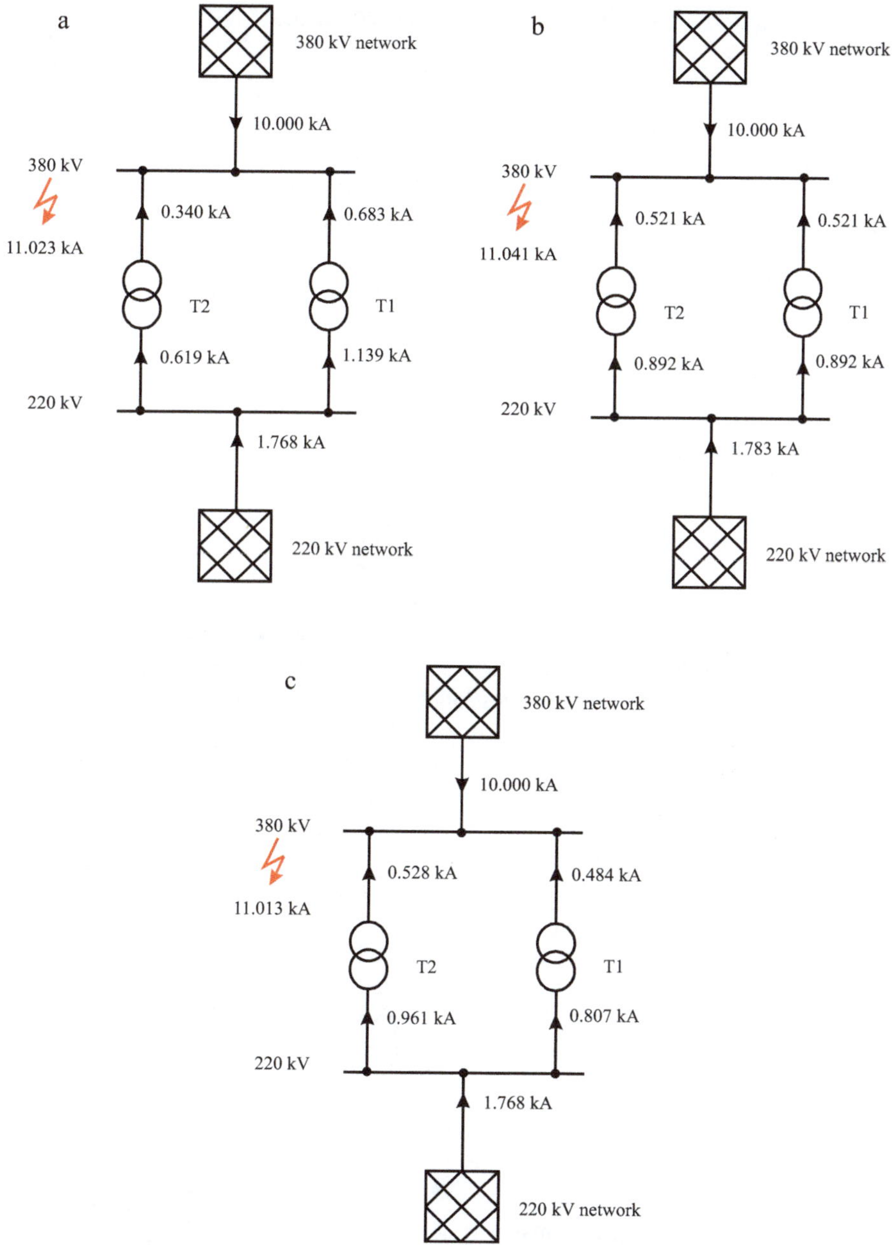

Fig. 15.10 Example of calculating the three-phase short-circuit current. (**a**) Method D, equivalent voltage source, different transformation ratio, (**b**) equivalent voltage source, same transformation ratio, (**c**) distributed voltages, different transformation ratio

15.2 Neutral Point Treatment on Both Sides

When selecting and operating a transformer between networks with different voltage levels, network operators must agree on the type and location of the neutral point earthing, e.g. when operating industrial networks [7]. This involves, in particular, the selection of the step-up transformer and the question of whether its neutral points should be earthed in any way or should not be earthed, because earthing the neutral points of a transformer on the HV and LV sides can generally be problematic [8].

To evaluate the various options for transformer neutral point treatment, an example is assumed in which a 110 kV network feeds a 10 kV network. Both voltage and current problems are addressed here.

15.2.1 Faults Considered

When considering the influence of the 110 kV network on the 10 kV network in the context of the neutral point treatment of the transformer, the focus is on unbalanced short circuits with earth contact in the higher-level network, which lead to consequences for voltages and currents in the lower-level network. The effect of such load cases on an unearthed secondary-side transformer neutral point or secondary-side delta winding of the transformer is described in detail in the literature, e.g., [9]. This means that earthing a transformer neutral point is generally only effective if there is a delta winding. If solidly (low-impedance) earthing is provided on both sides (high and low voltage), a compensating winding in a delta connection must be present. Both-sided earth fault compensation should be avoided, and only effective earthing provides adequate protection.

Unbalanced faults with earth contact in the 110 kV network primarily include line-to-earth faults (k1) and line-to-line faults with earth contact (kE2E). According to Table 9.2, the ratio of the zero sequence voltage $U_{(0)}$ to the voltage $E_{(1)}$ positive sequence voltage of the parallel system can be determined. For the simplification, the resistance is neglected in the following. For the zero sequence voltage, the following generally applies to the fault cases considered:

$$\underline{U}_{(0)} = \mathrm{j}X_{(0)} \cdot \underline{I}_{(0)} \tag{15.67}$$

With

$X_{(0)}$ reactance of the zero sequence system,
$I_{(0)}$ current of the zero sequence system.

Depending on the short circuit, the current in the zero sequence system can be determined based on the voltage in the positive sequence system. In this case, the reactances in the positive and negative sequence systems ($X_{(1)} = X_{(2)}$) are equated.

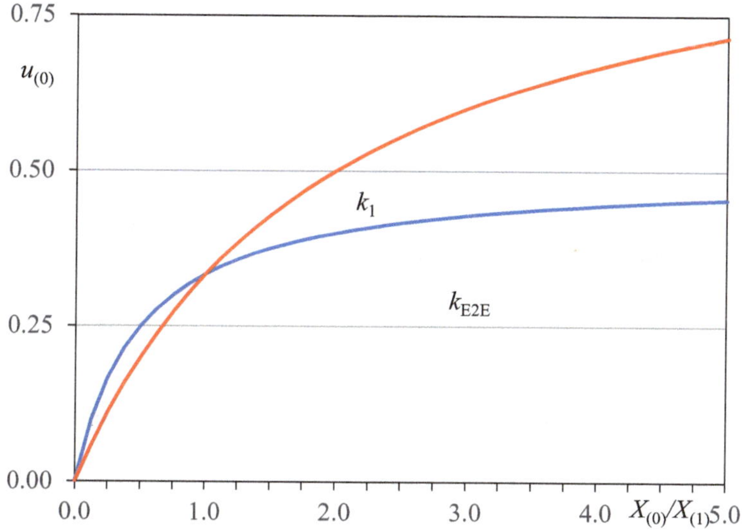

Fig. 15.11 Zero sequence voltage $u_{(0)} = |\underline{U}_{(0)}/\underline{E}_{(1)}|$ at the fault location relative to the positive sequence voltage $E_{(1)}$ for a line-to-earth (k1) and a line-to-line short circuit with earth contact (kE2E) as a function of the reactances of the positive and zero sequence systems

Line-to-line short circuit with earth contact:

$$\underline{U}_{(0)} = -X_{(0)} \cdot \frac{X_{(1)}}{X_{(1)} + X_{(0)}} \cdot \frac{\underline{E}_{(1)}}{X_{(1)} + \frac{X_{(1)} \cdot X_{(0)}}{X_{(1)} + X_{(0)}}} \tag{15.68}$$

$$u_{(0)} = \left| \frac{\underline{U}_{(0)}}{\underline{E}_{(1)}} \right| = \frac{1}{2 + X_{(1)}/X_{(0)}} \tag{15.69}$$

Line-to-earth short circuit:

$$\underline{U}_{(0)} = X_{(0)} \cdot \frac{\underline{E}_{(1)}}{2X_{(1)} + X_{(0)}} \qquad u_{(0)} = \left| \frac{\underline{U}_{(0)}}{\underline{E}_{(1)}} \right| = \frac{1}{1 + 2X_{(1)}/X_{(0)}} \tag{15.70}$$

Of these two fault cases, the line-to-earth fault causes a greater shift in the zero sequence voltage than the line-to-line fault, as illustrated in Fig. 15.11 for $X_0/X_1 > 1$. The ratio $X_0/X_1 > 1$ means that the three-phase short-circuit current is greater than the line-to-earth current, which is the case for network faults, except those directly at the terminals of step-up transformers (e.g., Dy).

15.2.2 *Transformation of the Zero Sequence Voltage*

The influence of the overlaid 110 kV network on the underlaid 10 kV network with a step-up transformer earthed on both sides is first illustrated using the transferred zero sequence voltage $\underline{U}_{(0)}$ with the aid of the symmetrical components. The further considerations apply under the assumption that no additional neutral point is earthed in the 10 kV network.

15.2.2.1 General

Figure 15.12a shows the network connection of a transformer Y(N)y(n)d with compensation winding, which is earthed on both sides, as a step-up transformer for a 10 kV network from a overlaid 110 kV network in the event of a line-to-earth fault on the HV side. The impedances $\underline{Z}_{E1}$ and $\underline{Z}_{E2}$ represent a resistance or a reactance depending on the neutral point treatment. The corresponding equivalent circuits in the positive, negative, and zero sequence systems are shown in Fig. 15.12b.

Figure 15.12c is useful for easily assessing the voltage distribution in the zero sequence system. In this case, the voltage $\underline{E}_0$ in the zero sequence system represents the voltage $\underline{E}_{(1)}$ coupled by the active system voltage in the event of a line-to-earth fault at the short-circuit location. The value $\underline{U}_{(0)}/\underline{E}_{(1)}$ can be taken from Fig. 15.11 for the steady-state case as a function of the ratio $X_{(0)}/X_{(1)}$. In contrast, the voltage $U_{(0)}$ describes the voltage transferred to the fault-free side of the transformer in the zero sequence system.

The following variables are shown in Fig. 15.12:

(1, 2, 0) values of the positive, negative, and zero sequence system,
$\underline{E}_{(1)}$ active voltage of the positive sequence system (110 kV),
$\underline{E}_{(0)}$ voltage of the zero sequence system (110 kV) in event of a line-to-earth fault,
$\underline{U}_{(0)}$ transferred voltage in the zero sequence system (10 kV) in the event of line-to-earth fault,
X_N, C_N network reactance, capacitance (110 kV),
C_Q network capacitance (10 kV),
X_{W1} reactance of the HV winding,
X_{W2} reactance of the LV winding,
X_{W3} reactance of the stabilizing winding,
$\underline{Z}_{E1}$, $\underline{Z}_{E2}$ earth impedance of the 110 and 10 kV networks.

15.2.2.2 Transfer Equations and their Evaluation

Equations (15.71)–(15.75) describe the relationships between the zero sequence voltage $U_{(0)}$ transferred to the 10 kV network and the active voltage $E_{(0)}$ of the 110 kV network in the event of a line-to-earth fault. According to Fig. 15.12c, the resistances of the transformers are neglected in relation to the short-circuit inductances, so that, for simplicity, the following applies:

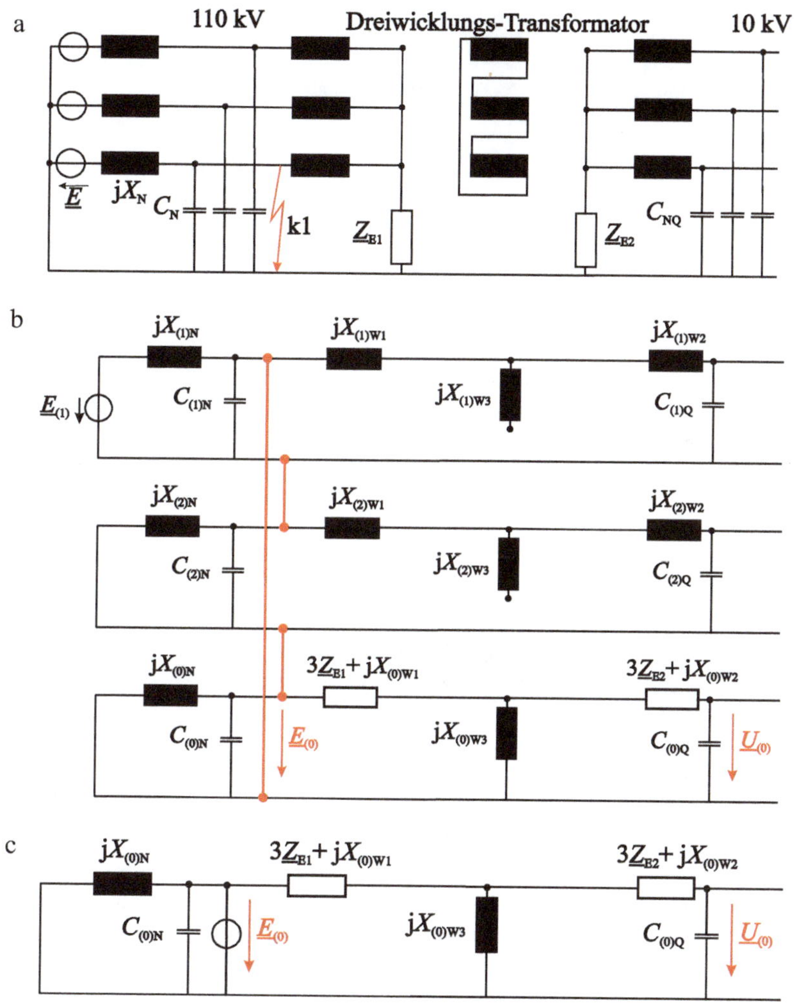

Fig. 15.12 Example illustrating zero sequence voltage $\underline{U}_{(0)}$ in the underlaid 10 kV network in the event of a line-to-earth fault in the overlaid 110 kV network. (**a**) Network circuit, (**b**) equivalent circuit of the component systems, (**c**) simplified equivalent circuit in the zero sequence system

$$\mathrm{j}X_{(0)\mathrm{W1}} = \mathrm{j}\omega L_{(0)\mathrm{W1}} \qquad \mathrm{j}X_{(0)\mathrm{W2}} = \mathrm{j}\omega L_{(0)\mathrm{W2}} \qquad \mathrm{j}X_{(0)\mathrm{W3}} = \mathrm{j}\omega L_{(0)\mathrm{W3}} \tag{15.71}$$

For further consideration, simplification is achieved by adding the inductance of the transformer earthing to the inductance of the secondary winding, if it is taken into account:

$$L_1 = L_{(0)W1} \qquad L_2 = L_{(0)W2} + 3 \cdot L_{E2} \qquad L_3 = L_{(0)W3} \tag{15.72}$$

Operating frequency steady state zero sequence voltage:

$$u_{(0)} = \frac{U_{(0)}}{E_{(0)}} = \frac{\omega_1^2}{1 + L_1/L_3} \cdot \frac{1}{\sqrt{\left(\omega_1^2 - \omega^2\right)^2 + 4\delta^2 \cdot \omega^2}} \tag{15.73}$$

Transient zero sequence voltage:

$$u_{(0)\omega 1} = u_{(0)} \cdot \sqrt{\frac{\left(\omega_1^2 - \omega^2\right)^2 + 4\delta^2 \cdot \omega_1^2}{\left(\omega_1^2 - \omega^2\right)^2 + 4\delta^2 \cdot \omega^2}} \tag{15.74}$$

Natural frequency:

$$\omega_1 = \sqrt{\frac{L_1 + L_3}{C_{(0)Q} \cdot (L_1 \cdot L_2 + L_2 \cdot L_3 + L_1 \cdot L_3)}} \tag{15.75}$$

Damping time constant when general resistance or neutral point earthing is considered as resistance that dampens the oscillation curve:

$$2\delta = \frac{R \cdot (L_1 + L_3)}{L_1 \cdot L_2 + L_2 \cdot L_3 + L_1 \cdot L_3} \tag{15.76}$$

The zero sequence voltage $U_{(0)}$ in the 10 kV network consists of a steady state component with the operating frequency of the network and a transient, decaying component (usually) of higher frequency. The voltage $U_{(0)}$ is related to the value $E_{(0)}$, for which the phase voltage of the active network, converted to the 10 kV level, is used.

Table 15.4 Related zero sequence voltages $u_{(0)}$ (steady state) and $u_{(0)\omega 1}$ (transient) in underlaid network in the event of a line-to-earth fault in the overlaid network and different earthing (E) of the neutral points of the step-up transformer with and without stabilizing winding. The values apply to a 10 kV cable network with a length of 40 km or resonance tuning

	Earthing of the overlaid network (110 kV)					
	Solidly earthed				Resonant earthed	
Earthing (E) of the underlaid network (10 kV)	$u_{(0)}$ [a]	$u_{(0)}$ [b]	$u_{(0)\omega 1}$ [a]	$u_{(0)\omega 1}$ [b]	$u_{(0)}$ [a]	$u_{(0)\omega 1}$ [a]
E1	0.25	1.00	0.25	1.00	0.02	0.02
E2	0.25	1.05	0.25	1.05	0.02	0.02
E3	0.30	1.20	0.30	1.20	0.02	0.02
E4	0.24	1.00	0.00	0.00	0.02	0.00
E5	0.24	1.00	0.00	0.00	0.02	0.00
E6	≈ 7	$\gg 7$	≈ 7	$\gg 7$	≈ 0.4	≈ 0.4

[a] Two-winding transformer with stabilizing winding (three-winding transformer)
[b] Two-winding transformer without stabilizing winding.

Table 15.5 Short-circuit voltages of the transformers in the positive sequence system, relative to the throughput power of $S_{rT} = 31.5$ MVA

Short-circuit voltages (%)	Stabilizing winding	
	Without	With
u_{k12}	12	12
u_{k23}	–	4
u_{k31}	–	16

Table 15.4 contains the relative zero sequence voltages $U_{(0)}$ and $U_{(0)\omega 1}$ calculated according to Eqs. (15.73) and (15.74), depending on the neutral point treatment of the transformer with and without stabilizing winding. The voltages are each related to the nominal system voltage (10 kV). The values in Table 15.4 are based on a 31.5 MVA three-winding transformer with stabilizing winding as the step-up transformer. Here, the two main windings are connected in star configuration, and the stabilizing winding is connected in delta configuration. The following earthing impedances $\underline{Z}_{E2}$ are considered for the neutral point earthing (E) of the underlaid 10 kV network:

- E1: $\underline{Z}_{E2} = 0$, solidly earthed,
- E2/E3: $\underline{Z}_{E2} = jX_{E2}$ ($I''_{k1} = 2000\,\mathrm{A}/500\,\mathrm{A}$), current limiting earthed,
- E4/E5: $\underline{Z}_{E2} = R_{E2}$ ($I''_{k1} = 2000\,\mathrm{A}/500\,\mathrm{A}$), current limiting earthed,
- E6: $\underline{Z}_{E2} = j/(3\omega C_{(0)Q})$, resonant earthed neutral point.

The impedance $X_{(0)W3} = \omega \cdot L_3$ represents the tertiary winding connected in a delta configuration, which is arranged as a concentric winding on the core, followed by the 10 and 110 kV windings. The lower the value of the stabilizing winding in relation to $X_{(0)W1}$, the better the separation between the primary and secondary sides of the transformer in the zero sequence system. In a transformer without a stabilizing winding Y(N)y(n), L_3 depends on the leakage flux that takes its path through the transformer tank. $L_3/(L_1 + L_2)$ is approximately 7–100. Such transformers cause strong coupling of the zero sequence system from the primary side.

The short-circuit voltages of the transformers are listed in Table 15.5. The values in the zero sequence system correspond to 85–90% of the values in the positive sequence system. The following considerations provide information about variants of the neutral point treatment according to Table 15.4.

15.2.2.3 Low-Impedance Neutral Point Earthing

When the primary neutral point of a transformer without a stabilizing winding is solidly earthed and the secondary neutral point is earthed via a reactance or resistor, Table 15.4 shows that the zero sequence voltage is coupled into the secondary network.

Values for $u_{(0)} > 1$ are to be expected if, in the case of reactance earthing, the inductances ($L_1 + L_2$) form a series resonant circuit with the cable capacitance $C_{(0)}$ of the underlaid network, whose natural frequency approaches the operating frequency of the network. In this case, $\omega_1^2 - \omega^2 \approx 0$ is approximately equal to the value given by Eq. (15.73). As a result, only the attenuation of the short-circuit path limits the voltage $U_{(0)}$. Such configuration must be avoided.

When the primary neutral point of a transformer with stabilizing winding is solidly earthed, the size of the secondary earth impedance has almost no influence on the transferred zero sequence voltage $u_{(0)}$. The value $u_{(0)}$ is primarily determined by the ratio L_1/L_3 according to Eq. (15.73) if there is no other earthed neutral point in the underlaid network. This configuration is possible.

Secondary-side resistance earthing of the transformer neutral point ensures that the transient oscillation process always decays quickly.

15.2.2.4 Earth Fault Compensation (Resonant Earthing) of the Overlaid Network

When an arc suppression coil is connected to the primary neutral point of a step-up transformer, with stabilizing winding and simultaneous low-resistance earthing of the secondary neutral point, only a negligible zero sequence voltage is transferred to the underlaid network. This circuit is uncritical with current-limited low-impedance neutral point earthing in the 10 kV network.

A transformer without a stabilizing winding only carries a earth-fault current corresponding to the dimensioning of the arc suppression coil (current through the inductance L_1 according to Fig. 15.12) if this zero sequence current—taking into account the transformation ratio—can continue to flow in the secondary-side network. If, as assumed in this case, the secondary neutral point is the only earthed neutral point in this network, the primary neutral point will not contribute to the earth fault current of the overlaid network, except for the small capacitive current through $C_{(0)}$ and the small inductive current through L_3.

The quality factor of the earthing of the 110 kV network therefore depends on the operating status of the 10 kV network. (This is why Table 15.4 does not contain a column for the primary-side connection of an arc suppression coil.) The effect of this earthing is minimal and unpredictable, which is why it must be avoided.

15.2.2.5 Earth Fault Compensation (Resonant Earthing) of the Underlaid Network

Earth fault compensation in the 10 kV network with solidly earthing in the 110 kV network at the same transformer is not permitted because, according to Table 15.4, this results in high overvoltages in the low-voltage network (in the event of a line-to-earth short circuit in the high-voltage network). Both-sided earth fault compensation is not recommended [8] because the effectiveness of the compensation depends on the switching status.

15.2.2.6 Isolated Neutral Point in the Underlaid Network

If the neutral point of the step-up transformer is insulated on the 10 kV network side, no inductively transferred zero sequence voltages occur in this network. However, it should be noted that capacitive coupled voltages could reach considerable values in 110 kV transformers if no other cables, lines, or capacitances are connected to the 10 kV busbar. A capacitive voltage divider is created from the high- and low-voltage side capacities of the transformer and the underlaid network.

15.2.3 Balancing Currents in the Underlaid Network

Due to the neutral point treatment of step-up transformers, balancing currents occur when the overlaid network is earthed with low impedance, $\underline{Z}_{E11} = \underline{Z}_{E12} = 0\,\Omega$ (with $\underline{Z}_{E1}$ being the earthing impedance of the neutral points on the high-voltage side). These currents remain uncritical and are therefore not considered in 110 kV networks, operated in a resonant earthed mode.

15.2.3.1 General

The balancing current flows in both the positive and negative sequence systems regardless of the type of neutral point treatment, but special attention must be paid to it in the zero sequence system in the case of neutral point earthing. The balancing current occurs when two transfer transformers are fed via two parallel lines from a remote switchgear or from separate networks and an earth fault occurs on one of these lines at the terminals of a transformer.

The neutral points on the primary sides of the step-up transformers are both directly earthed. Figure 15.13 shows the network connection examined in more

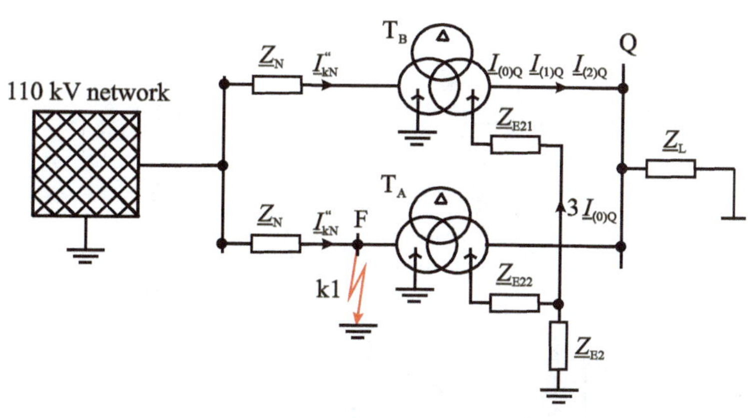

Fig. 15.13 Example illustrating the balancing currents in parallel transformers T_A and T_B (each with stabilizing winding) in the event of a line-to-earth short circuit (k1) in the overlaid 110 kV network at the terminals of a transformer. $\underline{I}''_{kN}$ initial symmetrical short-circuit current of the network, $\underline{I}_{(0)Q}$, $\underline{I}_{(1)Q}$, $\underline{I}_{(2)Q}$ component currents in the zero, positive, and negative sequence systems (underlaid network), $\underline{Z}_E$ earthing impedances, $\underline{Z}_L$ load ($\rightarrow \infty$)

detail below. Here, the impedance of the 110 kV network is divided between the two network branches, so that it is assumed that the two transformers are fed from different parts of the network. In the examples listed, it is assumed that the 10 kV network is unloaded ($\underline{Z}_L \rightarrow \infty$).

The network circuit shown in Fig. 15.13 can be converted into a component equivalent circuit for the line-to-earth fault in the overlaid network, resulting in Fig. 15.14. This assumes that the negative sequence system is identical to the positive sequence system, but without an active voltage source. When representing the zero sequence system, it is assumed that the underlaid 10 kV network is earthed exclusively via the neutral points of the transformers and that there is therefore no

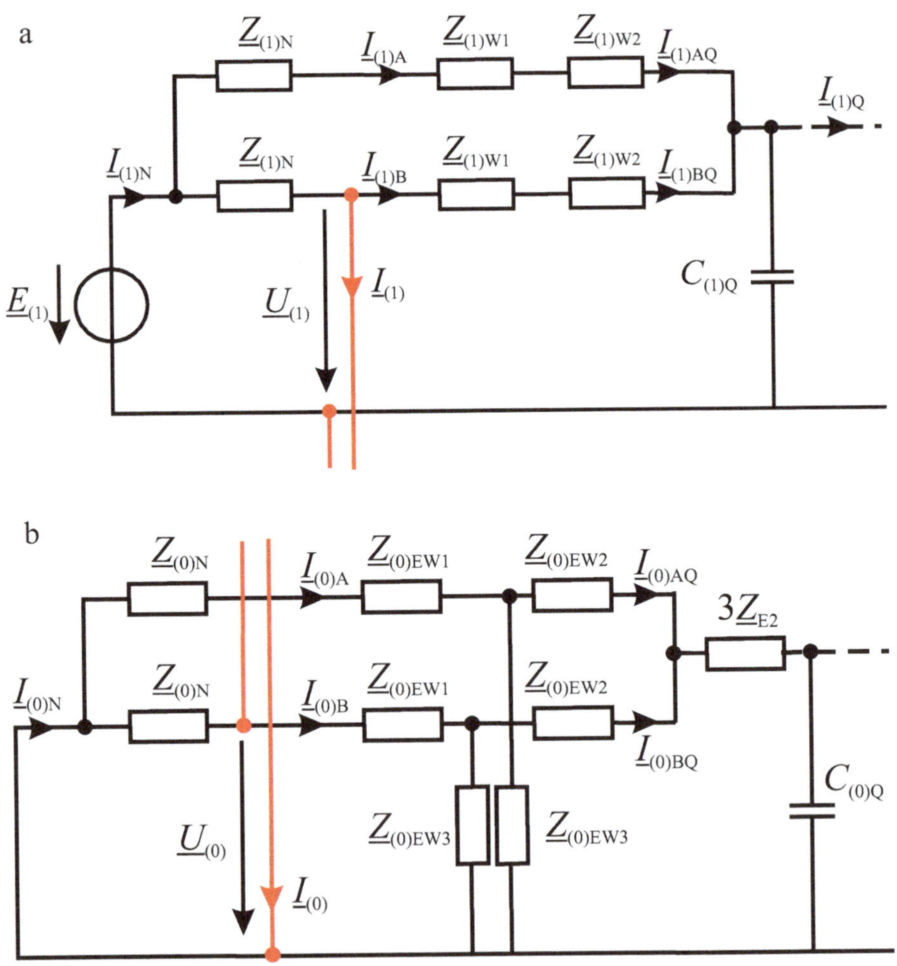

Fig. 15.14 Component equivalent circuits for determining currents in the event of a line-to-earth fault in the overlaid network. (**a**) Positive sequence system, (**b**) zero sequence system

other earthing point. When representing the transformers, the index W is used to describe the windings (see legend for Fig. 15.12).

In the positive and negative sequence system, the impedances $\underline{Z}_{(1)\mathrm{W1}}$ and $\underline{Z}_{(1)\mathrm{W2}}$ are combined to form the resulting transformer impedance $\underline{Z}_{(1)\mathrm{T}}$, since the transformer with balancing winding is operated as a two-winding transformer and this winding is therefore not accessible. For this reason, the balancing winding is not included in the representation of the positive and negative sequence system (Fig. 15.14a).

In accordance with Fig. 15.12b, the following designations are used in the zero sequence system, which are entered in Fig. 15.14b. The following relationships apply, whereby a distinction is made between the neutral point impedance and the neutral impedance of the transformer. For winding 1, it is assumed that the transformer neutral points on the high-voltage side are always solidly earthed:

$$\text{Winding 1}: \quad \underline{Z}_{(0)\mathrm{EW1}} = 3\underline{Z}_{\mathrm{E11}} + \underline{Z}_{(0)\mathrm{W1}} = 3\underline{Z}_{\mathrm{E12}} + \underline{Z}_{(0)\mathrm{W1}} = \underline{Z}_{(0)\mathrm{W1}} \tag{15.77}$$

$$\text{Winding 2}: \quad \underline{Z}_{(0)\mathrm{EW2}} = 3\underline{Z}_{\mathrm{E21}} + \underline{Z}_{(0)\mathrm{W2}} = 3\underline{Z}_{\mathrm{E22}} + \underline{Z}_{(0)\mathrm{W2}} \tag{15.78}$$

$$\text{Winding 3}: \quad \underline{Z}_{(0)\mathrm{EW3}} = \underline{Z}_{(0)\mathrm{W3}} \tag{15.79}$$

Depending on the following examples, the impedances shown in Eqs. (15.77)–(15.79) take on different values, depending on the earthing conditions on the secondary side ($\underline{Z}_{\mathrm{E22}}$, $\underline{Z}_{\mathrm{E21}}$). In contrast, the transformers on the primary side are always solidly earthed ($\underline{Z}_{\mathrm{E11}} = \underline{Z}_{\mathrm{E12}} = 0$).

To calculate the currents in the event of a fault, it is advisable to represent the impedance systems in such a way that the values for $\underline{Z}_{(1)}$ and $\underline{Z}_{(0)}$ can be determined directly, Fig. 15.15. While the positive sequence system remains unchanged, Fig. 15.15a, the zero sequence system changes, Fig. 15.15b, depending on the earthing of the transformer neutral points on the low-voltage side.

In principle, the component currents can be determined according to Sect. 9.2.2.4, Eq. (15.80), whereby all impedances are related to the voltage level of the underlaid network and the capacitance of $C_{(1)\mathrm{Q}}$ of the underlaid network is neglected.

$$\underline{I}_{(0)} = \underline{I}_{(1)} = \underline{I}_{(2)} = \frac{E_{(1)}}{2\underline{Z}_{(1)} + \underline{Z}_{(0)}} \tag{15.80}$$

With

$$\underline{Z}_{(1)} = \frac{\underline{Z}_{(1)\mathrm{N}} \cdot \left(\underline{Z}_{(1)\mathrm{N}} + 2\underline{Z}_{(1)\mathrm{T}}\right)}{2\underline{Z}_{(1)\mathrm{N}} + 2\underline{Z}_{(1)\mathrm{T}}} \tag{15.81}$$

The component currents of the positive and negative sequence systems flowing through the busbar Q can be determined according to Eq. (15.85) using the designations shown in Fig. 15.14.

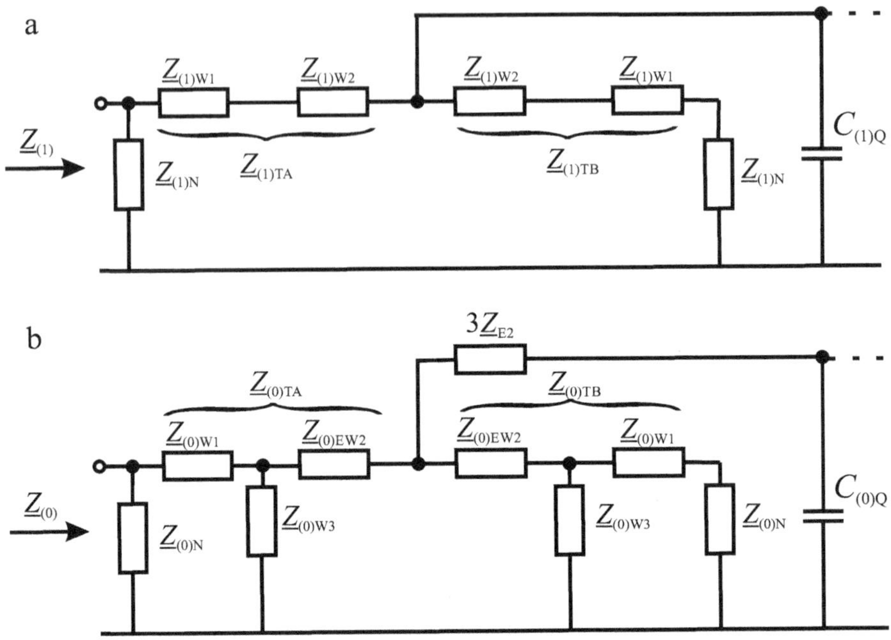

Fig. 15.15 Equivalent circuit for determining the impedances in the event of a line-to-earth fault at location F according to Fig. 15.13 (**a**) positive sequence system, (**b**) zero sequence system

$$\underline{I}_{(1)\text{AQ}} = \underline{I}_{(2)\text{AQ}} = \frac{\underline{Z}_{(1)\text{N}}}{2\underline{Z}_{(1)\text{N}} + 2\underline{Z}_{(1)\text{T}}} \cdot \underline{I}_{(1)} \tag{15.82}$$

In the event of an earth fault, the balancing currents $\underline{I}_{(1)\text{Q}}$ and $\underline{I}_{(2)\text{Q}}$ flow through the busbar of the underlaid network in the positive and negative sequence systems, which primarily result from the ratio of the network impedances to the transformer impedances. Taking into account the conditions according to Eqs. (9.76) and (15.82), the following results for the balancing currents flowing through the busbar Q:

$$\underline{I}_{\text{R}} = 2\underline{I}_{(1)\text{AQ}} \qquad \underline{I}_{\text{S}} = \underline{I}_{\text{S}} = -\underline{I}_{(1)\text{AQ}} \tag{15.83}$$

The zero sequence impedance of the step-up transformers depends on the design and can be specified as a function of the positive sequence impedance as follows, Sect. 4. 8.2, Table 4.7:

- Y(N)yd.: $\underline{Z}_{(0)\text{T}} \approx 2.4 \cdot \underline{Z}_{(1)\text{T}}$ (with balancing winding)
- Y(N)y(n)d: $\underline{Z}_{(0)\text{T}} \approx 1.5 \cdot \underline{Z}_{(1)\text{T}}$ (with balancing winding)
- Y(N)y: $\underline{Z}_{(0)\text{T}} \gg 7 \cdot \underline{Z}_{(1)\text{T}}$ to 100 (without balancing winding)
- Y(N)y(n): $\underline{Z}_{(0)\text{T}} \approx 1.0 \cdot \underline{Z}_{(1)\text{T}}$ (without balancing winding).

In the following, various examples of neutral point treatment are considered, so that the zero sequence impedance $\underline{Z}_{(0)}$ changes accordingly.

15.2.3.2 Isolated Neutral Point of the Transformer in the Underlaid Network

If the neutral points of both transformers on the 10 kV network side are isolated ($\underline{Z}_{E22} = \underline{Z}_{E21} = \rightarrow \infty$), the simplified zero sequence system shown in Fig. 15.16 results, since the impedances $\underline{Z}_{(0)EW2} = \rightarrow \infty$ apply to both transformers with and without balancing windings.

The zero sequence impedance $\underline{Z}_{(0)T}$ of the three-winding transformer is determined according to Fig. 15.16 from the sum of the impedance values of windings 1 and 3.

$$\text{With balancing winding}: \underline{Z}_{(0)} = \frac{\underline{Z}_{(0)N} \cdot \underline{Z}_{(0)T}}{\underline{Z}_{(0)N} + \underline{Z}_{(0)T}} \tag{15.84a}$$

$$\text{Without balancing winding}: \underline{Z}_{(0)} \approx \underline{Z}_{(0)N} \text{ due to } \underline{Z}_{(0)T} \approx \infty \tag{15.84b}$$

In the illustration in Fig. 15.17, the current of the positive sequence system $I_{(1)A}$ is related to the rated current of a transformer I_{rT}. It can be seen that, for a short time corresponding to the duration of the fault in the 110 kV network, almost twice the rated transformer current can flow in the phase affected by the fault according to Eq. (15.83). Due to the neutral point treatment, the current in the zero sequence system on the secondary side is zero.

15.2.3.3 Low-Impedance Neutral Point Earthing in the Underlaid Network

In the case of low-impedance neutral point earthing of the 10 kV network, the following three cases can be distinguished:

- Only the neutral point of one of the two transformers is earthed ($\underline{Z}_{E2} = \underline{Z}_{E22} = 0$, $\underline{Z}_{E21} \rightarrow \infty$). Under these conditions, the zero sequence system for a transformer with stabilizing winding is shown in Fig. 15.18. The corresponding two-winding transformer can be derived from the condition $\underline{Z}_{(0)W3} \rightarrow \infty$. The result is that, in addition to the currents in the positive and negative sequence systems, a small current flows in the zero sequence system (busbar Q), which depends on the existing earth capacitance $C_{(0)Q}$ of the 10 kV network.

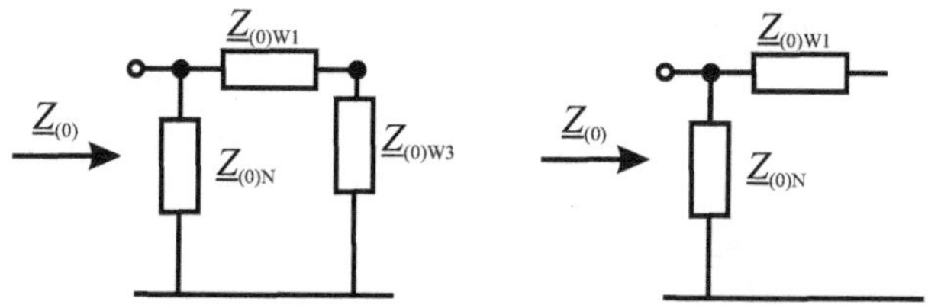

Fig. 15.16 Equivalent circuit of the zero sequence system for isolated transformer neutral points. Left: with balancing winding, right: without balancing winding

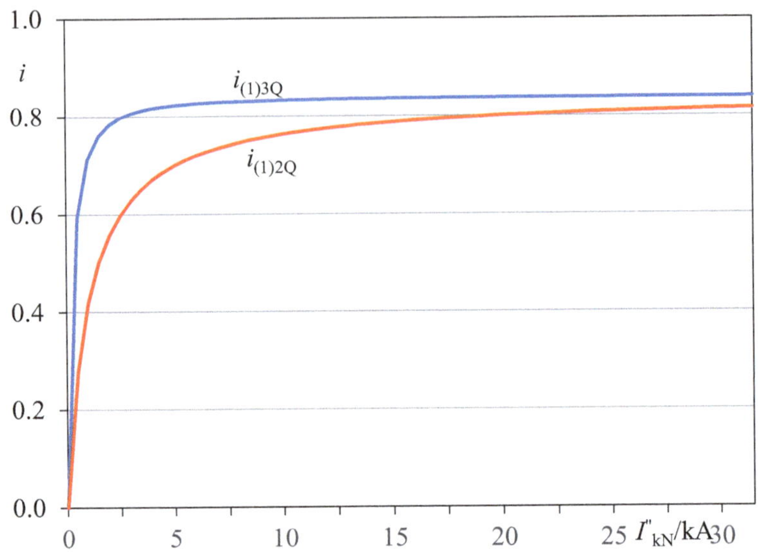

Fig. 15.17 Balancing currents in the positive sequence system in the event of a fault according to Fig. 15.14 as a function of the initial symmetrical short-circuit current of the overlaid network, in each case relative to the rated current of a transformer. $i_{(1)3Q}$ transformer with balancing winding (three-winding transformer), $i_{(1)2Q}$ transformer without balancing winding (two-winding transformer)

- The neutral points of both transformers are directly connected to each other and are earthed via a common impedance ($\underline{Z}_{E2} = \underline{Z}_{E21} = \underline{Z}_{E22} = 0$). In addition to the currents in the positive and negative sequence systems, a current of the zero sequence system flows via the transformer neutral points and via the busbar Q of the 10 kV network (but not via the earthing impedance $\underline{Z}_{E2}$). Figure 15.19a shows a slightly modified representation of the zero sequence system compared to Fig. 15.15b, which shows that almost no current flows through this impedance and can therefore be neglected, since the 10 kV network has no additional earthing point and the capacitance $C_{(0)Q}$ is not taken into account in this context. Since the individual impedances of the windings can be combined to form a resulting zero sequence impedance of the transformer, a simplified representation according to Fig. 15.19b is derived to determine the partial currents $\underline{I}_{(0)AQ}$ und $\underline{I}_{(0)BQ}$

The following therefore applies to the zero sequence impedance for determining the resulting current $\underline{I}_{(0)}$:

$$\underline{Z}_{(0)} = \frac{\underline{Z}_{(0)N} \cdot \left(\underline{Z}_{(0)N} + 2\underline{Z}_{(0)T}\right)}{2\underline{Z}_{(0)N} + 2\underline{Z}_{(0)T}} \tag{15.85}$$

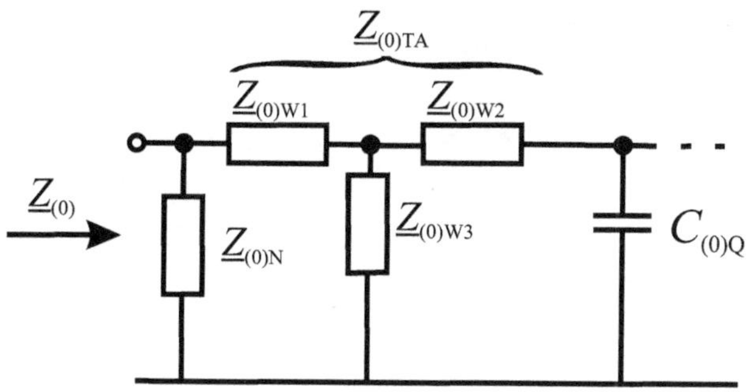

Fig. 15.18 Zero sequence system for a transformer with stabilizing winding in the case of only one earthed neutral point on the low-voltage side

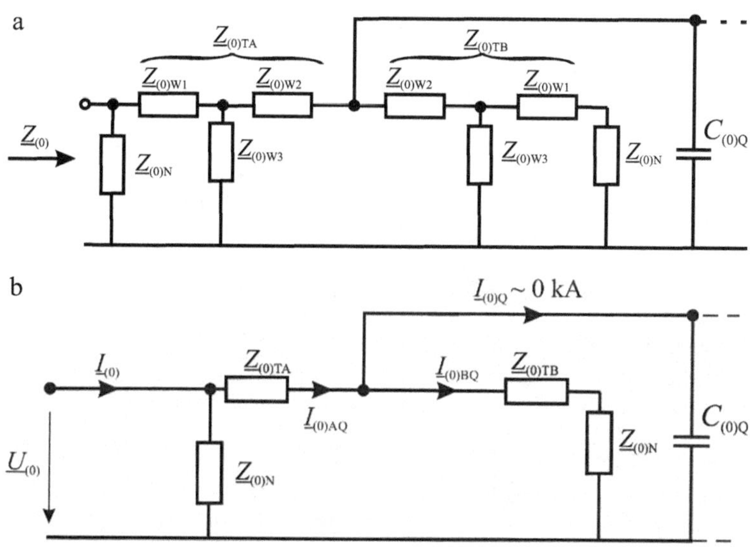

Fig. 15.19 Zero sequence system according to Fig. 15.15 in the case of transformers earthed on both sides. (**a**) Detailed zero sequence system, (**b**) simplified representation

Figure 15.20 shows the relative currents occurring in the positive sequence ($i_{(1)Q}$) and zero sequence ($i_{(0)Q}$) systems on the secondary side of the transformers for this neutral point treatment. While the currents $i_{(1)Q} = i_{(2)Q}$ remain smaller than the transformer rated current even with increasing network short-circuit current, this is no longer true for the zero sequence current $i_{(0)Q}$. Values of $i_{(0)Q} < 1$ apply in the case of a three-phase network short-circuit current of max. 2.5 kA and up to $i_{(0)Q} \sim 2.5$ for values of max. 30 kA, depending on the transformer type. In this case, a two-winding transformer Yy results in a higher load. The phase currents $\underline{I}_R = 2\underline{I}_{(1)Q} + \underline{I}_{(0)Q}$ and $\underline{I}_S = \underline{I}_R = I_{(0)Q} - \underline{I}_{(1)Q}$ flow through the busbar when a line-to-earth fault

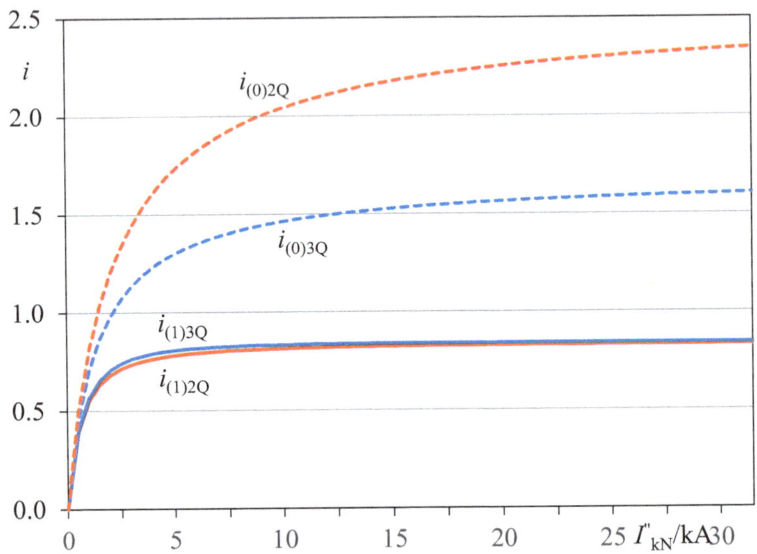

Fig. 15.20 Balancing currents in the positive and zero sequence systems in the case of a fault according to Fig. 15.16 as a function of the initial symmetrical short-circuit current of the overlaid network, in each case relative to the rated current of a transformer. $i_{(1)3Q}$, $i_{(0)3Q}$ positive and zero sequence currents in a transformer with stabilizing winding (busbar Q), $i_{(1)2Q}$, $i_{(1)2Q}$ positive and zero sequence currents in a transformer without stabilizing winding (busbar Q)

occurs in phase R in the 110 kV network. If the ohmic components are neglected, they are in the same direction. They can be considerably high and unnecessarily trigger the system protection.

- Both transformer neutral points are earthed via an impedance ($\underline{Z}_{E2} = 0$). In this case, the zero sequence currents occurring on the secondary side can generally be neglected because the parallel connection $\underline{Z}_{E21}//\underline{Z}_{E22}$ significantly determines the dimensioning of the impedances with regard to current limitation in the event of a line-to-earth fault on the low-voltage side. In the event of a fault in the overlaid network, the zero sequence current is limited by the series connection $\underline{Z}_{E21} + \underline{Z}_{E22}$.

15.2.4 Result

When the two neutral points of a transformer (high-voltage and low-voltage sides) are earthed between two networks, the neutral systems are coupled with each other. An unbalanced fault with earth contact in the overlaid network can appear to the underlaid network as its own fault or even cause a resonance increase in voltages and currents.

The coupling remains low:

- with a low-impedance earth on the side of the underlaid network,
- when using a transformer with a stabilizing winding,
- with a transformer whose stabilizing winding is constructed as the bottom winding on the core.

The coupling is large:

- with earth fault compensation on both sides of the transformer,
- with earth fault compensation on the side of the underlaid network and solidly earthed on the side of the overlaid network.

The coupling of the zero sequence voltage from the overlaid network into the underlaid network means in any case that, in the event of a line-to-earth fault in the overlaid network, the voltage in the underlaid network drops more sharply on one phase (namely the phase affected by the fault) than without this coupling in the zero sequence system. This can lead to undervoltage tripping of motors or to unwanted tripping of motor protection switches [9].

In the case of transformers without stabilizing windings, not only is the influence of the overlaid network on the underlaid network high, but the quality of the earthing of the overlaid network also depends on the zero sequence currents that can flow in the underlaid network. This type of operation should be avoided. In the case of parallel transformers, the neutral points of these transformers should not be connected to each other. Otherwise, balancing currents may occur between the transformers and, under certain circumstances, activate the system protection.

15.3 Conclusions

In the case of parallel transformers with different transformation ratios, calculating the short-circuit current using the equivalent voltage source method at the short-circuit location leads to significant deviations compared to calculations using actual voltages or the superposition method. These deviations do not occur if the transformation ratios are identical. In order to still use the proven method of the equivalent voltage source at the short-circuit location, it is necessary to apply the arithmetic mean of the transformation ratios in the case of parallel transformers according to [2].

In the case of both sides being earthed for the neutral points of transformers, line-to-earth faults on the high-voltage side result in mutual interference between both networks, depending on the network connection. Transformers with a stabilizing winding allow low-impedance earthing on both sides of the transformer star points. In contrast, arrangements of transformers without stabilizing windings or with earth fault compensation on one or both sides are not recommended.

References

1. Balzer G (2016) Short-circuit calculation with fullsize converters according to IEC 60909. In: 21st conference of electric power supply industry, CEPSI 2016, Bangkok, Thailand, October 2016, B1
2. Balzer G, Wasserrab A, Busarello L (2015) Nachbildung paralleler Transformatoren. EW (Zeitschrift für Energiewirtschaft) 10, pp 64–68
3. IEC 60909-0:2002 (2002) Short-circuit current calculations in three-phase a.c. systems—Part 0: calculation of currents. IEC, Geneva
4. Balzer G, Remde H (1983) High- and low-voltage side earthing of transformers linking public power supply systems and industrial systems. CIRED-1983, b.01.1-0.1.6
5. Balzer G, Remde H (1985) Probleme bei der beidseitigen Erdung von Transformatoren. Brown Boveri Technik 72(7), pp 349–354
6. Balzer G, Remde H (1985) Problems arising with transformers having both neutral points earthed. Elektron (J S Afr Inst Electr Eng) 1985, pp 25–30
7. Clerici E, Lionetto PF, Tommazollini GF, Vergani A (1981) Suggested items to be considered when studying the interface between industrial and public distribution systems. In: CIRED 1981, vol 1(5), pp 279–284
8. Gröber R, Komurka J (1973) Übertragung der Nullspannung bei zweiseitig geerdeten Transformatoren. Technische Mitteilung, Forschungsgemeinschalt für Hochspannungs- und Hochstromtechnik e.V., Mannheim/Deutschland
9. Bayer W, Buchmann G (1981) Effect of faults in high-voltage power supply systems on industrial networks supplying induction motors. In: CIRED 1981, vol 1, pp 290–293

Chapter 16
Representation of Autotransformers in Short-Circuit Current Calculations

Autotransformers are generally used as coupling transformers between two extra-high voltage levels, and autotransformers are generally treated as separate winding transformer in short-circuit current calculations in accordance with Sect. 4.8. In the following, particular emphasis is placed on determining the equivalent circuit diagrams and the current distribution when determining the line-to-earth short-circuit currents [1].

16.1 General

Figure 16.1 shows the basic design of an autotransformer. In this autotransformer, the two windings (high and low voltage) are connected in series. Based on a separate winding transformer, an additional galvanic connection is established between the two windings with the number of turns w_1 and w_2, as shown in Fig. 16.1 (left), so that it is characteristic of autotransformers that both inductive and galvanic coupling of the windings exists. In terms of design, an autotransformer can consist of the following windings:

- Main winding or parallel winding (voltage drop $\underline{U}_2$), Fig. 16.1,
- additional winding or series winding (voltage drop $\underline{U}_X$), Fig. 16.1,
- stabilizing winding (voltage drop $\underline{U}_3$), Fig. 16.2.

The main difference to a separate winding transformer is that the total winding power S_E is lower than the throughput power S_D, relative to a separate winding transformer. The total winding power is defined as the arithmetic mean of the high voltage and low voltage winding power, which is therefore a measure of the material costs of the windings and thus of the costs. Taking into account the designation according to Fig. 16.1, the following relationships result for transformers consisting of three single-phase units.

G. Balzer, *Short-Circuit Currents in Three-Phase Systems*,
https://doi.org/10.1007/978-3-658-50710-7_16

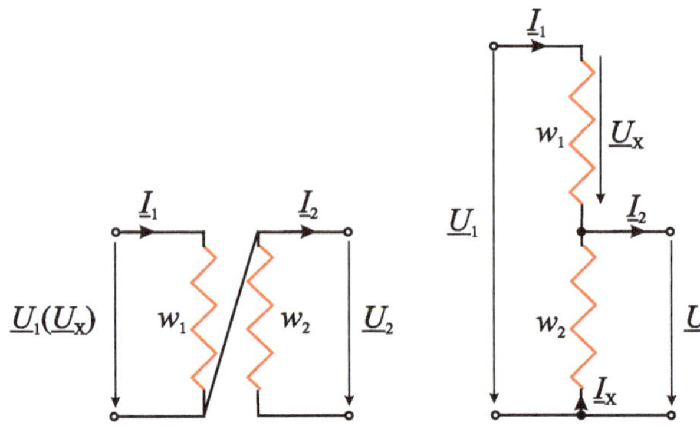

Fig. 16.1 Equivalent circuit diagram of an autotransformer with an internal winding connection derived from a separate winding transformer (w turns)

- Separate winding transformer

$$S_E = \frac{3}{2} \cdot (U_1 \cdot I_1 + U_2 \cdot I_2) \tag{16.1}$$

$$S_D = 3 \cdot U_1 \cdot I_1 = 3 \cdot U_2 \cdot I_2 \tag{16.2}$$

With

U_1, I_1 voltage und current of the high voltage winding,
U_2, I_2 voltage und current of the low voltage winding.

Taking into account Eqs. (16.1) and (16.2), it follows that for a separate winding transformer, the total winding power is equal to the throughput power, i.e. $\underline{S}_E = \underline{S}_D$. This means that the total winding power of a separate winding transformer always corresponds to the throughput power and must be designed for this purpose. The same applies to autotransformers:

- Autotransformer

$$S_E = \frac{3}{2} \cdot [I_1 \cdot U_X + U_2 \cdot I_X] = \frac{3}{2} \cdot [I_1 \cdot (U_1 - U_2) + U_2 \cdot (I_2 - I_1)] \tag{16.3}$$

$$\underline{S}_D = 3 \cdot \underline{U}_1 \cdot \underline{I}_1 = 3 \cdot \underline{U}_2 \cdot \underline{I}_2 \tag{16.4}$$

With

U_X, I_X voltage und current of the high voltage winding.

Combining Eqs. (16.3) and (16.4) yields the following for the power ratio, with the transformation ratio $t = U_1/U_2$:

$$\frac{S_E}{S_D} = \frac{1}{2} \cdot \frac{I_1 \cdot U_1 + I_2 \cdot U_2 - 2 \cdot I_1 \cdot U_2}{S_D} = 1 - \frac{U_2}{U_1} = 1 - \frac{1}{t} \tag{16.5}$$

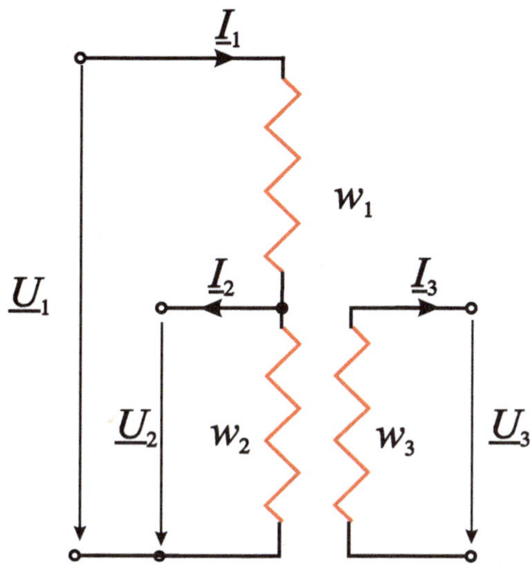

Fig. 16.2 Autotransformer with stabilizing winding and different windings and numbers of turns w. w_2 number of turns (parallel winding), w_1 number of turns (series winding), w_3 number of turns (stabilizing winding)

Equation (16.5) shows that the power consumption of an autotransformer, and thus the manufacturing and material costs, are always lower than those of a comparable separate winding transformer with the same throughput power. Due to the lower material costs of autotransformers compared to separate winding transformers, the short-circuit voltages are also generally lower [2]. The transformation ratio should not deviate too much from one in order to achieve the cost-saving benefit. Autotransformers are therefore used in Germany to couple extra-high voltage grids (e.g., 380/220 kV) when both sub-grids are solidly earthed. However, this financial advantage is offset by an electrical disadvantage: Due to the conductive connection between the input and output windings, the autotransformer can only be used if the two connected networks have the same neutral point treatment, as a line-to-earth fault in the overlaid network automatically leads to a short-circuit current in the underlaid network.

In general, an autotransformer is equipped with a stabilizing winding (connected in delta), so that Fig. 16.2 can be extended; in this case, the voltage group is Y(d)y [3].

16.2 Equivalent Circuits

An autotransformer can be viewed on the basis of two circuit diagrams, which differ in terms of the description of the windings and the associated electrical quantities [4]. These two representations are:

- Winding configuration: In this case, the voltages and currents are assigned to the individual windings so that the respective leakage reactances can be considered.

- Circuit configuration: This form of representation corresponds to the approach that is consistent with operational use. This means that the voltages and currents accessible at the terminals are used.

The conversion of the two representations for a two-winding transformer is derived below. Figure 16.1 can be used to derive a T equivalent circuit for autotransformers. According to Sect. 4.8.2, the dependence of the voltages and currents of a separate winding transformer is given by Eq. (4.200) if the current designations given there are used, whereby in this case too, the ohmic losses that occur are not taken into account:

$$\begin{pmatrix} \underline{U}_X \\ \underline{U}_2 \end{pmatrix} = j\omega \cdot \begin{pmatrix} w_1^2 \cdot (\Lambda_H + \Lambda_{\sigma 1}) & w_1 \cdot w_2 \cdot \Lambda_H \\ w_1 \cdot w_2 \cdot \Lambda_H & w_2^2 \cdot (\Lambda_H + \Lambda_{\sigma 2}) \end{pmatrix} \cdot \begin{pmatrix} \underline{I}_1 \\ -\underline{I}_X \end{pmatrix} \tag{16.6}$$

With

U_1	voltage of the high voltage winding,
I_2	current of the low voltage winding,
U_1, I_X	voltage and current according to Fig. 16.1,
Λ_H	magnetic conductance of the magnetizing flux,
$\Lambda_{\sigma 1,\,\sigma 2}$	magnetic conductance of the leakage flux,
w	number of turns of the windings.

Due to the special configuration of an autotransformer, the internal current and voltage values can be replaced by the external values [5]:

$$\underline{I}_X = -\underline{I}_1 + \underline{I}_2 \qquad \underline{U}_X = \underline{U}_1 - \underline{U}_2 \tag{16.7}$$

After substitution into Eq. (16.6), the result is:

$$\begin{pmatrix} \underline{U}_1 - \underline{U}_2 \\ \underline{U}_2 \end{pmatrix} = j\omega \cdot \begin{pmatrix} w_1^2 \cdot (\Lambda_H + \Lambda_{\sigma 1}) & w_1 \cdot w_2 \cdot \Lambda_H \\ w_1 \cdot w_2 \cdot \Lambda_H & w_2^2 \cdot (\Lambda_H + \Lambda_{\sigma 2}) \end{pmatrix} \cdot \begin{pmatrix} \underline{I}_1 \\ \underline{I}_1 - \underline{I}_2 \end{pmatrix} \tag{16.8}$$

By transforming, the voltages $\underline{U}_1$ and $\underline{U}_2$ are determined to be.

$$\begin{aligned} &\underline{U}_1 = j\omega \cdot \left\{2 \cdot w_1 \cdot w_2 \cdot \Lambda_H + w_1^2 \cdot (\Lambda_H + \Lambda_{\sigma 1}) + w_2^2 \cdot (\Lambda_H + \Lambda_{\sigma 2})\right\} \cdot \underline{I}_1 \\ &-j\omega \cdot \left\{w_1 \cdot w_2 \cdot \Lambda_H + w_2^2 \cdot (\Lambda_H + \Lambda_{\sigma 2})\right\} \cdot \underline{I}_2 \\ &\underline{U}_2 = j\omega \cdot \left\{w_1 \cdot w_2 \cdot \Lambda_H + w_2^2 \cdot (\Lambda_H + \Lambda_{\sigma 2})\right\} \cdot \underline{I}_1 - j\omega \cdot w_2^2 \cdot (\Lambda_H + \Lambda_{\sigma 2}) \cdot \underline{I}_2 \end{aligned} \tag{16.9}$$

The result is in the following matrix form

$$\begin{pmatrix} \underline{U}_1 \\ \underline{U}_2 \end{pmatrix} = \mathrm{j}\omega \cdot \begin{pmatrix} \underline{Z}_{11} & \underline{Z}_{12} \\ \underline{Z}_{12} & \underline{Z}_{11} \end{pmatrix} \cdot \begin{pmatrix} \underline{I}_1 \\ \underline{I}_2 \end{pmatrix} \tag{16.10}$$

With the abbreviations:

$$\begin{aligned} \underline{Z}_{11} &= \mathrm{j}\omega \cdot \{2 \cdot w_1 \cdot \ w_2 \ \cdot \Lambda_\mathrm{H} + w_1^2 \cdot \ (\Lambda_\mathrm{H} + \Lambda_{\sigma 1}) + w_2^2 \cdot \ (\Lambda_\mathrm{H} + \Lambda_{\sigma 2})\} \\ \underline{Z}_{12} &= \underline{Z}_{21} = \mathrm{j}\omega \cdot \{w_1 \cdot \ w_2 \ \cdot \Lambda_\mathrm{H} + w_2^2 \cdot \ (\Lambda_\mathrm{H} + \Lambda_{\sigma 2})\} \\ \underline{Z}_{22} &= \mathrm{j}\omega \cdot w_2^2 \cdot \ (\Lambda_\mathrm{H} + \Lambda_{\sigma 2}) \end{aligned} \tag{16.11}$$

Taking into account the equivalent circuit diagram of a separate winding transformer according to Fig. 4.35, the transformed voltages $\underline{U}_{2\mathrm{t}}$ and currents $\underline{I}_{2\mathrm{t}}$ shown in Fig. 16.3 can be introduced, so that according to Fig. 16.1 at no load ($\underline{I}_2 = 0$) the following applies:

$$\frac{\underline{U}_1}{\underline{U}_2} = \frac{w_1 + w_2}{w_2} = t \tag{16.12}$$

The following then applies to the transformed values:

$$\underline{U}_2 = \frac{\underline{U}_{2\mathrm{t}}}{t} \qquad \underline{I}_2 = t \cdot \underline{I}_{2\mathrm{t}} \tag{16.13}$$

Inserting the transformed values gives the results for the matrix:

$$\begin{aligned} \begin{pmatrix} \underline{U}_1 \\ \underline{U}_{2\mathrm{t}} \end{pmatrix} &= \mathrm{j}\omega \cdot \begin{pmatrix} \underline{Z}_{11} & t \cdot \underline{Z}_{12} \\ t \cdot \underline{Z}_{12} & t^2 \cdot \underline{Z}_{22} \end{pmatrix} \cdot \begin{pmatrix} \underline{I}_1 \\ \underline{I}_{2\mathrm{t}} \end{pmatrix} \\ &= \mathrm{j}\omega \cdot \begin{pmatrix} \underline{Z}_{11} & \underline{Z}_{12\mathrm{t}} \\ \underline{Z}_{12\mathrm{t}} & \underline{Z}_{22\mathrm{t}} \end{pmatrix} \cdot \begin{pmatrix} \underline{I}_1 \\ \underline{I}_{2\mathrm{t}} \end{pmatrix} \end{aligned} \tag{16.14}$$

For the T equivalent circuit of the autotransformer, Fig. 16.3 is the result with a changed current direction $\underline{I}_{2\mathrm{t}}$ compared to Fig. 16.1, so that the matrix can be used with positive elements $\underline{Z}_{12\mathrm{t}}$ and $\underline{Z}_{22\mathrm{t}}$.

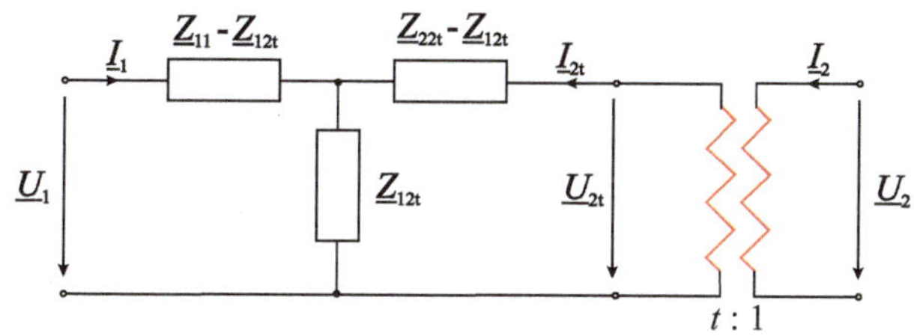

Fig. 16.3 T equivalent circuit of an autotransformer

For the elements of the T equivalent circuit, the following relationships are the result according to Eq. (16.15):

$$\begin{aligned}
\underline{Z}_{11} - \underline{Z}_{12t} &= j\omega \cdot w_1 \cdot (w_1 \cdot \Lambda_{\sigma 1} - w_2 \cdot \Lambda_{\sigma 2}) \\
\underline{Z}_{22t} - \underline{Z}_{12t} &= j\omega \cdot w_1 \cdot (w_1 + w_2 \cdot)\Lambda_{\sigma 2} \\
\underline{Z}_{12t} &= j\omega \cdot \left\{ (w_1 + \ w_2)^2 \ \cdot \Lambda_{\mathrm{H}} + w_2 \cdot (w_1 + \ w_2)\Lambda_{\sigma 2} \right\}
\end{aligned} \tag{16.15}$$

Since the magnetic conductance of the magnetizing reactance is significantly greater than the corresponding values of the leakage reactance, branch $\underline{Z}_{12t}$ of the T equivalent circuit can be neglected for the short-circuit current calculation. The resulting short-circuit reactance is then as follows:

$$\underline{Z}_{\mathrm{k}} = \underline{Z}_{11} - 2 \cdot \underline{Z}_{12t} + \underline{Z}_{22t} = j\omega \cdot w_1^2 \cdot (\Lambda_{\sigma 1} + \Lambda_{\sigma 2}) \tag{16.16}$$

A comparison with Eq. (4.205) shows that the short-circuit reactance of an autotransformer corresponds to the reactance of a separate winding transformer.

Three-winding transformers are generally used in extra high voltage networks, for which a component circuit can be derived in accordance with Fig. 4.42, as shown in Fig. 16.4. The windings w_1 and w_2 with the internal connection represent the parallel and series windings. Figure 16.4a thus shows an equivalent circuit representing the windings of the autotransformer (winding configuration). In contrast, the equivalent circuit shown in Fig. 16.4b is used for the calculation, since the two terminals of the windings 2 and 1 are available for the corresponding networks and thus represent the network circuit (circuit configuration). According to Fig. 16.4, the following transmission ratios apply:

$$t = (w_1 + w_2)/w_2 \qquad t_3 = (w_1 + w_2)/w_3 \tag{16.17}$$

The circuit shown in Fig. 16.4b, which is useful for applications involving symmetrical components, for example, can be derived from Fig. 16.4a, since $\underline{U}_1$ is separate from $\underline{U}_2$ here. Since the equivalent circuits consist of three reactances and two transmission ratios, in total of five equations are required for a clear determination. These equations can be derived from two open-circuit and three short-circuit measurements. According to [4, 6], the relationships according to Eq. (16.16) result for the reactances, corresponding to Fig. 16.4b, if the variables of the circuit configuration are derived from the variables of the winding configuration. Here, all reactances are related to the reference voltage U_2:

$$\begin{aligned}
X_{\mathrm{b}1} &= \left(\tfrac{t-1}{t}\right)^2 \cdot X_{\mathrm{a}4} - \frac{t-1}{t} \cdot X_{\mathrm{a}2} & X_{\mathrm{b}2} &= \frac{t-1}{t} \cdot X_{\mathrm{a}2} \\
X_{\mathrm{b}3} &= X_{\mathrm{a}3} + \frac{1}{t} \cdot X_{\mathrm{a}2} & t &= \frac{U_1}{U_2} = \frac{w_1 + w_2}{w_2}
\end{aligned} \tag{16.18}$$

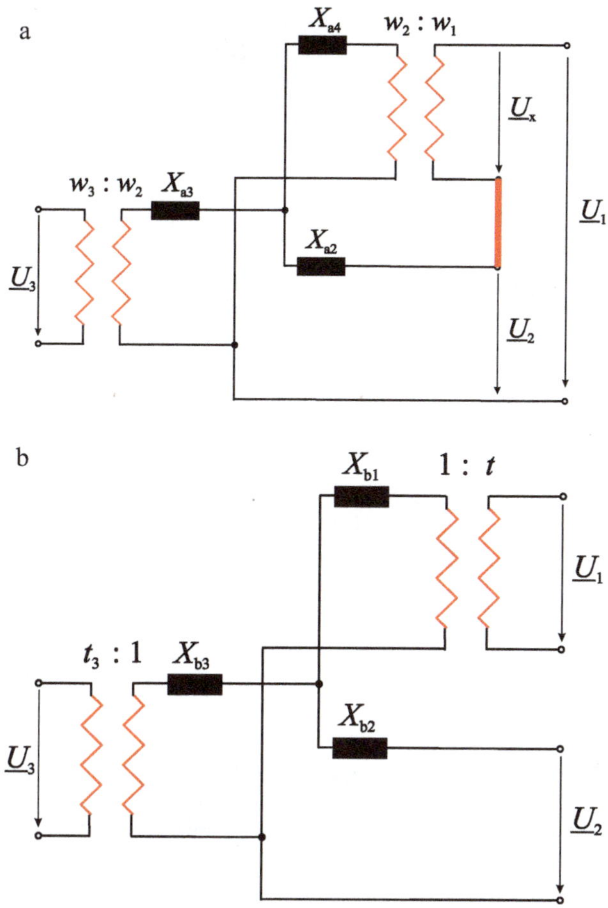

Fig. 16.4 Equivalent circuit of an autotransformer with stabilizing winding (reference voltage of the reactances is the voltage level U_2 in each case). (**a**) With internal connection (winding configuration, index a), (**b**) network circuit for network coupling (circuit configuration, index b)

With

U_1 high voltage,
U_2 low voltage.

If, in general, the neutral point of the autotransformer is directly earthed (earthing impedance $\underline{Z}_E = 0\ \Omega$), taking into account the stabilizing winding (delta), the equivalent circuit in the zero sequence system (Fig. 16.4) is obtained, which corresponds to Fig. 4.43, but with an earthing impedance $\underline{Z}_E = 0\ \Omega$. The zero sequence system diagram shows that in the event of line-to-earth fault, e.g. in the overlaid network ($\underline{U}_1$), a fault current is transmitted to the underlaid network ($\underline{U}_2$).

If, in contrast to Fig. 16.5, a neutral point impedance is taken into account, the zero sequence impedance of the autotransformer must be extended according to [4], Fig. 16.6. In this case, the neutral point impedance must be assigned to not only one

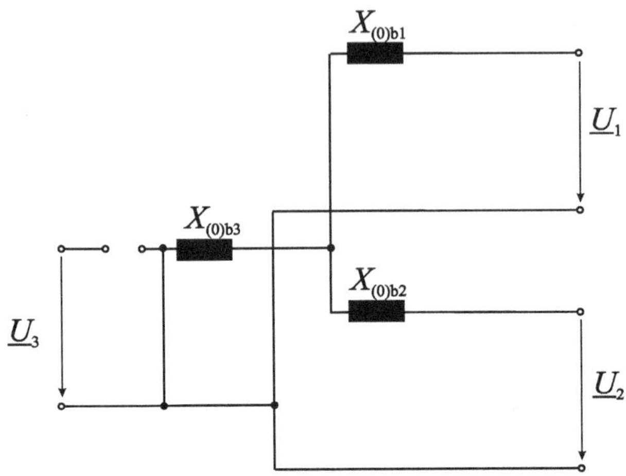

Fig. 16.5 Equivalent circuit of a earthed autotransformer in a zero sequence system (reference voltage U_2)

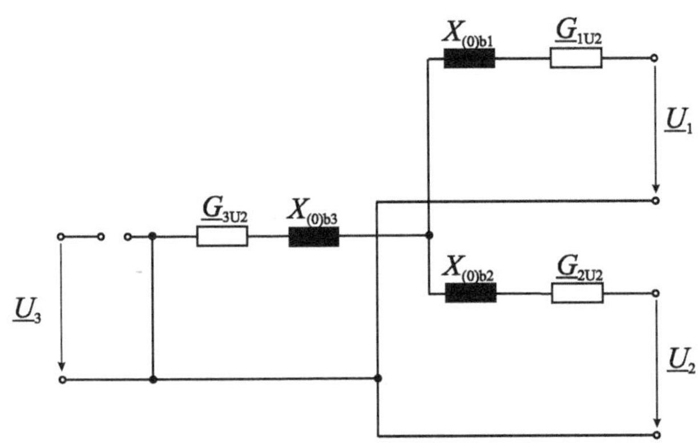

Fig. 16.6 Equivalent circuit of an autotransformer in a zero sequence system, taking into account an earth impedance $\underline{Z}_{E,}$ according to Eq. (16.19)

winding, but also two windings (high voltage and low voltage), since the neutral-point current causes a different voltage drop. According to [4], the following applies to the transformed neutral point impedances $\underline{G}$, each related to the high voltage $\underline{U}_1$ (for $\underline{Z}_E \neq \infty$) or to the low voltage $\underline{U}_2$ with the transformation ratio t according to Eq. (16.19):

$$\begin{aligned} \underline{G}_{1U1} &= -3\underline{Z}_E \cdot (t-1) & \underline{G}_{2U1} &= +3\underline{Z}_E \cdot t \cdot (t-1) & \underline{G}_{3U1} &= +3\underline{Z}_E \cdot t \\ \underline{G}_{1U2} &= -3\underline{Z}_E \cdot \frac{(t-1)}{t^2} & \underline{G}_{2U2} &= +3\underline{Z}_E \cdot \frac{(t-1)}{t} & \underline{G}_{3U2} &= +3\underline{Z}_E \cdot \frac{1}{t} \end{aligned} \tag{16.19}$$

If the neutral point of an autotransformer is not earthed, a star-delta configuration should be implemented from the star configuration in the zero sequence system as shown in Fig. 16.6. Converting the star configuration with $\underline{Z}_E \rightarrow \infty$ does not make sense, as in this case all branches have an infinite value. However, due to galvanic coupling, the zero sequence impedance of the transformer represents a finite value [7]. Considering Eq. (16.20), the reactances in the zero sequence system for a free neutral point according to [4] are obtained when the high voltage U_1 is the reference voltage:

$$\begin{aligned} X_{(0)b12} &= t \cdot X_{(0)b1} + \frac{1}{t} \cdot X_{(0)b2} + \frac{(t-1)^2}{t} \cdot X_{(0)b3} \\ X_{(0)b23} &= -\left[\frac{t^2}{t-1} \cdot X_{(0)b1} + \frac{1}{t-1} \cdot X_{(0)b2} + (t-1) \cdot X_{(0)b3}\right] \\ X_{(0)b31} &= \frac{t}{t-1} \cdot X_{(0)b1} + \frac{1}{t \cdot (t-1)} \cdot X_{(0)b2} + \frac{t-1}{t} \cdot X_{(0)b3} \end{aligned} \tag{16.20}$$

The circuit in the zero sequence system is referred to as a resonance triangle, since the sum of the reactances equals zero ($X_{(0)b12} + X_{(0)b23} + X_{(0)b31} = 0$).

16.2.1 Example of the Application of the Component Equivalent Circuit

Data for the autotransformer Y(d)y:

Nominal system voltage	U_n/kV: 380/220/30
rated voltage	U_r/kV: 400/230/30
rated power (winding)	S_r/MVA: 1000/1000/200
Short-circuit voltage	u_{k12}/%: 15.5
	u_{k23}/%: 11.2
	u_{k13}/%: 15.2

For simplicity, resistances are not taken into account. Using the short-circuit voltages and the throughput power (in each case the smallest power between the windings involved), the reactances of the equivalent circuit in the interconnected system can be determined as shown in Fig. 4.42 (reference voltage level $U_n = 380$ kV).

$$X_{12} = \frac{15.5}{100} \cdot \frac{(400\ \text{kV})^2}{1000\ \text{MVA}} = 24.8\ \Omega \qquad X_{23} = \frac{11.2}{100} \cdot \frac{(400\ \text{kV})^2}{200\ \text{MVA}} = 89.6\ \Omega$$

$$X_{31} = \frac{15.2\%}{100\%} \cdot \frac{(400\,\text{kV})^2}{200\,\text{MVA}} = 121.6\,\Omega$$

According to Table 4.7, the reactances (star configuration) for the equivalent circuit can be determined.

$$X_1 = \frac{X_{31} + X_{12} - X_{23}}{2} = \frac{121.6 + 24.8 - 89.6}{2} = 28.4\,\Omega$$

$$X_2 = \frac{X_{12} + X_{23} - X_{31}}{2} = \frac{24.8 + 89.6 - 121.6}{2} = -3.6\,\Omega$$

$$X_3 = \frac{X_{23} + X_{31} - X_{12}}{2} = \frac{89.6 + 121.6 - 24.8}{2} = 93.2\,\Omega$$

Based on the 220 kV voltage level, the following reactances (index t) result:

$$X_{1\mathrm{t}} = 9.39\ \Omega \qquad X_{2\mathrm{t}} = -1.19\ \Omega \qquad X_{3\mathrm{t}} = 30.81\ \Omega$$

Due to the design of winding 2, the associated reactance X_2 is negative. Since autotransformers consist of three single-phase units, the zero sequence reactances are equal to the positive sequence reactances in a solidly earthed transformer ($\underline{Z}_{\mathrm{E}} = 0\,\Omega$). In contrast, in an isolated transformer ($\underline{Z}_{\mathrm{E}} \rightarrow \infty$), the reactances of the resonance triangle are determined according to Eq. (16.20), with the transformation ratio t:

$$t = \frac{U_1}{U_2} = \frac{400}{230} = 1.739$$

$$X_{(0)12} = t \cdot X_1 + \frac{1}{t} \cdot X_2 + \frac{(t-1)^2}{t} \cdot X_3$$

$$= 1.739 \cdot 28.4\Omega - \frac{3.6\Omega}{1.739} + \frac{0.739^2}{1.739} \cdot 93.2\Omega = 76.598\,\Omega$$

$$X_{(0)23} = -\left[\frac{t^2}{t-1} \cdot X_1 + \frac{1}{t-1} \cdot X_2 + (t-1) \cdot X_3\right]$$

$$= -\frac{1.739^2}{0.739} \cdot 28.4\Omega + \frac{3.6\Omega}{0.739} - 0.739 \cdot 93.2\Omega = -180.231\,\Omega$$

$$X_{(0)31} == \frac{t}{t-1} \cdot X_1 + \frac{1}{t \cdot (t-1)} \cdot X_2 + \frac{t-1}{t} \cdot X_3$$

$$= \frac{1.739}{0.739} \cdot 28.4\Omega - \frac{3.6\Omega}{1.739 \cdot 0.739} + \frac{0.739}{1.739} \cdot 93.2\Omega = 103.633\,\Omega$$

Based on the 220 kV voltage level, the following reactances (index t) result:

$$X_{(0)12\mathrm{t}} = 25.33\ \Omega \qquad X_{(0)23\mathrm{t}} = -59.59\ \Omega \qquad X_{(0)31\mathrm{t}} = 34.26\ \Omega$$

The corresponding equivalent circuits can be shown using the reactances calculated in the positive and zero sequence systems, Fig. 16.7.

The equivalent circuit of a earthed autotransformer in the component equivalent circuit thus corresponds to the equivalent circuit of a three-winding transformer Y (d)y with solidly earthed neutral points on both sides [8]. Section 16.3.2 provides an example of a line-to-earth fault taking into account an earthed autotransformer. In contrast, Sect. 16.3.3 shows the procedure for an insulated neutral point of the autotransformer.

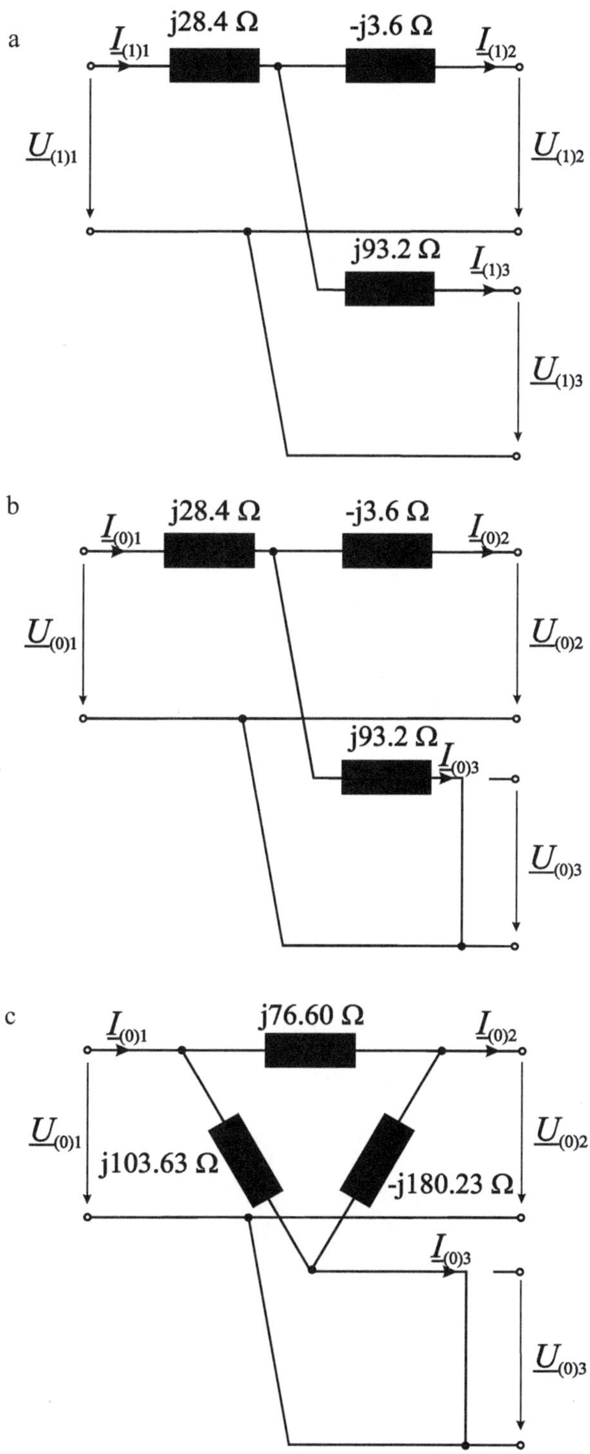

Fig. 16.7 Equivalent circuits of an autotransformer (all values refer to the high voltage side U_1). (**a**) Positive sequence system, (**b**) zero sequence system (neutral point solidly earthed), (**c**) zero sequence system (neutral point isolated), 1 high voltage winding, 2 low voltage winding, 3 stabilizing winding

16.3 Calculation of Currents in the Event of a Line-to-Earth Short Circuit

It is standard practice to earth the neutral point of autotransformers so that the neutral sequence systems of the connected voltage levels are coupled and currents and voltages are transformed in the event of a line-to-earth short circuit [7]. The following section examines the influence of the neutral point treatment of an autotransformer (with 30 kV stabilizing winding) on the line-to-earth short-circuit current.

16.3.1 Network Circuit

According to Fig. 16.8, the current is calculated for a short circuit on the 220 kV side, with feeder from the overlaid 380 kV network. In this figure, the neutral point of the autotransformer is assumed to be solidly earthed. For simplicity, only reactances are taken into account; the stabilizing winding is unloaded. All reactances are related to the 220 kV level, since the line-to-earth short circuit is calculated at this voltage level and the equivalent voltage source is applied at the short-circuit location. To simplify the calculation of the short-circuit current, the correction factor K_T for transformers is not applied.

The following equipment data is used for the calculation (**the impedances are referenced to the 220 kV level**):

380 kV network (A)

Initial symmetrical short-circuit current: 45 kA (contribution from the 380 kV network)

$$X_{(1)A} \approx \frac{1.1 \cdot U_n/\sqrt{3}}{I''_{k3A}} \cdot \frac{1}{t^2} = \frac{1.1 \cdot 380/\sqrt{3}}{45} \cdot \left(\frac{230}{400}\right)^2 \Omega = 1.773\,\Omega$$

Zero sequence reactance ratio: $X_{(0)A}/X_{(1)A} = 3.0 \rightarrow X_{(0)A} = 5.319\ \Omega$.

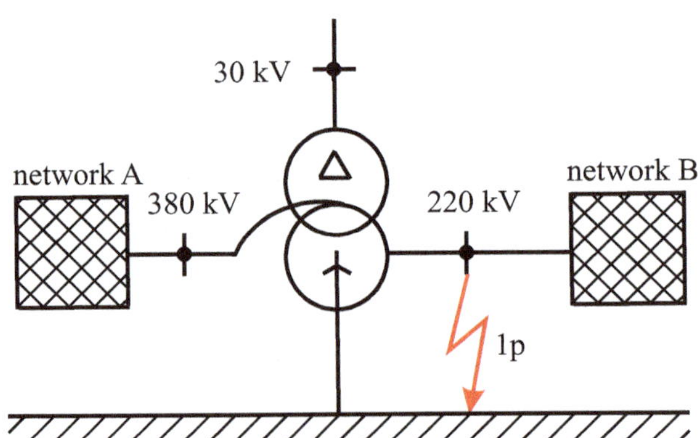

Fig. 16.8 Network circuit for determining the line-to-earth short-circuit current in the 220 kV network (earthed autotransformer)

220 kV network (B)

Initial symmetrical short-circuit current: 50 kA (contribution from the 220 kV network)

$$X_{(1)\mathrm{B}} \approx \frac{1.1 \cdot U_{\mathrm{n}}/\sqrt{3}}{I''_{\mathrm{k3B}}} = \frac{1.1 \cdot 220/\sqrt{3}}{50}\,\Omega = 2.794\,\Omega$$

Zero sequence reactance ratio: $X_{(0)\mathrm{Bt}}/X_{(1)\mathrm{Bt}} = 2.0 \rightarrow X_{(0)\mathrm{B}} = 5.589\ \Omega$.

Autotransformer

Information according to Sect. 16.2 “Example,” transformation ratio $t = 400/230$.

According to Sect. 8.3, impedance correction factors must be introduced when calculating short-circuit currents for transformers, which may take into account a higher active voltage because of inductive load flow before a short circuit occurs. This impedance correction must be introduced in the positive, negative, and zero sequence systems. In this example, the correction is neglected to simplify the calculation. Applying the correction would reduce the impedance values of the autotransformer by a maximum of 4.4%, so that the current components would increase accordingly.

16.3.2 Earthed Neutral Point of a Transformer

If the neutral point of the transformer is solidly earthed ($\underline{Z}_{\mathrm{E}} = 0$ applies to the earthing impedance), Fig. 16.9 can be used to calculate the line-to-earth short-circuit current, assuming that the only voltage in the network is the equivalent voltage source at the short-circuit location. Since no network is connected to the transformer's stabilizing winding, this transformer terminal is not considered for this reason.

For the line-to-earth short-circuit current at the fault location, according to Eq. (3.3), assuming $\underline{Z}_1 = \underline{Z}_2$, the following applies:

$$\underline{I}''_{\mathrm{k1}} = 3 \cdot \underline{I}_{(1)} = 3 \cdot \underline{I}_{(0)} = \frac{1.1 \cdot \sqrt{3} \cdot U_{\mathrm{n}}}{2 \cdot \underline{Z}_{(1)} + \underline{Z}_{(0)}} \approx -\mathrm{j}\frac{1.1 \cdot \sqrt{3} \cdot U_{\mathrm{n}}}{2 \cdot X_{(1)} + X_{(0)}} \tag{16.21}$$

The calculated reactances are included in Fig. 16.9 for the calculation of the line-to-earth short-circuit current.

16.3.2.1 Determination of the Positive and Negative Sequence Reactance

Positive sequence reactance (380 kV network, transformer)

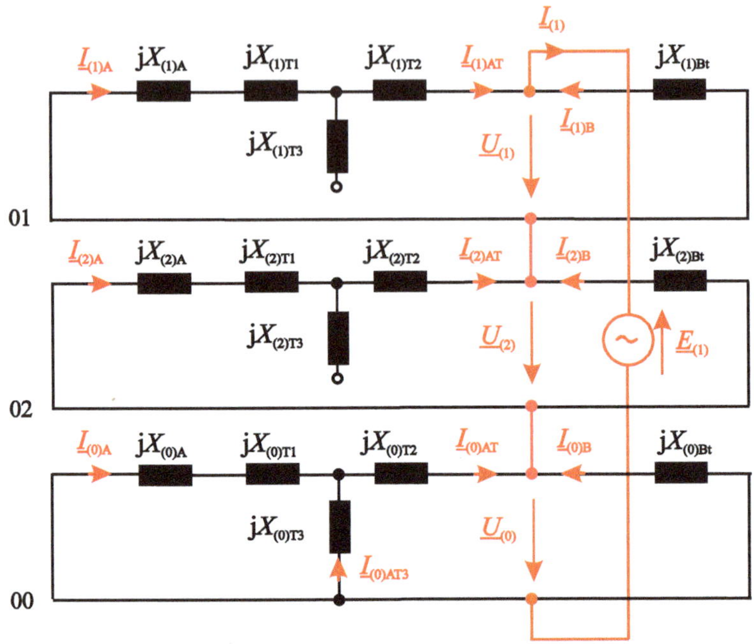

Fig. 16.9 Equivalent circuit for calculating the line-to-earth short-circuit current (neutral point of the autotransformer earthed)

$$X_{(1)\mathrm{AT}} = X_{(1)\mathrm{A}} + X_{(1)\mathrm{T1}} + X_{(1)\mathrm{T2}} = (1.773 + 9.390 - 1.190)\,\Omega = 9.973\,\Omega$$

Positive sequence reactance (220 kV network): $X_{(1)\mathrm{B}} = 2.794\,\Omega$.

Resulting positive sequence reactance:

$$X_{(1)} = \frac{X_{(1)\mathrm{AT}} \cdot X_{(1)\mathrm{B}}}{X_{(1)\mathrm{AT}} + X_{(1)\mathrm{B}}} = \frac{9.973 \cdot 2.794}{9.973 + 2.794} = 2.183\,\Omega$$

16.3.2.2 Determination of the Zero Sequence Reactance

Zero sequence reactance (380 kV network, transformer)

$$X_{(0)\mathrm{AT}} = \frac{\left(X_{(0)\mathrm{A}} + X_{(0)\mathrm{T1}}\right) \cdot X_{(0)\mathrm{T3}}}{X_{(0)\mathrm{A}} + X_{(0)\mathrm{T1}} + X_{(0)\mathrm{T3}}} + X_{(0)\mathrm{T2}}$$
$$= \frac{(5.319 + 9.390) \cdot 30.814}{5.319 + 9.390 + 30.814}\,\Omega - 1.190\,\Omega = 8.766\,\Omega$$

Zero sequence reactance (220 kV network): $X_{(0)\mathrm{B}} = 5.589\,\Omega$.

Resulting zero sequence reactance:

$$X_{(0)} = \frac{X_{(0)AT} \cdot X_{(0)B}}{X_{(0)AT} + X_{(0)B}} = \frac{8.766 \cdot 5.589}{8.766 + 5.589}\,\Omega = 3.413\,\Omega$$

16.3.2.3 Calculation of Component Currents and Voltages

The component currents are determined taking into account Fig. 16.9 and the reactances determined. The component currents can be calculated from Eq. (16.19).

$$\underline{I}_{(1)} = \underline{I}_{(0)} = \underline{I}''_{k1}/3 \approx -\mathrm{j}\frac{1.1 \cdot \sqrt{3} \cdot U_n}{2 \cdot X_{(1)} + X_{(0)}} = -\mathrm{j}\frac{1.1 \cdot 220/\sqrt{3}}{2 \cdot 2.183 + 3.413}\,\mathrm{kA} = -\mathrm{j}17.96\,\mathrm{kA}$$

$$\underline{I}_{(1)AT} = \underline{I}_{(2)AT} = \underline{I}_{(1)A} = \underline{I}_{(2)A} \approx \frac{X_{(1)B}}{X_{(1)AT} + X_{(1)B}} \cdot \underline{I}_{(1)}$$

$$= \frac{2.794 \cdot (-\mathrm{j}17.96)}{9.973 + 2.794}\,\mathrm{kA} = -\mathrm{j}3.93\,\mathrm{kA}$$

$$\underline{I}_{(1)B} = \underline{I}_{(2)B} = \underline{I}_{(1)} - \underline{I}_{(1)AT} = -\mathrm{j}14.03\,\mathrm{kA}$$

$$\underline{I}_{(0)AT} \approx \frac{X_{(0)B}}{X_{(0)AT} + X_{(0)B}} \cdot \underline{I}_{(0)} = \frac{5.588 \cdot (-\mathrm{j}17.96)}{8.766 + 5.588}\,\mathrm{kA} = -\mathrm{j}6.99\,\mathrm{kA}$$

$$\underline{I}_{(0)B} = \underline{I}_{(0)} - \underline{I}_{(0)AT} \approx -\mathrm{j}10.97\,\mathrm{kA}$$

$$\underline{I}_{(0)A} \approx \frac{X_{(0)T3}}{X_{(0)A} + X_{(0)T1} + X_{(0)T3}} \cdot \underline{I}_{(0)AT} = \frac{30.814 \cdot (-\mathrm{j}6.99\,\mathrm{kA})}{5.319 + 9.390 + 30.814} = -\mathrm{j}4.73\,\mathrm{kA}$$

$$\underline{I}_{(0)AT3} = \underline{I}_{(0)AT} - \underline{I}_{(0)A} \approx -\mathrm{j}2.26\,\mathrm{kA}$$

The component currents are included in Fig. 16.10; all currents and voltages are referred to the 220 kV voltage level. Due to the application of the equivalent voltage source method at the short-circuit location, the actual voltages in the positive sequence system must be calculated from the difference between the calculated and measured values. The voltages at the transformer terminals are calculated below.

$$\underline{U}_{(1)A} = 1.1 \cdot \frac{U_n}{\sqrt{3}} - X_{(1)A} \cdot \underline{I}_{(1)A} = 139.719\,\mathrm{kV} - \mathrm{j}1.773 \cdot (-\mathrm{j}3.932)\,\mathrm{kV}$$

$$= 132.75\ \mathrm{kV}$$

$$\underline{U}_{(2)A} = -X_{(1)A} \cdot \underline{I}_{(1)A} = -\mathrm{j}1.773 \cdot (-\mathrm{j}3.932)\,\mathrm{kV} = -6.97\ \mathrm{kV}$$

$$\underline{U}_{(0)A} = -X_{(0)A} \cdot \underline{I}_{(0)A} = -\mathrm{j}5.319 \cdot (-\mathrm{j}4.734)\,\mathrm{kV} = -25.18\ \mathrm{kV}$$

$$\underline{U}_{(1)B} = 1.1 \cdot \frac{U_n}{\sqrt{3}} - X_{(1)B} \cdot \underline{I}_{(1)B} = 139.719 - \mathrm{j}2.794 \cdot (-\mathrm{j}14.031)\,\mathrm{kV} = 100.51\ \mathrm{kV}$$

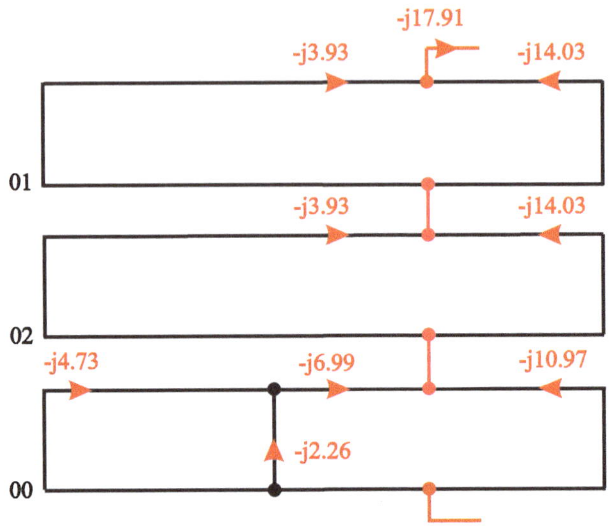

Fig. 16.10 Currents of the reduced component system, values in kA and inductive (reference 220 kV voltage level)

$$\underline{U}_{(2)\mathrm{B}} = -X_{(1)\mathrm{B}} \cdot \underline{I}_{(1)\mathrm{B}} = -\mathrm{j}2.794 \cdot (-\mathrm{j}14.031)\,\mathrm{kV} = -39.21\,\mathrm{kV}$$

$$\underline{U}_{(0)\mathrm{B}} = -X_{(0)\mathrm{B}} \cdot \underline{I}_{(0)\mathrm{B}} = -\mathrm{j}5.589 \cdot (-\mathrm{j}10.969)\,\mathrm{kV} = -61.30\,\mathrm{kV}$$

16.3.2.4 Calculation of Network Currents and Voltages (Index "t": Related to the 380 kV Level)

The network currents in the RST system can be determined using the transformation according to Eq. (9.76), and all currents are labeled as inductive. The currents and voltages are determined according to Eq. 16.22 in accordance with [3] using the general quantity $\underline{G}$ and $\underline{a}$ or $\underline{a}^2$ according to Eqs. (9.4) and (9.5). When calculating the currents in the 380 kV voltage level, it is assumed that no phase shift occurs due to the use of the autotransformer between the high voltage and low voltage sides.

$$\begin{pmatrix} \underline{G}_\mathrm{R} \\ \underline{G}_\mathrm{S} \\ \underline{G}_\mathrm{T} \end{pmatrix} = \begin{pmatrix} 1 & 1 & 1 \\ 1 & \underline{a}^2 & \underline{a} \\ 1 & \underline{a} & \underline{a}^2 \end{pmatrix} \cdot \begin{pmatrix} \underline{G}_{(0)} \\ \underline{G}_{(1)} \\ \underline{G}_{(2)} \end{pmatrix} \tag{16.22}$$

Since the component currents in the positive and negative sequence systems are equal, these equations simplify to:

$$\underline{I}_\mathrm{R} = \underline{I}_{(0)} + 2 \cdot \underline{I}_{(1)} \quad \text{and} \quad \underline{I}_\mathrm{S} = \underline{I}_\mathrm{T} = \underline{I}_{(0)} - \underline{I}_{(1)}$$

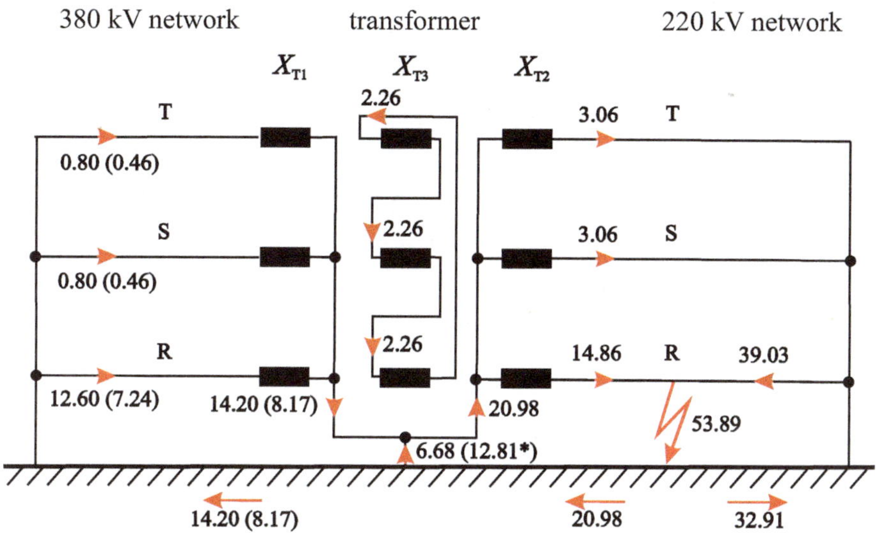

Fig. 16.11 Current distribution in the event of a line-to-earth short circuit in the 220 kV network (data in kA, with stabilizing winding), currents are related to 220 kV and are inductive; values in brackets refer to the 380 kV voltage level (the reactances of the networks are not shown). * resulting current via the common neutral point

The conversion to the 380 kV voltage level is performed using the transformation ratio $t = 400/230$. Figure 16.11 shows the determined current distribution in the two voltage levels in the event of a line-to-earth fault at the 220 kV level.

Feeder from the 220 kV network (B), right of the short-circuit location

$$\underline{I}_{RB} = -j10.97\text{ kA} - j28.06\text{ kA} = -j39.03\text{ kA} \qquad \underline{I}_{RBt} = -j22.44\text{ kA}$$

$$\underline{I}_{SB} = \underline{I}_{TB} = -j10.97\text{ kA} + j14.03\text{ kA} = +j3.06\text{ kA} \qquad \underline{I}_{SBt} = \underline{I}_{TBt} = +1.76\text{ kA}$$

Current via earth: $\underline{I}_{EB} = 3 \cdot I_{(0)B} = -j32.91\text{ kA} \qquad \underline{I}_{EBt} = -j18.92\text{ kA}$

Feeder from the 380 kV network (AT), left of the short-circuit location

$$\underline{I}_{RAT} = -j6.99\text{ kA} - j7.86\text{ kA} = -j14.86\text{ kA} \qquad \underline{I}_{RATt} = -j8.54\text{ kA}$$

$$\underline{I}_{SAT} = \underline{I}_{TAT} = -j6.99\text{ kA} + j3.93\text{ kA} = -j3.06\text{ kA} \qquad \underline{I}_{SATt} = \underline{I}_{TATt} = -j1.76\text{ kA}$$

Current via earth: $\underline{I}_{EAT} = 3 \cdot I_{(0)AT} = -j20.98\text{ kA}$ $\underline{I}_{EAt} = -j12.06\text{ kA}$

Current at short-circuit location

$$\underline{I}''_{k1} = 3 \cdot \underline{I}_{(1)} = -j3 \cdot 17.96 = -j53.89\text{ kA} \qquad \underline{I}''_{k1t} = -j30.99\text{ kA}$$

Feeder into the 380 kV network

$$\underline{I}_{\text{RA}} = -\text{j}4.73\ \text{kA} - \text{j}7.96\ \text{kA} = -\text{j}12.60\ \text{kA} \qquad \underline{I}_{\text{RAt}} = -\text{j}7.24\ \text{kA}$$

$$\underline{I}_{\text{SA}} = \underline{I}_{\text{TA}} = -\text{j}4.73\ \text{kA} + \text{j}3.93\ \text{kA} = -\text{j}0.80\ \text{kA} \qquad \underline{I}_{\text{SAt}} = \underline{I}_{\text{TAt}} = -\text{j}0.46\ \text{kA}$$

Current via earth: $\underline{I}_{\text{EA}} = 3 \cdot \underline{I}_{(0)\text{A}} = -\text{j}14.20\ \text{kA} \qquad \underline{I}_{\text{EAt}} = -\text{j}8.17\ \text{kA}$

Current of the neutral point

The current via the transformer neutral point cannot be derived from the equivalent circuit because the neutral point must be assigned to both the 380 and 220 kV levels. It must therefore be derived from the difference between the currents, taking into account the transformation ratio.

$$\underline{I}_{\text{STt}} = -\text{j}3 \cdot \underline{I}_{(0)\text{AT}} - \text{j}3 \cdot \underline{I}_{(0)\text{A}} \cdot \frac{230}{440} = -\text{j}12.81\,\text{kA}$$

The representation in Fig. 16.11 (related to 220 kV) follows:

$$\underline{I}_{\text{ST}} = 3 \cdot \underline{I}_{(0)\text{AT}} - 3 \cdot \underline{I}_{(0)\text{A}} = -\text{j}6.78\,\text{kA}$$

Current of the stabilizing winding

The current in the delta stabilizing winding corresponds to the component current in the zero sequence system with $I_{(0)\text{AT3}} = -\text{j}2.26$ kA (reactance $X_{(0)\text{T3}}$; current value is related to 220 kV).

Transformer voltage of the 380 kV level (no shifting vector group of the autotransformer is assumed).

$$\underline{U}_{\text{RA}} = \underline{U}_{(0)\text{A}} + \underline{U}_{(1)\text{A}} + \underline{U}_{(2)\text{A}} = 100.60\ \text{kV} \qquad U_{\text{RAt}} = 174.95\ \text{kV}$$

$$\underline{U}_{\text{SA}} = \underline{U}_{(0)\text{A}} + \underline{a}^2 \cdot \underline{U}_{(1)\text{A}} + \underline{a} \cdot \underline{U}_{(2)\text{A}} = -88.07\ \text{kV} - \text{j}121.00\ \text{kV}$$

$$\underline{U}_{\text{TA}} = \underline{U}_{(0)\text{A}} + \underline{a} \cdot \underline{U}_{(1)\text{A}} + \underline{a}^2 \cdot \underline{U}_{(2)\text{A}} = -88.07\ \text{kV} + \text{j}121.00\ \text{kV}$$

$$U_{\text{SA}} = U_{\text{TA}} = 149.66\ \text{kV} \qquad U_{\text{SAt}} = U_{\text{TAt}} = 260.27\ \text{kV}$$

Transformer voltage of the 220 kV level

$$\underline{U}_{\text{RB}} = \underline{U}_{(0)\text{B}} + \underline{U}_{(1)\text{B}} + \underline{U}_{(2)\text{B}} = 0.00\,\text{kV}$$

$$\underline{U}_{\text{SB}} = \underline{U}_{(0)\text{B}} + \underline{a}^2 \cdot \underline{U}_{(1)\text{B}} + \underline{a} \cdot \underline{U}_{(2)\text{B}} = -91.96\,\text{kV} - \text{j}121.00\,\text{kV}$$

$$\underline{U}_{\text{TB}} = \underline{U}_{(0)\text{B}} + \underline{a} \cdot \underline{U}_{(1)\text{B}} + \underline{a}^2 \cdot \underline{U}_{(2)\text{B}} = -91.96\,\text{kV} + \text{j}121.00\,\text{kV}$$

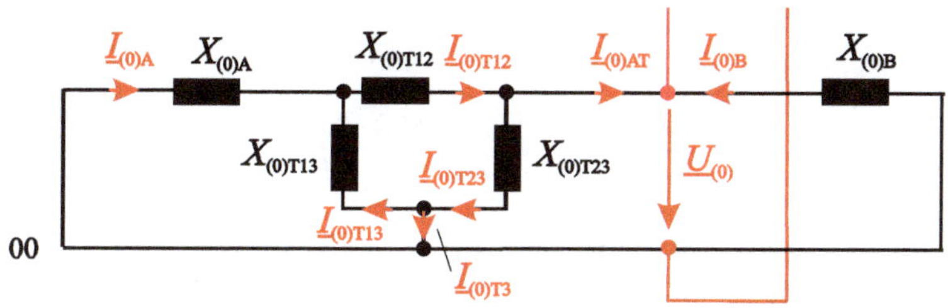

Fig. 16.12 Zero sequence system of the equivalent circuit for calculating the line-to-earth short-circuit current, neutral point isolated

$$U_{\mathrm{SB}} = U_{\mathrm{TB}} = 151.98\,\mathrm{kV}$$

Earth-fault factor

The zero sequence voltage at the fault location (220 kV) corresponds to the component voltage $\underline{U}_{(0)\mathrm{B}} = 61.30\,\mathrm{kV}$. The earth fault factor at the short-circuit location can be determined from the voltage of the conductors not affected by the fault to the unaffected voltage. Due to the application of the voltage factor c in the short-circuit current calculation, the following results:

$$c_{\mathrm{E}} = \frac{\underline{U}_{\mathrm{SB}}}{1.1 \cdot U_{\mathrm{n}}/\sqrt{3}} = \frac{151.98}{1.1 \cdot 220/\sqrt{3}} = 1.08$$

This means that the voltage increase of the conductor not affected by the fault is only slight compared to the unaffected operating voltage due to the use of the autotransformer with an earthed neutral point. The calculation shows that, as a result of the neutral point earthing of the autotransformer, a line-to-earth fault, in this case on the low voltage side (220 kV), is transformed to the high voltage side (380 kV).

16.3.3 Isolated Neutral Point of the Transformer

With an isolated neutral point of the autotransformer, only the zero sequence system of the equivalent circuit changes, while the other component systems are unaffected by the neutral point earthing. This results in the zero sequence system shown in Fig. 16.12, taking into account Fig. 16.7c.

While the positive and negative sequence impedances remain unchanged, the resulting zero sequence reactance must be recalculated.

16.3.3.1 Calculation of the Zero Sequence Reactance

Zero sequence reactance (380 kV network, transformer)

$$X_{(0)\mathrm{ATx}} = \frac{X_{(0)\mathrm{A}} \cdot X_{(0)\mathrm{T13}}}{X_{(0)\mathrm{A}} + X_{(0)\mathrm{T13}}} + X_{(0)\mathrm{T12}} = \frac{5.319 \cdot 34.264}{5.319 + 34.264}\,\Omega + 25.325\,\Omega = 29.930\,\Omega$$

$$X_{(0)\mathrm{AT}} = \frac{X_{(0)\mathrm{ATx}} \cdot X_{(0)\mathrm{T23}}}{X_{(0)\mathrm{ATx}} + X_{(0)\mathrm{T23}}} = \frac{-59.590 \cdot 29.930}{29.930 - 59.590}\,\Omega = 60.133\,\Omega$$

Zero sequence reactance (220 kV network): $X_{(0)\mathrm{B}} = 5.589\,\Omega$.

Resulting zero sequence reactance:

$$X_{(0)} = \frac{X_{(0)\mathrm{AT}} \cdot X_{(0)\mathrm{B}}}{X_{(0)\mathrm{AT}} + X_{(0)\mathrm{B}}} = \frac{60.133 \cdot 5.589}{60.133 + 5.589}\,\Omega = 5.114\,\Omega$$

16.3.3.2 Calculation of Component Currents and Voltages

The line-to-earth short-circuit current at the fault location (220 kV) is then calculated as follows:

$$\underline{I}_{(1)} = \underline{I}_{(0)} = \underline{I}_{\mathrm{k1}}''/3 \approx -\mathrm{j}\frac{1,1 \cdot \sqrt{3} \cdot U_{\mathrm{n}}}{2 \cdot X_{(1)} + X_{(0)}} = -\mathrm{j}\frac{1.1 \cdot 220/\sqrt{3}}{2 \cdot 2.183 + 5.114}\,\mathrm{kA} = -\mathrm{j}14.74\,\mathrm{kA}$$

Considering the zero sequence impedance, the various component currents and voltages are determined in accordance with the calculation for the earthed autotransformer.

$$\underline{I}_{(1)\mathrm{AT}} = \underline{I}_{(2)\mathrm{AT}} = \underline{I}_{(1)\mathrm{A}} = \underline{I}_{(2)\mathrm{A}} \approx -\mathrm{j}3.23\ \mathrm{kA} \qquad \underline{I}_{(1)\mathrm{B}} = \underline{I}_{(2)\mathrm{B}} = -\mathrm{j}11.51\ \mathrm{kA}$$
$$\underline{I}_{(0)\mathrm{AT}} = -\mathrm{j}1.25\ \mathrm{kA} \qquad \underline{I}_{(0)\mathrm{B}} = -\mathrm{j}13.49\ \mathrm{kA}$$

Based on the zero sequence current of the transformer $\underline{I}_{(0)\mathrm{AT}}$, the remaining partial currents of network A can be derived in the zero sequence system, which are shown in Fig. 16.13 together with the other component currents. The following applies to the voltages at the transformer terminals, related to the 220 kV level:

$$\underline{U}_{(1)\mathrm{A}} = 1.1 \cdot \frac{U_{\mathrm{n}}}{\sqrt{3}} - X_{(1)\mathrm{A}} \cdot \underline{I}_{(1)\mathrm{A}} = 139.719\ \mathrm{kV} - \mathrm{j}1.773 \cdot (-\mathrm{j}3.226)\ \mathrm{kV} = 134.00\ \mathrm{kV}$$
$$\underline{U}_{(2)\mathrm{A}} = -X_{(1)\mathrm{A}} \cdot \underline{I}_{(1)\mathrm{A}} = -\mathrm{j}1.773 \cdot (-\mathrm{j}3.226)\ \mathrm{kV} = -5.72\ \mathrm{kV}$$
$$\underline{U}_{(0)\mathrm{A}} = -X_{(0)\mathrm{A}} \cdot \underline{I}_{(0)\mathrm{A}} = -\mathrm{j}5.319 \cdot (-\mathrm{j}2.180)\ \mathrm{kV} = -11.60\ \mathrm{kV}$$

$$\underline{U}_{(1)\mathrm{B}} = 1,1 \cdot \frac{U_{\mathrm{n}}}{\sqrt{3}} - X_{(1)\mathrm{B}} \cdot \underline{I}_{(1)\mathrm{B}} = 139.719 - \mathrm{j}2.794 \cdot (-\mathrm{j}11.514)\,\mathrm{kV} = 107.55\ \mathrm{kV}$$
$$\underline{U}_{(2)\mathrm{B}} = -X_{(1)\mathrm{B}} \cdot \underline{I}_{(1)\mathrm{B}} = -\mathrm{j}2.794 \cdot (-\mathrm{j}11.514)\,\mathrm{kV} = -32.17\ \mathrm{kV}$$
$$\underline{U}_{(0)\mathrm{B}} = -X_{(0)\mathrm{B}} \cdot \underline{I}_{(0)\mathrm{B}} = -\mathrm{j}5.589 \cdot (-\mathrm{j}13.486)\,\mathrm{kV} = -75.37\ \mathrm{kV}$$

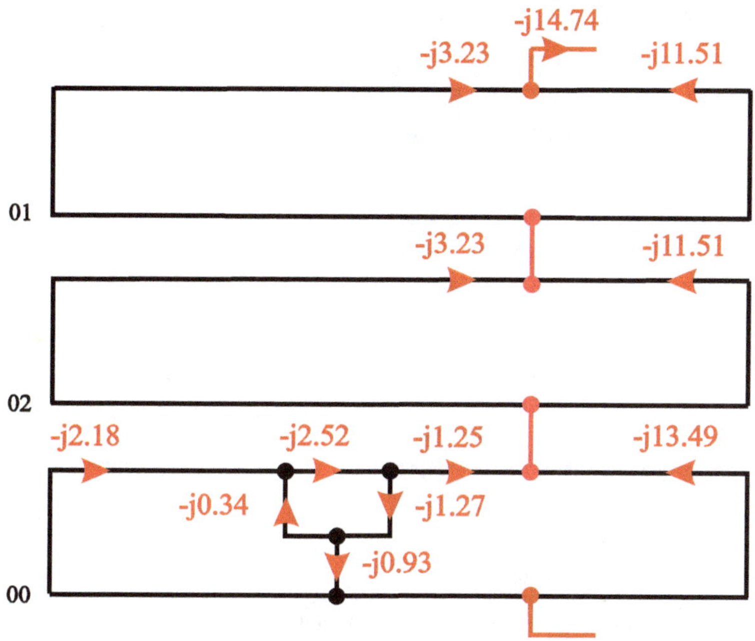

Fig. 16.13 Currents of the reduced component system, reference 220 kV voltage level, values in kA and inductive

16.3.3.3 Calculation of Network Currents and Voltages (Index "T": Related to the 380 kV Level)

Feeder from the 380 kV network (AT), left of the short-circuit location

$$\underline{I}_{\text{RAT}} = -\text{j}1.25\ \text{kA} - \text{j}6.45\ \text{kA} = -\text{j}7.71\ \text{kA} \qquad \underline{I}_{\text{RATt}} = -\text{j}4.43\ \text{kA}$$

$$\underline{I}_{\text{SAT}} = \underline{I}_{\text{TAT}} = -\text{j}1.25\ \text{kA} + \text{j}3.23\ \text{kA} = +\text{j}1.97\ \text{kA} \qquad \underline{I}_{\text{SATt}} = \underline{I}_{\text{TATt}} = +\text{j}3.43\ \text{kA}$$

Current via earth

$$\underline{I}_{\text{EAT}} = 3 \cdot \underline{I}_{(0)\text{AT}} = -\text{j}3.76\ \text{kA} \qquad \underline{I}_{\text{EAt}} = -\text{j}2.16\ \text{kA}$$

Current at the short-circuit location

$$\underline{I}''_{\text{k1}} = 3 \cdot \underline{I}_{(1)} = -\text{j}3 \cdot 14.74 = -\text{j}44.22\ \text{kA} \qquad \underline{I}''_{\text{k1t}} = -\text{j}25.43\ \text{kA}$$

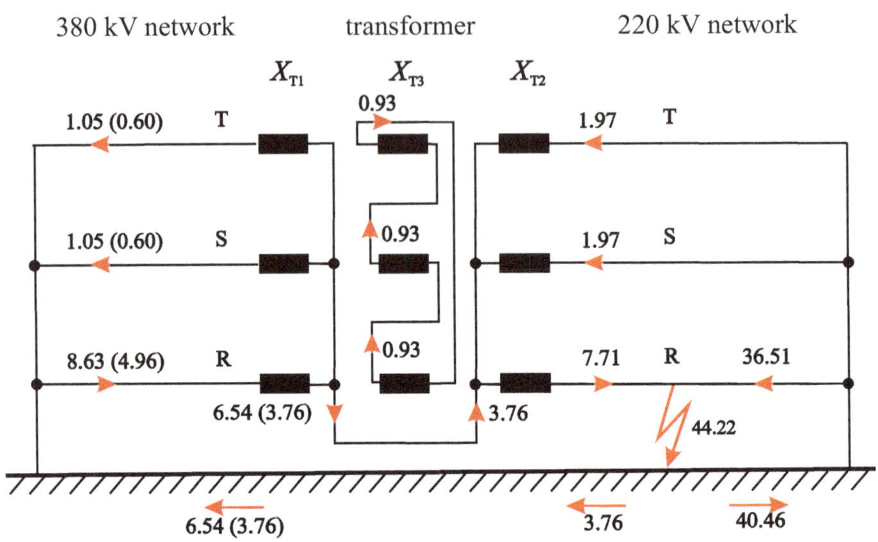

Fig. 16.14 Current distribution in the event of a line-to-earth fault in the 220 kV network (figures in kA, with stabilizing winding), currents are based on 220 kV and are inductive; values in brackets refer to the 380 kV voltage level, deviations due to rounding errors

Feeder into the 380 kV network

$$\underline{I}_{RA} = -j2.18\text{ kA} - j6.45\text{ kA} = -j8.63\text{ kA} \qquad \underline{I}_{RAt} = -j4.96\text{ kA}$$

$$\underline{I}_{SA} = \underline{I}_{TA} = -j2.18\text{ kA} + j3.23\text{ kA} = +j1.05\text{ kA} \qquad \underline{I}_{SAt} = \underline{I}_{TAt} = +j0.60\text{ kA}$$

Current via earth

$$\underline{I}_{EA} = 3 \cdot \underline{I}_{(0)A} = -j6,54\text{ kA} \qquad \underline{I}_{EAt} = -j3,76\text{ kA}$$

Current of the neutral point
Since the transformer's neutral point is isolated, no current can flow via the neutral point.

Current of the stabilizing winding
The current in the delta stabilizing winding corresponds to the component current in the zero sequence system with $\underline{I}_{0AT3} = -j0.93$ kA (reactance $X_{(0)T3}$; current value is based on 220 kV). Figure 16.14 shows the currents determined in the RST system, where rounding errors occur during addition.

Transformer voltage of the 380 kV level (no shifting vector group of the autotransformer is assumed)

$$\underline{U}_{RA} = \underline{U}_{(0)A} + \underline{U}_{(1)A} + \underline{U}_{(2)A} = 116.68\,\text{kV} \qquad U_{RAt} = 202.93\,\text{kV}$$

$$\underline{U}_{SA} = \underline{U}_{(0)A} + \underline{a}^2 \cdot \underline{U}_{(1)A} + \underline{a} \cdot \underline{U}_{(2)A} = -75.73\,\text{kV} - \text{j}121.00\,\text{kV}$$

$$\underline{U}_{TA} = \underline{U}_{(0)A} + \underline{a} \cdot \underline{U}_{(1)A} + \underline{a}^2 \cdot \underline{U}_{(2)A} = -75.73\,\text{kV} + \text{j}121.00\,\text{kV}$$

$$U_{SA} = U_{TA} = 142.75\,\text{kV} \qquad U_{SAt} = U_{TAt} = 248.26\,\text{kV}$$

Transformer voltage of the 220 kV voltage level

$$\underline{U}_{RB} = \underline{U}_{(0)B} + \underline{U}_{(1)B} + \underline{U}_{(2)B} = 0.00\,\text{kV}$$

$$\underline{U}_{SB} = \underline{U}_{(0)B} + \underline{a}^2 \cdot \underline{U}_{(1)B} + \underline{a} \cdot \underline{U}_{(2)B} = -113.06\,\text{kV} - \text{j}121.00\,\text{kV}$$

$$\underline{U}_{TB} = \underline{U}_{(0)B} + \underline{a} \cdot \underline{U}_{(1)B} + \underline{a}^2 \cdot \underline{U}_{(2)B} = -113.06\,\text{kV} + \text{j}121.00\,\text{kV}$$

$$U_{SB} = U_{TB} = 165.60\,\text{kV}$$

Earth fault factor

The zero sequence voltage at the fault location (220 kV) corresponds to the component voltage $U_{(0)B} = 75.37$ kV. The earth fault factor at the short-circuit location can be determined from the voltage of the conductors not affected by the fault to the unaffected voltage. Due to the application of the voltage factor c in the short-circuit current calculation, the following results:

$$c_E = \frac{U_{SB}}{1.1 \cdot U_n/\sqrt{3}} = \frac{165.60}{1.1 \cdot 220/\sqrt{3}} = 1.19$$

This means that the voltage increase of the conductor not affected by the fault relative to the unaffected operating voltage is greater due to the use of the autotransformer with an unearthed neutral point than in the earthed case. The calculations show that, in this case, the neutral point earthing of the transformer has only a minor influence on the voltage levels of the fault-free conductors in the event of a line-to-earth fault (increase from 151.98 to 165.60 kV), which is a consequence of the low-impedance earthing of the faulty network (220 kV network). Even with an isolated transformer neutral point, the overlaid network contributes to the short-circuit current, but this contribution is reduced.

16.4 Conclusions

The autotransformer has a cost advantage over a comparable separate winding transformer, especially when the transformation ratio is close to 1. However, a disadvantage is that even with an isolated neutral point of the transformer, a line-to-earth fault, e.g., on the high voltage side, is transformed to the low voltage side and vice versa, so this must be taken into account when designing the system protection.

For the coupling of high voltage networks, e.g., 380 and 220 kV, autotransformers are often used because their throughput power is greater in relation to their total winding power and are therefore more cost-effective than a comparable separate winding transformer. When calculating short-circuit current with a stabilizing winding and an earthed neutral point, they can be treated like a three-winding transformer. In contrast, in the case of autotransformers with an isolated neutral point, the zero sequence system of the transformer must be converted into a separate delta equivalent circuit when determining the line-to-earth short-circuit current.

References

1. Balzer G, Neumann C, Balzer C (2025) Schalt- und Ausgleichsvorgänge in elektrischen Netzen, 2nd Edn. Springer Verlag, Berlin
2. IEC/TR 60909-2 (2008) Edition 3.0 (2008) Short-circuit current calculations in three-phase a.c. systems—part 2: Data of electrical equipment for short-circuit current calculations. IEC, Geneva
3. Kuchler R (1966) Die Transformatoren. Grundlagen für ihre Berechnung und Konstruktion, 2nd Edn. Springer-Verlag, Berlin
4. Funk G (1962) Der Kurzschluß im Drehstromnetz. R. Oldenbourg Verlag, München
5. Heuck K, Dettmann K-D, Schulz D (2010) Elektrische Energieversorgung, 8th Edn. Vieweg +Teubner Verlag, Wiesbaden
6. Schlosser K (1962) Vergleich der Ersatzschaltungen des Voll- und des Spartransformators bei symmetrischer Belastung. BBC-Nachrichten, vol 44, pp 3–8
7. Schlosser K (1965) Spartransformatoren mit freiem Sternpunkt. BBC-Nachrichten, pp 470–479
8. Schlosser K (1962) Die Ersatzschaltungen des Voll- und des Spartransformators bei Fehlerberechnungen. BBC-Nachrichten, pp 123–132

Chapter 17
Voltage Flashover between Three-Phase Systems with Different Nominal Voltages

In the case of overhead lines with different voltage levels on a single tower, e.g., 380 and 110 kV, the conductors of the 380 and 110 kV systems can approach each other at the flashover distance under extreme boundary conditions. These processes place increased stress on connected equipment and, in particular, on the installed surge arresters in the lower voltage network, depending on the effectiveness of the neutral point earthing (earth fault factor) [1]. In these cases, equipment of the lower-level network is briefly exposed to the operating voltage of the higher-level network. In principle, a flashover is possible if insulators and lines are defective, lines swing against each other, or come into contact with each other due to external influences (flying objects).

The following section determines the steady-state voltage distribution and short-circuit current in the lower-level network when the line conductors of the two three-phase systems are in galvanic contact. Additional information, e.g., the transient process, is presented in [2].

17.1 Calculation of Steady-State Voltage and Current Distribution

Figure 17.1 shows the equivalent circuit for further investigation, in which a voltage flashover occurs between the line conductor R of the two systems 380 and 110 kV, resulting in a short-circuit current $\underline{I}_F$. This assumes that the 380 kV network is always solidly earthed, while the lower-level 110 kV network is connected to earth via an impedance $\underline{Z}_E$, e.g. with an arc suppression coil, or with low impedance.

If two line conductors touch each other as shown in Fig. 17.1, an equivalent component circuit in the positive, negative, and zero sequence system can represent the two networks. This assumes that the internal voltages of the two networks do not influence each other, which is particularly true for a 380 kV network with a higher

G. Balzer, *Short-Circuit Currents in Three-Phase Systems*,
https://doi.org/10.1007/978-3-658-50710-7_17

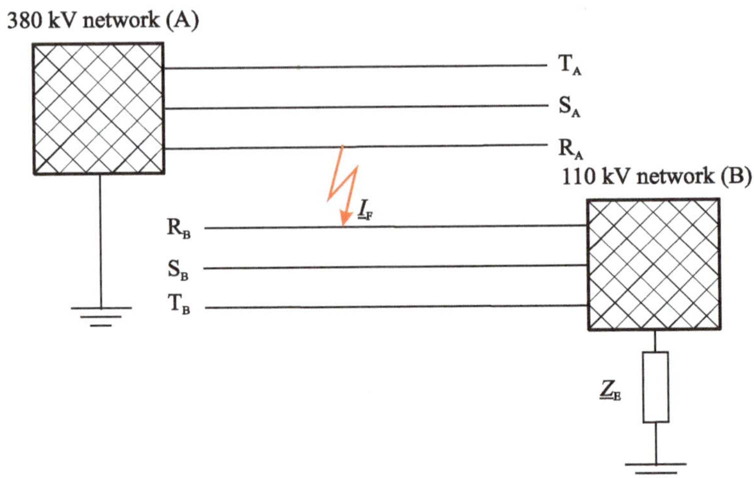

Fig. 17.1 Equivalent circuit for determining the steady-state voltage distribution as a function of the neutral point impedance $\underline{Z}_E$ of the lower-level network

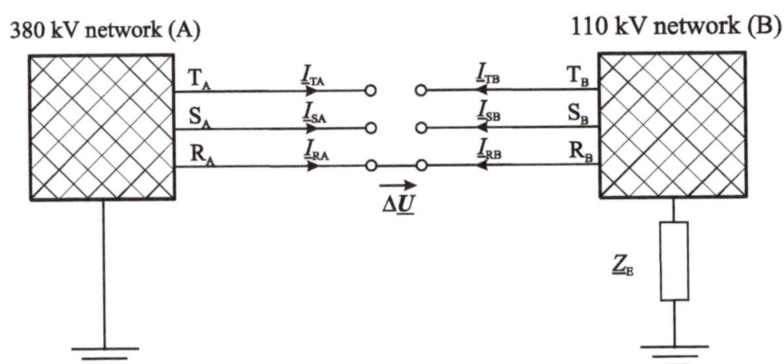

Fig. 17.2 Substitute circuit for deriving the component equivalent circuit in the event of a voltage flashover in a line conductor

three-phase short-circuit current and a 110 kV network with an earth fault factor of $c_E > 1.4$. This means that the lower voltage level is not supplied directly from the higher-level network.

The calculations shown refer exclusively to the unaffected short-circuit current. The use of surge arresters in the lower-level networks influences the voltage conditions, and due to the energetic overload of the arresters, they form a short circuit and thus change the earthing conditions [2].

The simulation of the fault shown in Fig. 17.1 corresponds to a two-phase line interruption, so that an equivalent component circuit can be derived in the positive, negative, and zero sequence systems, assuming the network circuit shown in Fig. 17.2.

The following equations apply to the derivation of the fault conditions between the two networks:

$$\underline{I}_{SA} = \underline{I}_{SB} = 0 \qquad \underline{I}_{TA} = \underline{I}_{TB} = 0 \tag{17.1}$$

$$\Delta\underline{U}_R = \underline{U}_{RA} - U_{RB} = 0 \tag{17.2}$$

The fault conditions (current/voltage) in the component system can be derived from the fault conditions in the original system.

$$\begin{pmatrix} \Delta\underline{U}_{(0)} \\ \Delta\underline{U}_{(1)} \\ \Delta\underline{U}_{(2)} \end{pmatrix} = \frac{1}{3} \cdot \begin{pmatrix} 1 & 1 & 1 \\ 1 & \underline{a} & \underline{a}^2 \\ 1 & \underline{a}^2 & \underline{a} \end{pmatrix} \cdot \begin{pmatrix} 0 \\ \Delta\underline{U}_S \\ \Delta\underline{U}_T \end{pmatrix} \tag{17.3}$$

Adding the rows gives the following result for the differential voltage:

$$\Delta\underline{U}_{(0)} = 0 \tag{17.4}$$

The following applies to the currents of the individual systems:

$$\begin{pmatrix} \underline{I}_{(0)A} \\ \underline{I}_{(1)A} \\ \underline{I}_{(2)A} \end{pmatrix} = \frac{1}{3} \cdot \begin{pmatrix} 1 & 1 & 1 \\ 1 & \underline{a} & \underline{a}^2 \\ 1 & \underline{a}^2 & \underline{a} \end{pmatrix} \cdot \begin{pmatrix} \underline{I}_{RA} \\ 0 \\ 0 \end{pmatrix} \tag{17.5}$$

$$\begin{pmatrix} \underline{I}_{(0)B} \\ \underline{I}_{(1)B} \\ \underline{I}_{(2)B} \end{pmatrix} = \frac{1}{3} \cdot \begin{pmatrix} 1 & 1 & 1 \\ 1 & \underline{a} & \underline{a}^2 \\ 1 & \underline{a}^2 & \underline{a} \end{pmatrix} \cdot \begin{pmatrix} \underline{I}_{RB} \\ 0 \\ 0 \end{pmatrix} \tag{17.6}$$

$$\begin{aligned} \underline{I}_{(1)A} &= \underline{I}_{(2)A} = \underline{I}_{(0)A} = \underline{I}_{RA}/3 = \underline{I} \\ \underline{I}_{(1)B} &= \underline{I}_{(2)B} = \underline{I}_{(0)B} = \underline{I}_{RB}/3 = -\underline{I} \end{aligned} \tag{17.7}$$

Equation (17.7) shows that all component currents are equal, resulting in a series connection of the individual systems as an equivalent circuit diagram, as shown in Fig. 17.3. To simplify the calculation of the current and voltage distribution, the following impedances are combined:

$$\underline{Z}_{(1)} = \underline{Z}_{(1)A} + \underline{Z}_{(1)B} = \underline{Z}_{(2)} \qquad \underline{Z}_{(0)} = \underline{Z}_{(0)A} + \underline{Z}_{(0)B} \tag{17.8}$$

The component currents are determined as follows:

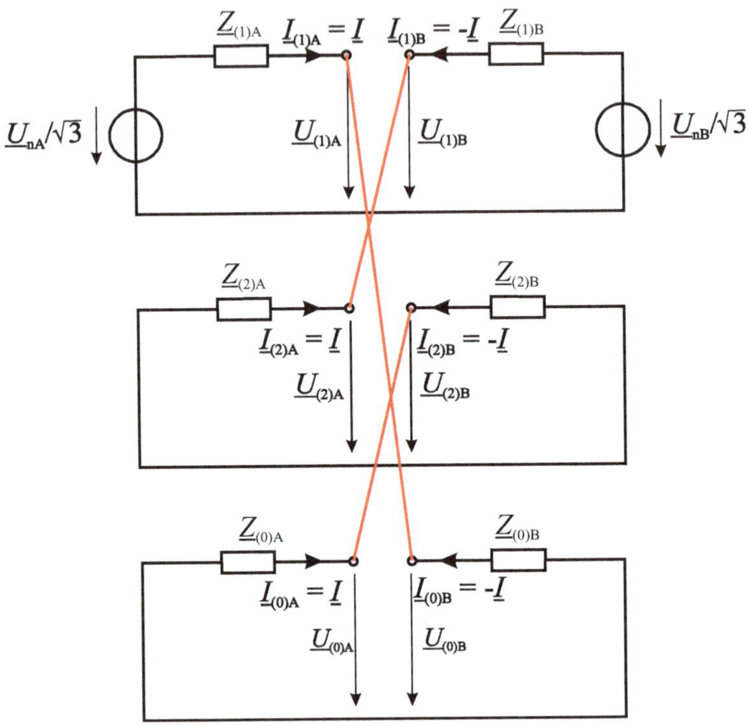

Fig. 17.3 Representation of the fault condition in the positive, negative, and zero sequence systems in the event of a voltage flashover between two three-phase systems A and B

$$\underline{I} = \underline{I}_{(1)\mathrm{A}} = \underline{I}_{(2)\mathrm{A}} = \underline{I}_{(0)\mathrm{A}} = -\underline{I}_{(1)\mathrm{B}} = -\underline{I}_{(2)\mathrm{B}} = -\underline{I}_{(0)\mathrm{B}} = \frac{\underline{E}}{2 \cdot \underline{Z}_{(1)} + \underline{Z}_{(0)}} \tag{17.9}$$

For the voltages in the symmetrical components of network B, taking into account the active voltage $\underline{E}$, which is calculated according to Eq. (17.10) from the difference between the two network voltages:

$$\underline{E} = (\underline{U}_{\mathrm{nA}} - \underline{U}_{\mathrm{nB}})/\sqrt{3} \tag{17.10}$$

$$\underline{U}_{(2)\mathrm{B}} = \frac{\underline{Z}_{(1)\mathrm{B}}}{2 \cdot \underline{Z}_{(1)} + \underline{Z}_{(0)}} \cdot \underline{E} \tag{17.11}$$

$$\underline{U}_{(1)\mathrm{B}} = \frac{\underline{Z}_{(1)\mathrm{B}}}{2 \cdot \underline{Z}_{(1)} + \underline{Z}_{(0)}} \cdot \underline{E} + \underline{U}_{\mathrm{nB}}/\sqrt{3} \tag{17.12}$$

$$\underline{U}_{(0)\mathrm{B}} = \frac{\underline{Z}_{(0)\mathrm{B}}}{2 \cdot \underline{Z}_{(1)} + \underline{Z}_{(0)}} \cdot \underline{E} \tag{17.13}$$

Under these conditions, the component current $\underline{I}$ can be determined according to Eq. (17.14) or Eq. (17.9) if, for the simplification, only reactances are considered and the reactances in the positive and negative sequence systems are equal, which is always the case with passive elements. In addition, the voltage factor $c_{max} = 1.1$ is used to calculate the maximum short-circuit current.

$$\underline{I} = -\mathrm{j}\frac{1,1\cdot(\underline{U}_{nA} - \underline{U}_{nB})/\sqrt{3}}{2\cdot\left(X_{(1)A} + X_{(1)B}\right) + X_{(0)A} + X_{(0)B}} \tag{17.14}$$

with

U_n voltages of the two networks A (380 kV) and B (110 kV); (line-to-line),

$X_{(1)}$, $X_{(0)}$ reactances in the positive and neutral sequence systems of the two networks.

In the case of an extra-high voltage network, it can generally be assumed that this network is solidly earthed, so that a reactance ratio of

$$X_{(0)A} = 3\cdot X_{(1)A} \tag{17.15}$$

can be assumed for the overlaid network. In contrast, various neutral point treatments are possible in a 110 kV network, so that, depending on the earth fault factor c_E, the following applies to the ratio $X_{(1)B}/X_{(0)B}$:

$$\frac{X_{(1)B}}{X_{(0)B}} = \frac{1}{2}\cdot\left[\frac{3}{\sqrt{4\cdot c_E^2 - 3}} - 1\right] \tag{17.16}$$

The earth fault factor c_E according to Eq. (17.16) indicates the ratio of the voltage of the conductor not affected by the fault during the short circuit to the voltage before the short circuit. According to Eqs. (9.85) and (9.86), the voltages $\underline{U}_S$ and $\underline{U}_T$ during a line-to-earth short circuit can be determined if only reactances are considered.

$$\begin{aligned}\underline{U}_{SB} &= \left[\underline{a}^2 + \frac{X_{(1)B} - X_{(0)B}}{2\cdot X_{(1)B} + X_{(0)B}}\right]\cdot\underline{E}_{(1)}\\ \underline{U}_{TB} &= \left[\underline{a} + \frac{X_{(1)B} - X_{(0)B}}{2\cdot X_{(1)B} + X_{(0)B}}\right]\cdot\underline{E}_{(1)}\end{aligned} \tag{17.17}$$

If the resistances are neglected, the two voltages are identical, so that the relationship corresponding to Eq. (17.16) can be derived from Eq. (17.18).

$$\frac{\underline{U}_{SB}}{\underline{E}_{(1)}} = c_E = \left[-\frac{1}{2} - \mathrm{j}\frac{1}{2}\cdot\sqrt{3} + \frac{\frac{X_{(1)B}}{X_{(0)B}} - 1}{2\cdot\frac{X_{(1)B}}{X_{(0)B}} + 1} \right] \tag{17.18}$$

The earth fault factor can take on various characteristic values, resulting in the following consequences for the ratio $X_{(1)A}/X_{(0)A}$:

- $c_E = 1$: The line-to-earth short-circuit current is identical to the three-phase current with $X_{(1)A} = X_{(0)A}$. In this case, the network is low-impedance earthed.
- $c_E = \sqrt{3}$: This is a network whose neutral points are isolated, or the network is operated in a compensated manner. Only an earth fault current flows in the event of a line-to-earth fault.

The positive sequence impedances of the two systems can be determined from the three-phase initial symmetrical short-circuit currents $\underline{I}''_{k3A}$, $\underline{I}''_{k3B}$, using Eq. (17.19), depending on the nominal system voltages U_n.

$$\underline{I}''_{k3A} = -\mathrm{j}\frac{1.1\cdot\underline{U}_{nA}/\sqrt{3}}{X_{(1)A}} \qquad \underline{I}''_{k3B} = -\mathrm{j}\frac{1.1\cdot\underline{U}_{nB}/\sqrt{3}}{X_{(1)B}} \tag{17.19}$$

According to Eq. (9.80), the fault current $\underline{I}_F$ (Fig. 17.1) is determined from three times the value of the component current $\underline{I}$, so that, taking Eq. (17.15) into account, the following applies:

$$\underline{I}_F = -\mathrm{j}\frac{1.1\cdot\sqrt{3}\cdot(\underline{U}_{nA} - \underline{U}_{nB})}{2\cdot\left(X_{(1)A} + X_{(1)B}\right) + X_{(0)A} + X_{(0)B}} = -\mathrm{j}\frac{1.1\cdot\sqrt{3}\cdot(\underline{U}_{nA} - \underline{U}_{nB})}{5\cdot X_{(1)A} + 2\cdot X_{(1)B} + X_{(0)B}} \tag{17.20}$$

The fault current $\underline{I}_F$ can be related to the three-phase short-circuit currents of the two networks A and B, resulting in the relative fault currents $\underline{i}_F$.

$$\frac{\underline{I}_F}{\underline{I}''_{k3A}} = \underline{i}_{FA} = \frac{3\cdot\left(1 - \frac{\underline{U}_{nB}}{\underline{U}_{nA}}\right)}{5 + \frac{X_{(1)B}}{X_{(1)A}}\cdot\left(2 + \frac{X_{(0)B}}{X_{(1)B}}\right)} = \frac{3\cdot\left(1 - \frac{\underline{U}_{nB}}{\underline{U}_{nA}}\right)}{5 + \frac{I''_{k3A}}{I''_{k3B}}\cdot\frac{U_{nB}}{U_{nA}}\cdot\left(2 + \frac{X_{(0)B}}{X_{(1)B}}\right)} \tag{17.21}$$

$$\frac{\underline{I}_F}{\underline{I}''_{k3B}} = \underline{i}_{FB} = \frac{3\cdot\left(\frac{\underline{U}_{nA}}{\underline{U}_{nB}} - 1\right)}{5\cdot\frac{X_{(1)A}}{X_{(1)B}} + 2 + \frac{X_{(0)B}}{X_{(1)B}}} = \frac{3\cdot\left(\frac{\underline{U}_{nA}}{\underline{U}_{nB}} - 1\right)}{5\cdot\frac{I''_{k3B}}{I''_{k3A}}\cdot\frac{U_{nA}}{U_{nB}} + 2 + \frac{X_{(0)B}}{X_{(1)B}}} \tag{17.22}$$

Equation (17.16) is used for the reactance ratio $X_{(0)B}/X_{(1)B}$ of the underlaid 110 kV network as a function of the earth fault factor.

Since the components of the underlaid network are exposed to the voltage of the overlaid network, the original voltages $\underline{U}_{RB}$, $\underline{U}_{SB}$, and $\underline{U}_{TB}$ are determined below based on the component voltages $\underline{U}_{(1)B}$, $\underline{U}_{(2)B}$, und $\underline{U}_{(0)B}$. According to Eqs. (9.24) and (9.42), the following applies:

$$\underline{U}_{RB} = \underline{U}_{(0)B} + \underline{U}_{(1)B} + \underline{U}_{(2)B} = \frac{2 \cdot X_{(1)B} + X_{(0)B}}{2 \cdot X_{(1)} + X_{(0)}} \cdot \frac{(\underline{U}_{nA} - \underline{U}_{nB})}{\sqrt{3}} + \frac{\underline{U}_{nB}}{\sqrt{3}} \tag{17.23}$$

$$\underline{U}_{SB} = \underline{U}_{(0)} + \underline{a}^2 \cdot \underline{U}_{(1)} + \underline{a} \cdot \underline{U}_{(2)} = \frac{X_{(0)B} - X_{(1)B}}{2 \cdot X_{(1)} + X_{(0)}} \cdot \frac{(\underline{U}_{nA} - \underline{U}_{nB})}{\sqrt{3}} + \underline{a}^2 \cdot \frac{\underline{U}_{nB}}{\sqrt{3}} \tag{17.24}$$

$$\underline{U}_{TB} = \underline{U}_{(0)} + \underline{a} \cdot \underline{U}_{(1)} + \underline{a}^2 \cdot \underline{U}_{(2)} = \frac{X_{(0)B} - X_{(1)B}}{2 \cdot X_{(1)} + X_{(0)}} \cdot \frac{(\underline{U}_{nA} - \underline{U}_{nB})}{\sqrt{3}} + \underline{a} \cdot \frac{\underline{U}_{nB}}{\sqrt{3}} \tag{17.25}$$

Equations (17.21) and (17.22) show that the fault current and thus also the voltages depend on the phase angle of the network voltages of the two systems. For this reason, the phase angle of the faulty conductors is taken into account in the calculation by using the following angles:

- Angel $\varphi = 0°$: The two systems are in phase,
- angel $\varphi = 180°$: The two systems are in phase opposition.

In addition, the ratio $\underline{i}_k$ of the three-phase short-circuit currents without voltage flashover is used as an additional parameter. Since only reactances are taken into account in this example, the following simplified formula applies:

$$\underline{i}_k = \frac{\underline{I}''_{k3B}}{\underline{I}''_{k3A}} = i_k \tag{17.26}$$

The voltages of the underlaid network are determined for the following parameters in the event of a fault:

- Short-circuit ratio: $0.25 \leq i_k \leq 1$
- Earth fault factor: $0.9 \leq c_E \leq 1.725$

Only real values are assumed for the earth fault factor, as no resistances are included in the calculation. The following applies to the voltage operators:

$$\underline{a} = -\frac{1}{2} + \mathrm{j}\frac{\sqrt{3}}{2} \qquad \underline{a}^2 = -\frac{1}{2} - \mathrm{j}\frac{\sqrt{3}}{2} \tag{17.27}$$

Figure 17.4 shows the fault current flowing between the two systems as a function of the system voltage. The figure shows the referenced fault current $i_{FB} = I_{FB}/I''_{k3B}$ as a function of the earth fault factor c_E of the underlaid network B.

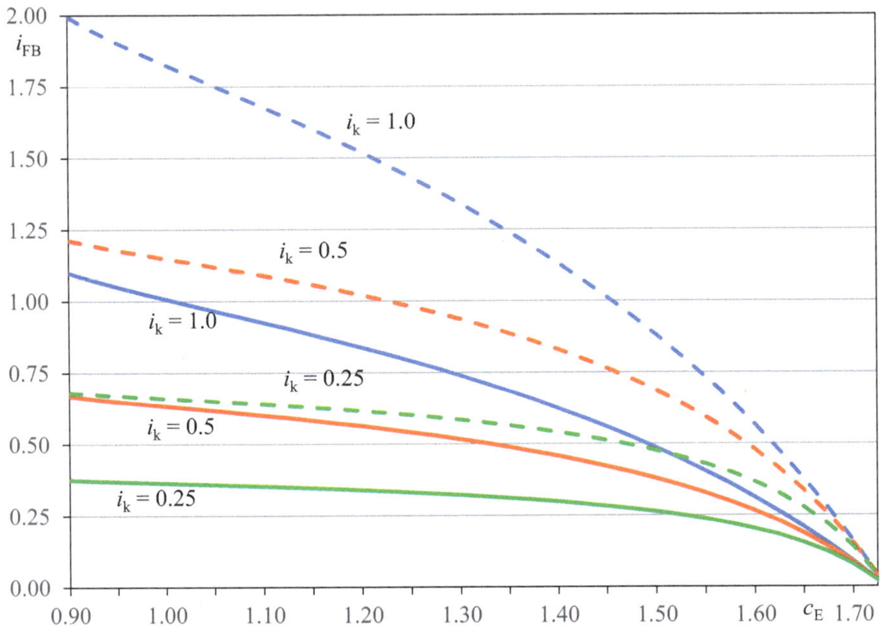

Fig. 17.4 Relative fault current i_{FB} in the underlaid network as a function of the earth fault factor c_E and the short-circuit current ratio i_k. Solid line: same phase angle of the systems, dotted line: opposite phase angle of the systems

If the voltages of both systems are in phase, the fault current can be at most equal to the three-phase short-circuit current of the underlaid network if the earth fault factor has a value of $c_E \approx 1.0$, provided that the three-phase short-circuit currents of the two systems are identical, $i_k = 1$. If the phase angles of the two voltage systems are opposite, the fault current can be greater than the three-phase short-circuit current of the underlaid network.

When the earth fault factor c_E reaches ≈ 1.7, the underlaid network is either isolated or compensated so that the fault current has a value of zero due to the zero sequence impedance $X_{(0)B} \rightarrow \infty$, or corresponds to the earth fault current.

The line-to-earth voltages of the underlaid voltage levels $\underline{U}_{RB}$, $\underline{U}_{SB}$, and $\underline{U}_{TB}$ and the neutral point-to-earth voltage $\underline{U}_{(0)A}$ are calculated using Eqs. (17.13) and (17.23)–(17.25). Here, the voltages are related to the nominal system voltage of the network B and, in addition, the relationships according to Eqs. (17.15) and (17.19) are used.

$$|\underline{u}_{RB}| = \left|\frac{\underline{U}_{RB}}{\underline{U}_{nB}/\sqrt{3}}\right| = \left|\frac{2 + \dfrac{X_{(0)B}}{X_{(1)B}}}{5 \cdot \dfrac{I''_{k3B}}{I''_{k3A}} \cdot \dfrac{U_{nA}}{U_{nB}} + 2 + \dfrac{X_{(0)B}}{X_{(1)B}}} \cdot \left(\frac{\underline{U}_{nA}}{\underline{U}_{nB}} - 1\right) + 1\right| \tag{17.28}$$

$$|\underline{u}_{SB}| = \left|\frac{\underline{U}_{SB}}{\underline{U}_{nB}/\sqrt{3}}\right| = \left|\frac{\frac{X_{(0)B}}{X_{(1)B}} - 1}{5 \cdot \frac{I''_{k3B}}{I''_{k3A}} \cdot \frac{U_{nA}}{U_{nB}} + 2 + \frac{X_{(0)B}}{X_{(1)B}}} \cdot \left(\frac{\underline{U}_{nA}}{\underline{U}_{nB}} - 1\right) + \underline{a}^2\right| \quad (17.29)$$

$$|\underline{u}_{TB}| = \left|\frac{\underline{U}_{TB}}{\underline{U}_{nB}/\sqrt{3}}\right| = \left|\frac{\frac{X_{(0)B}}{X_{(1)B}} - 1}{5 \cdot \frac{I''_{k3B}}{I''_{k3A}} \cdot \frac{U_{nA}}{U_{nB}} + 2 + \frac{X_{(0)B}}{X_{(1)B}}} \cdot \left(\frac{\underline{U}_{nA}}{\underline{U}_{nB}} - 1\right) + \underline{a}\right| \quad (17.30)$$

$$\left|\underline{u}_{(0)B}\right| = \left|\frac{\underline{U}_{(0)B}}{\underline{U}_{nB}/\sqrt{3}}\right| = \left|\frac{\frac{X_{(0)B}}{X_{(1)B}}}{5 \cdot \frac{I''_{k3B}}{I''_{k3A}} \cdot \frac{U_{nA}}{U_{nB}} + 2 + \frac{X_{(0)B}}{X_{(1)B}}} \cdot \left(\frac{\underline{U}_{nA}}{\underline{U}_{nB}} - 1\right)\right| \quad (17.31)$$

Figure 17.5 shows the voltage conditions in the underlaid network B as a function of the earth fault factor c_E and the short-circuit current ratio i_k of the two networks. The line voltages (line-to-earth) of the underlaid network (network B) are plotted, each relative to the unaffected network voltage.

The faulty conductor then assumes the highest voltage at the same phase angle of the network voltages ($\varphi = 0°$) when the earth fault factor has a value of $c \approx 1.7$. This means that the neutral points of the underlaid network are either isolated or connected to a compensation coil (Petersen coil) so that the network is operated in a compensated manner. In contrast, this applies to the fault-free conductors S and T when the phase angle is opposite at $\varphi = 180°$. The figure shows the additional parameter i_k according to Eq. (17.26). Figure 17.6 also shows the neutral point voltage $u_{(0)B}$ (relative to the line-to-earth voltage).

In the event of a voltage flashover, the conductor voltages sometimes assume higher voltages than the network voltage, with the voltage increase rising in the event of a high earth fault factor. If the zero sequence impedance in the underlaid network is high ($c_E = \sqrt{3}$, compensated or isolated neutral point treatment), the line-to-earth voltages assume at least the voltage value of the overlaid network.

The different voltages can be illustrated using Fig. 17.7 for an isolated or compensated underlaid network. It can be seen that, in the case of phase opposition of the faulty voltages (Fig. 17.7b), the fault-free line-to-earth voltages ($\underline{U}_{SB}$; $\underline{U}_{TB}$) always assume the highest value and are equal in magnitude [3]. This also applies to the neutral point voltage of the underlaid network. With the help of Fig. 17.7a, b, the voltage values in Table 17.1 can be explained for an earth fault factor of $c_E = \sqrt{3}$.

Depending on the current ratio i_k, the fault current can be equal to the value of the three-phase short-circuit current. For discrete earth fault factors, the current and voltage values are listed in Table 17.1, which distinguishes between the following values:

- $c_E = 1$: The three-phase short-circuit current is equal to the line-to-earth short-circuit current,

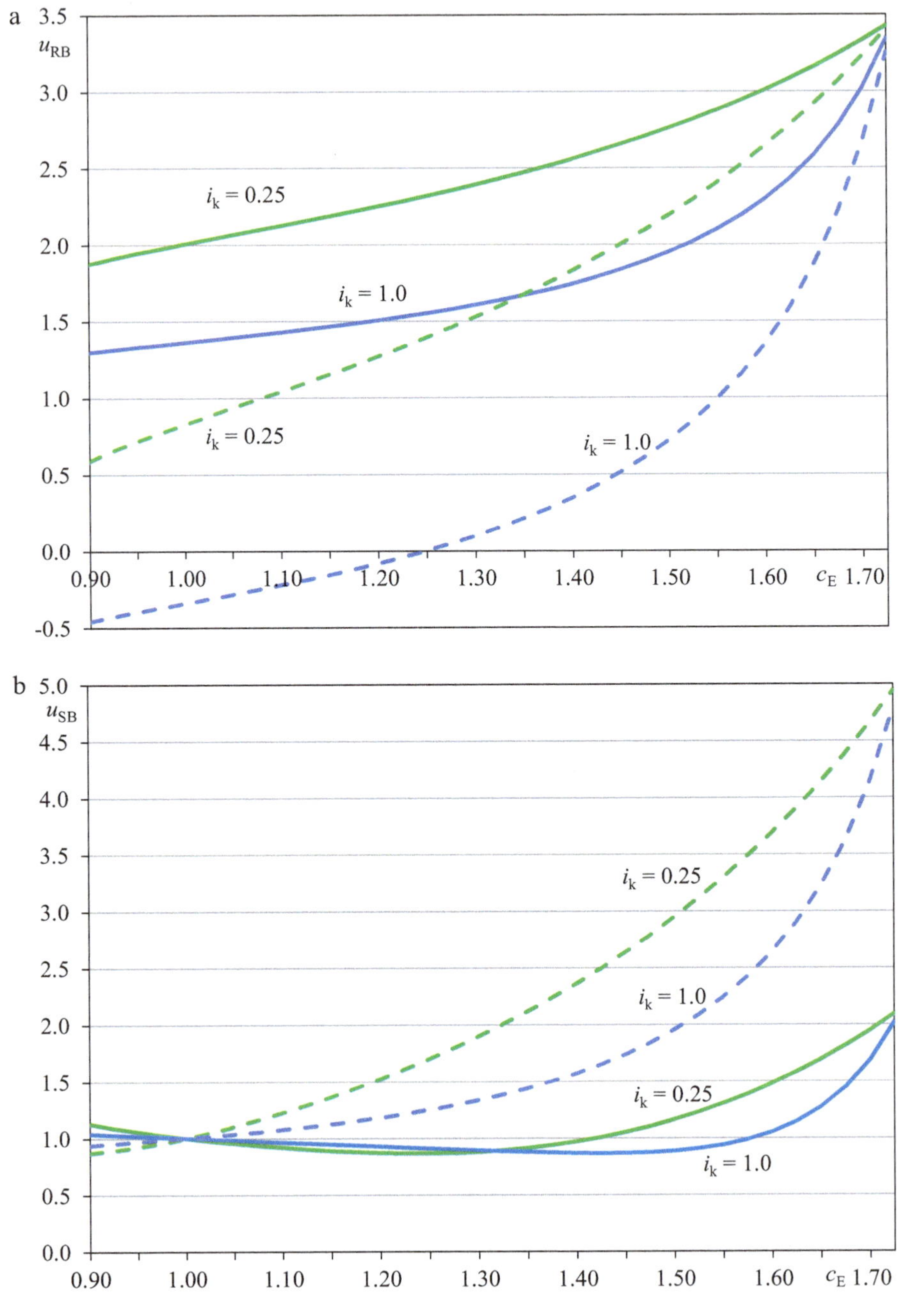

Fig. 17.5 Voltage conditions in the underlaid network B (line-to-earth) as a function of the earth fault factor c_E and the short-circuit current ratio i_k. (**a**) Faulty conductor, (**b**) fault-free conductor, solid line: same phase angle of the systems, dotted line: opposite phase angle of the systems

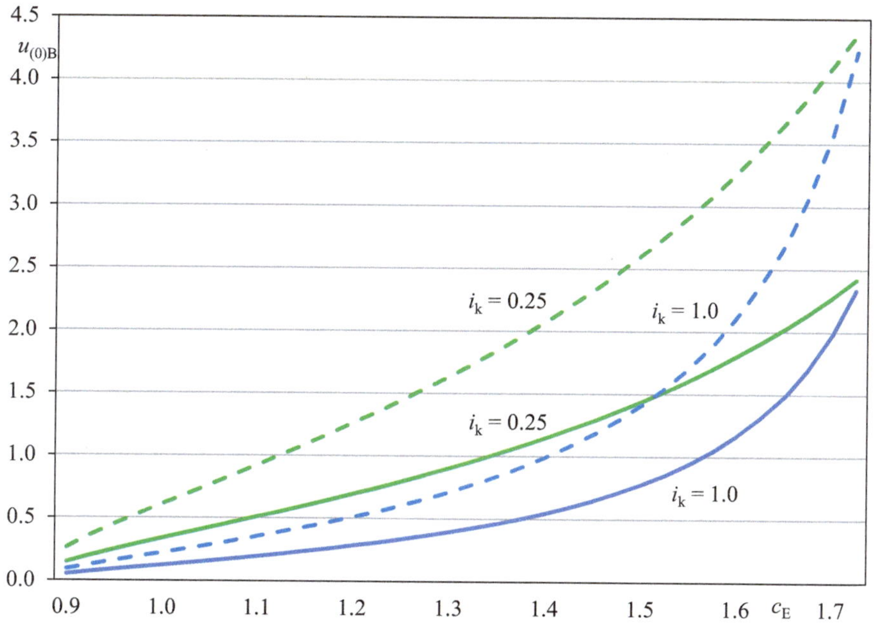

Fig. 17.6 Voltage conditions of the neutral point voltage in the underlaid network B as a function of the earth fault factor c_E and the short-circuit current ratio i_k. Solid line: same phase angle of the systems, dotted line: opposite phase angle of the systems

- $c_E = 1.4$: typical value for low-resistance neutral point earthing,
- $c_E = \sqrt{3}$: compensated or isolated neutral point earthing.

In networks with isolated or compensated transformer neutral points, only capacitive current flows, but this is not taken into account in the calculation according to Eq. (17.21).

17.2 Conclusions

In principle, it is possible that galvanic contact between different three-phase systems on an overhead line tower can occur, e.g., due to a conductor break. In this case, the potential of the lower voltage level is raised to the potential of the higher voltage level. The fault current depends on the neutral point treatment of the lower-level network, assuming that the higher-level network is earthed with low impedance.

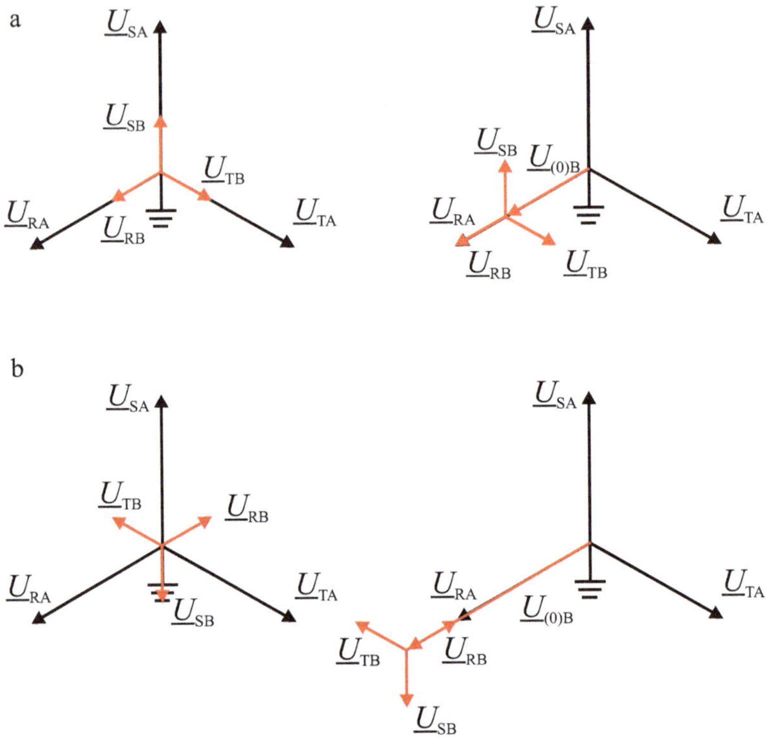

Fig. 17.7 Circuit diagram showing the displacement voltage during a voltage flashover as a function of the voltage angle between the two systems, undisturbed operation (left), disturbed operation (right). (**a**) Voltage angle $\varphi = 0°$, (**b**) voltage angle $\varphi = 180°$

Table 17.1 Related line-to-earth voltages and neutral point voltages (reference values 1 p. u. = $U_{nB}/\sqrt{3}$) and related short-circuit currents i_{FB}

c_E	i_k	φ	u_{RB}/p.u.	u_{SB}/p.u.	$u_{(0)A}$/p.u.	i_{FB}
1.0	0.25	0°	2.01	1.00	0.34	1.01
1.0	0.25	180°	0.83	1.00	0.61	1.83
1.0	1.0	0°	1.36	1.00	0.12	0.36
1.0	1.0	180°	0.34	1.00	0.22	0.66
1.4	0.25	0°	2.56	0.97	1.14	0.62
1.4	0.25	180°	1.83	2.36	2.07	1.13
1.4	1.0	0°	1.74	0.87	0.54	0.30
1.4	1.0	180°	0.35	1.57	0.99	0.54
√3	a ((no index))	0°	3.46	2.14	2.45	0.00
√3	a ((no index))	180°	3.46	5.03	4.45	0.00

c_E earth-fault factor
φ voltage angle between the voltages $\underline{U}_{RA}$ and $\underline{U}_{RB}$
i_{FB} related fault current, Fig. 17.4
i_k ratio of the three-phase short-circuit currents, Eq. (17.25)
[a] Independent of the short-circuit current ratio i_k

Isolated or compensated neutral point treatment of the underlaid network results in high voltages with a low fault current, e.g., <1 kA. In contrast, low-impedance neutral point treatment of the network results in lower voltages with a higher fault current, e.g., >10 kA.

References

1. Balzer G (1989) Verhalten von Metalloxid-Ableitern bei einem Spannungsübertritt zwischen zwei Drehstromsystemen. Elektrizitätswirtschaft, Jg 88(7): pp 407–413
2. Balzer G, Neumann C, Balzer C (2025) Schalt- und Ausgleichsvorgänge in elektrischen Netzen, 2nd edn. Springer Verlag, Berlin
3. Balzer G, Solbach HB (1990) Behavior of metal-oxide surge arrester during flashover between two three phase systems with different nominal system voltages. In: Cigre Session 1990, August 26–September 1, Paris, pp 33–207

Index

G. Balzer, *Short-Circuit Currents in Three-Phase Systems*,
https://doi.org/10.1007/978-3-658-50710-7

FSC
www.fsc.org
MIX
Papier | Fördert
gute Waldnutzung
FSC® C083411

Kolibri 360 GmbH
Ferdinand-Jühlke-Straße 7
99095 Erfurt, Deutschland
produktsicherheit@kolibri360.de